Principles of Acarology

Er cof annwyl am fy Nhad a Mam
a roddodd gymaint i mi

Principles of Acarology

Gwilym O. Evans

former Professor of Agricultural Zoology,
University College, Dublin

C·A·B International

C·A·B International
Wallingford
Oxon OX10 8DE
UK

Tel: Wallingford (01491) 832111
Fax: (01491) 8335058
Email: cabi@cabi.org
Website: http://www.cabi.org

A catalogue record for this book is available from the British Library

ISBN 0 85198 822 9

First published 1992

Transferred to digital print on demand, 2003

Typeset by Leaper & Gard Ltd, Bristol
Printed and bound by Antony Rowe Ltd, Eastbourne

Contents

Preface

This book has its origin in a series of lectures on the functional morphology and general biology of the Acari which I gave over a period of 25 years for courses in acarology organized by Dr P.W. Murphy for the University of Nottingham. The growth in knowledge of all branches of acarology during this period has been remarkable and this has been accompanied by a great increase in the literature on acarological topics, as evidenced by the annual list of titles published in the Arachnida section of the *Zoological Record*. The major advances are to be seen in studies on fine structure, physiology and behaviour whilst new approaches to phylogenetic classification have generated increased interest in and work on acarine taxonomy. Much of this new information is contained in papers widely scattered in very many scientific journals, a number of which are not readily available to the student or researcher. I have attempted to include a summary of these recent contributions to our knowledge of acarine structure and function in the first 11 chapters of the book. The object is to provide an introductory text for the non-specialist and student as well as a source of reference. The list of references at the end of the volume is far from being a full catalogue of important works on the Acari – that would be impossible in a volume of this size. References to older papers have been omitted if the information they contain has been adequately summarized in later and more comprehensive works. The classification of the Acari, which is introduced in Chapter 12, is in a state of flux with traditional concepts being challenged by the results of new methods of classification. It is hoped that methodology will remain the tool and not become the master of the systematist.

Many colleagues and friends have given me assistance and encouragement but I should like to record, in particular, my sincere thanks to Gerd Alberti, Anne S. Baker, Alex Fain, Evert E. Lindquist, Don Macfarlane and Paul W. Murphy for reading one or more chapters of the book in draft form and making many valuable suggestions, the majority of which I have followed. Gerd Alberti, Alex Fain and Harald Witte have given me permission to reproduce some of their published illustrations, and Don

Macfarlane has generously allowed me to use a number of his unpublished drawings of Oribatida. Through the kindness of Don A. Griffiths, I have received permission from the Central Science Laboratory, MAFF, Slough, to publish the following SEM micrographs: Figs 5.1B; 5.4A, B; 5.11A, B; 6.15A–D; 6.16A–C. I am grateful to Philip J.A. Pugh for Figs 4.4, 4.5 and 4.15, to Elsevier Science Publishers, Amsterdam, for allowing me to reproduce Figs 9.9 and 9.14 and to Chapman & Hall, London, for permission to use Figs 10.2, 10.3 and 10.5. J.G. Wernz and G.W. Krantz have kindly allowed me to reproduce Fig. 6.3. I wish to express my thanks to Barry Cregg, University College, Dublin, for his technical assistance in preparing a number of SEM micrographs.

The labour of writing this book has been lightened by the constant support, patience and assistance of my wife, Angela, to whom I express my thanks and appreciation. Finally, to quote from A.D. Michael's Preface to the *British Oribatidae*, 'if the study gives to others half the pleasure it has given to me, I shall be well rewarded for my labours'.

Gwilym Owen Evans
Storrington
March 1992

Introduction

The Acari, comprising mites and ticks, form one of the largest and most biologically diverse groups of the Arachnida. They are worldwide in distribution and rival the insects in the extent to which they have successfully colonized terrestrial and aquatic habitats. They abound in forest and grassland soils and in temporary accumulations of organic debris. Many species spend part or all of their life cycle on trees and shrubs while some are cavernicolous and others have become adapted for life in thermal springs. Aquatic forms live in salt and fresh water. Unlike other arachnids, which are free-living, a large number of acarines have developed intimate associations with other animals. The mite associations range from commensalism to parasitism while many species living in temporary habitats practise phoresy, using a variety of other arthropods as vehicles for dispersal. Ticks are parasitic on a wide range of vertebrate hosts. This evolutionary diversification of the Acari into a number of different modes of life has been accompanied by a variety of structural and life-history adaptations. The former affect, for example, body shape and mouthparts and the latter are seen in strategies influencing survival and reproduction. An estimated number of approximately 30 000 described species of Acari, distributed among more than 1700 genera, was given by Radford (1950) but the number of undescribed species could exceed this total by up to twentyfold.

The highest population densities and species richness of free-living mites occur in the organic strata of soils where they form the numerically dominant component of the arthropod mesofauna and may constitute up to 7% of the total weight of the invertebrate fauna. They comprise saprophagous, microphytophagous and predatory species and are exceptional among the Arachnida in including forms which ingest solid food of plant and animal origin. The Oribatida, commonly referred to as beetle or moss mites, are generally the most abundant of the acarine taxa in soils rich in decaying organic matter. For example, in soils of deciduous woodlands in southern England, they form over 70% of the mite fauna (Evans *et al.*, 1961). This dominance of the oribatid mites is less obvious or replaced by the Prostigmata in some tropical and subtropical soils with relatively low levels of organic matter (Loots & Ryke, 1966). The majority of the oribatid species feed on the decaying organic material and/or on the microorganisms associated with its decay but the part they play in the functioning of soil systems is not fully understood. Their comminution of decaying plant tissues and production of faecal pellets, which increase the possible surface area for decomposition by microorganisms, are undoubtedly beneficial within the system but their role in translocating organic matter, both vertically and laterally, is probably of minor importance. The species richness of mites in soil habitats can be extremely high; for example, Purvis (1982) records up to 77 species per core (5 cm diameter × 3.25 cm depth) in a dune-grass tussock site in Ireland.

Some free-living mites live in stored food where they often multiply rapidly and attain pest status. In species composition, the acarine community of stored products, and also human habitations, is very similar to that occurring in the nests of birds and mammals and it has been suggested that nests rather than the soil form the natural habitat of these mites. The order Astigmata, which contains the species of major economic importance, utilizes a wide range of food including grain, fishmeal and substances containing sugars (e.g. dried fruit) but of the large number of astigmatic species associated with these materials only relatively few can be considered to be important pests (Hughes, 1976). In addition to infesting food, certain of the species also live in human habitations where they are found in damp situations favouring the growth of fungi upon which they feed. Members of one family, commonly referred to as house dust mites, are of considerable medical importance in being the causative agents of atopic asthma and rhinitis, the allergen(s) being present in their faecal pellets.

The majority of the species of mites living on the aerial parts of higher plants, especially on trees, feed mainly on microflora or prey on other small arthropods or nematodes living on the plant. However, some of the Prostigmata, chiefly the spider mites, false spider mites and eriophyoid mites, use their specialized mouthparts to feed on the vascular tissues of higher plants and by their activity can cause severe losses to field and protected crops. Many species of the Eriophyoidea are responsible for the production of plant galls, while others transmit plant viruses. Considerable success has been achieved in using predatory mites of the family Phytoseiidae in the biological control of spider mites, especially those infesting glasshouse crops.

The evolution of animal parasitism in the Acari is unique among the Arachnida. It is in reference to the parasitic acarines that the distinction is made between mites and ticks. Both common names can be traced back to Old English (before 1150 AD) and have become well established in the scientific literature. This distinction is also made in other European languages, for example *milben* and *zecken* in German and *acare* and *tique* in French. However, the common names do not coincide with the taxonomic division of the Acari into its two major taxa; in fact some groups of mites are more closely related to ticks than they are to other mite taxa. Ticks parasitize vertebrates, including man and domesticated animals, and are vectors of a wider range of disease organisms than any other group of arthropods. They are essentially haematophagous and, in the engorged state, some species attain a length of 20–30 mm. Parasitism in mites has evolved independently in a number of groups. The term parasite is often used rather loosely in reference to mites associated with both invertebrate and vertebrate hosts and in many instances the mite–host relationship is unknown or a matter of conjecture. Most microhabitats provided by skin, hair and feather have been occupied by mites, which display an amazing range of structural and biological adaptations. On mammals, those living in or on the skin often cause mange accompanied by pruritus, while on birds some are commensals feeding on debris and oil secretions on feathers and others are parasites living inside quills or in the skin. A few species of Trombiculidae, which are parasitic in the larval phase only, are vectors of the rickettsial disease known as scrub typhus or tsutsugamushi.

Mites have a long fossil history but even the earliest forms can be readily accommodated in existing family taxa. The first described fossil mite, *Protacarus crani*, was found in the Devonian Rhynie Chert of Scotland which appears to have been formed by the rapid inflow of hot silicified material into a bog (Hirst, 1923). The species was assigned provisionally to the prostigmatic family Eupodidae, but Dubinin (1962) considered that the material examined by Hirst represented five species of which four, including *P. crani*, were classified in the early derivative group Endeostigmata and one in the Tydeidae. More recently, 13 specimens representing four species, all of the division Actinotrichida (=Acariformes), have been recovered from Devonian deposits (376–379 mya), predominantly terrestrial in origin, in New York State (USA). Two of the species belong to the Endeostigmata and the other two to the Oribatida (Norton *et al.*, 1988). One of the oribatid species was considered to be a highly plesiomorphic member of the Enarthronota. Previously, the oldest recorded specimens of Oribatida were of Jurassic age and comprised members of the somewhat highly derived taxa Hydrozetidae and Achipteriidae (Krivolutsky & Druk, 1986). According to Krivolutsky (1979), actinotrichid mites were 'rather numerous in the Mesozoic era' and the rate of evolution of the group 'is one of the slowest in the animal kingdom'. The Anactinotrichida (=Parasitiformes) do not appear in the fossil record until the Upper Eocene (Bernini, 1991). A rich source of fossil mites is amber, a fossil resin from prehistoric coniferous trees (Vitzthum, 1940–43). One of the oldest described is a larval erythraeoid mite from amber of the Cretaceous in Canada, but representatives of the Anactinotrichida also occur, for example *Dendrolaelaps fossilis* (Mesostigmata), from Tertiary amber in Mexico (Hirschmann, 1971).

Preparation of Mites for Microscopic Study

The small size and often delicate nature of mites makes it necessary to use special techniques for preparing them for study under the microscope. The following account deals mainly with methods of preparation for light microscopy. Techniques for transmission and/or scanning electron microscopy are described by Crooker *et al.* (1985), Alberti *et al.* (1981) and Walzl and Waitzbauer (1980). Methods of collecting and culturing free-living and parasitic mites are discussed by Evans *et al.* (1961) and sectioning techniques are given in Rhode (1964), Woodring (1970) and Crooker *et al.* (1985).

Much information on the ornamentation of the cuticle and form of cuticular organules of larger mites can be obtained by using the carbon-block method of Grandjean (1949a) in conjunction with a binocular microscope and incident light. A small area of a porous carbon block is moistened with 70% alcohol siphoned through a capillary tube from a small reservoir (Figs 1 and 2). The region of the block saturated with alcohol becomes very black. For examination, the specimen is moved a short distance from the alcohol near the mouth of the delivery tube but as soon as it shows a tendency to dry out, it is returned to the alcohol.

The identification and detailed study of the external morphology of

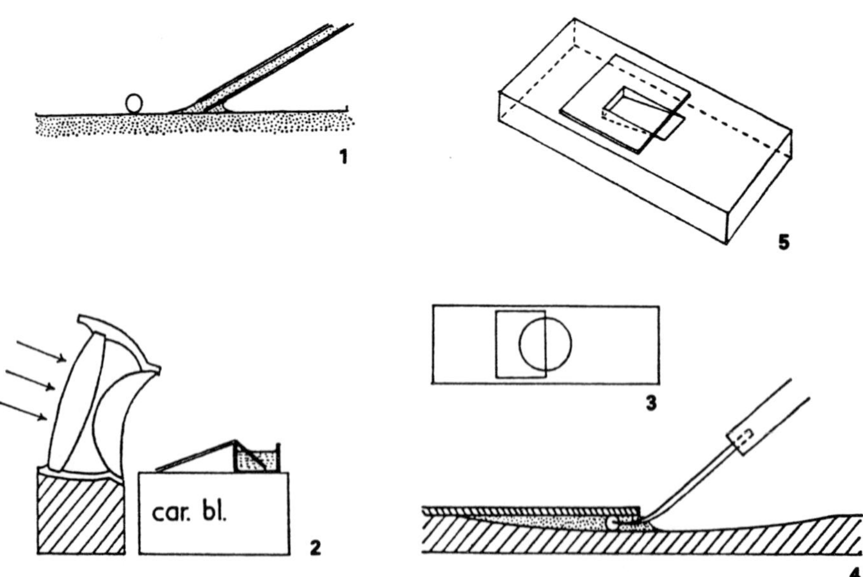

Figs 1–5. Grandjean's carbon block method for the examination of small arthropods by: incident light (*car. bl.*, carbon block) (1, 2); glass (3) and perspex cavity slides (5). Manipulation of a specimen in a cavity slide (4). (From Evans *et al.*, 1961.)

mites require the use of a compound microscope and some form of microscope-slide preparation either of a temporary or a permanent nature. The basic requirements are the same in both cases and comprise methods of killing and fixing the specimen and the removal of non-cuticular structures so as to render it more or less transparent with the cuticular structures intact. In some species, bleaching of the darkly coloured cuticle may also be necessary.

Killing and fixation

Mites that have been collected alive are best killed by pouring onto them a small quantity of boiling water (Michael, 1884). This causes the extension of the trophic and ambulatory appendages which facilitates the detailed study of these structures. Usually, mites are extracted from soil and organic materials by funnel-heat desiccation or flotation methods and the 'catch' is usually preserved and fixed in 70–80% alcohol (industrial methylated spirit) to which a small amount of glycerol may be added to prevent the drying out of the specimens during storage. Oudemans' fluid (alcohol 87 parts, glycerol 5 parts and glacial acetic acid 8 parts) may be used as an alternative to alcohol for terrestrial mites, and Viets' fluid (glycerol 10 parts, acetic acid 3 parts and distilled water 6 parts) is suitable for freshwater species. Specimens collected in an aqueous solution of picric acid should be transferred to alcohol for storage. For histological preparation, more specialized fixation fluids are necessary (see Romeis, 1968).

Maceration and bleaching

The most widely used macerant is lactic acid in aqueous solutions ranging in concentration from 50 to 95% (v/v) depending on the degree of sclerotization of the mite, the lower concentrations being suitable for weakly sclerotized forms. In the absence of a phase-contrast or interference-contrast microscope, the addition of a stain (lignin pink, acid fuchsin or methylene green) to the macerant is often advantageous when dealing with small delicate species that tend to over-clear. Lactic acid saturated with iodine is a useful selective stain for distinguishing birefringent setae in actinotrichous mites when polarizing equipment is unavailable (Grandjean, 1935c). After staining, the specimen is transferred to warm stain-free lactic acid before study. Birefringent or anisotropic setae appear brown in colour while isotropic sensilli are colourless.

The cuticle of some strongly sclerotized species, especially among Oribatida, remains opaque after maceration in lactic acid and it is necessary to render it translucent for microscopic study. One of the most convenient methods is to bleach the specimen in sodium perborate after maceration (Balogh, 1959). The specimen is placed in a small quantity of water on a slide to which a minute amount of solid sodium perborate is added. The slide is then gently heated. Caustic potash may be added to hasten the bleaching process.

Mites engorged with blood are difficult to macerate without some damage to the specimen. It is advisable to puncture the body with a fine needle to facilitate entry of the macerant and light pressure may be applied to the specimen to force out some of the gut contents. Lactic acid is not entirely suitable and more drastic agents such as lacto-phenol (lactic acid 50 parts, phenol crystals 25 parts and distilled water 25 parts) and 5–10% aqueous solution of sodium hydroxide may be required. For macerating Halacaridae, Newell (1947a) recommends the gentle pressing out of some of the gut contents through a body incision before digestion in trypsin solution (0.2 g powdered trypsin dissolved in 10 ml 0.5% Na_2CO_3) in an atmosphere saturated with toluene. Mites are left in the solution for 24–72 hours and then transferred to distilled water before mounting on slides.

Temporary preparations

For routine identification, mites may be placed in a drop of the appropriate concentration of aqueous lactic acid on a slide, oriented and the coverslip applied. The preparation is then warmed on a hotplate or over a naked flame to hasten maceration. The specimen is ready for examination when the maceration process is completed. After examination the mite can be stored in 70% alcohol plus glycerol (up to 5%). Temporary preparations are also recommended for the detailed study of the external morphology of mites. The method, however, requires the use of a cavity slide and not a normal microscope slide so as to avoid compression of the specimen and to facilitate its orientation to the desired position for study (Grandjean, 1949a). A square coverslip is placed over half to two-thirds of the cavity and sufficient lactic acid is introduced to fill the portion of the cavity below the coverslip (Figs 3 and 4). The specimen, which has been gently macerated in warm or cold lactic acid in a glass container, is transferred into the lactic acid in the cavity and oriented under the coverslip by means of a fine tungsten needle (made by dipping the tip of a piece of 100 or 200 μm tungsten wire in boiling sodium nitrite). The viscosity of the lactic acid is usually sufficient to maintain the specimen in the desired position but its orientation may be changed by gently moving the coverslip or manipulating it with the tungsten needle. A wedge-shaped cavity in sheet perspex was found by Evans and Browning (1955) to be more versatile than the form of cavity usually available in glass slides (Fig. 5). After study, the specimen is stored in 70% alcohol (with added glycerol, if desired), preferably in a small glass specimen tube (6 mm diameter and 25 mm long) plugged with a porous cork made from plastic, botanical pith or polyporus. The tube, together with a label bearing the specific name and details of habitat and locality, is placed in a larger specimen tube (12 mm diameter and 25 mm long) filled with alcohol and plugged with cotton wool. The larger tube is then inverted in a large flat-bottomed storage jar containing 70% alcohol, which is closed by a well-fitting lid to minimize evaporation.

Permanent preparations

Permanent slide preparations are desirable for building up a readily accessible reference collection as well as providing an alternative to temporary lactic acid preparations for routine identifications. A number of media have been introduced for permanent preparations and these have replaced Canada balsam and Euparal, so favoured by early workers on mites. The most widely used are those based on gum-chloral or polyvinyl alcohol (PVA). The Faure-type gum-chloral medium (gum arabic 30 g, chloral hydrate 50–150 g, glycerol 20 g and distilled water 50 ml) is water soluble and the specimen can be easily recovered from the medium should the preparation deteriorate during storage. The higher concentration of chloral hydrate is recommended for Tetranychidae (spider mites). The coverslip of gum-chloral mounts should be sealed with a proprietary ringing compound such as Glyceel, especially when the slides are stored in humid conditions. One of the most satisfactory PVA media is a modified form of Heinze-PVA by Boudreaux and Dosse (1963) and contains polyvinyl alcohol 10 g, chloral hydrate 100 g (changed from 20 g after publication), 85–92% lactic acid 35 ml, 1% aqueous solution of phenol 25 ml, glycerol 10 ml and distilled water 40–60 ml. Specimens are more difficult to recover from this medium but this can be accomplished by soaking the preparation in lacto-phenol.

Freshly killed or alcohol-preserved specimens may be placed directly into these media without pre-treatment although in the case of the gum-chloral medium it is advisable to macerate and clear heavily sclerotized species before mounting. Vitzthum's fluid (phenol 9 parts, chloral hydrate 1 part and distilled water 1 part) is commonly used as a clearing fluid. Specimens macerated in lactic acid should be washed in alcohol if they are to be placed directly into gum-chloral media. Slides prepared by both types of media should be partially hardened in an oven at about 50°C for 24–36 hours before use. However, care must be taken that PVA slides are not overheated or shrinkage of the medium and damage to the specimen will occur.

The above techniques are also applicable to the majority of unfed ticks although the larger size of adults of some species, in comparison with mites, makes them unsatisfactory material for preparing whole mounts. It has been traditional to study entire specimens of alcohol-preserved nymphal and adult ticks under a binocular dissecting microscope and to make temporary or permanent preparations of only dissected parts of the specimen for detailed study under the compound microscope.

Literature

Comprehensive texts on the functional anatomy and classification of the Acari are relatively few. Among the earliest are the classic works of A.D. Michael, the *British Oribatidae* (Michael, 1884, 1888a) and *British Tyroglyphidae* (Michael 1901, 1903) published by the Ray Society. Their publication was a landmark in the history of acarology and they remain a valuable source of information and an inspiration for students of acarology. Subsequently, no comparable text in scope or detail appeared until H. Graf Vitzthum's two major works: 'Acari = Milben' in W. Kukenthal and Th. Krumbach's *Handbuch der Zoologie* (Vitzthum, 1931) and 'Acarina' published in *Bronn's Klassen und Ordnungen des Tierreiches* at intervals from 1940 to 1943 (Vitzthum, 1940–43). The latter is a monumental work and provides an extensive review of the knowledge of the morphology, biology and classification of mites and ticks up to about 1940. Its thousand or so pages indicate the extensive nature of the work. This was followed by M. André's contribution on the 'Ordre des Acariens' in P-P. Grassé (ed.) *Traité de Zoologie* (Vol. VI) published in 1949. Three further general texts on acarology have been published since this date. *An Introduction to Acarology* by E.W. Baker and G.W. Wharton in 1952 provided a much desired English text at a period when acarology was developing as a subject of considerable importance in the fields of medicine, veterinary science and agriculture. Its treatment was largely taxonomic with notes on the biology of selected taxa but with little or no information on anatomy and physiology. On the other hand, the publication in 1959 of the *Mites or the Acari* by T.E. Hughes dealt in some detail with the functional anatomy of the Acari. A general text by T.A. Woolley entitled *Acarology, Mites and Human Welfare* appeared in 1988.

Publications on the descriptive external morphology and classification of the Acari have formed, and will continue to form, a major part of acarological literature although their earlier dominance has decreased in recent years with the increasing interest in other fields of the science. The works of A. Berlese, A.C. Oudemans and F. Grandjean have been outstanding in their contributions to these aspects of acarology. Of particular significance have been Berlese's great work *Acari, Myriapoda et Scorpiones hucusque in Italia reperta*, which was published at intervals from 1882 to 1903, and the *Kritisch Historisch Oversicht der Acarologie* compiled by Oudemans and appearing between 1926 and 1937. The latter provides a critical survey of acarology from 850 BC to 1850 AD. However, it is the contributions made by Grandjean to the descriptive morphology of the actinotrichid mites that have probably had the greatest influence on acarology through the detail, accuracy and originality of his work. His papers on Acari have been reprinted under the title *Oeuvres Acarologiques Complètes*, Vols. I–VII (Lochem, The Hague, 1972–1976) and should be required reading for any serious student of mites. Other valuable contributions, many of which are mentioned in the final chapter of this work, refer to the production of key works for the identification of the Acari on a worldwide or a regional basis. In this context, the classic works of G.H.F. Nuttall and his colleagues (Nuttall *et al.*, 1908b; Nuttall & Warburton, 1911, 1915), M. Sellnick (1929, 1960), Sig Thor (Thor, 1931, 1933; Thor & Willmann, 1947), K. Viets (1925) and C. Willmann (1931) are worthy of mention.

A valuable source of information on current work and developments in acarology is provided by the *Proceedings of the International Congresses of Acarology*. The first Congress was held in Fort Collins, USA, in 1963 and since then meetings have taken place at approximately four-year intervals at different venues. The papers presented at each of the first four Congresses were published as the *Proceedings* of the specific (1st–4th) International Congress of Acarology but, subsequently, the volumes have been given unconnected titles: *Recent Advances in Acarology* (published in 1979), *Acarology VI* (1984) and *Progress in Acarology* (1989). In addition, a number of publications dealing with specific acarological topics have appeared in the last ten years and these include the *Morphology, Physiology and Biology of Ticks* (Ellis Horwood Ltd, Chichester, 1986), *Spider Mites, their Biology, Natural Enemies and Control* (Vols 1A & 1B, published by Elsevier, Amsterdam, 1985) and the proceedings of the first symposium organized by the European Association of Acarologists (EURAAC) entitled *The Acari, Reproduction, Development and Life-History Strategies* (Chapman & Hall, London, 1991).

Although papers on the Acari are published in a wide range of zoological, entomological and related periodicals, three are devoted entirely to works on mites and ticks. The first of these, *Acarologia* (published in France), was founded by the late Marc André in 1959. It specializes in the publication of papers on the descriptive morphology and classification of the Acari. The second, the *International Journal of Acarology*, first appeared in 1975 and is published by the Indira Publishing House in the USA under the direction of V. Prasad. It publishes papers on general and applied acarology. The third publication, *Experimental and Applied Acarology*, has been produced since its foundation in 1985 by Elsevier, Amsterdam. Papers appearing in scientific periodicals on the morphology, classification and biology of the Acari are listed in the Arachnida part of the *Zoological Record* (published yearly by Biosis and the Zoological Society of London). Important aspects of each paper are indicated in subject, geographical, palaeontological and systematic indexes. The extensive systematic index listing new taxa, new synonymies, localities and hosts is an invaluable tool for the systematic acarologist.

Terminology

Although acarological terminology has much in common with that of entomology, many of the terms referring to descriptive morphology are applicable only to the Acari and largely stem from the work of F. Grandjean and his disciples, especially L. Van der Hammen. The plethora of terms (and their synonymy) is daunting for the uninitiated. A glossary of acarological terminology has been compiled by Van der Hammen (1980). His treatment of terms relating to the Actinotrichida is extensive and thorough but some of the general terminology, as well as that relating specifically to the Anactinotrichida, is incomplete (Athias-Henriot, 1982). The less familiar terms used in this work are explained in the text or illustrated and appear in the Subject Index. The selected term, in the case of synonymy, is accompanied, when first cited, by the alternative term(s) in parenthesis. Many of the terms relating to the external morphology of the Acari are defined or illustrated in the relevant sections of Chapter 12 (Classification) and not in the chapters on functional morphology.

The names Archoribatida and Euoribatida used in this work are equivalent, respectively, to the Macropylina and Brachypylina, and the Lower Oribatida and Higher Oribatida.

1 Integument and Moulting

The body covering or integument in the acarines is complex in structure and diverse in function. It provides a protective exoskeleton and sites for muscle attachment as well as lining the tracheal system, the ducts of glands and parts of the genital system and alimentary tract. Its high degree of impermeability plays an important role in restricting water loss from the body. Although showing considerable flexibility, the sclerotized non-cellular component of the integument is limited in the extent to which it can expand to allow for the increase in size of the animal. In order to accommodate this, it is shed periodically and replaced during the moulting cycle.

Integument

The structure of the acarine integument is essentially similar to that of other arthropods, particularly insects, and comprises a cellular epidermis and a non-cellular cuticle which is secreted by the epidermis. Epidermal cells form a single layer which is bordered basally by a basement membrane or lamina (Fig. 1.1).

Cuticle and cerotegument

The cuticle consists of a thick procuticle, which may be differentiated into an endocuticle (=hypostracum) and exocuticle (=ectostracum), a thin epicuticle (=epiostracum) and a superficial wax-cement or secretion-layer, the tectostracum or cerotegument. (Some authors follow Grandjean (1951a) in referring to the region of the cuticle comprising the procuticle and the inner and outer epicuticle, but excluding the wax and cement layers, as tegument. This term should not be confused with the same name given to the metabolically active body covering of the Cestoda and Trematoda, which is entirely different in structure and

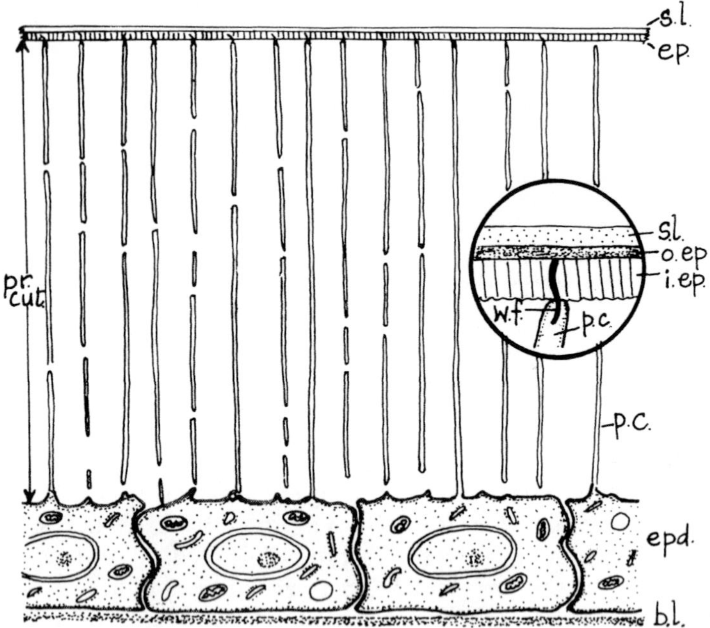

Fig. 1.1. Diagrammatic section of the acarine integument (modified from Amosova, 1979). *b. l.* = basement lamina; *ep.* = epicuticle; *epd.* = epidermis; *i. ep.* = inner epicuticular layer; *o. ep.* = outer epicuticular layer; pr. cut. = procuticle; *p. c.* = pore canal; *s. l.* = secretion layer (cerotegument); *w. f.* = wax filament.

function.) The relative thickness of the procuticle and epicuticle in, for example. *Tetranychus urticae*, is 0.25–2.0 μm and 0.05–0.15 μm, respectively (Mothes & Seitz, 1982). Wharton *et al.* (1968) indicate the presence of a distinct deposition zone (Schmidt layer) in *Echinolaelaps echidnina*. This zone occurs between the epidermis and the stabilized cuticle and is granular in appearance. The most extensive study of the acarine cuticle has been carried out by Alberti *et al.* (1981).

The procuticle is constructed of microfibrils of chitin, a structural polysaccharide, embedded in a protein matrix. The microfibrils are laid down in layered sheets or lamellae and within each sheet the microfibrils are parallel to one another (Fig. 1.2). When microfibrils in successive sheets are aligned at regular changing angles, the resulting helicoidal arrangement produces a parabolic patterning of the cuticle. Unidirectional orientation of the microfibrils in successive sheets gives a preferred layer which does not have a lamellar appearance (Neville, 1975). In most arthropods, exocuticles are helicoidal (lamellate) while endocuticles may show varying proportions of helicoidal and unidirectional layers. Details of the structure of the endocuticle do not appear to be known for the Acari but, in other Arachnida, helicoidal and unidirectional layers are present in the endocuticle of the spider *Tegenaria agrestis* whereas only helicoidal layers occur in the scorpion *Leiurus quinquestriatus*. In

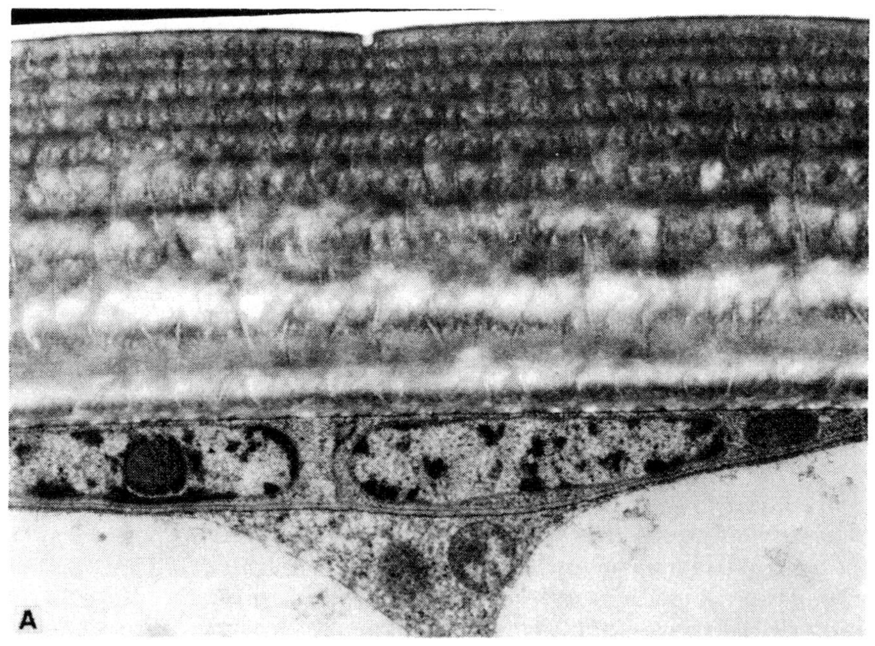

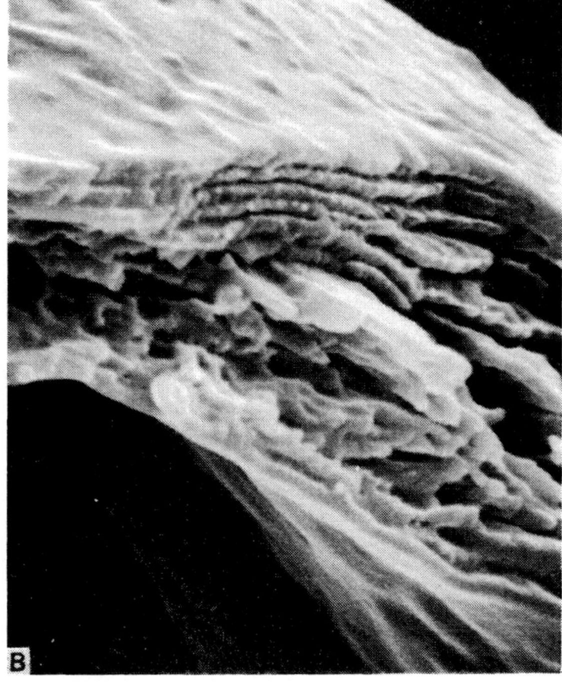

Fig. 1.2. A, TE micrograph of a section of the integument of *Eulohmannia ribagai* (Oribatida) showing lamellar structure and pore canals of the cuticle (from Alberti *et al.*,1981); **B**, cuticular lamellae in an SE micrograph of the fractured base of the gnathosoma of *Phytoseiulus persimilis* (Mesostigmata).

Haemaphysalis leporispalustris three to five and up to 20 lamellae occur, respectively, in the endo- and exocuticle of the sclerotized integument (Nathanson, 1967), whereas 4–16 procuticular lamellae have been reported in *Tetranychus urticae*. Exocuticle occurs in the hard parts of the body, such as the gnathosoma, body sclerites and the cuticle of the legs. Endocuticle is stated to be absent from the gnathosomatic base and legs in ticks (Hackman, 1982).

The epicuticle comprises two main layers: a homogeneous innermost layer, the inner epicuticle, containing lipids, phenols and protein, and an electron-dense outer epicuticle which lies immediately below the wax layer (Weis-Fogh, 1970). Some authors consider the outer epicuticle, as conceived in this work, to constitute, in ticks, a very thin (2 nm) outer layer which can be compared to the lipid monolayer of the insect epicuticle, and below it a dense cuticulin layer (Amosova, 1979). The outer epicuticle is the first part of each new cuticle to be secreted and covers almost the entire integumentary surface of the acarine. When newly secreted it is resistant to enzymes in the moulting fluid. Chitin is said to be absent from the epicuticle but this needs confirmation (Neville, 1975). The presence of vertically oriented electron-translucent, non-staining microfibrils in the outer epicuticular layer of *Boophilus microplus* may indicate the presence of chitin microfibrils as in insects (Filshie, 1976). The secretion-layer (Sekretschicht of Alberti *et al.*, 1981) overlying the epicuticle is typically formed of two components. The outermost of these, when present (as in the Argasoidea), is the cement layer which is probably produced by the dermal glands and consists of tanned proteins and lipids. The function of this layer is to protect the underlying wax layer. The latter is considered to be secreted by way of pore canals which extend from the epidermis to the base of the epicuticle but do not penetrate it (Fig. 1.1, inset). The canals follow a spiral course through the helicoidally arranged lamellae of the cuticle. Distally, the pore canals are connected to the surface of the epicuticle by fine wax canals. Wax filaments emerge from the wax canals and are thought to act like a wick of a candle in transporting wax to the surface of the epicuticle (Neville, 1975). Some authors, for example Brody (1970) and Krantz (1978), consider the wax and cement layers to be part of the epicuticle but the distinction between the epicuticle and the secretion-layer has long been made in acarological literature as signified by the use of the term epiostracum for the former and tectostracum or cerotegument for the latter (Vitzthum, 1940–43; Grandjean, 1951a).

Hardening, often accompanied by darkening in colour, of the outer part of the procuticle is assumed to be due to orthoquinone tanning of proteins as in insects and involves the use of orthoquinones to form cross-linkages between free imino or amino groups of protein molecules. This produces a highly resistant and insoluble protein called sclerotin and the hardening process is referred to as sclerotization. The term 'chitinized' is often used to refer to various types of hard cuticle in arthropods but this terminology is misleading since the hardening of arthropod cuticle does not primarily involve chitin at all (Barrington, 1967). Hughes (1959) considers that some form of sulphur tanning, probably due to the preferential reaction of quinones and sulphydril groups, may occur in the 'white type' of cuticle of Acaridae. A sulphur-tanned keratin-like substance has been found in the eggshells of *Panonychus ulmi* by Beament (1951). Exocuticle is fully tanned and stiff but the endocuticle is untanned and pliant. A mesocuticle, ultrastructurally resembling endocuticle in texture but more electron dense, may also be defined. It lies between the endo- and exocuticle and is impregnated with stabilized lipid as is the exocuticle but not the endocuticle. The mesocuticle is not tanned and stains with acid fuschin.

Protein sclerotization, although the most common process, is not the only method of cuticular hardening in the Oribatida. In representatives of the Ptyctima and Enarthronota, cuticular hardening by the deposition of calcium salts has been demonstrated (R.A. Norton, personal communication). Calcite ($CaCO_3$) appears to be the mineral involved in the Ptyctima and whewellite (CaC_2H_2O), a form of calcium oxalate, in the Enarthronota. The whewellite is probably derived from crystals originally precipitated by the fungal food of the mites. Chambers and canals (?pore canals) in the procuticle are possible centres of calcification and in the Enarthronota have been found to be filled with electron-opaque matter (Alberti *et al.*, 1981). Calcification produces a rigid, inflexible, birefringent cuticle. This appears to be the only record of cuticular hardening by calcification in the Arachnida.

The degree of sclerotization of the body varies considerably and ranges from forms without apparent areas of differentiaton of the cuticle through those with isolated shields or plates widely separated by flexible cuticle to those in which the body is extensively armoured (Fig. 1.3E–H). Immature forms are usually less sclerotized than adults, and there is a tendency during ontogenetic development for a progressive fusion of shields to form more extensive areas of sclerotization (Fig. 1.3A–D).

The structure of the cuticle shows considerable diversity in the degree of development and form of its constituent layers and has been studied extensively by Alberti *et al.* (1981). In general, the epicuticle is uniform whereas the procuticle and the secretory layer (cerotegument) show functional adaptations. In forms ingesting large quantities of fluid such as haematophagous parasites, there is usually a reduction in the areas of sclerotized cuticle and an extensive folding of the epicuticle of the soft cuticle of the body so that its surface appears corrugated. The unfolding of the epicuticular layer together with the stretching of the basal layers of the cuticle allow for the rapid distension of the body. The nonsclerotized cuticle connecting sclerotized areas (intersegmental cuticle) consists of epicuticle and lamellate endocuticle with pore canals.

The synthesis of new cuticle occurs during the slow phase of the feeding cycle in larvae, nymphs and females of ixodoid ticks (Lees, 1952; Balashov, 1960). The growth in the thickness of the cuticle takes place chiefly in the endocuticle. On the last day of the 6–7-day feeding period

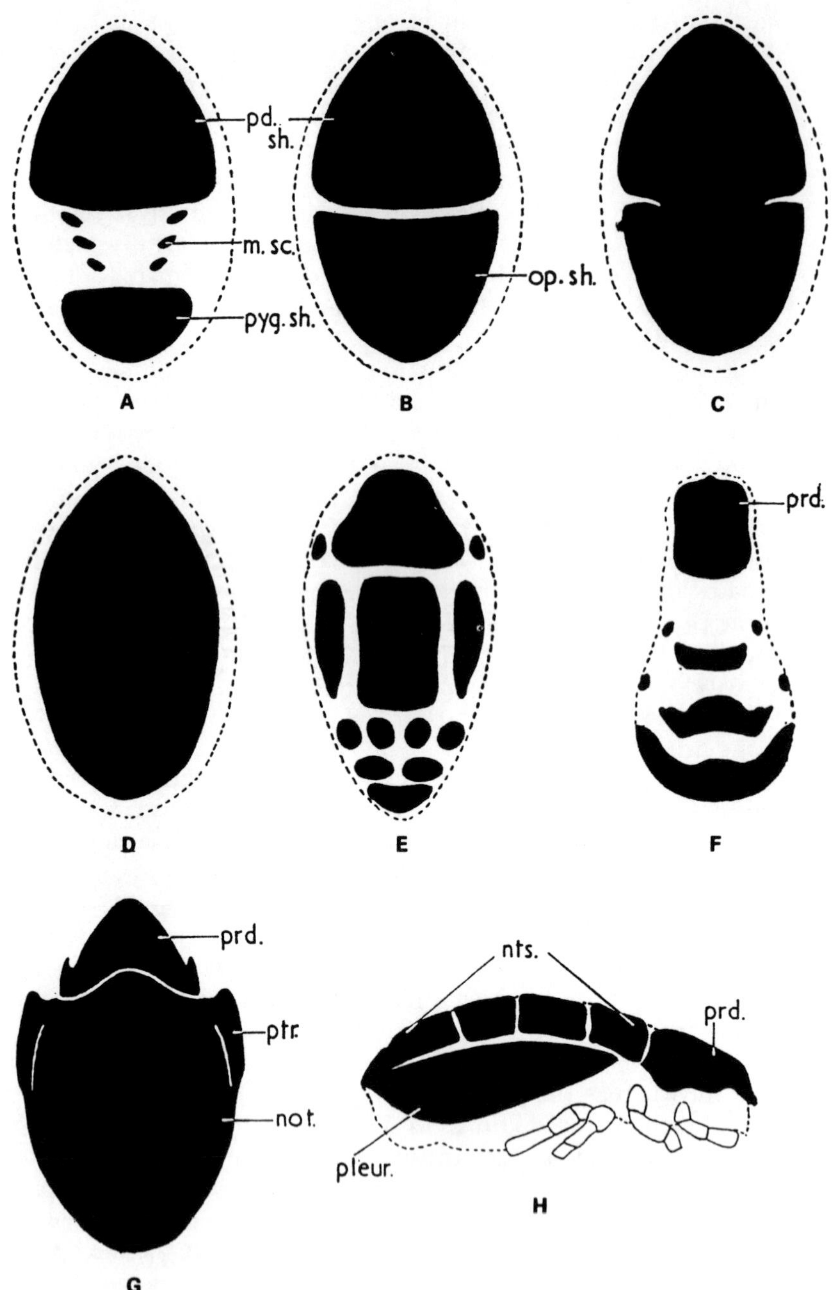

Fig. 1.3. A range of types of dorsal sclerotization in mites. Black areas denote tanned cuticle. **A–D**, protonymph (**A**) and adults (**B–D**) of a parasitid mite (Mesostigmata); **E**, dorsal sclerites of a stigmaeid mite (Prostigmata); **F–H**, adults of some Oribatida – **F**, Palaeostomata; **G**, Circumdehiscentiae; **H**, Enarthronota. *m. sc.* = mesonotal scutellum; *not.* = notogaster; *nts.* = notaspis; *op. sh.* = opisthonotal shield; *pd. sh.* = pronotal shield; *pleur.* = pleuraspis; *prd.* = prodorsum; *ptr.* = pteromorpha; *pyg. sh.* = pygidial shield.

in *Ixodes ricinus*, for example, there is a rapid distension of the idiosoma which coincides with the cessation of cuticle synthesis. The stretching of the cuticle to accommodate rapid engorgement results in the thickness of the procuticle becoming less than half that of the partially fed tick. Cuticular synthesis apparently only occurs in the larvae of argasoid ticks which resemble ixodoids in being slow feeders. According to Arthur (1962), in the rapid feeding nymph and adult argasoid no cuticular synthesis takes place.

The synthesis of new cuticle in a single active instar of an arthropod was termed neosomy by Audy *et al.* (1963) and the phenomenon is also known to occur in some larval Trombiculidae and Hydrachnidae and some females of the Pyemotidae, Pterygosomatidae and Teinocoptidae in which the body becomes greatly distended either as the result of feeding or of viviparity (Audy *et al.*, 1972). Neosomy is seen at its most spectacular in the larva of the trombiculid *Vatacarus ipoides*, which infests the air sacs of amphibious sea snakes of the genus *Laticauda*. During engorgement the transverse rows of body setae become borne on conspicuous papillae to the extent that the engorged larva bears little resemblance to the unengorged form. The papillae enable the larvae to move in the mucus covering the air sacs. New external cuticular structures such as the papillae of *Vatacarus* are referred to as neosomules.

Heavily armoured species of the Oribatida and Mesostigmata have a uniformly thick cuticle with the endo- and exocuticular areas of the procuticle clearly defined and having numerous parallel lamellae. Unusual modifications of the cuticle occur in the Enarthronota, in which the procuticle is transformed into a meshwork, and in some members of the Trombidioidea, which have a unique system of cuticle-derived fibrils situated within the epidermis (Fig. 1.4A,B). In those water mites in which the cuticle is flexible and relatively thin, such as *Hydrodroma* and *Protzia*, the procuticle is largely constructed of macrofibrils which show cross-striation and resemble collagen fibrils (Alberti *et al.*, 1981). Some species of Tetranychidae have the ridges of the cuticle provided with numerous lobes which are considered to be evaporative structures in feeding mites (absent in diapausing females) or to function in increasing the rigidity of the cuticle (Alberti & Crooker, 1985).

The surface of sclerotized areas of the cuticle can be variously ornamented. In the Mesostigmata and Ixodida in particular, the most common form encountered is the reticulate-areolate pattern which is considered to represent the outlines of the surface of epidermal cells that lie beneath the cuticle (Wigglesworth, 1961). Other types of surface are described as foveate, punctate, granulate, rugose and tuberculate (Fig. 1.5). The terminology for these surface structures has not been standardized in descriptive morphology of the Acari and the terms are used rather loosely. The system used for describing the surfaces of seeds by Murley (1951) appears to be applicable and if adopted could avoid the confusion that exists at present.

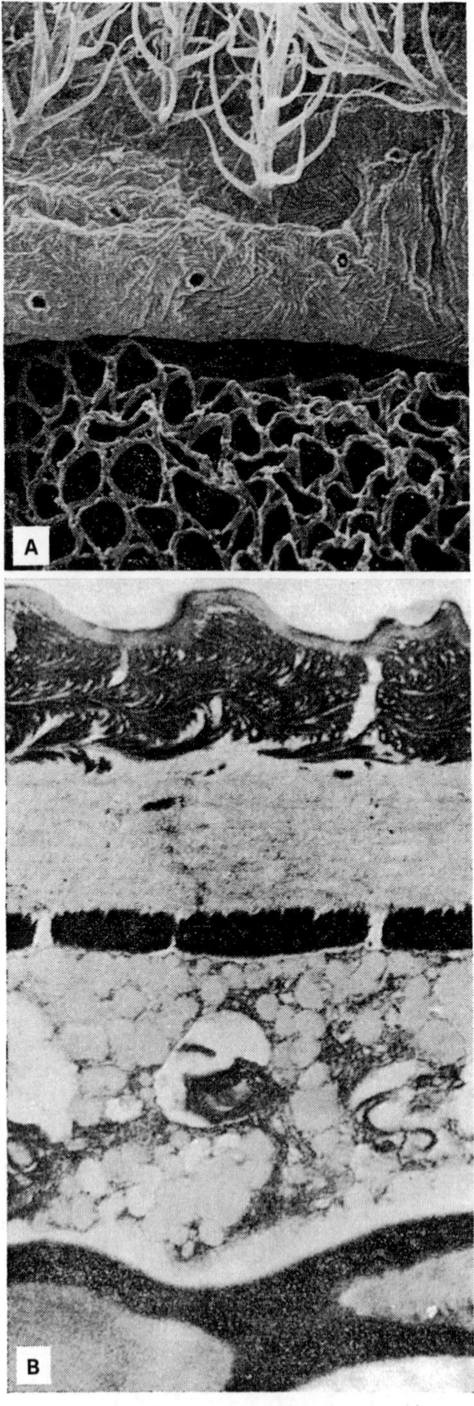

Fig. 1.4. *Allothrombium fuliginosum* (Prostigmata). **A**, cuticle after KOH and ultrasonic treatment showing the cuticle-derived network of fibrils; **B**, section of the integument with the cuticle-derived network shown at the bottom of the micrograph. (After Alberti *et al.*, 1981.)

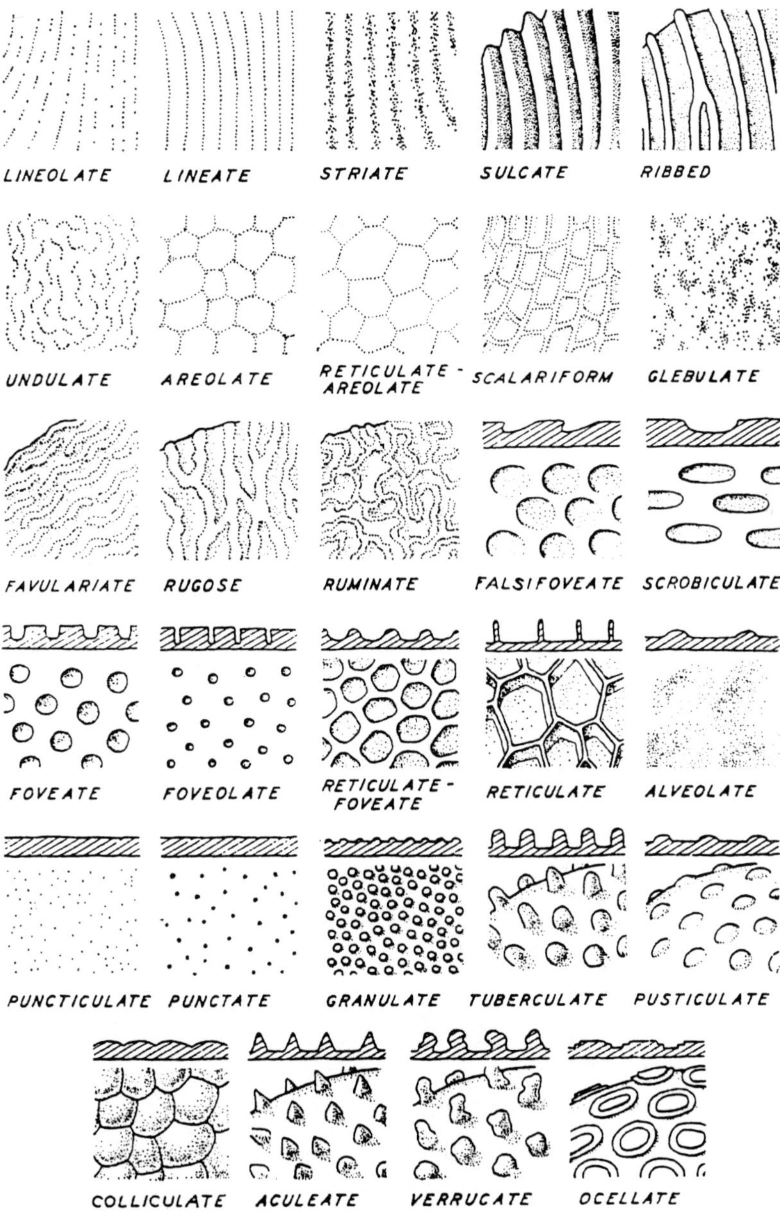

Fig. 1.5. Terminology for different types of surfaces of seeds which can be used for descriptions of cuticular ornamentation in the Acari. (From Murley, 1951.)

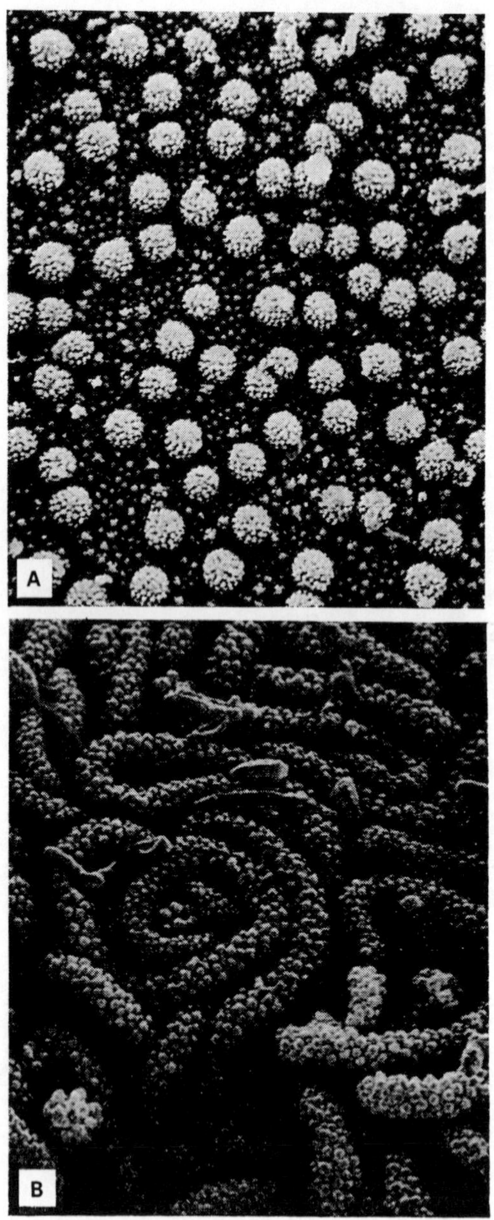

Fig. 1.6. SE micrographs of the secretion layer (cerotegument) of the oribatids *Damaeus onustus* (**A**) and *Belba pulverulenta* (**B**). (After Alberti *et al.*, 1981.)

A highly developed secretion-layer or cerotegument occurs in the Oribatida (Fig. 1.6A,B), Mesostigmata (*Epicrius*) and Prostigmata (*Caeculus, Bryobia* and *Halacarellus*) and gives the body a characteristic ornamentation. It is particularly thick in *Plesiodamaeus cratifer* and in some species of *Pelops*. Differences in the distribution of the cerotegument over the body surface are evident in the Euoribatida and it is possible that secretions of different glands such as the single pair of podosomatic glands opening under the pteromorphae (Woodring & Cook, 1962a) and those opening by way of the areae porosae of the notogaster may be involved in its production, Alberti *et al.* (1981) state that the cerotegument of *Damaeus, Belba, Hydrozetes, Galumna*, etc. is strongly differentiated and consists of several components (up to four or five in *Galumna*). The cerotegument of a number of species of Oribatida has been shown to have a respiratory function by trapping an air film and forming a plastron (Crowe & Magnus, 1974; Pugh *et al.*, 1987c). This aspect of its structure and function is dealt with in the chapter on respiration.

Of particular interest is the development of cuticular outgrowths and depressions of the cuticle for the protection of the legs which occurs among the Oribatida and Uropodina. The prodorsum of the former may be provided with blade or flange-like processes (lamellae and tutoria) which protect the distal parts of the anterior legs when these are withdrawn to the body. The antero-lateral region of the notogaster may be produced into immovable or movable wing-like expansions (pteromorphae) which in their well-developed state bend ventrally to protect the posterior legs. In the Uropodina, on the other hand, a platform-like structure, the scabellum, is often present below the vertex to accommodate the first pair of legs and ventral depressions or pedofossae positioned external to legs II to IV provide protection for these appendages. Leg joints are often protected by tecta, flange-like extensions of a podomere which overhang the arthrodial membrane. Invaginations and inflexions of the cuticle to form apodemes for muscle attachment commonly occur as part of the endoskeleton.

Dermal glands

Certain cells of the epidermis are specialized for specific functions. Amongst the most conspicuous are those forming the dermal gland and its duct, which opens on the surface of the cuticle via a pore. In ticks, each gland typically consists of a pair of large glandular cells and a few smaller cells forming the glandular duct (Fig. 1.7). Its activity is restricted to the period between the beginning of feeding and apolysis (or a short time after). Dermal glands in argasoid ticks are of one type and secrete the cement layer of the cuticle a few hours after moulting. In ixodoids, where the cement layer is lacking, the products of the dermal glands according to Lees (1947) 'appear to add nothing of functional significance

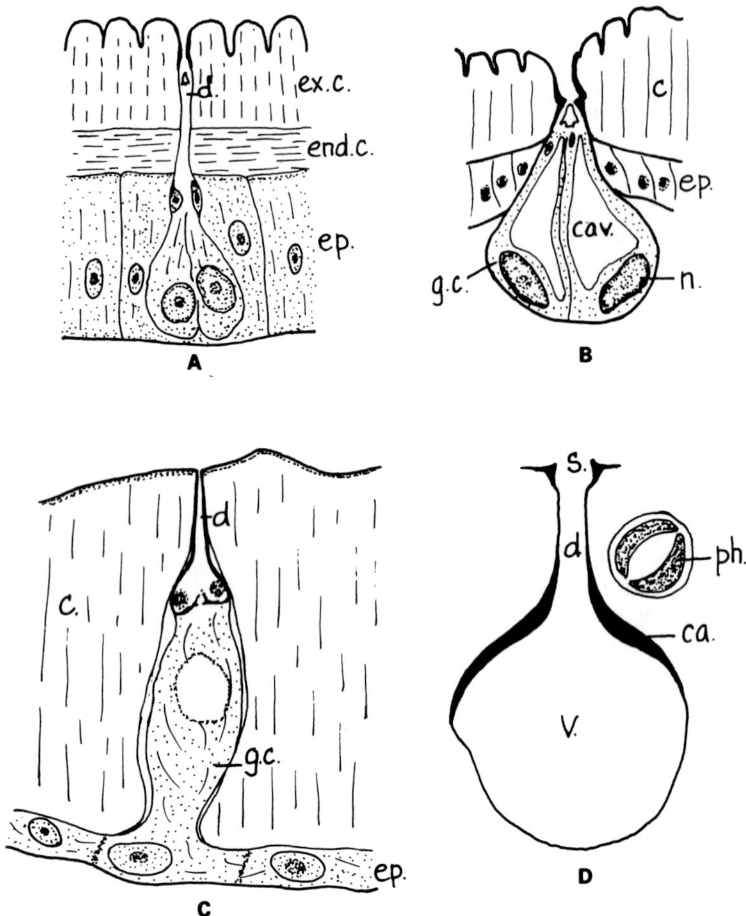

Fig. 1.7. A–B, type I (**A**) and type II (**B**) dermal glands in ixodid ticks (after Filippova, 1985 and Balshov, 1968); **C**, dermal gland in *Allothyrus australasiae* (Holothyrida); **D**, diagrammatic representation of a simple dermal gland in the Mesostigmata (based on Athias-Henriot, 1969). *c.* = cuticle; *ca.* = calyx; *cav.* = cavity of gland cell; *d.* = duct of gland; *end. c.* = endocuticle; *ep.* = epidermis; *ex. c.* = exocuticle; *n.* = nucleus of gland cell, *ph.* = valves at mouth of duct ('phragmides'); *s.* = solenostome; *v.* vesicle.

to the substance of the cuticle', whereas Balashov (1968) suggests that they may participate in the formation of the wax layer. Three types of dermal glands are recognized by Lees in *Dermacentor andersoni* and each has a different cycle of activity. The large glands, type A (=type II of Balashov, 1968), in nymphs attain their maximum size at the end of engorgement. They show degenerative changes just before the onset of moulting when yellow greasy droplets appear in the cytoplasm, which becomes yellowish and reticulated. The gland undergoes involution and the yellow greasy residue, the end-product of the degenerating gland cells, passes through the duct to the surface of the cuticle. The greasy material is completely insoluble in cold chloroform and xylol and only slightly soluble in hot chloroform. The smaller dermal glands, type B (= type I of Balashov, 1968), begin to accumulate cytoplasm during feeding but undergo rapid involution after the tick has dropped off the host and within a few days have disappeared whereas type A still persists. The products of degenerating type B glands also pass onto the surface of the cuticle. Lees (1947) found that a group of about eight nuclei remaining near the old duct of a type B gland stained more deeply with haematoxylin than other epidermal nuclei. Soon after apolysis these nuclei remain grouped around a new cuticular duct that is forming and subsequently commence to secrete cytoplasm. Thus, each group of nuclei plays a part in forming one gland cell, which Lees refers to as type C. The round or pear-shaped multinucleate gland cells attain their maximum size before the moulting fluid is secreted and it is suggested that they may secrete, or participate in secreting, the moulting fluid. Only one type of dermal gland, resembling type B, is present in *Ixodes ricinus* but the glands in the nymph are not provided with ducts so that their degenerative products are absorbed. However, in the adult the characteristic yellow greasy material is 'thrown out on to the surface of the cuticle'. Yalvac (1939) and Schulze (1942) considered the dermal glands in the hard ticks to function as sense organs with an additional glandular function, and referred to them as 'Drüsensinnesorgane'. Glands type A to C are without sensory innervation and the sensory function attributed to them is rejected by Lees (1948).

Schulze (1942) attributed a sensory function to duct-like structures penetrating the cuticle of ticks and called them *sensilla hastiformia*. They formed one of the three types of cuticular structures which he referred to as 'krobylophores'. Each duct is a wide-based, open canal into which the cytoplasm of an epidermal cell projects. The cytoplasm extends only about half the length of the canal, which then continues to the surface as a narrow tube. The associated nuclei lie in the epidermis. No innervation of the structure was found by Lees (1948) or Foelix and Axtell (1972) and it is considered to be a gland. Structures with the same characteristic configuration also occur in the Notostigmata and Holothyrida (Fig. 1.7C).

Dermal or cuticular glands appear to be common in the Mesostigmata and have been described in the Dermanyssina and Sejina (Athias-Henriot, 1970a; Krantz & Redmond, 1987) and in the Uropodina (Woodring & Galbraith, 1976). They occur on the body and legs. The secretion of the glands in *Uroactinia agitans* is completely dissolved in alcohol–xylol and is probably a lipid. Our knowledge of the dermal glands in the Dermanyssina, Parasitina and Sejina is based largely on optical sections of macerated specimens and, therefore, refers only to sclerotized parts of the gland complex. Athias-Henriot (1970a) refers to the cuticular glands as 'crobylophores' and states that they are characterized by having a 'calyx', i.e. a lightly sclerotized corolla encircling the junction between the cell and the duct of the gland (Fig. 1.7D). The duct, or ducts in the case of compound glands, may open to the exterior in a depression or infundibulum and the opening (solenostome) may be closed by a pair of valves ('phragmides') as in the Sejina. The distribution of 'crobylophores' in the Mesostigmata is referred to as adenotaxy and Athias-Henriot (1979a) recognizes two basic conditions; *protoadenic*, in which the glands are idionymic, i.e. they can be homologized with corresponding glands in other species of the same group, and *euneoadenic*, in which each gland with a calyx is not idionymic and their number is always greater than in the protoadenic type. The former condition occurs in the Dermanyssina, Parasitina and Zerconina and the latter in the Antennophorina, Epicriina, Heterozerconina and Uropodina. No glands having the characteristic form of the 'sensilla hastiformia' appear to have been described, to date, in the Mesostigmata but it is likely that they represent a primitive form of gland in the Anactinotrichida and will probably be found to occur in the more primitive members of the order.

A pair of lateral opisthosomatic glands (=expulsory vesicles, oil glands, latero-abdominal glands, latero-opisthosomatic glands) are conspicuous structures in the Astigmata and in the Oribatida (except the Palaeosomata and Enarthronota). They contain a highly refractive liquid which may be colourless, yellow, brown or, as in *Australoglyphus geniculatus*, red. Each gland in *Dermatophagoides farinae* comprises a single cell with a cuticle-lined lumen (Brody & Wharton, 1970). Its duct opens by a crescent-shaped orifice provided with a cuticular flap that is probably capable of opening and closing the gland. Contractile elements or muscle fibres associated with the gland are considered to be involved in the intermittent expulsion of the gland contents, which comprise both volatile and non-volatile components. Michael (1901) suggested that the secretion of the lateral opisthosomatic glands served to lubricate the surface of the mite while Brody and Wharton (1970) favoured a repugnatorial function. More recently, the glands in acaroid mites are considered to be the site of secretion of alarm, aggregation and sex pheromones (Kuwahara *et al.*, 1980; My-Yen *et al.*, 1980; Leal *et al.*, 1989).

The small paired areas of porose cuticle ('areae porosae') occurring on the notogaster of adult Euoribatida and thought at one time to be respiratory structures are openings of dermal glands whose products may contribute to the formation of the secretory layer overlying the epicuticle (Woodring & Cook, 1962a; Alberti *et al.*, 1981). In immature euoribatids, the dermal glands are much smaller and more numerous than in the adult and are not precursors of adult dermal glands. Some appear to be associated with setal sockets and others open to the exterior on very small, lightly sclerotized porose areas.

Three other types of specialized epidermal cells occurring in the integument of the Acari form the cuticular structures of a wide range of sensory organules. These so-called enveloping cells of the organule comprise the trichogen, tormogen and thecogen cells and are discussed in Chapter 3, dealing with sense organs.

Coloration

Many mites and ticks show coloration of the body. In pale species, this is often due to pigments of ingested food in the alimentary tract such as the green of phytophagous mites and the reddish-brown of haematophagous forms. This is referred to as extrinsic or exogenous coloration. In *Ixodes ricinus*, the colour of the contents of the mid-gut and its diverticula changes from pale straw to slate grey during the feeding cycle. Some pale and pigmented prostigmatic forms have a distinct white median stripe in the posterior half of the idiosoma indicating the contents of the post-ventricular mid-gut. In addition to this type of coloration, many forms are brightly coloured as the result of pigments located in the integument (intrinsic or endogenous coloration). The pigments are lodged in the cells of the epidermis and they may be red, orange, yellow, green or purple. The most brightly coloured forms occur among the Trombidiidae, Bdellidae and the water mites. Little is known of the chemistry of the pigments but the red pigmentation in water mites of the genera *Eylais*, *Hydrodroma* and *Limnochares* is due to carotenoids (Czezuga & Czerpak, 1968) and the orange and red pigments of diapausing females of many Tetranychidae are hydroxyketo-carotenoids (Veerman, 1974).

Grandjean (1951d) described a metallic coloration of the cuticle of a species of *Carabodes* when viewed in reflected light. The ridges on the notogaster were blue-green and the intervals between them appeared a pale rose. The colours were lost when the mite was submerged in liquids. The coloration was attributed to a physico-chemical peculiarity of the cuticle and considered to be similar in origin to the metallic coloration of the ground beetle, *Carabus*.

Moulting

The moulting process in the Acari has not been studied extensively. Although no glands analogous to the Y-organ of the Crustacea or the prothoracic glands of the Insecta have been identified with certainty, the results of experimental work with ecdysteroid hormones indicate that these are involved in the initiation and control of the moulting process (Solomon *et al.*, 1982). Ellis and Obenchain (1984) suggest that cells associated with the fat body are the natural source of ecdysteroids in ticks.

The moulting cycle appears to resemble that in insects and has been studied in some aspects, in the mite *Sancassania bohari* by Woodring (1969) and in detail in the tick *Hyalomma asiaticum* by Balashov (1963) and Amosova (1979). The onset of moulting in *H. asiaticum* is indicated by epidermal mitosis, which spreads from the anterior to the posterior end of the body. At the same time, the epidermis begins to separate from the cuticle during the process known as apolysis (Jenkin & Hinton, 1966). The separation is brought about by lysis of the procuticular

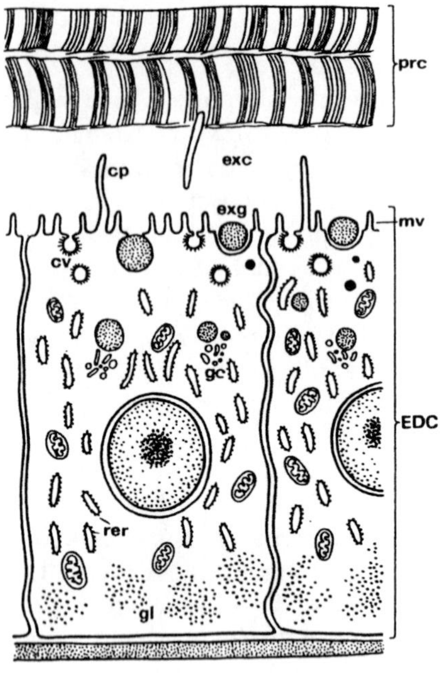
prc
cp
exc
exg
mv
cv
gc
EDC
rer
gl

A

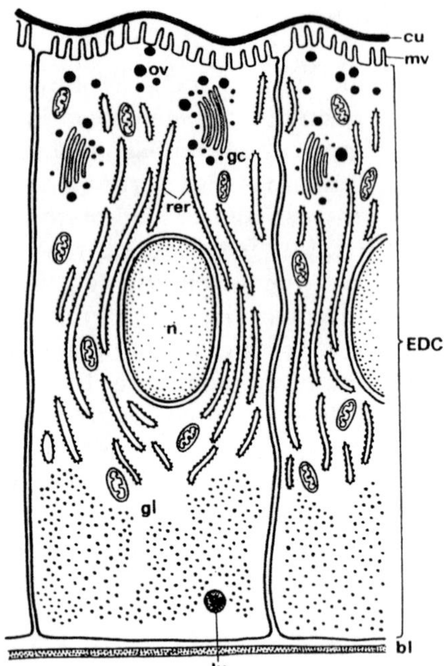
cu
mv
ov
gc
rer
n
EDC
gl
bl
ly

B

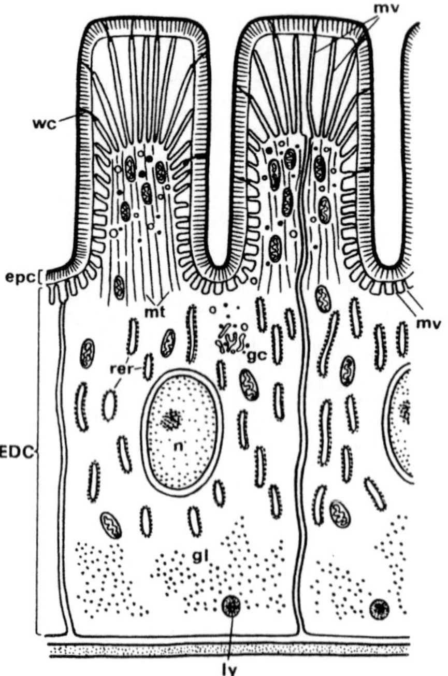
mv
wc
epc
mt
mv
rer
gc
EDC
n
gl
ly

C

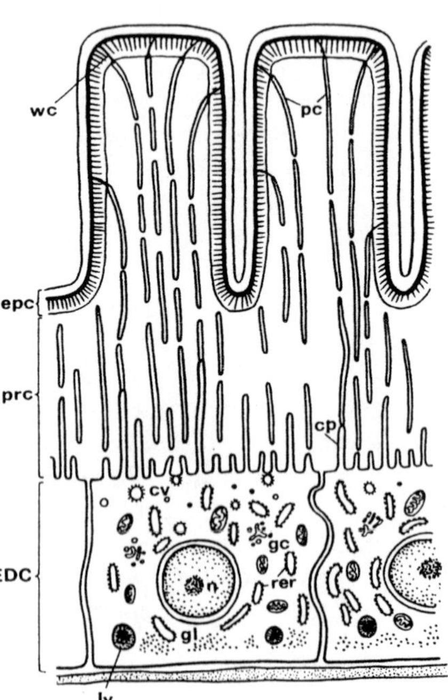
wc
pc
epc
prc
cp
cv
EDC
gc
rer
n
gl
ly

D

lamellae nearest to the epidermis and results in the formation of an exuvial cavity or space (Fig. 1.8A). The secretion of the outer epicuticular layer by the epidermal cells precedes the activation of enzymes of the moulting fluid, which occupies the exuvial cavity. Simultaneously with the digestion of the old procuticle or the endocuticle, there is an uptake of the lysed components of the cuticle by the epidermal cells. Formation of new cuticle commences with the deposition of the outer epicuticle (Fig. 1.8B). The material first appears as discrete, dense particles on the plasma membrane plaques of the tips of the microvilli, but these then gradually elongate to form a continuous layer. During the deposition of the inner epicuticle, the epidermal cells elongate and show high synthetic activity. Coincidental with the deposition of the epicuticle folds, which are so characteristic of extensible cuticle, begin forming. During the first stage of this process, regular invaginations of the apical surface of the epidermis covered by epicuticle are seen but later the folds assume their typical form (Fig. 1.8C). The microvilli of the epidermal cells play an important role in the formation of the folds. Procuticle is laid down below the epicuticle and during this phase the epidermal cells decrease in size (Fig. 1.8D). Cytoplasmic projections at the surface of the epidermal cells participate in the formation of the pore canals. The sclerotization and differentiation of the procuticle begins after ecdysis, i.e. after the shedding of the remnants of the old cuticle, the exuvium.

During the latter part of the life of an instar, the body may become distended followed by the cessation of feeding and locomotion. This inactive state is made up of a pre-apolytic period when the cuticle is still attached to the epidermal layer (cuticular phase) and a post-apolytic period when the cuticle becomes detached from the epidermal layer (exuvial phase) (Hinton, 1971a). The exuvial phase is always a pharate phase, i.e. the developing instar is enveloped by the cuticle of the previous instar. A quiescent episode is a feature of the developmental cycle of most Acari with the apparent exception of the Mesostigmata and argasid ticks (Rockett & Woodring, 1972). In the Oribatida, for example, about one-third of the total developmental time may be spent in this state (Luxton, 1981) with the period of pre-ecdysial inactivity less in the early than in the later instars (Lebrun, 1970).

During the moulting cycle in the Astigmata and ixodoid ticks, leg tissue progressively dedifferentiates and is regressed into a mound of

Fig. 1.8. Schematic representation of the structural changes in the integument of an ixodoid nymph during the moulting cycle. **A**, procuticle separation from the epidermis (apolysis); **B**, formation of the outer epicuticle; **C**, deposition of inner epicuticle and formation of cuticular folds by invaginations of the apical surface of the epidermis; **D**, deposition of procuticle and formation of pore canals (after Amosova, 1979). *cp.* = cytoplasmic process; *cu.* = outer epicuticle; *cv.* = coated vesicle; *exc.* = exuvial space; *EDC.* = epidermal cell; *epc.* = epicuticle; *exg.* = ecdysial granule; *gc.* = Golgi complex; *gl.* = glycogen; *ly.* = lysosome; *mt.* = microtubules; *mv.* = microvilli; *n.* = nucleus; *ov.* = dense vesicles (contents are precursors of protein epicuticle); *pro.* = procuticle; *rer.* = rough endoplasmic reticulum; *wc.* = wax canal.

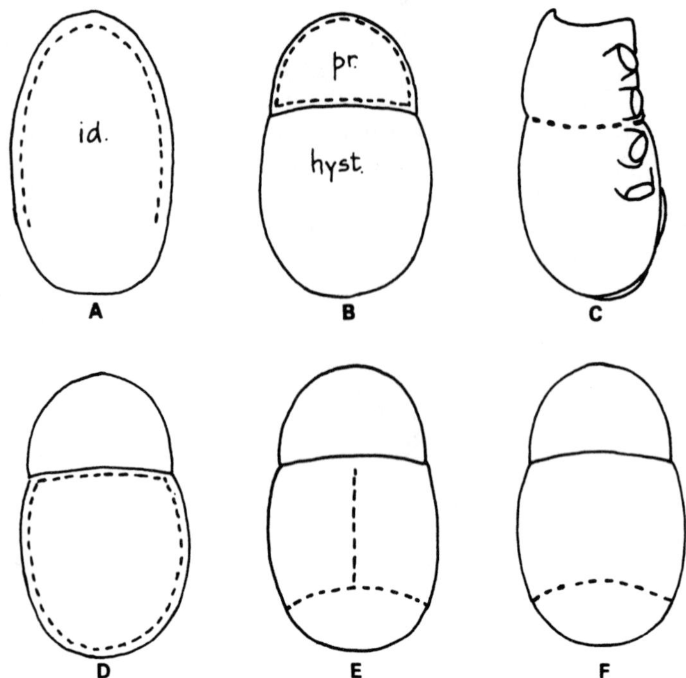

Fig. 1.9. Types of ecdysial cleavage (dehiscence) in the Acari (lines of dehiscence indicated by broken lines). **A**, anterior idiosomatic as in the Parasitiformes; **B**, prodorsal; **C**, amphidehiscence; **D**, circumdehiscence; **E**, inverted T; **F**, transverse. *hyst.* = hysterosoma; *id.* = idiosoma; *pr.* = prodorsum.

tissue ('coxal bud') in the coxal region. Subsequent elongation and differentiation of the coxal bud is in a ventro-median direction and this results in the limb of the succeeding instar developing in the ventral exuvial space. This contrasts with the condition in the Prostigmata and Oribatida in which the limbs of the next instar form within the old limb cavities and are withdrawn into the ventral exuvial space before ecdysis. During apolysis in the Mesostigmata and argasid ticks, however, the acarine usually remains active, or if appearing inactive normally responds to touch up to the time of ecdysis. The limbs apolyse in place as in the Prostigmata and Oribatida but are not withdrawn until ecdysis (Rockett & Woodring, 1972).

The emergence of an instar from the cuticle of the preceding instar is facilitated by the splitting of the old cuticle along a definite line (or lines) of weakness known as the line of dehiscence or the ecdysial cleavage line. The rupture of the cuticle is largely achieved by hydrostatic pressure resulting from muscular contractions of the body, although in some species putative egg bursters may also play a part (Grandjean, 1954a; Evans *et al.*, 1963). The initial cleavage of the old cuticle in the Mesostigmata and Ixodida occurs in the anterior region of the idiosoma

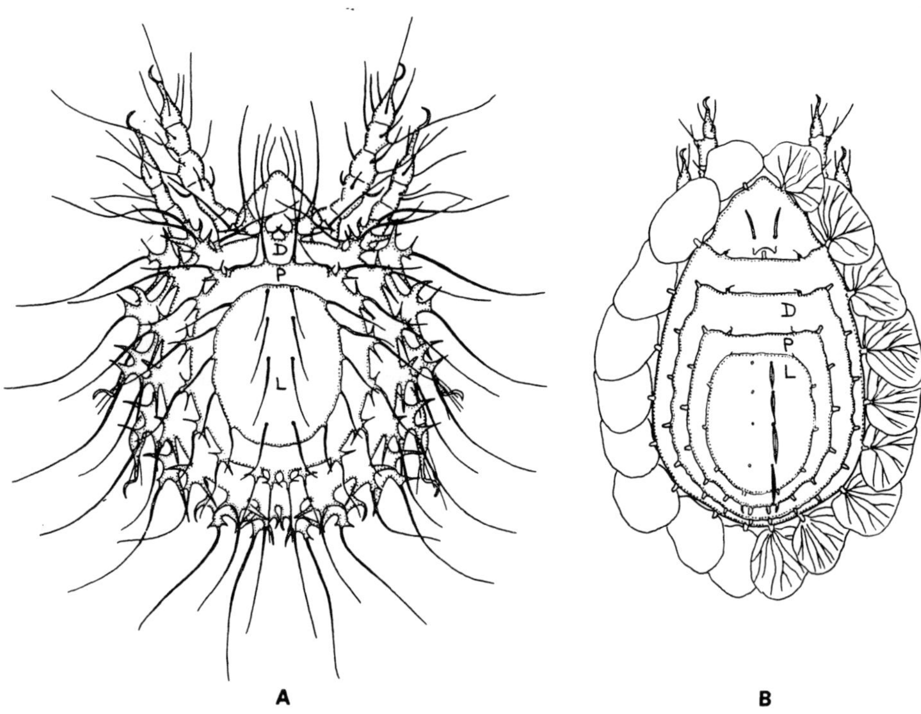

Fig. 1.10. A, B, eupherederm condition in the tritonymphs of *Cepheus* (**A**) and *Ommatocepheus* (**B**). *L.P.D.* = larval, protonymphal and deuteronymphal scalps, respectively.

and then progresses in a posterior direction along the lateral margins (Fig. 1.9A). The extent of the lateral prolongation of the cleavage line may vary between instars as, for example, in *Uropoda orbicularis* in which the lines do not extend posterior to the legs for the eclosion of the deuteronymph but extend to the posterior part of the idiosoma for the emergence of the adult so that the dorsal disc has only a narrow connection with the remainder of the exuvium. In the Actinotrichida, splitting of the cuticle takes place in the anterior (proterodehiscence) or posterior (hysterodehiscence) part of the body. Proterodehiscence occurs in the Prostigmata and may take the form of prodorsal dehiscence in which the cleavage line follows the abjugal furrow and the exuvium of the prodorsum is partly or completely excised (Fig. 1.9B), or amphidehiscence in which the cleavage line is around the girth of the mite between the second and third pair of legs (Fig. 1.9C). The line of dehiscence in actinotrichid regressive stases extends laterally from the anterior margin of the body. Hysterodehiscence is characteristic of the Sarcoptiformes. In the Astigmata and Euoribatida, the initial split occurs at the posterior end of the body and progresses along the lateral margins of the opisthosoma and, in the Euoribatida, often also along its anterior margin. This type of ecdysial cleavage is referred to as circumdehiscence

(Fig. 1.9D). In the Archoribatida, on the other hand, the line of dehiscence may be transverse and restricted to the posterodorsal region of the opisthosoma (Fig. 1,9E) or inverted T-shape in which a median split runs anteriorly from a posterodorsal transverse slit (Fig. 1.9F). Rarely, as in the larva of *Hypochthonius rufulus*, the line of dehiscence is T-shaped, i.e. the transverse component is anterior (Luxton, 1981). The emerging instar normally leaves the exuvium in a backward direction in hysterodehiscent forms and in an anterior direction in proterodehiscent species.

The occurrence of dehiscence has been most extensively studied in the Oribatida by Grandjean, who used it as a major criterion in his classification of the order (Grandjean, 1953). For example, the taxon Circumdehiscentiae (Euoribatida) contains species in which the line of dehiscence is circumgastric, i.e. around the margin of the opisthosoma. When the line is complete a portion of the old cuticle (the scalp) is separated from the remainder of the exuvium and may be carried on the back of the emerging instar. When the three nymphal instars, and in some cases also the adult, carry the scalp of the preceding instar in contact with their backs, the condition is referred to as *eupherederm* and occurs, for example, in the Liodidae and Belbidae (Fig. 1.10A,B). Thus, a eupherederm tritonymph will carry three scalps and the adult may transport four scalps, one larval and three nymphal. If the scalps carried by the nymphal instars are supported by dorsocentral setae at some distance from the cuticle of the dorsum, the condition is known as *apopherederm* and is characteristic of the Oribatellidae. In *opsiopherederm* forms, such as the Hermanniellidae, the nymphs are nude (without scalps) but the adult carries the scalp of the tritonymph. Finally, the forms in which a scalp is not retained or is not produced owing to the line of dehiscence being incomplete anteriorly are termed *apherederm*.

2 Segmentation, Musculature and Legs

Segmentation

The segments forming the soma of extant Arachnida are grouped into two tagmata, an anterior prosoma (=cephalothorax) and a posterior opisthosoma (=abdomen). The tagmata are either broadly joined (latigastric) as in Pseudoscorpiones and Opiliones or connected by a narrow pedicel (caulogastric) as in, for example, the Araneae and Amblypygi (Fig. 2.1A–C). In latigastric forms, the prosoma and opisthosoma are more or less distinctly separated by a disjugal furrow (*disj*) (Van der Hammen, 1989). The prosoma comprises six segments and an unsegmented precheliceral lobe including the acron. Each of the six segments (I–VI) bears a pair of appendages comprising the chelicerae (I), pedipalps (II) and four pairs of legs (III–VI). The tergites of the prosoma show varying degrees of fusion. They may form a single shield, carapace or peltidium, covering the entire dorsal surface of the prosoma (e.g. Araneae, Scorpiones, Amblypygi) or three scutal elements comprising a propeltidium covering the first four segments, a mesopeltidium of segment V and a metapeltidium of segment VI. A subdivision of the mesopeltidium (Palpigradi) and the metapeltidium (some Uropygi), each into two sclerotized elements, may also occur (Fig. 2.1D). This reduction in the sclerotization of the dorsal regions of segments V and VI allows for greater flexibility of the soma in the metapodosomatic region. In the Acari and also the Ricinulei, the appendages of the cheliceral and pedipalpal segments form a trophic-sensory unit termed the gnathosoma (=capitulum), which is movably articulated to the remainder of the body, the idiosoma (Fig. 2.1D). Further, the fourth pair of legs is generally suppressed or reduced to vestiges in the larval phase, which is hexapod as opposed to the octopod nymphal and adult phases.

The segmentation of the arachnid opisthosoma is variable, with the maximum number of 13 segments occurring in the Scorpiones (Millot, 1949). The opisthosoma in many of the groups shows external evidence of segmentation by the presence, in each segment, of a dorsal tergite and

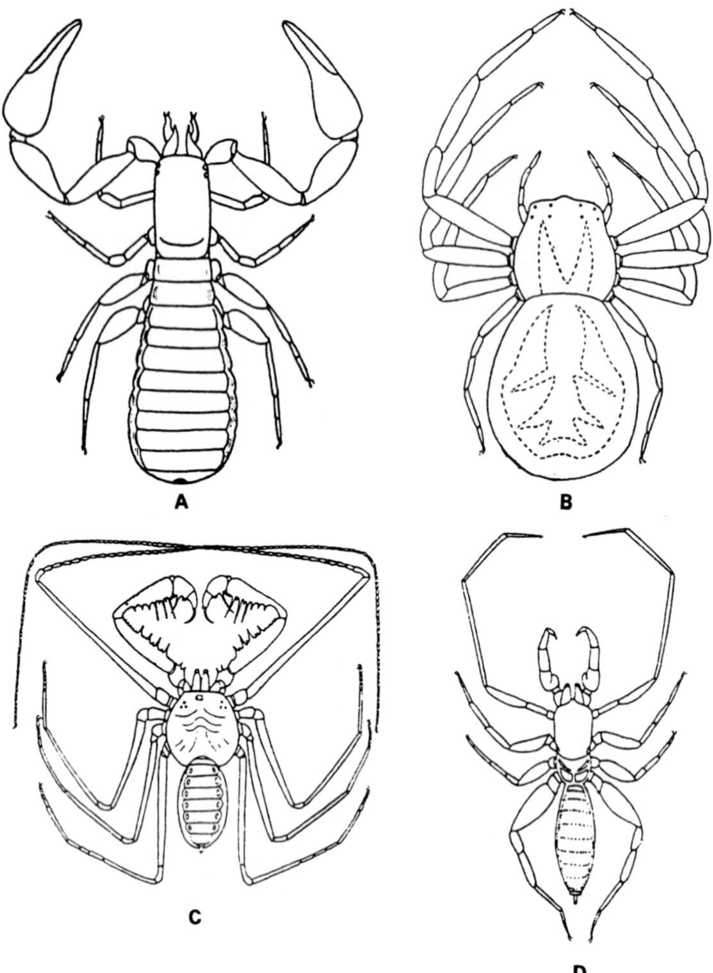

Fig. 2.1. Representatives of caulogastric and latigastric Arachnida. **A**, Pseudoscorpiones (=Chelonethida); **B**, Araneae; **C**, Amblypygi; **D**, Uropygi.

ventral sternite but there is a tendency in some taxa, for example the Araneae and Acari, for such conspicuous evidence to be restricted to some taxa or to be lacking. In the case of the Acari, the distribution of cuticular organules such as setae and slit organs (lyrifissures, poroids, cupules) are often used as indicators of segmentation. The number of segments incorporated in the opisthosomatic region of the Acari is a continual source of debate amongst acarologists and this is not surprising when much of the evidence is conjectural. Embryological data are few but in those taxa where metamerism of the opisthosoma occurs at some stage of development, for example ticks (Aeschlimann, 1958; Anderson, 1973) and some Tarsonemina (Brucker, 1900; Reuter, 1909), five or six segments have been reported, the higher number probably incorporating the telson which is considered by some authors to be a primary germinal disc 'acting perhaps as a growth zone for opisthosomal segments' (Aeschlimann, 1984). The same complement of opisthosomatic segments, based on tergal, chaetotactic and other cuticular criteria, appears to be characteristic of larval actinotrichid mites and may possibly be a derived or apomorphic state for the Acari as a whole (Grandjean, 1939a; Van der Hammen, 1970, 1989; Sitnikova, 1978; Lindquist, 1984).

The Arachnida in general are epimorphic, i.e. the larva emerging from the egg possesses the definitive number of segments, but anamorphosis is considered to occur in the Actinotrichida where up to three segments may be added to the posterior opisthosoma during the nymphal phase. The three segments are the adanal, anal and peranal which appear, respectively, in the protonymph, deuteronymph and tritonymph. In the more derived taxa, anamorphosis may be suppressed, as in the Tarsonemina and Tetranychoidea, or reduced by the retention of the protonymphal (some Eupodoidea) or deuteronymphal (many Oribatida) complement of segments in succeeding instars. Whether the increase in the number of opisthosomatic segments during ontogeny in the Actinotrichida is actually due to anamorphosis (the increase in the number of segments by the splitting of the pre-anal zone of proliferation) or to some other method is not certain. Van der Hammen (1989) is of the opinion that the phenomenon is the result of the reappearance in the course of post-embryonic development of segments that had been suppressed during embryonic development, and he referred to the condition as hysteromorphosis.

The larvae of the majority of the Parasitiformes and Actinotrichida show a posteroventral curvature of the opisthosoma which is referred to as the caudal bend. This results in the location of the anus on the ventral side of the opisthosoma and is considered to lead to a reduction in body size (Sitnikova, 1978). The curvature is particularly marked in the anamorphic Sarcoptiformes in which the anal opening is large to facilitate the ejection of a faecal pellet resulting from the ingestion of solid food. In the more derivative groups of the Actinotrichida in which anamorphosis is suppressed or reduced, the relatively smaller anus of these forms, which ingest food in a liquid or finely particulate state, is usually located subterminally. The large anus in the Notostigmata is terminal and regarded as the retention of a primitive condition.

Anactinotrichida

Among the Anactinotrichida, the Notostigmata show some external evidence of segmentation but the number of segments comprising the opisthosoma is undecided; for example, 13 segments are defined by Van der Hammen (1970) and 11 by Sitnikova (1978). The total somatic complement of either 17 or 19 segments represents the largest number of

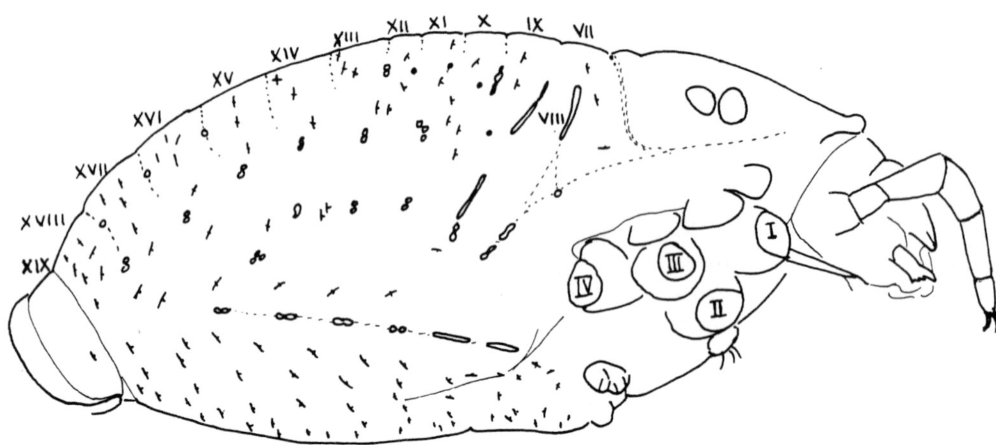

Fig. 2.2 Segmentation of the opisthomsoma of *Opilioacarus texanus* (Notostigmata) based on the concept of Van der Hammen (1970).

body segments that have been noted in the Acari (Fig. 2.2). Embryological evidence of only five segments (plus telson) in the opisthosoma of the Ixodida points to the presence of 11 somatic segments if, as is usually accepted, the Anactinotrichida are epimorphic. Raw (1957) interpreted the number of festoons along the posterior margin of the body in some hard ticks (Ixodoidea) to indicate the presence of six opisthosomatic segments but this criterion is not reliable.

In the absence of metamerism in the embryo or of definite external evidence of segmentation in the larva and succeeding instars, opisthosomatic segmentation in the Mesostigmata and Holothyrida is speculative. Some attempts have been made to determine the number of body segments on the basis of setal patterns and the distribution of other cuticular organules in the more derivative members of the Mesostigmata such as the Dermanyssina. The relatively low number of these organules on the body of the free-living members of this taxon make it possible, on a positional basis, to determine the ontogenetic development of the chaetotaxy. However, conclusions reached on somatic segmentation from setal patterns are varied and generally subjective. For example, Zachvatkin (1952) considered that 14 segments and a telson are represented in *Rhodacarus* of which eight segments occur in the opisthosoma, while Evans and Till (1979) were of the opinion that only 11 segments (plus telson) could be defined on the basis of chaetotaxy in the Dermanyssina although this number would be increased to 12 segments if, as assumed by Zachvatkin, a pregenital segment (VII) is present (Fig. 2.3A,B).

A characteristic feature of the dorsal sclerotization of the larva in the Mesostigmata is the presence of an anterior or pronotal (podonotal) shield associated with a remarkably constant chaetotaxy. In succeeding instars, the anterior shield remains separate or fuses with sclerotized elements of the dorsum of the opisthosoma, the opisthonotum, to form an entire dorsal shield. The pronotal shield is possibly homologous with the arachnid peltidium and may define the posterior margin of the prosoma. According to Van der Hammen (1970, 1989), however, a large part of the prodorsum has disappeared in the Mesostigmata and the pronotal shield incorporates dorsal elements of five opisthosomatic segments. This is highly speculative and the anterior division of the pronotal shield in some euedaphic Dermanyssina, which is considered by him to represent 'a dorsodisjugal furrow with advanced position', could well be a secondary division of the pronotal shield allowing increased

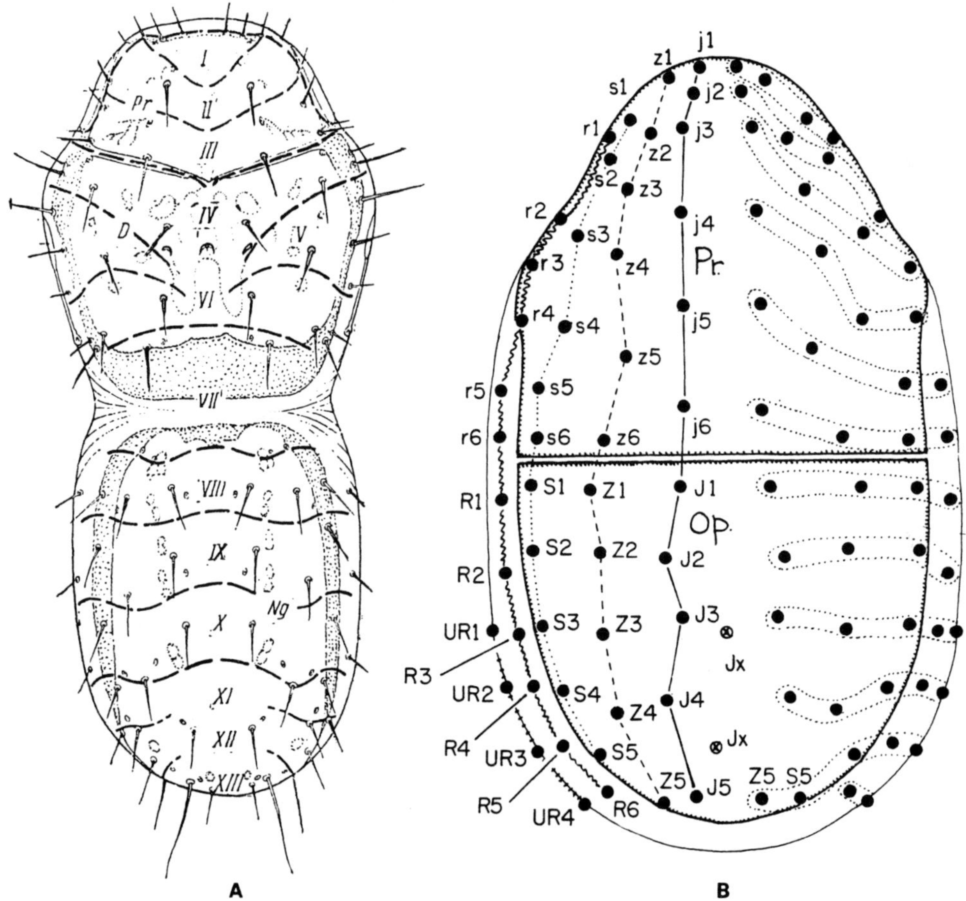

Fig. 2.3. A, segmentation of the soma of *Rhodacarus* (Mesostigmata) according to Zachvatkin (1952); segment XIV forms part of the anal shield and is not indicated in the figure. **B**, terminology for the dorsal idiosomatic chaetotaxy in the holotrichous Dermanyssina (Mesostigmata) following Lindquist and Evans (1965). *j-J.* = dorsocentral series; *Op.* = opisthonotal shield; *Pr.* = pronotal shield; *r-R* = marginal series; *s S* = lateral series; *UR* = *submarginal series; z-Z mediolateral series.*

flexibility of the soma for moving in confined spaces in the soil (Evans, 1982).

The number and distribution of setae (chaetotaxy) putative stress-sensitive organules or poroids (poroidotaxy) and dermal gland pores (adenotaxy), especially those occurring on the dorsum of the idiosoma, are used in the classification of the Dermanyssina and Parasitina and various systems of terminology have been proposed for them. The most widely used system for the chaetotaxy of the dorsum of the idiosoma recognizes the division of the body into a pronotal and an opisthonotal region, each partially or entirely covered by a sclerotized shield. Each longitudinal row of setae is thereby divided into a pronotal and an opisthonotal subseries that are numbered independently. The symbols for the rows of setae are mainly based on Sellnick (1944) and comprise dorsocentral (*j.J.*), mediolateral (*z.Z.*), lateral (*s.S.*), marginal (*r.R.*) and submarginal (*UR*) series, with the setae of the pronotum being denoted by lower case letters and those of the opisthonotum by capital letters (Fig. 2.3B). Each subseries is numbered consecutively from anterior to posterior. Sellnick's system which referred only to the opisthonotal chaetotaxy was extended for the pronotum by Hirschmann (1957) but a modified form of his system by Lindquist and Evans (1965) is usually preferred, and this is shown in Fig. 2.3B. The Sellnick symbols have also been used by Lee (1981) for the idiosomatic chaetotaxy of the Oribatida and presents an alternative system to that of Grandjean (1934d, 1939a) for this group (see p. 471).

Athias-Henriot (1975a,b) adopted the symbols *g* for gland pores and *i* for the poroids. In protoadenic forms (Dermanyssina, Parasitina and Zerconina), the maximum number of dorsal idiosomatic gland pores (*gd*) is nine pairs (*gd*1–9) of which five pairs are located in the prosomatic region and four pairs in the opisthosomatic region (Fig. 2.4). This contrasts with the larger number of gland pores in euneoadenic species (Heterozeronina, Uropodina, Epicriina and Antenophorina). Ventral gland pores are designated by *gv*. The poroids of the idiosoma dorsum in protoadenic forms are usually divided into three longitudinal series: dorsal (*id*) in the pronotal region and dorsomedian (*idm*) and dorsolateral (*idl*) in the opisthosomatic region. The gland pores, poroids and muscle sigilla are combined to form organogerous territories.

Actinotrichida

The segmentation of the Actinotrichida is also debatable. If it is accepted that the opisthosoma of the larva is made up of a maximum of six segments (including anal lobe), then the adult number of somatic segments would range from 12 in forms without anamorphosis to 15 in species with complete anamorphosis, i.e. with three segments added in the nymphal phase (Fig. 2.5). Van der Hammen (1970) included a pregenital segment VII in his calculation and proposed a somatic comple-

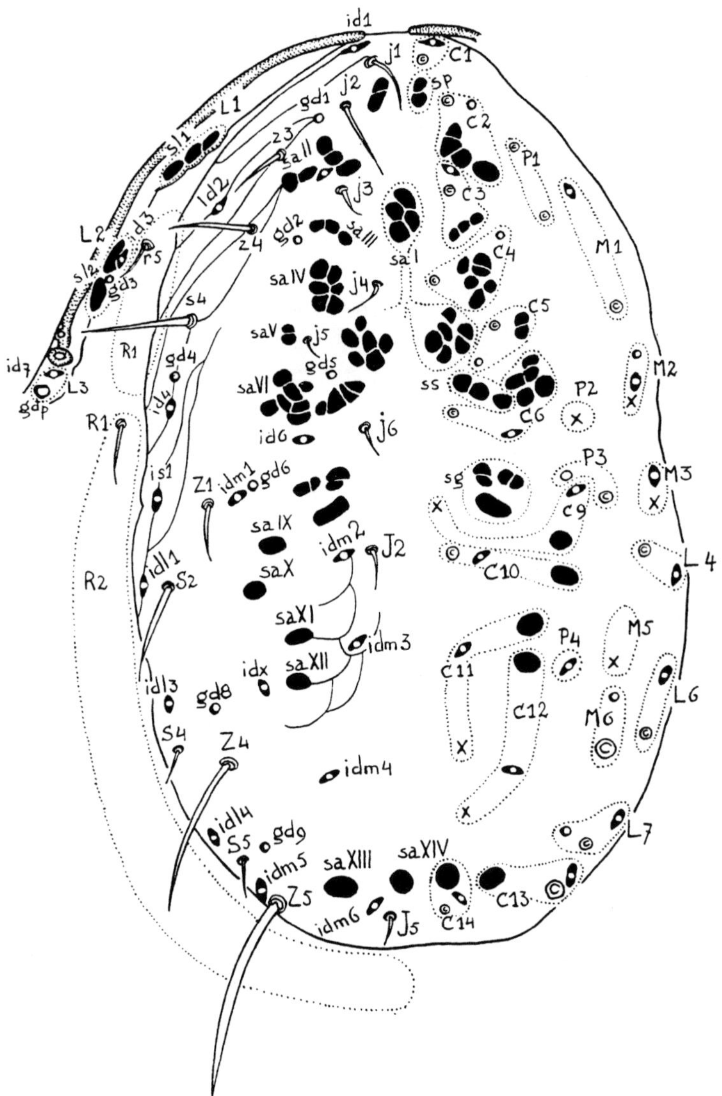

Fig. 2.4. *Adenotaxy, poroidotaxy and sigillotaxy of the dorsum of the idiosoma in Amblyseius* (Mesostigmata). *C.* = paraxial series of organogerous territories; *gd.* = dorsal gland pores; *id.* = dorsal poroid; *idm.* = dorsomedian poroid; *idl.* = dorsolateral poroid; *L.* = antiaxial series of organogerous territories; *M.* = intermediate series of organogerous territories; *P.* intercalary territories; *sa* I–XIV = sigilla. (From Athias-Henriot, 1975b.)

ment of from 13 to 16 segments. Sitnikova (1978), on the other hand, considered that the pregenital segment (VII) is lost in the Actinotrichida and that the maximum number of somatic segments, excluding the anal lobe, is 12 (comprising segments I–VI and VIII–XIII) with the rare exception of 13 in some primitive members of the Oribatida (e.g. Parhypochthoniidae) which have a peranal segment (XIV).

The terminology for the segments and the cuticular organules of the hysterosoma, i.e. the region of the soma comprising segments V, VI and the opisthosoma, usually follows the system proposed by Grandjean (1934d, modified 1939a) in which the segments from anterior to posterior are designated by the capital letters C,D,E,F,H,PS (pseudanal),

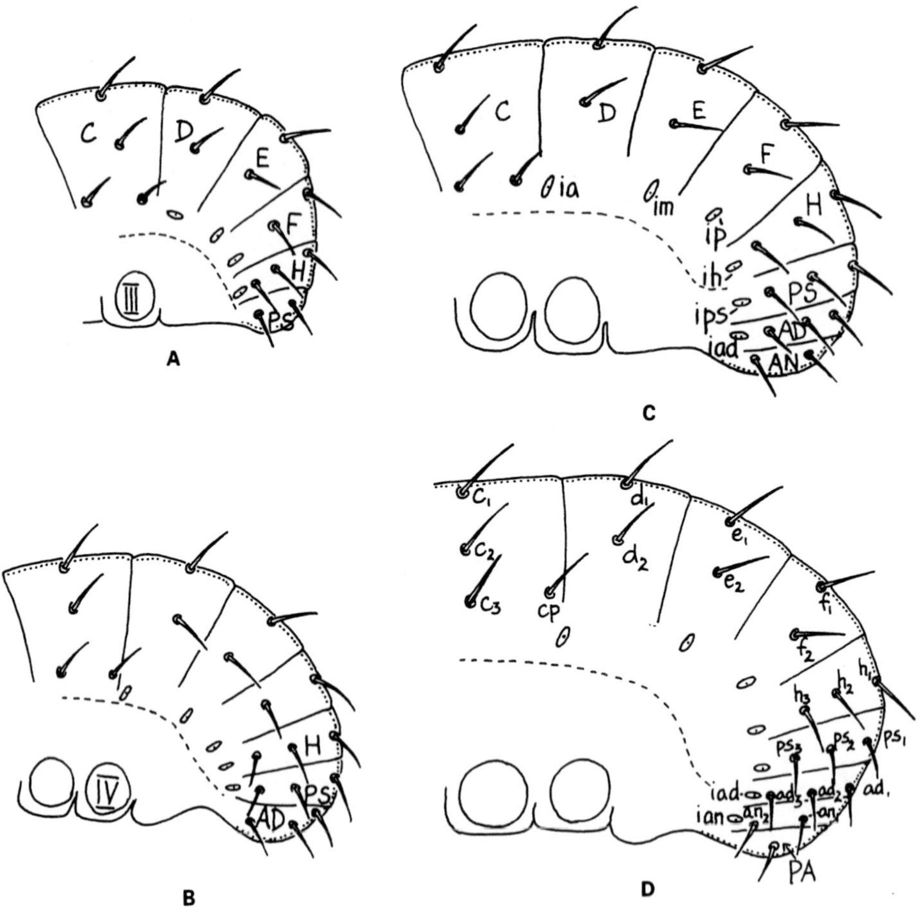

Fig. 2.5. Diagrammatic representation of the segmentation of the hysterosoma in the larva (**A**), protonymph (**B**), deuteronymph (**C**) and tritonymph (**D**) of a primitive oribatid showing addition of segments during ontogeny, together with the notation for the dorsal setae and cupules, based on Grandjean (1939a). *AD* = adanal segment; *AN* = anal segment *PA* = peranal segment; *PS* = pseudanal segment. For other abbreviations see text.

AD (adanal), AN (anal) and PA (peranal) (Fig. 2.5). Van der Hammen (1970) considers all these segments to be opisthosomatic and for segment C to be formed by the fusion of two segments (VII and VIII), whereas Sitnikova (1978), who uses a different terminology for the posterior opisthosomatic segments based on the earlier system of Grandjean (1934d), favours an alternative hypothesis that C and D belong to podosomatic segments V and VI with E (VIII) being the apparent first opisthosomatic segment. An interesting aspect of Sitnikova's terminology is the establishment of an anal lobe (*An*), bearing the anus, in the larva so that additional segments are added in front of the lobe as one would expect if the mite is anamorphic. A dorsal suture located between setal rows *d* and *e* and continuing latero-ventrally until joining the ventral flexible cuticle in the oribatid *Elliptochthonius profundus* (Norton, 1975) would appear to indicate that both segments C and D are the tergites of segments V and VI (Fig. 2.6A). However, this so-called podo-opisthosomatic articulation may be an adaptation for moving in confined

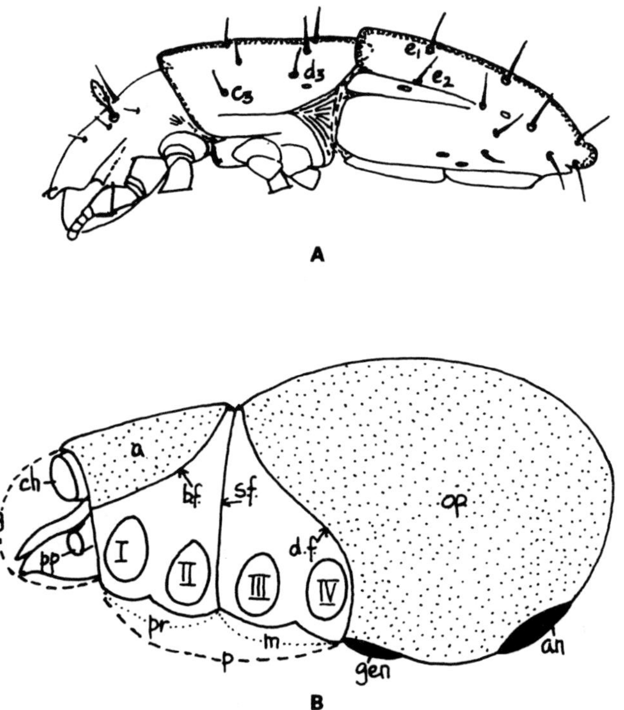

Fig. 2.6. A, lateral view of *Elliptochthonius profundus* (Oribatida) (based on Norton, 1975); **B**, schematic lateral view of an actinotrichid mite showing divisions of the soma, based on Grandjean (1970a). *a.* = aspidosoma; *an.* = anal plate; *b. f.* = abjugal furrow; *ch.* = chelicera; *d. f.* = disjugal furrow; *g.* = gnathosoma; *gen.* = genital plate; *m.* = metapodosoma; *op.* = opisthosoma; *p.* = podosoma; *pp.* = pedipalp; *pr.* = propodosoma; s.f. = sejugal furrow; *I–IV* = acetabula of legs I–IV. *g* + *a* + *pr* = proterosoma; *op* + *m* = hysterosoma; *g* + *a* = epiprosoma; *p* + *a* + *g* = prosoma.

spaces in mineral soil and a functionally similar modification to that occurring in some euedaphic Mesostigmata, such as *Rhodacarus* and *Protogamasellus*. Norton (1975) referred to the condition in which protero-hysterosomatic and podo-opisthosomatic articulations are present as *trichoidy*.

The setae of a specific segment carry the symbol of that segment in lower case and the numbering of the setae of each transverse row begins with the nearest to the mid-line, as shown diagrammatically in Fig. 2.5. Putative paired slit organs (= lyrifissures, cupules) are considered to be segmental structures and are designated by the letter *i*. A maximum of seven pairs may be present laterally on the opisthosoma in the adult and consist of *ia* in the region of *c* setae, *im* near *e* setae, *ip* at the level of *f* setae, *ih* near *h* setae, *ips* near *ps* setae, *iad* near setae *ad* and *ian* at the level of setae *an*. Four pairs of cupules, *im*–*ips*, are present in the larva, the remainder, when present, being added during the nymphal phase.

The soma of the Actinotrichida has been so markedly modified in the course of evolution that it is difficult to recognize the limits of the prosoma and opisthosoma, especially from the dorsal surface of the body. Grandjean (1970a) is of the opinion that the most significant development has been a marked shortening of the dorsal region of the ancestral podosoma (segments III–VI) and the advancement of the dorsal region of the opisthosoma into the podosomatic region so that the dorsal and ventral parts of the disjugal furrow are usually not continuous in the vertical plane (Fig. 2.6B). According to his hypothesis, the reduction in the dorsal part of the podosoma has been so drastic that this region has been virtually obliterated in extant forms resulting, for example, in the dorsal region of the anterior of the idiosoma consisting only of the dorsal parts of the 'precheliceral', cheliceral and pedipalpal segments which together form the aspidosoma. His proposed terminology for the divisions of the soma in the Actinotrichida is shown in Fig. 2.6B. Whether the reduction in the podosoma has been so extensive is impossible to ascertain on the basis of our present knowledge of actinotrichid embryology and morphology. Grandjean's (1970a) proposed terminology has been accepted, in general, by acarologists working on actinotrichid mites.

Subdivisions of the actinotrichid soma, superimposed on the original prosoma–opisthosoma tagmata, are defined by furrows and are referred to as pseudotagmata by Van der Hammen (1989). The circumcapitular furrow separates the gnathosoma from the idiosoma and the sejugal furrow, located between segments IV and V, divides the proterosoma from the hysterosoma. The gnathosoma is movably articulated to the idiosoma but movable articulation between the proterosoma and hysterosoma or the aspidosoma and opisthosoma of adults, at least in the Oribatida, is found only among the Palaeosomata, Enarthronota, Parhyposomata, Mixosomata and Euptyctima. In the Mixosomata, for example, the sejugal region is composed of a band of soft cuticle and the protero–hysterosomatic articulation is referred to as dichoidy (Fig. 2.7A). This contrasts with ptychoidy in adult Euptyctima and all stases of some

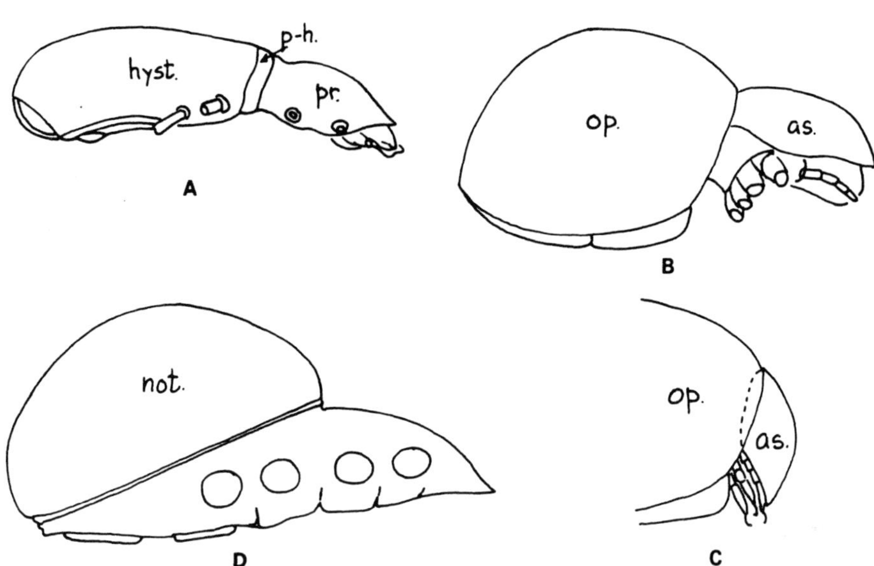

Fig. 2.7. Various types of articulation between the proterosoma and hysterosoma of the Oribatida. **A**, dichoidy in Eulohmanniidae; **B**, **C**, ptychoidy in Phthiracaridae with the aspidosoma open (**B**) and semi-closed with legs partly withdrawn (**C**); **D**, holoidy in Euoribatida. **as.** = aspidosoma; *hyst.* = hysterosoma; *not.* = notogaster; *op.* = opisthosoma; *pr.* = proterosoma; *p-h.* = protero-hysterosomatic articulation.

Enarthronota (Protophoroidea and Mesoplophoroidea) in which a dorsal hinge between the aspidosoma and opisthosoma allows the former to be closed, like a hinged lid of a box, on to the opisthosoma at the same time as the withdrawal of the legs into the body (Fig. 2.7B,C). No movement occurs between the proterosoma and hysterosoma in the Desmonomata (=Nothronota) and in adult Euoribatida (=Brachypylina). The articulation has been lost dorsally and ventrally although the limits of the two body divisions can be determined dorsally from the dorsosejugal discontinuity (*dsj*) between the prodorsum and notogaster (Fig. 2.7D) and ventrally by the position of the sejugal apodeme. This condition is referred to as holoidy. A third furrow, the abjugal, separates the aspidosoma (prodorsum) from the propodosoma. Other subdivisions of the soma which are used in descriptive morphology are: the propodosoma and metapodosoma forming, respectively, the anterior and posterior parts of the podosoma separated by the sejugal furrow; the epiprosoma (the aspidosoma plus gnathosoma) and the stethosoma comprising the prosoma without the gnathosoma.

Idiosomatic Musculature

The Acari have a well-developed musculature which may be divided into four basic groups comprising gnathosomatic muscles, idiosomatic muscles, visceral muscles and leg muscles. Only the muscles of the idiosoma are discussed in this section while the remainder are considered in the section on legs and in the chapters dealing with the structure of the reproductive, feeding, digestive and secretory organs.

Muscle cells or fibres (myofibres) are spindle-shaped to cylindrical and formed of a specialized cytoplasm, the sarcoplasm, surrounded by a membrane termed the sarcolemma (Fig. 2.8A). Embedded in the sarcoplasm are the contractile elements, the myofibrils, each of which is divided into sarcomeres by the so-called Z-lines. The whole fibre consists either of a single specialized cell or a syncytium. Such fibres when grouped together form a primary bundle or fasciculus. A muscle is a group of primary bundles. In the Acari, all the muscles are striated, i.e. they have a cross-striated appearance due to the regular alternation of bands with strongly birefringent (anisotropic or A-bands) and weakly birefringent (isotropic or I-bands) properties. A lighter H-zone is situated in the middle of the A-band (Fig. 2.8B). In ticks, the Z-line or disc is flanked by narrow I-bands and the greater part of the sarcomere is occupied by the A-band (El Shoura, 1989). The myofibril consists of two proteins, actin and myosin, whose molecules are arranged, respectively, into thin and thick filaments (myofilaments). The A-band is interpreted as being composed of overlapping thick and thin filaments, the H-zone of thick filaments and the I-band of thin filaments (Huxley, 1958; Barrington, 1967). During the complex process of contraction, the actin and myosin filaments are believed to slide past each other as a result of the actin filaments moving inwards (Fig. 2.8C). It is this movement that causes the shortening of the muscle fibre. The movement is reversed during relaxation. Striated muscles give increased speed of contraction and hence response, in comparison with smooth muscle although visceral striated muscle fibres are considered to perform slow contractions similar to those of the smooth muscles of vertebrates (Smith, 1968). The muscles may be inserted into the inner surface of the cuticle, attached to apodemes (cuticular projections into the body) or connected to an endosternite. In some muscles, for example the antagonistic flexor and extensor muscles of the movable digit of the chelicerae and the claw-complex of the ambulatory appendages, a sclerotized cord or tendon intervenes between the muscle and the point of insertion.

The method of attachment of muscles to the cuticle has been extensively studied in insects but little detailed information is available for the Acari. In insects, the muscle fibres are attached to the cuticle by way of epidermal cells whose cytoplasm has numerous microtubules oriented in parallel with the long axis of the adjacent muscle fibres (Lai-Fook, 1967). These bundles of microtubules are apparently involved in

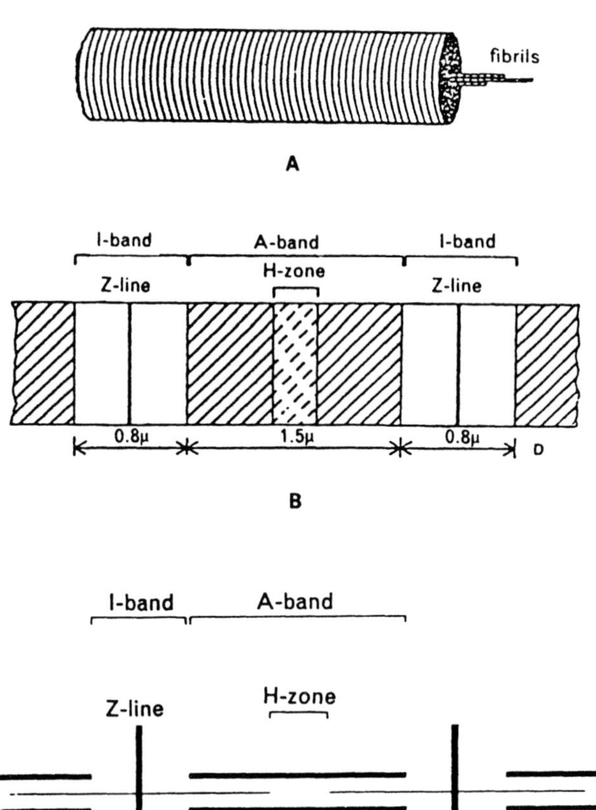

Fig. 2.8. Muscle fibre (**A**) and a myofibril (**B**) of rabbit poas muscle showing band pattern at resting length; (**C**) relationship between actin filaments (thin) and myosin filaments (thick) in a striated muscle. (After Barrington 1967.)

transferring stress across the epidermal cells. The basal surface of such an epidermal cell bears processes which interdigitate with matching projections of the muscle cell so that their plasma membranes are in apposition. The plasma membranes are linked by desmosome junctions. The outer surface of the cell, adjacent to the cuticle, is anchored to the cuticle by extracellular fibrils (tonofibrillae) which penetrate the procuticle and attach to the outer epicuticular layer. The same type of myoepidermal junction occurs in the ixodoid tick *Hyalomma dromedarii* (El Shoura, 1989). Muscle attachment in the astigmatic mite, *Sancassania mycophagus,* however, appears to differ from that in other arthropods (Kuo *et al.*, 1971). There are no bundles of microtubules extending from the myo-epidermal junctions on the inner surface of the epidermal cells to the outer surface adjacent to the cuticle and the stress across these cells is transferred by desmosome-like structures which form intraepidermal cell bridges. It is suggested that the absence of microtubules from the epidermal cells of this mite may be related to the flattened nature of the cells so that fibrillar material of the desmosome complex may be sufficient to transmit the tension between muscle and cuticle across the cell.

On the basis of the arrangement of striated fibres, two main types of voluntary muscles are recognized, namely parallel-fibred and pinnate muscles (Alexander, 1968). In the parallel-fibred muscle, all fibres typically run the whole length of the muscle and are attached at their ends by a tendon or an apodeme, whereas in the pinnate form the fibres are much shorter than the muscle and converge on a central tendon. The idiosomatic and leg muscles are examples of parallel-fibred muscles and the strong levator and depressor muscles of the movable digit of the chelicera are usually pinnate.

The most comprehensive studies of the musculature of the Acari include the following, which should be consulted for specific details: Notostigmata (With, 1904), Holothyrida (Thon, 1906); Ixodida (Robinson & Davidson, 1914; Ruser, 1933; Douglas, 1943); Mesostigmata (Michael, 1892; Young, 1970; Woodring & Galbraith, 1976; Akimov & Yastrebtsov, 1988); Prostigmata (Blauvelt, 1945; Mitchell, 1962a, 1964; Mathur & LeRoux, 1965; Gorgol & Yastrebtsov, 1987); Astigmata (Nalepa, 1884, 1885; Michael, 1901; Zachvatkin, 1941; Woodring & Carter, 1974); Oribatida (Akimov & Yastrebtsov, 1989).

Dorsoventral muscles

These are usually the strongest paired muscles of the idiosoma and as their name signifies they originate on the dorsal body wall and insert on the ventral wall (Fig. 2.9A). They are usually attached to extensive sclerotized shields or to isolated scutella but in some members of the Ixodida, Astigmata and Prostigmata, for example, sclerotized areas on the hysterosoma are lacking and the sites of muscle attachment may be marked by dimples (small depressions) as in the Parasitengona or longitudinal grooves or depressions as in the immature instars and females of ixodoid ticks. Numerous dorsoventral muscles occur in ticks, divided into marginal, genital and anal series in the Ixodoidea, and a relatively high number occur in the Parasitengona – 15 pairs in *Allothrombium lerouxi* and 17 in *Blankaartia ascoscutellaris*. In the Dermanyssina, on the other hand, seven pairs of dorsoventral muscles are usually present comprising two pairs in the podosomatic region originating laterally on the dorsal shield and inserting on the internal lateral surface of the sclerotized cuticle between coxae I and II and III and IV, and five pairs (reduced to three pairs in *Varroa* and four pairs in, for example, *Veigaia* and *Ameroseius*) in the opisthosomatic region, originating on the dorsal or opisthonotal shield and, with the exception of the first pair, inserting on scutella and sclerites on the opisthogaster. The first pair is attached by tendons to ' podemes behind coxae IV (Fig. 2.9B). Additional dorsoventral

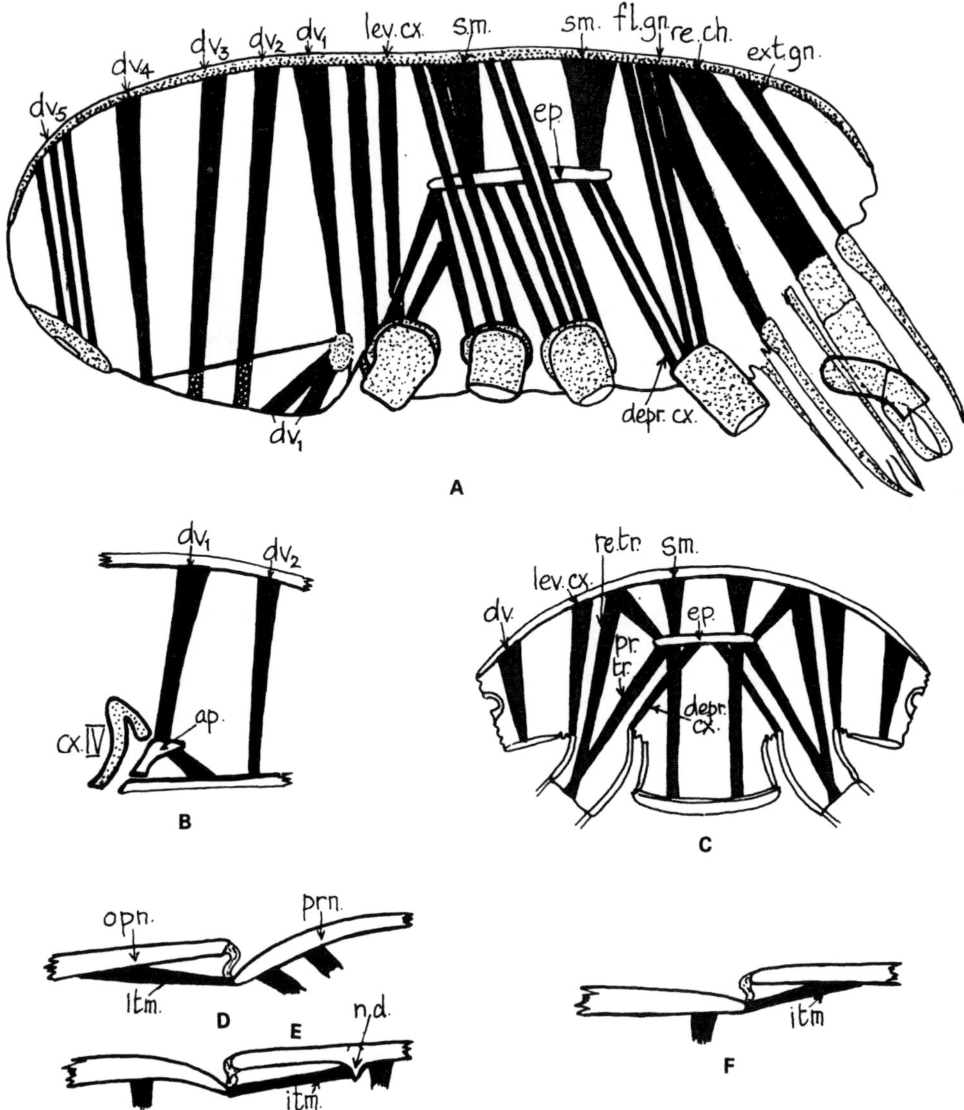

Fig. 2.9. **A**, diagrammatic representation of the idiosomatic musculature of a mesostigmatic mite; **B**, insertion of the first pair of opisthosomatic dorsoventral muscles on an apodeme posterior to coxae IV; **C**, diagrammatic transverse section of the prosoma of a mesostigmatic mite at the level of legs III showing musculature; **D–E**, intertergal muscles in *Veigaia* (**D**), *Dendrolaelaps* (**E**) and *Zercon* (**F**). (Based on Akimov & Yastrebtsov, 1989.) *ap.* = apodeme; *depr. cox.* = depressor muscles of the coxa; *dv.* = dorsoventral muscles; *ep.* = endosternite; *ext. gn.* = extensors of gnathosoma; *fl. gn.* = flexors of gnathosoma; *it. m.* = intertergal muscles; *lev. cx* = levator muscles of the coxa; *nd.* = scleronodule; *opn.* = opisthonotal shield; *pr. tr.* = promotor muscle of the trochanter; *prn.* = pronotal shield; *re. ch.* = retractor muscles of the chelicera; *re. tr.* = remotor muscle of the trochanter; *sm* = suspensory muscle of endosternite.

muscles occur in the Zerconidae and originate on the inner surface of the four fossae situated near the posterior margin of the opisthonotal shield and insert on the opisthogastric shield. The dorsoventral muscles in the uropodid *Uroactinia agitans* and the tick *Argas persicus* are equally spaced around the lateral margins of the body.

The pliable cuticle connecting the dorsal and ventral sclerotized regions of the idiosoma allows the compression of the idiosoma during the contraction of the dorsoventral muscles with the result that there is an elevation in hydrostatic pressure which is responsible, for example, for cheliceral protraction and the extension of those articles of the legs and pedipalps which do not have antagonistic muscles. Contraction of these muscles in weakly sclerotized species may also have the effect of changing the shape of the idiosoma through infolding of the cuticle. In the Mesostigmata and Holothyrida, the dorsoventral muscles of the podosomatic region which have their insertion near the stigmata probably play a part in ventilating the tracheal system.

Suspensory muscles of the endosternite

Certain of the extrinsic muscles of the first movable podomere of the legs, and in some taxa (Mesostigmata) also the second podomere, usually originate on the ventral surface of a horizontal plate, probably derived from connective tissue, called the endosternite, which may be suspended by muscles arising from the cuticle of the dorsum of the idiosoma and attaching to its dorsal surface (Fig. 2.9C). In the Ixodoidea, however, the abductor and adductor muscles inserting on the coxae originate from the dorsal scutum and grooves on the dorsum of the unsclerotized body wall while in the Argasoidea they arise from small muscle discs on the dorsal surface of the idiosoma. The cervical and lateral grooves of the dorsal scutum in the ixodoids mark the attachment sites of the muscles. The well-developed endosternite in the majority of the Mesostigmata is a thin transparent structure having an X configuration due the presence of two anterior and two posterior arms. It lies in the podosomatic region below the ventriculus and the longer anterior arms pass on either side of the synganglion. In the majority of the Dermanyssina and Parasitina, three paired groups of suspensory muscles are present. The endosternite in *Spinturnix myoti* is markedly reduced. In the Argasidae, five pairs of mesial intercoxal muscles and the superior genital muscles originate on a small endosternite which does not have suspensory muscles. The endosternite in the Bdellidae is a conspicuous structure which in some species shows a distinct lumen between its thick upper and under walls (Michael, 1896). It lies beneath the ventriculus of the mid-gut and the hind part of the synganglion. An endosternite with suspensory muscles is also present in the Oribatida but in the Astigmata it is often reduced, according to Michael (1901), 'to its smallest possible dimensions' at the junction of the dorsal endosternite and the trochanteral adductor muscles. This condition is similar to the so-called muscle-knot at the junction of the anterior and ventral retractor muscles of the gnathosoma with the muscles inserting on the wall of the progenital cavity and on the postero-lateral cuticle between legs II and III in some Erythraeidae (Witte, 1978).

Intertergal muscles

These muscles have been described in mites with two dorsal scutal elements and are known for the Dermanyssina, Parasitina, Zerconina and Cheyletidae (Gorgol & Yastrebtsov, 1987; Akimov & Yastrebtsov, 1988). They are responsible for the movement of the scuta and may also cause bending of the body at the juncture between the shields (Fig. 2.9D–F). In the Parasitidae, three muscle groups are attached to the anterior end of the opisthonotal shield and the posterior edge of the pronotal shield. In the Rhodacaridae and Digamasellidae scleronodules on the pronotal shield serve as attachment points for the intertergal muscles (Shcherbak & Akimov, 1979). The number of scleronodules indicates the number of muscle groups; for example, three in *Rhodacarus* and four in *Rhodacarellus* and *Dendrolaelaps.*

Longitudinal muscles

These muscles, which shorten the body, are particularly well developed in the Astigmata and Prostigmata. According to Nalepa (1884, 1885), four pairs of longitudinal muscle-bands may be present in adult Acaroidea. They arise in the scapular region, pass close to the outer wall of the hysterosoma before inserting by broad tendons into the cuticle above the third pair of legs. In the deuteronymph (hypopus) of *Sancassania boharti,* three pairs of propodosomatic retractor muscles occur which originate on the dorsal hysterosomatic shield at the level of the third coxae and insert on the prodorsal shield. They probably function, in the same way as the dorsoventral muscles, to increase hydrostatic pressure (Woodring & Carter, 1974).

The longitudinal (horizontal) muscles in the Trombidioidea consist of a dorsal and a ventral series according to whether they have both attachments on the dorsal or on the ventral surface of the idiosoma. The dorsal series in *Blankaartia,* for example, comprises six muscles forming two rows down the dorsum of the idiosoma (see Mitchell, 1962a for details). Their attachment sites are termed dorsalia, which may be shared with some of the dorsoventral muscles. Tension on the dorsal series of muscles produces transverse integumentary folds at the level of dorsalia I, II and III. Five pairs of dorsal longitudinal muscles occur in *Tetranychus urticae* comprising three parallel series orginating from a dorsal apodeme and inserting on the dorsal integument, and two shorter pairs also arising from the dorsal apodeme but extending somewhat laterally before inserting on the dorsal integument (Blauvelt, 1945).

Sigilla and sigillotaxy

The attachment sites of muscles on the sclerotized integument of the idiosoma of the Mesostigmata are called sigilla (*sing*, sigillum) and are usually conspicuous in macerated specimens examined by phase-contrast or interference-contrast microscopy (Fig. 2.4). Their number and distribution have been extensively studied by Athias-Henriot (1970b). She recognizes a maximum of 19 pairs of sigilla groups in orthoadenic forms which she has attempted to relate to the musculature. The terminology and identity of the muscle groups require revision following the more recent studies of idiosomatic musculature in the Dermanyssina (e.g. Akimov & Yastrebtsov, 1988). However, sigillotaxy, in combination with poroidotaxy and adenotaxy, has proved to be a valuable taxonomic tool. The sigilla, poroids and gland pores of the dorsum are combined to form a number of 'organogerous areas' ('territoires organogères') and these have been described in detail for the Phytoseiidae (Athias-Henriot, 1975b).

Legs

The acarine locomotory appendages, normally comprising three pairs in the prelarval and larval phases and four pairs in the nymphal and adult phases, typically consist of seven podomeres (= segments, articles) – namely, from proximal to distal, coxa, trochanter, femur, genu (or patella), tibia, tarsus and apotele (the primitive terminal podomere which according to Grandjean (1952c) is represented in extant species *only by the basilar piece and claws*). Division of the trochanter, femur and tarsus, each into two articles, may occur and, exceptionally, the tarsi may be divided into a number of 'false' articles as in *Tarsolarkus* (Grandjean, 1952b, 1954c). The podomeres are connected by flexible cuticle (arthrodial membrane), usually have a condylar articulation and are moved by antagonistic muscles or by the alternate action of muscles and hydrostatic pressure. The podomeres are typically provided with bristle-like mechanosensory organules called setae and may have other cuticular structures such as tube-like chemosensory sensilli termed solenidia and stress-sensitive slit organs.

The types of articulation between podomeres are various and have been extensively studied by S.M. Manton in the course of her work on arthropod locomotion (see Manton, 1977) and by Van der Hammen (1989) in the Arachnida. Two types of joints are evident: the eudesmatic joint in which movement at the joint is accomplished by muscles, and the adesmatic joint in which there is no muscle attachment and any movement at the joint is effected mainly by stress created at the joint when the appendage makes contact with the substrate during ambulation. The suture forming the adesmatic joint in the Anactinotrichida usually passes

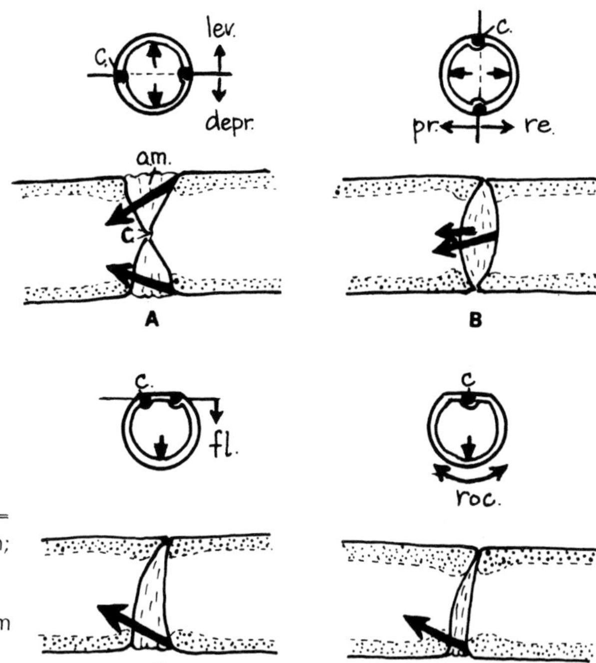

Fig. 2.10 Types of eudesmatic leg joints. **A**, horizontal pivot joint; **B**, vertical pivot joint; **C**, dicondylar hinge joint; **D**, monocondylar rocking joint. *a.m.* = arthrodial membrane; *c* = condyle; *depr* = depressor action; *fl.* = flexion; *lev.* = levator action; *pr* = promotor action; *re* = remotor action; *roc* = rocking movement. Long arrows denote direction of movement resulting from muscular action; short arrows indicate insertion points of muscles.

through one or more slit organs thereby forming a peripodomeric stress-sensitive organ as, for example, on the femur and tarsus of the Mesostigmata. Articulation of eudesmatic joints is basically dicondylar, i.e. two condyles on the distal margin of a podomere engage depressions, cotyles or acetabula, on the proximal margin of the succeeding podomere. The orientation and location of the condyles and cotyles on the distal margin of the podomere determine to a great extent the plane of movement at the joint. Paired condyles lying in a horizontal or vertical plane passing approximately through the diameter of the podomere so that considerable arthrodial membrane is present on either side of the articulation form pivot joints (Fig. 2.10A,B). Pivot joints allow the action of antagonistic muscles which produce levator/depressor movements when the plane of movement is horizontal and promotor/remotor movements when the plane is vertical. Condyles situated dorsally at the junction of two podomeres form a hinge joint which is normally flexed by muscles and extended by hydrostatic pressure (Figs 2.10C & 2.11A,B). The joint is characterized by having an expanse of arthrodial cuticle ventrally. One of a pair of articulation points may be lost and this can result in a rocking movement at the monocondylar joint; the rocking movement takes place about the longitudinal axis of the leg with the 'rock' occurring alternately in opposite directions (Fig. 2.10D). The function of a joint as a component of the locomotory mechanism of an appendage is often difficult to establish and requires a knowledge of both the disposition of muscles and the nature of articulations, including podomere and arthrodial membrane geometry. 'To do no more than describe a joint as a hinge or pivot type invites thorough misunderstanding of the locomotory function of that joint' (Manton, 1978).

Anactinotrichida

The podomeric composition of the leg in the majority of this group conforms to the typical form except in the Notostigmata in which there is a subdivision of the trochanters of legs III and IV into two articulating podomeres in the tritonymph and adult. The pivot joint between

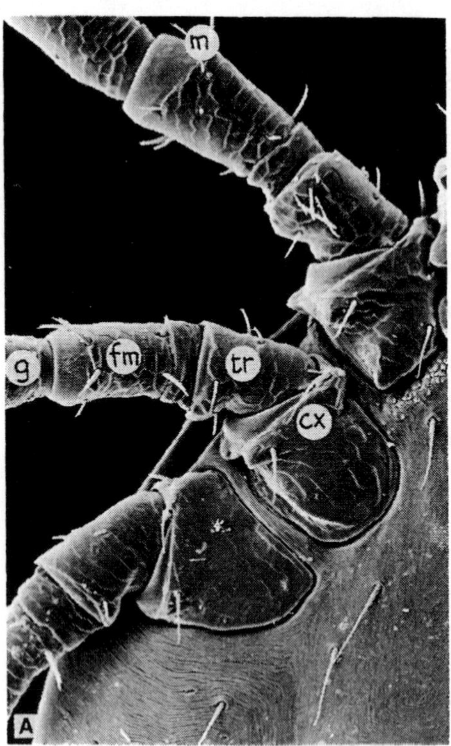

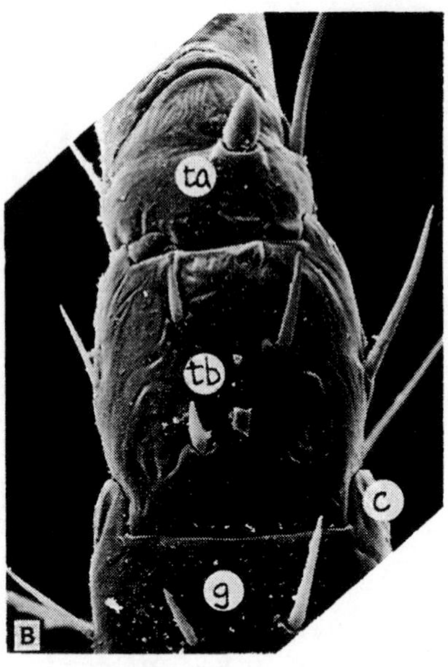

Fig. 2.11. SE micrographs of leg joints in the Anactinotrichida. **A**, basal podomeres of legs of larval ixodid tick showing vertical pivot joint between coxa and trochanter and the wide expanse of ventral arthrodial membrane at the flexure joint between femur and genu; **B**, hinge joints between genu-tibia and tibia-tarsus in a uropodine mite. *c.* = condyle; *cx.* = coxa; *m.* = arthrodial membrane; *tr.* = trochanter; *fm.* = femur; *g.* = genu; *ta.* = tarsus; *tb.* = tibia.

trochanter I and trochanter II, as these podomeres are designated, and the antagonistic muscles inserting by tendons dorsally and ventrally on the proximal margin of trochanter II produce a levator/depressor movement at the joint. The subdivision of the trochanters is thought to be a plesiomorphic (primitive) retention of a specialized state that is also present in all active instars of the Ricinulei and Solifugae (Grandjean, 1936a, Lindquist, 1984).

Joints and musculature

The coxa is present as a movable podomere in all anactinotrichids although the mobility is much reduced in the case of coxae II–IV in the Ixodida. Oudemans (1936a, 1937) divided the opisthogoneate arthropods into two groups on the basis of the mobility or immobility of the coxae: the Soluticoxata (movable coxae), which included the acarine groups Notostigmata, Holothyrida and Mesostigmata, and the Fixicoxata, which contained the Trombidi-Sarcoptiformes and also the Ixodida and Spinturnicidae! The coxae lie in cavities or acetabula situated ventrally or ventrolaterally on the podosoma and may be surrounded by a holopodal shield or discrete endo- and exopodal shields as in the Mesostigmata. In the Notostigmata, the first three coxae are protected by suprapedal lobes. The coxae are separated ventrally by the sternal region which may be sclerotized to form a sternal shield. A condylar articulation between the body and the coxa appears to be lacking, at least when the appendage is used only for locomotion. The first pair of legs, however, is often sensory rather than ambulatory in function and the articulation of the joints often differs from those of the walking legs. For example, the body–coxa articulation of the anteriorly directed first legs in the Dermanyssina and Parasitina is ventral and monocondylic (Evans & Till, 1979). Extrinsic muscles originating ventrally on the endosternite and on the internal surface of the dorsal shield insert, respectively, on the external ventral surface and on the proximal dorsolateral surface of the coxa and produce levator–depressor (or abductor–adductor) movements of the coxae as well as limited rotation (Fig. 2.9C). The Ixodida are an exception in that both the extrinsic muscles of the coxae orginate on the dorsal scutum or the lateral groove of the dorsum of the idiosoma in the Ixodoidea (Ruser, 1933) and on dorsal muscular discs in the Argasoidea. In the Mesostigmata, the coxae usually have a plate-like apodeme which extends for a distance into the body. Three sets of trochanteral muscles and one set of femoral muscles originate on this apodeme in the uropodid *Uroactinia agitans* (Woodring & Galbraith, 1976). The flattened and medially contiguous coxae I occurring in most of the Uropodoidea cover and protect the gnathosoma when in their resting position. They are articulated to the posterior rim of the gnathopodal cavity.

With some exceptions, the trochanter has a vertical dicondylar articulation with the coxa and promotor–remotor movements are produced by extrinsic and intrinsic muscles inserting on its anterior

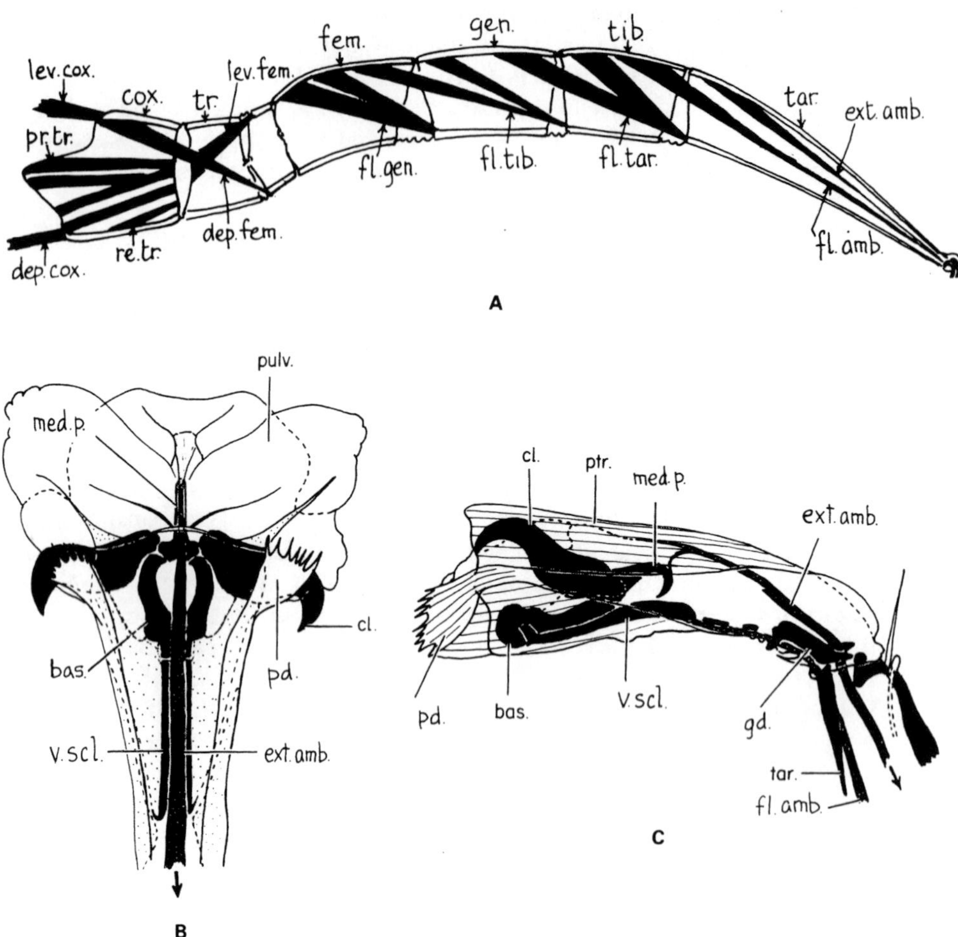

Fig. 2.12. **A**, musculature of an ambulatory appendage of a mesostigmatic mite; **B, C**, structure of the ambulacrum in *Haemogamasus hirsutus* (Mesostigmata) in the extended (**B**) and retracted state (**C**) (based on Evans & Till, 1965). *bas.* = basilar piece; *cl* = claw; *cox.* = coxa; *dep. cox.* = coxal depressors (adductors); *dep. fem.* femoral depressor; *ext. amb.* = ambulacral extensors; *fem.* = femur; *fl. amb.* = ambulacral flexors; *fl. gen.* = genual flexors; *fl. tib.* = tibial flexors; *fl. tar.* = tarsal flexors; *gd.* = tendon guide; *gen.* = genu; *lev. cox.* = coxal levators (abductors); *lev. fem.* = femoral levators; *med. p.* = median piece; *pd.* = paradactylus; *pr. tr.* promotors of the trochanter; *ptr.* = pretarsus or ambulacral stalk; *pulv.* = pulvillus or caruncle; *re. tr.* = remotors of the trochanter; *tar.* = tarsus; *tend. ext.* = tendon of extensor muscles of ambulacrum; *tib.* = tibia; *tr.* = trochanter; *v. scl.* = ventral sclerite.

lateral and posterior lateral faces (Fig. 2.12A). The extrinsic muscles in the Dermanyssina and Parasitina originate on the endosternite and the dorsal shield (Akimov & Yastrebtsov, 1988). The intrinsic promotor and remotor muscles arise within the coxa. Exceptions appear to occur in those members of the Uropodoidea in which the legs are folded into pedofossae for protection. In *U. agitans*, the coxa–trochanter articulation point is monocondylar and this allows trochanteral rotation in anterior and posterior directions by the action of the trochanteral muscles (Woodring & Galbraith, 1976). Posterior rotation leads to a defensive leg curl while anterior rotation leads to a transition from the leg curl to the walking position.

The trochanter–femur joint is eudesmatic and dicondylar with movement taking place about a horizontal axis, i.e. at right-angles to the vertical axis of the coxa–trochanter articulation. In the Mesostigmata, levator muscles originate on the ventral surface of the coxa (or on the ventral surface of the trochanter) and insert on the proximal dorsal of the femur while the antagonistic depressor muscles originate on the dorsal surface of the coxa and insert on the proximal ventral of the femur (Fig. 2.12A). A distinct peripodomeric suture subdivides the podomere into a short proximal basifemur and a long distal telofemur. The suture passes through one or more slit organs (or lyrifissures) and may be entire or interrupted by one or more narrow sclerotized bridges connecting the basi- and telofemur. It has not been established if the division of the femur represents a primitive state or a secondary division of a primitive entire femur.

The articulations between femur–genu, genu–tibia and tibia–tarsus are dorsal eudesmatic hinge joints with flexion operated by muscles originating, respectively, on the dorsal surface of the femur, femur/genu and genu/tibia and inserting by tendons on the proximal ventral margin of the genu, tibia and tarsus. Extension of the podomeres is by hydrostatic pressure. The hinge joints are dicondylar or monocondylar, the latter permitting rocking at the joint. As in the case of the femur, the tarsi are normally subdivided by a complete or interrupted peripodomeric suture into a short basitarsus and a long telotarsus. Subdivision is lacking in tarsus I of the Holothyrida. The suture usually passes through one or more slit-organs. The basi-telotarsal joint is adesmatic. In the Notostigmata, the basi- and telotarsus of legs II–IV are completely separated by flexible cuticle and some movement at the suture by the indirect action of ambulacral muscles and by hydrostatic pressure may be possible. The extent of dorsal extension is limited by a pair of dorsal protuberances on the basitarsus opposing cavities on the proximal dorsal surface of the telotarsus (Van der Hammen, 1989). A small sclerite occurs ventrally in the flexible cuticle of the suture in many Mesostigmata and probably forms part of a mechanoreceptor. In the Sejina and Antennophorina, the sclerite on leg IV is larger and bears a pair of setae (Evans, 1969). A similar setigerous sclerite occurs in the suture on legs II and III of the larva of the Ixodoidea and is retained, with a tendency for its distal margin to fuse with the telofemur, in the nymph and adult. A secondary distal subdivision of the telotarsus into a short acrotarsus or apicotarsus occurs in legs II–IV of the Notostigmata and in legs I of the Allothyridae (Holothyrida) and some Mesostigmata.

The ambulacrum comprises the peduncle-like extension of the tarsus, the ambulacral stalk or pretarsus, as well as the apotele. In the Mesostigmata, for example, two pairs of sclerites, located laterally and ventrally, give stability to the ambulacral stalk. The ventral sclerites extend almost the length of the stalk and abut the basilar piece in much the same way as the condylophores of some actinotrichid ambulacra. The claw-complex comprises a basilar piece and paired claws. The latter have a condylar articulation with the lateral cotyles (cotyloid cavities) of the basilar piece. Flexor muscles of the ambulacrum have their origin on the dorsal surface of the tibia and insert by a single tendon on the ventral surface of the basilar piece while the extensors arise from the dorsal surface of the tibia and tarsus and insert by a common tendon either on its dorsal surface and/or on the membraneous pad, the pulvillus, associated with the claw-complex (Fig. 2.12A,C). The pulvillus appears to be most highly developed in the free-living Mesostigmata, in which it is provided with weakly sclerotized veins and can function as a sucker. In many Dermanyssina, the pulvillus and the claws can be retracted into the ambulacral stalk (Evans & Till, 1965). The tendon of the extensor (or retractor) muscle in *Haemogamasus hirsutus* is connected to the pulvillus and a median sclerite, and does not appear to be attached to the basilar sclerite (Fig. 2.12C). A pair of lateral denticulate processes arising from the ambulacral stalk, the paradactyli or pretarsal opercula, are not retracted with the pulvillus and claws. Two pairs of setae are present on the ambulacral stalk of the Notostigmata (Grandjean, 1936a) and one pair in the Holothyrida (Van der Hammen, 1989) but there is none in the Ixodida and Mesostigmata. The ambulacra of the first pair of legs are often reduced or lost. Reduction is seen in the size of the pulvillus and claws. Antagonistic muscles are lacking in those species in which the ambulacrum is completely lost as, for example, in the Macrochelidae.

The direction adopted by the legs is determined by the orientation of the coxae and trochanters in relation to the longitudinal axis of the idiosoma. Usually, legs I and II are directed forwards and legs III and IV backwards although in some Mesostigmata leg III may be directed laterally. The typical sequence in the relative lengths of the legs in descending order is I, IV, II, III or, more rarely, IV, I, II, III. Legs I are often very long in the Mesostigmata; for example, in the genus *Podocinum* they are two to three times the length of the idiosoma with the tarsus lacking an ambulacrum but terminating in two long tactile setae. One or more podomeres of legs II and sometimes legs IV are spurred in the males of the Mesostigmata and used to grasp the female during sperm transference. The spurs are usually hypertrophied setae. In males of the Parasitidae, femur II is often crassate. In the Uropodoidea, a longitudinal flange or tectum on the femora is usually present and protects the distal

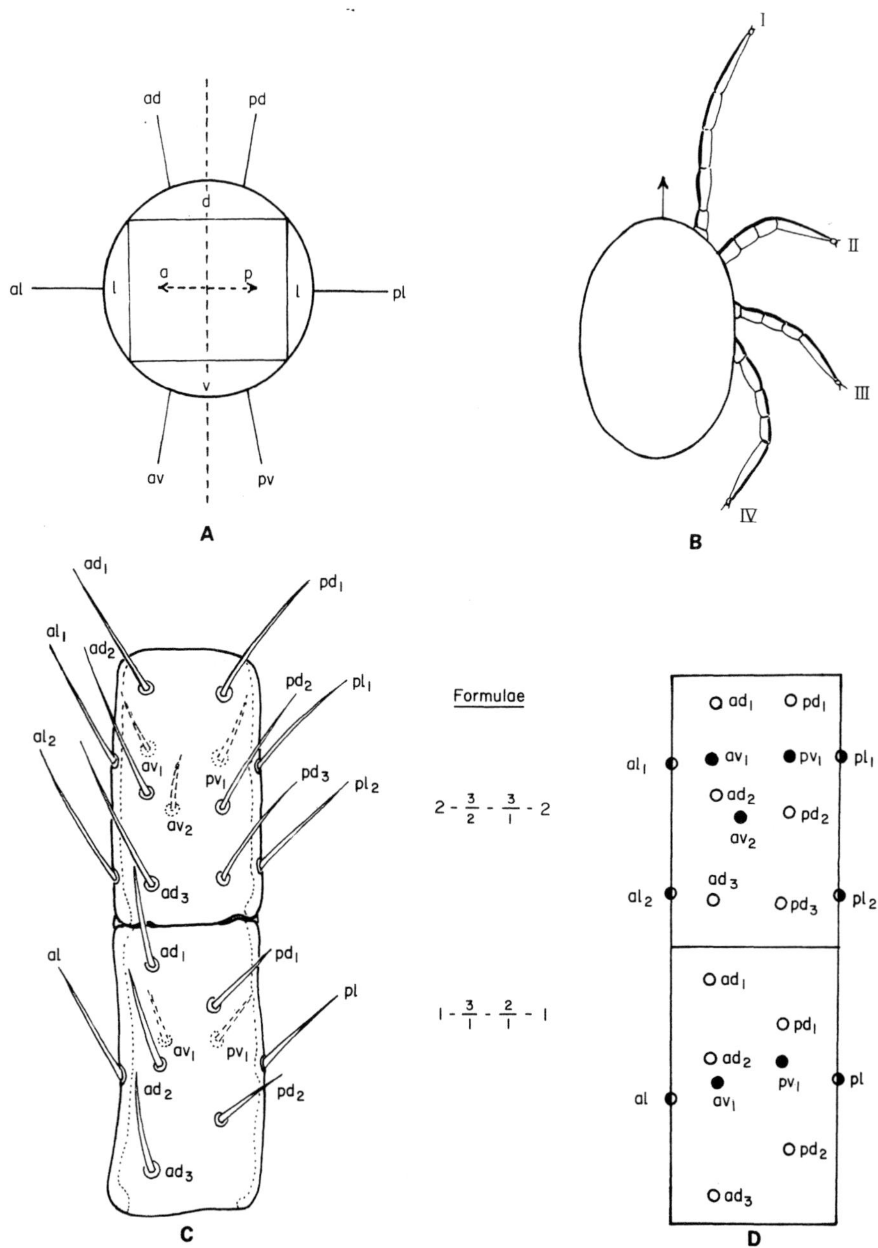

Fig. 2.13. Terminology for the leg chaetotaxy of the Mesostigmata. **A**, diagram of a podomeric whorl or verticil of setae; **B**, designation of anterior faces (bold lines) of podomeres used in defining setal patterns; **C**, example of terminology applied to the chaetotaxy of two leg podomeres; **D**, diagrammatic representation of podomeric chaetotaxy of C (*open circles*, dorsal setae; *half-open circles*, lateral setae; *black circles*, ventral setae). (After Evans & Till, 1979.) *a* = anterior segment of podomere; *ad* = antero-dorsal seta; *al* = antero-lateral seta; *av* = antero-ventral seta; *d* = dorsal face of podomere; *l* = lateral face; *p* = posterior segment of podomere; *pd* = postero-dorsal seta; *pl* = postero-lateral seta; *pv* = postero-ventral seta; *v* = ventral face.

podomeres when the legs are folded in the pedofossae.

Chaetotaxy

The podomeres bear a number of tactile setae which are basically arranged in whorls or verticils, i.e. in rings around the circumference of the podomere. The recognition of individual whorls is difficult or impossible in nymphs and adults of the Notostigmata, Holothyrida and Ixodida because of the large numbers of setae on the majority of the podomeres. The Mesostigmata, however, are an exception and the relatively few setae on the podomeres makes it possible to plot the ontogenetic development of individual setae and to use the numbers and distribution of setae on a podomere as taxonomic characters (Evans, 1963a, 1969, 1972). A complete whorl in this order consists of six setae and representatives of from one to four whorls appear to be present on the podomeres, excluding the tarsus of the first pair of legs (Fig. 2.13A). In defining setal patterns, each podomere is considered to have four setae-bearing faces: two lateral (*l*), one dorsal (*d*) and one ventral (*v*). The lateral faces are designated antero-lateral and postero-lateral on the basis of the division of each podomere into an anterior and posterior region: the anterior and posterior faces of the podomere being those which are directed, respectively, in an anterior and posterior direction when the leg is extended laterally at right-angles to the longitudinal axis of the idiosoma (Fig. 2.13B). Thus, the setae on the dorsal surface of the podomere are designated antero-dorsal (*ad*) and postero-dorsal (*pd*); those on the ventral surface antero-ventral (*av*) and postero-ventral (*pv*); and those on the lateral faces antero-lateral (*al*) and postero-lateral (*pl*). Owing to the difficulty of assigning setae on some podomeres to their correct whorl, the setae on each face are simply numbered from distal to proximal. Thus, setae with the same numerical signature do not necessarily belong to the same podomeric whorl. The setae of each longitudinal row are numbered from distal to proximal on each podomere (Fig. 2.13C,D). The number of setae in each row is represented by the formula: *al-ad/av-pd/pv-pl* (see formulae for Fig. 1.13C). In cases where the non-cylindrical form of the podomere makes it difficult to distinguish anterior and posterior members of the dorsal and ventral series a simpler formula, *al-d/v-pl*, is used. The presence of a number of chemosensory sensilli on the tarsus of leg I necessitates a different system of terminology for this podomere but, at present, a generally acceptable system is not available. The above system of terminology has also been applied to the leg chaetotaxy of larval Ixodoidea (Edwards & Evans, 1967) and Argasoidea (Edwards, 1975).

Actinotrichida

Unlike the Anactinotrichida, the coxae of extant actinotrichids are never free but are considered to be incorporated into the ventral body wall of the podosomatic region. According to Grandjean (1952a), the ventral or coxisternal region of the podosoma of the Archoribatida (=Macropylina) is composed of four epimera each carrying a pair of legs and separated by epimeric (or epimeral) furrows (Fig. 2,14A). The epimeric furrows are the limits of the four segments of the podosoma of which the epimera are

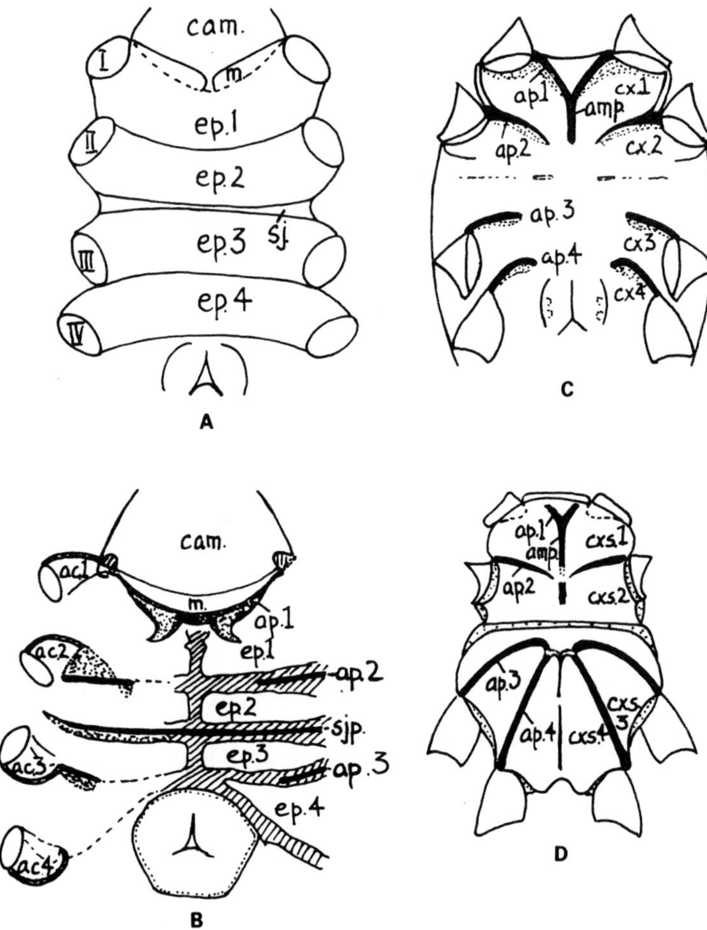

Fig. 2.14. Coxisternal region of selected groups of actinotrichid mites. **A**, primitive oribatid mite; **B**, euoribatid (brachypyline) mite; **C**, astigmatic mite; **D**, male tarsonemoid mite. (**A** and **B** based on Grandjean, 1952a.) *ac.* 1–4 = 'acetabula' of legs; *amp.* = anteromedian or prosternal apodeme; *ap.* 1–4 = apodemes; *cam.* = camerostome; *cx.* 1–4 = coxal fields; *cxs* 1–4 = coxisterna; *ep.* 1–4 = epimera; *m.* = mentotectum; *sj.* = sejugal furrow; *sjp.* = sejugal apodeme.

the ventral and ventrolateral tegument. The furrow situated between epimera 2 and 3 is the sejugal furrow. In the majority of cases, the epimeric furrows are prolonged inside the body to form apodemes. Apodeme I is often associated with a tectum, the *mentotectum*, whose role is to protect the ventral articulation between the gnathosoma and the podosoma. In the Euoribatida, the epimeral region is quite flat and darker pigmented bands, the epimeric, sejugal and sternal borders, separate the epimera (Fig. 2.14B). The basal articulation of the trochanter with the coxal region in the Archoribatida is exposed but in many Euoribatida the base of the trochanter is enclosed in a cup-like cavity, the so-called acetabulum. A well-defined coxisternal region is also present in the Astigmata and in the Tarsonemina. In the former, the use of the term epimera by Zachvatkin (1941) differs from that of Grandjean (1952a) and refers to the apodemes forming the anterior margin of a coxal field, while the term epimerite is given to a linear sclerite marking the posterior edge of a coxal field (Fig. 2.14C). The epimerite is often fused with the neighbouring apodeme while epimera I often fuse in the mid-line with a median apodeme to form a Y-shaped sternum. The coxisternal plates (coxisterna) are usually fused in the Tarsonemina and the medial fusion of the anterior plates (the anterior sternal shield of Schaarschmidt, 1959) is indicated by the prosternal or anteromedian apodeme (Fig. 2.14D). The coxisternal plates of the metapodosomatic region are also consistently

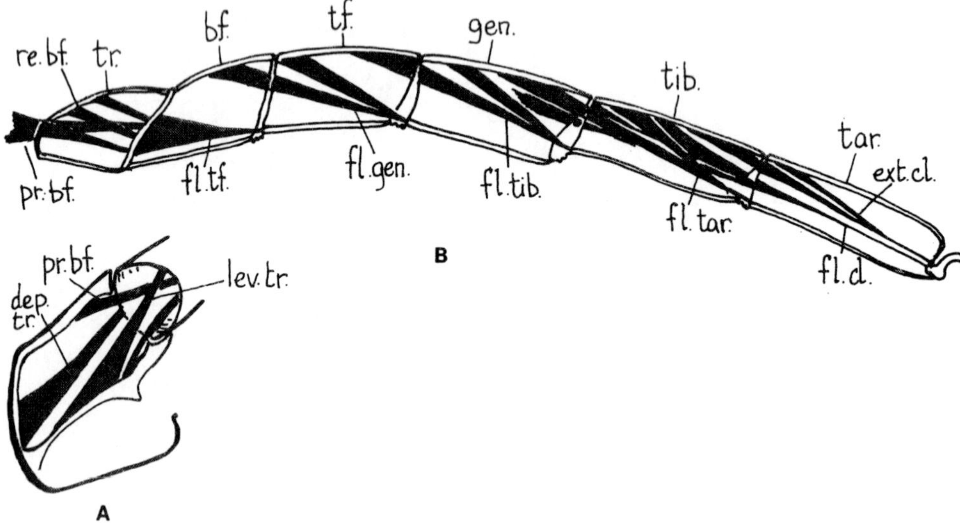

Fig. 2.15. Diagrammatic representation of the musculature of an ambulatory appendage of a trombidiid mite (Prostigmata). **A**, coxisternum of first leg showing origin and insertion of muscles controlling the movement of the trochanter; **B**, musculature of the movable podomeres of the leg. (Based on Mitchell, 1962a.) *bf.* = basifemur; *dep. tr.* = depressor of trochanter; *ext. cl.* = extensor of claw; *fem.* = femur; *fl. cl.* = flexor of claw; *fl. gen.* = flexors of genu; *fl. tar.* = flexors of tarsus; *fl. tf.* = flexors of telofemur; *fl. tib.* = flexors of tibia; *gen.* = genu; *lev. tr.* = levator of trochanter; *pr. bf.* = basifemoral promotor; *re. bf.* = basifemoral remotor; tar. = tarsus; *tf.* = telofemur; *tib.* = tibia.

consolidated into one posterior plate in the adults of the Tarsonemidae and Podapolipidae (Lindquist, 1986). In many other Prostigmata, the coxal fields are usually discernible either as distinct flat plate-like structures incorporated in the body wall and representing the ventral walls of the coxae or by areas differentiated from surrounding cuticle by their surface ornamentation or lack of it (Fig. 2.16 below).

The confusion referred to above in the general application of the term epimeron in acarine morphology also applies to the definition of the coxa. Dugès (1834) considered the skeletal divisions of the sternal region in the Oribatida to represent fixed coxae or more properly their ventral surfaces. Michael (1884), however, rejected the presence of fixed coxae, considering them to be free podomeres (equivalent to the trochanters of Grandjean) and used the term epimera for the 'chitinous, rod-like or blade-like pieces sunk in the sternal cuticle of soft-skinned Acarina and forming a rigid skeleton for the support of the legs'. Van der Hammen (1989), on the other hand, suggested that the coxae of the Chelicerata, when present, have evolved from epimera which are the sternites of the prosoma. He further considered that the coxa has never been developed in the Actinotrichida, which have retained to a great extent the primitive condition of the 'coxisternal region', hence their classification with the Palpigradi, which he considered to share the same condition, in his subclass Acoxata (also referred to by him as the Epimerata). This hypothesis has received little support (Lindquist, 1984). An attempt to clarify the structure and terminology of the coxisternal region of the Oribatida has been made in a preliminary note by Piffl (1991).

Joints and muscles

Although considerable detailed information is available on the external morphology of the coxisternal region and the legs, there are relatively few detailed works on musculature and on locomotory function at the leg joints. Movement at the body–trochanter joint, at least in the Prostigmata and Astigmata, is primarily depressor–levator (abduction–adduction) with some rocking of the trochanter, while that at the trochanter–femur joint is mainly promotor–remotor. Flexure occurs at the joints between femur–genu, genu–tibia and tibia–tarsus.

Trochanter

This is the first movable podomere of the legs in the Actinotrichida and is variously articulated to its acetabulum. In the Parasitengona, the trochanter has an anterior basal monocondylar articulation with the acetabulum and the body–trochanter joint allows the trochanter (and the rest of the leg) to be raised or lowered around the axis defined by the anterior articulation and the tight membrane forming the posterior part of the joint (Mitchell, 1957a, 1962a). The depressor and levator muscles of the trochanters I–III in *Blankaartia ascoscutellaris* originate within the plate-like coxae with the depressors inserting anteroventrally and the levators anterodorsally on the base of the trochanter (see Fig. 2.15A). These insertion sites also allow some rocking of the joint as well as the main depressor–levator action. The muscles originating within coxal plate IV are differently arranged from those of coxal plates I–III and their action in conjunction with the coxa–trochanter joint produces a counter-clockwise rotation accompanying depression and a clockwise rotation with elevation. The body–trochanter articulation in the Tarsonemina and Astigmata is dicondylar with the condyles of the acetabulum being located anteriorly and posteriorly so that the plane of movement at the joint is primarily horizontal permitting abduction–adduction of the leg (Woodring & Carter, 1974). In the deuteronymph of *Sancassania boharti*, abductors of each trochanter originate on one of the posterior coxisternal apodemes and insert on the dorsal proximal edge of the trochanter while the adductors arise on the ventral surface of the endosternite and attach onto the ventral proximal rim of the podomere. According to Grandjean (1970b), horizontal dicondylar articulations appear to characterize the joint between the acetabulum and trochanter in the Oribatida and this suggests that abduction–adduction of the leg at this joint also occurs in this group. This contrasts with the description by Akimov and Yastrebtsov (1989) of a vertical pivot joint between the body and trochanter in *Nothrus palustris* and the consequent promotor–remotor movement at the joint. Obviously, further study of the musculature and joints is required in this group. The trochanters and femora in taxa with well developed acetabula may be strongly curved so as to direct the legs forward (I and II) or backward (III and IV).

Femur

This podomere in the Actinotrichida is considered to have been divided primitively into two podomeres but a regressive tendency became established early in the evolutionary history of the group which has led to the progressive suppression of the division (Grandjean, 1952c). According to Grandjean (1954c), three categories of actinotrichid mites can be recognized on the basis of femur characteristics:

A. Adults and immatures with the femur divided into a basi- and telofemur on all legs, as in the majority of the Parasitengona.
B. All or certain of the legs with the femur entire in the larva and protonymph but divided in the succeeding instars, as in the Palaeosomata, Endeostigmata and some Bdelloidea.
C. All active instars with the femur on all legs entire, as in the Tetranychoidea, Raphignathoidea, Astigmata and Euoribatida.

The apparent unique occurrence of divided femora in larval Parasitengona is interpreted by Lindquist (1984) as a 'secondary, within-group transformation reversal'. The instar and the leg in which the division of the femur first appears varies between taxa; for example, in the Palaeosomata the division first appears on leg I in the deuteronymph and on legs II–IV in the tritonymph, while in some Endeostigmata (*Alycus* and *Petralycus*) only femur IV is divided and this first occurs in the deuteronymph. The majority of the Bdelloidea have a division of legs I–III in the protonymph but that of leg IV appears in the deuteronymph. The articulation between the basi- and telofemur may be a eudesmatic hinge joint, as on all legs of the Parasitengona and leg IV of *Alycus*, or represented by a peripodomeric suture forming at the most an adesmatic joint as in, for example, the Bdelloidea and Eupodoidea. In some actinotrichids the division is indicated by only sutural vestiges as on the femora of legs III and IV in the males of some Tarsonemina.

Articulation of the femur with the trochanter in the Prostigmata and Astigmata is usually dicondylic and more or less vertical so permitting promotor–remotor movements at the joint. In the former, the dorsal condyle situated on the proximal edge of the trochanter may not lie directly above the ventral condyle but distal to it so that the axis of articulation is oblique (Figs 2.15B & 2.16). The promotors (=protractors) of the basifemora in *Blankaartia ascoscutellaris* originate dorsally on the trochanter (legs II–IV) or on the coxa (leg I) and insert on the anterior base of the podomere while the remotors (=retractors) arise from the coxae and trochanter and attach to the posterior base of the femur (Mitchell, 1962a). In the Astigmata, the femoral promotor and remotor muscles originate, respectively, on the anterior face of the trochanter and the anterior coxisternal apodeme and insert on the anterior proximal edge and the posterior proximal edge of the femur (Woodring & Carter, 1974). When a telofemur is present as in the Para-

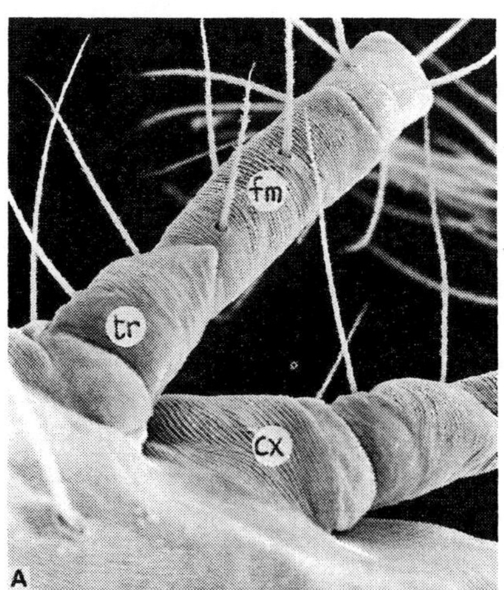

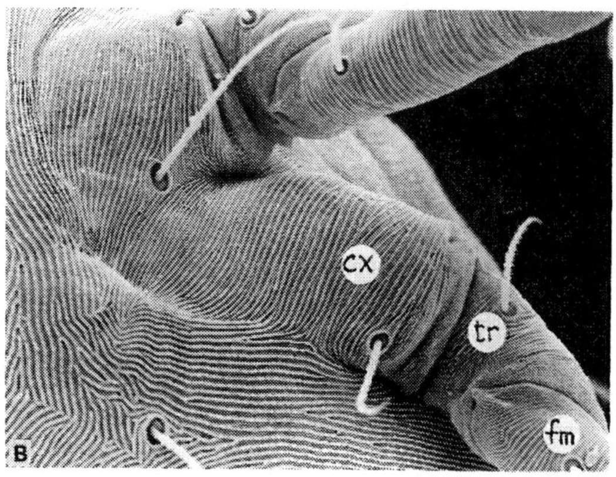

Fig. 2.16. SE micrograph of the basal podomeres of the posterior pair of legs in *Tetranychus urticae* (Prostigmata) in dorsal (**A**) and ventral (**B**) view showing the articulation of the coxa (cx) and trochanter (tr), and trochanter and femur (fm). Note the oblique vertical condylar joint between trochanter and femur and the nodular surface of the arthrodial membrane.

sitengona, a dorsal eudesmatic hinge joint is present between the basi- and telofemur and flexion is obtained through muscles originating on the mid-dorsal wall of the basifemur and inserting on the ventral proximal edge of the telofemur.

The only detailed study of the musculature and joints in the Oribatida is by Akimov and Yastrebtsov (1989) for *Nothrus palustris*. The pivot joint between the trochanter and femur is stated to be horizontal with depressor–levator (flexion–extension) movements effected by antagonistic muscles originating on the epimeric apodeme and inserting on the dorsal and ventral proximal edge of the femur.

Genu, tibia and tarsus

As in the Anactinotrichida, hinge joints are present between the femur and genu, the genu and tibia, and tibia and tarsus with flexion by muscular action and extension by hydrostatic pressure. In forms with two femora, additional flexion occurs at the eudesmatic basi–telofemoral joint (see Fig. 2.15B above), In *B. ascoscutellaris*, the genual, tibial and tarsal flexors originate, respectively, on the dorsal walls of the telofemur, genu and tibia, but in *Allothrombium*, according to Mathur and LeRoux (1965), the genual, tibial and tarsal flexors each comprising two slips arise from the dorsal walls of basifemur/telofemur, telofemur/genu and genu/tibia, respectively. The genual and tibial flexors in the Astigmata arise in the femur and the tarsal flexors in the genu and tibia.

Ambulacrum

The ambulacrum in the Oribatida consists of the apotele (the basilar piece and the claws, the latter formed of modified setae) and a short ambulacral stalk (Fig. 2.17A), A pair of condylophores articulate proximally with the tarsus and distally with lateral cotyles on the basilar piece (Grandjean, 1941a). The origin of the condylophores is debatable. Grandjean (1941a) suggests that they represent the extremity of the tarsus, in which case the internal articulation between the condylophores and the basilar piece would coincide with the articulation between the tarsus and the apotele. The base of the median or empodial claw, unlike the bases of the lateral claws, is partially or entirely fused with the basilar piece. Movements of the claws are effected through the action of flexor muscles originating in the tibia and extensor muscles arising in the tibia and tarsus which connect by tendons, respectively, to the ventral and dorsal surfaces of the basilar piece. In the Astigmata, the base of the empodial claw is also fused with the basilar piece and has a similar articulation with the condylophores (Fig. 2.17B,C). The claw is associated with a membraneous pad or caruncle which in parasitic forms is often enlarged and supported by secondary sclerites. According to Grandjean (1943c), the tendon of the extensor muscles in *Acarus* attaches to the caruncle while the flexor tendon inserts on the basal region (actually to the fused basilar piece) of the empodial claw. Atyeo (1979), in his comparative study of the ambulacra of the Astigmata, considers the terms ambulacrum, pretarsus and apotele to be synonymous (with a preference for the term pretarsus). Hé refers to the caruncle as the conjuctiva which with the claw-complex forms the ambulacral disc. The flexible extension of the tarsus through which the condylophores and tendons run is termed the ambulacral shaft. The apotele in the Prostigmata is basically similar in

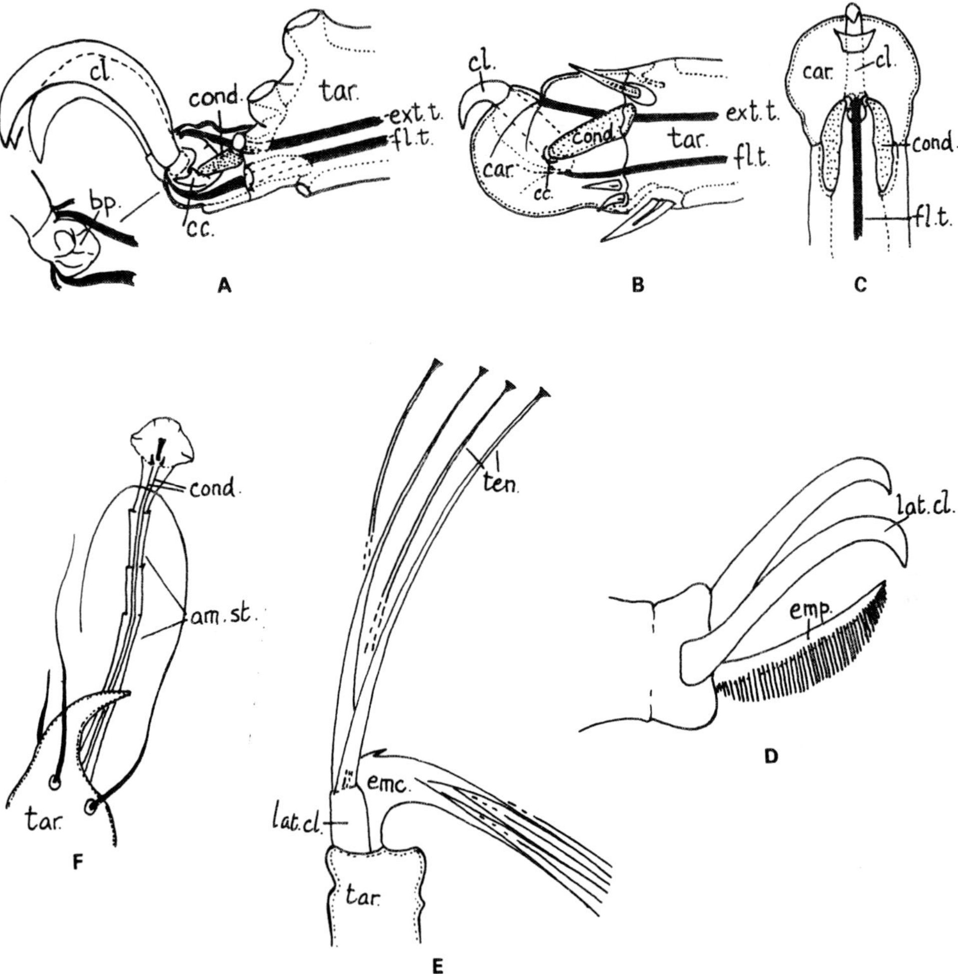

Fig. 2.17. Ambulacra. **A**, oribatid mite, lateral view (based on Grandjean, 1941a); **B**, **C**, acarid mite (Astigmata), lateral (**B**), ventral (**C**) showing articulation between cotyles of basilar piece of empodial claw and condylophores (based on Grandjean, 1943c); **D**, Tydeidae (Prostigmata); **E**, Tetranychidae (Prostigmata); **F**, Psoroptidae (Astigmata). *am. st.* = ambulacral stalk; *bp.* = basilar piece; *car.* = caruncle; *cc.* = cotyle (cotyloid cavity); *cl.* = claw; *cond.* = condylophore; *emc.* = empodial claw; *emp.* = empodium; *ext. t.* = tendon of extensor muscles; *fl. t.* = tendon of flexor muscles; *lat. c.* = lateral claw; *tar.* = tarsus; *ten.* = tenent hair.

structure to that in the Oribatida but may show considerable modification, particularly amongst parasitic and phytophagous forms.

The number of claws ranges from one to three. The tridactyl condition is common in adults of the Euoribatida but the immatures are invariably monodactyl, having only the median claw. The bidactyl form is either the result of the suppression of one of the lateral claws, as in *Nothrus*, or more usually the reduction of the median or empodial claw to a vestige or its modification into a pad-like empodium (Fig. 2.17D) or rayed claw. In the Tetranychoidea, the ancestral form of the ambulacrum is considered to comprise a pair of uncinate claws and a median pad-like empodium, all furnished with tenent hairs, and this condition has been retained in the Tuckerellidae and the majority of the Tenuipalpidae but considerably modified within the Tetranychidae (Lindquist, 1985). For example, in *Tetranychus* the lateral claws are reduced to short rod-like structures from which arise a pair of specialized tenent hairs, and the empodium may be claw-like with three pairs of hair-like processes arising from it proximoventrally or, in the case of the marked reduction in the size of the empodial claw, represented largely by the proximoventral processes (Fig. 2.17E). The ambulacrum in the free-living Astigmata and some Oribatida has one claw which is homologous with the median claw of the tridactyl condition. In certain epizooic Astigmata, the ambulacrum is variously modified (Atyeo, 1979). For example, in mites of the family Freyanidae and Epidermoptidae associated with birds, the functional claws are lacking and the ambulacral disc or caruncle is in the form of a large circular or subcircular bell-like structure with the condylophores, basilar piece and vestiges of the claws forming supporting sclerites, while in the Psoroptidae the relatively small caruncle is borne on a long annulated stalk and associated with a proximal claw-like process of the tarsus (Fig. 2.17F). The apotele may be completely lost on the first pair of legs, when they assume a mainly sensory function as, for example, in the tydeid genus *Pronematus*.

In *Anystis* and *Balaustium*, the long tendon of the extensor muscles extends well into the tibia and some of the muscles originate in the telofemur. During contraction of the ambulacral extensor muscles in these taxa, the tarsus is raised but the extent is limited by cuticular protuberances on the dorsal distal surface of the tibia engaging corresponding depressions on the proximal edge of the tarsus.

Fusion of two or more podomeres to form a compound podomere is not uncommon in the Actinotrichida. Reference has already been made to the union of the ancestral basi- and telofemora in the more derived taxa of the group, a condition which Grandjean (1952c, 1954a) refers to as an anergastic union. Further examples of the fusion of podomeres are seen in the Tarsonemina; for example, in the Tarsonemidae the tibia and tarsus of the first pair of legs and the femur and genu of the third pair of legs are fused in the female to form, respectively, a tibiotarsus and femorogenu, but they remain as separate podomeres in the larva and male. Further, leg IV in the male has usually only four podomeres (trochanter, femorogenu, tibia and tarsus) but this complement is reduced to three in *Acarapis* through the loss of the tarsus while in the female only three podomeres are consistently present and comprise trochanter, femorogenu and tibiotarsus (Lindquist, 1986). Union of podomeres in the Astigmata is seen in the presence of a femorogenu on legs I and II in *Fusacarus* (Grandjean, 1953).

As in the Uropodina, cuticular projections may occur on some podomeres of the Euoribatida which protect the arthrodial membranes or other podomeres. Proximal collar-like tecta referred to as *crispins* may be present on the genu, tibia and tarsus (Passalozetidae) and ventral longitudinal tecta may also be present, especially on the femora. In some genera of the Scutacaridae, the inner edges of trochanters IV have a sclerotized flange which covers and protects the legs when they are withdrawn, with legs III, into lateral recesses formed between the enlarged posterior coxisternal plate and the lateral plates (Ebermann, 1991).

Chaetotaxy

The cuticular organules commonly referred to as setae are classified into at least four types within the Actinotrichida on the basis of their optical, chemical and structural properties (Grandjean, 1935a). The three types of 'true' setae, comprising ordinary setae, eupathidia and famuli, are anisotropic while the fourth type, the solenidion, is isotropic. The characteristics and possible function of these sensilli are discussed in Chapter 3. The true setae of the legs are regarded as having been primitively arranged in whorls or *verticils* as in the Anactinotrichida and the nomenclature for individual setae is based on the systems developed by F. Grandjean for the Oribatida (Grandjean, 1940b, 1941c) and the Astigmata (Grandjean, 1939c). In the Oribatida, whorls of five setae are recognizable in the majority of instances and comprise: dorsal (*d*), anterolateral (*al′*), anteroventral (*v′*), posteroventral (*v″*) and posterolateral (*pl″*) setae (Fig. 2.18A). The symbols ′ and ″ refer, respectively, to the paraxial and antiaxial position of the setae on a podomere. In the Palaeosomata (Fig. 2.18B), a whorl is composed of seven setae: dorsal (*d*), anterior dorsolateral (*ls′*), anterior ventrolateral (*li′*), anterior subtibial (*st′*), posterior subtibial (*st″*), posterior ventrolateral (*li″*) and posterior dorsolateral (*ls″*). When a whorl of six setae is present, as on tibiae I and II of certain Desmonomata, the five-setae nomenclature is used for the anterior face and the seven-setae for the posterior face, thus *d, l′, v′, st″, li″, ls″*.

The dorsal and ventral members of the basal whorls on the tarsi are aligned with those of the genu and tibia but the dorsal setae do not correspond so that a separate nomenclature is used for the tarsal setae (Fig. 2.19A). The base of the ambulacrum is surrounded by two pairs of setae, the dorsal *prorals* (*p*) and the ventral *unguinals* (*u*). Dorsally behind the prorals are the paired *iterals* (*it*), *tectals* (*tc*) and *fastigials*

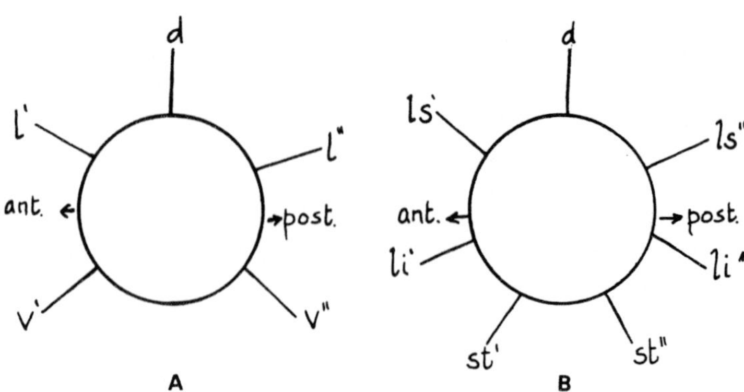

Fig. 2.18. Terminology for the leg chaetotaxy of actinotrichid mites. **A**, nomenclature of the five-setae whorl (verticil) in the Oribatida. **B**, the same of a seven-setae whorl (see text for abbrevations).

(*ft*), and ventrally behind the unguinals are an unpaired ***subunguinal*** (*s*), and paired *anterolaterals* (*a*), *primiventrals* (*pv*) and primilaterals (*pl*). Tarsus I is also provided with a famulus (ε), which is not included when calculating the setal complement of the podomere. All the above setae, except the iterals, are larval setae, or protonymphal in the case of legs IV, and are termed fundamental setae, as opposed to accessory setae which may appear after the larval instar on legs I–III and after the protonymphal instar on legs IV. Accessory setae, on all tarsi, include the *iteral* setae and the posterior setae located behind the fundamental setae, which often occupy the position of setae *pl* and *pv* when these are lost. The iterals usually first appear in the nymphs or adults. A seta appearing in one stase exists also in subsequent stases so that in the adult one finds all the setae present in the larva as well as those which appear later in ontogeny. The prorals and subunguinals and, to a lesser extent, the unguinals, iterals and anterolaterals, may be in the form of eupathidia (chemosensory sensilli). Eupathidia are designated by the Greek letter zeta (ζ).

The solenidia occur only on the genua, tibiae and tarsi. They are designated by Greek letters: σ (sigma) for those on the genua, φ (phi) for the tibiae and ω (omega) for the tarsi (Fig. 2.19B). When more than one solenidion is present on a podomere, the first appearing during ontogeny is given the subscript 1 and the second the subscript 2, etc.; thus, ω_1, ω_2 for the solenidia on the tarsi. The largest number of solenidia occur in the more primitive taxa as, for example, in the Palaeosomata in which there are four solenidia on tarsus I and three or four on tibia I. The podomeric complement of solenidia can be expressed as a solenidial formula comprising a set of three numbers for the genu, tibia and tarsus of each leg, commencing with leg I. The most widespread formula for the Euoribatida is (1-2-2)(1-1-2)(1-1-0)(0-1-0). Many Desmonomata have the formula (1-2-3)(1-1-2)(1-1-0)(1-1-0).

On leg I of the free-living Astigmata, a circle of eight setae occurs

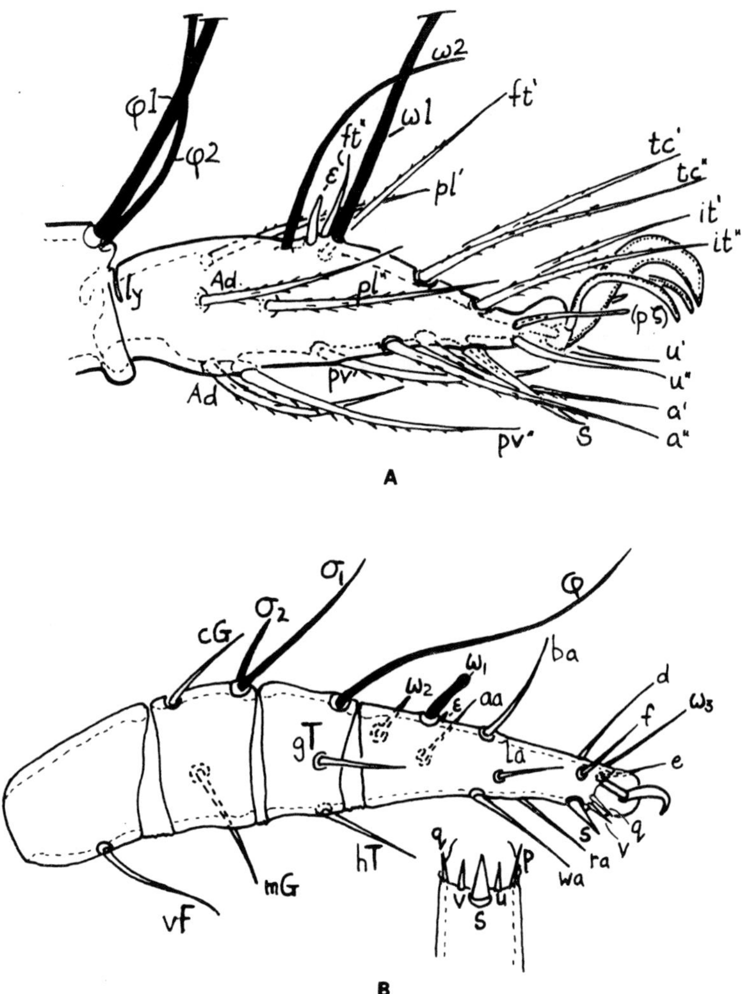

Fig. 2.19. **A**, nomenclature for the setae, famuli and solenidia of tarsus I (and end of tibia I) in *Edwardzetes edwardsii* (Oribatida) (after Grandjean, 1940a); **B**, the same for the femur, genu, tibia and tarsus of leg I of an acarid mite (Astigmata) (see text for abbrevations).

distally on the tarsus and consists of a dorsodistal group *d*, *e* and *f*, and a ventrodistal group *p,q,u,v* and *s*, usually taking the form of short, blunt spines (Fig. 2.19B). These setae are also present on the tarsi of legs II–IV. The median region of tarsus I has a whorl comprising *ba* (dorsal in position), *wa* (ventral), *la* (paraxial) and *ra* (antiaxial). These four setae are also found on tarsus II but only *r* and *w* occur on tarsi III and IV. The basal group comprises two solenidia, namely omega$_1$ (ω_1) and the more proximal omega$_2$ (ω_2), a famulus closely associated with omega$_1$ and seta *aa*. The famulus, omega$_2$ and *aa* are restricted to tarsi I but omega$_1$ also occurs on tarsi II. Setae of the trochanter, femur, genu and tibia are designated, respectively, by the letters R, F, G and T; thus, on the first leg, the seta on the trochanter is *pR*, and on the femur *vF*, while the two setae on the genu and tibia are, respectively, *cG* and *mG*, and *gT* and *hT*.

The appearance of specific setae during ontogeny, the number of setae per podomere and the form of the setae, famuli and solenidia show a remarkably high degree of intraspecific stability, especially in free-living forms, and are used as important taxonomic characters throughout the Actinotrichida.

Autotomy and Limb Regeneration

Autotomy appears to be a rare phenomenon in the Acari and has only been reported in the Notostigmata and the erythraeid *Lucasiella plumipes* (Coineau & Legendre, 1975; Vitzthum, 1940–43). A plane of autotomy is probably present in the Notostigmata since self-amputation of a leg usually occurs between the coxa and trochanter. It is rare to collect these delicate mites with all their long appendages intact. Coineau & Legendre (1975) have shown that regeneration of the autotomized leg of *Phalangiacarus brosseti* occurs between apolyses with the regenerated limb appearing as a loop-shaped structure beneath the cuticle. After the next ecdysis the regenerated leg occupies its usual functional position although slightly shorter than normal. These authors suggest that the 'horse-shoe glands' considered by With (1904) to be coxal glands in *Opilioacarus segmentatus*, are in reality loop-shaped regenerating first pair of legs. Autotomy of the long fourth leg probably also occurs in *L. plumipes*. The genu and tibia of the leg is provided with a dense covering of long setae which gives a plume-like effect to this region of the appendage. The leg appears to break readily at the joint between the trochanter and the basifemur.

Experimental amputation of the legs of a range of other Acari by Rockett & Woodring (1972) has provided some interesting data on limb regeneration. Complete regeneration, irrespective of the site of amputation, only occurs in ticks. This was achieved after one apolysis in *Amblyomma* and after two apolyses in *Argas*. Among the Mesostigmata, the uropodoid *Uroactinia* regenerated a maximum of two distorted podomeres beyond the site of amputation whereas *Macrocheles* showed no regenerative capacity and the site of amputation usually formed the terminal end of the limb in the succeeding instar. However, in a female phytoseiid mite found in the field, Evans (1987) described the regeneration of a full complement of podomeres, except the ambulacrum, after the apparent loss of the tibia, tarsus and ambulacrum in a preceding instar. The regenerated podomeres lacked movable articulation, setae and sensilli. Irrespective of the number of apolyses or the site of amputation, representatives of the Astigmata and Oribatida displayed podomeric reduction and/or distortion distal to the point of incision whereas those of the Prostigmata showed no regenerative capacity and the lack of coagulation of the haemolymph at the site of amputation resulted in the majority dying of exsanguination or desiccation. Coagulation is rapid and a good clot is formed in ixodid ticks, Mesostigmata, Astigmata and Oribatida but the process is very slow in argasid ticks.

Woodring (1969) observed the absence of mitosis during the elongation of a new leg in *Sancassania boharti* which led him to suggest that this mite probably has a constant cell number after the larval instar. A similar lack of post-larval mitosis appears to be characteristic of the Tetranychidae. Little or no limb regeneration occurs in these taxa. In ticks, on the other hand, which exhibit complete limb regeneration after one or two apolyses, extensive mitosis takes place in the post-larval instars during the first phase of apolysis (Balashov, 1963). On the basis of this evidence, Rockett and Woodring (1972) concluded that the degree of limb regeneration is probably determined by the presence or absence of post-larval mitosis.

3 Circulation, Nervous System and Sense Organs

In the arthropods, the circulatory, nervous and neurosecretory (or neuroendocrine) systems are usually closely associated and this is particularly evident in those Acari in which the central nervous system is enclosed within a perineural sinus of the circulatory system. Considerable advances have been made during the last 25 years in determining the fine structure and function of these systems, particularly in the Ixodida. Many of these advances are due to the availability and application of new techniques in the fields of electron microscopy and electrophysiology.

Circulation

The circulatory system in the Acari is lacunar. The fluid-filled haemocoel forms the basic unit of the system and the fluid within it surrounds the tissues of the internal organs and enters the cavities of the appendages. The haemocoel lacks an epithelial lining so that tissues of the internal organs are separated from the haemolymph by thin basement membranes only. The haemolymph not only functions in the transport of nutrients, chemical messengers such as hormones, and waste products but also plays an important mechanical role in supporting tissues and in transferring energy in the form of hydrostatic pressure. In many acarines, the haemocoel contains a loose network of parenchyma cells which appear as a foamy tissue in living forms, and a fairly dense fat body in the areas of the synganglion, coxae and tracheae. The volume of the haemocoel is markedly reduced in some mites, for example Trombiculidae, and Mitchell (1970) considers the mesenteron to have a gastrovascular function as suggested by Mégnin (1876) for the Mesostigmata. In support of this hypothesis Mitchell points to the close connection between the germinal portions of the ovary and the digestive–absorptive mid-gut epithelium, and between the synganglion and oral glands and the anterior mid-gut caeca.

The haemolymph is a relatively clear fluid containing haemocytes (amoebocytes, leucocytes) which are stellate, triangular, rectangular or oval in shape. Following the system of nomenclature proposed by Jones (1962), at least five classes of haemocytes have been recognized in ticks, namely prohaemocytes, plasmatocytes (amoebocytes), granulocytes, spherulocytes and oenocytes. Our knowledge of the functions of these cells is incomplete (Binnington & Obenchain, 1982). Prohaemocytes have a high nucleus to cytoplasm ratio and lack both granules and vacuoles. They possibly serve as germinal cells, particularly for plasmatocytes. The latter have a low nuclear–plasma ratio and their surface usually forms cytoplasmic projections or filopodia. Jones (1950a) considers that 'amoebocytes' are probably involved with or are responsible for the dissolution of soft limb tissue during apolysis in certain prostigmatic mites, while Brinton and Burgdorfer (1971) consider them to be the chief haemocytes in the haemolymph of ixodid ticks and to perform the function of macrophages. Granulocytes are extremely polymorphic, have numerous round or oval granules in their cytoplasm and are believed to

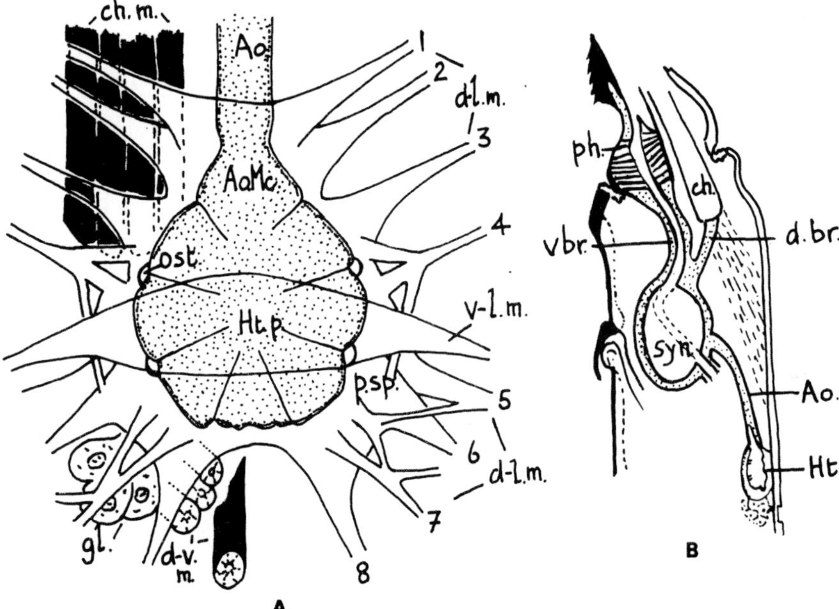

Fig. 3.1. **A**, diagrammatic representation, in ventral view, of the heart, pericardial septum and suspensory muscles of *Amblyomma tuberculatum* (Ixodida); **B**, diagrammatic longitudinal section of the anterior region of an ixodid tick showing the aorta, periganglionic sinus and the division of the anterior arterial sinus into a dorsal branch surrounding the nerves to the eyes and chelicerae and a ventral branch surrounding the stomodaeal and pedipalpal nerves. (Based on Binnington & Obenchain, 1982.) *Ao.* = aorta; *Ao. Mc.* = aortic-myocardial cone; *ch.* = chelicera; *ch. m.* = cheliceral retractor muscles; *d. br.* = dorsal branch of arterial sinus; *d-l. m* = dorsolateral suspensory muscles; *d-v. m.* = dorsoventral muscles of body; *gl.* = glandular tissue of foveae dorsalis; *Ht.* = heart; *Ht. p.* = pulsatory part of heart; *ost.* = ostium; *ph.* = pharynx; *p. sp.* = pericardial septum; *syn.* = synganglion; *v.br.* = ventral branch of arterial sinus.

participate in haemolymph clotting. According to Balashov (1968), oenocytes in engorged female and nymphal ticks appear as giant cells on connective tissue strands and fibres of the fat body. There is no evidence of their relationship to oenocytes of insects.

Circulation in the majority of the smaller Acari results entirely from the action of body musculature and the movement of the internal organs. In ticks and some parasitiform mites, however, circulation also involves a dorsal heart and arterial vessels. The heart in ticks lies mid-dorsally and is surrounded by a pericardial sinus which, according to Obenchain & Oliver (1976), is formed by a perforate septum of connective tissue supported by the terminal processes of dorsolateral and ventrolateral suspensory muscles (Fig. 3.1A). The septum is loosely attached to the dorsal body wall along its lateral and posterior edges and to the wall of the aortic–myocardial cone anteriorly. The heart is divided into two regions on the basis of the orientation of muscle fibres. The anterior region (aortic–myocardial cone), which appears to provide the force that pumps the haemolymph anteriorly, has its striated muscle fibres orientated longitudinally. This region of the heart is considered to be a modified part of the anterior aorta. The posterior region, on the other hand, has a radiating band of striated muscles and is provided with two pairs of ostia in both argasoid and ixodoid ticks.

Diastole is achieved by the contraction of the dorsolateral suspensory muscles and during this phase haemolymph is drawn into the pulsative region of the heart from the pericardial sinus by way of the ostia situated ventrolaterally. The contraction of the intrinsic musculature of the wall of the heart pumps the haemolymph past a septal valve into the anterior aorta which leads into a periganglionic sinus surrounding the synganglion (Fig. 3.1B). The haemolymph then flows laterally through arterial vessels surrounding the pedal nerves and anteriorly through sinuses surrounding the nerves to the gnathosomatic appendages, pharynx and oesophagus. Haemolymph flows back from the lacunae of the appendages into the body haemocoel. It probably enters the pericardial sinus by flowing around the lateral and posterior margins of the septum and by filtration through its perforations (Binnington & Obenchain, 1982). During diastole, the heart is subtriangular (Argasoidea) or pentagonal (Ixodoidea) in outline but assumes a spherical shape during systole.

A heart with two pairs of ostia has been described by Thon (1906) in the Holothyrida. Winkler (1888) records a heart with one pair of ostia and an anterior aorta in the Parasitina but anatomical studies on related groups of the Mesostigmata such as the Dermanyssina and Uropodina have not located a heart.

Nervous and Neurosecretory Systems

The loss of segmentation in the Acari is clearly seen in the form of the central nervous system, which comprises a fused circumoesophageal

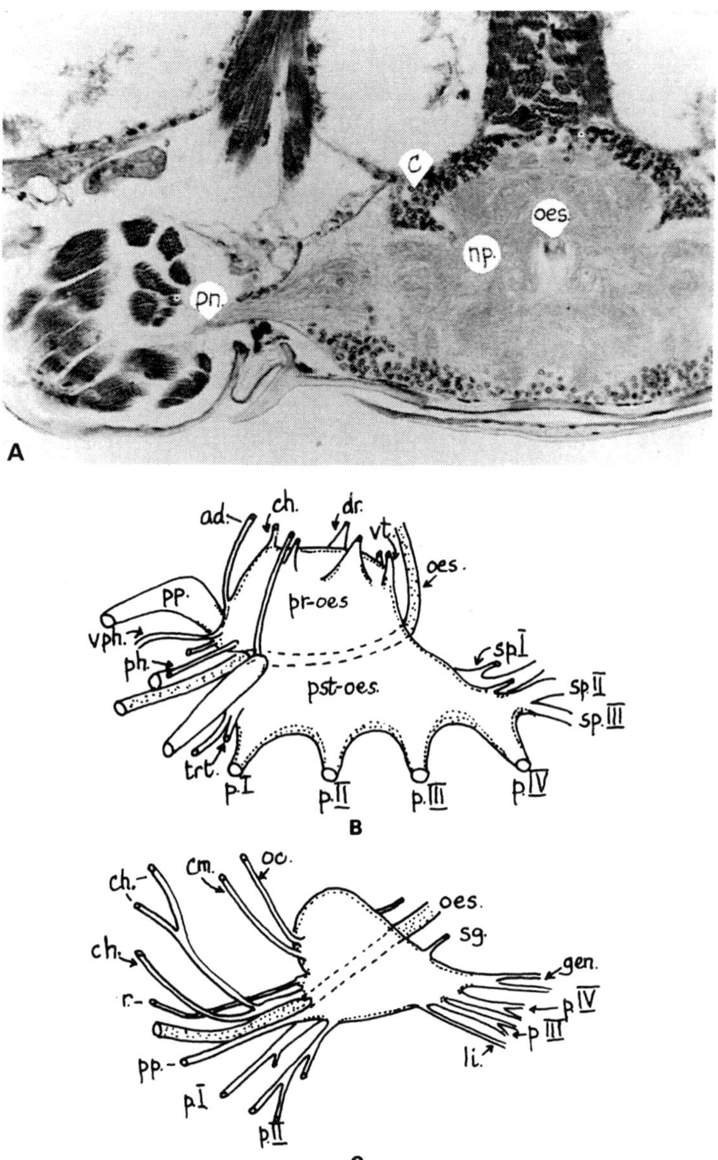

Fig. 3.2. **A**, transverse section of the synganglion of *Spinturnix myoti* (Mesostigmata); **B**, synganglion of *Echinolaelaps echidninus* (Mesostigmata) (based on Jakeman, 1961); **C** synganglion of *Allothrombium lerouxi* (based on Moss, 1962). *ad.* = anterior dorsal nerve; *c.* = cortex; *ch.* = cheliceral nerve; *cm.* = nerve to crista metopica; *gen.* = nerve to genital tract; *li* = lateral idiosomatic nerve; *np.* = neuropile; *oes.* = oesophagus; *oc.* = ocellar nerve; *p. I–IV* = pedal nerves; *pr-oes.* = pre-oesophageal region of the synganglion; *pst-oes.* = post-oesophageal region of the synganglion; *ph.* = pharyngeal nerve; *pp.* = pedipalpal nerve; *pn.* = pedal nerve; *r.* =rostral nerve; *sg.* = somatogastric nerve; *sp I–IV* = splanchnic nerves; *trt.* = tritosternal nerve; *vph.* = ventral pharnygeal nerve; *vt.* = ventricular nerve.

ganglionic mass with peripheral nerves extending to various parts of the body. This synganglion or brain is basically similar in all major taxa and is located in the prosomatic region of the body. In ticks, which have been the most extensively studied, the synganglion is enclosed within a periganglionic blood sinus and only the opisthosomatic nerves penetrate the sinus at their point of origin from the synganglion. The nervous tissue, surrounded by a neural lamella (neurilemma), consists of two zones, an outer cortex (rind layer) and an inner neuropile (Fig. 3.2A). The neural lamella provides mechanical support for the underlying tissues and resists positive hydrostatic pressure. It is considered to be permeable to nutrients and ionic exchange. Coons *et al.* (1974) have shown that the ultrastructure of the lamella differs between ticks and mites. In the former it consists of repeated layers of homogeneous, finely granular material with the areas between the layers occupied by fibres of a collagen-like protein. The lamella of mites (*Macrocheles*), on the other hand, forms a single homogeneous layer lacking fibrillar structures (Coons & Axtell, 1971). In some Mesostigmata and Ixodida a layer of glial cells (neuroglia) lies beneath the neural lamella (Coons *et al.*, 1974; Obenchain, 1974). Axons containing neurosecretory vesicles are present in the lamella.

The cortex consists of perineurium, comprising a thin layer of glial cells, and neuronal cell bodies. Perineurium glial cells in ixodoid ticks contain glycogen deposits and form a continuous layer surrounding the neuronal cell bodies, while in argasid ticks the glial cells and their extensions contain intracellular spaces or vacuoles and are considered to be analogous to the glial–lacunar system of insects. Soluble nutrients may be stored and passed through the vacuoles to underlying neurones. This difference between hard and soft ticks may reflect their different feeding behaviour (Coons *et al.*, 1974). The neurones of the cortical region in ticks are of three main types: motor or association-motor neurones (type I); neurosecretory cells (type II); and olfactory globuli cells (type III). The neurones of one type are usually situated in bilaterally symmetrical groups. Type I neurones are cells with a large nucleus and relatively small cytoplasmic volume. They are present in all ganglia and the perikarya contain mitochondria, Golgi complexes, neurotubules, lysosome-like bodies and a few RER cisternae (Ivanov, 1979). Type II cells have a larger cytoplasmic volume and contain neurosecretory granules. Type III cells have intensely staining nuclei and form a paired mass in the first pedal ganglia. Glial cells investing the neurones in ticks have sheaths which extend into invaginations on the surface membranes of the perikarya. The neurones in the synganglion are mainly unipolar with pyriform (most common), ellipsoid or polyhedral perikarya.

The neuropile, occupying a large central area of the synganglion, appears to be formed of numerous ganglionic areas which are defined by invaginations of the cortex and linked by commissures (Jakeman, 1961). It is largely formed of nerve fibres (axons and dendrites) which are partially ensheathed by thin glial cells (Coons *et al.*, 1974). The glial cells of the neuropile originate in the neuropile or are extensions of glial cells of the cortex.

The oesophagus passes through the synganglion along its central axis and divides it into two regions (Figs 3.2B,C & 3.4 below): the pre-oesophageal (supra-oesophageal) and post-oesophageal (suboesophageal). The former is considered by Ioffe (1963) to include the protocerebrum, which gives rise to optic, cheliceral and pedipalpal ganglia. The post-oral commissure and stomodaeal bridge are considered by Obenchain (1974)

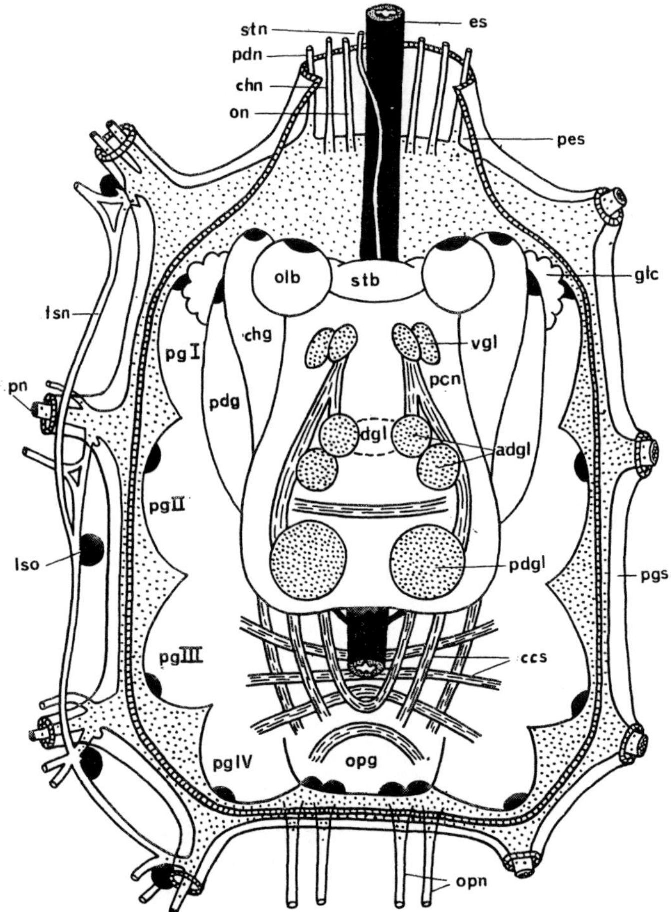

Fig. 3.3. Diagrammatic representation of the synganglion of *Dermacentor variabilis* (Ixodida) lying within the periganglionic sinus, dark areas indicate innervation of ganglia (from Ivanov, 1979). *adgl.* = anterior dorsal glomeruli; *ccs.* = commissure-connective system; *chg.* = cheliceral ganglion; *chn.* = cheliceral nerves; *dgl.* = diffuse glomerulus; *es.* = oesophagus; *glc.* = globuli cells; *lsn.* = lateral sympathetic nerve; *lso.* = lateral segmental organ; *olb.* = optic lobes; *on.* = optic nerves; *opg.* = opisthosomatic ganglion; *opn.* = opisthosomatic nerves; *pcn.* = protocerebral neuropile; *pdg.* = pedipalpal ganglion; *pdgl.* = posterior dorsal glomeruli; *pdn.* = pedipalpal nerves; *pes.* = perioesophageal sinus; *pgI–pgIV,* = pedal ganglia; *pgs.* = periganglionic sinus; *pn.* = pedal nerves; *stb.* = stomodaeal bridge; *stn.* = stomodaeal nerve; *vgl.* = ventral glomeruli.

to be equivalent to the arthropodan tritocerebrum. In *Dermacentor variabilis,* associated centres in the pre-oesophageal region are represented by a number of bilaterally symmetrical glomerular structures of which the anterodorsal, posterodorsal and ventral glomeruli are connected by nerve fibre trunks (Fig. 3.3). In the absence of antennae, the deuterocerebrum is thought to be absent. Pedal and posterior opisthosomatic ganglia are the main components of the post-oesophageal region whose complex of nerve fibres and trunks forms a five-level commissure-connective system (Ioffe, 1963). Lateral 'sympathetic' nerves arise from the bases of the nerve trunks of the first pair of pedal ganglia and connect all the pedal ganglia on each side of the synganglion. Ganglionic swellings, forming the lateral segmental organs, occur on the lateral 'sympathetic' nerves near to the point of origin of each pedal nerve trunk.

The peripheral nerves, at least in ticks, are described as tunicated, the nerve fibres being ensheathed by mesaxons from glial cells. The nerves have mixed sensory and motor axons (Tsvileneva, 1964). Nerves from the pre-oesophageal region of the synganglion serve the eyes or other types of photoreceptors and the gnathosoma (including the labrum, pharynx, chelicerae and pedipalps), while those from the post-oesophageal region serve the legs, the alimentary tract and the genital organs. The majority of the nerves are paired but the oesophageal nerve is a notable exception in being unpaired. The presence of only motor and simple association-motor neurones in the post-oesophageal part of the synganglion of ticks indicates that they have retained a simple bineural (reflex) arc (Balashov, 1968).

Acetylcholine and compounds such as dopamine and noradrenaline are considered to be possible neurotransmitters in ixodoid ticks (Smallman & Shunter, 1972; Megaw & Robertson, 1974).

Neurosecretion

In addition to the neurones which are concerned with the propagation of nerve impulses and with the secretion of transmitter substances, the synganglion also has cells containing secretory products and which are referred to as neurosecretory cells. They differ from typical neurones by their larger perinuclear cytoplasm. Their perikarya contain numerous electron-dense secretory granules as well as Golgi complexes, mitochondria, ribosomes, neurotubules and other cytoplasmic organelles. The secretory products of the neurosecretory cells (neurosecretions or neurohormones) pass down the axons, which end in close association with the vascular system and form compound structures called neurohaemal organs. These organs store and release the neurosecretion into the haemolymph where it is carried around the body to produce a specific physiological effect at a site well removed from the point of its release.

Major neurosecretory pathways and tracts in ticks have been studied in detail in *Dermacentor variabilis* by Obenchain and Oliver (1975) and to some extent in a species of *Ornithodoros* and in *Boophilus microplus* by Gabe (1955) and Binnington and Tatchell (1973), respectively. Up to 18 neurosecretory centres with from one to 16 pairs of neurosecretory cells have been observed in *D. variabilis* and these are indicated diagrammatically in Fig. 3.4. Binnington and Obenchain (1982) state that

> neurosecretion in ticks apparently follows various axon pathways some to discrete neuroendocrine complexes which may be the equivalent of the corpus cardiacum–corpus allatum complex in insects, in part, some to more diffuse terminals associated with the perineural layer–neutral lamella of the synganglion, and some others via peripheral nerves to the lateral segmental organs and probably also directly to various effector sites (as in the salivary glands).

Less information is available on neurosecretion in mites. Woodring

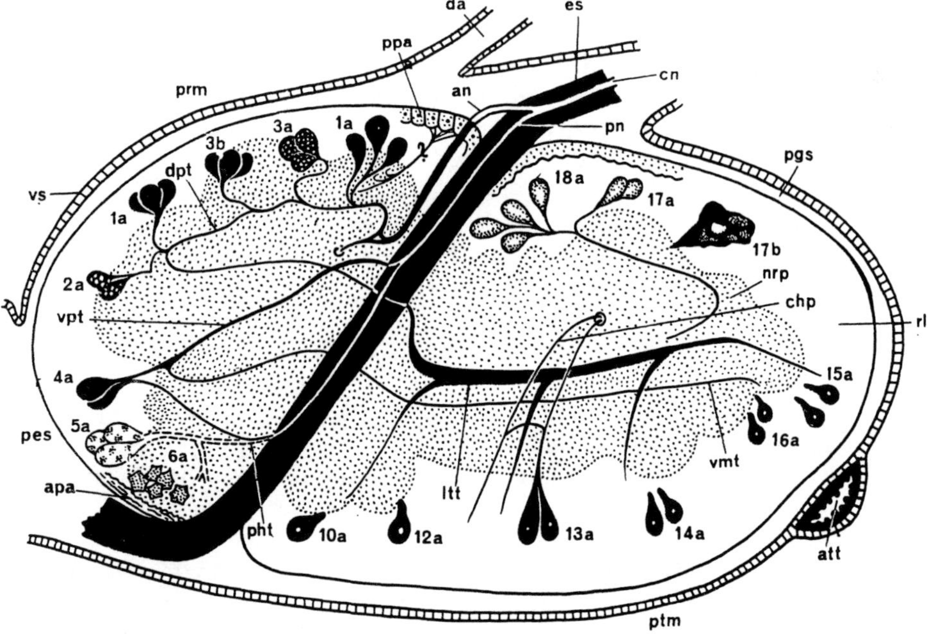

Fig. 3.4. Diagrammatic representation of a sagittal section of the synganglion of *Dermacentor variabilis* (Ixodida) showing major neurosecretory centres (from Ivanov, 1979). *an.* = accessory nerve to retrocerebral organ complex; *apa.* = anterior perineural association; *att.* = anterior tracheal trunk; *chp.* = chiasmatic neurosecretory pathway pedal ganglion II; *cn.* = compound nerve to retrocerebral complex; *da.* = dorsal aorta; *dpt.* = dorsal protocephalic neurosecretory tract; *es.* = oesophagus; *ltt.* = lateral neurosecretory tract; *nrp.* = neuropile; *pes.* = peri-oesophageal sinus; *pgs.* = periganglionic sinus; *pht.* = pharyngeal neurosecretory tract; *pn.* = primary nerve to retrocerebral organ complex; *ppa.* = peri-oesophageal ridge perineural association; *prm.* = pre-oesophageal nerve mass; *ptm.* = post-oesophageal nerve mass; *rl.* = rind layer of synganglion; *vmt.* = ventromedian neursoecretory tract; *vpt.* = ventral protocephalic neurosecretory tract; *vs.* = vascular sheath. 1a–18a = neurosecretory centres; 1a = protocerebral; 2a = anterodorsal cheliceral; 3a, 3b = dorsolateral cheliceral; 4a = anteroventral cheliceral; 5a = frontal stomodaeal; 6a = ventral stomodaeal; 10a = globular; 12a–15a = ventrolateral pedal, 16a = ventral opisthosomal; 17a, 17b = dorsal opisthosomal; 18a = dorsal post-oesophageal.

and Galbraith (1976) described a glandular structure comprising three or four large elongate cells arising from the base of the pedipalpal nerve in *Uroactinia agitans* which they considered to be part of a neuroendocrine system. Putative ductless glands have also been described by Jakeman (1961) and Coons and Axtell (1971) in the Dermanyssina. The most detailed studies to date have been carried out by Akimov *et al.* (1986) on the bee parasite *Varroa jacobsoni* and by Severino *et al.* (1984) on *Dermanyssus gallinae*. The former plotted 20 neurosecretory centres in the synganglion. The cell group situated within the motor nerve of the fourth leg had the histological characteristics of the lateral segmental organ of ticks.

Attempts to relate a neurosecretory cycle with a specific physiological process, such as moulting, diapause and reproductive cycle, have been made by a number of authors using histochemical methods. In ticks, for example, Ioffe (1965) related the volume of secretory material in neurosecretory cells of the female of *Dermacentor pictus* to the development of ovipositional diapause, while Gabby and Warburg (1976) used the criterion of the numbers of neurosecretory cells present to correlate the activity of the neurosecretory systems with physiological conditions in *Ornithodoros tholozani*. In mites, Hughes (1964) observed the disappearance of neurosecretory granules from cells in the pre-oesophageal region of the synganglion of *Acarus siro* during nymphal apolysis and suggested that the neurosecretions may be involved in the moulting process. More recently, Hänel (1986) has found that 'important periods' in the life of *V. jacobsoni* correspond to patterns of 'filling in' of groups of neurosecretory cells; for example, cell group 1 in the medial protocerebral area of the synganglion appears to be involved in moulting and oviposition; secretions of cell group 2, lying between the two ganglionic masses, seem to be necessary during the entire reproductive process while those of cell group 8, situated in the opisthosomatic ganglion, are probably responsible for the initiation of reproduction. These observations suggest some relationship between neurosecretory activity and the physiological state of the acarine but as Binnington and Obenchain (1982) remark 'Confirmation of relationships suggested by histochemical observations must await demonstration of the physiological effects of extracted hormones.'

Sense Organs

Sensory perception is accomplished by means of a variety of sense organs which are located at the peripheral endings of nerve cells. The most common form of sensory receptor is the sensillus* comprising a cuticular

*There is considerable confusion in the literature concerning the spelling of this word. The author follows those who consider it to be the diminutive form of *sensus* so that the spelling is *sensillus* (plural *sensilli*)

structure by or through which a stimulus is received and conveyed to processes of nerve cells. The majority of sensilli are exteroceptors which detect external stimuli but some act as proprioceptors and register stresses in the cuticle resulting from muscular activity or changes in hydrostatic pressure. The minute size of the sensilli in mites and ticks limits the extent to which their function can be ascertained by experimentation so that their role is often inferred from their structure and location.

On the basis of ultrastructural characteristics Altner (1977) and Altner and Prillinger (1980) classified arthropod sensilli into the following three main categories:

1. **Sensilli without a pore system** (NP-sensilli): these include mechanosensitive setae with a flexible socket but without innervation of the shaft as well as peg-like sensilli with a rigid socket (Fig. 3.5C).

2. **Sensilli with terminal pore system** (TP-sensilli): these appear as hair-like sensilli, pegs or cuticular elevations in which the terminal pore or pores are located near the tip of the sensillus (Fig. 3.5A,B). They have either a flexible or a rigid socket. Those with a flexible socket, which allows flexion of the sensillus when in contact with a substrate, are also provided with a mechanosensitive unit which is stimulated in the same way as in a mechanosensitive seta. Thus, this type of organule performs the dual function of mechanoreception and chemoreception (contact chemoreceptors).

3. **Sensilli with wall pores** (WP-sensilli): these sensilli have a number of pores which are not restricted to the apical region. They may be single-walled (sw) with or without pore tubule systems (Fig. 3.5D) or have a double wall (dw) with slits and spoke channels (Fig. 3.5E). Single-walled and double-walled sensilli with wall pores are indicated, respectively, by the symbols (sw/WP) and (dw/WP). They usually have a rigid socket and are olfactory receptors.

Acarologists engaged in the descriptive morphology of the Actinotrichida use an alternative system of terminology for the sensilli of the body and appendages which was proposed by Grandjean (1935a, 1970a). This system is based on the optical properties (when examined in polarized light) and structure of the sensillus after maceration in lactic acid. Only the cuticular structures of the organule are considered and examination is by light microscopy. Grandjean found that the majority of the sensilli of the body and appendages in the Actinotrichida are birefringent and he considered this to be due to their having a core or layer of anisotropic material, termed actinochitin (later changed to actinopilin), which resists boiling in lactic acid and is insoluble in basic hypochlorites. (It was on the basis of this property that Grandjean first distinguished the taxon Actinochitinosi (=Actinotrichida) from the Anactinochitinosi (=Anactinotrichida excluding the Opilioacariformes), which do not have actinopilin in their sensilli.) The axis of actinopilin is surrounded by a thin isotropic layer which forms the barbules and other processes of ornamented setae (Fig. 3.6A). He further distinguished different types of birefringent sensilli or 'true setae': those with an axis of actinopilin but no protoplasmic core comprising ordinary or normal setae and trichobothria (Fig. 3.6A,B), and those with a protoplasmic core consisting of eupathidia (=acanthoides) and famuli (Fig. 3.6C–F). Other sensilli, mainly on the distal segments of the two anterior pairs of legs, are isotropic, lack actinopilin and have distinct lumina. These sensilli, termed solenidia, may be piliform, baculiform (same diameter at the extremity as at the base) or ceratiform (thinner at the extremity than at the base). They appear to be transversely striated on their inner surface with the striae forming a helical structure (Fig. 3.6G–I). Grandjean (1943a)

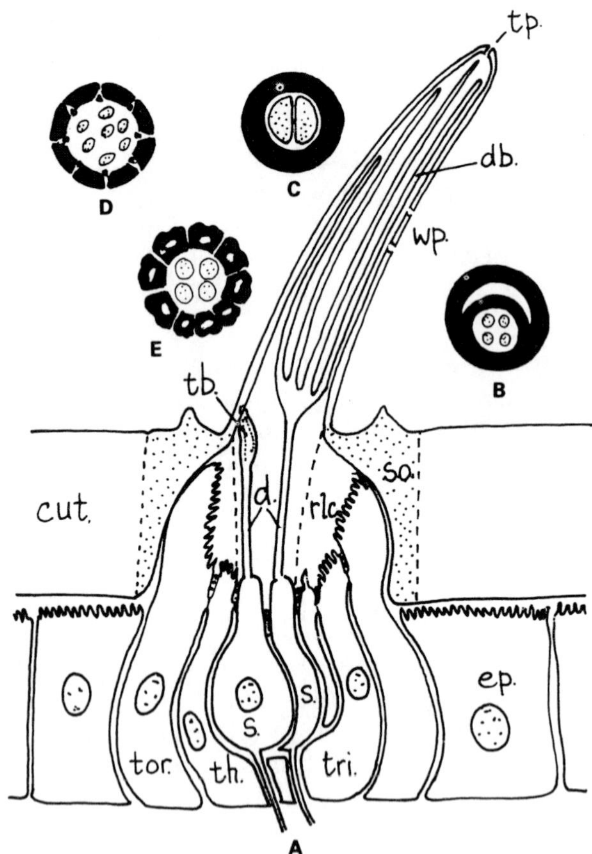

Fig. 3.5. **A**, diagrammatic representation of a peg-like sensillus. **B–E**, transverse sections of different types of sensilli: **B**, with terminal pore (TP); **C**, without wall pores (NP); **D**, with single wall, wall pores and plugs (sw/WP); **E**, double-walled with wall pores (dw/WP). (Based on Altner & Prillinger, 1980.) *cut.* = cuticle; *d.* = dendrites; *db.* = dendritic branches; *ep.* = epidermis; *s.* = sensory cell; *rlc.* = receptor lymph cavity; *so.* = socket; *tb.* = tubular body; *th.* = thecogen cell; *tor.* = tormogen cell; *tp.* = terminal pore; *tri.* = trichogen cell; *wp.* = wall pores.

showed that a seta may become a euphathidion during ontogenetic development but, so far as is known, an ordinary seta cannot develop into a solenidion. There is no information on the chemical composition and molecular structure of actinopilin or on its functional and evolutionary significance. According to Zachvatkin (1952), birefringence of body setae also occurs in some other Arachnida, which he grouped with the Actinochitinosi in the taxon Actinochaeta. It has not been established whether this optical property is due to the general occurrence of actinopilin or to some other substance or substances. For example, as noted elsewhere, calcification produces a birefringent body cuticle in some Oribatida.

Comparatively little ultrastructural work has been carried out on the sensilli of the Actinotrichida which can be used as a basis for reconciling the two systems. The majority, if not all, the setae and trichobothria are poreless mechanoreceptors. The eupathidia are probably terminal pore sensilli with a gustatory/mechanosensory function and their location on the pedipalps and legs would support this. Many of the thin-walled baculiform and ceratiform solenidia have wall pores which are visible in

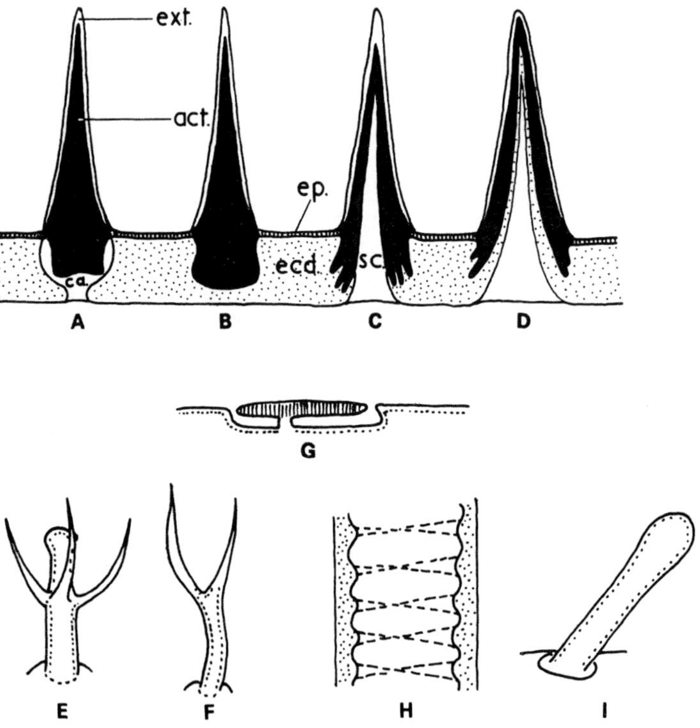

Fig. 3.6.A–D, schematic sections of some different types of birefringent setae based on Grandjean (1947) (actinopilin shown in black). **E–F**, examples of famuli; **G**, recumbent solenidion (rhagidial organ); **H**, optical section of solenidion showing striae; **I**, erect solenidion. *act.* = actinopilin; *ca.* = socket; *ecd,* = procuticle; *ep.* = epicuticle; *ext.* = external isotropic layer; *sc.* = setal canal.

scanning EM micrographs and there seems little doubt that these are olfactory sensilli. The function of the widely occurring aporous famulus, which is restricted to the tarsi of the first legs and usually associated with a solenidion, is unknown.

Five main categories of sensory receptors occur in the Acari: mechanoreceptors, chemoreceptors, thermoreceptors, hygroreceptors and photoreceptors.

Mechanoreceptors

The sensilli of this group show considerable diversity in form but share the property of being activated by mechanical deformation of some part of the receptor by contact with solid objects, by movement in air and water or by stresses in the cuticle.

The most widely distributed and numerous mechanoreceptor is the articulated sensory seta or hair ('le poil ordinaire' of F. Grandjean). It is the product of the epidermis and comprises a bristle-like shaft secreted by a trichogen cell. The shaft is movably articulated to the body by means of a membraneous socket which is formed by a tormogen cell. Associated with the base of the seta are one or two neurones whose dendrites are enveloped for varying distances along their course by a cuticular dendritic sheath or scolopale secreted by a thecogen cell. These three non-neural auxiliary cells surround the sensory cell(s). The latter does not contribute in any way to the cuticular structures of the organule.

The sensory cell is a bipolar receptor cell and excitation is conducted by way of an axon to the central nervous system. Its dendritic process, the receptor region of the cell, comprises two segments: an inner segment in the form of a cylindrical process of the perikaryon provided with numerous mitochondria, and an outer segment which arises from the inner one as a modified cilium (Altner & Prillinger, 1980). In mechanoreceptor sensilli, the outer dendritic segment terminates in a tubular body which comprises densely packed microtubules (neurotubules) embedded in an electron-dense material of unknown chemical composition. The dendritic sheath surrounding the tubular body connects it to the articulating membrane of the socket (see Fig. 3.5A above). The tubular body is a specific feature of cuticular mechanoreceptors in the Arthropoda and is considered to have a special function in the transduction process (Thurm, 1964).

Tactile setae are the most numerous type of sensory receptor in the Acari and display a wide range in shape and length (Fig. 3.7A–J). In many groups of mites, the number and arrangement of the setae (chaetotaxis or chaetotaxy) of the body and of the legs are used widely as taxonomic criteria and various systems of terminology for tactile and chemosensory sensilli have been introduced (see Chapter 2).

A second type of mechanoreceptor encountered in mites is the

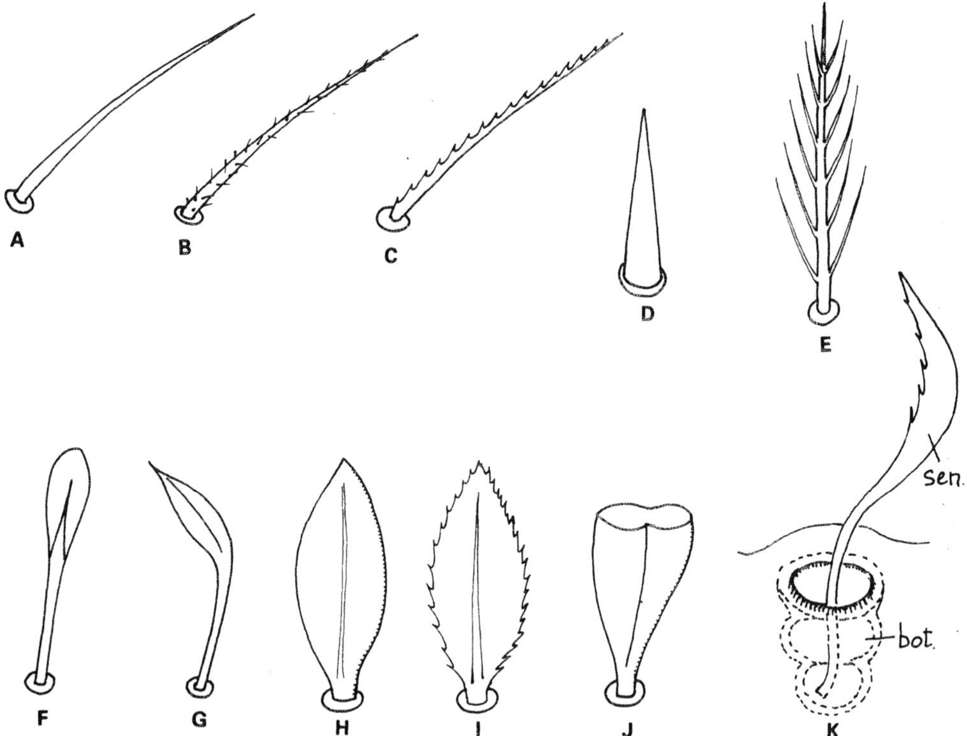

Fig. 3.7. Examples of different types of setae in mites. **A**, simple; **B**, pilose; **C**, serrate; **D**, spiniform; **E**, bipectinate; **F**, spatulate; **G**, falcate; **H**, lanceolate; **I**, lanceolate-serrate; **J**, cuneiform; **K**, bothridium (*bot.*) and sensillus (*sen*) of an oribatid mite.

trichobothrium which has a relatively large cup-like socket in relation to its long, slender shaft. The latter has a swollen base and usually arises at right-angles to the socket. The fine structure of this organule in the Acari does not appear to have been studied. It is considered to respond to air movements. Trichobothria occur in small numbers on the legs and body of some Actinotrichida, particularly the Bdelloidea. They are absent in the Anactinotrichida with the exception of two lateroventral trichobothria on the femora in *Simacarus* (Notostigmata) and a trichobothrium-like sensillus on the tarsus of the fourth leg of some nymphal Veigaiidae and Parasitidae (Athias-Henriot, 1969; Leclerc, 1989).

Characteristic paired trichobothria-type sense receptors referred to as 'pseudostigmatic organs' are present on the prodorsum of the Oribatida, except in a few mainly aquatic and marine genera, and in most Heterostigmata. Each receptor comprises a deep cup or cone-shaped bothridium from which arises a sensillus (Fig. 3.7K). In the Oribatida, the internal

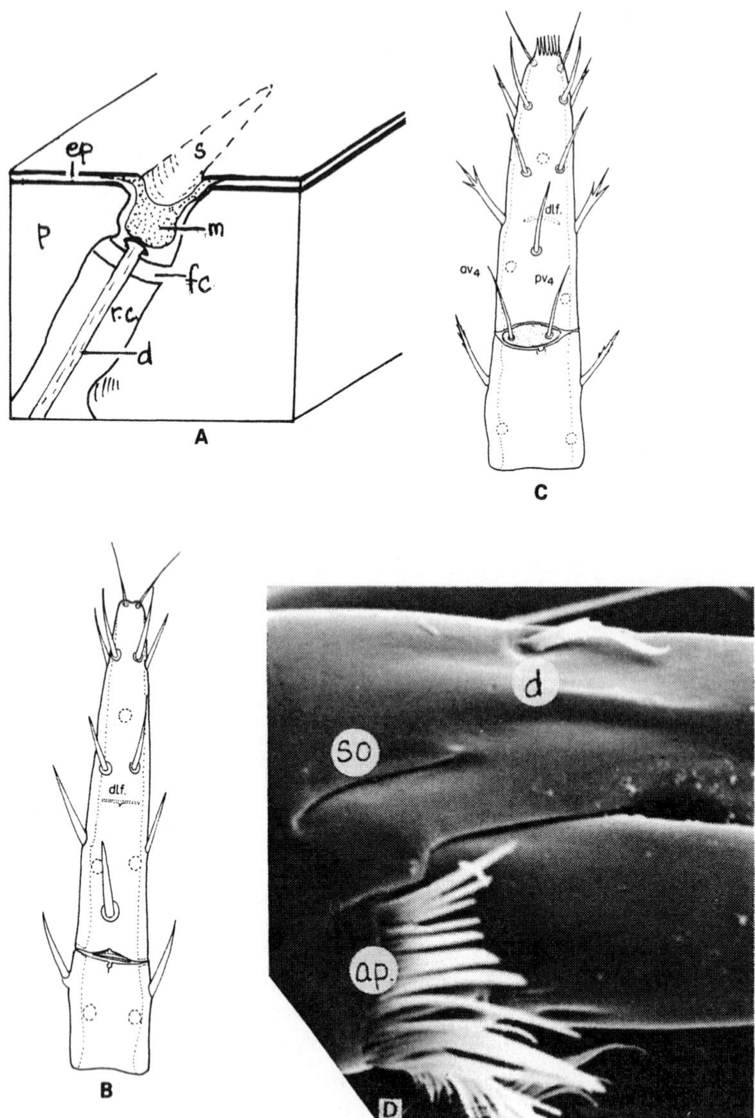

Fig. 3.8. **A**, diagrammatic representation of the structure of a slit organ from the dorsodistal surface of tarsus I in *Amblyomma variegatum* (Ixodida) (based on Hess & Vlimant, 1984); **B**, **C**, peripodomeric suture and dorsal slit organ on tarsus IV (ventral) of a dermanyssoid (**B**) and a sejid mite (**C**) showing the form of the sclerite located lying ventrally in the suture; **D**, antiaxial slit organ on the chelicera of a mesostigmatic mite. *ap.* = arthrodial processes; *d.* (in **A**) = dendrites; *d.* (in **D**) = dorsal seta; *dlf.* = dorsal slit organ; *ep.* = epicuticle; *fc.* = fibrous cup; *m.* = covering membrane; *p.* = procuticle; *ps.* = peripodomeric suture; *rc.* = receptor canal; *s.* = surface slit or depression; *so.* = slit organ.

wall of the bothridium is often plicate, chambered or provided with a helicoidal structure. Fine tubes (brachytracheae) and clusters of sacculi may also be associated with the bothridium but the function of these structures is unknown. The sensillus is diverse in form (setiform, pectinate, clubbed, flattened, etc.). However, Aoki (1973) records a remarkable uniformity in the length (short) and shape (clavate) in arboricolous species. In *Ommatocepheus ocellatus*, the bothridium appears to completely enclose the sensillus. According to Pauly (1956), the sensillus is sensitive to shock and slightly sensitive to air movements, but it has also been suggested that it functions as an organ of balance. Further work is required to elucidate the fine structure and function of this sensory organ. The sensilli in the Heterostigmata are usually capitate and when present occur only in the adult female, except in the Tarsocheylidae where they appear in both sexes. The bothridia are generally located below the lateral margins of the prodorsal shield and the entire organ may be covered by a hood-like expansion of the anterior tergite of the opisthosoma. They appear in this group to be derived from ordinary setae homologues during ontogeny (Lindquist, 1986).

Some sensilli on the terminal article of the palp and the first pair of legs in ticks and mesostigmatic mites, and probably also in actinotrichid mites, are mechanogustatory receptors and are considered in the discussion of chemoreceptors.

Slit organs or lyrifissures measure strains or loads induced by muscular activity, substrate vibrations and haemolymph pressure and are widely distributed in the Arachnida. The receptor is in the form of an elongate slit or hole in the cuticle and is covered by a thin membrane which mainly consists of the epicuticular layer. The only detailed accounts of the structure of slit organs in the Acari are those by Penman and Cone (1974a) in *Tetranychus urticae* and Hess and Vlimant (1984, 1986) in ticks. The distal tarsal slit organ (DTSSO) on the dorsal surface of the tarsi of legs II–IV of *Amblyomma variegatum* is situated in the distal third of the podomeres. It comprises a receptor canal with a fibrous cup which bears the endings of the dendrites of two sensory cells (Fig. 3.8A). The receptor canal extends from the haemolymphatic cavity to the covering membrane of the surface slit and the dendrites are enclosed in a single dendritic sheath whose enlarged tip forms a peduncle which is often continuous with the covering membrane. Forces which result in the deformation of the slit are transmitted to the fibrous cup and are focused to the tip of the dendrites of two sensory neurones, each provided with structurally different tubular bodies, where transduction of the stimulus takes place. Two transverse slits located in the procuticle but without communication with the exterior are covered and partially filled out by covering membrane. They are considered to represent mechanical filters absorbing oblique strains. The sensillus appears to have structural similarities with both arachnid slit organs and insect campaniform organs. Hess and Vlimant (1984, 1986) suggest three possible functions for this type of mechanoreceptor:

1. coordinating walking movements by analysing strains in the tarsal cuticle set up by the weight of the body;
2. gravity perception;
3. detection of substrate/air vibrations.

In regard to the last mentioned, ticks are known to react to substrate vibrations and sound has been found to be a host-detection cue in *Ornithodoros concanensis* by Webb (1979). Subsequently, Hess and Vlimant (1986) have discovered an additional nine different structural types of intracuticular mechanoreceptors distributed on the legs of Ixodida. These may be single organules, as in the case of the distal tarsal sensillus referred to above, or composite structures comprising a ventral and two or more lateral sensilli forming an incomplete ring-like suture (peripodomeric suture) in the basal region of the femur and tarsus. Argasoids have a greater number of intracuticular mechanosensory leg sensilli than ixodoids and this 'indicates differences in the physiology of their host-seeking behaviour'.

Similar types of intracuticular mechanoreceptors appear to be present on the legs of the Mesostigmata and comprise single as well as composite structures (Grandjean, 1935b; Evans & Till, 1979). The composite type, usually comprising three lyrifissures, forms a ring-like scissure in the basal region of femora I–IV and tarsi II–IV. The membrane of the ventral region of the basitarsal scissure of leg IV is often provided with a microsclerite which seems to be a component of the ventral sensillus (Fig. 3.8B). In the nymphs and adults of the suborders Sejina and Antennophorina this so-called intercalary sclerite is provided with a pair of setae (Fig. 3.8C) and this is also the case on tarsi II to III of larval ixodoids (Evans, 1969; Edwards & Evans, 1967).

Slit organs are more numerous in the Anactinotrichida than in the Actinotrichida. With the exception of the Notostigmata, in which the opisthosomatic sensilli are arranged in transverse rows, and the composite peripodomeric sensilli of the legs of the Mesostigmata and Ixodida, slit organs are usually isolated and do not form groups of parallel-arranged slits as in many other archnids. In the Mesostigmata, for example, slit sensilli occur on the chelicerae (Fig. 3.8D), pedipalps, legs and on various regions of the idiosoma. The term cupules is used by some authors for a range of putative mechanoreceptors on the opisthosoma of actinotrichid mites.

Mechanoreceptors having the typical appearance of insect campaniform sensilli occur in ticks (and possibly also in mites). They consist of a socket closed off by a dome of cuticle which has an indentation across the middle (Waladde & Rice, 1982). The sensillus is innervated by a single neurone.

Waladde and Rice (1977, 1982) consider that a terminal papilla on each inner cheliceral digit of the cattle tick *Boophilus microplus* functions as a mechanoreceptor. Two dendrites, of which one possesses a more dense arrangement of microtubules than the other, innervate the papilla (see Fig. 3.11A below). A similar differentiation of dendrites occurs in the pair innervating the shaft of the inner digit where it joins the cheliceral hood membranes. These authors suggest, on the basis of ultrastructural criteria, that the mechanoreceptive neurones of the cheliceral papillae are stimulated as the cheliceral digits cut into the tissues of the host and monitor the interaction between the chelicerae and the integument while dendrites innervating the inner digit shaft may monitor stresses in the cuticle of the shaft as the result of the action of the muscles that pull the digits laterally.

Long spindle-shaped sensory cells attached to the arthrodial membranes of the legs of ***Erythraeus*** are considered to be proprioceptors (Gossel, 1935). According to Waladde and Rice (1982), the stretch receptor described by Beadle (1973) in ticks 'is actually the basal region of a typical trichoid sensillum'.

Schulze (1942) attributed a sensory function to certain duct-like structures occurring in the cuticle of ixodoid ticks. He divided these sensilli into two groups: trichoid sensilli and krobylophores (or *Schopforgane*). The former type includes the 'sensillum auriforme' which appears to be similar in form to a campaniform sensillus. The krobylophores are represented by three types of structures: sagittiform, hastiform and laterniform sensilli. Schulze considered these sensilli to respond to both mechanical and chemical stimuli and to be secretory. The hastiform sensilli (sensilla hastiformia) are the most numerous and these are now considered to be integumentary glands (Douglas, 1943; Lees, 1948).

Chemoreceptors

Chemoreceptors respond to changes in the chemical environment of the organism and the sensory cells are associated with cuticular structures formed by the non-neural enveloping cells (trichogen, tormogen and thecogen). Chemosensory sensilli differ from mechanoreceptors in having the dendritic processes of two or more sensory neurones passing into the lumen of the sensillus. The dendritic outer segments of the sensory cells are bathed, at least for some of their distance, in a fluid secreted by the enveloping cells into a receptor lymph cavity (see Fig. 3.5 above).

Olfactory sensilli are usually concentrated on the dorsal surface of the tarsi, and in some taxa also on the genua and tibiae, of the first pair of legs, which in many groups are predominantly sensory and not ambulatory in function; however, sensilli may also occur on other legs. Gustatory or taste receptors are chiefly associated with the gnathosoma, especially the tarsi of the pedipalps (Fig. 3.9A), and with the tarsi of the first pair of legs. Our knowledge of chemoreceptors in the Acari is mainly based on ultrastructural and behavioural studies although in recent years electrophysiological techniques have been used to determine function. The most detailed work has been carried out on the Ixodida by Foelix and Axtell (1971, 1972), Foelix and Chu-Wang (1972), Roshdy *et al.* (1972), Axtell

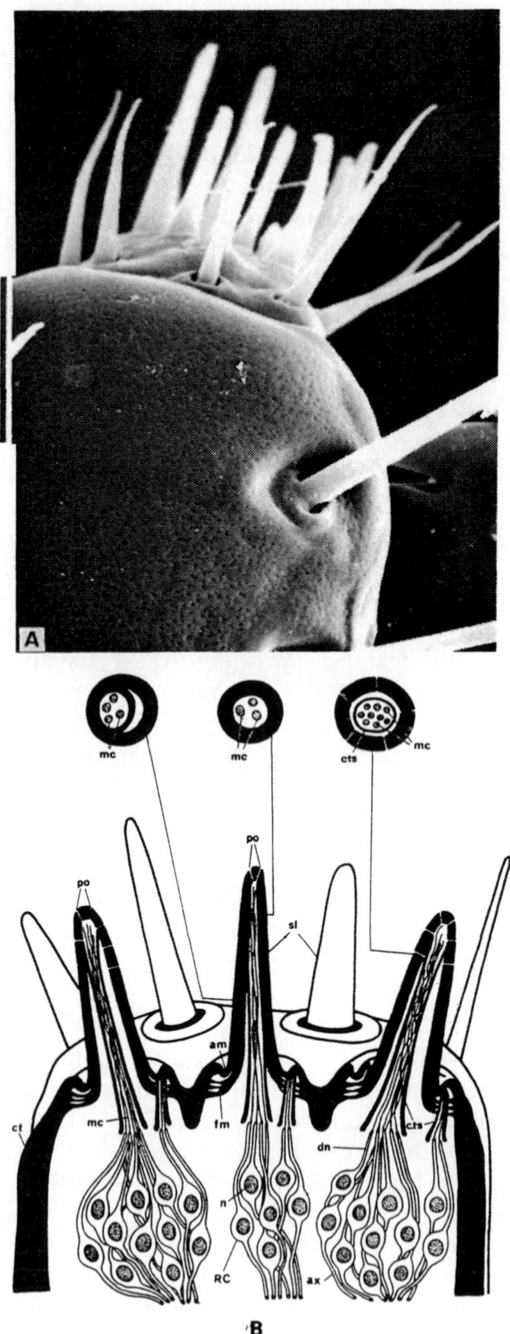

Fig. 3.9 **A**, SE micrograph of the pedipalp-tarsal sensilli in an ixodoid tick. **B**, diagrammatic representation of the structure of the pedipalp-tarsal sensilli (palpal organ) in an amblyommid tick (from Ivanov & Leonovich, 1979). Bar = 10 μm. *am.* = articulating membrane; *ax.* = axon; *ct.* = cuticle; *cts.* = cuticular sheath; *dn.* = dendrites; *fm.* = fibrous membrane; *mc.* = dendritic branches; *n.* = nucleus; *po.* = wall pores; *RC.* = sensory cell; sl = sensillus.

et al. (1973), Waladde (1982) and Hess and Vlimant (1986), and on the mesostigmatic families Macrochelidae, Ascidae and Phytoseiidae, respectively, by Coons and Axtell (1973), Egan (1976) and Jagers op Akkerhuis *et al.* (1985). Comparatively little information is available on the function and ultrastructure of the chemoreceptors in the Actinotrichida (Bostanian & Morrison, 1973; Mills, 1974a).

Most of the detailed studies have been conducted on the chemosensory organules of the tarsus of the pedipalp, the sensory field of the tarsus of leg I and the chelicerae.

Pedipalps

Lees (1948) in his study of the sensory physiology of *Ixodes ricinus* showed by experimentation that the sensilli of the palptarsus ('palpal organ') are probably contact chemoreceptors. This is supported by subsequent ultrastructural and, to some extent, electrophysiological investigations.

The group of ten sensilli occurring on the distal end of the palptarsus of adult ixodoid ticks are thick-walled and, with the exception of a single central sensillus, receive dual innervation, i.e. several unbranched dendrites enter the lumen of the sensillus while two dendrites terminate in tubular bodies and make contact with the articulating membrane of the socket (Foelix & Chu-Wang, 1972; Ivanov & Leonovich, 1979). The dendrites in the lumen of the sensillus are chemoreceptors and communicate with the environment through a single pore or a number of subterminal pores (Fig. 3.9B). Thus, nine of the sensilli have dual mechano-chemosensory function. The shafts of four of the ten TP-sensilli each have two unconnected lumina (type A) of which the circular central lumen has four dendrites and the outer crescent-shaped lumen none (Fig. 3.9B). Both lumina are filled with electron-dense fluid. In the remaining six sensilli (type B), including the sensillus that lacks a mechanosensitive unit, there is only a central lumen which contains 6 to 12 dendrites enclosed in a thin scolopale (Fig. 3.9B). The communication between these dendrites and the exterior is by means of several narrow canals traversing the wall of the sensillus a short distance below its tip. The dendritic ciliary structure comprises 11 double microtubules without central elements (11 + 0) which differs from the (9 + 0) condition in *Argas.* Type A sensilli tend to be longer and more slender than those of type B. The remaining sensilli of the tarsus appear to be tactile in function. According to Waladde and Rice (1982), the end of the palpal organ in *Boophilus microplus* can be retracted to form a depression into which the sensilli are withdrawn and can be everted by hydrostatic pressure to expose the sensilli when required.

Type A sensilli of *Hyalomma asiaticum* have been found by Balashov *et al.* (1976) to be sensitive to gustatory irritants such as salt solutions and the sensilli are considered to orientate ticks to gradients of chemical stimuli such as secretions of skin glands, and to play an important role in

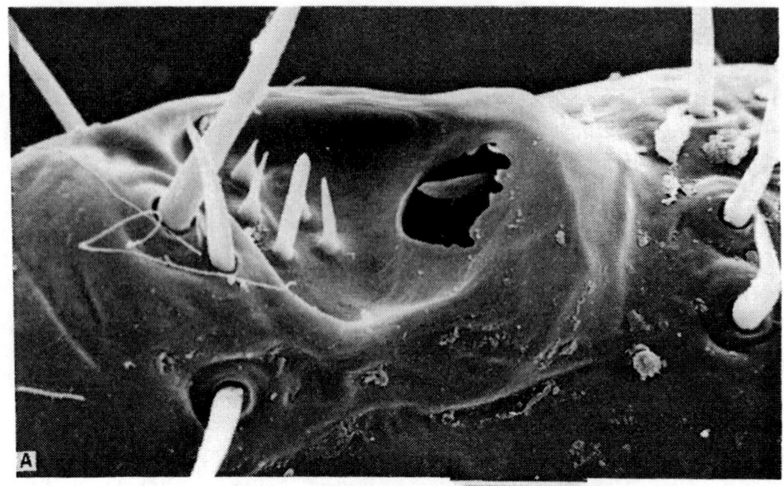

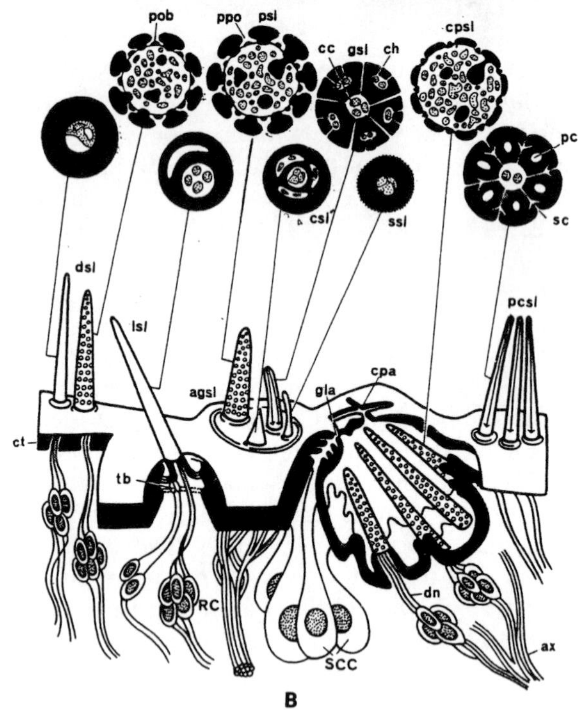

Fig. 3.10. **A**, SE micrograph of tarsus I of *Ixodes ricinus* (Ixodida) in dorsal view showing position of the capsule and anterior pit of Haller's organ. **B**, diagrammatic representation of the sensilli of Haller's organ and surrounding area in an amblyommid tick with sections of the different types of sensilli (only one sensillus of each type is shown in the anterior pit) (from Ivanov & Leonovich 1979). Bar = 10 μm. *agsl.* = sensilli of anterior pit; *ax.* = axon; *cc.* = central cavity; *ch.* = canal; *cpa.* = capsular opening; *cpsl.* = capsular sensilli; *csl.* = conical sensillus; *ct.* cuticle; *dn.* = dendrites; *dsl.* = distal sensillus; *gla.* = gland duct opening; *gsl.* = grooved sensillus; *lsl.* = lateral sensillus; *pc.* = peripheral cavity; *pcsl.* = postcapsular sensilli; *pob.* = plugged body; *ppo.* = plugged pore; *psl.* = wall pore sensillus; *RC.* = sensory cell; *sc.* = slit cavity; *SCC.* = gland cells; *ssl.* = slender sensillus; *tb.* = tubular body.

the behaviour connected with attachment to the host. Type B sensilli do not react to salt or glucose solutions and do not have the functional characteristics of typical taste receptors. It is suggested that they probably perceive pheromones. The palps of male ticks are important in mating and it has been demonstrated that when the palpal organ of male ixodid ticks is cauterized, the males will not copulate or approach females (Feldman-Muhsam & Borut, 1971). Thus, the sensilli of the palpal organ appear to respond to a range of stimuli but our knowledge of the functional response of individual sensilli of the organ is lacking.

The distal palptarsal sensilli in the mesostigmatic predator *Phytoseiulus persimilis* also appear to be of the TP-type with dual mechano-chemosensory function (Jagers op Akkerhuis *et al.*, 1985). In section, the sensilli show two lumina as in type A sensilli of *Amblyomma.* Bostanian and Morrison (1973) record sensilli with a terminal pore on the palps of *Tetranychus urticae* and consider these to be contact chemoreceptors; otherwise there does not appear to be any detailed account of the ultrastructure of palpal eupathidia in the Actinotrichida. An enlarged palptarsal eupathidion in tetranychine spider mites functions as a spinneret.

Legs

Most of the work on the ultrastructure and function of the sensilli of the legs has been concentrated on those of the tarsi of the first pair of legs which in the majority of the Acari function as important sensory appendages.

On the dorsal surface of the fore-tarsus in the Ixodida two sensory areas comprising an anterior pit and a posterior capsule have long been recognized and attempts have been made by experiment to elucidate the function of their constituent sensilli (Fig. 3.10A). These two structures form the components of Haller's organ. The capsule is a relatively deep cuticular depression containing a number of sensilli (Fig. 3.10B). It is covered except for a variously shaped opening by the cuticular extension of the rim of the cavity. The sensilli are thin-walled and lack sockets. The lumen of each is filled by branches of dendrites and the shaft is perforated by large pores. Each pore is plugged by a circular body which has filamentous extensions in direct connection with a dendritic branch. Such pore-plugs are unknown in other arthropods. These sw/WP sensilli have been shown to be olfactory sensilli, and Hess and Vlimant (1986) refer to the presence of receptor cells specific to methyl salicylate (a component of the aggregation–attachment pheromone) associated with these sensilli. Glandular structures comprising groups of two-celled glands open by separate ducts into the capsule of Haller's organ in *Amblyomma americanum* (Foelix & Axtell, 1972) but in *Hyalomma asiaticum* the ducts fuse into a common gland duct (Ivanov & Leonovich, 1979). The function of these glands is unknown.

The chemosensory sensilli of the anterior pit are less uniform in structure but they are all thick-walled and usually have sockets. In the six anterior pit sensilli of *H. asiaticum*, for example, one is a sw/WP sensillus with numerous large, plugged pores and innervated by branched dendrites, two are dw/WP sensilli having small cuticular pores that open into longitudinal grooves along the shaft and give the sensillus a spoke-wheel appearance in section, two are relatively short and slender NP sensilli with unbranched shaft innervating units and one is conical and poreless (Fig. 3.10B). Receptor cells of extracapsular sw/WP sensilli react to 2,6-DCP (dichlorophenol) and to *o*-nitrophenol (a component of the aggregating–attachment pheromone). Ammonia-sensitive cells have been found within the dw/WP sensilli in *Rhipicephalus sanguineus* by Haggart and Davis (1980).

In addition to the sensilli of Haller's organ, aporous tactile setae (NP), terminal pore sensilli (TP) with mechano-gustatory function, double-walled porose olfactory sensilli (dw/Wp) and a single-walled porose olfactory sensillus (sw/WP) also occur on the foreleg tarsus (Fig. 3.10B). The majority of the aporous sensilli are situated in the proximal half of the podomere while the terminal pore and wall-pore receptors are more or less restricted to the distal half.

The most detailed information on the sensilli of the Ixodida is provided by Hess and Vlimant (1986). They recognize a maximum of 13 structural types of sensilli and their distribution has been plotted on the foreleg tarsi of five species of ixodoid ticks and one argasid tick. The maximum number of structural types only occurs in the prostriate ticks while the metastriates and the argasoids have 12 and 11 types, respectively. The numbers of sensilli ('sensory hairs' – SH) and the corresponding numbers of mechanosensory (MSU) and shaft innervating (SIU) units encountered in the different species are presented. In *Amblyomma variegatum* and *Ornithodoros moubata*, for example, the numbers of SH, MSU, and SIU on a foreleg tarsus are 64, 96, 129–185, and 30, 28 and 94–128, respectively. Of the 12 different types (subgroups) of sensilli present on the tarsus of leg I in *Amblyomma variegatum* only three types (two NP and one TP) occur on 'other' podomeres. It was concluded that these three sensilli play an important role in taste reception and possibly also in detecting temperature differences, substrate vibrations and tactile stimuli. Of all the ticks studied ixodoids were found to have a better olfactory sense than argasoids and this probably reflects the difference in importance of olfaction in host-finding between the two groups. Although the number of capsular sensilli ranges between four and seven, the total number of neurones innervating the sensilli is the same in all the species examined and the lower number of sensilli in the smaller metastriate ticks is considered to be due to the fusion of some of the sensilli. So far as taste-sensory organs (terminal pore sensilli) are concerned, metastriates appear to be the best equipped.

Jagers op Akkerhuis *et al.* (1985) in their ultrastructural study of the dorsal sensilli of the distal region of the first tarsus of *Phytoseiulus persimilis* found NP, sw/WP and dw/WP sensilli to be present. The aporous sensilli are long, tapering and provided with sockets whereas the porose sensilli are short, blunt-tipped and with or without sockets. The single-walled WP organule with branching dendrites has funnel-shaped pore canals but unlike similar sensilli in ticks these do not have cuticular plugs. One type of porose sensillus without a complete second lumen has only one dendrite extending into the shaft and the pores are connected to the receptor lymph cavity by mushroom-shaped canals. This type of sensillus has not previously been recorded in the Acari but a similar structure is known for the insects. Recent electrophysiological work has been conducted on the sensory field of the first tarsus in *P. persimilis* and responses were obtained to two of the four known components of a volatile kairomone emitted by Lima bean plants infested with *Tetranychus urticae* and used by the predator in distant prey location (de Bruyne *et al.*, 1991).

On the tarsus of the first leg in *Tetranychus urticae*, Bostanian and Morrison (1973) report the presence of mechanosensory setae (NP), thick- and thin-walled sensilli with longitudinal stripes and a single apical pore (TP-sensilli), and numerous multispiculate 'setae' with each spicule having an apical pore. The TP sensilli were considered to be contact chemoreceptors and the multispiculate setae to have mechanoreceptor and olfactory functions but further study is required to establish their ultrastructure and function. Working on the same species, Mills (1974a) investigated the fine structure of some of the anterodorsal idiosomatic setae. Setae v_2, sc_1, sc_2 and c_3 are each innervated by two sensory cells whose dendrites terminate in tubular bodies. The setal cilia have 8 + 0 pattern of doublet microtubules. A mechanosensory function was suggested for these setae. Setae c_1, however, appeared to be innervated by more than the two dendrites terminating in tubular bodies and the setal cavity contained a cellular process which might represent the dendrites of another sensory cell(s). A mechano-chemosensory function was attributed to this organule. The small amount of ultrastructural work which has been carried out on cuticular organules in the Actinotrichida is surprising in view of the detailed knowledge of other aspects of the external morphology of this group.

The location of some of the dorsal sensilli of the distal podomeres of the anterior legs in cuticular depressions is relatively uncommon in mites. Ampullaceous sensilli in which the sensillus is sunk deep into the cuticle and connected to the surface by a long tube, occur on the first tarsi of the Notostigmata (Grandjean, 1936a) and on the tibia of the foreleg in the Ereynetidae. Recumbent isotropic sensilli (solenidia) in elongate shallow depressions of the cuticle, called rhagidial organs, are present on at least the tarsi and tibiae of the first and second legs of the Eupodoidea (see Fig. 3.6G above). In the Mesostigmata, three small sensilli placed in a capsule-like depression have been described on the first tarsus of the bat mite *Spinturnix myoti* by Evans (1968), and in the Holothyrida some of the chemosensory sensilli of the first tarsus are encapsulated in much the same way as in the Ixodida. The function and fine structure of these organules are unknown. The cuticular depressions and cavities protect the more delicate thin-walled sensilli from mechanical damage and they probably also have the role of isolating a sensillus with a specific function or concentrating a group of sensilli with similar function, such as the olfactory sensilli of the capsule of Haller's organ in ticks.

A close association of a solenidion (which is generally elongated and tapered like a seta) and a shorter tactile seta, often within a combined base, occurs on tarsi I and II within the Tetranychidae and the pair are referred to as duplex setae. A similar type of association between a famulus and a solenidion is found on tarsus I in the Oribatida and forms 'coupled setae'.

Chelicerae

Reference has already been made to the presence of mechanosensory structures on the chelicerae of ixodid ticks but in addition Waladde and Rice (1977) have located organules of probable chemosensory function on the inner and outer digits of ***Boophilus microplus.*** The inner digit has two sensory pits designated ps1 and ps2, and the outer digit has an opening under a small spur (Fig. 3.11A). Pit ps1 is innervated by a single dendrite and ps2 by 11 dendrites divided into three groups comprising five, four and two dendrites, with each group situated in a separate scolopale or dendritic sheath. Electrophysiological studies with contact chemicals have indicated that neurones of the inner digit sensilli respond to cattle blood, plasma, saline, and saline with adenosine diphosphate (ADP) or reduced glutathione (GSH). These sensilli by responding to haematophagostimulants undoubtedly have an important role to play in feeding behaviour. Thirteen dendrites distributed in four dendritic sheaths appear to communicate with the exterior by way of the opening beneath the spur on the outer digit. A chemosensory function has been suggested for this structure.

In addition to the function of detecting phagostimulants in the host's blood, evidence has been obtained from experiments on *Dermacentor variabilis* and *D. andersoni* that the sensilli of the cheliceral digits are also important in copulation (Sonenshine *et al.*, 1984; Sonenshine, 1986), Inner digit sensilli of males respond strongly to washings of tick vagina and to 20-hydroxyecdysone (a possible component of genital sex pheromone). Males deprived of their cheliceral digits fail to copulate or to form spermatophores.

The fixed digit of the chelicera of the Dermanyssina has two trichoid sensilli, the dorsal seta and pilus dentilis (Fig. 3.11B) and an apical 'pit sensillus' (Fig. 3.11C) which bears a resemblance to the pit sensillin ticks (Fig. 3.11B). The role of these organules is not known but it is possible that the pilus has a mechanoreceptive function and the apical sensillus a gustatory function. The apical pit sensillus and pilus dentilis in the Uropodina appear to be replaced by subapical antiaxial sensilli of very different form (Fig. 3.11D). In the oribatid *Hermannia gibba*, Walzl (1987) has found that the teeth of the fixed and movable digits of the chelicerae contain the dendrites of sensory cells. The dendrite(s) of a tooth terminates in an apical pit and each tooth possibly functions as a TP sensillus with a gustatory function. This structure of the cheliceral teeth is most interesting in view of Grandjean's (1947) hypothesis that the fixed digit in Oribatida is of setal origin. The seta *chb* on the fixed digit is a mechanoreceptor.

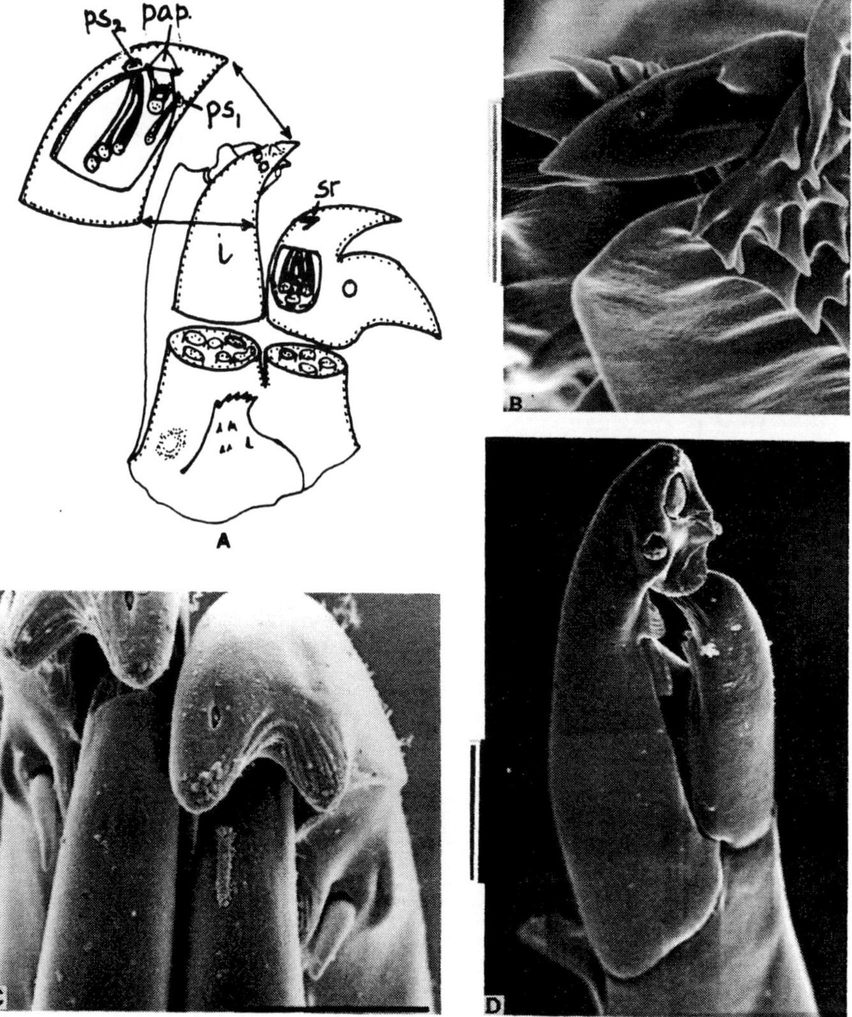

Fig. 3.11. A, cut-away diagram of the cheliceral digits of *Boophilus microplus* (Ixodida) showing details of dendrites and location of the outer digit spur and inner digit sensory papilla and pit sensilli (after Waladde & Rice, 1977). **B**, SE micrograph of the pit sensillus on the cheliceral digit of *Ixodes ricinus* (Ixodida). **C**, SE micrograph of the apical 'pit sensillus' on the fixed cheliceral digit of a species of *Paragamasus* (Mesostigmata). **D**, SE micrograph of the distal sensilli on the fixed digit of a species of *Dinychus* (Mesostigmata). Bar. = 10 μm. *i.* = inner digit; *o.* = outer digit; *pap.* = papilla; *ps.* = pit sensillus; *sr.* = spur.

Thermoreceptors and hygroreceptors

Lees (1946) observed the behavioural response of the sheep tick, *Ixodes ricinus*, to temperature and humidity after the removal of the tarsal segments of the first pair of legs, or the partial or complete exclusion of Haller's organ using cellulose paint. He concluded that the sensilli of the pit of Haller's organ were hygroreceptive while the short (less than 50 μm), straight, thick-walled sensilli on the dorsal and lateral faces of the proximal half of the first tarsi were probably thermoreceptive. Although these experiments appear to have established the location of humidity receptors in the pit and temperature sensilli on the proximal region of the tarsus, they did not identify which of the pit sensilli are hygroreceptive or prove conclusively that the short sensilli act as thermoreceptors.

Subsequent ultrastructural and electrophysiological studies of the tarsal sensilli of the forelegs by Waladde *et al.* (1981) and Hess and Loftus (1984) have succeeded in isolating organules with thermosensitive receptor neurones. The former found that one receptor cell of a median dorsal sensillus (located just behind the capsule) on the first tarsus of *Rhipicephalus appendiculatus* responded to sudden changes in temperature. Spontaneous receptor cell activity was markedly increased by the application of cold stimuli that lowered the ambient temperature by 3–4°C. Warm stimuli in the form of human breath and radiant heat reduced or blocked receptor activity. It was suggested that the cold receptors in the sensillus assist the tick in recognizing temperature changes brought about by thermal gradients originating from the host or by shadows cast by the host during grazing.

Hess and Loftus (1984) demonstrated antagonistic thermal receptors in each of a pair of long, tapering NP sensilli located distally on the tarsus of the first leg of *Amblyomma variegatum*. One of the pair of receptor cells responded to a rapid drop in temperature (cold receptor) and the other to a rapid rise in temperature (warm receptor). In addition to the two thermosensitive neurones whose unbranched dendrites extend into the shaft, the sensillus also has two other units which are mechanosensitive. In *Boophilus*, *Ixodes ricinus* and *Ornithodoros moubata* one or both of these long NP sensilli possess only one shaft innervating unit which suggests that either the cold or the warm temperature-sensitive unit is lacking (Hess & Vlimant, 1986). The sensilli of the proximal region of the first tarsus have the fine structure of TP or NP receptors. The latter, which predominate, are without shaft-innervating units and are mechanosensitive only. Thermo-hygrosensitivity appears to be reduced in argasoids in comparison to ixodoids.

Oribatid mites react to humidity gradients by congregating in preferred zones but no sensilli have been identified that specifically function as hygroreceptors (Madge, 1964).

Photoreceptors

The perception of light has been attributed to a number of sense receptors. 'Eyes' when present are never faceted and appear as circular or elliptical areas of differentiated cuticle often backed by pigment. Among the Anactinotrichida, two or three pairs of strongly pigmented ocelli with well-developed lenses occur in the Notostigmata and they form closely set structures on either side of the prosoma, while a pair of lens-like ocelli has been described in the holothyrid *Australothyrus ocellatus* by Van der Hammen (1983). In the Ixodida, a single pair may be present on the antero-lateral margins of the scutum in metastriate ticks while one to three pairs of photoreceptors may occur laterally in the anterior region of the body in the argasoids (Binnington, 1972). Eyes are absent in the Mesostigmata. In the Actinotrichida, ocelli are commonly encountered in the Prostigmata but more rarely in the Oribatida and Astigmata.

Grandjean (1958) postulated the primitive number of ocelli on the proterosoma of the Actinotrichida to be three pairs, comprising two lateral and one median. The median pair of the primitive complement is probably represented, for example in *Bimichaelia, Brachychthonius* and *Caeculus*, by a hemispherical protuberance lying under an anterior projection, the naso, of the prosoma. This median structure is often bilobed and considered to be a 'double eye' (un 'oeil double'). In some species of Caeculidae, the upper surface of the naso has a clear patch which Coineau (1970) refers to as the 'cornée supérieure'. The function of this structure is not known. In some forms such as *Cyta*, a median ocellus is situated on the antero-dorsal surface of the prosoma while in other Bdellidae a reduced naso overlying a median eye occurs in the Spinibdellinae; a median eye without cuticular differentiation (cornea) has been proved histologically to be present in the Bdellinae and Odontoscirinae (Alberti, 1975). Two pairs of lateral ocelli are frequently present in the Prostigmata and may be sessile or stalked as in some species of Parasitengona or situated on distinct ocular plates as in marine and freshwater forms. A maximum of one pair occurs in the Astigmata and is found on the prodorsum in, for example, adults of *Carpoglyphus* and deuteronymphs (hypopodes) of *Calvolia* and *Histiogaster*, and on the dorso-anterolateral corners of the hysterosoma in deuteronymphs of *Bonomoia* and *Copronomoia*. The structure of the so-called eyes of the Rhagidiidae, Eupodidae and Tydeidae has not been studied in detail and they appear as aggregates of pigment granules in the integument, probably associated with sensory cells, and apparently without differentiated surface cuticle.

The structure of the acarine visual system has been studied in most detail in the Notostigmata, metastriate ticks, Tetranychidae, Caeculidae and Oribatida. The ocellus typically consists of a cuticular corneal lens, biconvex or convex–concave, and an underlying photoreceptor unit comprising retinula cells whose axons form the optic nerve. The microvilli of a retinula cell form one or more rhabdomeres. A rhabdome is

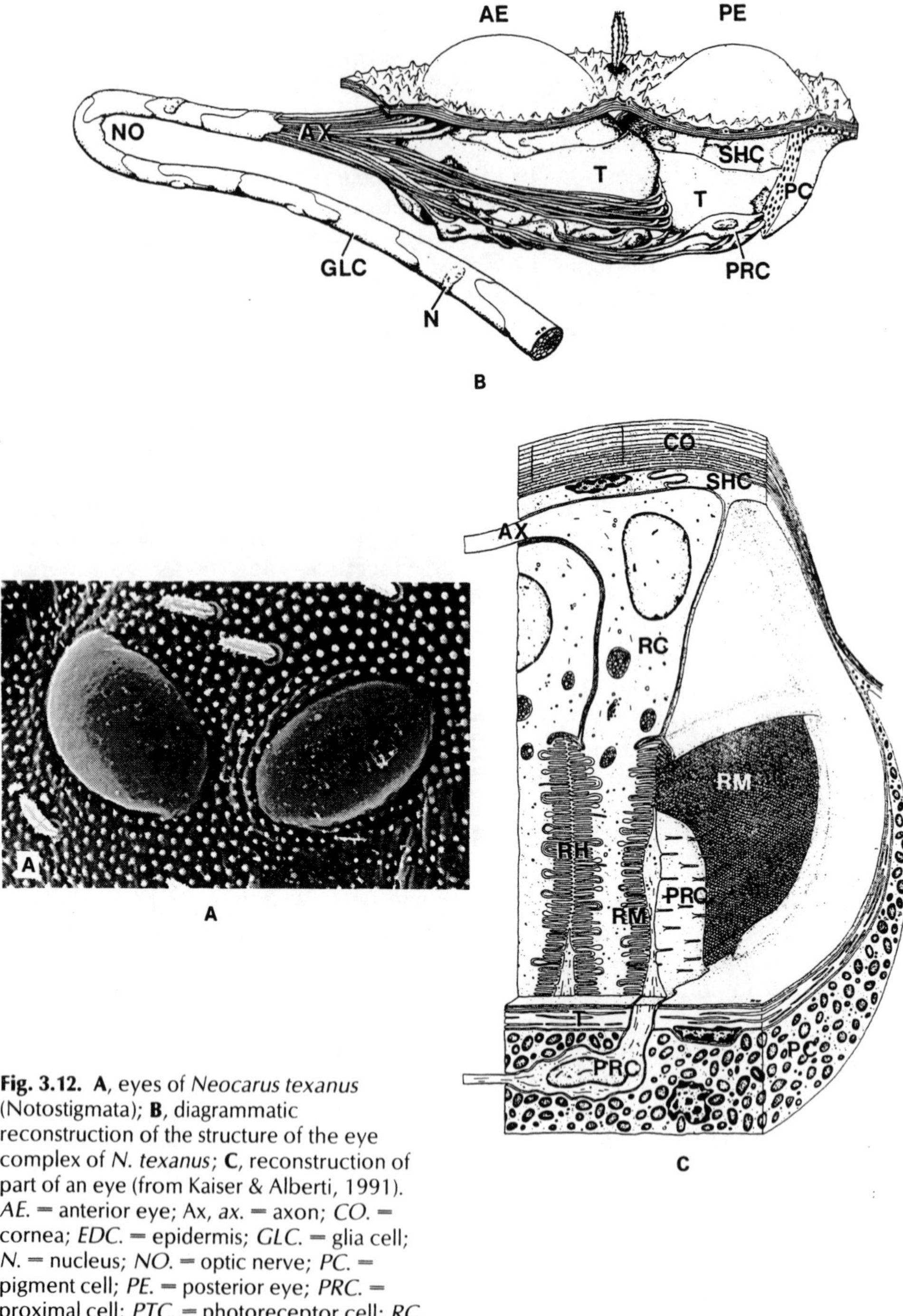

Fig. 3.12. **A**, eyes of *Neocarus texanus* (Notostigmata); **B**, diagrammatic reconstruction of the structure of the eye complex of *N. texanus*; **C**, reconstruction of part of an eye (from Kaiser & Alberti, 1991). *AE.* = anterior eye; Ax, *ax.* = axon; *CO.* = cornea; *EDC.* = epidermis; *GLC.* = glia cell; *N.* = nucleus; *NO.* = optic nerve; *PC.* = pigment cell; *PE.* = posterior eye; *PRC.* = proximal cell; *PTC.* = photoreceptor cell; *RC.* = retinula cell; *RH.* = rhabdom; *RM, rbm.* = rhabdomere; *SHC.* = sheath cell; *T.* = tapetum.

produced by contact between the rhabdomeric microvilli of adjacent retinula cells. Pigment cells are typically present in actinotrichid eyes but are not always present or are poorly represented in anactinotrichids. A tapetum, a reflecting layer, may be present.

Under the cuticular convex–concave cornea (Fig. 3.12A) of each eye of *Neocarus texanus* (Notostigmata), modified cornea-producing epithelial cells called sheath cells send extensive processes into the retina (Kaiser & Alberti, 1991). The retinula cells, 20 in the anterior eye and 14 in the posterior eye, form respectively ten and seven disc-like rhabdomes. The retinula cells are inversely oriented and have flat processes directed away from the light which bear the rhabdomeric microvilli. Each rhabdome encloses the dendritic process of a neurone whose perikaryon is located outside the retina (Fig. 3.12B,C). These neurones are called proximal cells. The retina of each eye is underlain by a crystalline tapetum (produced by three cells) which surrounds it like a thin-walled cup except for nearly circular holes through which dendritic processes of the proximal cells pass. The two eyes on each side of the body are situated in a common capsule of pigment cells whose cytoplasm is densely filled with electron-dense pigment granules. The optic nerve, one on each side of the body, is formed from the axons of the retinula and proximal cells and contains afferent fibres only. The presence of these cells together with the inverse retinae and tapeta form a unique combination within the Acari and probably reflect a primitive condition. The resolution of the eyes is considered to be rather low.

The lens in ticks is a hemispherical cuticular thickening separated from the photoreceptor cells by a layer of epidermal cells (Fig. 3.13A,B). The distribution of pore canals in the lens cuticle differs from that in the surrounding cuticle, and in *Amblyomma americanum*. Phillis and Comroy (1977) state that the pores are thinner and organized in parallel bundles. In *Hyalomma asiaticum*, the photoreceptor section of the eye comprises 20 to 30 large unipolar sensory cells situated with their longitudinal axes parallel to the surface of the body cuticle. Their thin axons contribute to the optic nerve. The tip of each of these elongate photoreceptor cells (retinula cells) has numerous microvilli which form the rhabdomere. All the rhabdomeres are in contact and therefore may be considered to constitute a rhabdome (Ivanov & Leonivich, 1979). The rhabdomeres are orientated at different angles from each other but are always positioned in the rhabdome so that the orientation of the microvilli is at right-angles to incident light. Thus, the orientation of the photoreceptor cells in ticks contrasts with that in the majority of other arthropods in which the longitudinal axes of the photoreceptor cells coincide with the direction of the light. Large irregularly shaped cells containing electron-dense granules ('pigment-like granules') of unknown function occur beneath the photoreceptor unit. The axons leave the eyes laterally. In contrast to the above, El Shoura (1988) working with *Hyalomma dromedarii* describes everse retinula cells which form several rhabdomes and the axons leave the eye basally. The tick eye is thought to

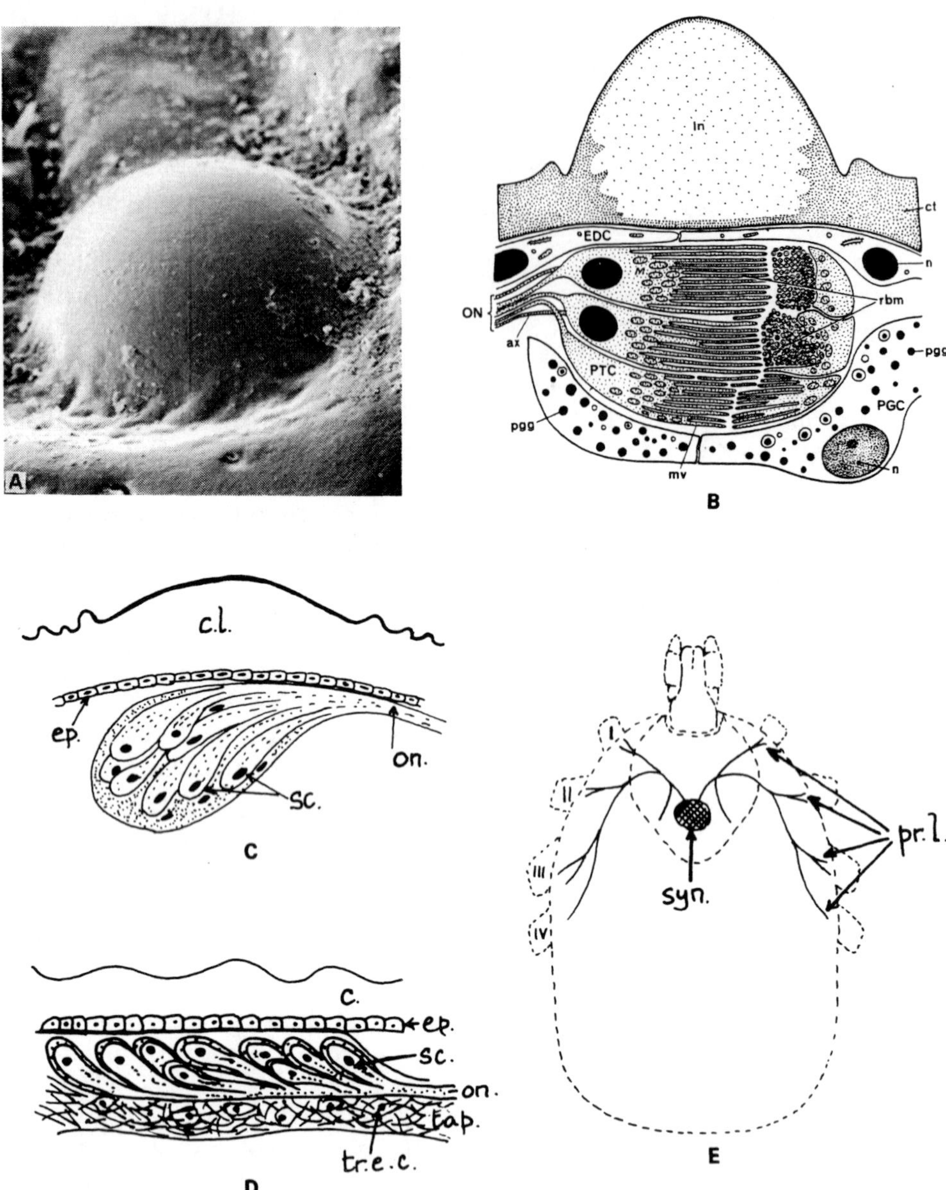

Fig. 3.13. A, SE micrograph of an ocellus of an amblyommid tick; **B**, diagrammatic representation of a section of an eye of *Hyalomma asiaticum* (Ixodida) (from Ivanov & Leonovich, 1979); **C**, schematic section of a lateral photoreceptor in *Ornithodoros gurneyi* (Ixodida); **D**, the same of a lateral photoreceptor in *Argas persicus*; **E**, branches of optic nerves to lateral groups of photoreceptors in an amblyommid tick (after Binnington, 1972). Bar = 10 μm. *ax.* = axon; *c.l.* = cuticular lens; *c, st.* = cuticle; *EDC.* = epidermal cell; *ep.* = epithelium; *In.* = lens; *on. ON.* = optic nerve; *PGC.* = pigment cell; *pgg.* = pigment granules; *PTC.* = photoreceptor cell; *pr. l.* = groups of sensory cells on lateral margin of prosoma; *rbm.* = rhabdomere; *sc.* = sensory cell; *syn.* = synganglion; *tap.* = tapetum; *tr. e. c.* = tracheal cell.

be capable of perceiving differences in light intensity and the direction of light. Species of *Hyalomma* and *Ornithodoros savignyi* are considered to use their visual system in host-seeking behaviour. Recently, Kaltenrieder (1990) has shown that the camel tick *Hyalomma dromedarii* exhibits positive scototaxis (e.g. it orientates towards a black disc in front of a white background) and its response to stationary and moving targets increases with target size.

Binnington (1972) has described anterior and lateral photoreceptors in argasoid ticks and these comprise groups of unipolar neurones lying, respectively, beneath the epidermis lateral to the gnathosoma and in the region of the supracoxal fold above coxae I–III. In *Ornithodoros gurneyi*, two pairs of lateral photoreceptors, each with a cuticular lens, are present in the region dorsal to coxae I–III. The neurones are enveloped by a neurilemma-like membrane and their axons form a branch of the optic nerve (Fig. 3.13C). A cuticular lens is absent in the pair of lateral photoreceptors situated above coxae II–III in *Argas persicus* but each photoreceptor appears as a white elongated spot as a result of reflection from a tracheated tapetum through the transparent cuticle. The 50–60 unipolar neurones of each photoreceptor lie between the integument and the boat-shaped tapetum (Fig. 3.13D). Removal or destruction of the photoreceptor of one side of the body in *A. persicus* interferes with phototaxis.

Some ticks, such as *Ixodes ricinus*, lack eyes but respond to light (Lees, 1948). Binnington (1972) has traced branches of the optic nerves of species of *Ixodes*, *Haemaphysalis* and *Aponomma* to groups of sensory cells in the epidermis of the lateral margins of the prosoma dorsal to legs I–IV (Fig. 3.13E). Their distribution corresponds with Lees' finding in *I. ricinus* that the anterior region of the soma is particularly responsive to increased light intensities. Thus, it appears probable that all ticks, regardless of the presence or absence of eyes, have an optic nervous system which responds to light. This property is likely to be present also in other Acari which lack ocelli but in which light plays an important role in regulating certain biological processes such as diapause.

In *Tetranychus urticae*, a pair of eyes is located on either side of the prodorsum with each pair comprising an anterior and posterior component (Fig. 3.14A). Structurally, each pair consists of three elements: the two lenses, an eye manifold (made up of one vitreous cell, six corneal cells, 15 retinula cells and six pigment cells) and an optic nerve (Mills, 1974b). The anterior eye has a biconvex lens surrounded by red oily pigment while the posterior eye has a convex lens. Both are covered with epicuticle having fine ridges which are considered to serve as an anti-reflection 'coating'. The anterior lens is secreted by the underlying vitreous cone (part of the vitreous cell) and is composed of 25–30 cuticular parabolic lamellae similar to those in the lenses of the ommatidia of insects. The posterior lens is produced by the corneal cells and its internal electron-dense layers may function as an optical interference filter.

Five of the retinula cells serve the anterior eye and have rhabdomeres, each with one or two sets of microvilli, which occupy indentations in the vitreous cone (Fig. 3.14B). The photoreceptors are cup-shaped. The remaining ten retinula cells have rhabdomeres beneath the posterior lens and nine of the cells with their rhabdomeres form a single rhabdome. The majority of these rhabdomeres make contact with the overlying corneal cells and three of them form a pillow-shaped photoreceptor. The cell bodies of the six pigment cells are situated near the vitreous cell. The pigmented processes of the cell bodies form a thin sheath under the posterior lens and penetrate upwards between the retinula processes which connect the rhabdomeres of the anterior eye with their cell bodies. Below the anterior lens, the sheath is relatively thick and surrounds the photoreceptive structures, except dorsally where the vitreous cone is situated. Two kinds of pigment are present in the processes of the pigment cells, namely spherical droplets, light-grey to black in TEM micrographs, which constitute the red oil that gives the eyes their brilliant colour in life, and black staining granules similar to those occurring in ordinary epithelial cells. Axons of the 15 retinula cells in an eye manifold are connected to the pre-oesophageal region of the synganglion by an optic nerve.

According to McEnroe and Dronka (1969), the anterior eye can act as a scanning point detector but does not form an image. Receptors for both green and ultraviolet (UV) are present in the anterior eye. The posterior eye with a simple convex lens is a non-directional receptor for near-UV

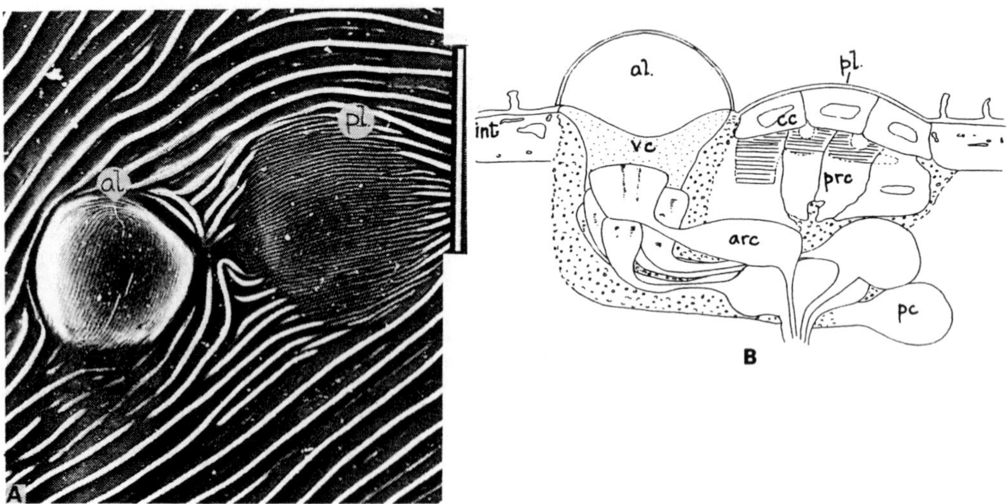

Fig. 3.14. **A**, SE micrograph of the lateral eyes of *Tetranychus urticae* (Prostigmata); **B**, schematic longitudinal section through the lateral eyes of *T. urticae* (based on Mills, 1974b). *arc.* = anterior retinular cells; *al.* = anterior lens; *int.* = integument; *cc.* = corneal cell of posterior eye; *pc.* = pigment cell; *pl.* = posterior lens; *prc.* = posterior retinular cells; *v.* = vitreous cell; *vc.* vitreous cone. Bar = 10 μm.

light only. Reaction to green light is dependent on the feeding status of the mite (McEnroe & Dronka, 1971). Young females which have not fed are green photopositive whereas mites that have fed for at least one day are green photonegative. Further, green photonegative mites maintained at low ambient relative humidity (RH) conditions change to photopositive. The switch in behaviour is due to a decrease in the volume of the mite and is probably perceived by a pressure receptor. However, the loss of the green photopositive response under a high RH regime is considered to be controlled by a hygroreceptor. Mites of both green photo-states exhibit a negative geotaxis in the dark but this can be broken by UV illumination from below. The changes in photo-behaviour are thought to be due to water stress and ambient RH. Green photonegative mites predominate as long as fresh food is available but green photopositive mites develop when the food source is destroyed by heavy feeding activity (Van der Geest, 1985).

The basic structure of the eyes in other prostigmatic mites which have been studied in any detail appears to be similar to that in *T. urticae* but significant differences do occur. For example, in *Microcaeculus* the lenses of the paired lateral eyes are all biconvex, there is no vitreous (crystalline) cone or special corneal cells, and the rhabdomeres of the retinula cells are either single or irregularly grouped (Wachmann, 1975). The lateral eyes in the water mite *Hydryphantes ruber* are also biconvex while the median eye is convex–concave (Mischke, 1981).

Primary eyes with primary receptor cells occur in only few taxa of the Oribatida, for example on the prodorsum of *Heterochthonius gibbus*, and probably represent a plesiomorphic condition (Balogh & Mahunka, 1985). However, in the Euoribatida new light-sensitive organs have been developed and externally appear as areas of clear or translucent cuticle situated on the anterior region of the notogaster. An area of translucent cuticle clearly differentiated from the surrounding darkly tanned cuticle and termed the lenticulus occurs on the anterior region of the notogaster of a number of species and is particularly evident, for example, in the Phenopelopidae, Hydrozetidae and Scutoverticidae. The lenticulus was considered to function as an eye or light-sensitive region by Piersig (1895) and Oudemans (1914, 1917) and its fine structure has been studied recently by Alberti and Fernandez (1988) in *Hydrozetes lemnae*. It comprises a lens-like cornea overlying extensions of epidermal cells, two pigment cells and a pair of lamellated bodies which are accompanied by glial cells and separated medially by two large fat body cells (Fig. 3.15A,B). The lamellated bodies, the presumed photosensitive area, consist of about 100 lamellae that are arranged vertically and oriented longitudinally. The pigment cells are situated postero-lateral to the lamellated bodies and also extend over their dorsal faces. Two bundles of dendrites connect the lamellated bodies with the synganglion. After entering the synganglion each bundle continues to a large lateral perikaryon from which an axon leads to a distinct neuropile. Both axons contribute to an optic chiasma in the centre of the synganglion. In some

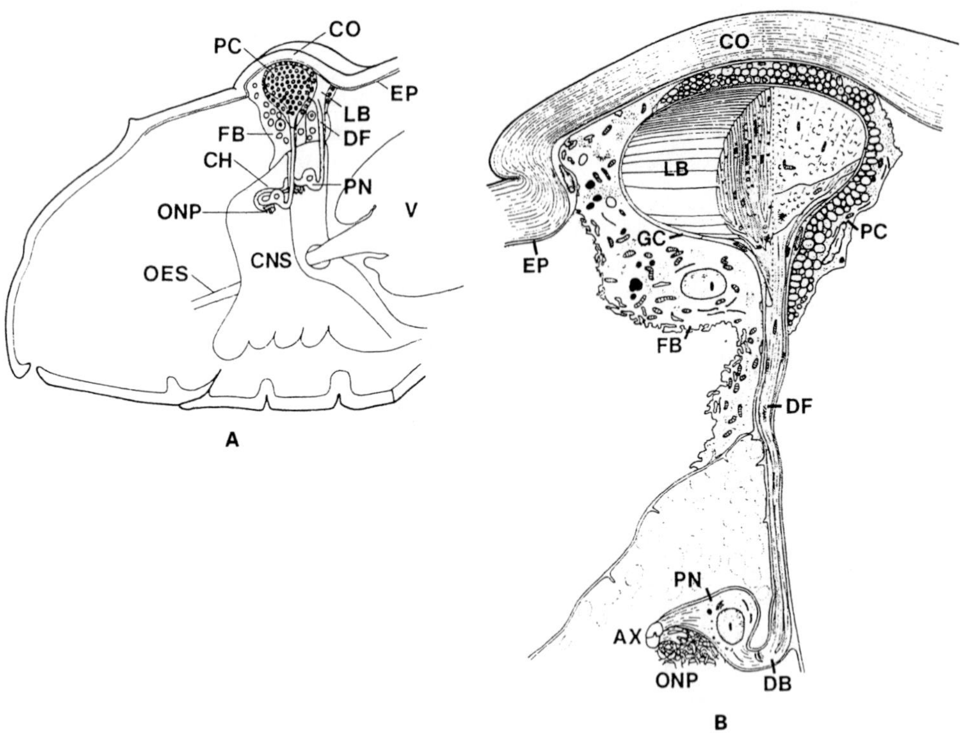

Fig. 3.15. **A**, lenticulus and associated components of *Hydrozetes lemnae* (Oribatida); **B**, schematic sagittal section through the lenticulus and its connection with the synganglion. (From Alberti & Fernandez, 1988.) AX = photoneurone axons; CH = chiasma opticum; CNS = synganglion; CO = cuticular cornea; DB = dendritic base; DF = dendritic fibres; EP = epidermis; FB = fat body; GC = glial cell; LEN = lenticulus; LB = lamellated body; OES = oesophagus; ONP = optic neuropile; PC = pigment cell; PN = perikaryon of photoneurone; V = ventriculus.

other Euoribatida, such as *Oribella* and *Chamobates*, a lenticulus is lacking but a clear spot of cuticle occurs in the same position. These also have lamellated bodies with similar fine structure to those of lenticulate forms (Alberti & Fernandez, 1990). However, the bodies are situated at the dorsolateral surface of the brain and not in close contact with the cuticle. Pigment cells and elongated dendrites are absent.

The presence of an 'eye' in the hysterosomatic region of the body of the Euoribatida is exceptional for the Acari, in which the eyes are typically located in the proterosomatic region of the prosoma. However, the structure of the receptor organ of the lenticulus is also exceptional in not being of the rhabdomeric type and there seems little doubt that it is a unique organ in the Acari and of secondary origin. Alberti and Fernandez (1988) suggest that after the reduction of the proterosomatic eyes light sensitivity was only possible by specialization of distinct neurones in the synganglion, whose dendrites moved to the surface of the central nervous system and finally came to lie close to the dorsal surface of the body. The development of a cuticular lens and of pigment cells from modified epidermal cells completed the required components of a somewhat complex eye.

Photoreceptors have been established by experiment on the pulvilli of the first pair of legs of the snake mite, *Ophionyssus natricis*, by Camin (1953a). They consist of two symmetrically placed pigmented spots on each pulvillus and when the leg is raised from the substrate the 'pulvillus folds up fan-like leaving these spots in a perpendicular plane on opposite sides of the folded pulvillus'. The octopod instars of the mite wave their first pair of legs from side to side during ambulation, similar to the antennae of insects, with the result that the pulvillar regions are constantly moving in and out of the shadow of the mite's body when it is facing away from a light source. Camin observed that the mites move in relatively straight pathways in darkness but displayed winding movements in light. They turned abruptly when coming from a shaded area into light and moved directly back into the darkness. This negative phototactic behaviour is particularly pronounced in replete females and causes them to move into dark crevices for oviposition after feeding under the scales of their host.

4 Respiratory Systems

Arachnids respire by means of invaginations of the integument which communicate with the exterior through openings referred to as stigmata or pneumostomes. These invaginations comprise two distinct types: elongate tube-like structures called tracheae, which usually ramify in the internal organs, and vestibules or atria whose anterior wall is developed into a series of parallel leaflets forming a book lung or phyllotracheae. Scorpions, Amblypygi and Uropygi possess only book lungs, and Pseudoscorpiones, Solifugae, Opiliones, Ricinulei and Acari only tracheae. The Araneae have forms with book lungs or tracheae, or both. In the Palpigradi and some Acari a respiratory system is lacking (Millot, 1949).

A respiratory system comprising tracheae opening to the exterior by paired stigmata (spiracles, ostia) is present in the nymphal and adult phases of the Anactinotrichida and in the majority of the terrestrial Prostigmata. It occurs only in the adults of the Euoribatida and is absent in the Astigmata, the Archoribatida and in some terrestrial and aquatic Prostigmata in which gaseous exchange between the acarine and its external environment is through the cuticle or by means of structures considered to have a respiratory function such as brachytracheae, porose areas of the cuticle and genital tracheae. A respiratory system is lacking in the larvae of the Anactinotrichida with the exception of some species of argasid ticks but is present at this instar in a number of species of Prostigmata (Tetranychidae, Bdellidae). Brachytracheae occur in the larvae of some Euoribatida. Prelarvae lack a respiratory system except in the Tetranychidae and Stigmaeidae in which components of an embryonic gaseous exchange mechanism are present in the prelarva within the egg shell. As in the Hexpoda, the stigmatic–tracheal system is derived from ectoderm and its cuticular lining is shed at ecdysis. Little is known of the chemistry and physiology of respiration in the acarines.

There is considerable diversity in the position of the stigmata and in the form of the tracheal system and these are considered to indicate that the respiratory system is of independent origin in the Anactinotrichida and Actinotrichida, and probably also at ordinal level within these groups

(Hughes, 1959; Witalinski, 1979a). The relative positions of the openings of the tracheal system were used by Kramer (1877) and Canestrini (1891) as the basis of their classifications of the Acari and this is reflected in the nomenclature adopted for the major taxa. The suffix 'stigmata' was used in much the same way as 'ptera' in the formation of ordinal names in the insects; the prefix in the case of the Acari being based on either the position of the stigmata, as in Mesostigmata and Prostigmata, or on some other characteristic such as the hidden nature of the stigmata in the oribatid mates (Cryptostigmata) and the absence of a stigmatic–tracheal system in the Astigmata.

Anactinotrichida

The stigmata and tracheae of the Anactinotrichida, with the exception of ixodid ticks, are hypothesized by Van der Hammen (1968a) to belong to a homonomous series. Of the four pairs of opisthosomatic stigmata (1–4) in the Notostigmata, only stigmata 2 and 3 are considered to be present in the Holothyrida and apparently only stigmata 2 in the Mesostigmata and argasid ticks (Coineau & Van der Hammen, 1979). However, it is now known that only one pair of stigmata is present in the Holothyrida (see below) so that all anactinotrichids other than the Notostigmata have one pair of stigmata in the nymphal and adult phases. In the absence of concrete data on the comparative segmentation of the soma of members of the four orders comprising the Anactinotrichida, it is not possible to relate the position of the stigmata in the Parasitiformes to that in the Opilioacariformes. The apparent location of the stigmata of the Mesostigmata and Holothyrida in the lateral prosomatic region and their association with peritremes would appear to support an independent origin for the stigmatic–tracheal system of these taxa from that of the Notostigmata.

Notostigmata

In adults, a pair of stigmata is located dorsally on each of the second to the fifth opisthosomatic segments (Fig. 12.2A). Few details are available on the postembryonic development of the tracheal system but in *Phalangiacarus brosetti* only the second and third pairs of stigmata are present in the protonymph whereas the posterior pair (fourth) first appears in the deuteronymph and the anterior pair (first) in the tritonymph (Coineau and Van der Hammen, 1979). According to With (1904), in *Opilioacarus segmentatus* three bifurcating tracheal trunks arise from each anterior stigma with one trunk extending anteriorly, one laterally and one posteriorly. Tracheae of the second and third stigmata divide into a strong anterior and thin posterior branch which extend, respectively, into the prosoma and opisthosoma. The tracheae of the fourth pair of stigmata comprise an inner and outer trunk, both of which are bifurcated. Spiral thickenings (taenidia) of the tracheae appear within a short distance of the stigmata.

Grandjean (1935b) noticed that each stigma in the Notostigmata was situated in a segmental row of 'lyrifissures' and this led him to postulate that the stigmata and tracheae originated by specialization of these organules. The detailed structure and function of the 'lyrifissures' are unknown but if they are mechanoreceptors, as is generally assumed, their modification from sensory to respiratory structures would be unique so far as the Arachnida (and probably the Arthropoda) is concerned! If the precursor of the stigmatic–tracheal system is indeed an integumentary organule, then the integumentary gland with its duct would appear to be the type requiring the least transformation.

Mesostigmata

One pair of stigmata is present in the nymphal and adult phases of this taxon and is located laterally in the region of coxae III–IV or, as in the majority of the Uropodoidea, in the region of coxae II–III (Figs 4.1A,B,D & 4.2A). Stigmata and peritremes are absent in the larva. The stigmata are displaced to the dorsal surface in some parasitic forms (Spinturnicidae, Rhinonyssidae) and in the intertidal species *Thinozercon michaeli*, in which the idiosoma is conspicuously flattened dorsoventrally. The stigma is usually subcircular or slit-like in shape and leads into a cup-shaped cavity, the stigmatic chamber, which opens basally by the tracheal orifice into the tracheal atrium (Fig. 4.2B). The tracheal opening is usually guarded by variously shaped cuticular processes or fimbriae, which form a barrier to the entry of particles of organic or mineral origin into the tracheal system. There is no closing mechanism. The stigmatic chamber is extended anteriorly into an open canal or groove, termed the peritreme, whose internal surface is provided with numerous micropapillae (microtrichia, acicula) which may extend over the entire surface or be restricted to the floor of the canal (Fig. 4.1C). The stigma and peritreme are typically located in a peritrematic shield which may lie free or show various degrees of fusion with the dorsal shield and/or the scutal elements, exopodal shield(s), bordering the leg acetabula. In *Varroa jacobsoni*, each peritreme is situated, for the greater part of its length, on the underside of a ventrolateral appendage of the body although the stigma is located in the normal position (Akimov *et al.*, 1988).

Arising from the tracheal atrium are a variable number of major tracheal trunks which in the Dermanyssina and Parasitina branch repeatedly to end in clusters of fine tracheoles. In *Haemolaelaps ambulans*, an anterior trunk branches into all parts of the gnathosoma, the synganglion and the legs while the branches of a posterior trunk extend to the alimentary tract and to the reproductive and excretory systems (Young, 1959). Tracheae from separate atria anastomose in the gnathoso-

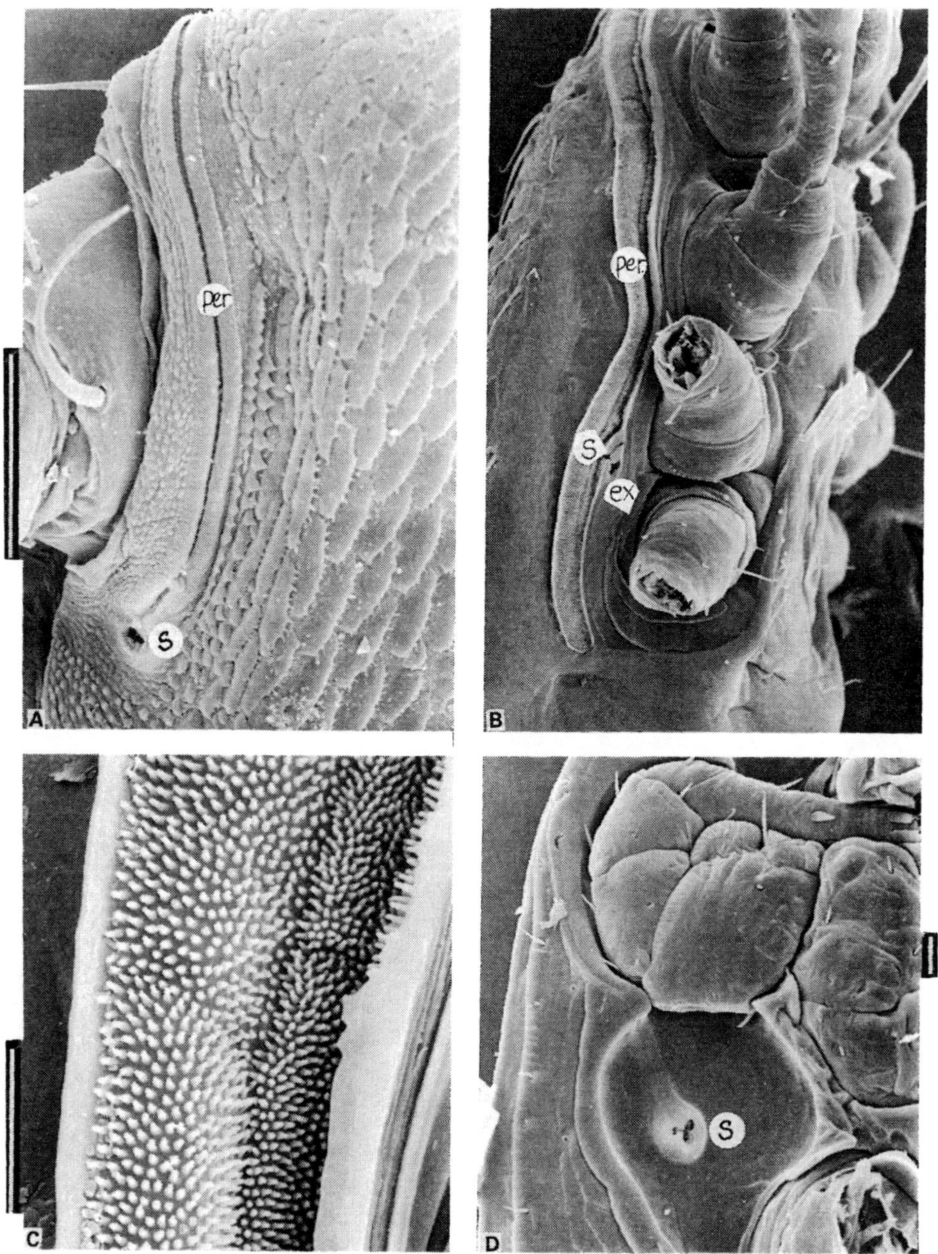

Fig. 4.1. A, B, D SE micrographs of stigmata and peritremes of three mesostigmatic mites: *Spinturnix myoti* (**A**), *Platyseius italicus* (**B**) and *Cilliba cassidea* (**D**); **C**, micropapillae of the open peritreme of *P. italicus*. Note the protection of the stigma by an overlapping part of the expodal shield in *P. italicus* and the location of the stigma in the pedofossa of leg III in *C. cassidea* whose peritreme runs as a narrow slit anterolaterally from the stigma. *ex.* = exopodal shield; *per.* = peritreme; *s.* = stigma.

matic and podosomatic regions and, presumably, make it possible for air entering one stigma to pass throughout the tracheal system. The tracheal atrium in *Echinolaelaps echidninus* also gives rise to two major trunks as well as smaller tracheae serving the dorsal region of the podosoma and the legs (Jakeman, 1961). The connection between the two sides of the tracheal system, however, is by a transverse trachea connecting the two atria (Fig. 4.3A). In some Higher Uropodina the major tracheal trunks are blunt ended and perforated with numerous fine pores terminally from which arise long unbranched tracheoles. Michael (1889), for example, describes a single blunt-ended trachea with tracheoles leading from each stigma in *Leiodinychus krameri* but Woodring and Galbraith (1976) describe four blunt-ended tracheae arising from each atrium in *Uroactinia agitans*.

In *Pergamasus viator*, the tracheae and tracheoles are lined by a cuticular lamina produced by the tracheolar cells which lie externally (Witalinski, 1979a). The lamina comprises an exocuticular component forming a helically coiled band (taenidia) and an epicuticular layer lining the tracheal lumen and extending between the coils of the taenidial band (Fig. 4.3B). Unlike the taenidia in ticks (and insects) in which the taenidial rings are clearly separated and protrude into the tracheal lumen, those in *P. viator* are close to each other and give the appearance, ultrastructurally, of shallow ridges.

The peritreme is a characteristic feature of the respiratory system of

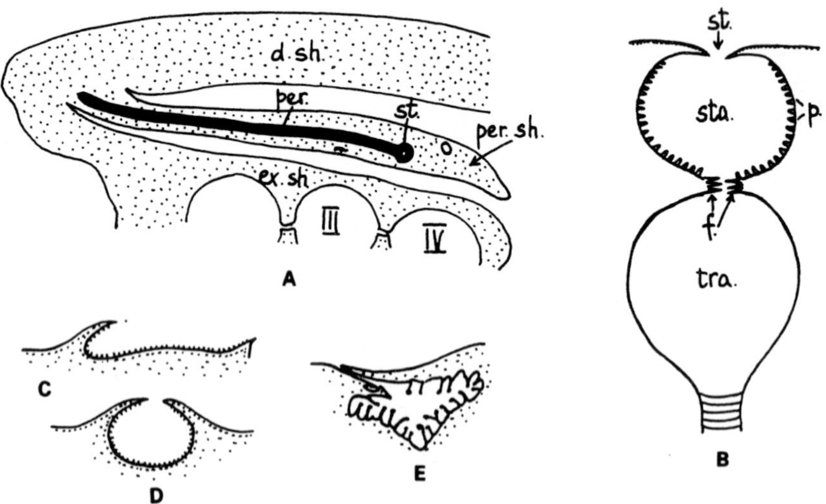

Fig. 4.2. A, schematic lateral view of the stigma and peritreme in the Dermanyssina. **B**. diagram of a section through the stigmatic and tracheal atria of a mesostigmatic mite. **C–E**, schematic sections of the peritrematic canal in *Platyseius* (**C**), *Pergamasus* (**D**) and *Thinozercon* (**E**). (**B** and **E** based on Pugh *et al.*, 1987a, b.) *d.sh.* = dorsal shield; *ex. sh.* = expodal shield; *f.* = fimbriae; *p.* = micropapillae; *per.* = peritreme, *per. sh.* = peritrematic shield; *st.* = stigma; *sta.* = stigmatic atrium; *tra.* = tracheal atrium.

the Mesostigmata and first appears in the protonymph as a short appendage of the stigma that scarcely extends beyond the level of the third coxa. In the majority of species, its length increases in both the deuteronymph and the adult. A notable exception occurs in the Zerconidae in which the adult peritreme is shorter than that of the deuteronymph. The adults of obligatory parasitic species usually show some reduction in the length of the peritreme in comparison with free-living forms, often as the result of the retention of the protonymphal form of the peritreme. Extreme reduction in the peritreme is particularly evident in the endoparasitic Dermanyssoidea such as the Rhinonyssidae but short peritremes also occur in the soil-dwelling species and are particularly evident in the Zerconidae and euedaphic Rhodacaridae. In the Epicriidae, a typical peritreme is lacking (Evans, 1955a). The peritreme is normally a straight channel extending anteriorly from the stigma to the vertical region of the idiosoma but in some Uropodina it follows a sinuous course (e.g. *Dinychus*) while in others the short peritreme is hooked (*Cilliba*). In some taxa which live in wet habitats, for example *Platyseius* and *Dinychus*, the peritreme is also developed posterior to the stigma (Fig. 4.1B).

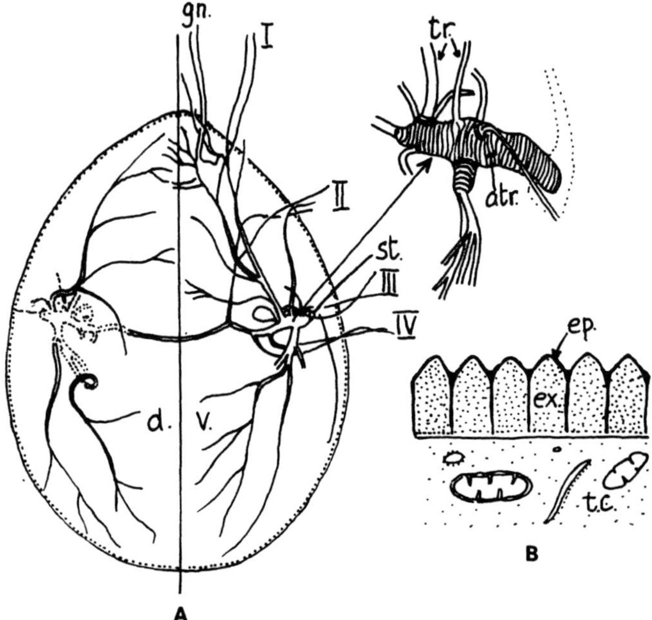

Fig. 4.3. A, tracheal system of *Echinolaelaps echidninus* (Mesostigmata) showing tracheae supplying dorsal (d) and ventral (v) regions of the body and enlarged view of the stigmatic and tracheal atria (after Jakeman, 1961). **B**, schematic section of the cuticular lamina of the tracheae and tracheoles of *Pergamasus viator* (based on Witalinski, 1979a). *atr.* = combined tracheal and stigmatic atria; *ep.* = epicuticular layer; *ex.* = exocuticular component; *gn.* = gnathosomatic tracheae; *st.* = stigma; *t. c.* = tracheolar cell; *tr.* = tracheal trunks; I–IV. tracheae to legs I–IV.

Hinton (1971b), in his classic paper on respiration in the semi-aquatic species *Platyseius italicus*, showed that the peritremes of the two stigmata formed an efficient plastron, i.e. a type of air store in which a thin film of air in communication with the stigma is held by a system of hydrofuge cuticular processes when the animal is submerged in water. In this species, the peritreme plastron provides a relatively large area for the extraction of dissolved oxygen from water when the animal is periodically submerged without reducing the impermeability of the cuticle, while in air the function of the plastron ceases and atmospheric oxygen enters the tracheal system by way of the stigmata. In order to act as an efficient respiratory structure, the plastron must resist wetting at hydrostatic pressures to which it is normally subjected in nature and resist loss of waterproofing from surface-active materials. Further, the ratio between the area of the plastron and the weight of the respiring tissue must be satisfactory, and the drop in oxygen pressure along the plastron should be small enough so that the whole of the plastron is effectively used. The basal limit for the area of the air–water interface of the plastron in relation to the weight of the animal has been determined as 1.5×10^4 μm^2 mg^{-1} (or 15×10^{-6} μm^2 μm^{-3} if body volume is used instead of weight) although Hinton pointed out that this is a somewhat arbitrary figure and that 'one might well refer to mites with 10^4 μm^2 mg^{-1} as plastron breathers'.

Three components of the peritreme contribute to the development of the plastron (Pugh *et al.*, 1987a). Firstly, the hydrofuge property of the hydrocarbon layer overlying the epicuticle of the stigmatic region and the peritreme prevents the entry of water into the tracheal system. This is an essential prerequisite for plastron function (Thorpe, 1950). Secondly, the hydrofuge micropapillae which line the inner surfaces of the peritrematic canal support an air film. Thirdly, the peritrematic canal increases the area of air film to form a viable plastron.

The broad, flat, open peritrematic canal of *Platyseius* (Fig. 4.2C) is not typical of the Dermanyssina and Parasitina and probably represents the most highly evolved and efficient peritreme plastron in the Mesostigmata. In the large majority of the species, the canal is subcircular in section with its lateral margins built up to form lip-like structures which enclose the canal except for a narrow slit to the exterior (Fig. 4.2B,D). The lips protect the micropapillae from abrasion, Hinton (1971b) suggested that the narrowing and lengthening of the stigma to form this type of channel or canal evolved as a mechanism for reducing the chances of the stigmatic opening becoming totally blocked by debris. The development of cuticular processes within the canal would probably increase the effectiveness of the structure as a filter. This type of stigmatic canal could form the structural basis for the operation of a different kind of selective pressure, namely one to increase the width of the slit and to modify the canal into a plastron.

In soil-dwelling forms Witalinski (1979a) considered that the respiratory function of the narrow peritremes was to act as temporary air stores during inundation. Recently, however, Pugh *et al.* (1987a), working with a number of intertidal Dermanyssina and Parasitina, found that the relationship between the peritrematic canal length and body volume in species living in the lower intertidal zone showed no significant difference from supralittoral species. All resembled the structure of the peritremes of soil-dwelling species and in no instance was there the development of elaborate peritremes as in *Platyseius*, although several species such as *Hydrogamasus salinus*, *Macrocheles superbus* and *Parasitus kempersi* were able to survive submergence in seawater for considerable periods. It was calculated that the maximum oxygen content of the peritrematic canals of *H. salinus* was only sufficient to maintain metabolism for about 45 minutes when submerged, considerably less than the four months observed under experimental conditions by Pugh and King (1985). This indicated that the canals were more than a simple air store. Pugh *et al.* (1987a) found that the ratio of the area of the water/air interface presented by the stigmatic orifices and peritrematic slits to body volume was above the basal limit of plastron function in all the species studied by them. They postulated that the air film supported by the peritrematic slits could form crude gas gills capable of maintaining normal respiration even if the width of the peritrematic slit is as narrow as 1.0 μm. The respiratory efficiency of the peritreme as a gas gill would, in their opinion, greatly improve if the air film was supported by the micropapillae of the floor of the canal instead of by the slit. This would involve the partial flooding of the peritrematic canal to the level of the tips of the hydrofuge micropapillae and could be achieved without flooding the tracheal system except at pressures greatly exceeding those generated by the maximal tidal range on the coast of their study area.

Among the Uropodina, Krantz (1974) has demonstrated the existence of an efficient peritreme plastron in the intertidal species *Phaulodinychus mitis*. Unlike *Platyseius italicus*, the opening of the broad-based peritreme in this species is narrow but extensive branching of the peritreme increases the ratio of the slit area to the weight of the body above the basal limit for an effective plastron. In the intertidal species *Thinozercon michaeli* and *Phaulodinychus repleta* the lips of the subcircular peritrematic canal overlap, and Pugh *et al.* (1987b) have shown that the peritrematic slits are not capable of forming a gas gill but probably function to exclude water from the peritrematic canal (Fig. 4.2E). These authors suggest that extra-peritrematic plastrons in the form of coarse sculpturing of the dorsal surface of the peritrematic shield in *T. michaeli* and pedofossae in *P. repleta* may trap a film of air which communicates with the respiratory system. The extremely narrow opening of the peritreme also appears to be a feature of many terrestrial Uropodidae such as *Cilliba* (Fig. 4.1D).

Holothyrida

The paired stigmata in this taxon are situated dorsolateral of coxae III–IV as in the majority of the Mesostigmata and the associated well-developed peritremes extend anterior and posterior to the stigmata. The peritrematic shield is fused with the lateral region of the dorsal shield (Fig. 4.4A,B). In the genera *Thonius* and *Holothyrus*, the stigma is slit-like in form and the entrance from the stigmatic into the tracheal atrium is guarded by closely set rows of fimbriae, which in *Holothyrus coccinella* comprise numerous parallel cuticular rods with short projections arranged in regular rows and cross-linked by lateral struts to form a sieve-like structure (Pugh *et al.*, 1991). In *Thonius braueri*, Travé (1983) records about 20 tracheal trunks arising from the atrium and these ramify into tracheae and tracheoles which spread throughout the body and appendages. The peritreme in *H. coccinella* is situated on the ventro-lateral surface of the idiosoma and consists of a broad, shallow, open canal surrounded by a raised lip. Numerous mushroom-shaped major papillae, each comprising a cap and pedicel, arise from the floor of the canal and tubular micropapillae project from each pedicel (Fig. 4.4C–F,H,J,K). The cap of each pedicel is covered by cerotegument (Fig. 4.4G). A duct, possibly that of an integumentary gland, opens into the majority of the interpapillar spaces (Fig. 4.4I). The ducts are similar to those opening on to the dorsal shield. In *Allothyrus australasiae*, the major pillar-like papillae are capless and some are covered by cuticular tongues developed from the rim of the lip surrounding the peritreme. The gland and its duct opening into the interpapillar spaces in this species bear a marked resemblance to the 'sensilla hastiformes' of ixodid ticks (see Fig. 1.7C, p. 12). The peritreme does not appear to function as a respiratory plastron but probably provides an alternative route for the entry of air into the tracheal system by way of the interpapillar spaces should the stigma be occluded by debris, while in *H. coccinella* it may also function as a mechanism for retarding water vapour transpiration (Pugh *et al.*, 1991).

A relatively large circular pore lying dorsal to the posterior extremity of each peritreme was considered by Thon (1905a) to be the opening to a system of air sacs which might function as an accessory respiratory organ (Fig. 4.4B). The pore leads by way of a duct provided with taenidia to a tubular blind sac or atrium whose walls are perforated. Each

Fig. 4.4. Stigmata and peritremes in *Holothyrus coccinella* (Holothyrida). **A**, latero-ventral view showing position of the peritreme (*p*) in relation to the dorsal (*D*) and ventral (*V*) shields; **B,C**, peritreme and stigma (*s*); **D**, caps (*a*) of major papillae in peritrematic groove (*g*); **E,F,H,J**, major papillae showing distribution of micropapillae (*m*); **G**, cerotegumental (*q*) covering of the cap of a major papilla (*M*); **I** paired valve (*u*) of duct opening between the bases of the major papillae; **K**, section showing major papillae arising from thick cuticle (*z*) which is pierced by a narrow duct (*d*). Scale bars in microns: A = 500, B = 200, C = 50, D,I,J = 2, E,K = 20, F = 10, G,H = 5. (From Pugh *et al.*, 1991.) *c.* = pedicel; *r.* = raised lip.

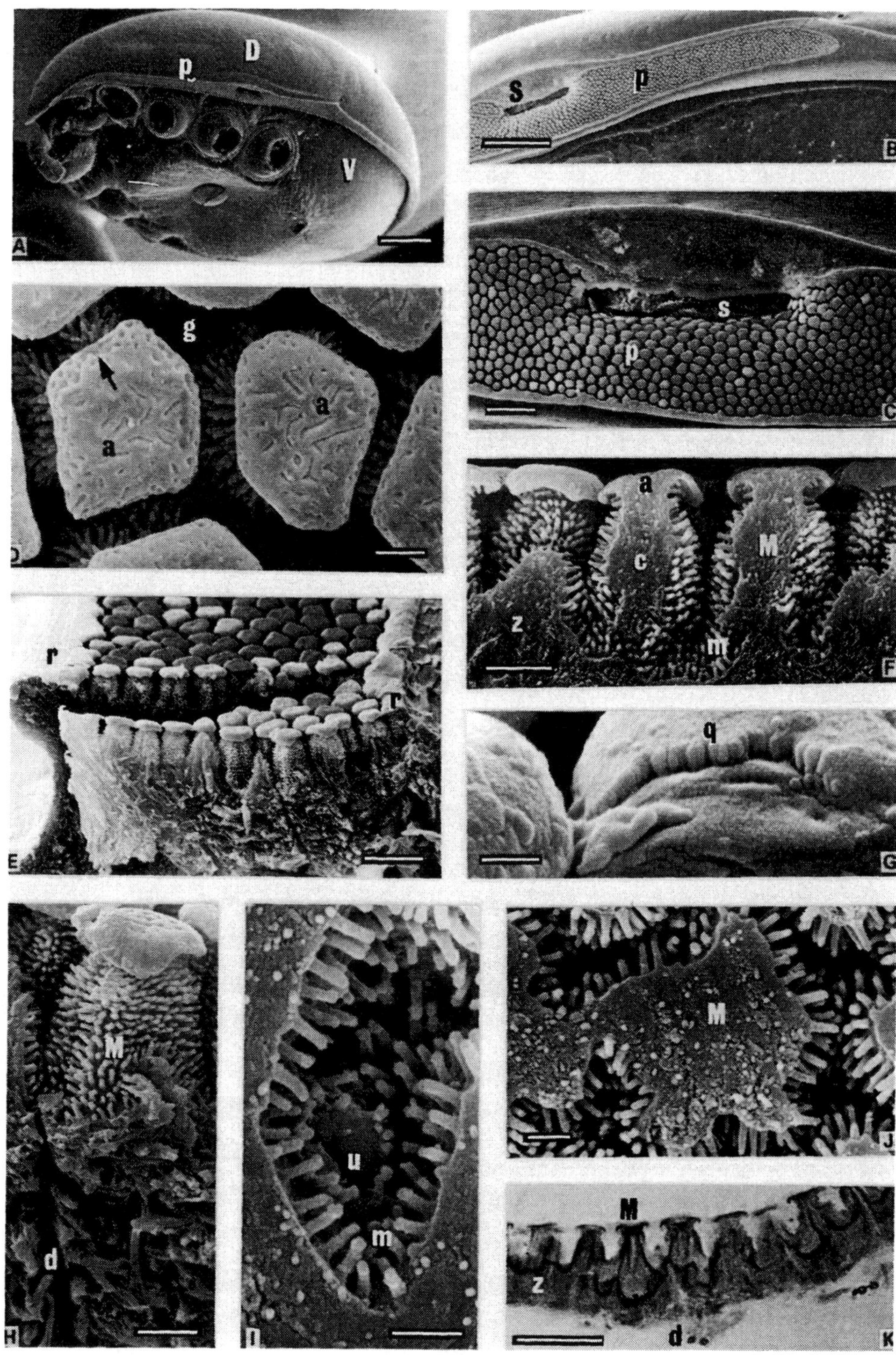

A
D
p
V
B
S
p
C
s
p
D
g
a
a
F
a
c
M
z
m
E
r
r
G
q
H
M
d
I
u
m
J
M
K
M
z
d

perforation leads into a membraneous saclet and the saclets are arranged radially around the atrium. Thon described a muscular closure device at the entrance to each saclet but Hughes (1959) was unable to find such a mechanism in serial sections of the organ. However, a valve-like structure is present at the saclet exit-duct into the atrium and is reminiscent of the form found in the 'coxal glands' of the Dermanyssina opening on coxae I. Hughes recorded the presence of a fine conglomeration of granular matter in the saclets and homogeneous staining material in the atrium. In *Allothyrus australasiae*, only some of the 'saclets' have granular contents while the remainder contain homogeneous staining material. The presence of 'secretory' material in the saclets and atrium has also been observed by Travé (1983), who named the structure Thon's organ and considered it to be glandular and not respiratory in function. He also suggested that the glands could be the source of the irritant secretion known to be produced by some species of holothyrids as a defence against predators. Whether Thon's organ has a secretory or an excretory function has not been established. A gland pore similar in appearance to that of Thon's organ also occurs behind each eye in the Allothyridae. Globules of a milky white fluid have been observed by D.C. Lee (personal communication) in limited areas anterior to the postocular pores and the openings of Thon's organ in *A. australasiae*. It is probable that the fluid originates from the gland pores but this requires confirmation.

Ixodida

Ixodoidea

The paired spiracles or stigmata of the Ixodoidea are situated on the ventrolateral surfaces of the body posterior to the coxae of the fourth pair of legs and each consists of a subcircular or comma-shaped raised sclerotized plate called the spiracular (or stigmatic) plate (Fig. 4.5A). The surface of this plate has an eccentric heavily pigmented area, the macula, which in part constitutes the upper lip (columella *sensu* Arthur, 1960) of a crescentic fissure termed the ostium. The macula is surrounded by a number of pores or aeropyles (Fig. 4.5B). Arthur (1956) states that the pores are closed by a membrane, but subsequent electron microscopic studies have confirmed the earlier work of Batelli (1891) and Nuttall *et al.* (1908a) who considered them to be open (Hinton, 1967; Sixl *et al.*, 1971). Underlying the surface of the spiracular plate is a labyrinth comprising chambers and pedicels (pillars) resting on a thickened base plate (Fig. 4.5C). The latter are usually provided with projections or flanges.

In *Ixodes ricinus*, the aeropyles are arranged in concentric rings and three types of labyrinthal chambers may be distinguished: primary atrial chambers (the goblets of Arthur, 1960); secondary atrial chambers; and peripheral chambers (Pugh *et al.*, 1988). The primary chambers in *I.*

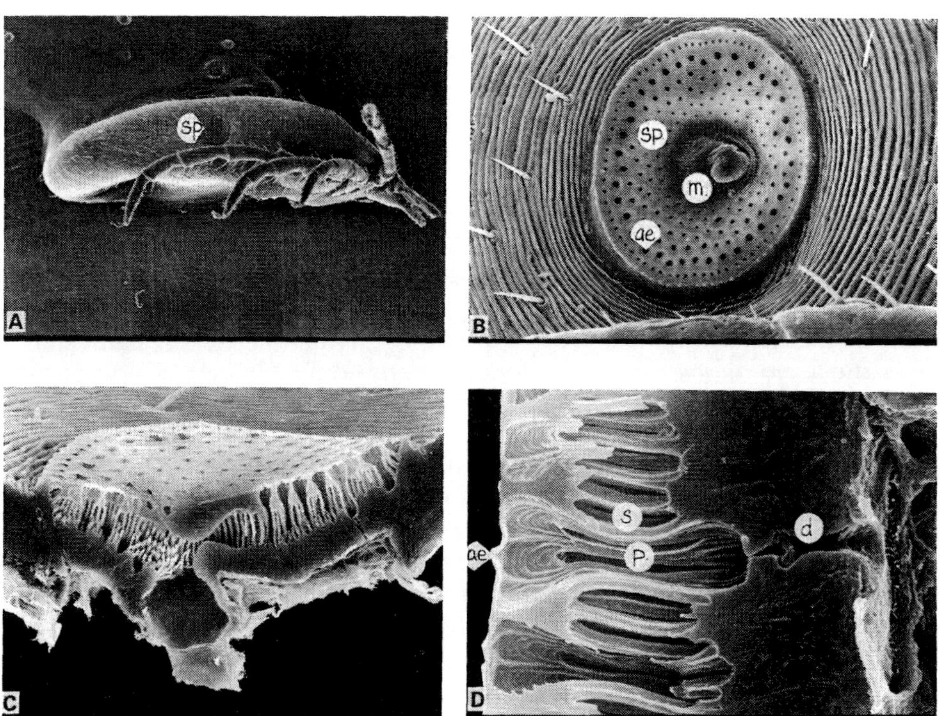

Fig. 4.5. A, lateral view of *Ixodes ricinus* (Ixodida) showing position of the spiracular plate (= sieve plate); **B**, surface structure of the spiracular plate; **C**, vertical fracture of the plate showing the labyrinth and subostial space; **D**, detail of labyrinth showing arrangement of chambers. Scale bar: A = 1000 μm; B, C = 100 μm; D = 10 μmm. (A–D from Pugh *et al.*, 1988.) *ae.* = aeropyles; *d.* = gland duct; *m.* = macula; *p.* = primary atrial chamber; *s.* = secondary atrial chamber; *sp.* = spiracular plate.

ricinus are situated beneath each of the aeropyles, except for those of the peripheral series, and the cuticle is exceedingly thin where it overlies the chambers (Fig. 4.5D). Narrow canals (sub-atrial ducts), containing fine cytoplasmic extensions from underlying epidermal cells, traverse the base plate and connect the chambers and the epidermis. This is reminiscent of the condition in *Allothyrus* and it is possible that the stigmatic plate with its elaborate labyrinthal chambers has evolved from papillae and interpapillar spaces of the open peritreme. The canals were at one time considered to have a sensory function, for example by Nordenskiöld (1908), but the cytoplasmic extensions are now thought to be the fine ducts of integumentary glands (Roshdy, 1974; Sixl & Sixl-Voigt, 1974). The communication between primary chambers is indirect by way of secondary atrial chambers which lack aeropyles. The latter comprise tubular structures surrounded by a number of pedicels and communicate directly with each other and, apparently, by means of slit-shaped fenestrations with the primary and the peripheral chambers. A single ring of chambers on the periphery of the spiracular plate forms the so-called peritreme. The use of the term peritreme for these chambers is inappropriate if the aeropyles are the openings (the tremata!) of the respiratory system. Further, the structure and function of the tick 'peritreme' is entirely different from that of the Holothyrida and Mesostigmata. To avoid confusion, the term should not be used in the descriptive morphology of the tick spiracle and in this work the 'peritreme' is referred to as peripheral aeropyles.

The upper lip of the macula is connected to the inner cuticular layer by the stalk of the columella, a septum of varying thickness which extends the length of the ostium and is continuous both anteriorly and posteriorly with the cuticle of the macula. Within the limits of the ostium, the columella and its stalk are hollowed out and lined with epithelium (Arthur, 1960). Below the macula and surrounding the stalk is a horseshoe-shaped cavity referred to as the subostial space. This space is bordered by the pedicels of the spiracular plate and has connections with the airspace of the labyrinth through the gaps between the pedicels. Below the subostial space is the tracheal atrium (atrial chamber) whose ventral surface extends into the atrium as a broad bulbous plicate lobe, the atrial valve, and separates the tracheal cavity from the subostial space (Figs 4.5C & 4.7A). Arthur (1960) described two muscles inserting on the walls of the atrial cavity in *Dermacentor*: a single band originating from the ventrolateral body wall below the spiracular plate and inserting on the lobe of the ventral wall, and a series of oblique broad bands arising from the dorsal surface of the body and inserting along the dorsal and dorsolateral walls. There appears to be some intergeneric variation in the atrial musculature; for example, Roshdy and Hefnawy (1973) found only one muscle inserting on the dorsal atrial wall in *Haemaphysalis longicornis.* There are no direct occluding muscles in the spiracle. Arising from the tracheal atrium are several tracheal trunks which repeatedly branch to end in bundles of fine tracheoles that enmesh all the internal organs. The tracheae have taenidia.

There is disagreement surrounding the function of the components of the spiracular plate in the respiratory process. Roshdy and Hefnawy (1973) exposed unfed female *H. longicornis* to cyanide gas and found that those in which the ostial lip had separated from the ostial edge as a result of prior exposure to carbon dioxide died within 30 min whereas those without exposure to carbon dioxide and with the ostial lip and edge firmly opposed showed no mortality after 24 h. They concluded from this experiment that the ostium is the functional opening of the respiratory system. The closing and opening of the ostium was considered to depend on haemolymph pressure inside the ostial lip, and on the atrial and dorsoventral body muscles. Thus, 'withdrawal of haemolymph from the lip to the body cavity probably permits slight contraction of the columella and opening of the ostium' while the reverse occurs when haemolymph enters the lip cavity. The aeropyles and labyrinth were not thought to be involved in gaseous exchange but to be a monitoring device for carbon dioxide concentrations.

The function ascribed to the ostium by Roshdy and Hefnawy (1973) contradicts the view of Hinton (1967) who suggested that the macula represents an ecdysial scar and that the ostium of the adult is the external opening of a collapsed ecdysial tube which becomes closed after the transition from nymph to adult. Thus, it is through the aeropyles of the spiracular plate and not the ostium that ambient air enters the tracheal system. Woolley (1972), supporting Hinton's findings, further suggested that the spiracular plate formed a respiratory plastron, but according to Pugh *et al.* (1988) this is not possible because there is insufficient area of pores to support respiration. Rudolph and Knülle (1979) found that carbon dioxide sensitivity of the spiracle could be induced in isolated preparations of the atrial valve although the actual site(s) of the receptors is not known. If the carbon dioxide sensitive atrial valve and not the ostium is the gating mechanism of the tick spiracle, then the effects of cyanide gas on *H. longicornis* observed by Roshdy and Hefnawy (1973) could have been due to the entry of the gas by way of the aeropyles and not the ostium.

If the aeropyles are accepted as constituting the functional openings of the spiracle, what is the role of the labyrinth in respiration? As long ago as 1908, Nuttall *et al.* (1908a) suggested that the spiracle acted as a filter for incoming air. They postulated that dilation of the atrial cavity by muscular action caused the closure of the ostium and the inspiration of air through the aeropyles whereas contraction of the dorsoventral body muscles effected the expulsion of air from the tracheal system by way of the 'spiracle' (?=aeropyles and ostium). Knülle and Rudolph (1982) expressed the opinion that the spiracle might act as a diffusion barrier to slow down the escape of water vapour and this interesting idea has been further developed by Pugh *et al.* (1988). These authors, working with *Ixodes ricinus*, consider that the water-saturated air within the tracheal system and non-saturated air beyond the spiracle 'causes humidity and hence diffusion gradients to form between the tracheal system and the atmosphere'. If the spiracle is to prevent the loss of water vapour from the tracheal system it must (i) reduce the diffusion gradients and (ii) move them as far as possible from the tracheal system. To achieve the first requirement, the mechanism must have a resistance to airflow so that air leaving the spiracle is slowed as much as possible, and have a means of spreading the airflow evenly over the surface of the spiracular plate. This is achieved by the elaborate structure of the surface plate and the underlying labyrinth. Air is expelled centrally into the labyrinth by way of the subostial space and the airflow is reduced to a minimum by the flanged pedicels forming a physical obstruction and inducing surface drag in moving air. The enlargement of the aeropyles radially across the spiracular plate, except in the region of the peripheral aeropyles ('peritreme'), enhances gaseous exchange towards the periphery and thereby effects a more even rate of water vapour loss across the entire sieve plate. The aeropyles are very closely spaced so that water vapour escaping from adjacent aeropyles is thought to experience a mutual interference effect and form a cloud of water-saturated air a short distance (20–30 μm) from the surface of the sieve plate. The small peripheral aeropyles are considered to retard lateral diffusion of water vapour and cause this cloud or disc of saturated air to be held over the surface of the plate. Thus, no diffusion gradients occur within the tracheal system but only beyond it. It has not been determined whether gaseous exchange is the result of passive diffusion or of a more active pumping mechanism.

Differences in the morphology of the spiracle from that observed in *Ixodes* occur in the amblyommid species *Aponomma latum. Dermacentor marginatus* and *Rhipicephalus sanguineus* and are evident in the spatial distribution of the aeropyles and in the form and arrangement of the chambers forming the labyrinth (P.J.A. Pugh, personal communication). The aeropyles are regularly spaced and not arranged in concentric rings, are much smaller and more closely packed. The atrial chambers are homomorphic in *A. latum* and most have a surface aeropyle, while they are heteromorphic in the other two species comprising three (*D. marginatus*) or four (*R. sanguineus*) types. The pedicels in all three species occupy more of the labyrinth and the development of their flanges is greater than in *I. ricinus*. It is postulated that an alternative mechanism of retarding transpiration of water vapour occurs in these three species and is achieved by restricted air movement in the labyrinth as the result of the pedicels and their flanges occupying more of the labyrinth space and by a high resistance to diffusion generated by numerous small aeropyles.

Water loss from the tracheal system is also reduced by the spiracles being open only periodically for gaseous exchange, as shown by Rudolph and Knülle (1982) in *Amblyomma variegatum*. When the tick is inactive, the spiracles are only opened 1–2 times per h for a total period of 4–7 min whereas when it is active the openings are increased to about 15 times per h.

The well-argued case presented by Pugh *et al.* (1988) for the structurally elaborate sieve plate of *Ixodes ricinus* to function as a trap for water vapour is convincing. Their observation that the ostium cannot function as an opening because it is too narrow and 'would thus generate a much higher resistance to airflow because of its crescentic shape' refutes the opinion of Roshdy and Hefnawy (1973) that it is the functional opening of the respiratory system in the ixodoid ticks. However, the function of the ostium remains somewhat of an enigma – if it is a collapsed ecdysial tube then the ostium should be absent in the nymph. Nuttall *et al.* (1908a) found this to be the case in the nymph of *Haemaphysalis punctata* but according to Roshdy (1974) a nymphal ostium is present in *H. longicornis* and in *Boophilus microplus*. Further fine structural and physiological studies are necessary to solve some of the outstanding problems regarding the functional morphology of the tick spiracle, particularly the role of the 'glands' opening into the atrial chambers.

Argasoidea

The ovoid or subcircular spiracles of the Argasoidea are considerably smaller than in the Ixodoidea and are situated posterolateral to coxae III or IV (Fig. 4.6). Each essentially comprises two elevated sections, a smaller anterior sieve plate (area porosa, cribriform plate, spiracular plate) and a posterior macula, which enclose a crescentic ostium (Fig. 4.7B). Unlike the condition in the Ixodoidea, the ostium is capable of opening and closing (Roshdy, 1961; P.J.A. Pugh, personal communication). The sieve plate is provided with aeropyles and the underlying labyrinth, lacking distinct atrial chambers, comprises an airspace surrounding simple columnar pedicels which do not create a significant resistance to airflow (Fig. 4.7D). The chambers do not have associated epidermal glands as in the primary chambers of the ixodoids. In *Argas (Persicargas) persicus*, the elongated atrial cavity is constricted about its middle by the bulging of the atrial wall (Robinson & Davidson, 1913b). A small muscle orginating on the dorsal body wall and inserting on the atrial wall near its base acts as a dilator of the atrium. Robinson and Davidson (1913b) consider that the projection of the atrial wall serves to close the atrium by being squeezed against the opposite wall and that this is effected by the contraction of the dorsoventral body muscles. A number of tracheal trunks (3 to 5) arise from the atrium. Each trachea has an inner, spirally thickened, cuticular sheath and an outer epithelium

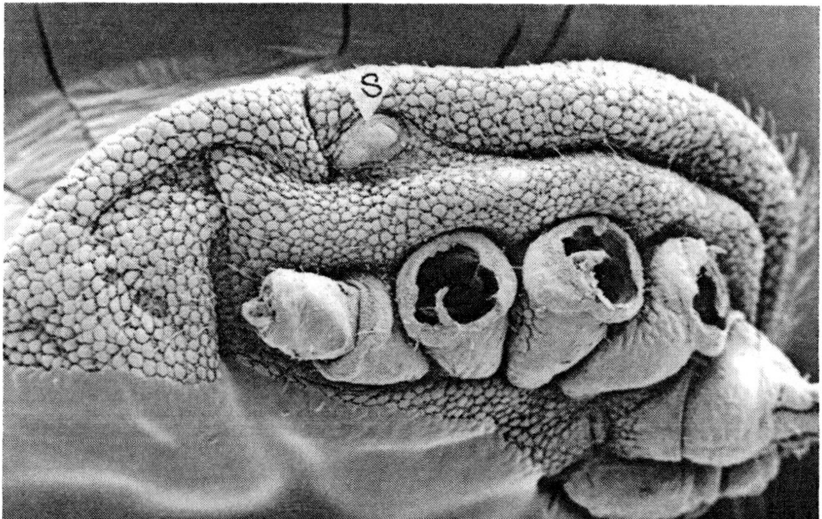

Fig. 4.6. SE micrograph of the lateral view of a species of *Ornithodoros* (Ixodida) showing position of the spiracle (s).

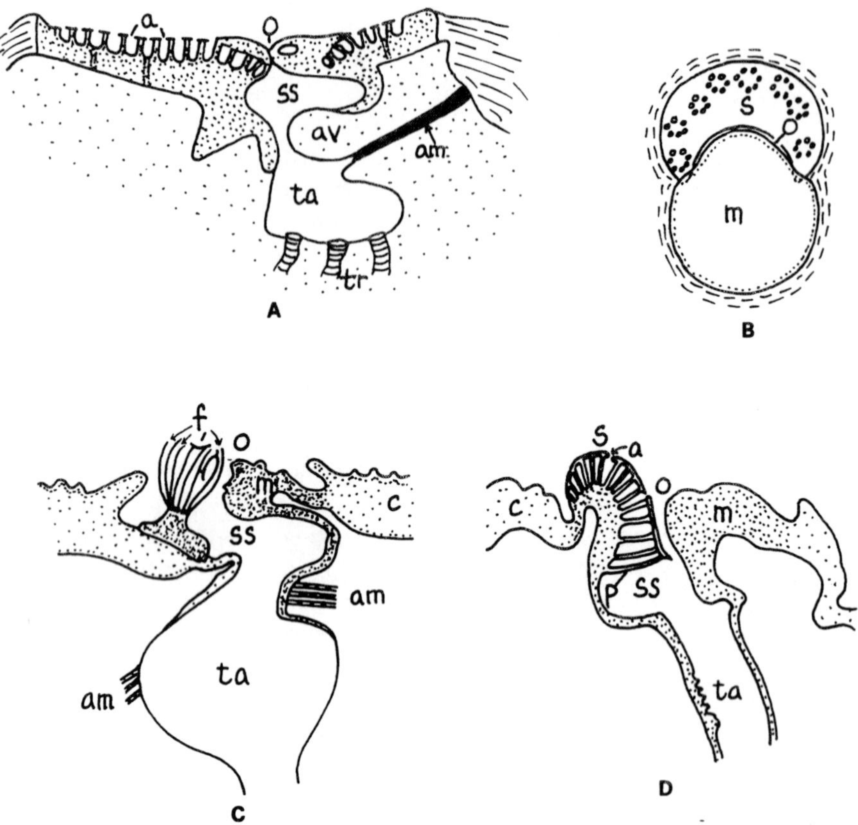

Fig. 4.7. **A**, schematic vertical section of the spiracular plate of *Amblyomma variegatum* showing the atrial valve and atrial muscle inserting on the wall of the atrium (after Rudolph & Knülle, 1979); **B**, spiracular plate of *Argus persicus*; **C**, diagrammatic vertical section through spiracle of *Nuttalliella namaqua*; **D**, semi-schematic vertical section through spiracular plate of *A. persicus* (after Roshdy *et al.*, 1983). *a.* = aeropyles; *am.* = atrial muscle; *av.* = atrial valve; *c.* = cuticle; *f.* fenestrae; *m.* = macula; *o.* = ostium; *p.* = pillar; *s.* = sieve plate; *ss.* = subostial space; *ta.* = tracheal atrium; *tr.* = tracheal trunks.

comprising flattened cells. According to P.J.A. Pugh (personal communication), argasoids inspire by way of the ostia and expire via the sieve plate and this is a xeric adaptation to establish undisturbed humidity gradients above the sieve plates.

A tracheal system in communication with ambient air through a pair of minute openings with coxae I and II has been described in some larval argasids of the genera *Ornithodoros* and *Argas* by Theodor and Costa (1960) and Roshdy *et al.* (1982). In *A. (P.) persicus*, for example, the stigma between coxae I and II leads to a narrow vestibule whose walls are provided with small spiniform projections. The walls of the wide atrial chamber leading from the vestibule have fine elongate processes which form a sieve-like structure in the lumen and together with the projections of the vestibule prevent foreign matter entering the respiratory system. A single tracheal trunk with taenidia arises from the atrial chamber and divides into tracheae which ramify into tracheoles over the body organs. A spiracular plate is absent and there is no valve in the vestibule or atrium.

Nuttallielloidea

The spiracles in *Nuttalliella namaqua* (Nuttallielloidea) are located posterolateral to coxae IV and each comprises a relatively small rectangular plate surrounded by strongly convoluted cuticle (Roshdy *et al.*, 1983). The plate consists of a macula and a larger fenestrated or latticed area enclosing the crescentic ostium (Fig. 4.7C). The fenestrated area is formed of pedicels enclosing interpedicellar spaces which communicate to the exterior by way of the fenestrae. There are no primary chambers (goblets) of the form present in the ixodoids. The subostial space is bordered by pedicels and the inner margin of the ostial lip, and thus appears to be in communication with the exterior through the interpedicellar spaces and the fenestrae although this is not stated by Roshdy *et al.* (1983). The atrial chamber has a valve-like projection and atrial muscles. Five tracheal trunks arise from the chamber. These authors consider the ostium to be a functional opening of the respiratory system with the atrial valve controlling airflow to and from the tracheal trunks. The fenestrated plate appears to be an analogous structure to the sieve plate of the argasoid spiracle.

Actinotrichida

There is greater variation in the occurrence and the form of the respiratory system in this taxon than in the Anactinotrichida. A primary system is absent in the Astigmata and poorly developed in the Oribatida, and it is within these orders that one finds, in particular, the development of putative secondary respiratory mechanisms such as porose areas of the cuticle overlying muscles of the appendages and short blind tubes (brachytracheae) leading internally from small orifices on the surface of the body and legs. The respiratory system is best developed in the terrestrial Prostigmata but even within this group there is considerable diversity in its form.

Prostigmata

The stigmata in this order are located on the gnathosoma or, more rarely, on the dorsal and anterolateral surface of the prodorsum. The diversity in the form of the components of the respiratory system and in the position

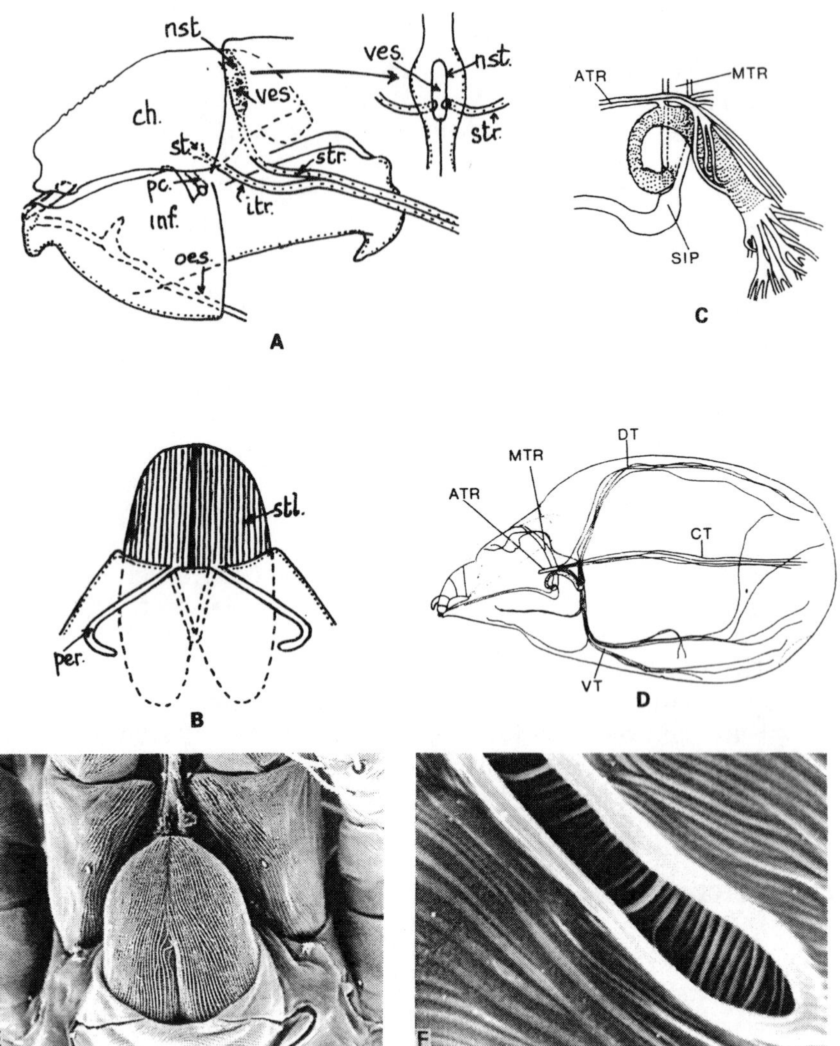

Fig. 4.8. **A**, diagrammatic section showing position of stigmata and major tracheal trunks in *Retetydeus viviparus* (Prostigmata) (based on Grandjean, 1938a); **B**, stylophore and peritremes of *Tetranychus*; **C**, sigmoid piece and major tracheal trunks of *Tetranychus urticae*; **D**, lateral view of respiratory system of *T. urticae*, female; **E**, SE micrograph of the stylophore and peritremes of *T. urticae*; **F**, SE micrograph of a section of the peritreme of *T. urticae*. (C, D after Alberti & Crooker, 1985.) *ATR.* = anterior tracheal trunk; *ch.* = chelicera; *CT.* = central tracheae; *DT.* = dorsal tracheae; *inf.* = subcapitulum; *itr.* = inferior tracheal trunk; *MTR.* = main or dorsoventral tracheal trunk; *nst.* = neostigma; *oes.* = oesophagus; *pc.* = podocephalic canal; *per.* = peritreme; *SIP.* = sigmoid piece; *st.* = stigma; *stl.* = stylophore; *str.* = superior tracheal trunk; *ves.* = vestibule; *VT.* = ventral tracheae.

of the stigmata appears to be due to the modifications of the gnathosoma, particularly the chelicerae, for different methods of feeding (Evans *et al.*, 1961). Grandjean (1938a), on the basis of the form of the respiratory system in a tydeid mite, *Retetydeus viviparus*, postulated that primitively the tracheal system opened by two pairs of stigmata, i.e. the mite was tetrastigmatic. One pair of stigmata was situated below the chelicerae (subcheliceral) and the other between the chelicerae (intercheliceral) in an infraparaxial position. Each stigma opened into a tracheal trunk. When the four stigmata are present there are always four distinct tracheal trunks but if the subcheliceral stigmata are lost, then their tracheal trunks become incorporated with the tracheal trunks of the intercheliceral stigmata. In extant species, the subcheliceral stigmata, when present, are always simple but the intercheliceral stigmata may remain simple or be replaced as functional openings of the respiratory system by a secondary orifice or paired orifices. The latter condition arises from the development of a pair of channels or vestibules (more rarely a single vestibule probably formed by the fusion of two channels), termed neostigmatic processes by Grandjean, ascending vertically from the stigmata along the paraxial faces of the chelicerae to open at the level of the dorsal surfaces of the chelicerae or on the dorsal surface of the gnathosoma. The dorsal openings of the neostigmatic processes are called neostigmata and the extension of each process with its slit-like neostigma onto the surface of the gnathosoma forms the peritreme. Neostigmata may be formed in species in which the proximal parts of the chelicerae are free or fused together. Following Grandjean's terminology, some authors refer to open gutters on the surface of the body as 'taenidia'. This is most confusing in view of the generally accepted use of the term taenidia to describe ring-like structures which project into the lumen of tracheae and glandular ducts and function to keep them distended. In this work the term peritreme is used for the open gutter associated with the stigma.

R. viviparus is an example of a species in which the intercheliceral stigmata lose their function as primary openings to the respiratory system. The subcheliceral stigmata, located between the chelicerae and subcapitulum, retain a simple form and give rise to a pair of tracheal trunks (Fig. 4.8A). The tracheal trunks associated with the intercheliceral stigmata, however, no longer open directly to ambient air through their stigmata but do so by way of a single air chamber or vestibule running up between the paraxial faces of the chelicerae to open in a dorsal position as the neostigma.

The retention of the subcheliceral stigmata and the development of neostigmata also occurs in the Tetranychidae and the respiratory system has been studied in detail in *Tetranychus urticae* (Blauvelt, 1945; André & Remacle, 1984; Alberti & Crooker, 1985). The pair of slit-like subcheliceral stigmata are situated between the stylophore and the subcapitulum. The intercheliceral stigmata, however, open into a pair of dorsoventral tracheal trunks or vestibules which rise to the dorsal surface through a cleft in the proximal part of the stylophore, the fused cheliceral bases of the chelicerae (Fig. 4.8B,C,E). The dorsoventral or main tracheal trunks, lying in close proximity to each other, are separated by a thin septum and enclosed by a common thick sclerotized supporting structure, the sigmoid piece. The latter continues below the point where the tracheae of the intercheliceral stigmata enter the dorsoventral trunks to curve and run forward before attaching to the base of the subcapitulum at the posterior extremity of the cheliceral stylet channels. The sigmoid piece holds the lower part of the dorsoventral trunks in position and protects them from possible damage by the in-and-out movements of the stylophore. The dorsoventral trunks extend dorsally onto the flexible integument encircling the stylophore as a pair of peritremes whose long slit-like openings form the neostigmata. The lateral walls of the peritreme are connected by transverse struts which are thickened at intervals to give the peritreme a chambered appearance (Fig. 4.8F). The struts not only give support to the walls of the peritreme but may also form a sieve-like structure to prevent debris entering the tracheal system. When the stylophore is retracted the peritremes are withdrawn into the flexible cuticular collar ensheathing its base and this prevents air entering the neostigmata. Blauvelt considers that this mechanism might explain the high degree of resistance of this species to certain toxic gases.

The tracheal system of *T. urticae* consists of a pair of accessory tracheal trunks leading from the subcheliceral stigmata and a larger pair, representing those associated with the intercheliceral stigmata, which leave the bases of the dorsoventral trunks. The accessory trunks tap the larger pair so that their air supply is probably supplemented from the neostigmata. Each of the trunks gives off a number of slender tracheae that may branch once or twice. These tracheae then go into bundles before separating into smaller groups and finally into individual tracheae. The groups of tracheae within the body run in dorsal, median and ventral planes (Fig. 4.8D). The tracheae have taenidia and their thin cuticular lining is surrounded by simple squamous epithelium. It is considered that alternate compression and relaxation of the flexible tracheal trunks by the movement of the stylophore assists the ventilation of the tracheal system. Blauvelt (1945) illustrates a well-developed respiratory system in the larva and notes that in this instar the accessory and 'main' tracheal trunks are about equal in size.

Simple intercheliceral stigmata, each with a tracheal trunk and occupying their primitive position, occur in the Nicoletiellidae and Rhagidiidae, but the subcheliceral stigmata are lost, Oudemans (1906, 1931a) placed prostigmatic mites with this type of stigmatic structure into the taxon Stomatostigmata. Unfortunately in his illustrations of the tracheal system of some of the species included in this taxon, the ducts of podocephalic glands are depicted as tracheae and this has led to some confusion.

The loss of the subcheliceral stigmata and the replacement of the intercheliceral stigmata by neostigmata occur in a range of prostigmatic taxa. In the Trombidioidea, the respiratory apparatus, when present, consists basically of paired tubular structures, running downwards between the chelicerae and backwards into the prosoma, from which a number of tracheae arise. Air enters the structures through neostigmata either in the form of simple orifices located between the chelicerae (Fig. 4.9A) or by slit-like openings of peritremes. Feider (1955) defined five categories of respiratory apparatus in the Trombidiidae represented in the Romanian fauna. In the first of three categories which lack peritremes and is represented, for example, by ***Rhinothrombium***, sigmoid pieces ('teaca chitinoasa', cheliceral sclerites) are not present in the respiratory apparatus and the tracheae appear to dichotomize from a short vestibule.

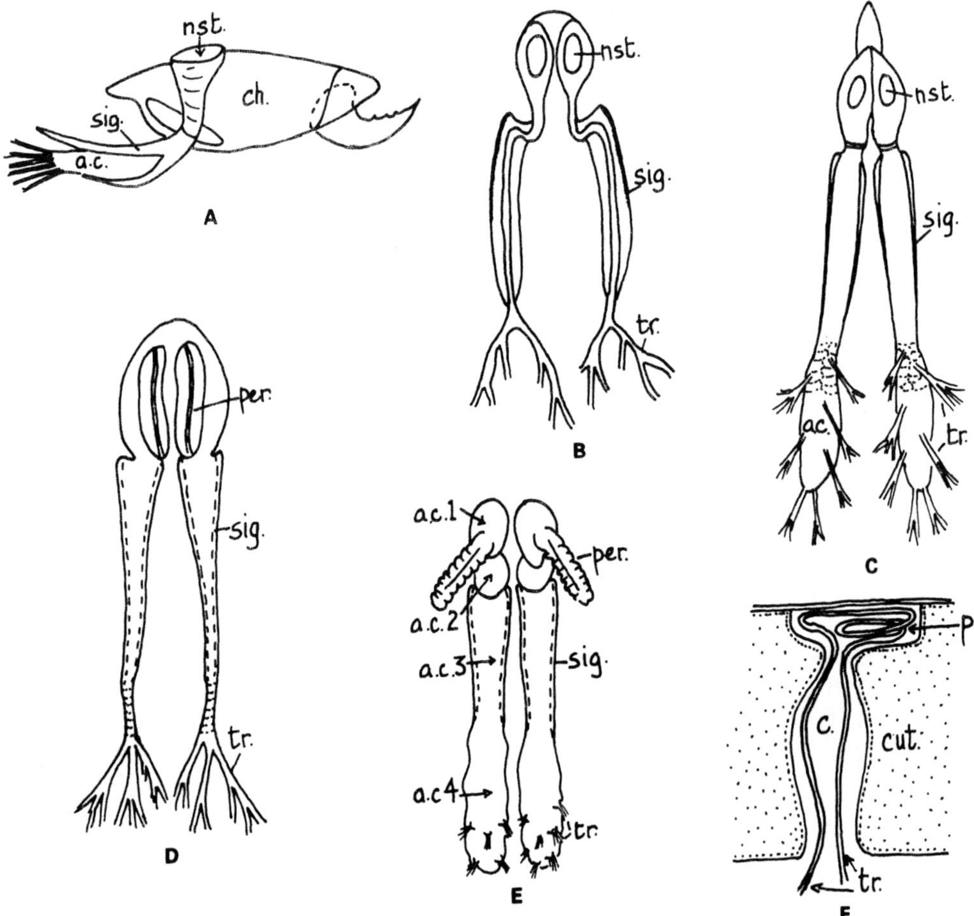

Fig. 4.9. **A**, chelicera, sigmoid piece and atrium of a trombidiid mite in paraxial view (based on Vitzthum, 1940–43); **B–E**, diagrammatic representation of the respiratory apparatus in *Georgia* (B), *Eutrombidium* (C), *Podothrombium* (D) and *Allothrombium* (E). **F**, cuticular pit and channel with coiled tracheole in *Arrenurus*. (B–D based on Feider, 1955; E after Moss, 1962). *a. c.* = atrial chamber; *c.* = channel; *ch.* = chelicera; *cut.* = cuticle; *nst.* = neostigma; *p.* = pit; *per.* = peritreme; *sig.* = sigmoid piece; *tr.* = tracheae.

In the other two categories sigmoid pieces ensheathe the vestibules but in one (*Georgia*), the tracheal trunks branch arborescently soon after leaving the vestibule while in the other (*Eutrombidium*) each trunk expands into a chamber which gives off numerous groups of tracheae (Fig. 4.9B,C). This difference in the form of the tracheal trunk posterior to the vestibule also distinguishes the remaining two categories in which the respiratory apparatus has peritremes and sigmoid pieces: *Podothrombium* having branching trunks (Fig. 4.9D) and *Allothrombium* having chambered trunks (Fig. 4.9E). The 'lateral tracheal trunks' referred to in Feider's illustrations of the respiratory systems are salivary ducts and not tracheae (Moss, 1962).

The respiratory system of *Allothrombium* has been described in detail by Henking (1882) and Moss (1962). In *A. lerouxi,* studied by the latter, the peritremes are supported by a thickened membrane 'that arches over the incised posterior bases of the chelicerae and joins them laterally'. In dorsal aspect, the pair of peritremes have the configuration of a horseshoe with its open end directed posteriorly. Moss considers that the peritremes are closed structures with diffusion of air taking place across the unsclerotized portions. However, Hinton (1971b) has shown that in *A. fuliginosum* the peritreme is a groove lined with micropapillae and with the neostigma only 0.05–1.0 μm in width. This confirms the observations of Henking (1882) and Vitzthum (1930) that the peritremes in this genus have longitudinal slits. Each peritreme leads into a small subspherical chamber which connects ventrally through a short, strongly sclerotized tube with a second chamber whose smooth walls are strengthened by a network of sclerotized ridges (Fig. 4.9E). A small flap overlies the opening of the tube where it leaves the first chamber and this is considered to be a possible closing mechanism. A third air chamber broadly connected to the second is the largest of the three and it runs ventrally and backward within the sigmoid piece. Cheliceral protractor muscles are attached to the sigmoid piece. Leading posteriorly from the third chamber to a point just in front of the synganglion is a weakly sclerotized blind sac which possibly represents a modified tracheal trunk. At intervals towards the posterior end of the sac, clusters of small tracheae with taenidia arise from tubercles and spread throughout the body and legs.

The intimate association between the sigmoid pieces and the neostigmatic processes in *A. lerouxi* is such that Moss (1962) found it difficult to determine whether the third chamber runs as a tube within the length of the hollow sigmoid piece or whether the chamber and the piece are one and the same. In other words do the sigmoid pieces, whose primary role is to act as protractor sclerites for the chelicerae, merely act to support the neostigmatic processes or do they actually form the vestibules of the neostigmatic processes? The first alternative is the more likely in species in which the sigmoid pieces retain their role as protractor sclerites, as in the case of the Trombidiidae, but the second alternative appears to apply in some of the forms in which the cheliceral protraction mechanism does not involve the sigmoid pieces. For example, Witte (1978) found that in larval Erythraeidae, which lack a respiratory system, the hook-like chelicerae are protracted by muscles attached to a pair of well-developed sigmoid pieces (capitular apophyses). In the deuteronymph and adult, however, the gnathosoma undergoes radical changes in which the sigmoid pieces are transformed into 'tracheae' and the basal segment of each chelicera becomes a styliform chelicera with a different protractor mechanism. The pair of 'tracheae' are the equivalent of neostigmatic processes and connect the tracheal sacs, from which numerous small tracheae arise, to the neostigmata.

A tracheate respiratory system is apparently absent in some adult Trombiculidae, for example in the genera *Trombicula* and *Blankaartia* (Brown, 1952; Mitchell, 1964). The reports of a tracheal system in these taxa by Hughes (1959) and Mitchell (1962a) are due to the erroneous description of podocephalic gland ducts as tracheae.

The respiratory apparatus of water mites, in common with many other aquatic arthropods, comprises a closed system in which oxygen from the water diffuses through the cuticle into a tracheole system. In most water mites, the tracheoles radiate from two regions of anastomosis lateral to the synganglion (Wiles, 1984). The network of tracheoles on one side of the synganglion may or may not be connected with that of the opposite side. There does not seem to be a connection between the tracheole system and the neostigmatic processes associated with the sigmoid pieces (Schmidt, 1935; Mitchell, 1972; Wiles, 1984). Thus, the so-called pro-stigmata are blind-ending structures. Wiles suggests that the region of anastomosis could be the result of the atrophy of a trachea that once led to a pro-stigma or a new structure evolved to supply the synganglion with 'copious quantities of oxygen'. The body of mites of the genus *Arrenurus* is completely sclerotized and they have evolved a novel way of providing for gaseous exchange (Mitchell, 1972). Their integument has a regular pattern of pits. Each pit lies beneath a thin layer of cuticle and is connected to the inside of the body by a narrow channel (Fig. 4.9F). In the majority of the pits, a tracheal loop extends through the channel and coils in the chamber of the pit so that part of each trachea is only separated from the water by a thin layer of cuticle. The cuticular pits develop after the emergence of the adult mite and it is only after their formation that the tracheae move into them. Apparently, teneral mites have a thin outer layer and it is during the first week or so of adult life that a second layer is deposited and in which the pits are formed.

An unusual form of neostigma and peritremes occurs in the Caligonellidae in which the chelicerae are completely fused together (Grandjean, 1946a). The neostigma is situated anterodorsally on the fused cheliceral bases and the peritremes extend over the surface of the chelicerae. A tube-like vestibule joins the neostigma to a pair of tracheal trunks which extend to the posterior end of the body and give off branches to the legs. In the related Raphignathidae the peritremes are not restricted to the chelicerae but run laterally over the membrane connecting the

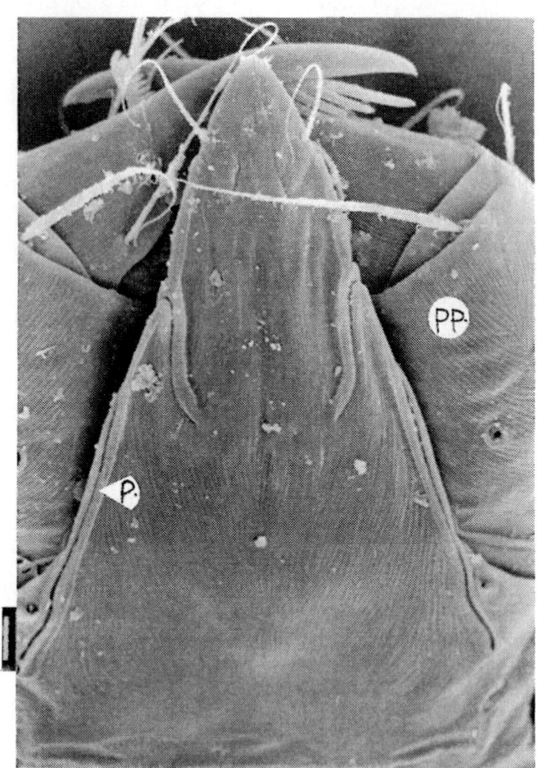

Fig. 4.10. SE micrograph of the peritreme (p.) of *Cheyletus eruditus* (Prostigmata). Bar = 10 μm. *pp.* = pedipalp.

gnathosoma and idiosoma. A tracheate respiratory system is absent in the Stigmaeidae and the descriptions of tracheae in this family refer to the ducts of podocephalic glands.

The restriction of the peritremes to the dorsal surface of the gnathosoma is also seen in the Cheyletidae. Hughes (1958) considered the peritremes to be covered by a thin membrane and that air only entered through a small opening near the posterior end of each peritrematic limb. Scanning EM studies, however, have shown this to be incorrect and the peritreme is a typical open gutter (Fig. 4.10).

Stigmata and a tracheal system are present in the females of the Tarsonemina but are absent in the males (hence the group name Heterostigmata). The stigmata are located on the dorso-anterolateral surface of the prodorsum except in those genera, for example *Nasutotarsonemus* and *Pseudotarsonemus*, in which they open on the ventral surface of an extensive hood-like prodorsal shield (Lindquist, 1986). Typically, a tracheal trunk arises from each stigma and, within a variable distance of its origin, is enlarged to form a tubular atrium. In the Tarsonemidae, each tracheal trunk may divide into two tracheae immediately beyond the atrium or remain undivided until further into the body (Fig. 4.11A). The tracheae do not enter the legs and in many species each trachea ends in a

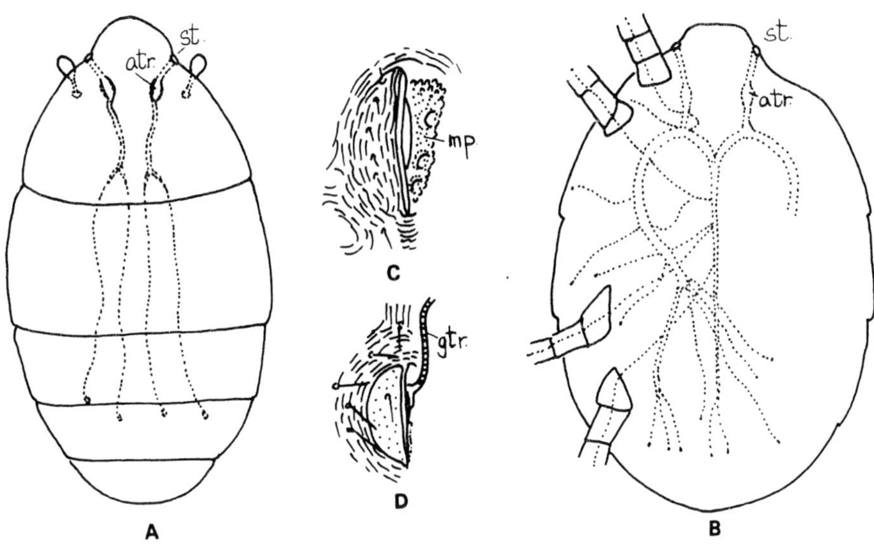

Fig. 4.11. **A**, stigma (*st.*), atrium (*atr.*) and tracheae of a tarsonemid mite; **B**, respiratory system of *Acarophenax* (based on Newstead & Duvall, 1918); **C**, genital vestibule with mamillated pouch (*mp.*) in *Bdella*; **D**, genital tracheae (*gtr.*) of *Trachymolgus* (Bdellidae). (C, D based on Grandjean, 1938b.)

small ovoid bulb. The tracheal system is more extensive in the Pyemotidae and Acarophenacidae. In the former each trunk divides into a bundle of fine tracheae immediately beyond the atrium but as in the Tarsonemidae the tracheae do not penetrate the leg cavities. Each tracheal trunk divides into two limbs in *Acarophenax* and the two median ones fuse a short distance from their origin and give rise to a group of fine tracheae, some of which penetrate the legs (Fig. 4.11B). The tracheae arising from the lateral limbs appear to terminate in small bulbs. Only one pair of stigmata opening anterolaterally on the prodorsum occur in this genus and not two pairs as described by Newstead and Duvall (1918).

In the bdellid subfamilies Cytinae and Spinibdellinae, genital tracheae are present in addition to the normal tracheal system opening at the base of the chelicerae (Fig. 4.11D), *Cyta* has one pair of tracheae in the protonymph but two pairs occur in the deuteronymph with the two tracheae on each side of the midline being contiguous at their point of origin (Grandjean, 1938b). The tracheae extend anteriorly from the region of the genital papillae. The two genital tracheae of *Trachymolgus* first appear in the deuteronymph. In the adult they extend as far as the level of the second pair of legs and have an unusual form. Each assumes a flattened, spatulate shape distally with the tracheal wall having a granular appearance. Grandjean refers to this type of trachea as a platytrachea. Genital tracheae are always absent in the Odontoscirinae and may be

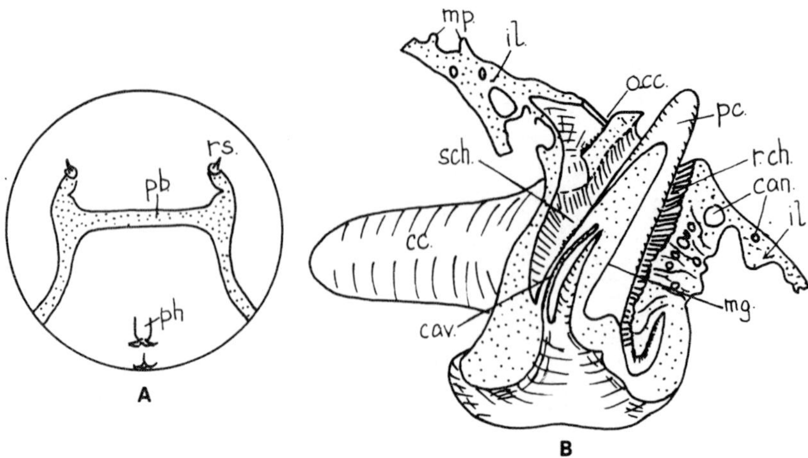

Fig. 4.12. A, diagrammatic representation of the egg of *Tetranychus lintearicus* (Prostigmata) showing position of 'residual setae' and punctate bands (after Grandjean, 1948); **B**, diagrammatic representation of the structure of a perforation organ and associated structures in *T. urticae* (based on Dittrich, 1971). *can.* = canals; *cav.* = cavity of perforation cone; *cc.* = centripetal cone; *il.* = intermediate lamella; *mg.* = median groove; *mp.* = micropillars; *o. cc.* = outer connecting canal; *pb.* = punctate band; *pc.* = perforation cone; *ph.* = pharynx; *r.ch.* = rough chamber; *rs.* residual seta; *s.ch.* = smooth chamber.

absent or nearly so in the Bdellinae. In the latter, however, the majority of species have their genital vestibule partially differentiated into a mamillated pouch or sac adjacent to the genital papillae of each side (Fig. 4.11C). This type of pouch as well as the genital tracheae of other taxa contain air and both are considered to have a respiratory function. The water/air interface of the peritrematic slit in *Bdella* is insufficient to maintain respiration during inundation but the peritreme allows gaseous exchange by way of the tracheal system if the neostigmata become occluded by debris (Pugh *et al.*, 1987a). This probably also applies to other Prostigmata with peritremes.

The calyptostasic prelarva of the gorse mite, *Tetranychus lintearicus*, was shown by Grandjean (1948) to possess, in addition to vestiges of the mouth and pharynx, a pair of birefringent 'residual setae' situated in deep pits and three bands of punctate cuticle forming an H configuration (Fig. 4.12A). Similar setae and punctate bands in *T. urticae* are considered by Dittrich and Streibert (1969) and Dittrich (1971) to be components of an embryonic respiratory system. EM studies show that the 'residual seta' is a slender cone-like structure, about 5 μm in length, with internal cavities and a median groove along its longitudinal axis. It lies in a sac-like cavity of an intermediate lamella, composed of laminae which show a substructure of parabolically bent fibrils, possibly of chitin (Fig. 4.12B). The intermediate lamella, probably representing the deutovum or prelarval cuticle, *sensu* Grandjean, is present at oviposition and completely surrounds the embryo. A lateral extension of the wall of the perforation organ forms a large cone-like structure, the 'centripetal cone', which extends towards the centre of the egg where it connects with the developing embryo. The seta and its pit, *sensu* Grandjean, form the perforation organ and the punctate bands are zones of conical protrusions or micropillars of the intermediate lamella which form air ducts; the micropillars probably maintain an open space between the lamella and the shell as the growing embryo presses against the shell. The perforation organs appear to have two functions, namely to penetrate the egg shell and allow the entry of respiratory gases and to conduct these gases towards the system of air ducts. The actual mechanism of penetration of the shell by the cone of the perforation organ is not known but it is suggested that cavities of the cone may conduct a lytic substance or a plastizing enzyme to the shell to soften it before penetration. Respiratory gases are conducted through the V-shaped chambers (*r.ch., s.ch.,* Fig. 4.12) into the connecting canal (*o.cc.*) and thence to the air duct system (the zones of micropillars). The gases are thought to pass through perforations in the intermediate lamella to the embryo. A similar respiratory apparatus has been found in the eggs of other Tetranychoidea and a modified form occurs in the Stigmaeidae (Suski, 1989).

According to Van Impe (1985), Dittrich considers that the prelarva, *sensu* Grandjean, is an embryonic developmental phase and not a stase, the residual setae are not regressive structures but embryonic organs adapted for gaseous exchange and the vestiges of the mouth and pharynx belong to the embryonic larva. Detailed embryological studies of prostigmatic species exhibiting various degrees of morphological regression in the prelarva are required to determine the origin of the deutovum. At present there is no conclusive evidence to seriously challenge Grandjean's concept of the prelarva.

Interestingly, this is not the first reference in the literature to the presence of an embryonic respiratory system. Henking (1882) referred to two funnel-shaped ingrowths, the urpores, on the apoderma of the prelarva of *Allothrombium* and each formed the point of attachment of a cord which linked the urpore to the embryo.

Astigmata

A tracheal system opening by paired stigmata is typically absent in the Astigmata, as the ordinal name implies. Hirst (1921) illustrated a system of paired tubes with taenidia, which he referred to as tracheae, in *Otodectes cynotis* and *Chirodiscoides caviae.* In the latter, each 'trachea' arises from an opening situated laterally between the first and second pair of legs and extends posteriorly to the level of the anus. These are now known to be ducts of glands. Similarly, Fain (1968a) has described 'pseudotracheae' in the glycyphagids *Grammolichus* and *Sclerolichus.* In *Grammolichus,* the short, round or cylindrical trachea-like structures arise from sclerotized pockets on the dorsal surface of the trochanters of the second pair of legs, whilst in *Sclerolichus* the pseudotracheae comprise two pairs of transversely striated tubes arising from the base of coxae II, with one pair terminating in the region of the genital orifice and the other in the anal region. The function of these structures is uncertain. A pair of genital tracheae opening into the genital atrium occurs in the female of *Goheria fusca* and although Hughes (1959) observed them to contain air there is some doubt as to their function.

Oribatida

The oribatids exhibit a number of structures which are considered to be involved in gaseous exchange and our knowledge of respiratory mechanisms in this group is largely based on the work of Michael (1884, 1888a) and Grandjean (1933, 1934a). In the atracheate primitive oribatids such as the Hypochthoniidae and Lohmanniidae, gaseous exchange through the sclerotized cuticle is considered to be facilitated by micropores which traverse the cuticle. The micropores in some cases are grouped together to form porose areas and these commonly occur on the surface of leg podomeres (Fig. 4.13A). It should be noted, however, that not all porose areas have a respiratory function and, in particular, those constituting the octotaxic system of *areae porosae* on the notogaster of the Euoribatida and considered by Grandjean to be respiratory organs are in fact openings of glands (Woodring & Cook, 1962a; Alberti *et al.*, 1981).

Grandjean (1934a) considers that the oribatid trachea has developed from a simple invagination of the cuticle in the form of a small thin-walled pocket ('poche trachèenne') which when connected to the exterior by a neck forms a tracheal sacculus ('saccule trachèen'). This type of structure is best seen on the legs and a single podomere may have more than one sacculus; for example, seven to nine sacculi in a longitudinal row are present on the femora of the adult of *Platyliodes scaliger* (Fig. 4.13B). The elongation of such a pocket is considered by Grandjean (1933) to result in the formation of a short trachea, which he termed a brachytrachea. The brachytracheae may be flat or cylindrical in transverse section. They can arise from the leg acetabula, bothridia of the prodorsal sensilli and various other external surfaces of the body and the legs, particularly on the femora and trochanters. The brachytracheae of the legs never extend beyond the podomere of their origin (Fig. 4.13C). Brachytracheae can occur in the larvae and nymphs as well as adults.

The long tracheae which originate in the leg acetabula of the adults of the Euoribatida (=Higher Oribatida, Brachypylina) are considered to have developed from brachytracheae (Fig. 4.13D). The trachea of the first acetabulum (trachea I or the cephalo-thoracic tracheae of Michael) is divided into two branches soon after leaving the stigma (*Galumna, Humerobates, Cepheus, Belba*, etc.). In *Hermanniella granulata* the two branches appear to be separate and each to arise from a stigma whilst in

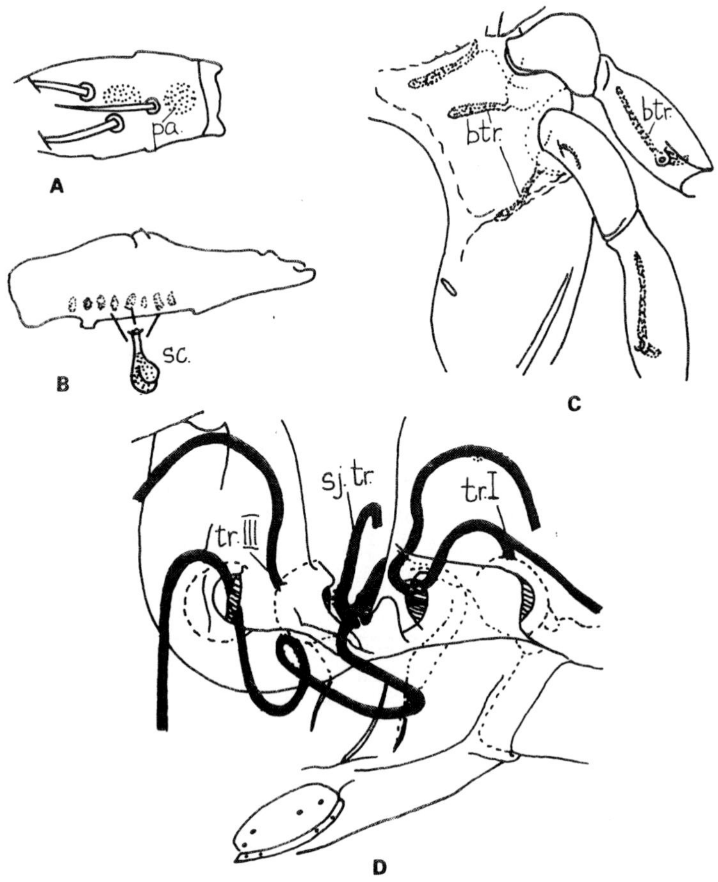

Fig. 4.13. **A**, porose areas on tarsus I ventral of *Phauloppia lucorum* (Oribatida); **B**, paraxial view of right femur I of *Platyliodes scaliger* showing distribution of sacculi; **C**, sejugal, acetabular and podomeric brachytracheae of the metapodosomatic region of *Neoliodes theleproctus*; **D**, lateral view of the tracheal system of an adult euoribatid mite. (Based on Grandjean, 1934a.) *btr.* = brachytrachea; *pa.* = porose area; *sc.* = sacculus; *sj. tr.* = sejugal trachea; *tr. I. tr. III* = tracheae of acetabula of legs I and III.

Neoliodes theleproctus trachea I is divided into four lateral branches in the adult and two branches in the three nymphs but only a brachytrachea in the form of a sac is present in the larva. No tracheae open into acetabula II except for a brachytrachea in *H. granulata*. A simple trachea arises from acetabulum III and it is often larger than the other tracheae, as in *Cepheus*. In *N. theleproctus* it is represented in the adult as well as in the larva and nymphs by a brachytrachea (Fig. 4.13C). Although Michael (1884) describes a large trachea ('great ventral') originating from acetabulum IV in *Cepheus* and other euoribatids, Grandjean (1933) states that he has never seen a trachea in this position in any of the species he has examined although brachytracheae occur in the nymphs and adults of

Neoliodes and a short, broad transversely directed tracheal pocket is present in *H. granulata.* In addition to the tracheae opening into the leg acetabula, adults of the higher oribatids also have a large trachea, usually comprising two branches as in trachea I, opening into the sejugal furrow which separates the proterosoma from the hysterosoma. The stigma is normally a slit that is difficult to see at the bottom of the sejugal furrow. Michael (1888) considered the stigma of this trachea to lie in acetabulum II.

The large tracheae associated with the acetabula and sejugal furrow lack taenidia, are usually of even diameter before terminating in a point or in an oval-shaped bulb reminiscent of the swollen bulb of a gland duct. They do not enter the appendages. Woodring (1973) found that the terminal bulbs are larger in fresh water than in terrestrial species and that in many the bulbs contained solid debris.

Long tracheae are also associated with the region of the pregenital atrium of certain Archoribatida (Lower Oribatida) such as *Cosmochthonius* and *Haplochthonius* (Grandjean, 1934b). In the former two pairs of so-called genital tracheae are present in the adult, the second pair first appearing in the tritonymph. The anterior pair passes laterally and then forwards to about the level of coxae II whilst the posterior pair runs backwards towards the anal region before turning to end almost in line with the level of their origin.

The cerotegument of species of Euoribatida living in water or in habitats susceptible to periodic flooding has been shown to have a respiratory function by trapping a film of air and forming a plastron which may or may not be in contact with the openings of the tracheal system (Crowe & Magnus, 1974; Alberti *et al.*, 1981, Krantz & Baker, 1982; Pugh *et al.*, 1987c). Contact between certain stigmata and the plastron is achieved in some species by means of channels or canals in the surface of the integument. Such channels were first described by Van der Hammen (1963a) in an intertidal species of *Fortuynia* and were considered to be air stores. The plastron–channel system has been studied in detail by Pugh *et al.* (1990) in *Fortuynia maculata,* a species living in crevices or in empty barnacle shells in the extreme upper marine littoral in Kenya. The plastron system consists of a ventral element extending from the camerostome to a point between coxae IV and the genital valves, and a posterior element in the form of a horseshoe band extending from above acetabula II around the lateral margins of the ventral shield with a connection to the ventral element behind acetabula IV (Fig. 4.14A). The cerotegument of the plastrons covering the cuticular tubercles is formed of two sheets connected by tapering pillars which interrupt the airfilm between the sheets (Fig. 4.14B). The airfilm is folded over the contours of the cuticular tubercles, which has the effect of increasing the surface area of the airfilm to gaseous exchange and strengthening the cerotegument. The channel system on each side of the body (Van der Hammen's organ) in *F. maculata* comprises three narrow, semi-closed, cerotegument-lined channels (the sejugal, prodorsal and posterior) on the lateral surfaces of

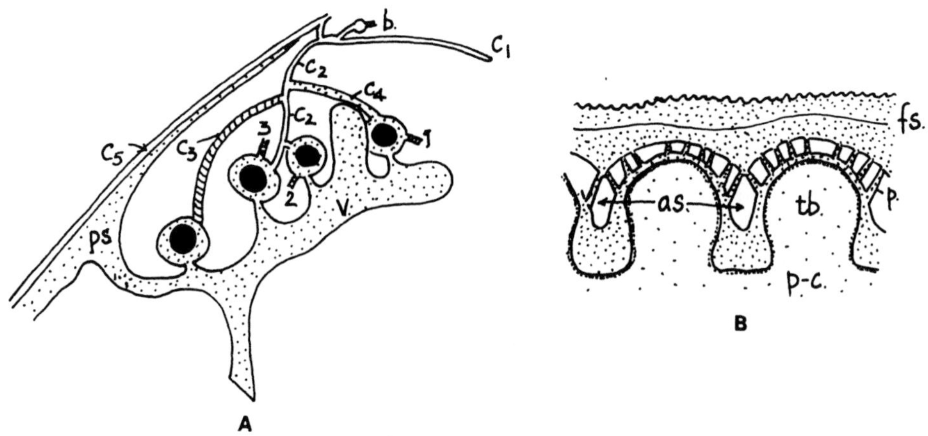

Fig. 4.14. **A**, diagrammatic representation of the plastron system and air-flow channels (Van der Hammen organ) in *Fortuynia maculata* (Oribatida); **B**, section of cuticular tubercles and cerotegument in *F. maculata* (after Pugh *et al.*, 1990). *as.* = air space; *b.* = brachytrachea associated with the bothridium; c_{1-5} = channels (c_1 = prodorsal, c_2 = sejugal, c_3 = posterior, c_4 = link channel of first acetabulum, c_5 = circum-notogastral); *p.* = pillar; *fs.* = fissure; *p-c.* = procuticle; *ps.* = posterior plastron; *tb.* = tubercle; *v.* = ventral plastron; 1–3 = tracheae.

the prosoma connecting acetabular cavities II, III and IV with the bothridium of the prodorsal sensillus. The sejugal and prodorsal channels are considered to be derived, respectively, from the sejugal and abjugal furrows of the actinotrichid soma. The cerotegumental lining of the channels is formed into micropapillae which are similar in form to those in the peritreme of the Mesostigmata. The posterior channel has, in addition, numerous transverse septa. A link channel, in the form of a narrow band of flat cerotegument-covered tubercles, joins acetabular I with the sejugal channel and a circum-notogastral channel unites the canal system of the right and left side of the idiosoma and the posterior plastron. The system of cuticular tubercles and overlying ventral cerotegument functions as a plastron during the periodic tidal immersion of the mite and communicates directly with tracheae III and the tracheae opening at the bottom of the sejugal furrow and indirectly with the brachytracheae of the bothridium. The sejugal channel allows an airflow between the brachytracheae of the bothridium and the plastron. It is suggested that the function of the posterior channel and its septa is to act as a safety valve by allowing water to enter the system under surge pressure and thereby preventing the remainder of the system becoming flooded. The septa act as baffles to restrict inundation.

Travé (1986) records similar structures in *Hydrozetes*, where the channel is usually associated with trachea I, and in *Dolicheremaeus dorni* (Otocepheidae), in which the first trachea does not open directly into the first acetabulum but into a canal which connects with a microtuberculated area in the pleural region. Trachea III opens into a vestibule

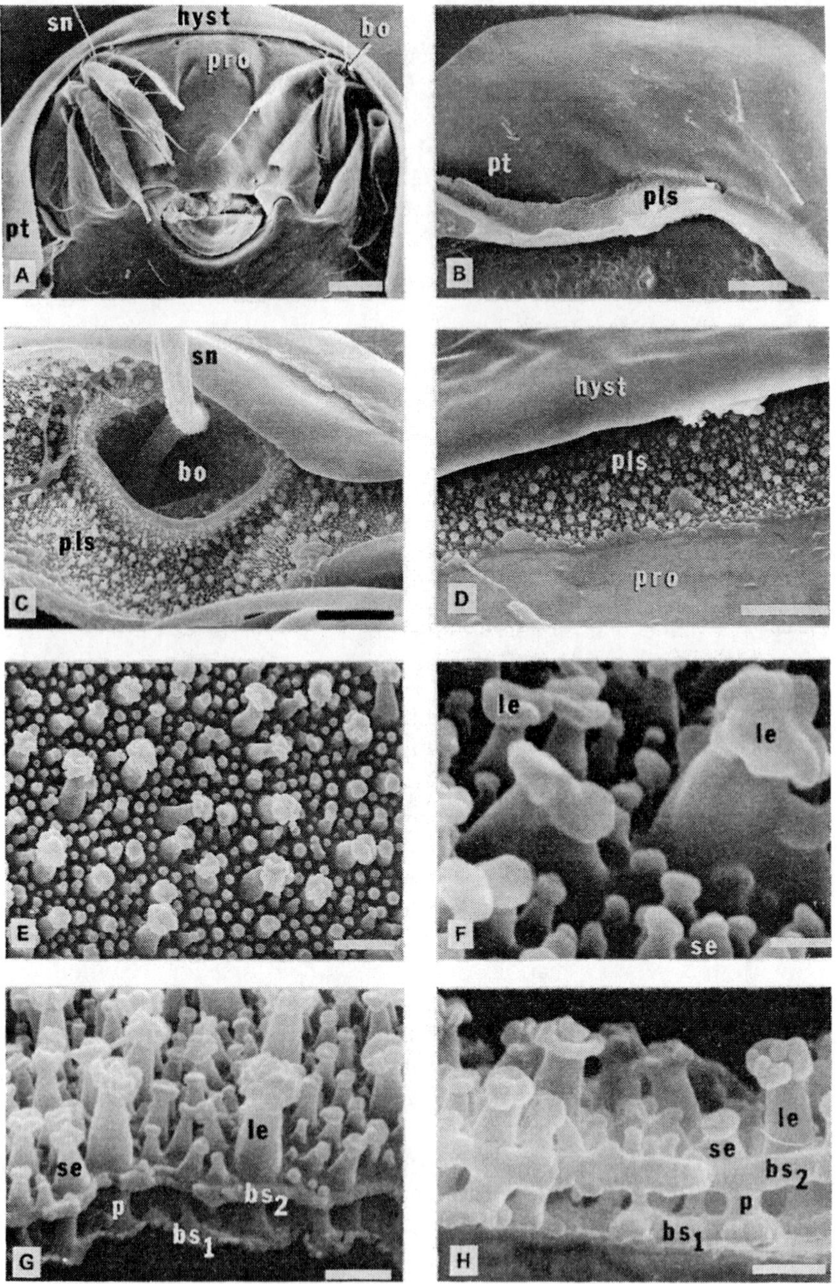

Fig. 4.15. SE micrographs of the cerotegument of *Euzetes globulus* (Oribatida). **A**, *en face* view of the prodorsum (*pro*) and hysterosoma (*hyst*); **B**, venter of detached pteromorph (*pt*) showing narrow band of plastron pile (*pls*); **C**, cerotegument around bothridium (*bo*) and sensillus (*sen*); **D**, plastron pile on prodorsum and under overhanging ledge of notogaster; **E**, close-up of pile in D; **F**, detail of pile shown in E; **G, H**, fractured edge of plastron pile showing two types of excrescences (*le, se*) attached to a double basal sheet (bs_1, bs_2) which are separated by pillars (*p*). (After Pugh *et al.*, 1987c.)

where it communicates with a smooth semi-tubular channel.

In the strongly sclerotized terrestrial species *Euzetes globulus*, the cerotegument is well protected from abrasion by being restricted to a horseshoe-shaped band extending from the posterior margin of the trochanters of the fourth pair of legs along the lateral margin of the body under the pteromorphae and on to the region of the prodorsum where it is overhung by the notogaster (Fig. 4.15A–D). It comprises a pile of short and long excrescences attached to a double basement layer (Fig. 4.15E–H). The function of the latter is not known but Pugh *et al.* (1987c) suggest that the two-tiered pile may be a means of increasing respiratory efficiency by enlarging the surface area of the water–air interface through folding or making the interface undulate. The plastron is in communication with the openings of the tracheal system.

The cerotegument of *Ameronothrus marinus*, an intertidal species that is capable of surviving immersion for periods of up to four months, is more extensive than in *E. globulus* and covers the notogaster, except for the region of the lenticulus, and much of the ventral shield. It consists of a flat uniform basal layer with a smooth surface towards the epicuticle and a series of granules connected to the basal layer by narrow pillars. Pugh *et al.* (1987c) suggest that this coarse, strongly developed cerotegument is a mechanism for protecting the thin airfilm it supports from mechanical stress generated by wave action. The cerotegumental plastron lacks a communication with the tracheal system and it is suggested that gaseous exchange takes place across the cuticle, which is not extensively sclerotized and has a leathery texture. The absence of a direct connection between the plastron and the tracheal system has also been reported in a species of *Hydrozetes* living in fresh water (Crowe & Magnus, 1974).

5 Mouthparts

Feeding is an essential part of the life of all Acari and involves a series of behavioural and mechanical processes leading to the ingestion of food in a fluid or particulate state. The events in this complex of processes include the location and selection of the food source by sensory organules of the anterior legs and pedipalps, and the processing and reception of the food by the mouthparts. Mites utilize a wide variety of animal and plant substances as food and this is reflected in the diversity of their feeding behaviour and in the structural adaptations of their mouthparts.

The mouthparts of the Arachnida display a variety of structures but typically comprise the appendages of the first two body segments – the chelicerae and pedipalps (usually contracted to palps), the labrum (=lingula), mouth and, in some taxa, the so-called labium (the sternite of the pedipalpal segment). An exception occurs in the Palpigradi in which the pedipalps are locomotory in function and do not appear to play a part in feeding, whilst in the Scorpiones and Opiliones endites or apophyses of the coxae of one or more pairs of legs are also involved in the feeding process (Snodgrass, 1948). The lobes of the coxae of the first two pairs of legs of the Scorpiones form the floor of a quadrate cavity for the reception of food material and its comminution by the chelicerae. In the Opiliones, the coxal endites of the first pair of legs or of the first and second pair of legs are used for passing food on to the mouth. Van der Hammen (1989) refers to the type of feeding which involves endites of the leg coxae as coxisternal feeding in contrast to the rostral feeding of other arachnids.

The pre-oral chelicerae constitute the trophic appendages and are variously modified for cutting, tearing, crushing or piercing while the post-oral palps are sensory appendages that may also play a role in seizing and holding the prey during feeding. One of the evolutionary trends among those arachnids displaying rostral feeding is the development of a pre-oral chamber or channel for the reception of food. In the Araneae, for example, the prey is lacerated and crushed by the chelicerae and the food is conveyed to the mouth by way of a pre-oral chamber formed from apophyses of the coxae of the palps (maxillae), the labrum and labium (Fig. 5.1A). Strong setae situated on the apophyses, labium and labrum act as a filter to prevent solid particles of the prey entering the mouth (Gerhardt and Kästner, 1937). The formation of this type of pre-oral chamber does not involve the fusion of the palpcoxae, which are still capable of independent movement. In the Ricinulei and Acari, on the other hand, the formation of a pre-oral channel or trough entails the enlargement and fusion or approximation ventrally of the palpcoxae and their apophyses to form a subcheliceral unit, the subcapitulum (=infracapitulum) or hypognathum, incorporating the labrum, mouth and pharynx (Fig. 5.1B). The subcapitulum, pedipalps and chelicerae together constitute a discrete sensory–trophic structure, the gnathosoma (=capitulum), which is usually accommodated in an anterior or anteroventral cavity of the prosoma and movably articulated to it. In the majority of the Acari, this cavity contains only the gnathosoma and is referred to as the camerostome, but in the Mesostigmata it also accommodates the coxae of the first pair of legs and is termed the gnathopodal cavity.

In view of the wide range of feeding habits of mites, it is not surprising to find a variety of modifications of the various components of the gnathosoma for particular types of feeding. This diversity in structure not only occurs between species but may also be seen in the active stases

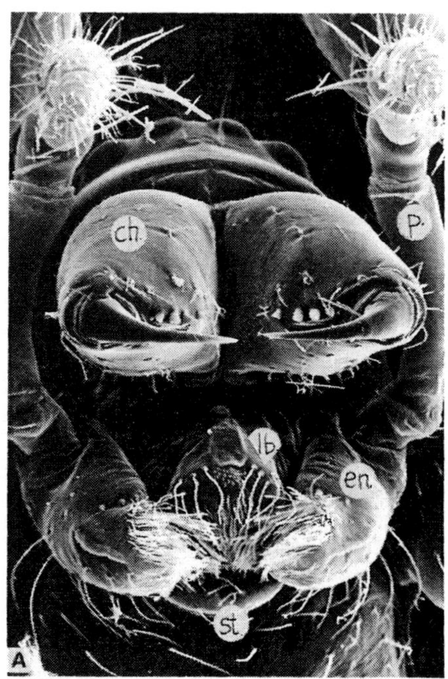

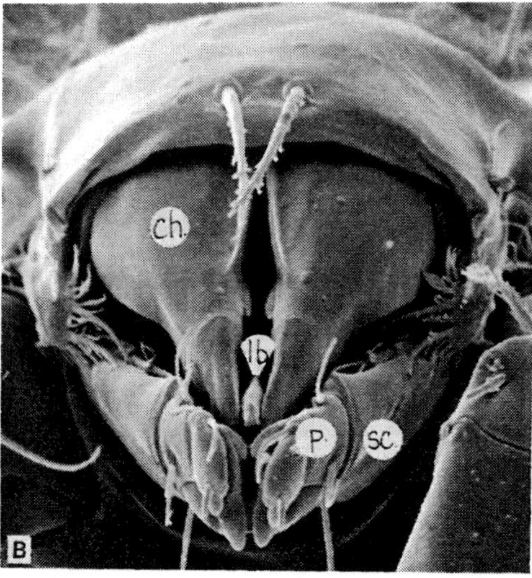

Fig. 5.1. SE micrographs of the mouthparts of a linyphiid spider (**A**) and the grain mite, *Acarus siro* (Astigmata) (**B**). *ch.* = chelicera; *en.* = endite of the coxa of the pedipalp (coxapophyses); *lb.* = labrum; *p.* = pedipalp; *sc.* = subcapitulum; *st.* = sternite ('labium').

Table 5.1. A list of selected works on the structure of the mouthparts of the Acari.

Taxon	Authors
General	Börner (1902); Snodgrass (1948)
Anactinotrichida	
Notostigmata	With (1904); Grandjean (1936b); Van der Hammen (1966, 1989)
Holothyrida	Thon (1906); Van der Hammen (1961, 1989)
Ixodida	
Argasoidea	Robinson & Davidson (1913b); True (1932); Bertram (1939); Balashov (1961)
Ixodoidea	Nuttall *et al.* (1905); Nordenskiöld (1908); Arthur (1951, 1957); Nikonov (1958); Gregson (1960a); Balashov (1965)
Mesostigmata	
Antennophorina	Oudemans (1928); Gorirossi & Wharton (1953); Gorirossi (1955b)
Dermanyssina	Stanley (1931); Hughes (1949); Gorirossi (1950); Bourdeau (1956); Van der Hammen (1964); Evans & Loots (1975); Evans & Till (1965, 1979); Akimov & Starovir (1976); Akimov & Yastrebtsov (1987); Akimov *et al.* (1988)
Parasitina	Winkler (1888); Gorirossi (1955c); Evans & Loots (1975)
Uropodina	Gorirossi (1955a); Woodring & Galbraith (1976)
Actinotrichida	
Astigmata	Nalepa (1884, 1885); Michael (1901); Hughes (1953, 1954a); Johnston (1965); Akimov (1985); Van der Hammen (1989)
Oribatida	Michael (1884); Grandjean (1957c); Akimov & Yastrebstov (1989); Van der Hammen (1989)
Prostigmata	
Endeostigmata	Grandjean (1939b); Theron (1979); Van der Hammen (1989)
Parasitengona	Henking (1882); Brown (1952); Mitchell (1962a,b); Mathur & LeRoux (1965); Witte (1978)
Raphignathae	Keifer (1959); Shevtchenko & Silvere (1958); Summers *et al.* (1973); Silvere & Setjein-Margolina (1976); Nuzzaci (1979a,b); Akimov & Yastrebstov (1981); André & Remacle (1984); Gorgol & Yastrebtsov (1987); Nuzzaci & de Lillo (1989, 1991)
Tarsonemina	Sachs (1951); Krczal (1959)

of a single species. The latter condition is exemplified by ***Balaustium murorum*** (Erythraeidae) in which the stout hook-like chelicerae of the parasitic larva are replaced by long styliform organs in the free-living deuteronymph and adult (Witte, 1978). This is accompanied by a reconstruction of the cheliceral region of the gnathosoma in the nymph in order to form a tubular frame to accommodate the styliform chelicerae. The diversity in gnathosomatic structure is rivalled by the diversity in terminology used in the descriptive morphology of the gnathosoma,

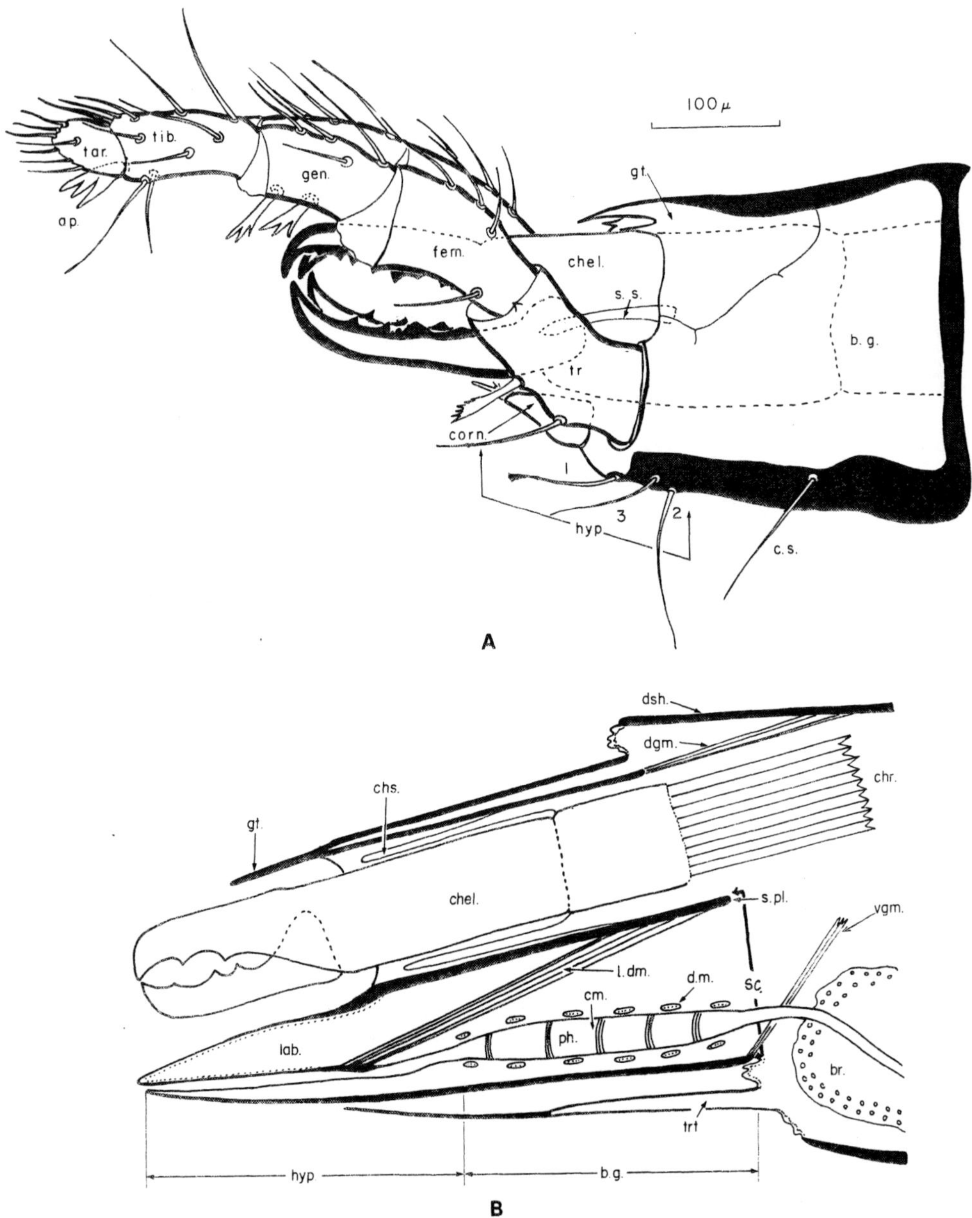

Fig. 5.2. **A**, lateral view of the gnathosoma of a parasitid mite (Mesostigmata); **B**, semidiagrammatic longitudinal section of the gnathosoma of a mesostigmatic mite. (After Evans & Till, 1979.) *ap.* = apotele; *b. g.* = gnathosomatic base; *br.* = synganglion; *c. s.* = palpcoxal seta; *chel.* = chelicera; *chr.* = cheliceral retractor muscles; *chs.* = cheliceral sheath; *cm.* = constrictor muscles of the pharynx; *corn.* = corniculus; *dgm.* = extensor muscles of the gnathosoma; *d. m.* = dilator muscles of the pharynx; *dsh.* = dorsal shield; *fem.* = palpfemur; *gen.* = palpgenu; *gt.* = gnathotectal process; *hyp.* = hypostome; *lab.* = labrum. *l. dm.* = labral muscles; *ph.* = pharynx; *sc.* subcapitulum; *s. s.* = salivary stylus; *tar.* = palptarsus; *tib.* = palptibia; *tr.* = palptrochanter; *trt.* = tritosternum; *vgm.* = flexor muscles of the gnathosoma; 1–3 = hypostomatic setae.

which has resulted in the same structure being referred to by many different terms. There is as yet no single generally acceptable terminology for the morphology of the acarine gnathosoma. The major contributions to the functional morphology of the acarine gnathosoma are listed in Table 5.1 and should be consulted for specific details.

Subcapitulum

The subcapitulum forms the inferior part of the acarine gnathosoma and its external walls are formed by the enlarged coxae of the pedipalp which meet and fuse ventrally. The role of the sternites of the cheliceral and pedipalpal segments in the formation of the subcapitulum has not been elucidated but it seems probable that they have either been markedly reduced or obliterated during the process of gnathosomatization. The mesial walls of the coxae are connected above the pharynx by the subcheliceral plate which forms the dorsal part of the subcapitulum. It provides attachment sites for certain of the dorsal dilator muscles of the pharynx, the labral muscles and, in some Actinotrichida, also for extrinsic muscles concerned with the movement of the gnathosoma (Fig. 5.2A,B). Snodgrass (1948) was of the opinion that this plate-like structure corresponded to the epistome (=clypeus) of mandibulate arthropods and hence should be so named in arachnids. The term epistome has also been used for the anterior projection of the dorsal wall of the gnathosoma in the Mesostigmata and in a restricted sense for a specific portion of the subcheliceral plate. The application of the same term to different structures has led to some confusion; for example, Woolley (1988) refers to the dorsal wall of the gnathosoma as 'epistome I' and to a part of the dorsal lobe in front of the mouth as 'epistome II'! The term subcheliceral plate is used in this work.

The subcheliceral plate is usually provided with a distinct median longitudinal ridge (capitular saddle) and the plate may extend beyond the posterior margin of the coxal region of the subcapitulum and into the idiosoma, particularly in those parasitiform mites with elongate cheliceral shafts. In the Mesostigmata, the plate is produced posteriorly either into two lateral arms or apodemes connected at their posterior extremity by a sclerotized bridge as, for example, in the Antennophorina and Uropodina, or into three elements comprising two lateral arms and a median apodeme (incorporating the median ridge) as in the Dermanyssina and Parasitina. The posterior of the dorsal pharyngeal dilator muscles originates on the lateral arms and the labral muscles arise from the posterior bridge or, in its absence, from the lateral arms and the median apodeme. Gorirossi (1955a) refers to the subcheliceral plate, in the sense used herein, as the 'horizonal shelf' and considers it to comprise three components:

1. an epistome consisting of a portion of the shelf from which the

pharyngeal muscles originate and the labrum is formed;
2. a subcheliceral plate forming an apodeme extending into the idiosoma and providing the surface upon which the chelicerae glide;
3. a tentorium – an apodeme connecting the epistome and subcheliceral plate with the mesial walls of the palpcoxae and extending anteriorly to form the dorsal wall of the hypostome.

In the Actinotrichida, Van der Hammen (1968b) uses the term 'cervix' for the plate connecting the palpcoxae and the term 'capitular apodeme' for its posterior extension from which the labral and certain of the pharyngeal dilator muscles originate. Akimov (1985), on the other hand, names the entire structure the epistome.

Anterior to the mouth, the subcapitulum contains the cavity or channel for the reception of food. In the Parasitiformes, this region of the subcapitulum is usually referred to as the hypostome and in ticks includes the harpoon-like structure of the gnathosoma. The mode of formation of the pre-oral region of the subcapitulum is problematical and it is not known to what extent the apophyses or endites of the palpcoxae and the ventral or ventrolateral region of the mouth are involved in its formation (Aeschlimann, 1984). It seems most likely that the pre-oral channel is formed during the embryonic back-growth of the mouth and labrum and a lateral forward shift of the rudiments of the anterior segments which Manton (1977) considers to be a basic theme shown by all arthropod groups. The walls of the pre-oral or food channel (hypopharynx of some authors) in the Mesostigmata and Holothyrida are continuous with the lateral walls of the pharynx (Fig. 5.3A–F). The channel is usually relatively narrow and V-shaped and in the Sarcoptiformes is bordered by the lateral lips (Grandjean, 1938e) and flanked by deep furrows along which the cheliceral digits travel during the retraction of the chelicerae (Fig. 5.4A). In the Astigmata, the lateral lips show considerable variety in degree of development. For example, in *Lepidoglyphus destructor* the paired lips are separated by a narrow groove and each terminates in a pad-like boss (Fig. 5.4A), while in *Sancassania* and *Rhizoglyphus* the lips are fused and form a solid ridge-like structure terminating in two bosses lying one above the other (Fig. 5.4B).

The walls of the pre-oral channel or the lateral lips are produced anteriorly into a pair of lobes or flaps of varying degrees of sclerotization and complexity. In the Mesostigmata and Holothyrida they form the hypostomatic processes (internal malae, hypopharyngeal processes), which are often fringed with setiform or rod-like projections or, as in *Asioheterozercon*, developed into a pair of large ventral lobes with a duct connecting their longitudinal groove to the pre-oral channel (Fig. 5.5). In the Oribatida and some Prostigmata, the lobes of the lateral lips bear the adoral setae of probable chemosensory function.

The labrum is always present as a lobe projecting above and beyond the mouth and lying in the pre-oral cavity (Figs 5.3E,F & 5.6B). It varies considerably in size from a short process in ixodoid ticks and many

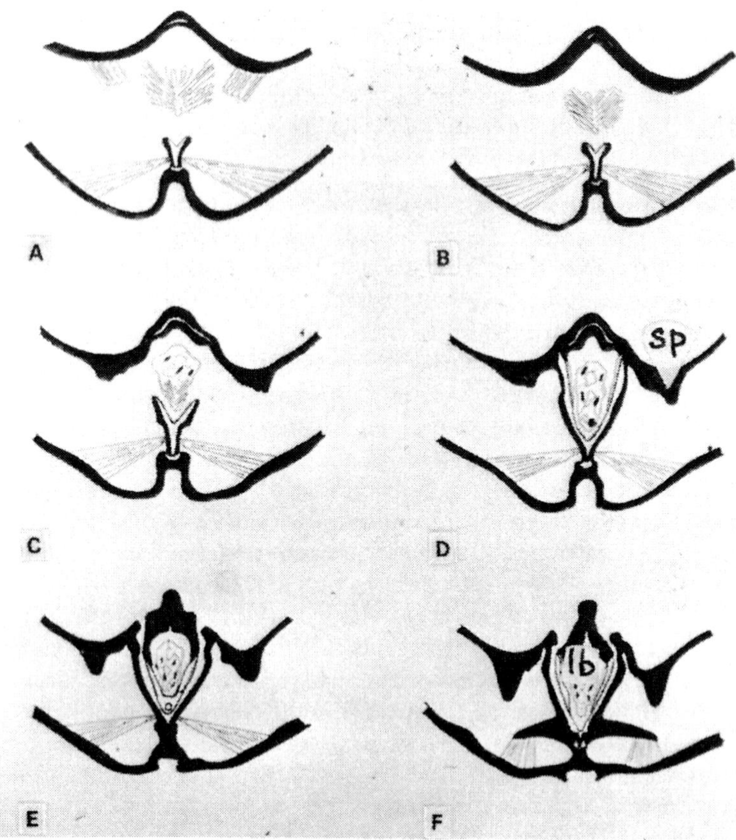

Fig. 5.3. Transverse sections of the gnathosoma of *Asioheterozercon audax* showing the formation of the pre-oral channel. **A–C**, post-oral sections showing the gradual dorsal extension of the lateral walls of the pharynx and the formation of a dorsal lip; **D**, section at the level of the mouth with the lateral walls of the pharynx extending to the subcheliceral plate; **E, F**, pre-oral sections illustrating the complete separation of the labrum from the lateral pharyngeal walls and subcheliceral plate. *lb.* = labrum; *sp.* = subcheliceral plate.

Prostigmata to a long and conspicuous structure extending more or less the length of the pre-oral cavity or channel in the Notostigmata, Holothyrida, Mesostigmata, Astigmata and Oribatida. At its origin above the mouth in the Mesostigmata and Holothyrida, the labrum comprises a ventral portion formed by the anterior prolongation of the dorsal wall of the pharynx and a dorsal more heavily sclerotized part originating from the subcheliceral plate. The subcheliceral plate component may continue anteriorly for a variable distance as an integral part of the labrum or, in some Mesostigmata, for example the Dermanyssina, as a separate hollow process overlying a distinct ventral lobe (Fig. 5.6A). Some authors refer to the portion originating from the cheliceral plate as the labrum and that from the prolongation of the dorsal wall of the pharynx as the epipharynx

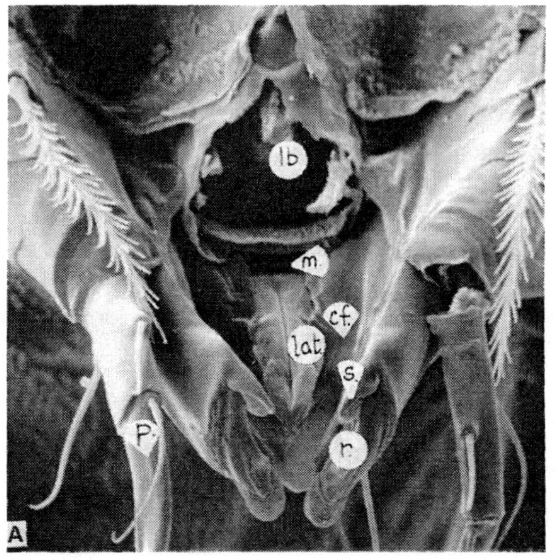

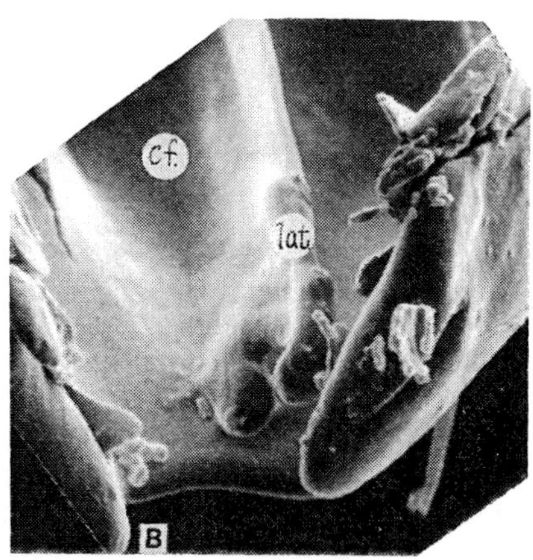

Fig. 5.4. **A**, SE micrograph of the pre-oral cavity of *Lepidoglyphus destructor* (Astigmata) showing the cheliceral furrows and form of the lateral lips; **B**, lateral lips in *Rhizoglyphus* forming a median ridge separating the cheliceral furrows. *c. f.* = cheliceral furrow; *lat.* = lateral lips; *lb.* = labrum; *m.* = mouth; *p.* = pedipalp; *r.* = rutellum; *s.* = strigilis.

(Hughes, 1949; Gorirossi, 1950; Akimov & Starovir, 1976). If this terminology is adopted, then the compound pre-oral lobe should be referred to as the labrum-epipharynx, but this rarely occurs and more commonly when the two components are not distinguished the lobe is called the epipharynx (Woodring & Galbraith, 1976) or the labrum (Van der Hammen, 1964). Evans and Loots (1975), on the other hand, referred to the pre-oral lobe as the labrum and the anterior extension of the subcheliceral plate as the supralabral process. In this confused state of the terminology, there is much to be said for adopting the opinion of Snodgrass (1948) that 'there is no apparent reason for calling the preoral lobe of any arachnid anything else than labrum'. A subcheliceral plate element may be incorporated in the pre-oral lobe in some Astigmata and Oribatida, at least at its origin above the mouth, and in these groups the entire lobe is generally referred to as the labrum (Van der Hammen, 1980; Akimov, 1985). The labral muscles, when present, originate on the postero-ventral region of the subcheliceral plate and insert, often on a sclerite, within the labrum, which is richly innervated from the labral nerve. The labrum is often provided with solid and/or hollow cuticular processes. Those on its dorsal surface in the Parasitidae, for example, may function as chemoreceptors (Fig. 5.6B,C). The elongate labrum in argasoid ticks divides the tube-like pre-oral cavity into dorsal and ventral sections which become independent channels. The ventral channel enters the buccal opening leading into the pharynx while the dorsal one opens

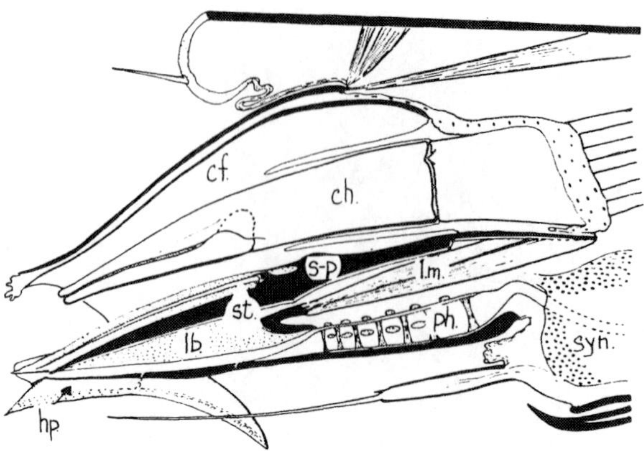

Fig. 5.5. Sagittal section of the gnathosoma of *Asioheterozercon audax* (Mesostigmata) showing the labella-like hypostomatic processes and paralabral stylus flanking the labrum. *c.f.* = cheliceral frame; *ch.* = chelicera; *hp.* = labella-like hypostomatic lobe with duct connecting pre-oral groove; *lb.* = labrum; *l. m.* = labral muscles; *ph.* = pharynx; *s-p.* = subcheliceral plate; *st.* = paralabral stylus; *syn.* = synganglion.

into the salivarium, a reservoir for saliva discharged into it by way of the salivary ducts (see Fig. 5.14F below).

In predatory Mesostigmata, variously shaped ridges or processes of the lateral walls of the pre-oral channel oppose processes on the labrum and are together considered to function as a sieve to prevent solid food particles entering the mouth (Evans & Loots, 1975). Muscles originating on the lateral and ventral walls of the subcapitulum of the Parasitiformes and inserting directly onto the walls of the channel or on their apodemes act as dilators of the pre-oral channel although the function of some of the shorter muscle pairs could be to provide stability to the channel rather than movement.

Near the origin of the labrum in free-living Parasitiformes, the walls of the pre-oral channel at their junction with the subcheliceral plate may be developed into flat denticulate processes which overhang the basal region of the labrum (Fig. 5.6B). These processes, which include a subcheliceral plate element, form the paralabra (hypopharyngeal styli, labella). Arising from each paralabrum in some Mesostigmata (for example Heterozerconidae and Antennophorina) and Holothyrida, is a long anteriorly directed stylus, the paralabral stylus, extending almost to the level of the tip of the labrum (Fig. 5.5). The styli may lie in dorsolateral grooves of the labrum as in *Asioheterozercon*. The function of the paralabra is not known but they may act to stabilize the labrum during its extension by hydrostatic pressure. Paralabra are usually absent or weakly developed in the Dermanyssina and the anterior sclerotized projection of the subcheliceral plate over the pre-oral lobe probably takes over the stabilizing role. Blade-like processes also arise from the dorsolateral wall of the pre-oral

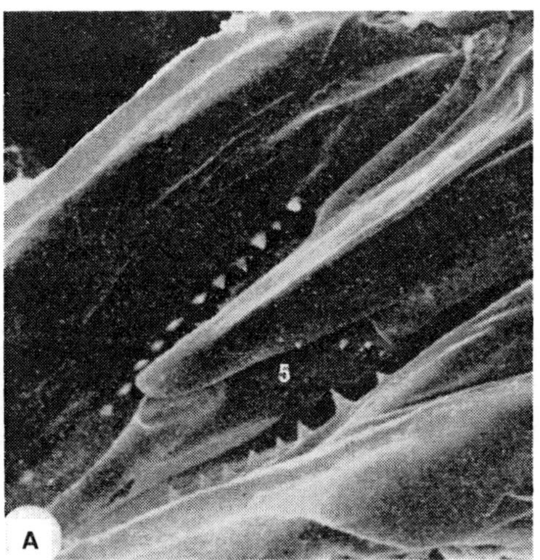

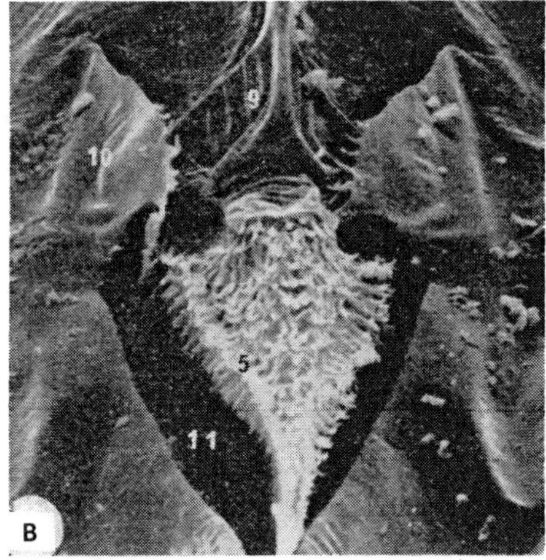

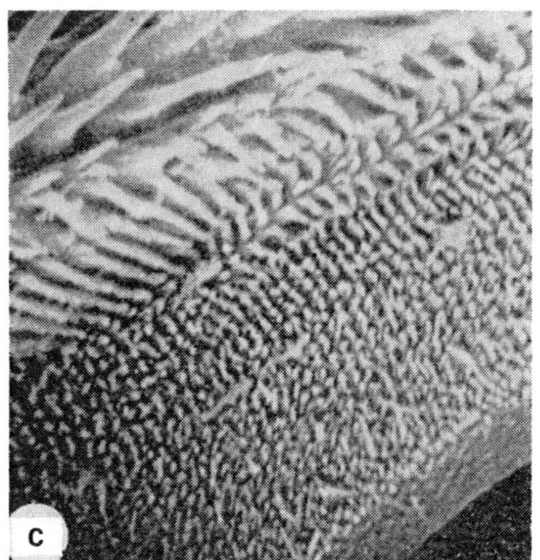

Fig. 5.6. **A**, SE micrograph showing the anterior extension of the subcheliceral plate as a process overlying the labrum in *Ornithonyssus sylviarum* (Mesostigmata); **B, C**, cuticular processes on the labrum of *Phityogamasus primitivus*. **B**, dorsal view of labrum lying in the pre-oral cavity. **C**, ventrolateral view of labrum. (After Evans & Loots, 1975.) 5 = labrum; 9 = ridge of subcheliceral plate (capitular saddle); 10 = paralabrum; 11 = pre-oral cavity.

cavity of some Astigmata and are shown abutting the base of the labrum of *Lepidoglyphus destructor* in Fig. 5.4A.

The anterior region of the subcapitulum on either side of the pre-oral channel or lateral lips is generally considered to be formed from the endites of the palpcoxae (=coxapophyses) and constitute the malae or malapophyses, the former term being used in the Mesostigmata and the latter in the Actinotrichida, Ventro- or dorsolaterally, the malae in the Notostigmata, Holothyrida and Mesostigmata are each provided laterally or dorsally with a horn- or chisel-like structure usually termed the corniculus (=external mala) and probably of setal origin (Fig. 5.2A). A second pair of less sclerotized corniculi-like structures lying internal to the dentate corniculi and termed With's organs occur in the Notostigmata. The corniculus in predatory forms is usually smooth and horn-like and in the free-living Dermanyssina is grooved dorsally to accommodate the salivary stylus. In species ingesting solid food such as the Notostigmata and the majority of the Ameroseiidae, the corniculi are dentate and more or less chisel-like.

An analogous structure is present on each malapophysis in the Oribatida and Astigmata. The process is terminal in the Oribatida and is of setal origin, being birefringent and usually having a basal root. It is called the rutellum (*pl.* rutella) and shows considerable diversity in form, ranging from a narrow seta-like to a massive dentate chisel-like structure (Grandjean, 1957c). Four main types are recognized:

1. *Primitive* or simple, such as that found in the Enarthronota, in which the narrow base is more or less cylindrical (Fig. 5.7A).
2. *Atelebasic*, as found in the Desmonomata and certain immature Euoribatida where the base is rather narrow but the distal region is differentiated into a stout dentate part and a thin paraxial lobe that may overlap its partner and completely hide the lateral lips when seen from below (Fig. 5.7B).
3. *Pantelebasic*, present in the majority of the Euoribatida, sub-triangular in outline from below with a broad base reaching the infrabuccal groove and paraxial edge hiding the lips, and differentiated into an anterior toothed part and a posterior smooth part (Fig. 5.7C).
4. *Suctorial*, leaf-like and semicircular in section and associated with species having long chelicerae with small digits as in the Phenopelopidae and Galumnidae (Fig. 5.7D).

A manubrial articulation in the form of a fissural break may occur near the base of the rutellum. Some authors consider that the lack of a basal root and of birefringence of the processes in the Astigmata are indicative of a non-setal origin and that they are probably formed from the malapophyses. They have been termed pseudorutella by Johnston (1965). Each structure in saprophagous and fungivorous species appears to comprise an anterior blade or pad-like anterior extension of the malapophysis and an internal-dorsal toothed or spike-like process(es). The entire structure is termed a rutellum by Griffiths (1977) and Akimov (1979) who, respectively, refer to the process(es) as a strigilis and prostomal teeth. Akimov (1979) states that a manubrial articulation between the rutellum and the 'hypostome dorsal' is common for all the acarid genera he has investigated. The term rutellum is used in the present work for both structures without implying homology with the similar named structure in the Oribatida. The rutella in the free-living Astigmata display a wide range in form which is associated with the processing of the different types of food (Akimov, 1985).

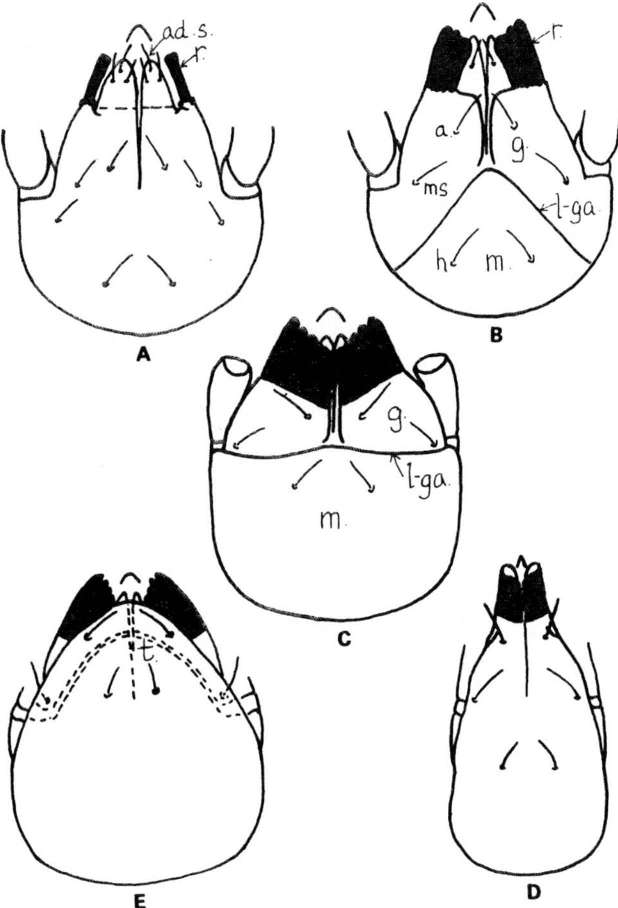

Fig. 5.7. Form of the rutella and venter of the subcapitulum in the Oribatida. **A**, simple rutella and anarthry. **B**, atelebasic rutella and stenarthry. **C**, pantelebasic rutella and diarthry. **D**, suctorial rutella and anarthry. **E**, diarthric state showing anterior tectum of the mentum. (Based on Grandjean, 1957c.) *a.* = anterior subcapitular seta; *ad. s.* = adoral setae; *gen.* = gena; *h.* = hysterostomatic seta (posterior subcapitular); *l-ga.* = labio-genal articulation; *m.* = mentum; *ms.* = median subcapitular seta; *r.* = rutellum; *t.* = tectum.

The ventral surface of the subcapitulum in the Anactinotrichida is formed from the coxae of the palps and their endites and, except in the Ixodida, is provided medially with a longitudinal groove extending from the posterior region of the subcapitulum to the anterior margin of the hypostome. The groove is termed the subcapitular or hypognathal groove (capitular groove, deutosternal groove, hypostomal groove). Some authors consider it to represent the deutosternum and protosternum, the latter forming the hypostomatic section of the groove (Gorirossi & Wharton, 1953). In the Mesostigmata, the groove or gutter is usually provided with anteriorly directed denticles arranged in transverse rows or in a single file. Closely associated with the groove is the tritosternum (=sternapophysis) which is a paired structure in the Notostigmata and an unpaired organ in the Mesostigmata. It is lacking in ticks and much reduced, when present, in the Holothyrida. The tritosternum arises ventrally from the perignathosomatic cuticle and its base is surrounded by thick pliable cuticle. It does not have an extrinsic musculature (Fig. 5.8A). The typical form in the Mesostigmata comprises a slender rectangular base from which arise a pair of pilose laciniae. The latter lie in the subcapitular groove when the gnathosoma is in its retracted state and their pili appear to engage the denticles (Fig. 5.8B). Thin muscle fibres originating at the base of the tritosternum and inserting on the laciniae are stated to occur in most of the Dermanyssina by Akimov and Yastrebtsov (1988). They are considered to be involved in moving the laciniae. The tritosternum is richly innervated by the tritosternal nerve. Four pairs of subcapitular setae are normally present on the gnathosoma venter in nymphal and adult Mesostigmata but only two pairs in the larva. In adult Ixodidae, on the other hand, only two pairs are present while species of Notostigmata may have 13 or more pairs.

The subcapitular groove and tritosternum are absent in the Actinotrichida. A division of the venter of the subcapitulum at the so-called labiogenal articulation into a posterior unpaired mentum (hysterostome) and a pair of anterior genae is encountered in the Oribatida (Grandjean, 1957c). The nature of the articulation is related, to some extent, to the form of the rutella and three conditions may be recognized:

1. *Anarthry* with the labiogenal articulation absent and rutella of the primitive or suctorial type (Fig. 5.7A).
2. *Stenarthry* having the labiogenal articulation running obliquely on either side from the mid-line to the lateral margin of the subcapitulum resulting in the mentum being triangular in outline, genae large and rutella atelebasic or pantelebasic (Fig. 5.7B).
3. *Diarthry* with a transverse labiogenal articulation running behind the palpal bases thus forming a large sub-rectangular mentum, genae small and rutella pantelebasic or suctorial (Fig. 5.7C).

The mentum may have an anterior tectum covering at least the bases of the rutella (Fig. 5.7E). Anarthry occurs in the primitive families of the Oribatida and in certain Euoribatida while stenarthry is characteristic of the Holostomata and Mixonomata and diarthry is most usually encountered in the Euoribatida. Deformation along the articulation permits lateral movement of the rutella which is an essential stage in the feeding mechanism of these mites (Dinsdale, 1974a). The labiogenal articulation is clearly an adaptation for a specific type of feeding and is restricted to the Oribatida. The terms genae and mentum are also applied in the descriptive morphology of the subcapitulum of those actinotrichids lacking a labiogenal articulation, each gena being the ventral surface of a malapophysis and a continuation of the mentum. Subcapitular setae are

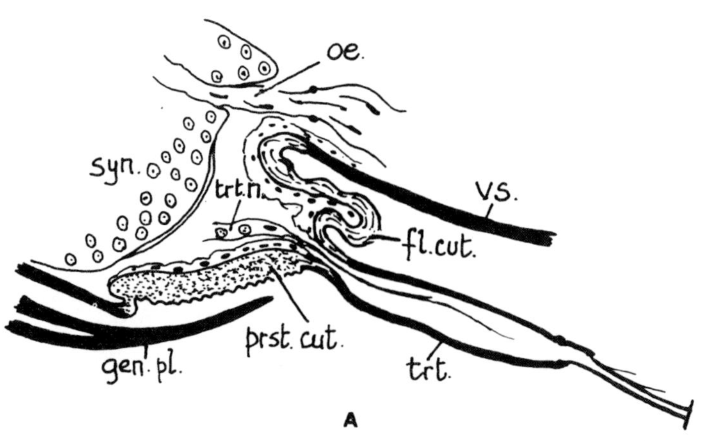

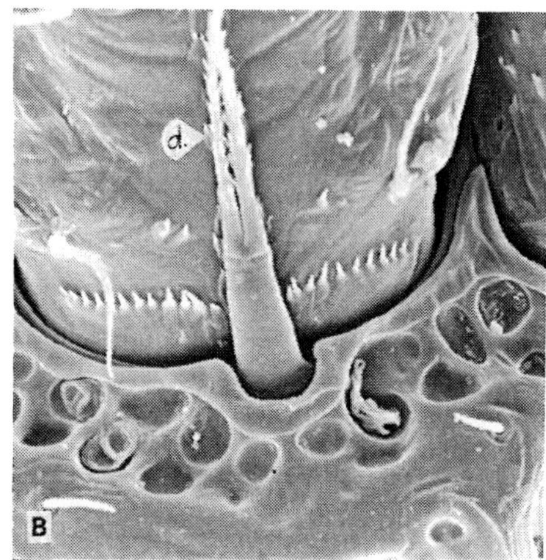

Fig. 5.8. **A**, sagittal section of the tritosternum in *Asioheterozercon audax* (Mesostigmata); **B**, SE micrograph of the tritosternum of the deuteronymph of *Protodinychus punctatus* lying along the capitular groove. *d.* = denticles of capitular groove; *fl. cut.* = flexible cuticle between gnathosomatic base and tritosternum; *gen. pl.* = genital plate of female; *oe.* = oesophagus; *prst. cut.* = presternal cuticle; *syn.* = synganglion; *trt.* = tritosternum; *trt. n.* = tritosternal nerve; *vs.* = ventral surface of subcapitulum.

normally present and in the Oribatida are designated anterior, median and posterior subcapitular (or infracapitular) setae. Additionally, a supracoxal seta (*e*) is normally present and situated above the base of the palp on each side of the subcapitulum.

Pedipalps

The pedipalps are articulated to the palpcoxal region of the gnathosoma and in the majority of the Acari form free limb-like appendages extending beyond the anterior margin of the subcapitulum. Rarely, as in certain Tarsonemina, they are extremely small and do not or scarcely extend beyond the gnathosomatic capsule. In the free-living Anactinotrichida, the palps comprise six articles or palpomeres: trochanter, femur, genu, tibia,

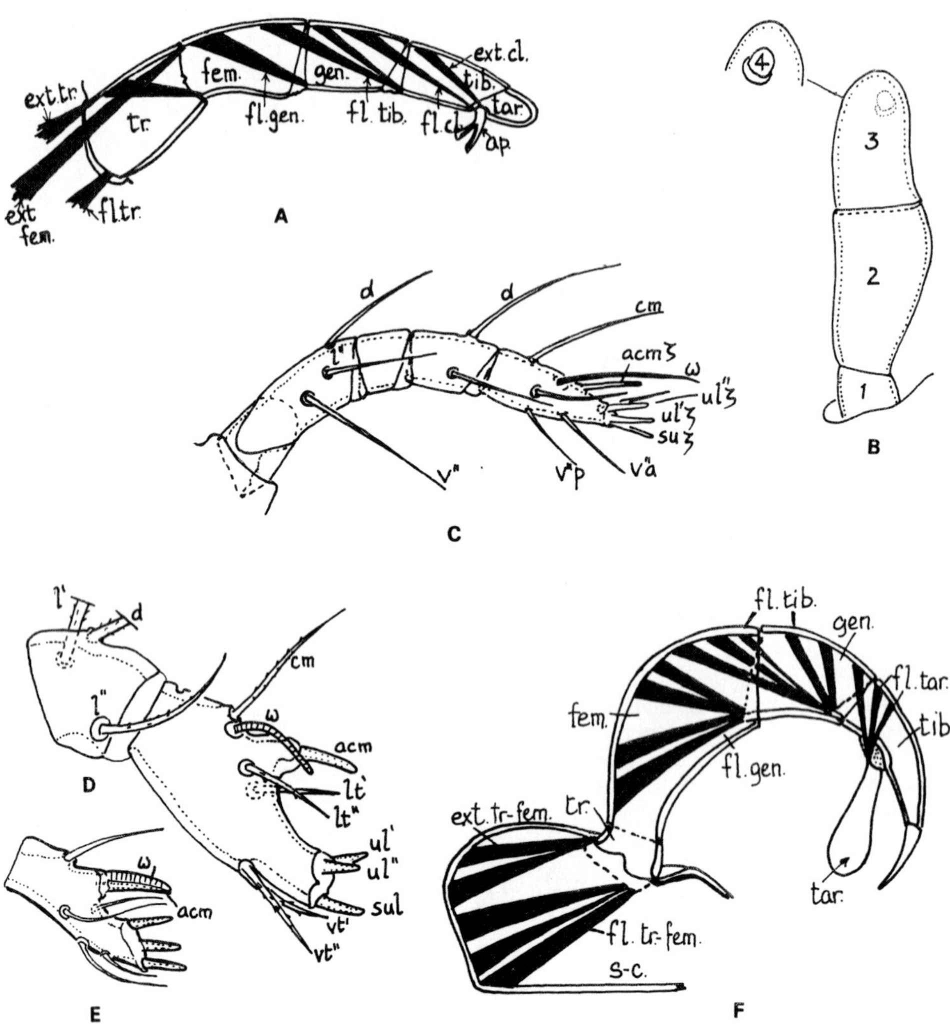

Fig. 5.9. **A**, lateral view of the pedipalp of a dermanyssoid mite (Mesostigmata) showing the musculature of the palpomeres; **B**, pedipalp of an ixodid tick; **C**, pedipalp of a phthiracaroid mite (Oribatida) with palpomeric chaetotaxy (see text for setal notation); **D**, chaetotaxy of the palptibia and tarsus *Neoliodes theleproctus* (Oribatida), antiaxial view; **E**, palptarsus of *Allogalumna* sp. illustrating the 'corne double'; **F**, pedipalp of a trombidioid mite (Prostigmata) showing musculature. *ap.* = apotele; *ext.* cl. = extensor of palpal claw (apotele): *ext. fem.* = extensors of palpfemur; *ext. tr.* = extensors of palptrochanter; *ext. tr-fem.* = extensors of palptrochanter-femur; *fem.* = palpfemur; *fl. gen.* = flexors of palpgenu; *fl. tar.* = flexors of palptarsus; *fl. tib.* = flexors of palptibia; *fl. tr.* = flexor of palptrochanter; *fl. tr-fem.* = flexors of palptrochanter-femur; *gen.* = palpgenu;, *s-c.* subcapitulum; *tar.* = palptarsus; *tib.* = palptibia; *tr.* = palptrochanter; 1–4 = articles 1–4.

tarsus and apotele (Fig. 5.9A). The nature of the articulation between the trochanter and the subcapitulum is often difficult to discern. Akimov and Yastrebtsov (1988) refer to the presence of a dicondylar dorsoventral joint in this position in the Mesostigmata and consider the muscles originating on the internal ventrolateral surface of the gnathosoma and inserting on the antiaxial and paraxial surfaces of the trochanter to function as promotors and remotors. Woodring and Galbraith (1976), on the other hand, are of the opinion that the muscles originating on the lateral wall of the gnathosoma and inserting on the dorsal and ventral proximal region of the palptrochanter in the Uropodoidea act as flexors and extensors. The position of the palptrochanter observed in relation to the subcapitulum in a series of SEM micrographs suggests that the segment is capable of flexion and extension as well as limited rotation. The pivot joint between the trochanter and femur is dicondylar and transverse with the depressor (flexor) muscles originating on the dorsal surface of the trochanter and inserting on the ventral proximal surface of the femur, and the levator (extensor) muscles originating on the ventrolateral wall of the subcapitulum and inserting on the dorsal proximal surface of the femur. The articulations between the femur–genu and genu–tibia are dorsal and monocondylar and the flexors of the genu and tibia originate on the dorsal surface of the femur and the genu and tibia, respectively. Extension is by hydrostatic pressure as in the case of the similarly named segments of the legs. A slit organ is present dorsally near the proximal margin of the genu and the tibia in the Mesostigmata. The ultimate palpomere, the apotele, is represented by a pair of terminal claws in the Notostigmata and by a tined claw-like structure (furcula) arising paraxially near the base of the tarsus in the Mesostigmata and some Holothyrida. It is lacking in the Ixodida. The apotele is flexed and extended by antagonistic muscles, the extensors originating on the dorsal posterior surface of the tarsus and inserting on the dorsal posterior edge of the apotele and the flexors originating on the dorsal posterior surface of the tibia and inserting on the ventral posterior edge of the apotele. In the Mesostigmata, the apotele or palpal claw (two, three or four-tined) appears to have a transverse bicondylar articulation with the tarsus.

Fusion of the tibia and tarsus of the palp to form a tibiotarsus occurs in some Uropodina and Antennophorina and probably also in the Ixodina. In the latter, the four palpomeres are usually referred to by tick workers as articles I (the proximal) to IV (the distal), but Van der Hammen (1983) considers the four articles in the Ixodoidea to be equivalent to the trochanter, femur, genu and tibiotarsus of the typical anactinotrichid palp (Fig. 5.9B). The four palpal articles in the Argasoidea are cylindrical or sub-cylindrical in shape and movably articulated except in the larvae of some Otobini (e.g. *Otobius lagophilus*), in which the second and third are immovably fused, and those of the Ornithodorini (e.g. *Ornithodoros savignyi*) that lack distinct articulations between the first and second, and the second and third articles (Edwards, 1975, Figs 5 & 9). The first articles in the nymph and male of *Nothoaspis reddelli* are exceptional in being enlarged and produced internally to form a flange (Keirans & Clifford, 1975; Keirans *et al.*, 1977). The flanges meet in the mid-line to conceal and protect the venter of the gnathosoma. In the Ixodoidea, the first article is movably articulated to the gnathosoma and usually to the second except, for example, in the genus *Dermacentor* (Arthur, 1960). Articles II and III are flattened and have a paraxial concavity which accommodates and protects the dorsal and dorsolateral region of the hypostome when the palps are apposed during the non-feeding period. The articulating membrane between these articles is very reduced dorsally and appears as a transverse suture. This results in little if any movement between them at the articulation. The fourth article lies in a cup-like hollow on the antero-ventral surface of the third and is articulated to it by a hinge joint. It is flexed by muscles originating to article II. The palp moves outwards at the joint between the first article and gnathosoma during feeding.

The palps of the Actinotrichida show considerable variety in their segmentation and form. An apotele is never present and the maximum number of palpomeres is five (trochanter, femur, genu, tibia and tarsus). This condition obtains in the majority of the Oribatida (Fig. 5.9C) although the number may be reduced in some taxa, for example to three in the Euphthiracaridae by the fusion of the femur, genu and tibia, and to two in the Epilohmanniidae by the fusion of the four proximal palpomeres. The palpomeres are cylindrical or subcylindrical in shape and the palp is essentially a sensory appendage. In the majority of families, the five palpomeres have the setal formula (0-2-1-3-9). In addition, there is invariably a tarsal solenidion (ω). Of the nine true setae on the tarsus (Fig. 5.9C,D), two, three or four situated at the distal end of the palpomere are eupathidia comprising an anteroculminal (*acm*), the ultimals (*ul′* and *ul″*) and the subultimal (*sul*) according to the terminology of Grandjean (1935c). When the anteroculminal lies immediately distal to and is continguous along its entire length with the solenidion, the pair form a *corne-double* (Fig. 5.9E). The five non-eupathidial setae are: the dorsalculminal (*cm*), the lateral (*lt′* and *lt″*) and the ventrals (*vt′* and *vt″*). The palptarsus in the Palaeosomata may have as many as 16 setae of which two or three are eupathidial.

The palps of the Prostigmata have from one to five palpomeres, which may be linear, simple sensory appendages as in, for example, the Endeostigmata, Tydeidae and Labidostommatidae, or two distal palpomeres may be modified for the purpose of grasping or tearing by the hypertrophy of a tibial seta to form a terminal claw-like structure and the displacement of the tarsus to a ventral position relative to the tibia, thereby forming the so-called thumb-claw condition (Fig. 5.9F). The raptorial function of the palp is particularly evident in the Cheyletoidea and terrestrial Parasitengona. In *Dinothrombium*, the tibial claw is also used for digging while in larvae of *Unionicola* parasitizing chironomids, the tibial claws are used to attach the mite to the cuticle of its host before the commencement of feeding (Baker, 1991). The function of the tibial claws in the phytopha-

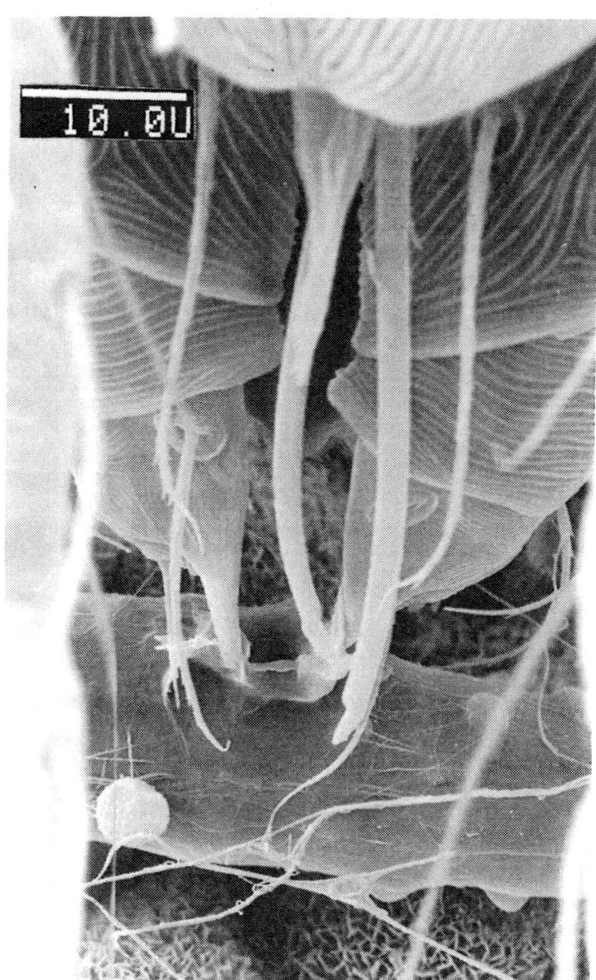

Fig. 5.10. SE micrograph of *Tetranychus urticae* (Prostigmata) showing the pedipalpal claws apparently tearing a hole in the hypha of a false mildew (Peronosporidae).

gous Tetranychidae does not appear to have been determined but in an unusual case of feeding by ***Tetranychus urticae*** on the contents of the hyphae of a species of false mildew (Peronosporidae) on the leaves of an ornamental plant, they appear to be involved in the tearing of the hyphal wall (Fig. 5.10). In the Cunaxidae, the tarsus has a claw-like termination and there may be apophyses along its inner margin. A reduction in the number of palpomeres to four, through the fusion either of the femur and genu or the tibia and tarsus, is not uncommon. A further reduction in the number of palpomeres to three comprising the trochanter, femorogenu and tibiotarsus occurs in the Tarsochelidae (Lindquist, 1986), while there is reduction to two (femorogenu and tibiotarsus) in the Tarsonemidae and to one in some species of Tenuipalpidae.

Little detailed information is available on the musculature of the pedipalps of the Prostigmata other than for the Parasitengona (Mitchell,

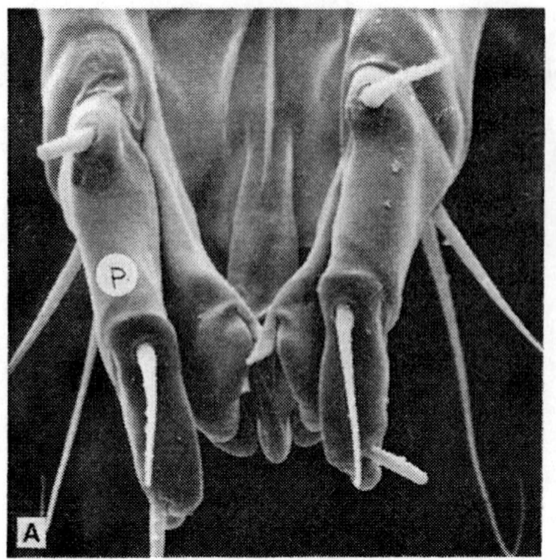

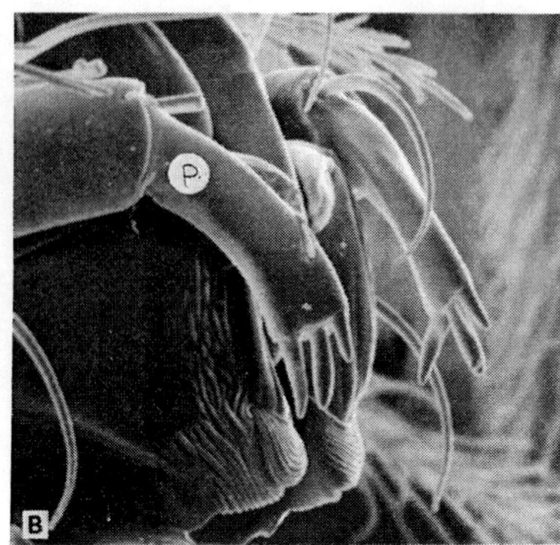

Fig. 5.11. SE micrographs of the pedipalps (p) of *Lardoglyphus* sp. (**A**) and *Lepidoglyphus destructor* (**B**).

1962a; Mathur & LeRoux, 1965), Tetranychidae (Akimov & Yastrebtsov, 1981) and Cheyletidae (Gorgol & Yastrebtsov, 1987). In *Blankaartia ascoscutellaris*, the trochanter moves more or less in a vertical plane by the action of depressor and levator muscles originating in the subcapitulum and inserting, respectively, on the ventral and dorsal surfaces of the base of the trochanter (Fig. 5.9F). The trochanter and femur cannot move independently. The flexor muscles of the genu and tibia insert on the ventral part of the base of the segments and originate, respectively, on the dorsal wall of the femur and the femur and genu. The tarsus is flexed by muscles mainly originating on the dorsal wall of the tibia and inserting ventrally on the base of the tarsus. Extension of the three terminal segments is by hydrostatic pressure.

The pedipalps are much reduced in the Astigmata and are held close to the subcapitulum (Fig. 5.11A,B). They normally comprise two articles although three occur in the deuteronymphs of the Schizoglyphidae (Mahunka, 1978). An adductor muscle rising from the subcapitular base and inserting on the basal article is described for *Sancassaria berlesei* by Akimov (1985). The degree of movement of the distal on the proximal article appears to vary within the group. The joint between the articles appears to be adesmatic in the majority of the taxa although a moderate amount of flexure of the distal article seems to be possible in the Glycyphagidae (Fig. 5.11B). In the Anoetidae, which feed in the liquid of decaying organic material, the palp has a lobed appearance and bears two flagella-like setae distally, one directed laterally and the other more or less posteriorly. Hughes (1953) considers the lobes to consist of thin sclerotized plates which extend the ventral surface of the palp laterally, whereas Scheucher (1957) refers to the transparent membraneous nature of the lobed structure. The palps together move obliquely upwards and outwards from the mid-line and by their sweeping action draw particles of food in the fluid medium towards the anterior region of the subcapitulum where, according to Hughes, they are apparently raked back into the pre-oral cavity by the forward and backward movements of the highly specialized chelicerae.

The pedipalps are provided with mechanoreceptive and chemoreceptive organules. The chemoreceptors probably have a gustatory function and occur on the tarsus or tibiotarsus, where they are often concentrated on or near the tip.

Chelicerae

The chelicerae are appendages that have become variously modified as trophic organs. In the least specialized form, each consists of three parts or articles: the first or basal article to which the cheliceral retractor muscles are attached; the second or middle article movably articulated to the basal one and ending in a hook-like fixed digit; and the third article or movable digit lying ventral to the fixed digit and hinged by a bicondylar articulation to the body of the chelicera (Fig. 5.12A). Some authors consider the basal article to represent the trochanter of the arachnid limb, the middle article to comprise the fused femur, genu, tibia and tarsus, and the movable digit to be the modified apotele or ambulacrum. According to Van der Hammen (1970), the coxa does not form part of the cheliceral shaft but is represented by the cheliceral sheath. This is debatable. The form of joints and the origin and insertion of muscles associated with them are the criteria used in arriving at these homologies and the assumption is made that specific intersegmental joints and their musculature remain constant irrespective of the change in the function of the limb. Some Uropodina are exceptional in having their long chelicerae consisting of four articles. Athias-Binche (1982) considers this to be due to a secondary division of the basal article in *Thinozercon*.

In the free-living Anactinotrichida, the three parts of the chelicera are clearly defined except in some Ameroseiidae (*Ameroseius* and *Brontispalaelaps*) in which the basal and middle articles are fused. The basal article has a condylar articulation with the middle article and this may be dicondylar, forming a horizontal pivot joint, as in the Notostigmata (Van der Hammen, 1966) or unicondylar and paraxial as in the majority of the Mesostigmata. The flexor muscle of the middle article is inserted on its postero-ventral margin and usually originates on the dorsolateral surface of the basal article (Fig. 5.12B), but in some Mesostigmata (*Heterozercon*) the powerful flexor muscle has its origin with the cheliceral retractor muscle group on the dorsum of the idiosoma. In the Uropodina,

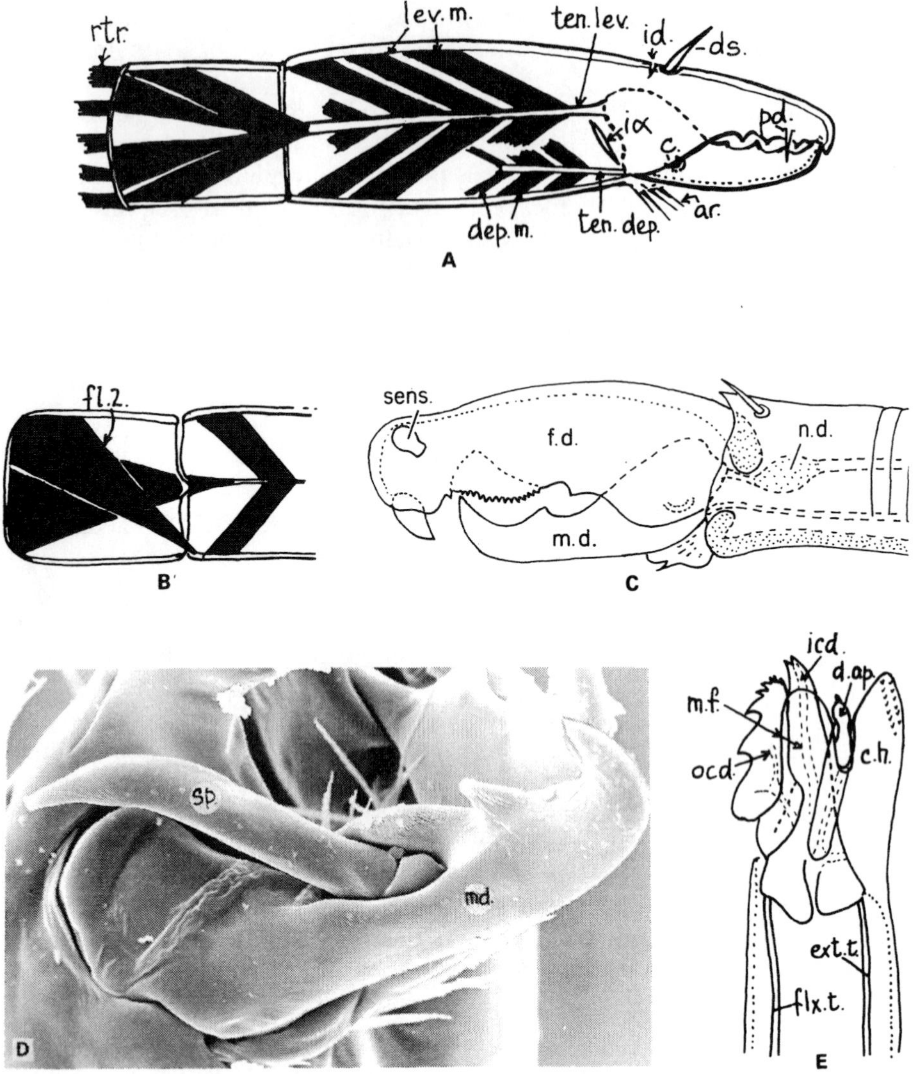

Fig. 5.12. **A**, musculature of the chelicera of a parasitid mite (Mesostigmata); **B**, flexor muscles of the middle article of the chelicera in *Varroa*; **C**, node on the tendon of the levator (extensor) muscle of the movable digit of the chelicera of a uropodid mite. **D**, chelicera of the male of a *Macrocheles* sp. illustrating the spermatodactyl; **E**, cheliceral digits of an ixodoid tick. *ar.* = processes at base of movable digit of chelicera; *c.* = condyle; *c. h.* = cheliceral 'hood'; *d. ap.* = dorsal digital appendage; *ext. t.* = tendon of extensor muscle; *dep. m.* = depressor muscle of the movable digit; *ds.* = dorsal seta; *f. d.* = fixed digit; *fl. 2.* = flexor muscle of middle article; *flx. t.* = tendon of flexor muscle; *ia.* = antiaxial lyrifissure; *icd.* = inner cheliceral digit; *id.* = dorsal lyrifissure; *lev. m.* = levator muscle of the movable digit; *m. f.* = membraneous fold or mantle; *m. d.* = movable digit; *n. d.* = node on tendon of levator muscle; *ocd.* = outer cheliceral digit; *pd.* = pilus dentilis; *sens.* = sensillus; *rtr.* = cheliceral retractors; *sp.* = spermatodactyl; *ten. dep.* = tendon of depressor muscle; *ten. lev.* = tendon of levator muscle.

Woodring and Galbraith (1976) consider the muscle group originating in the basal article and inserting on the proximal edge of the middle article to function as cheliceral rotators. According to Akimov and Yastrebtsov (1988), the flexors of the middle article in the Dermanyssina are characteristic only of specialized parasitic Dermanyssoidea. The levator muscle, when present, has its origin in the basal article and is attached by a tendon to the dorsal margin of the middle article. The flexion of the middle article by the action of the depressor muscles is particularly marked in some Holothyrida, for example in *Thonius braueri*, to the extent of its forming almost a right-angle with the basal article. The cheliceral shaft (basal and middle articles) has areas of strongly tanned cuticle which provide sites for muscle attachment, particularly for those that operate the movable digit. These may be in the form of dorsal and/or lateral longitudinal apodemes.

The cheliceral retractors have their origin on the dorsal wall of the idiosoma, usually in the pronotal region, and insert around the sclerotized posterior rim of the basal article. However, in those species with elongate cheliceral shafts, such as some Uropodina and parasitic Dermanyssoidea, the retractor muscles of the deuteronymph and adult have their origin on the posterior region of the dorsal idiosomatic shield while those of their less extensively sclerotized larva and protonymph originate on a posterior mid-dorsal extension of the pronotal shield (Evans & Till, 1965). The middle article is usually longer than the basal one although some Holothyrida are exceptional in having the basal part much the longer. The fixed digit forms the dorsal member of the chela, the anterior pincer-like part of the chelicera. It is usually a rigid structure but in some Uropodoidea (e.g. *Uroactinia*) it is capable of a certain amount of movement to provide a wider gape of the chela. The basal article is nude, except for one or two setae in the majority of the Notostigmata, while the middle article in its distal half has tactile, stress-sensitive and chemosensory sensilli. A maximum of five setae has been reported on this article in the male of *Opilioacarus vanderhammeni* by Juvara-Bals and Baltac (1977) as opposed to the usual three pairs in female notostigmatics. The full complement of sensilli has only been ascertained in a few free-living parasitiform taxa. Two 'lyrifissures', one situated dorsally (*id*) and the other antiaxially (*iα*), occur in the Notostigmata, Holothyrida and Mesostigmata. A dorsal 'seta' is usually present near the mechanoreceptor *id* in the Mesostigmata and a second dorsal seta may be present in some Holothyrida. A *pilus dentilis* (cheliseta) situated on the antiaxial face of the anterior half of the digit occurs in some Mesostigmata as well as up to three specialized sensilli of probable chemosensory function. The fixed digit is usually provided with one or more teeth in addition to the terminal hook.

The third article forming the movable digit moves in the vertical plane on its dicondylar articulation. This is achieved by the action of antagonistic depressor and levator muscles attached by strong tendons to the base of the digit (Fig. 5.12A). The muscles are often pinnate, i.e. the fibres are much shorter than the muscle and converge from either side on a central tendon. Such pinnate muscles are conspicuous in the chelicerae of the Mesostigmata, Astigmata and Oribatida. In comparison with a parallel-fibred muscle of similar shape, a pinnate muscle has more shorter fibres, its tendon moves through a shorter distance and it exerts more force (Alexander, 1968). Pinnate muscles also have an advantage over parallel-fibred muscles when a muscle must contract within a confined space such as the cavity of a rigid cheliceral shaft. A parallel-fibred muscle swells as it contracts and could not do so if surrounded by rigid walls whereas pinnate muscles do not swell and can contract in a confined space. The levator muscles have their origin on the dorsolateral and ventrolateral surfaces of the first and second articles and are connected by way of a common tendon to the dorsal surface of the basal region of the movable digit. The smaller depressor muscles originate within the middle article and are connected by a tendon to the ventral surface of the movable digit. In some Uropodoidea, the levator tendon is thickened to form a distinct node near its point of attachment to the digit and Evans (1972) has suggested that the movement of the node in relation to the base of the 'fixed' digit may increase the gape of the digits and provide a snapping action of the chela on contraction of the levator muscles (Fig. 5.12C). In predatory forms, the movable digit is provided with teeth but there appear to be no sensilli present. The digit in males of the Dermanyssina is modified for spermatophore transference by the development of a spermatodactyl (spermatostyle, spermatophoral process) which may be free distally (Fig. 5.12D) or almost entirely fused with the digit. Setiform processes associated with the arthrodial membrane at the base of the digit in the Mesostigmata are often arranged in a coronet or form one or two brush-like processes.

Modifications of the cheliceral shaft and the digits for specialized feeding, particularly in parasitic species ingesting tissue fluids of their host, are widespread in the Parasitiformes. The trends in the haematophagous species of the Dermanyssoidea are discussed in Chapter 6. In the Ixodida, the chelicerae are highly modified as cutting organs and have a form that is unique among the Acari. The shaft, somewhat bulbous proximally to accommodate the flexor and extensor muscles of the inner digit, lacks a distinct division into two articles. Tendons connect the muscles to the base of a triangular inner digit (? movable digit) to which an outer digit is articulated on its external surface (Gregson, 1960a; Balashov, 1968). The tendons in *Ixodes* lie in tubular channels within the shafts. Both digits have strong cutting teeth along their external margins and move in a horizontal plane (Fig. 5.12E). Attached movably to the outer side of the inner digit is a dorsal bifid appendage. Two other structures occur at the distal end of the cheliceral shaft – a thin membraneous fold or mantle arising dorsally from the base of the inner digit and extending laterally around the digits so that only their cutting teeth are exposed, and a so-called 'hood' arising from the mesal tip of the shaft. The hood is considered by Robinson and Davidson (1913a) to be an

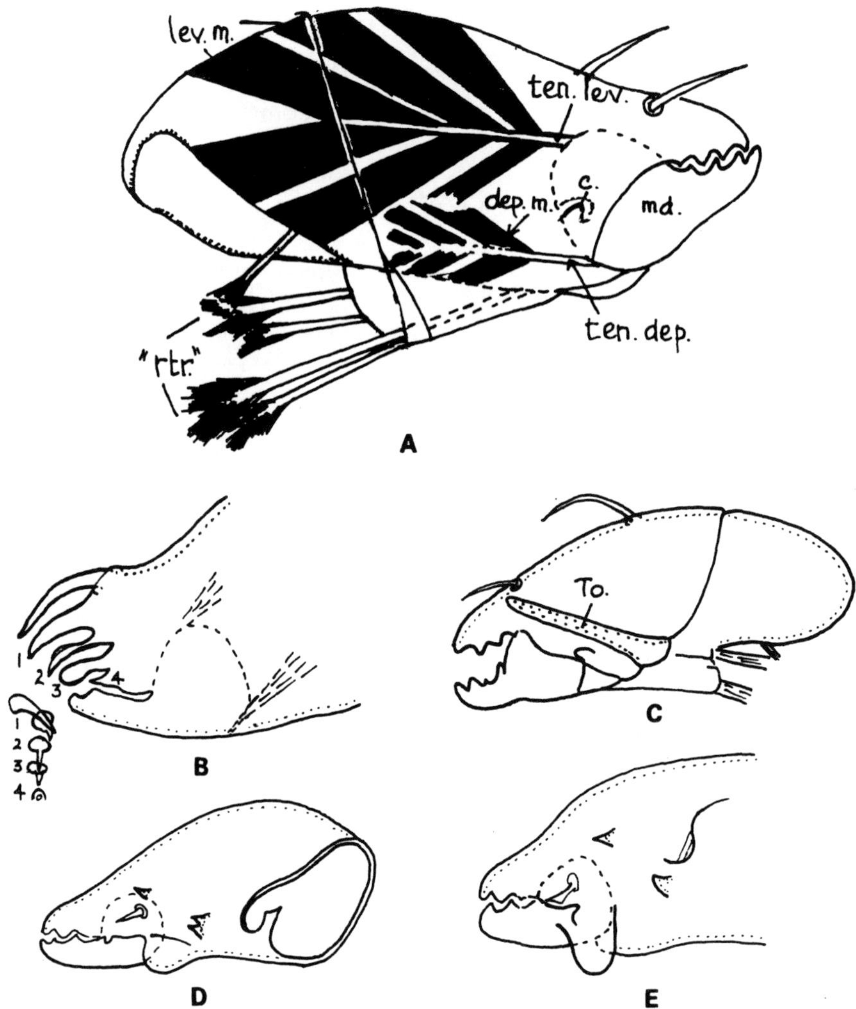

Fig. 5.13. **A**, form and musculature of the chelicera of an oribatid mite; **B**, chelicera of *Pterochthonius angelus* (Oribatida) (based on Grandjean, 1947); **C**, paraxial view of the chelicera of a euoribatid mite showing the position of Tragardh's organ; **D, E**, paraxial views of the chelicerae of *Tyrolichus* (**D**) and *Suidasia* (***E***) (based on Akimov, 1985). *To.* = Tragardh's organ. Other abbrevations as in Fig. 5.12.

extension of the soft 'articulating chitin of the joint between the shaft and digit', whereas Van der Hammen (1989) refers to it as the fixed digit. An oval membrane-covered area of weaker sclerotization and unknown function occurs dorsally on the cheliceral shaft about one-third along its length.

The chelicera of the Actinotrichida has the appearance of a bipartite structure in that it lacks a distinct basal article of the type found in the

Anactinotrichida. The basal article (trochanter of some authors) is considered to have been reduced to a ventral or ventrolateral element which in the Oribatida forms the site of insertion of the tendons of the retractor muscles (Fig. 5.13A). In the chelate condition occurring in the majority of Oribatida, Astigmata and some Prostigmata, for example the Rhagidiidae and Labidostommatidae, the movable digit has a bicondylar articulation with the shaft and opposes a fixed digit. The movable digit is operated by antagonistic depressor and levator muscles of the pinnate type originating within the shaft and connected by tendons to the base of the digit. The chelicerae show considerable diversity in form in the Prostigmata and modifications of the chelate form are seen in the reduction of the fixed digit and the development of the movable digit into a sickle-, hook- or stylet-like structure (see Fig. 5.15A below). The development of styliform digits is often accompanied by the partial or complete fusion of the cheliceral bases along their mesial faces to form a stylophore and this condition occurs in the Tetranychoidea, Raphignathidae, Caligonellidae, Tarsonemina and in some species of the Stigmaeidae and Tydeidae. The fused mesial walls of the cheliceral bases in the Tetranychoidea form a fenestrated median vertical septum which divides the stylophore cavity into two chambers. In the Tarsonemina and Caligonellidae, the stylophore is integrated with the subcapitulum to form a gnathosomatic capsule. The bases of the cheliceral stylets in the Eriophyoidea do not form a stylophore but are attached to tendons of retractor muscles originating on the posterior region of the dorsal shield (Nuzzaci, 1979b). According to Witte (1978), the shaft (proximal article) of each compact hook-like chelicera of larval Erythraeidae is transformed into a styliform chelicera following histolysis of the larval gnathosoma during the moult from larva to protonymph and is retained by the succeeding stases. In the oribatid genus *Gustavia*, the chelicera is very specialized and has only a single styliform digit, probably the movable digit, provided with recurved denticles in its distal third.

The digits of the chelate chelicerae of the Actinotrichida have a birefringent core and, on this basis, Grandjean (1947) postulated that they have a setal origin. The movable digit is homologized with the apotele or ambulacrum of a walking leg in which the paired claws and empodium also have a birefringent core while the fixed digit in the Oribatida is regarded as having been derived from up to four birefringent setae. In support of his hypothesis, Grandjean presents a number of examples of oribatid chelicerae showing gradations from a form in which the movable digit opposes a vertical group of four seta-like projections, as in *Pterochthonius angelus* (Fig. 5.13B), to the more usual type in which the fixed digit is entire.

The cheliceral shaft or body of the chelicera in the Oribatida bears a maximum of two dorsal setae (*cha* and *chb*) and in almost all the Holosomata and Euoribatida a paraxial Tragardh's organ. This organ usually appears as a long finger-like sclerotized protuberance arising from about the mid-length of the shaft basally and directed antero-dorsally

(Fig. 5.13C). According to Van der Hammen (1968b), the cuticle of the organ in *Hermannia convexa* is a continuation of the arthrodial membrane of the cheliceral sheath. A second paraxial protuberance, termed lower Tragardh's organ, may also be present as in the Phenopelopidae. The function of Tragardh's organ has not been established. A number of membraneous projections or oncophyses on the ventral and paraxial faces of the body of the chelicera may also be present. The paraxial face of the body of the chelicera in the Astigmata bears a single seta, usually short and spine-like, near the origin of the fixed digit and one or two short cuticular processes with a single or bifid pointed tip (Fig. 5.13D,E). In addition, a large process may occur at the articulation between the movable digit and the cheliceral body. None, one, two or many setae may be present on the body of the chelicera in the Prostigmata. Cheliceral setae are usually found in those taxa with chelate or hook-like chelicerae such as the Rhagidiidae, Nicoletiellidae, Bdellidae, Anystidae and Trombidiidae but one or two pairs of setae may also be present on the fused cheliceral bases of the specialized styliform chelicerae of the Tarsonemina. Slit organs are typically absent on the chelicerae of the Actinotrichida; the presence of a single antiaxial lyrifissure (*ia*) being of rare occurrence.

Cheliceral Frame and Sheaths

The chelicerae are usually enclosed in a distinct frame (the 'cadre mandibulaire' of Grandjean) or gnathotheca whose floor is formed by the subcheliceral plate. The lateral and dorsal walls of the cheliceral frame are constructed either by the extension of the frontal body wall of the proterosoma onto the subcapitulum or by the dorsal extension of the sclerotized walls of the pedipalpal coxae to enclose the chelicerae. In the former condition as seen in the Sarcoptiformes, the extended body wall is pliable and folded upon itself to accommodate the movement of the gnathosoma on the proterosoma (Fig. 5.14A). The dorsal part of the extended body wall may be sclerotized in the Actinotrichida and is referred to as the tegulum. A membrane separates the body cavity of the cheliceral region from the exterior and appears to be the extension of the body wall forming the cheliceral frame. It is attached ventrally to the subcheliceral plate. This membraneous tube-like structure is divided along part of its length by a median inter-cheliceral septum connected ventrally to the capitular saddle. The membrane continues anterior to the septum as two separate tubular sheaths in much the same way as the legs arise from the crotch of a pair of trousers. Each sheath surrounds the basal region of the chelicera like a sleeve and is attached to the proximal third of the body of the chelicera. The line of attachment of the sheath is designated *acx*. Each sheath is folded on itself and permits independent protection and retraction of the chelicerae while maintaining the seal between the body cavity and the exterior. The cheliceral sheaths are

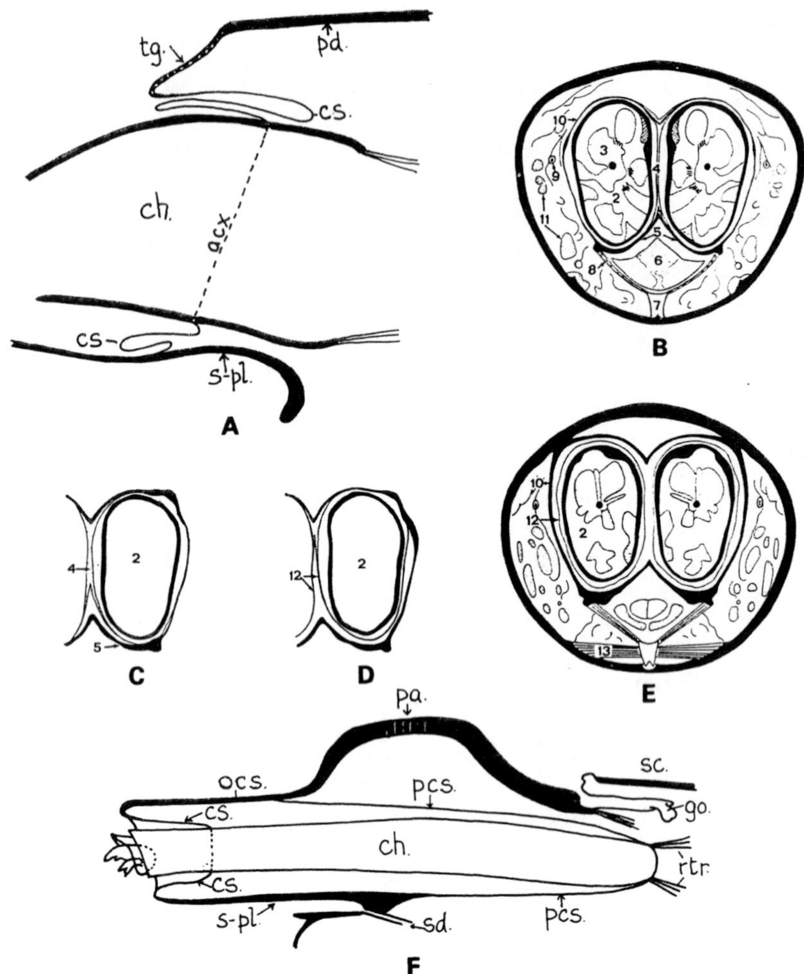

Fig. 5.14. **A**, lateral view of the chelicera and cheliceral sheath (*c.s.*) in *Sancassania* (Astigmata); **B–E**, transverse sections of the gnathosoma of *Phityogamasus primitivus* (Mesostigmata) showing the division of the intercheliceral septum and formation of individual cheliceral sheaths (after Evans & Loots, 1975); **F**, a schematic longitudinal section of the chelicera, cone sheath (*cs.*) and posterior cheliceral sheath (*pcs.*) in *Dermacentor andersoni* (Ixodida) (based on Gregson, 1960a). *acx.* = line of attachment of the cheliceral sheath; *ch.* = chelicera; *go.* = Gené's organ; *ocs.* = outer cheliceral sheath; *pa.* = porose area; *pd.* = prodorsum; *rtr.* = cheliceral retractor muscles; *sc.* = scutum; *sd.* = salivary duct; *s-pl.* = subcheliceral plate. 2 = chelicera; 3 = cheliceral muscles; 4 = intercheliceral septum; 5 = subcheliceral plate; 6 = labral muscles; 7 = pharynx; 8, 13 = pharyngeal dilator muscles; 9 = salivary duct; 10 = lateral walls of cheliceral frame; 11 = pedipalpal muscles; 12 = cheliceral sheath.

considered by Van der Hammen (1968b) to represent the coxal regions of the chelicerae, but there is no convincing evidence for this supposition.

The cheliceral frame in the Mesostigmata is formed, at least in part, from the mesial walls of the palpcoxae, which extend dorsally to surround the chelicerae in much the same way as the enlarged coxae of the first pair of legs in the Uropodoidea surround and protect the gnathosoma. The dorsal wall of the frame, the gnathotectum, is usually produced anteriorly as a gnathotectal process (or processes) of varying complexity which overhangs the retracted chelicerae and in some parasitic species also gives protection to the hypostome (see Fig. 5.2A above). The gnathotectal process has been referred to in the literature as the epistome (Berlese, 1900), tectum or tectum capituli (Snodgrass, 1948) and supracheliceral limbus (Van der Hammen, 1964). The frame is usually provided medially with a dorsal keel and a ventral ridge which more or less define separate channels for the movement of the chelicerae within the frame (Fig. 5.14E). The tubular sheath surrounding each cheliceral shaft is attached to the wall of its channel in the posterior region of the cheliceral frame except internally where it is fused with its opposite sheath to form a distinct intercheliceral septum. The septum is attached dorsally and ventrally to the keel and ridge, respectively (Evans & Loots, 1975). Anteriorly, the septum divides vertically and the remainder of each sheath separates from the frame and continues as a tube-like sheath which becomes attached to the cheliceral shaft in the arthrodial cuticle at the distal margin of the basal article (Fig. 5.14B–E). Each sheath is folded on itself and allows the chelicerae to move in unison or independently. The dorsal extension of the palpcoxae to form a frame surrounding the chelicerae also occurs in the deuteronymph and adult of the prostigmatic Erythraeidae. These have long slender chelicerae but, in the larva, the hook-like chelicerae are relatively short and stout (Witte, 1978).

The formation of the cheliceral cavity and the form of the cheliceral sheaths do not appear to have been fully elucidated in the strongly sclerotized gnathosoma of the Ixodoidea (Arthur, 1960; Gregson, 1960a). The roof of the cavity, the tectum, is formed from the body wall and the cuticle is thick and strongly sclerotized. In transverse section, the tectum is united with the vertical sclerotized walls separating the palps from the mesial region of the gnathosomatic base. These walls are connected below the chelicerae by the subcheliceral plate thereby forming a cavity containing the chelicerae. The cheliceral sheaths are considered by Snodgrass (1948) to be double-walled tubular folds of the gnathosomatic integument which extend individually around the cheliceral shafts. The outer sheath of each chelicera is produced beyond the dorsal wall of the gnathosoma and subcheliceral plate as a sclerotized tubular sleeve whose dorsal and lateral surfaces may bear numerous denticles arranged in oblique rows. Each inner cheliceral sheath is considered to be formed by the inflected distal margin of the outer sheath; this margin is invaginated between the outer sheath and the cheliceral shaft and attached to the expanded base of the chelicera. Some authors (Zebrowski, 1926; Bertram, 1939; Gregson, 1960a) refer to the presence of two inner sheaths associated with each chelicera. These comprise the cone sheath, which is an anterior invagination of the outer sheath forming a membraneous portion attached to the cheliceral shaft a short distance posterior to the digit, and a posterior cheliceral sheath originating dorsally from the inner region of the tectum and ventrolaterally from the subcheliceral plate (Fig. 5.14F). The posterior sheath is attached to the cheliceral shaft posteriorly. The form and function of each sheath is difficult to interpret but it is probable that the cone sheath prevents blood and debris passing into the cheliceral cavity during feeding.

The cheliceral sheaths are also complex in the Holothyrida (Thon, 1906) but further work is required to determine their form and function. The roof of the cheliceral frame is weakly sclerotized, unlike the gnathotectum of the Mesostigmata and Ixodoidea, and the gnathosomatic integument is much folded upon itself. In *Allothyrus australasia*, the integument appears to form two deep loops to which retractor muscles, originating on the dorsal shield, are inserted on their posterior margin. The cuticle of the internal walls of the dorsal loop have numerous blunt denticle-like projections.

Movements of the Chelicerae and Gnathosoma

Chelicerae

Protraction of the chelicerae in mites and ticks, other than in the Prostigmata, is by hydrostatic pressure maintained by the dorsoventral idiosomatic muscles and acting within the cavities of the cheliceral sheaths following the relaxation of the cheliceral retractor muscles. Retraction in the Anactinotrichida and Sarcoptiformes is by extrinsic muscles originating on the dorsal surface of the idiosoma, usually in the prosomatic region, and inserting on the posterior region of the body of each chelicera. In the Anactinotrichida, retractor muscles are inserted more or less around the posterior rim of the body of each chelicera while in the Oribatida they originate on the posterior margin of the prodorsum, insert postero-ventrally on each chelicera and may be represented by three distinct muscle groups which confer considerable mobility on the chelicerae during their retraction (Fig. 5.13A).

In those Prostigmata which have been studied in detail, protraction of the chelicerae is usually by muscular action via lever-like sclerites, the sigmoid pieces (capitular apophyses), in contact with the cheliceral shafts or, when a stylophore is present, with the cheliceral stylets. The cheliceral protractor muscles in those members of the Parasitengona whose chelicerae have a stout shaft and hook- or claw-like movable digits

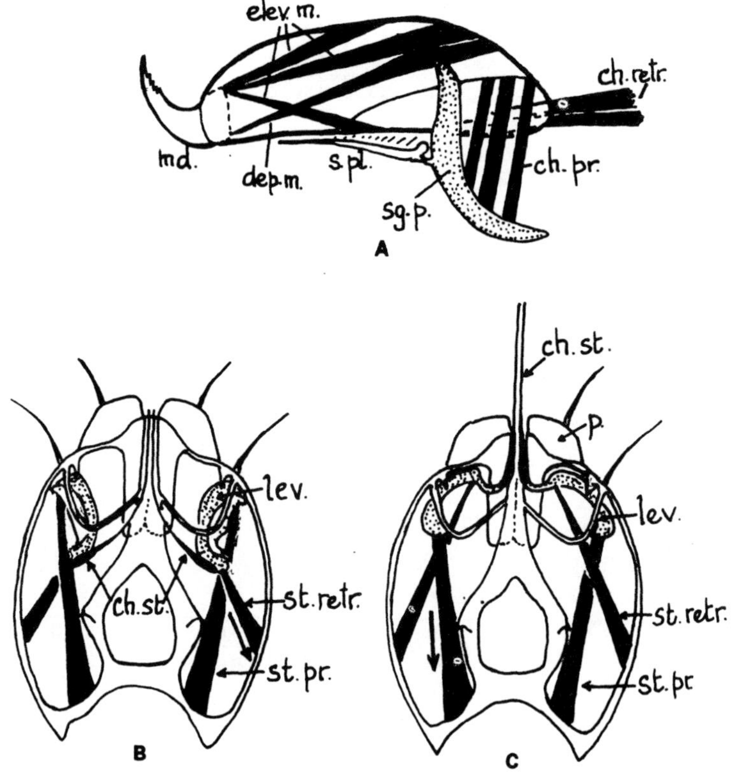

Fig. 5.15. **A**, paraxial view of the chelicera, sigmoid piece and associated musculature in a trombidioid mite (Prostigmata); **B, C**, cheliceral stylets of *Pyemotes scolyti* (Prostigmata) in the retracted (**B**) and protracted (**C**) state (based on Krczal, 1959). *ch. pr.* = cheliceral protractor muscles; *ch. retr.* = cheliceral retractor muscles; *ch. st.* = cheliceral stylets; *dep. m.* = depressor muscles of the movable digit; *elev. m.* = elevator muscles of movable digit; *lev.* = stylet lever; *md.* = movable digit; *p.* = pedipalp; *s. pl.* = subcheliceral plate; *sg. p.* = sigmoid piece; *st. pr.* stylet protractor muscles; *st. retr.* = stylet retractor muscles.

(Fig. 5.15A) extend fan-wise from the inner lateral wall of the sigmoid piece to the dorsal extension of the posterior region of the shaft (Mitchell, 1962a; Mather & LeRoux, 1965; Witte, 1978). Contraction of the muscles causes a lowering of the posterior half of the shaft, a simultaneous raising of its anterior half and the anterior movement of the apical part of the sigmoid piece. The latter is combined with a sclerotized bridge arising from the inner wall of the shaft and effects the protraction of the chelicerae. Retraction of the chelicerae in these forms is by powerful muscles originating on the base of the *crista metopica*, a median sensillary apodeme situated on the prodorsum. Protraction of the chelicerae in the deuteronymph and adult Erythraeidae, in which the chelicerae are in the form of long stylets, does not involve a lever mechanism. The protractor muscles are inserted on the medial and lateral faces of the posterior end of the chelicerae. The median protractors originate on the tracheal trunk, a structure which is probably formed from the larval sigmoid piece, and the lateral protractors arise from the wall of the subcapitulum (Witte, 1978). The retractor muscles, however, originate on the hind part of the *crista metopica* and insert on the posterior third of the chelicerae.

The lever of each cheliceral stylet in the Tarsonemina and Stigmaeidae is provided with protractor and retractor muscles which insert on opposite ends of the lever (Sachs, 1951; Krczal, 1959; André, 1977). Both muscles originate on the walls of the gnathosomatic capsule in the Tarsonemina and the levers may function in a vertical (*Acarapis*) or a horizontal (*Steneotarsonemus*) plane (Lindquist, 1986). In the Pyemotidae, whose gnathosomatic structure is shown diagrammatically in Fig. 5.15B,C, the lever is attached to the posterior end of the stylet. The protractor–retractor mechanism of the cheliceral stylets in the Tetranychoidea is still a matter of disagreement, particularly in relation to the mode of their retraction (Akimov & Yastrebtsov, 1981; André & Remacle, 1984). The recurved movable digit of each chelicera comprises a basal spatulate sclerite and a long slender terminal portion. The basal sclerite is suspended between and pivots on two knob-like apodemes, the one originating from the exterior cuticle and the other developed from the median vertical septum dividing the stylophore. Akimov and Yastrebtsov (1981) show stylet protractors and retractors inserting on the basal sclerites. In contrast, André and Remacle (1984) consider retractor muscles to be lacking and retraction to be a passive process which they liken to the return action of a 'brake cable'. The stylophore is provided with protractor and retractor muscles.

According to Desch (1988), each chelicera in *Demodex folliculorum* has a vertical sclerotized plate-like base termed the fulcrum. Protraction is achieved by muscles which originate on the dorsal wall of the gnathosomatic capsule and insert along the dorsal surface of the fulcrum. The retractor muscles also originate on the dorsal wall of the capsule but insert ventrally on the posterior margin of the fulcrum. Fibrous cuticular material appears to link the fulcra to each other and to the stylophore portion of the gnathosomatic capsule. This flexible material probably serves as the pivot point of the fulcra in the protraction and retraction of the chelicerae.

In the Eriophyoidea, the cheliceral bases are connected to a sclerotized knob-like structure, the motivator, by a large plate-like tendon arising mesially on the anterodorsal surface of the motivator and inserting on the dorsal surface of each chelicera (Nuzzaci, 1979a,b). Silvere and Setjein-Margolina (1976) consider the motivator to act as a point of rotation in the alternating protraction of the chelicerae. Thus, during the contraction of the retractor muscle of one chelicera, the motivator is turned in such a way as to move the other chelicera anteriorly. The origin of the motivator is not known but some authors consider it to be apodematic.

Gnathosoma

In the Anactinotrichida, the gnathosoma is protracted by hydrostatic pressure but its retraction, flexion and extension are accomplished by paired extrinsic muscles originating on the dorsal surface of the idiosoma and inserting on the gnathosomatic base. These muscles, probably derived from coxal muscles of the pedipalps, typically comprise extensors inserted dorsally or dorsolaterally on the gnathosomatic base and flexors inserted on the ventral edge (Fig. 5.16A). The latter may originate anterior to the extensors on the dorsal shield and this greater angle of attachment for the flexors enables the gnathosoma to be strongly deflected and projected posteriorly. This is particularly evident in males of the Mesostigmata which use the chelicerae to pick up sperm or spermatophores from their genital orifice and in females of the Ixodoidea during egg laying. The flexors and extensors in synchronous contraction act as gnathosomatic retractors. Woodring and Galbraith (1976) describe an additional pair of muscles inserting on the posterior lateral edge of the gnathosomatic base in *Uroactinia agitans* and give the origin of the ventral (flexor) muscles as the 'anterior endosternite'. A third pair of muscles is also described for the 'gamasid mites' by Akimov and Yastrebtsov (1988) and these are stated to originate on the dorsal shield and insert on the dorsal posterior edge of the subcheliceral plate, but this requires confirmation. Lateral movement of the gnathosoma is considered to be limited but in *Opilioacarus segmentatus* considerable lateral as well as vertical movement is possible (Grandjean, 1936a).

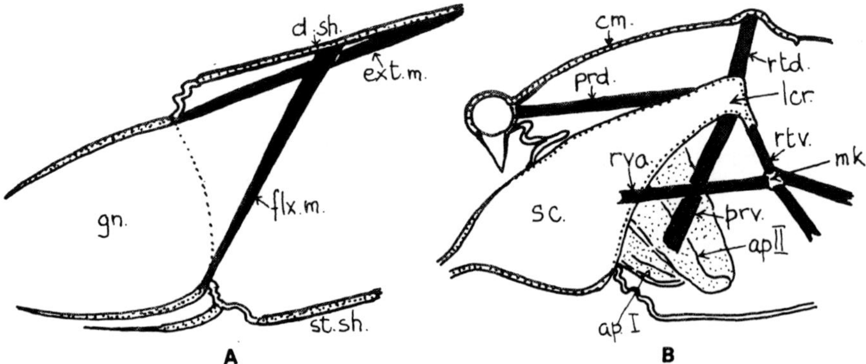

Fig. 5.16. **A**, diagrammatic representation of the extensor and flexor muscles of the gnathosoma in a mesostigmatic mite; **B**, semischematic representation of the protractor and retractor muscles of the subcapitulum in an erythraeid mite (based on Witte, 1978). *apI, apII* = apodemes of legs I and II; *cm.* = crista metopica; *d. sh.* = dorsal shield; *ext. m.* = extensor muscles; *flx. m.* = flexor muscle; *gn.* = gnathosoma; lcr. = latero-coxal region; *mk.* = muscle knot; *prd.* = dorsal protractor muscle; *prv.* = ventral protractor muscle; *rtd.* = dorsal retractor muscle; *rtv.* = ventral retractor muscle; *rva* = anterior retractor muscle; *sc.* = subcapitulum; *st. sh.* = sternal shield.

Protraction in the Sarcoptiformes is by hydrostatic pressure but retraction involves flexor and extensor muscles, which have been described in the Astigmata by Akimov (1985) and in the Oribatida by Akimov and Yastrebtsov (1989). In the prostigmatic Trombidiidae and Erythraeidae, on the other hand, protraction is performed by muscular action (Mitchell, 1962a; Witte, 1978). Two pairs of protractor muscles are present in the Erythraeidae comprising a dorsal and a ventral series (Fig. 5.16B). The dorsal protractors originate from the anterior end of the crista metopica and insert on the proximal end of the latero-coxal region of the pedipalp while the ventral protractors originate on the epimeral sclerite of leg II and insert immediately in front of the dorsal protractors. The retractor muscles consist of three pairs of which the dorsal group originates in the region of the posterior sensory field of the crista metopica and inserts dorsally on the proximal end of the subcapitular wall. The other two pairs, the anterior and the ventral retractors, originate from a 'muscle-knot' (endosternite) and are connected to the lateral coxal region of the subcapitulum. Gnathosomatic protractors do not appear to be universally present in the Parasitengona since they are stated to be absent in the Hydrophantidae, in which the flexors and extensors originate on the prodorsal shield and insert, respectively, on the ventrolateral and dorsolateral angles of the posterior margin of the subcapitulum (Mitchell, 1957a). In the Cheyletidae, the gnathosomatic levator muscles originate on the dorsolateral and dorsomedian regions of the prodorsal shield and insert dorsolaterally on the gnathosoma while the depressors originate on the ventrolateral walls of the idiosoma and medially on the apodemes of legs II and insert ventrolaterally on the gnathosoma (Gorgol & Yastrebtsov, 1987).

6 Feeding and Adaptations of the Mouthparts

The Acari display a wider range of feeding habits than any other group of the Arachnida and are exceptional in the extent to which they ingest fluid and particulate food of plant origin, such as tissues and cell contents of higher plants, pollen, fungi and algae, and in the way that many have adopted a parasitic mode of life. Although most of the Parasitiformes have retained the predatory lifestyle so characteristic of the Arachnida as a whole, some feed on living plant material or practise omnivory while others are facultative or obligatory blood-feeding parasites. A similar exploitation of a variety of food sources is found in the Actinotrichida. The Oribatida ingest decaying matter of plant and animal origin, bacteria, fungi, algae and lichens while some include nematodes in their diet. Free-living Astigmata also feed on living or decaying higher plant material and on fungi but, unlike the Oribatida, many species are parasitic and feed on the tissues of their vertebrate hosts. It is among the Prostigmata, however, that one encounters the greatest variety of feeding habits, with phytophagy being particularly widespread. Much of the information on the food preferences of mites is fragmentary and often inferred from the structure of the mouthparts or based on feeding behaviour under laboratory conditions where the confinement of the mite with a food source in a restricted arena can lead to biased results. The microscopic examination of gut contents has yielded considerable data on the type of food ingested by mites consuming solid food, such as soil-inhabiting Sarcoptiformes. This method is most suitable for analysing the less readily digestible material which remains in a recognizable form in the mid-gut food pellet. It is of limited use in identifying easily digestible materials and the method is not applicable to the study of feeding in predatory species that ingest their food in a liquid or a finely particulate state. In the case of predators, analyses of prey preferences using electrophoresis have provided useful information (Murray & Solomon, 1978; Lister, 1984; Pugh & King, 1985; Dicke *et al.*, 1988).

The range of feeding habits in the Acari may be broadly classified into four main categories:

1. Zoophagy (or carnivory) – consuming parts of living animals.
2. Phytophagy (or herbivory) – feeding on living plant materials, including fungi (mycetophagy) and algae.
3. Omnivory – utilizing both plants and animals as a source of food.
4. Saprophagy (or detritivory) – ingesting decaying plant or animal matter.

The placement of taxa in any one category is often difficult. For example, oribatid mites consuming decaying plant material (saprophagy) also ingest and utilize bacteria and fungi (microphytophagy) involved in the decomposition of the plant tissues. Further, many soil-dwelling mites considered at one time to be obligatory predators have been found to feed and complete their life cycle on fungi under laboratory conditions and this suggests that fungal components may be used by them as an alternative or supplementary food source in nature. Mites appear to be more catholic in the utilization of food materials than one would presume from the literature where statements are often based on speculation or inferred from morphology rather than on direct observation of feeding behaviour.

Notostigmata and Holothyrida

Relatively little is known of feeding in these two orders. Examination of the gut contents of *Opilioacarus* has yielded a mixture of pollen grains, fungal spores and fragments of arthropod appendages (With, 1904; Van der Hammen, 1989). It has not been established whether the presence of arthropod fragments is indicative of predation (in the sense of capturing and consuming living prey) or the result of consuming detritus containing pieces of exuviae. The ingestion of solid food is to be expected from the form of the corniculi (rutella) and from the large size of the terminal anus for the expulsion of solid faeces.

There are few observations on the feeding habits of holothyrids in nature. Lee and Southcott (1979) state that *Allothyrus australasiae* feeds on dead woodlice (Oniscidae) and 'hoppers' (Talitridae). In a laboratory arena, *Thonius braueri* has been observed to consume fragments of solid food (J. Travé, personal communication). A morsel of food was grasped by the digits of the long geniculate chelicerae and conveyed to and deposited in the hypostomatic region of the gnathosoma. The use of the chelicerae was reminiscent of that in some Opiliones.

Mesostigmata

Zoophagy, including predation and parasitism, is the predominant type of feeding in the Mesostigmata although phytophagy and omnivory are also represented.

Predation

The majority of Dermanyssina and Parasitina are general predators that show little or no preference for the type of prey, feeding on any small animals that they are able to overcome. They are usually rapidly moving mites that, while in motion, tap the substrate with their palps while constantly waving their elevated first pair of legs. These appendages are richly provided with tactile and chemosensory sensilli. The predators locate their prey by random contact or by chemical cues produced by the prey. General predators are fluid feeders and are commonly found, for example, in soil or in accumulations of organic debris such as compost and nests of birds and mammals; they are represented by members of the Ascidae, Cyrtolaelapidae, Laelapidae, Macrocheliidae, Ologamasidae, Pachylaelapidae, Parasitidae, Rhodacaridae and Veigaidae. Their prey includes collembolans, nematodes, enchytraeids, insect larvae and other mites (Karg, 1971). Carnivory, however, does not appear to be obligatory in some species, as exemplified by the ascid *Protogamasellus mica* which, under laboratory conditions, successfully develops from larva to adult on fungal diets although in continuous culture under this regime produces few adults and lower populations than those obtained in cultures on nematodes and collembolans (Walter & Lindquist, 1989). To what extent soil-inhabiting predators actually consume microflora under field conditions has not been ascertained.

Some members of the family Eviphididae are exclusively predators of nematodes (nematophages) as, for example, European species of the genus *Alliphis* (Karg, 1983). Nematodes appear to be the preferred food of many other Dermanyssina, particularly euedaphic species of the genera *Rhodacarus*, *Rhodacarellus* and *Gamasellodes*. This is also the case for a number of uropodoid mites such as *Uroobovella marginata* and *U. rackei* and *Uropoda orbicularis* and *U. sellnicki* (Faasch, 1967; Karg, 1989).

Many predatory mites do not feed on nematodes but take only arthropods as prey. This is particularly true of arboreal Phytoseiidae, which feed mainly on phytophagous mites such as spider mites (Tetranychidae), eriophyid mites and tarsonemid mites although they may consume microflora when prey is unavailable or of low density. *Phytoseiulus persimilis*, which is widely used as a biological control agent in pest management of Tetranychidae, feeds almost exclusively on spider mites. Prey searching and location behaviour of phytoseiids on leaves of plants has been extensively studied in the laboratory. The thigmotactic searching behaviour displayed by phytoseiids along leaf ribs and leaf edges is referred to as edge-orientated searching and is an essential component of prey-searching behaviour (Sabelis, 1981). Edge-walking patterns of *Amblyseius potentillae* on a rose leaf (Sabelis & Dicke, 1985) showed that when the predator arrived, by way of the stem, on the lower side of the leaf its searching path followed the main rib to almost the tip before meandering and reaching the leaf edge (Fig. 6.1A). Arrival on the upper side of the leaf usually resulted in the predator directly reaching and following the edges and not the less pronounced main rib (Fig. 6.1B) or roaming on the surface until reaching the edge (Fig. 6.1C). Phytoseiids are thought to deposit pheromones to mark the leaf area that has been searched and avoid recrossing these areas (Hislop & Prokopy, 1981).

The location of spider mites within high-density prey patches on a leaf area is probably by chance through physical contact. Berry and Holtzer (1990) refer to the type of ambulatory search involved at high prey densities by *Neoseiulus fallacis* as 'the random walk type' in which turning rate, walking rate and turning angle of the predator change (Fig. 6.1D). This type of behaviour helps to maintain the predator in a restricted area where prey is abundant. Edge-walking, on the other hand, is more linear and non-random and used when prey density is low (Fig. 6.1E). The location of distant prey colonies and the arresting of a predator within the colony are probably effected by kairomones whose presence has been demonstrated in experiments using Y-tube olfactometers (Sabelis *et al.*, 1984). For example, *P. persimilis* responds positively to the odour originating from leaves infested with *Tetranychus urticae* while *Amblyseius findlandicus* and *A. potentillae* respond to the odour of leaves infested with *Panonychus ulmi*. The kairomone(s) is associated with the faeces and eggs, but not the silk, of *T. urticae* and is present to a major extent on the lower epidermis of leaves on which they have been feeding. The four kairomone components emitted by Lima bean plants infested by *T. urticae* and used in distant prey locations by *P. persimilis* have been identified by Dicke *et al.* (1990) as: methyl salicylate, (E)-β-ocimene, linapool and (3E)-4-8-dimethyl-1,3,7-nonatriene. Dicke and Groeneveld (1986) have shown that carotenoid-deficient *A. potentillae* respond to kairomones of more prey species than do those that have carotenoids available. They suggest that this may be reflected in a diversification of the prey diet under carotenoid deficiency.

Some prey have developed defences against attack by predators which can reduce the probability of capture. The springtail *Tullbergia granulata* produces a distasteful chemical when attacked (D.E. Walter, personal communication) and many phytophagous and saprophagous mites (*Ctenoglyphus, Cosmochthonius, Phyllozetes*) have enlarged setae which when erected prevent the predator reaching the body of the prey. Soft-bodied mites of the families Eupodidae, Nanorchestidae and Terpnacaridae and hemi-edaphic Collembola evade predators by jumping.

Predatory Dermanyssina and Parasitina usually grasp the prey by means of their chelicerae (Fig. 6.2A,B) and/or palps, although when the prey is large and active the first two pairs of legs may also be used (Lee, 1974; Usher & Bowring, 1984). In *Pergamasus brevicornis*, Zukowski (1964) noted that the chelicerae draw the prey towards the hypostome thereby impaling it on the gnathotectal processes and corniculi. This contrasts with the method in the Rhodacaroidea where the prey is not impaled on these structures (Lee, 1974). This difference in behaviour may be attributed to the position of the openings of the salivary ducts in relation to the anterior extremity of the gnathosoma and their effective discharge into the prey. The salivary styli in the predatory Dermanyssina are located in a dorsal groove of the corniculi and the salivary ducts open at the anterior end of the hypostome whereas in the majority of the Parasitina, to which *Pergamasus* belongs, the ducts open lateral to the chelicerae at a distance from the tip of the hypostome. Thus, impaling the prey on the corniculi in the latter would facilitate the delivery of salivary secretions into the prey but this action would be unnecessary in those taxa where the secretions are delivered to the anterior extremity of the gnathosoma.

The cheliceral digits cut or tear an opening into the body of the prey which is rapidly immobilized and the internal organs are lacerated by the combined action of the digits and the backward and forward movements of the chelicerae. Digestive enzymes contained in the salivary secretions are thought to partially digest the prey and the resulting fluid food flows between and around the chelicerae to the hypostomatic region of the mite. The fluid produced from the prey is often in excess of the amount that can be directly imbibed by the mite and overflows towards the base

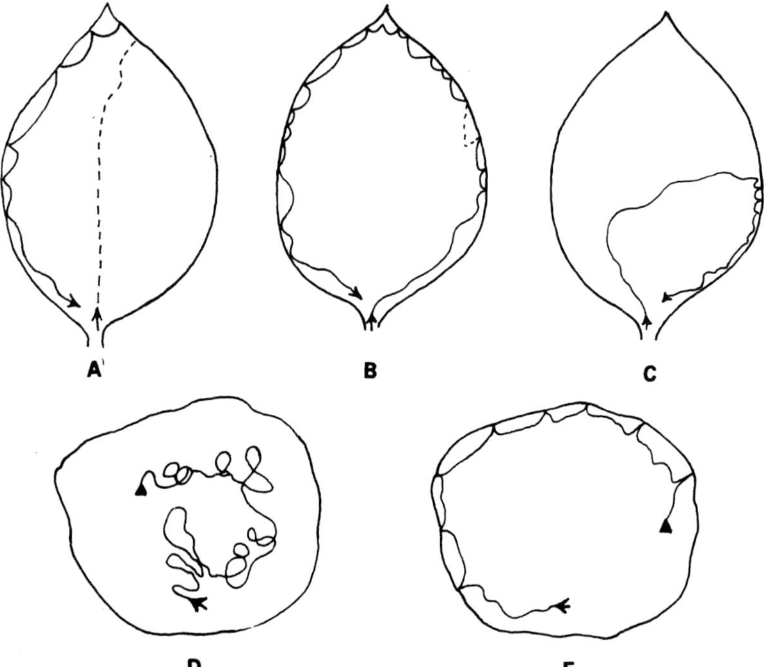

Fig. 6.1. A–C, walking patterns of *Amblyseius potentillae* (Mesostigmata) on rose leaves when arriving by way of the stem on the lower side (**A**) and upper side (**B** and **C**) of the leaf (after Sabelis & Dicke, 1985); **D, E**, random-walk pattern of *Neoseiulus fallacis* at higher (**D**) and lower (**E**) prey densities in an experimental arena (arrow and triangle indicate, respectively, the start and end of the walk) (after Berry & Holtzer, 1990).

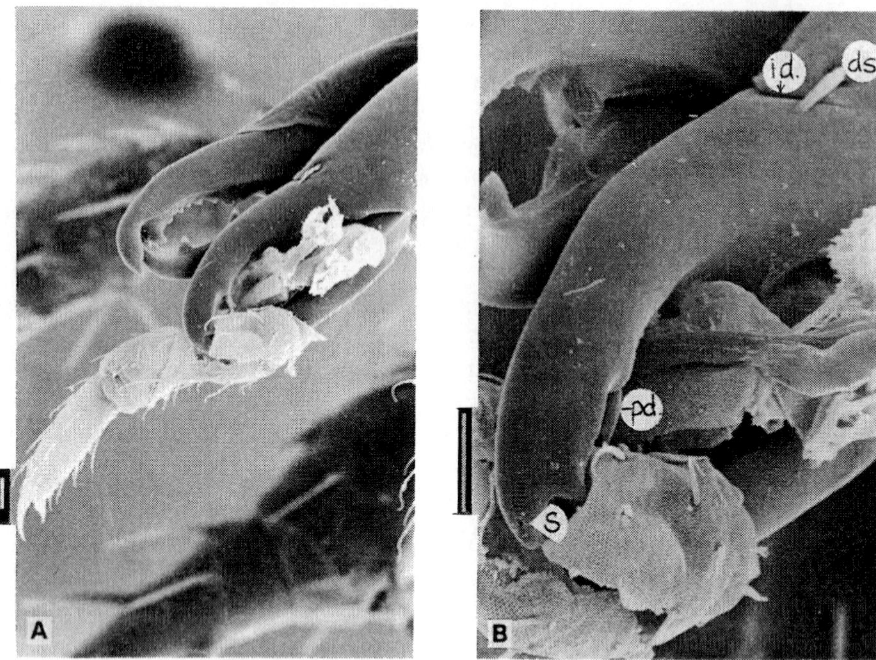

Fig. 6.2. **A**, **B**, SE micrographs of the cheliceral digits of a female *Hypoaspis aculeifer* grasping fragments of an onychiurid collembolan. Scale bar = 10 μm. *ds.* = dorsal seta; *id.* = dorsal slit organ; *pd.* = pilus dentilis; s. = apical sensillus.

of the gnathosoma. The flow of fluid is restricted laterally by the coxae of the first pair of legs, which lie with the gnathosoma in the gnathopodal cavity, and posteriorly by the folded cuticle of the presternal region. The folding of the presternal cuticle to form a furrow(s) occurs when the gnathosoma is retracted or flexed ventrally by the action of the gnathosomatic retractor muscles. The excess prey fluid is redirected from the presternal region to the hypostome in the form of an anteriorly directed stream between the tritosternum (base and laciniae) and the subcapitular groove (Wernz & Krantz, 1976) (Fig. 6.3A,B). The tritosternum lacks extrinsic musculature and its position relative to the venter of the subcapitulum is probably controlled by movements of the gnathosomatic base on the thick, flexible cuticle connecting it to the base of the tritosternum. Thus, it is suggested that extension of the gnathosoma by hydrostatic pressure is accompanied by a relaxation of the flexible cuticle and the subsequent lowering of the tritosternum whilst retraction or flexion of the gnathosoma by muscular action causes the raising of the tritosternum and brings the laciniae in contact with the subcapitular groove. The function of the two narrow muscles originating from the posterior internal surface of the base of the tritosternum and inserting on the laciniae has not been determined (Akimov & Yastrebtsov, 1988).

Food is drawn from the pre-oral channel into the alimentary tract.

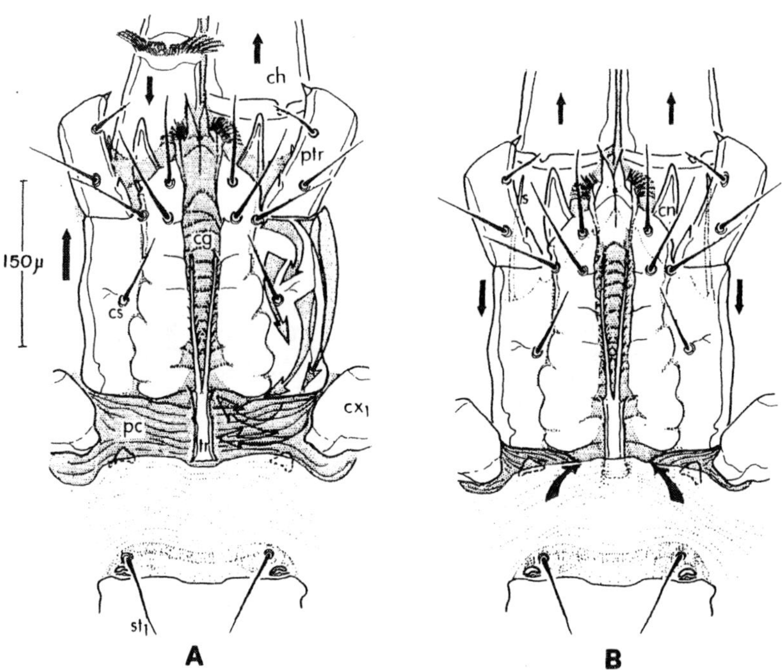

Fig. 6.3. The role of the tritosternum and capitular groove in fluid transport during feeding in *Parasitus coleoptratorum* according to Wernz & Krantz (1976). **A**, pattern of fluid disposition (denoted by stippling) when the gnathosoma is in the extended state with possible path of overflow fluids indicated by stippled arrows; **B**, pattern of fluid disposition when the gnathosoma is in the retracted state. *ch.* = chelicera; *cg.* = subcapitular groove; *cn.* = corniculus; *cs.* = palpcoxal seta; cx_1 = coxa I; *pc.* = presternal cuticle; *ptr.* = palptrochanter; *s.* = salivary stylus; st_1 = sternal seta.

According to Zukowski (1964), retraction of the labrum by muscular action results in the widening of the pharynx into which the fluid food is sucked. The labrum then returns to its original position thereby closing the entrance into the pharynx. The alternate contraction and relaxation of the pharyngeal constrictor and dilator muscles forces the food along the pharyngeal tract. The labrum and walls of the pre-oral channel in predatory Parasitina and the great majority of the Dermanyssina possess well-developed processes, which Evans and Loots (1975) consider to act as a strainer or sieve to prevent small fragments which may be suspended in the prey fluid from entering the alimentary tract (Fig. 6.4A). The predatory Phytoseiidae are exceptional in lacking these processes and resemble the condition in the microphytophagous Ameroseiidae (Fig. 6.4B).

The dimensions and dentition of the cheliceral digits are considered by Karg (1971) to give an indication of the diet of soil-inhabiting Dermanyssina and Parasitina. Thus, the chelicerae of species specializing in feeding on nematodes are characterized by short, stout digits with a few, often offset teeth, a type commonly found in the Eviphididae

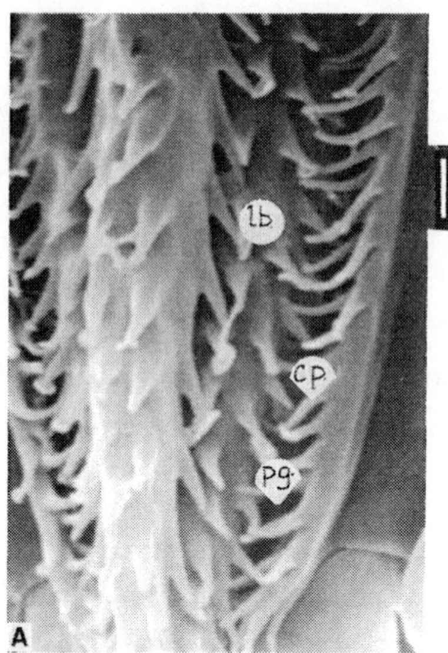

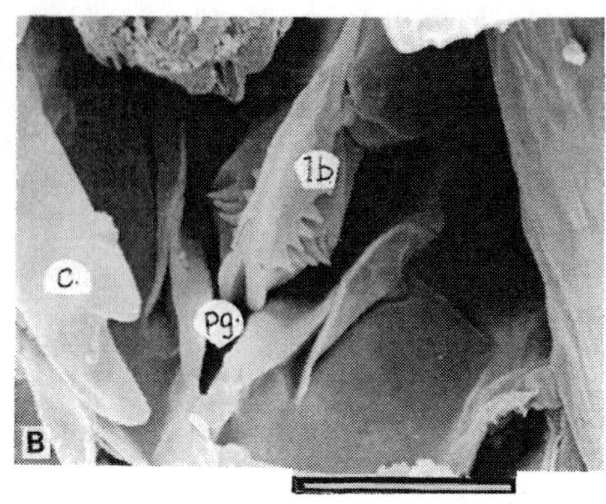

Fig. 6.4. **A**, cuticular processes of the labrum and lateral walls of the pre-oral groove forming a sieve-like structure in *Hypoaspis aculeifer* (Mesostigmata), scale bar = 1 μm; **B**, labrum and pre-oral groove in *Ameroseius echinatus*, scale bar = 10 μm. *c.* = corniculus; *cp.* = cuticular processes; *lb.* = labrum; *pg.* = pre-oral groove.

(Fig. 6.5A). Those catching rapidly moving prey such as collembolans and mites have longer, more slender digits with backwardly directed teeth (Fig. 6.5B), while polyphagous predators with diets including nematodes, mites, collembolans and insect larvae tend to have strong digits with a row of small, closely set teeth on the fixed digit, as in *Hypoaspis aculeifer* (Fig. 6.5C). Later, Karg (1983) illustrated other forms of chelicerae for a range of species in which nematodes from a major part of the diet: for example, those with a few large teeth on the movable digit opposing a closely set row of relatively smaller teeth on the fixed digit, as in some *Lasioseius* (Ascidae, Fig. 6.5D), or with the distal teeth on the movable digit opposed by a restricted area of saw-like teeth on the fixed digit, as in *Cheiroseiulus* (Fig. 6.5E). Scanning electron microscopy of cheliceral digits indicates that such criteria as the shape and spatial relationship of the teeth as well as the configuration of the opposing cutting surfaces should be considered in attempting to relate cheliceral structure to diet. Caution should be exercised, however, in attributing feeding habits on the basis of cheliceral structure alone, particularly since so much of the published information on predator diet is based on laboratory and not field data. The capture and ingestion of a prey species by a hungry mite in the restricted arena of a laboratory experiment is not necessarily a reliable indication of the preferred food of the predator under natural conditions.

Grooming of the mouthparts and the distal podomeres of the first pair of legs usually occurs after the remains of the prey have been discarded.

According to Lee (1974), this involves 'rubbing by part of an appendage (often the pretarsus), or nibbling by the chelicerae, of another part of the mite'. He observed in species of Rhodacaridae s.lat. that the chelicerae and palps groom each other and the distal ends of the first pair of legs. Particles of debris adhering to the tritosternal laciniae are picked up by the denticles of the subcapitular groove and are eventually swept towards the hypostomatic processes and to the dorsal region of the labrum and probably removed by the action of the arthrodial brush at the base of the movable digit of each chelicera.

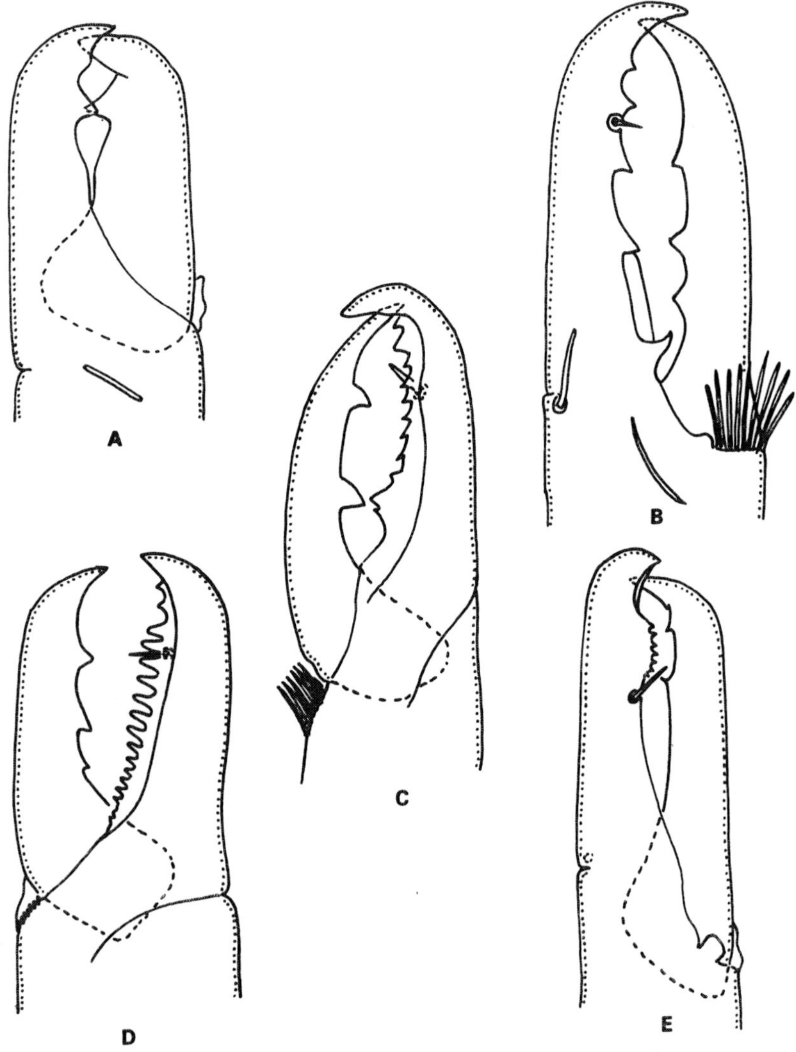

Fig. 6.5. A range of types of cheliceral digits and dentition in the predatory Mesostigmata (Dermanyssina and Parasitina). **A**, *Scarabaspis inexpectatus*; **B**, *Paragamasus* sp.; **C**, *Hypoaspis aculeifer*; **D**, *Lasioseius berlesei*; **E**, *Cheiroseiulus reniformis*.

Parasitism

The majority of the parasitic Mesostigmata occur in the Dermanyssoidea and are associated with other arthropods, reptiles, birds and mammals. The adoption of a parasitic mode of life by members of this taxon realized an evolutionary potential which has produced a vast assemblage of morphologically diverse forms (Evans & Till, 1966; Radovsky, 1969). It is considered that the parasitic forms have evolved from closely related free-living predatory species; in the case of parasites of birds and mammals, species that inhabited their nests. The nest habitat provided the basis for a close association between the mite and the nest occupant. Initially, the association may have been one of commensalism with the mite feeding on small ectoparasites and on blood exuding from abrasions on the body of the 'host'. Such a generalized type of feeding is characteristic of certain species of *Androlaelaps* and *Laelaps,* while in the nest-inhabiting species of the genus *Haemogamasus* a remarkable range of feeding habits occurs, including general predators (*H. pontiger*), polyphagous species with parasitic tendencies (*H. ambulans*) and obligate haematophagous species (*H. liponyssoides*).

Obligatory blood-feeding has been established in the Dermanyssidae, Hirstionyssinae, Hystrichonyssidae, Macronyssidae, Rhinonyssidae, Spinturnicidae and probably also occurs in the Halarachnidae and Spelaeorhynchidae. The adoption of haematophagy has been accompanied by a range of structural adaptations of the feeding organs and by a reduction in idiosomatic sclerotization to allow for the ingestion of a large volume of liquid food (Evans & Till, 1965). A distinct trend in macronyssid, dermanyssid and hystrichonyssid mites is for the elongation of the basal (Hystrichonyssidae) or the second (Dermanyssidae, Macronyssidae) article of the cheliceral shaft and the development of a concavity of the paraxial face of the shaft so that when the shafts of the two chelicerae are in contact they form a channel along which blood can pass to the hypostomatic region of the gnathosoma. In *Dermanyssus,* the shafts are held together by a coupling device (Radovsky, 1969). Modifications of the chelicerae of feeding instars in comparison with those of predatory dermanyssoids are seen in the narrowing of the digits (or their reduction to inconspicuous barbs at the tip of the shaft in the Dermanyssidae), reduction or complete loss of teeth, pilus dentilis, dorsal seta and arthrodial processes and the tendency for the development of membraneous structures associated with the digits (Evans & Till, 1965). The sensillus on the tip of the fixed digit is retained. The specialized digits in the Macronyssidae are used in puncturing the skin and disrupting the vascular bed so that a pool of blood is formed in the tissues. Blood is drawn directly from the pool with the chelicerae acting as a sucking tube. Lavoipierre and Beck (1967) consider this type of feeding in *Chiroptonyssus robustipes* to be intermediate between that of vessel-feeders and pool-feeders. The long, slender form of the chelicerae in the Dermanyssidae suggests that members of this family are true vessel-feeders. The weight of dermanyssid and macronyssid species increases 2–10 times during blood-feeding. The membraneous cheliceral digits of many of the blood-feeding endoparasitic Rhinonyssidae which inhabit the respiratory tract (most often the nasal cavities) of birds would appear to be too delicate to cut the skin or mucosa but no other structure has been implicated. According to Dahme and Popp (1963), *Orthohalarachne letalis* (Halarachnidae) uses its chelicerae as saw-blades to tear open the pulmonary tissue of its pinniped host.

Adaptations of the hypostomatic region of the gnathosoma are seen in the trends towards the increase in size of the pre-oral channel for the reception of blood, the reduction in the processes of the labrum and the walls of the pre-oral channel, and the modification of the corniculi into lobe-like or semi-membraneous structures (Evans & Till, 1965; Evans & Loots, 1975). Of particular interest is the absence of a sieve or strainer mechanism (formed from labral and hypostomatic wall processes) within the pre-oral food channel of those species which feed only on blood. The loss of such a mechanism in forms in which there is little likelihood of the ingestion of solid particles with their fluid food would appear to support the function attributed to it in predatory forms. The number of denticles in the subcapitular groove is markedly reduced in obligatory blood-ingesting species and they usually form a single file.

In members of the genera *Ixodorhynchus* and *Ixobioides* (Ixodorhynchidae), whose feeding habits are not known, the corniculi are each provided with one or two backwardly projecting hooks which probably serve to anchor the mite to its reptilian host. Members of the Spinturinicidae, on the other hand, use the strong claws of their legs for attaching to the wing and tail membranes of their chiropteran hosts.

There are no details of the feeding habits of certain species of Heterozerconidae, Paramegistidae, Diplogyniidae and Schizogyniidae, which have been found on snakes and, in the case of *Ophiomegistus*, also on lizards.

A number of mesostigmatic mites occur on other arthropods, especially insects and millepedes, but in the majority of cases their trophic behaviour is unknown. They comprise non-feeding phoretics and commensals as well as parasitic species. Some are found in depressions on the surface of the cuticle of the host, and in species of *Lasioseius* occurring on millepedes the shape of the mite corresponds closely to the spiracular depressions which they occupy (Evans & Sheals, 1959). The mouthparts of the putative parasitic forms provisionally classified in the Iphiopsidae or Laelapidae show a similar range of morphological adaptations to those of the dermanyssoids parasitizing vertebrates (Evans, 1955b; Kethley, 1978). Of particular interest are species of Varroidae associated with honey bees, of which *Varroa jacobsoni* is the most important economically. Originally a parasite of the Asian bee (*Apis cerana*), the species has been introduced almost worldwide into colonies of the European honey bee *Apis mellifera* (Griffiths & Bowman, 1981; De Jong *et al.*, 1982). It lives in the nest (or hive) of honey bees and feeds on the brood. A female enters a brood cell just before capping to lay eggs and she and her immature offspring feed on the haemolymph of larval and pupal instars of the bee. The chelicera has only a vestige of the fixed digit and it is the movable digit with its two strong teeth that is used to tear the host cuticle. The immatures usually develop into a 'family' of one male, which never leaves the cell, and several females (Ifantidis, 1983). When the brood cell is opened, the fertilized females leave and use adult bees as a means of transportation. They are capable of penetrating the soft cuticle between overlapping segments of the abdomen of the bee to feed on haemolymph.

Phytophagy

With the exception of some Phytoseiidae which feed on the leaf sap of higher plants (Porres *et al.*, 1975), mesostigmatic mites ingesting tissues of living plants are microphytophagous, feeding mainly on fungi and pollen. In the Dermanyssina, relatively few taxa are exclusively microphytophagous while many are omnivorous, feeding on microflora and animal tissues. Obligate microphytophagy appears to be restricted to species occurring in the families Ameroseiidae and Ascidae. Among the Ameroseiidae, species of *Ameroseius, Kleemannia* and *Epicriopsis* are predominantly mycophagous, although some species living in inflorescences probably feed on pollen or nectar (Krantz & Lindquist, 1979); *Neocypholaelaps*, for example, lives in inflorescences, utilizes pollen as the major food source and uses apoid bees as a main means of dispersal (Evans, 1963c). Modifications of the mouthparts for mycophagy in the *Ameroseius* group affect the cheliceral digits and hypostomatic region. The chelicerae tend to be relatively short, stout structures with the basal and middle articles often fused. The fixed digit is provided with a row of two to four strong teeth in its posterior half which opposes an edentate movable digit whose concave paraxial face forms a cavity bordered ventrally by a strongly sclerotized built-up rim (Fig. 6.6A). A hyaline membraneous appendage arising antiaxially from the fixed digit occurs in some of the taxa. This form of chelicera is probably capable of chopping fungal hyphae or puncturing fungal spores or pollen held in the concavity of the movable digit. The specialization of the hypostome is seen in the development of the corniculi into flat dentate structures, bearing a remarkable resemblance to the rutella of some Oribatida, and in the simplification of the hypostomatic processes (Fig. 6.6B). The characteristic 'strainer' formed by processes of the labrum and walls of the pre-oral channel in fluid-feeding predators is lacking in these particulate feeders. Whether the corniculi have a 'rutellate' function in removing macerated hyphal material from the cheliceral digits or in cutting fungal hyphae protruding on either side of the digits as they pass between the corniculi has not been established. Other adaptations associated with particulate feeding are seen in the tendency for the enlargement of the muscular

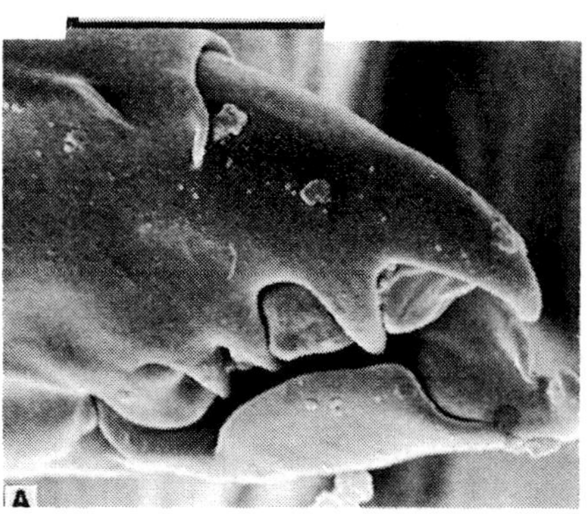

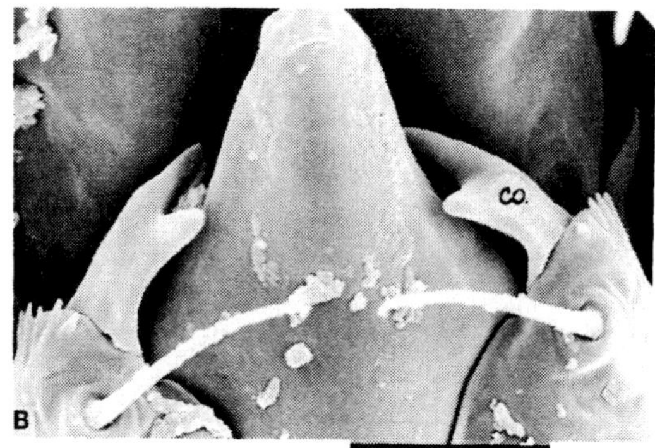

Fig. 6.6. SE micrographs of the cheliceral digits (**A**) and corniculi (**B**) of *Ameroseius echinatus* (Mesostigmata). Scale bar = 10 μm. *co.* = corniculus.

pharynx and anal orifice and for a reduction in the number of denticles in the subcapitular groove. Although undigested fragments of hyphae and amorphous matter can be seen in the post-colon of some species, a food pellet is never formed.

The modifications of the chelicerae and corniculi in the pollen-feeding *Neocypholaelaps* differ from those of *Ameroseius*. Both digits are less strongly sclerotized with the fixed digit having, at the most, a single sub-apical tooth ('Gabelzahn'). A large hyaline flap arises antiaxially from the fixed digit in some females (Evans, 1963c). This type of chelicera appears to be capable of piercing the wall of the pollen grain. The corniculi are relatively narrow, convergent structures, movably articulated to the hypostome and enveloped in membranes. The corniculi and associated membranes probably form an extension of the hypostome for the reception of food and provide support for the pollen grain manipulated by the palps, but the actual method of feeding is not known.

Few members of the Ascidae appear to be exclusively microphytophagous. Species of *Rhinoseius* occurring in flowers of plants pollinated by hummingbirds and transported from plant to plant in the nares of birds, appear to qualify since they probably feed exclusively on nectar and/or pollen (Lindquist & Evans, 1965). Their mouthparts show similar adaptations for microphytophagy as those encountered in the Ameroseiidae. Members of the genus *Hoploseius* occur in the tissues on the underside of bracket fungus and apparently use the rounded dentate extension of the fixed digit of the chelicera as a scraper for removing sporophores (Lindquist, 1963). They are phoretic on drosophilid flies frequenting the fungus.

The occurrence of large dentate cutting-type corniculi in conjunction with an enlarged anal orifice also occurs in the Ichthyostomatogasteridae (Sejina). These structures are particularly pronounced in the genus *Asternolaelaps* and suggest that its members consume particulate matter (Evans, 1954; Athias-Henriot, 1972).

Omnivory

Reference has already been made to the utilization of microflora as a supplement to prey by some predatory Dermanyssina. Omnivory is most widespread in the Phytoseiidae and in many species the ingestion of plant substances such as contents of pollen grains, sap and honeydew occurs mainly during periods of low prey density. However, some species feed primarily on plant sap or pollen and only facultatively on prey, while others are exclusively zoophagous (Kennett *et al.*, 1979; Tanigoshi, 1982). The mouthparts of phytoseiids exhibit structures which are more typical for microphytophagy than for predation. For example, the fixed digit of the chelicera may have a well-developed hyaline appendage (Fig. 6.7A) as in *Neocypholaelaps*, the corniculi are slender, slightly convergent and approximate (Fig. 6.7B) while the walls of the pre-oral channel are smooth and a 'strainer' mechanism of the type found in predators is lacking. The labrum may be of an unusual form in being divided distally into finger-like processes.

Feeding on both prey and fungi also occurs in some species of Ascidae, for example in the genera *Proctolaelaps* (Karg, 1971) and *Lasioseius* (Walter & Lindquist, 1989). The chelicerae usually have a row of closely set teeth on the fixed digit which fits into a groove along most of the length of the movable digit. This type of digital structure also occurs in some species of *Amblyseius*. The corniculi in certain species of *Proctolaelaps* and *Protogamasellus* are dentate and the anal opening is

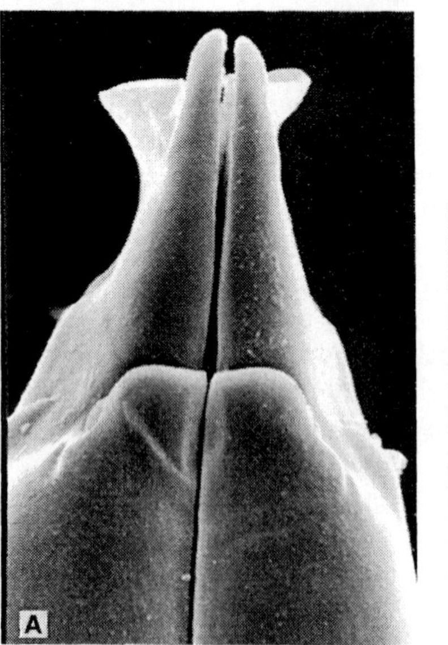

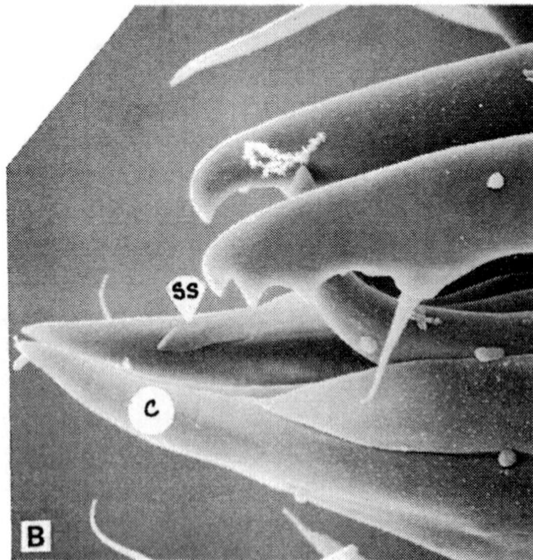

Fig. 6.7. **A**, dorsal view of the cheliceral digits of *Amblyseius* sp. (Mesostigmata) showing digital membranes; **B**, convergent corniculi (*c*) and salivary styli (*ss*) of a predatory phytoseiid mite.

enlarged. An unusual type of feeding has recently been described for the laelapid *Pneumolaelaps longanalis* by Royce and Krantz (1989). This species inhabits the nests of bumblebees in the Pacific Northwest region of the USA and appears to feed on bee larvae as well as on the coating of nectar and surface pollenkitt from pollen grains collected and processed by the bees. Sugars applied to the pollen grain by the bee are thought to serve as a feeding stimulus for the mite.

Feeding in the Uropodina has received little study but omnivory has been established in two relatively common species, *Uropoda orbicularis* and *Uroobovella marginata*. Both appear to prefer nematode and arthropod prey but *U. orbicularis* also feeds on fungal hyphae (Faasch, 1967) and *U. marginata* will feed on the parenchyma of cucumber plants (Karg, 1968). The mushroom-shaped apical appendage of the fixed digit of the chelicera in *Uroactinia* probably acts as a scraper to collect fungal hyphae or spores from substrates. The complex sensory organs on the fixed digits of the long chelicerae of many uropodoids suggest that they may be involved in locating and selecting food as well as in its manipulation.

Ixodida

Ticks are obligate external parasites of terrestrial vertebrates and the active instars ingest blood and tissue fluids of their hosts. Heavy infestations can lead to the death of the host through exsanguination but economically ticks are of greater importance as reservoirs and vectors of pathogenic organisms which are transmitted to vertebrates directly during attachment and feeding or indirectly by contact with tick coxal gland fluid and faeces (Balashov, 1968). Argasoid ticks are essentially 'habitat' ticks which select hosts inhabiting nests, burrows, caves and human dwellings that afford protection and relatively stable microclimatic conditions (Nuttall, 1911). Although larvae may remain attached to the host for several days, nymphs and adults engorge rapidly on the host, usually at night, and are rarely carried away from the nest or burrow. The majority are multihost ticks. Ixodoids, on the other hand, are 'field' ticks (except for relatively few species that have reverted to nest-burrow parasitism) and take only one meal of blood per instar (larva, nymph and adult) either on the same host, on two different hosts (larva and nymph remaining attached to the same host, adult on a different host) or on a maximum of three hosts during their life cycle.

Waladde and Rice (1982) identify a sequence of nine main behavioural events in tick feeding which cover host location (appetence and engagement), attachment (exploration, penetration and attachment) and engorgement (ingestion, engorgement, detachment and disengagement). This sequence is adopted in the following account of tick feeding behaviour.

Appetence and engagement

Ticks display appetence either by hunting for hosts or by preparing an ambush. Hunters, such as some species of *Hyalomma* and *Ornithodoros*, probably use their sight to run in the direction of the host or even pursue it, while a number of other ixodoid and argasoid species may be attracted over some distance to CO_2 produced by the host (Nevill, 1964; Garcia, 1969). In contrast, ixodoid ambushers remain quiescent on the vegetation and adopt a 'resting' posture (Fig. 6.8B) until alerted by a range of stimuli of an approaching animal. They display appetence by adopting a 'questing' attitude (Fig. 6.8A) in which the first pair of legs with their battery of sense organs are waved in the air towards the direction of the stimuli. In *Ixodes ricinus*, for example, the tendency for the tick to remain near the tips of the vegetation is assisted at first by their avoidance of the high humidity near the roots. The unfed tick in which the water balance is normal avoids the higher humidities of the base of vegetation and comes to rest in moist or dry air at the tips of vegetation (Lees, 1948). After an unsuccessful period of waiting in this position, the tick becomes desiccated and the avoiding response, a taxis, is replaced by a kinesis, the tick being active in dry air but coming to rest in moist air. Walking downwards, the tick reaches the humid vegetation mat where it takes up water and is ready for another period of activity at the tips. Stimuli which alert ticks to the imminent approach of the host include

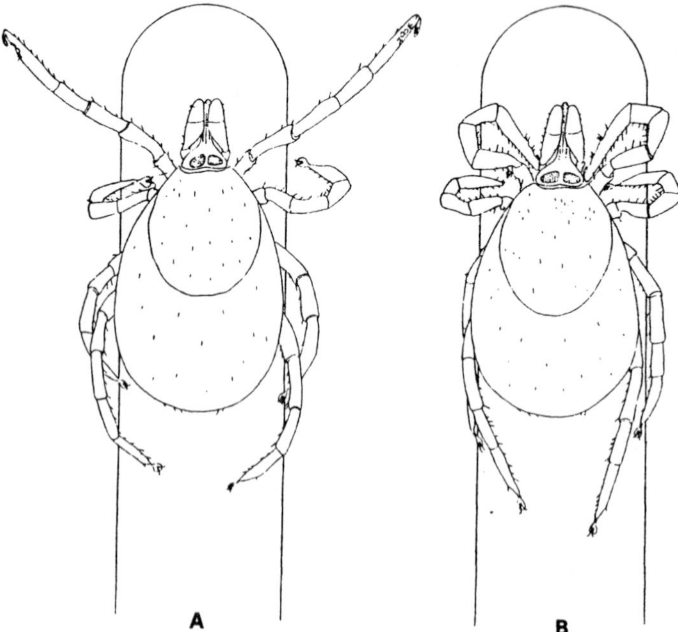

Fig. 6.8. Ixodid tick in the questing (**A**) and resting (**B**) posture (after Lees, 1948).

vibrations, shading and odours carried by air currents, while close stimuli such as warmth and touch are important in engagement. Ticks climb onto the host by attaching to the pelage by their legs, which are provided with strong claws.

Exploration, penetration and attachment

After engagement, the tick walks on the surface of the host in search of a suitable site for feeding. Thermal and chemical stimuli are probably involved in the establishment of a feeding site and are detected by sensory receptors located on the tarsi of the first pair of legs and on the terminal article of the palp, those on the latter being contact receptors (Lees, 1948). The stimuli may be non-specific, as in the case of *Ixodes ricinus*, which will attach to almost any part of the body of the host, or specific, as in *Rhipicephalus evertsi* on bovids, where the immature instars display a strong preference for the ears and the adults attach around the anus. Males of several species of *Ixodes*, especially nest and burrow-inhabiting forms, and larvae of some species of *Ornithodoros* do not feed whereas males in *Boophilus, Hyalomma, Rhipicephalus* and *Dermacentor* require a small meal before mating. Unmated females of ixodoid ticks appear not to engorge completely and may remain on the host for a longer period than usual but once mating occurs they begin rapid ingestion leading to engorgement, detachment and disengagement. There is some evidence that underfed females, which drop off their host, may be able to reattach to a new host and successfully complete engorgement (Arthur, 1962).

Prior to penetration, the tick grips the skin of the host by means of the ambulacra of its legs and lifts its body at an angle to the surface. Penetration of the skin of the host is achieved by the outward cutting movements of the cheliceral digits and a deeper lesion may be formed by the insertion of the digits into the skin by a forward movement of the chelicerae. The gradual insertion of the hypostome is accompanied by rocking movements. Favourable inputs from the digital receptors of the chelicerae as they come into contact with the body fluids of the host are thought to trigger the tick to complete the penetration event (Waladde and Rice, 1982). Ixodoids remain attached to the host for long periods and, with the exception of some species of *Ixodes* (*I. holocyclus, I. trianguliceps* and *I. pseudorasus* – group I in Fig. 6.9), utilize cement produced by the salivary glands for attachment. The cement is secreted in the form of a milky-white fluid as the hypostome is inserted into the skin and hardens almost immediately to form a cement sheath or tube of a latex-like consistency around the hypostome. Its form varies between genera (Moorhouse, 1969). In the Brevirostrata (*Haemaphysalis, Rhipicephalus, Boophilus* and *Dermacentor*), the short mouthparts hardly penetrate the epidermis and the cement, in addition to penetrating the lesion, forms an external cone around that part of the hypostome not

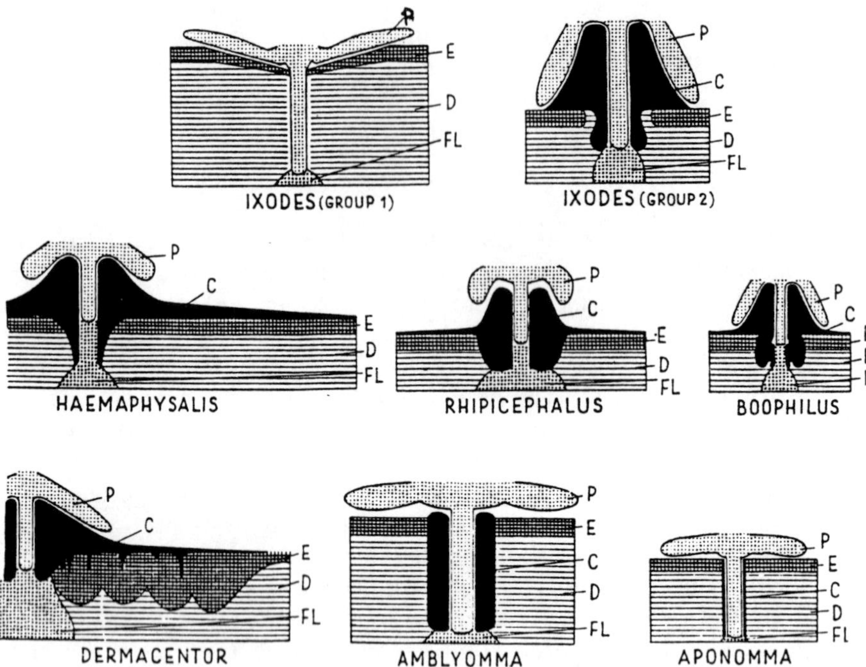

Fig. 6.9. Schematic diagrams of sections through the host skin to show form of attachment of female ixodoid ticks of various genera during the final stages of feeding (after Moorhouse, 1969). P = palps; C = cement; E = epidermis (Malpighian layer); D = dermis; FL = feeding lesion.

inserted into the skin (Fig. 6.9). Such an external cone is not formed in the Longirostrata (*Amblyomma, Aponomma* and *Hyalomma*), in which the hypostome is inserted deeply into the dermis and the cement lines the channel formed by the mouthparts. An external cement cone is also formed in *Ixodes tasmani* and *I. japonensis* in which only a portion of the hypostome enters the host. The hypostomatic teeth and the denticles of the outer cheliceral sheath become embedded in the cement cone and help to secure attachment. The cement cone appears to be essentially proteinaceous with some lipid and carbohydrate (Moorhouse & Tatchell, 1966).

Ingestion and engorgement

The cutting action of the cheliceral digits results in some damage to tissues and capillaries adjacent to the site of penetration (Lavoipierre & Riek, 1955; Gregson, 1960b). Salivary secretions are considered to cause dilation of the capillaries and extensive haemorrhaging in the feeding lesion leading to the formation of a pool of blood. A focus of haemorrhage develops soon after attachment in argasoids, which become fully engorged in a matter of minutes or at the most within a few hours, whereas in ixodoids haemorrhaging tends to be delayed and in ***Dermacentor andersoni*** occurs suddenly about 2½ h after attachment (Kemp *et al.*, 1982).

The outer cheliceral sheath and the hypostome in attached ixodoid ticks form a tube around the pre-oral channel which is further enclosed by the cement sheath. Feeding consists of alternating periods of drawing fluid from the inflammatory focus or blood pool at the tip of the hypostome (up to 30 s duration) and injecting saliva into the host's tissues (1 s or less in duration). The periods of sucking and injecting are separated by intervals lasting a few seconds to several minutes which prevents disturbance of fluids moving in opposite directions along the feeding channel (Balashov, 1968). During ingestion the pharyngeal dilator muscles contract thereby pulling apart the walls of the pharynx and creating a negative pressure in the pharynx which is transmitted along the pre-oral channel. This action draws blood and tissue fluids into the pharynx. The pharyngeal contents are forced along the oesophagus into the ventriculus by the contraction of the circular muscles of the pharynx after relaxation of the dilators. An anterior pharyngeal 'valve' closes just before the constriction of the pharynx. Its mode of functioning has not been fully elucidated but it seems likely that closure is brought about by the pharyngeal walls being forced against a dorsal tooth or wedge projecting into the lumen (Kemp & Tatchell, 1971).

The nymphs and adults of the Argasoidea do not utilize cement for attachment to the host and denticles are not present on the outer cheliceral sheath. In some slow-feeding argasoid larvae, a plug of host tissue has been observed at the attachment site but none is known to produce cement. The hypostome is a more delicate structure than in ixodoids but the heavier cheliceral digits are capable of cutting the host's skin more rapidly than those of ixodoids. The labrum is a larger structure than in the hard ticks and forms a flap-like lobe covering the opening into the pharynx. It is considered to separate the salivary secretions flowing out of the dorsal salivarium from host fluids sucked in through the ventrally directed pharyngeal opening (Robinson & Davidson, 1913a). The downward movement of a rod-like thickening of the labrum into the opening of the pharynx and feeding channel prevents regurgitation of the pharyngeal contents.

The ingested food consists of tissue fluids, lymph and blood. In the Ixodoidea, the ratio of whole blood to extravascular fluid is usually greater during the later stages of the feeding cycle and in older instars. During the first phase of feeding in members of this group lymph and lysed tissues as well as small quantities of blood issuing from ruptured capillaries are imbibed while in the second and third phases, respectively, non-blood fractions (plasma and infiltrating leucocytes) and whole blood form the predominant components of the food. Feeding proceeds over a period of days; for example, most ixodoid larvae feed for 3–5 days, nymphs for 3–6 days and females for 6–12 days. Phagostimulants which regulate ingestion have been investigated in *Argas, Ornithodoros* and *Boophilus* (Kemp *et al.*, 1982). Adenosine triphosphate (ATP) and glutathione (GSH) are considered to induce feeding. For example, *Boophilus microplus* attached to feeding membranes showed significant increases in the sucking component of feeding electrograms when ATP or GSH was added to the blood diet (Waladde & Rice, 1982). Electrophysiological work on this species has shown that neurones of the inner digit pit sensilli respond to saline with added ATP or GSH. In argasoids, Galun and Kindler (1968) suggested that at least two kinds of receptors are located on the mouthparts, one specific for GSH and nucleotides and the other for amino acids. Glucose appeared to act as a synergist at the receptor sites.

During the short feeding period of argasoid nymphs and adults, the weight of the tick increases 5–12 times but the relative volume of ingested blood usually decreases with each moult and is least in males (Balashov, 1968). The increase in body weight in larvae is considerably greater, being of the order of 15–20 times. Large quantities of blood are ingested over a period of several days by ixodoid ticks and this is apparent in the enormous enlargement of the body during attachment. In species such as those of *Amblyomma* and *Hyalomma*, the unfed weight of the female is about 10–15 mg while in the fully engorged state the weight lies between 600 and 1200 mg. These absolute weight increases in ixodoids do not take into account the amount of blood digested and assimilated or the quantity egested in a slightly modified form with the faeces. It is evident that the exact amount of blood ingested greatly exceeds that obtained by weighing the fully engorged tick.

Detachment and disengagement

The nature of the satiation stimuli (and their receptor sites) which lead to the detachment of the tick do not appear to have been determined. According to Arthur (1962), two mechanical processes seem to be involved in the actual detachment of the engorged ixodoid ticks from the skin of its host, namely the retraction of the chelicerae and the loosening of the hypostome from the cement sheath by the movements of the gnathosoma through muscular action. As a result of these actions, the engorged tick hangs loosely from the body of the host. Detachment appears to be preceded by a period of intense ingestion of blood. Thus, on day-grazing cattle, *Ixodes ricinus* and *Haemaphysalis bispinosa* take the last and largest ingested blood meal at night before detachment and disengagement in the morning (Balashov, 1968). Wharton and Utech (1969) consider the diurnal detachment–disengagement rhythm in the one-host species *Boophilus microplus* to be mainly stimulated by changes in light intensity. Female ticks 4.5 mm or more in length before nightfall rapidly increase to about 8–11 mm during the night before dropping off in maximum numbers shortly after daylight. Those smaller than 4.5 mm increase only slightly in length and complete their engorgement the following night. In the absence of the required stimulus, engorged ticks may stay on the host for several days. For example, in the Leningrad area completely engorged *I. ricinus* on cattle not pastured for two days because of the sudden onset of cold weather and snow remained attached until the cattle were driven to pasture on the third day, when they detached immediately (Balashov, 1968).

Prostigmata*

The feeding habits of members of the Prostigmata exhibit a greater diversity than is encountered in any other order of the Acari, especially in the extent to which they utilize materials of living higher plants. As well as phytophagy, many species are zoophagic, either as predators or animal parasites and parasitoids, while omnivory is not uncommon. Some phytophagous species are major pests of field and protected crops and a few parasitic species are vectors of disease to man.

Predation

This type of feeding behaviour is common and widespread within the Prostigmata and it is probable that only within the higher taxa Tetranychoidea and Eriophyoidea are predatory species not represented. The predators are fluid feeders except for some species of 'Endeostigmata' which are particulate feeders. The majority appear to prey on other small arthropods. Our knowledge of feeding behaviour within many major taxa is unknown or fragmentary and, as is common in the Acari, only those species of economic importance have been studied in any detail. Nematophagy appears to have been established only in the Alicorhagiidae (Walter, 1988), an intertidal species of *Bdella* (Pugh & King, 1985) and some species of Cunaxidae (Walter & Kaplan, 1991). The chelicerae are strongly chelate in the Alicorhagiidae, Rhagidiidae, Nicoletiellidae (Fig. 6.10A) and Cytinae although the size of the digits is much reduced in relation to the long shaft in other Bdellidae (Fig. 6.10B). In the majority of the predatory forms, the chelicerae are specialized for piercing or ripping rather than for cutting and show a reduction in the size or the loss of the fixed digit and the modification of the movable digit to form a movable hook or claw-like structure (Fig. 6.10C) or a piercing

*The complex of taxa forming the 'Endeostigmata' was originally included in the Prostigmata by Grandjean (1937) but is currently considered by some authors to be paraphyletic and to have given rise independently to the Prostigmata and the Astigmata–Oribatida lineages (OConnor, 1984). For convenience, the feeding habits of the group are considered here.

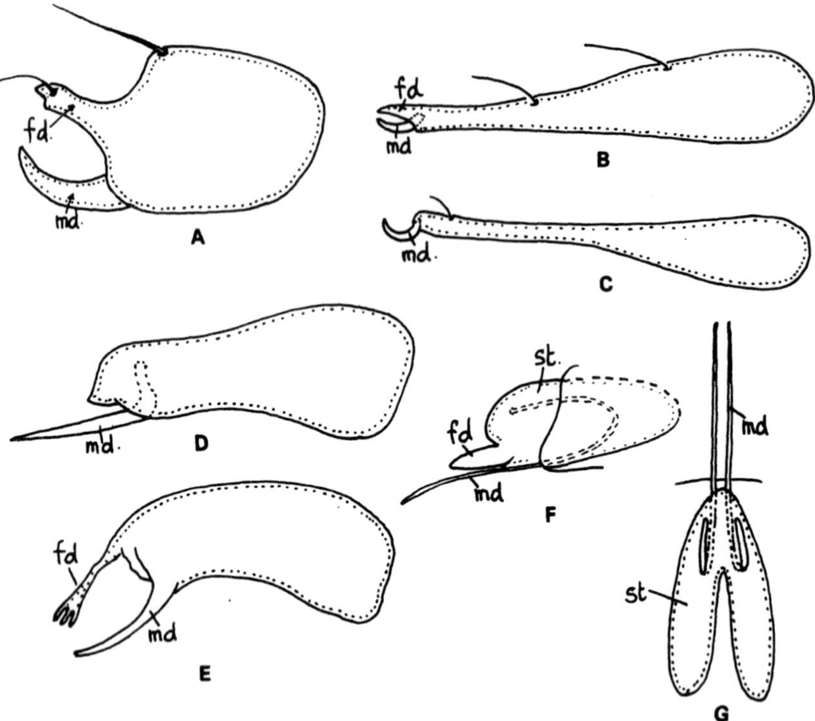

Fig. 6.10. Semidiagrammatic representation of some types of chelicerae in the Prostigmata. **A**, *Labidostomma* (Nicoletiellidae); **B**, *Bdella* (Bdellidae); **C**, *Cunaxa* (Cunaxidae); **D**, *Tydeus* (Tydeidae); **E**, *Penthaleus* (Penthaleidae); **F**, *Tetranychus* (Tetranychidae); **G**, *Brevipalpus* (Tenuipalpidae). *fd.* = fixed digit; *mf.* = movable digit; *st.* = stylophore.

stylet (Fig. 6.10D,E). Predatory forms with hook-like movable digits and separate cheliceral bases are represented, for example, by the deuteronymphs and adults of the majority of terrestrial and aquatic Parasitengona and by the Anystidae and Cunaxidae (Smiley & Knutson, 1983). Those with stylet-like movable digits of the chelicerae occur in the Cheyletidae, Erythraeidae, Raphignathoidea, Tarsonemidae and Tydeidae. The entire chelicera is styliform in the Erythraeidae but in the other taxa the cheliceral bases may be medially approximate or fused (Fig. 6.10F,G) while in the Cheyletidae, Caligonellidae and Tarsonemidae (see Fig. 5.15B,C, p. 159) the combined bases are also fused with the subcapitulum. Zoophagy is obligatory in the Cheyletidae but phytophagy occurs in some Erythraeidae, commonly occurs in the Raphignathoidea and Tydeidae and is predominant in the Tarsonemidae.

Soil-dwelling predators feed on the eggs and active instars of other small arthropods, including other mites, collembolans and dipteran larvae. An indication of the wide range of arthropods providing prey for the Bdellidae is given by Alberti (1973). Species of *Cheyletus* occurring in stored food products actively predate acarid mites and in some instances also feed on the young stages of storage moths and beetles and on psocids (Hughes, 1976). *Cheyletus eruditus* appears to have a preference for *Acarus siro* and *Lepidoglyphus destructor* and has been introduced into covered grain stores for the biological control of *A. siro* in Czechoslovakia (Pulpán & Verner, 1965). When prey is in short supply, cheyletids become cannibalistic. Arboreal species, particularly those in orchards, prey on phytophagous mites and insects. For example, *Zetzellia mali* and *Agistemus fleschneri* (Stigmaeidae) feed predominantly on spider mites of the genera *Panonychus* and *Tetranychus* and on the rust mite *Aculus schlechtendali* in apple orchards while *Pronematus ubiquitous* (Tydeidae) and *Paracheyletia bakeri* (Cheyletidae) prey on *Eutetranychus* spp. on citrus (Laing & Knop, 1983). '*Anystis agilis*' has been observed to prey on a number of species of mites and insects on broom in England (Baker, 1967). The oribatid mite *Chamobates borealis* appears to be an important prey species and nymphs of psocids, psyllids and thysanopterans as well as all stages of aphids were also included in the diet. On citrus, *A. agilis* feeds on the citrus thrips (*Scolothrips citri*) and tetranychid mites. All instars of *Anystis* are extremely cannibalistic in culture. *Bdellodes lapidaria* and *Neomolgus capillatus* are efficient predators of the lucerne flea, *Sminthurus viridis*, and have been introduced from Europe into Australia to control this collembolan which has attained pest status after introduction from Europe (Wallace, 1974). Species of *Acaronemus* and possibly some species of *Dendroptus* are exceptional among the Tarsonemidae in being free-living predators, the former feeding on the eggs of tenuipalpids and tetranychids and the latter feeding on eriophyids within their galls (Lindquist, 1986). Predacious mites living on plants, for example members of the families Stigmaeidae, Tydeidae and Phytoseiidae, as well as fungivorous species of the Astigmata and Oribatida, utilize leaf domatia (specialized chambers in the vein axils on the underside of the leaves of many plant species) for reproduction and development and for protection from environmental extremes (O'Dowd & Wilson, 1989).

The majority of the predatory Prostigmata appear to capture their prey by random contact using the first pair of legs and pedipalps as tactile and chemosensory organs. The prey is usually held by the pedipalps during the penetration of its body by the chelicerae. However, in predacious Halacaroidea, such as members of *Agauopsis* and *Thalassarachna*, spiniform setae on the telofemora, tibiae and genua of the first pair of legs are used to grasp the prey (Newell, 1947a). Two species of *Agauopsis* studied by MacQuitty (1984) feed on copepods and penetration of the cuticle was achieved by the mite's long rostrum. Members of the Bdellidae, Cunaxidae and Cheyletidae, on the other hand, ambush their prey. Some bdellids immobilize the prey by means of a stringy substance produced by the supra-oesophageal organ (receptaculum cibi) and extruded by way of the mouth. Bdellids and cunaxids may also utilize silk in capturing their prey, the former fastening the prey to the ground with silken threads and the latter spinning a trap of two different kinds of silk (Grandjean, 1938b; Alberti, 1973; Alberti & Ehrnsberger, 1977). The silk is produced by the podocephalic gland complex (probably gland 3) opening on the dorsal surface of the subcapitulum.

Parasitism

The parasitic forms encountered in the Prostigmata occur almost exclusively in five taxa, namely the Parasitengona, Pterygosomatoidea, Cheyletoidea, Tydeoidea and Tarsonemoidea. Parasitism in the Parasitengona is exceptional among the Acari in that it is limited to the larval stase (protelean parasitism) except in some water mites, for example the Unionicolidae in which adults and nymphs have secondarily acquired a parasitic habit on molluscs or sponges and some species of *Balaustium* which occasionally bite man (Newell, 1963). The larvae feed on vertebrate or arthropodan hosts to which they attach by means of the strong hook-like movable digits of their chelicerae. Subsequent to attachment, a 'feeding tube', the stylostome or histiosiphon, may be formed in

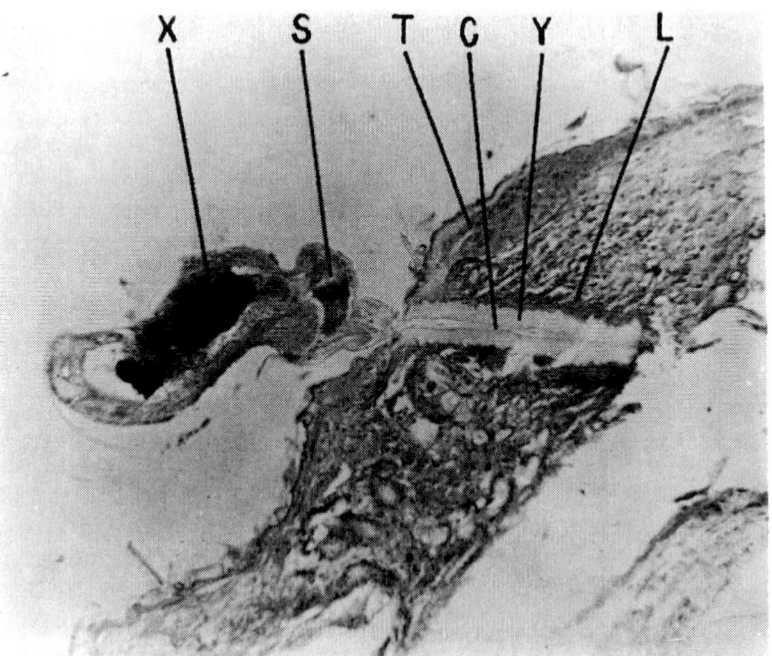

Fig. 6.11. Longitudinal section of *Eutrombicula splendens* (Prostigmata) and its feeding tube (stylostome or histiosiphon) in the skin of a mouse (from McLaughlin, 1968). *C* = canal; *L* = cell mass; *S* = salivary gland of mite; *T* = thickened epidermis; *X* = excretory material in the mite; *Y* = hyaline mass.

the host integument or skin and usually remains in the host after the parasite has engorged and disengaged (Fig. 6.11). The formation of the stylostome has been most extensively studied in larvae of the family Trombiculidae parasitizing mammals. During the feeding process, the larva (commonly referred to as a 'chigger') penetrates the stratum corneum of the host epithelium with its chelicerae and this is followed by the secretion of saliva which breaks down the deeper epidermis (Brown, 1952). A feeding tube, starting in the epidermis and extending well into the dermis, is then formed and comprises a narrow central canal bordered by a thick, cylindrical hyaline mass around which there is a massive accumulation of host cells (Jones, 1950b; Cross, 1964). The hyaline mass in *Eutrombicula splendens* contains neutral and acid mucosubstances with protein forming at least three-quarters of the hyaline material (McLaughlin, 1968). At its epidermal end, it is formed, at least in part, from the keratin in the epidermis but it has not been decided whether the main hyaline mass is derived from the skin of the host or from the salivary secretions of the mite. The chigger ingests tissue fluid and disintegrated tissue cells and nuclei.

The feeding tube in some larval Trombidiidae parasitizing insects extends into the haemocoel and is often extensively branched, as in *Teresothrombium susteri* feeding on a beetle (Feider, 1956). The stylostome is considered in this family to be formed from the salivary secretions of the larva and the fluid food passes to the mouth along its axial canal. Microtrombidiine and eutrombidiine larvae occurring on insects have a circular, denticulated, hyaline disc or coronet surrounding the subcapitulum and chelicerae at their area of contact with the surface of the cuticle. It probably plays a role in attaching the mite to its host and remains attached to the gnathosoma when the parasite is prematurely detached. Robaux (1974) is of the opinion that a stylostome is not developed in those larvae having a coronet. In water mite larvae parasitizing arthropods, the stylostome may be branched, as in Corixidae parasitized by species of *Hydrachna* (Davids, 1973) or unbranched and with a club-like or globular termination as in chironomids parasitized by *Unionicola aculeata* (Hevers, 1978). Davids (1973) suggested that the stylostome formed in the haemocoel of corixids during feeding by *Hydrachna conjecta* is a coagulation product of the host haemolymph in reaction to the saliva of the parasite.

Larvae of terrestrial Parasitengona, except the Trombiculidae, are external parasites of insects or arachnids and once attached to the integument of the host remain at the attachment site until engorged. A comprehensive list of their hosts is given by Welbourn (1983). Larvae of the two largest families, the Trombidiidae and Erythraeidae, chiefly parasitize Diptera (mainly Culicoidea and Muscoidea), Orthoptera, Coleoptera, Homoptera (primarily Aphididae), Lepidoptera (Noctuidae) and Arachnida (Acari, Araneae, Opiliones). Some species, particularly of *Allothrombium, Eutrombidium* and *Teresothrombium*, are considered to have potential for the control of insect pests. The larvae of Trombiculidae number about 2500 described species and, with few exceptions, parasitize vertebrates (Wharton & Fuller, 1952). The majority are external parasites but some occur subcutaneously as in the case of *Hannemania hylae* on tree frogs in North America (Ewing, 1922) while others live in the respiratory system, for example species of *Vatacarus* of marine iguanids and sea snakes (Vercammen-Grandjean, 1965). A number of species attack man and cause a dermatitis characterized by a small congestive elevation of the skin surrounded by a wheal. The most important species are *Trombicula (Neotrombicula) autumnalis* (the harvest mite) in Europe, *Trombicula (Eutrombicula) alfreddugesi, T. (E.) splendens* and *T. (E.) batatas* in the Americas, *T. (E.) wichmanni* in southern Asia and Australia and *T. (E.) hirsti* in Australia. Three species, *Leptotrombidium akamushi* in Japan and *L. deliensis* and *L. fletcheri* in southern Asia are vectors of *Rickettsia tsutsugamushi*, the causative agent of scrub typhus or Tsu Tsugamushi disease (Audy, 1968).

The larvae of a range of aquatic Parasitengona parasitize aerial insects whose immature instars are aquatic and not only use the host for food but also as a vehicle for dispersal. Mitchell (1957c) divides the larvae of water mites into aerial and aquatic forms. Aerial larvae have morphological features of terrestrial parasitengonid larvae (spherical membraneous body, relatively weak apodemes and very long body setae) and walk on the surface of the water. This type of larva occurs in the Hydrovolziidae, Eylaioidea and Hydrophantoidea. After making contact with a host, the slow-moving larvae of the Hydrovolziidae climb up the host's appendages and attach to its body, as in for example, *Hydrovolzia gerhardi*, attacking the terminal nymphs or adults of a mesoveliid bug. Most larvae of the Eylaioidea and Hydryphantoidea parasitize aquatic Diptera and after locating the pupa cluster in the region of its breathing horns to await the emergence of the imago. Aquatic larvae, on the other hand, are greatly modified in comparison with terrestrial parasitengonid forms, particularly for movement through water. They are flattened dorsoventrally with the dorsal and ventral surfaces each covered by a single shield, most setae are modified or lost (any long setae project parallel to the long axis of the body and do not reduce the 'streamlining') and the legs show modifications for swimming. This group is represented by the families Hydrachnidae, in which the very large gnathosoma of the unfed larva may be nearly as big as its idiosoma, Hygrobatoidea and Arrenuroidea. The hydrachnid larvae are not powerful swimmers and usually attach anywhere on an active host, such as an hemipteran nymph or adult, or an adult beetle. The swift swimming hygrobatoid and arrenuroid larvae appear to locate a pupa or terminal nymph of their host by random contact and rest on the instar until the emergence of the imago to which they usually attach in a specific location. Their hosts include Diptera and Odonata. Parasitism by larval *Arrenurus* can have a marked effect on populations of mosquitoes of the genera *Anopheles* and *Mansonia* (Smith, 1983).

The large superfamily Cheyletoidea includes parasitic as well as free-living predatory species. The parasitic forms are found in the families Cheyletidae (on birds and mammals), Cloacaridae (on reptiles), Demodicidae (on mammals), Harpyrhynchidae (on birds), Myobiidae (on mammals), Ophioptidae (on snakes), Psorergatidae (on mammals) and Syringophilidae (in quills of the body and flight feathers of birds). The majority are skin parasites feeding on tissue fluids and are often associated with mange (scaling, hair loss and pruritus). For example, *Cheyletiella parasitivorax* can cause mange in domesticated and wild rabbits while *C. yasguri* and *C. blakei*, respectively, produce similar signs in dogs and cats (Van Bronswijk & de Kreek, 1976). Humans in contact with dogs infected by *C. yasguri* sometimes develop an itching dermatitis. A putative mutualistic relationship between *Ornitocheyletia hallae* and the ascomycete *Micromonospora chalceae* on pigeons has been described by Haarløv and Mørch (1975). The oedemata and the hyperkeratinosis of the epidermal cells of the skin of the host resulting from the bite of the cheyletid provides a favourable culture medium for the fungus while the hyperkeratinized layer provides a microhabitat for the mite. It is suggested that the fungus breaks down complex proteins by means of its proteolytic enzymes and that the metabolic products are utilized by the mites.

Some species of Psorergatidae, such as *Psorobia ovis* infesting sheep, live in the corneous layer of the skin and the mange they produce is characterized by very strong keratinization and loss of hair. The small vermiform, poorly sclerotized, Demodicidae live in the pilosebaceous complex of the skin of mammals. *Demodex folliculorum* in man occurs in the superficial parts of the follicle of simple hairs, i.e. above the sebaceous glands, and feeds on cells of the follicular epithelium. It is considered to be a low-grade pathogen (Desch & Nutting, 1972) but *D. canis* infesting dogs produces a mange which may be either of a squamous type characterized by hair loss and dry scaly dermatitis with slight induration, or of a pustular type generally associated with secondary infection and characterized by chronic moist dermatitis and purulent exudate. The Cloacaridae are highly specialized parasites usually found attached to the mucosa of the rectum of adult turtles or, in the case of *Theodoracarus testudinis*, in the cellular tissues and muscles of the chest and legs of *Testudo graeca ibera*. The mode of transference of the mites from one host to another is not known but it has been suggested that it occurs during copulation (Camin *et al.*, 1967; Fain, 1968c). The gnathosoma is represented by a pair of dagger-like sclerites.

Among the Tydeoidea, parasitic species occur in the family Ereynetidae (Fain, 1965a). Members of the Lawrencarinae are nasicolous parasites of frogs and toads while those of the Speleognathinae occur in the nasal cavities of birds and mammals. The species of Ereynetinae, on the other hand are mainly free-living but those of *Riccardoella* live in the mucus of the mantle cavity of terrestrial molluscs and *Ereynetoides malayi* has been found in the nasal cavities of a bird. Little is known of the feeding habits of the nasicolous species.

The Pterygosomatidae are parasites of lizards, except *Pimeliaphilus podapolipophagus* which parasitizes cockroaches. The lizard parasites are usually found beneath the scales of their host and feed on blood and tissue fluids (Lawrence, 1935). The chelicerae have a reduced fixed digit in the form of a seta-like or branched process and the dentate face of the movable digit is directed laterally as in ticks. It functions to cut the skin of the host and probably also acts as a holdfast.

The tarsonemids parasitic on insects are highly host specific and of particular interest is *Acarapis woodi* infesting the prothoracic tracheal trunks of adult honey bees and the causal agent of Acarine or Isle of Wight disease. The larval and adult mites pierce the tracheal walls by means of their stylet chelicerae and feed on haemolymph. Heavy infestations impair the passage of air through the tracheae and this coupled with the degeneration of tissues around the feeding sites progressively weakens the bees. The pathogenicity of *Acarapis* infestations appears to have been exaggerated and the mortality attributed to Acarine disease earlier in this century is considered by Bailey (1982) to have been probably caused by chronic bee-paralysis virus. The symptoms of this viral disease correspond to those described for Acarine disease. Another inhabitant of the respiratory system of Apidae, *Bombacarus buchneri*, belongs to the Podapolipidae, a sister group of the Tarsonemidae. This species inhabits the large tracheal sacs of the first abdominal segment of species of *Bombus* and also feeds on haemolymph in much the same way as *Acarapis* (Evans *et al.*, 1961).

The biotic relationship of some heterostigmatic mites, such as members of the genera *Acarophenax, Iponemus* and *Pyemotes*, and their insect hosts cannot be satisfactorily defined as predacious or parasitic and Lindquist (1983) has argued for designating it a form of *parasitoidism*. In his modified concept of arthropodan parasitoidism, he lists seven characteristics of which the following three combined are the most essential:

1. The host is ultimately destroyed.
2. Only one host is required for development of one or more progeny of the maternal parasitoid.
3. Only one stage functions trophically on attacked hosts (the adult female in the case of heterostigmatic mites).

The mites of the above three genera have styliform chelicerae which pierce the soft body wall of their 'hosts'. Females of *Acarophenax* and *Pyemotes* feed on larval and pupal instars of insects (Coleoptera, Hymenoptera and Lepidoptera) whereas *Iponemus* engorges on the eggs of ipine bark beetles. They all exhibit physogastry in the engorged state. According to Herfs (1926), insertion of the chelicerae of *Pyemotes* into caterpillars evokes an initial curling or stiffening reaction which is followed in a few hours by paralysis. *Pymotes tritici* (*P. ventricosus* of some authors, see Krczal, 1959) is the most common species of the genus infesting stored grain and hay. A mixture of three neurotoxins, which have a contracting–paralysing action on insects, has been isolated from this species by Tomalski *et al.* (1989). The toxins are localized in podosomatic glands which flank the gut and appear to be connected with the cheliceral stylets through a series of ducts. People coming into contact with heavily infested materials are often bitten by the pest and develop a dermatitis accompanied by pruritus, headache, fever and perspiration. This has given rise to the common names hay, straw or grain itch mites for pyemotid species attacking humans.

Phytophagy

The ingestion of living plant material is common and widespread in the Prostigmata. The Tetranychoidea and Eriophyoidea feed almost exclusively on higher plants but in other taxa, such as the Tarsonemina, Eupodoidea, Tydeoidea and Raphignathoidea, fungivorous species are well represented or predominate among the plant-feeding forms. Pollen is a source of food for some mites, for example in the tydeid genus *Pronematus* (see Krantz & Lindquist, 1979) and in all active stases of the erythraeid *Balaustium florale* (Grandjean, 1946b). The latter seizes the pollen grain between its pedipalps and brings it up against the extremity of the subcapitulum. The pollen grain is pierced by the chelicerae, its contents sucked out and the shrivelled case rejected. Marine mites of the subfamily Rhombognathinae are thought to be algivorus and probably use their fine chelicerae to pierce algal cells (Newell, 1947a). The extent to which phytophagous mites consume living green plant tissue has not been studied extensively. Newell (1963) observed a larval *Balaustium* feeding on the leaf tissue of a species of *Veratum* in California and it has been suggested that the green gut contents of arboreal tydeids possibly indicate the ingestion of leaf tissue.

The chelicerae of phytophagous Prostigmata are adapted for piercing and, as in the case of many of the predatory forms, there is a tendency for the reduction of the fixed digit, modification of the movable digit into a stylet-like structure and fusion of the cheliceral bases. Exceptions occur among the Endeostigmata in which some of the rutellate forms feeding on fungi, for example *Terpnacarus* and *Alycus*, have chelate–dentate chelicerae (Theron, 1979). The gnathosomatic structure and probable feeding mechanisms in these taxa appear to resemble those of the Astigmata–Oribatida lineage. The chelicerae of the Tydeidae (Fig. 6.10D) with their non-retractable styliform movable digits and reduced fixed digits are among the least specialized of the piercing-type of chelicerae within the Prostigmata and are capable of dealing with animal and fungal food. Thus, it is not surprising to find predatory, fungivorous, algivorous and omnivorous species within this family. Omnivory, which is probably more widespread than has been reported in the literature, occurs, for example, in *Homeopronematus anconai* which feeds on *Aculops lycopersici*, eggs of *Brevipalpus* and, in the laboratory, on pollen, leaf tissue and fungi (Hessein & Perring, 1986, 1988). Relatively unspecialized chelicerae also occur in the phytophagous members of the Eupodoidea in which the chelicerae are characterized by edendate movable digits opposing a distorted fixed digit. Species of Penthaleidae, for which details of trophic behaviour are available, feed on higher plants; two species, *Penthaleus major*, a pest of a range of pasture and forage crops in Australia, South Africa and North America, and *Halotydeus destructor*, attacking vegetable and leguminous crops in Australia and South Africa, are of economic importance. The movable digits of the chelicerae in these species (Fig. 6.10E) are longer and more styliform than in the Eupodidae and Penthalodidae which are considered to be predominantly microphytophagous, feeding on fungi and possibly also on algae and lichens (Krantz & Lindquist, 1979; Goddard, 1982).

The Raphignathoidea are almost exclusively predacious but members of the genus *Eustigmaeus* have been observed to feed and reproduce on mosses (Gerson, 1972). A wide range of feeding habits occurs in the Tarsonemina. Species of *Tarsonemus* appear to be mainly fungivorous, those of *Steneotarsonemus* feed on monocotyledonous plants and those of *Daidalotarsonemus* are thought to utilize epiphytic algae and lichens on trees as their main source of food. Some members of *Dendroptus* take over eriophyoid galls and feed on erineal plant tissue within the galls (Beer, 1963). A number of species of Tarsonemidae are pests of crops and ornamentals and are discussed by Lindquist (1986). The styliform movable digits of the chelicerae in the Tarsonemina and Raphignathoidea are protracted and retracted by muscles acting through the sigmoid pieces. The contiguous or fused cheliceral bases in the Raphignathoidea have lost their capacity for independent movement and in the Tarsonemina, the stylophore is integrated with the subcapitulum. The cheliceral stylets in these groups are known to pierce the host tissue but whether this is their sole function or whether they also form a channel for the passage of fluids as in the Tetranychoidea does not appear to have been determined.

The Tetranychidae or spider mites feed mainly on the underside of leaves although some species prefer the upper side or feed on both sides. By their feeding activity, the mites sometimes cause damage to specific parts of the plant, for example cotyledons, tips of shoots, fruit spurs and flowers. The mites penetrate the plant by means of their long cheliceral stylets (movable digits) and suck out the cell contents. The chloroplasts disappear and the remaining cellular material coagulates to form an amber-coloured mass. The spongy mesophyll cells, and in some cases also the cells of the lowest palisade layer, are damaged during feeding from the under surface of the leaf while cells in all the palisade layers and sometimes a few of the adjacent spongy mesophyll cells are damaged from upper surface feeding. It is estimated that *Tetranychus urticae* is capable of exhausting about 18–22 cells per minute (Liesering, 1960). A series of stylet punctures on a leaf, usually in the form of a circle, results in the formation of small primary chlorotic spots. With continued feeding,

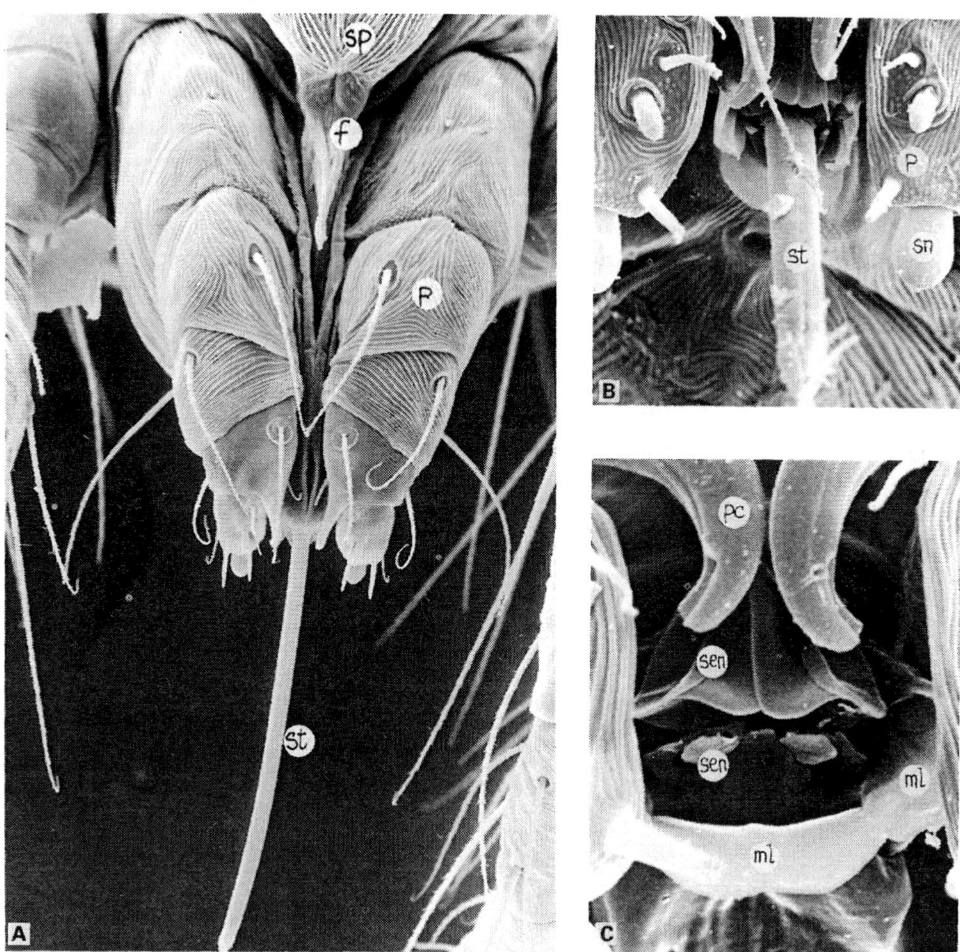

Fig. 6.12. SE micrographs of the chelicerae and subcapitulum of *Tetranychus urticae* (Prostigmata). **A**, stylophore and cheliceral digits in the protruded state and the dorsal subcapitular (or rostral) groove of the conical subcapitulum flanked by the pedipalps; **B**, frontal view of the extremity of the subcapitulum showing the sensilli, exit of the interlocked cheliceral stylets and membraneous lobes; **C**, higher magnification of the extremity of the subcapitulum with the cheliceral stylets withdrawn (note the somewhat blunt, channelled, open tip of the palptibial claw. *f.* = fixed digits of the chelicerae; *ml.* = membraneous lobes; *p.* = pedipalp; *pc.* = pedipalpal claw; *sen.* = sensilli; *sn.* = spinneret; *sp.* = stylophore; *st.* = stylets.

the primary spots integrate to form irregular spots which on prolonged exposure become white or greyish in colour. Bronzing of the leaf is usually associated with damage to mesophyll cells. The depth reached by the stylets ranges from 70 to 120 μm, depending on the length of the stylets and the characteristics of the plant (Tomczyk & Kropczynska, 1985). The stylets may pass between the cells of the epidermis as in the case of *Panonychus ulmi* feeding on apple and plum leaves (Avery & Briggs, 1968) or through the walls of the epidermal cells as, for example, in *Tetranychus mcdanieli* feeding on apple leaves (Tanigoshi & Davis, 1978). There is some evidence that certain substances (toxins or growth regulators) are injected into the plant tissue on which the spider mite is feeding but little is known of their composition or role (Avery & Briggs, 1968; Storms, 1971).

The function of the various components of the mouthparts of tetranychoid mites has received considerable attention since the classic work of Blauvelt (1945) but there remains some disagreement in relation to the actual part played by the cheliceral stylets in the feeding process. During protraction, the stylets are interlocked by a tongue-and-groove mechanism and form an efficient piercing organ (Fig. 6.12A). A longitudinal groove on the mesal face of each stylet forms a channel when the stylets are interlocked. In *Tetranychus urticae*, the subterminal opening of the channel is formed by the apposition of a notch in each stylet. The channel is considered by the majority of authors to function as a salivary channel through which products of the salivary glands are conveyed to the feeding puncture (Hislop & Jeppson, 1976; Alberti & Crooker, 1985; Nuzzaci & de Lillo, 1989, 1991). André and Remacle (1984), however, conclude that it is a food channel through which 'cell contents are sucked up' and that saliva mixes with the food in the 'prelabial cavity' before it is drawn into the pharynx. The recent description of a 'salivary pump' mechanism for controlling the flow of saliva along the stylet channel supports the hypothesis that the channel functions as a salivary and not a food channel (Nuzzaci & de Lillo, 1989, 1991).

Anterior to the pedipalpal base, the lateral walls of the conical subcapitulum in the tenuipalpid *Cenopalpus pulcher* extend dorsally to form a deep, relatively narrow, dorsal subcapitular or rostral groove which communicates with the median salivary duct, the cheliceral-stylet grooves, the labrum and the pre-oral groove (Nuzzaci & de Lillo, 1989). The median salivary duct and the cheliceral-stylet grooves which flank it are located on the dorsal surface of the subcheliceral plate. Distally, the duct opens into a vestibule whose lateral walls are formed by flexible plates, termed salivary plates, extending from the subcheliceral plate (Fig. 6.13A). Two narrow channels extending from the vestibule and lying on either side of a median ridge of the subcheliceral plate connect the vestibule to the salivary channel of the stylets. The sclerotized pre-oral groove is connected to the pharyngeal chamber and is covered dorsally, for the greater part of its length, by the labrum. The ventral lobe of the labrum is so formed that it fits the contour of the pre-oral groove (Fig. 6.13B) and together with its dorsal lobe forms a double valve which is capable of closing the food channel (combined pre-oral groove and labrum). The fixed digit of each chelicera comprises a dorsal and vertical

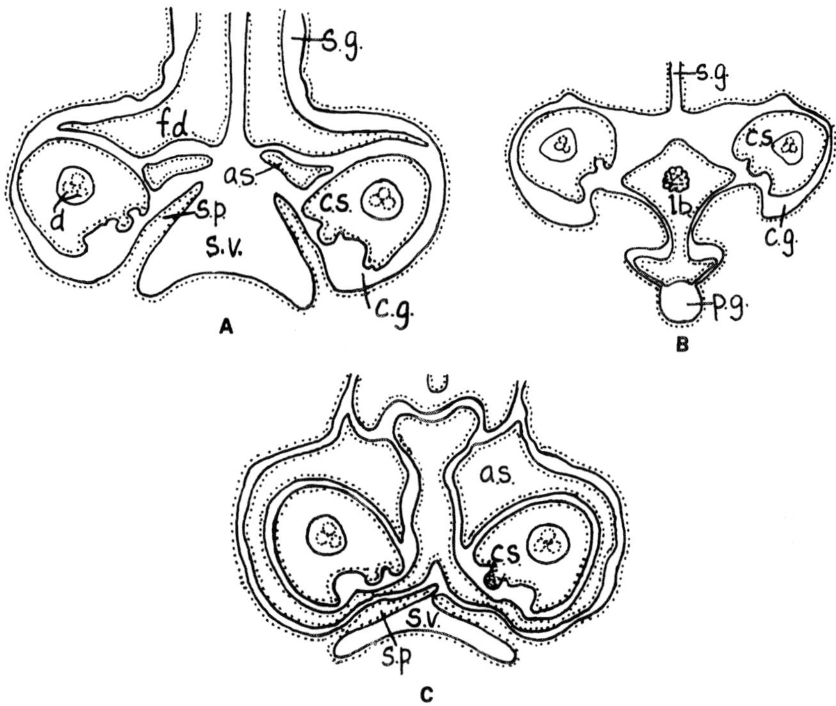

Fig. 6.13. Schematic transverse sections of the cheliceral digits and subcapitulum of *Cenopalpus pulcher* (Prostigmata) (based on Nuzzaci & de Lillo, 1989). **A**, through the region of the median salivary vestibule during the resting or suction stage when the cheliceral digits are retracted; **B**, showing form of the labrum pre-oral groove and cheliceral groove; **C**, through the same region as A but during the phase when the cheliceral stylets are protracted and saliva is being injected into the host plant. *a. s.* = accessory sclerites; *c. s.* = cheliceral stylet; *c. g.* = cheliceral stylet groove; *d.* = dendrites; *f. d.* = fixed digit; *lb.* = labrum; *p. g.* = pre-oral groove; *s. g.* = dorsal subcapitular or rostral groove; *s. p.* = salivary plate; *s. v.* = median salivary vestibule.

laminar element, which during protraction slides into the dorsal subcapitular groove, and a ventral horizontal plate-like portion that slides into the cheliceral-stylet groove. Beneath the fixed digits and between the movable digits where they are joined to the stylophore, lie a pair of accessory sclerites.

It is postulated by Nuzzaci and de Lillo (1989) that during the protraction of the stylets, the accessory sclerites exert pressure on the flexible salivary plates and cause a discharge of saliva from the duct into the channel within the interlocked stylets (Fig. 6.13C). During this phase, the apex of the subcapitulum, provided with ventral and lateral membraneous lobes and sensory organules, is pressed to the surface of the plant surface (Fig. 6.12B,C). After retraction of the stylets, plant juices flow to the surface of the plant and are sucked through the food channel by means of the decompression of the pharyngeal pump. A small orifice, the rostral fossette, on the ventral surface of the subcapitulum communicates with the pharynx through a short cuticle-lined tubule. The walls of the tubule are tricuspid-shaped in cross-section and are considered to form a valve that closes and opens, respectively, with the decompression and compression of the pharyngeal pump. Air bubbles which might be introduced with the food may be eliminated through the rostral fossette without disturbing the flow in the food channel. The basic structure and function of the mouthparts in *Tetranychus urticae* appear to be essentially the same as in *C. pulcher* although there are differences in detail (G. Nuzzaci, personal communication).

Eriophyoid mites feed only on the tissue of plant hosts (Jeppson *et al.*, 1975). The saliva of some species contains growth regulators which when injected into the plant tissue invoke a response in the form of discoloration of the plant or changes in the growth pattern of its affected cells. These are seen in the russeting of mature leaves after heavy feeding and in such growth modifications, initiated on embryonic plant tissue, as the excessive 'hairiness' (erineum) of the leaf surface (Fig. 6.14A) and the formation of galls. The gall-forming species exhibit a high degree of host specificity, particularly on broad-leaved plants. Most galls are hollow structures that are formed by hypertrophied growth around the area punctured by the mite. The majority are found on leaves but flower-, stem- and petiole-galls also occur. They tend to be localized and in their common form appear as hollow projections growing out of the upper surface of the leaf with an opening on the leaf under-surface which allows the mite to enter or leave the cavity of the gall (Fig. 6.14B,C). The common names used to describe the types of galls are based on their shape – for example nail galls, finger galls and bladder galls. Once the mite has initiated changes in plant cells by the action of regulators in its saliva, its feeding is no longer necessary for the development of the gall and it can move on to a new feeding site. Thus, many early formed galls are not occupied by mites until later in the season when the protected brood within the gall feeds on turgid cells and papillae. Gall-forming species, however, are in the minority, most eriophyoids being vagrants on the surface of foliage or living in the galls of other species. Some of the foliar vagrants are capable of living on exposed areas of the leaf surface but others occupy protected niches afforded by the plant, such as grass sheaths and adpressed bases of conifer needles.

Eriophyoid mites, on the whole, cause little or no mechanical damage to plant tissue by their feeding except possibly members of the Aberoptinae (Eriophyidae) which are suspected of leaf mining. In fact, it is to their advantage to keep the host plant alive and so preserve the viability of the succulent tissues on which they feed. Heavy infestations on crop and ornamental plants can cause severe damage which sometimes leads to the death of the plant (Keifer *et al.*, 1982). Usually, however, the damage is less severe and has more of a cosmetic than a harmful effect as, for

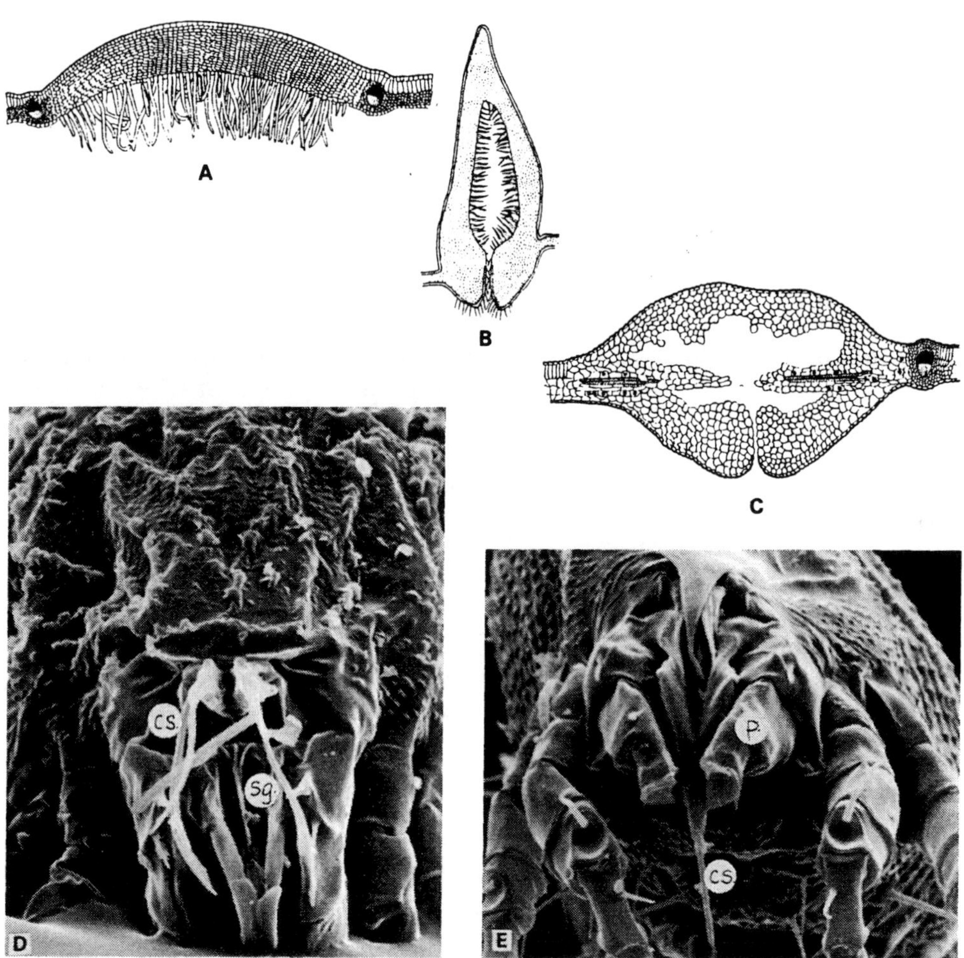

Fig. 6.14. A, perineum produced by an eriophyoid mite; **B**, section of a 'nail gall' on lime produced by *Eriophyes gallarumtiliae* (Prostigmata); **C**, section of a 'blister gall' on walnut produced by *Aceria tristriatus*; **D, E**, SE micrographs of two species of eriophyoid mites in frontal view showing form of the chelicerae, dorsum of the subcapitulum and pedipalps (chelicerae and stylets and dorsal subcapitular groove exposed in D). *cs.* = cheliceral stylets; *sg.* = groove for cheliceral stylets; *p.* = pedipalp.

example, in the production of malformed fruit and the discoloration and the distortion of leaves. Unlike the tetranychoid mites, some species of Eriophyidae are effective vectors of plant viruses (Jeppson *et al.*, 1975).

The mouthparts of the eriophyoids are adapted for piercing the epidermal cells of the host plant and comprise a complex of styli. Keifer (1959) recognizes five stylets in the complex, but Nuzzaci (1979a,b) describes nine. These stylets contain sensory elements (presumably dendrites). The cheliceral bases do not form a protractable stylophore but are much reduced in size. Arising from each cheliceral base is a pair of styli which Nuzzaci (1979b) considers to represent the fixed (dorsal) and movable (ventral) digits of the chelicera (Fig. 6.14D,E). Keifer (1959), on the other hand, recognizes only one pair of cheliceral stylets but observes that if 'the chelicerae are split, which often happens, additional threads will be present'. The styliform labrum (oral stylet of Keifer) is movably hinged to the subcheliceral plate. It is relatively short in the Eriophyidae but particularly long in the Diptilomiopidae. A further two pairs of stylets (the outer and inner subcapitular or infracapitular stylets) appear to be formed from structures associated with pedipalpal apodemes. The outer subcapitular stylets are the auxiliary stylets of Keifer (1959). The subcapitular walls extend dorsally to form a sheath-like structure for the accommodation of the stylets and the labrum. A curved spine, the cheliceral retainer, occurs dorsally on the basal segment of the pedipalp and curves centrally over the chelicerae. The spines of both palps are together thought to keep the chelicerae in position.

The stout pedipalps form a support on either side of the mouthparts and their blunt ends are considered to act as suction pads in anchoring the gnathosoma to the substrate (Fig. 6.14E). Insertion of the stylets into plant tissue is then effected by the force of suction created by the telescoping or deflection of segments in the pedipalpal supports by muscular action. Such pads are not present in some species, for example *Rhyncaphytoptus platani* and *Diptacus hederiphagus*, in which the pedipalps are bent backwards during feeding. These mites attach to the plant surface by their anal region and arch their bodies when feeding. It is thought that this arching of the body plays a part in inserting the stylets (Hislop & Jeppson, 1976). Saliva is injected into the plant along the cheliceral stylets with the assistance of the labrum while plant cell contents are ingested through a canal formed by the labrum and the subcapitular stylets. It is probable that all the styliform elements are introduced into the plant cell during feeding. In the majority of the eriophyoids, the stylets are short and are capable of piercing only epidermal cells but, in the Diptilomiopidae, the exceptionally long stylets reach the outermost cells of the parenchymal layer. The gnathosoma is directed anteriorly in the Eriophyidae but bent downwards in the Diptilomiopidae.

Astigmata

Members of this order are exceptional among the Acari in the extent to which they have successfully exploited a range of habitats, often of a temporary nature, created by other animals. These include nests, burrows, stored food, dung, compost, carrion and various microhabitats on the bodies of animal hosts (OConnor, 1982b). Success in exploiting temporary habitats depends, in part, on the ability of the animal to disperse when physical and/or biotic conditions become unfavourable. This could be a problem in such small animals as mites where walking would be ineffective and random dispersal by air currents hazardous when the food resource is not widely distributed. The Astigmata, however, have solved the problem through the evolution of a particular instar in the life cycle, the deuteronymph (hypopus), for phoresy (see section on Dispersal, Chapter 11). The free-living species are predominantly microphytophagous or saprophagous while those associated with other animals may be either commensals feeding on exudates or epidermal and other debris on the body of the host, or parasites ingesting host tissues and tissue fluids. The majority of the free-living species and commensals and some of the parasites utilize solid food which, like the Oribatida, they comminute before ingestion by means of the combined action of the chelicerae and subcapitular structures.

Some of the saprophagous–fungivorous species may attack living tissues of higher plants; for example, ***Tyrophagus neiswanderi*** feeds on the foliage of greenhouse cucumber plants (Johnston & Bruce, 1965) and species of ***Rhizoglyphus*** have been implicated in damage to healthy bulbs of ornamentals (Michael, 1903; Hodson, 1948), but these are exceptional. Stored food supports populations of free-living Astigmata which are similar in species composition to those occurring in nests of birds and small mammals. On this basis it has been suggested that nests, rather than the soil, form the natural habitat of the majority of these species, the primary source of food in both habitats being relatively dry organic matter as distinct from the saturated detritus in the upper soil layers (Woodroffe, 1953). Under suitable conditions, many of the species multiply rapidly to form extremely dense populations and attain pest status. Among the most important species are ***Acarus siro*** and ***Lepidoglyphus destructor*** in grain (Zdárková, 1967; Cusack *et al.*, 1975), ***Tyrophagus putrescentiae*** in stored food with a high fat and protein content (Hughes, 1976), and ***Lardoglyphus konoi*** in dried fish and salted fish products (Sasa, 1964). Some species of ***Tyrophagus*** are unusual among the Acaridae in feeding on living animal tissues as well as being saprophagous and/or fungivorous. For example, ***T. putrescentiae*** has been found preying on the eggs of a soil-inhabiting nematode (***Diabrotica undecimpunctata howardi***) in peanut and corn agroecosystems in the USA (Brust & House, 1988) and ***Tyrophagus zachvatkini*** feeds on injured and moulting arthropods as well as anhydrobiotic nematodes (Walter *et al.*, 1986).

Free-living species of the family Pyroglyphidae are typically inhabitants of the nests of birds and mammals where they feed on animal detritus and fungi. Some are external 'parasites' of birds and are thought to feed on debris from feathers, while a few live within the quills and ingest the pith of the feather (OConnor, 1982b). Of considerable interest are the pyroglyphids living in human habitations which are commonly referred to as house-dust mites. They occur in dust on the floors of living rooms but the largest populations occur in mattresses, which form the primary breeding sites for the three cosmopolitan species ***Dermatophagoides pteronyssinus, D. farinae*** and ***Euroglyphus maynei***. The total number of mites extracted from mattresses may be as high as 5000 per g of dust (Gridelet-de-Saint-Georges, 1976). The main diet of the house-dust pyroglyphids consists of human skin scales (500–1000 mg produced per person per day) and, to a lesser extent, fungi occurring in the dust. It has been shown that atopic individuals with asthma react positively, in skin tests and broncho-provocation, to extracts of pyroglyphids living in house dust, even at high dilutions (Wharton, 1976). The reaction is of the reaginic type (= IgE immunoglobulins) and the faecal pellets of the mite are the chief source of the allergens. The three species referred to above are the most widespread and important as sources of airborne allergens responsible for the onset of asthma and rhinitis in atopic individuals. *D. pteronyssinus* is the most widely distributed and is particularly abundant in the humid coastal climates of Western Europe and North America, whereas *D. farinae* prefers the drier continental climates of central Europe and central USA. Mites living in very damp human habitations and feeding on moulds, such as *Lepidoglyphus destructor* and *Tyrophagus putrescentiae*, may also be important sources of allergens (Wraith *et al.*, 1979).

The feeding habits of the majority of the astigmatic species associated with vertebrates have not been studied in detail and our knowledge is largely based on those of economic importance which attack man and domesticated animals. Species of the families Sarcoptidae and Psoroptidae are skin parasites of mammals and produce mange. The fertilized female of *Sarcoptes scabiei*, for example, burrows a sinuous channel, up to 2 cm in length, in the corneous layer of the skin where she lays her eggs. In man, the infestation is accompanied by pruritus, produced mainly by allergy, and scratching to relieve the itching is accompanied by scaling and thickening of the skin and secondary infection (Fain, 1978). Species of the genus *Psoroptes* are responsible for mange in sheep and cattle and for ear canker in rabbits. The mites cause severe dermatitis with subsequent scab formation and were thought at one time to pierce the epidermis during feeding. More recent work has indicated that *Psoroptes ovis*, the causative agent of sheep scab, feeds superficially on the outer epidermis, ingesting mainly lipids (Sinclair & Kirkwood, 1983). However, haemoglobin has been found in *P. cuniculi* and *P. ovis* in experimental infestations of the relatively thin skin of the ear of rabbits by Rafferty and Gray (1987). These authors suggest that when first established on rabbits, both 'species' feed on loose stratum corneum and lipid material. As feeding proceeds, antigenic material in the saliva or faeces of the mites causes an inflammatory response in the host skin. Skin breakages in the inflamed areas due to host scratching or to the mouthparts of the mite result in leakage of erythrocytes and serum fluids. The mites feed on this exudate and other surface material. Physical maceration and destruction of the erythrocyte cell wall appears to occur at the labellar-like rutella of the subcapitulum. Both 'species', which appear to be morphologically identical and antigenically similar, cross-mate and produce viable offspring and are capable of surviving and producing reproductive colonies on cattle and sheep (Wright *et al.*, 1983). Thus, it is highly probable that they are conspecific and the observed differences between them in infectivity may be explained on the basis of different strains or races.

Mites of the families Listrophoridae, Atopomelidae, Chirodiscidae and Myocoptidae attach to single hairs of their mammalian hosts either by a clasper formed from membraneous flaps extending from the coxal fields of the first pair of legs (Listrophoridae) or by means of modified anterior legs (Atopomelidae and Chirodiscidae) or posterior legs (Myocoptidae). Members of the first three families are thought to feed on sebaceous secretions and detritus although Hughes (1954a) considers that *Listrophorus* also ingests scales rasped off the hair surface. The Myocoptidae feed at the skin surface while still attached to the hair and may cause a thickening of the skin as well as loss of hair.

The astigmatic mites classified in the superfamilies Pterolichoidea and Analgoidea are associated with birds. Some, such as the Knemidokoptidae and Epidermoptidae, are skin parasites and by their feeding and/or burrowing activity produce pathological effects. Members of the genus *Knemidokoptes*, for example, burrow deeply into the corneous layer of the epidermis and may provoke hypertrophy of this layer leading to a condition described as scaly leg or scaly face according to the site of infestation; species of *Epidermoptes* cause depluming mange in domestic fowl and cage birds. A few are endoparasites and occur within quills of feathers (Dermoglyphidae and Syringobiidae), in lungs and air sacs (Cytoditidae) and under the skin or on the surface of muscles (Laminosioptidae). The large majority, loosely referred to as feather mites, live as external commensals on the wing and tail feathers of their host and feed on the oily secretion covering the feathers, fungal spores trapped in the oil, and feather debris (OConnor, 1982b).

The Anoetoidea differ from other Astigmata in the form of their mouthparts and in their feeding behaviour. They mainly inhabit wet situations rich in rotting organic matter, where they filter microscopic food material by means of the action of their modified palps and specialized non-chelate chelicerae, in which both digits are usually flattened laterally and provided with denticles (Hughes, 1953; Scheucher, 1957). In the other free-living taxa, the subcapitulum is characterized by the presence of a pair of subcapitular rutella, usually with one or more tooth-like projections paraxially, which in conjunction with the chelicerae and labrum mechanically process the food before it enters the mouth (Fig. 6.15A). Although there have been a number of studies of the morphology of the gnathosoma of astigmatic mites, especially of the Acaridae (Akimov, 1985), our knowledge of the function of the various components is incomplete.

The chelicerae move independently during feeding with the chela of the protracting chelicera taking an upper position in relation to that of the retracting chelicera (Fig. 6.15B). Protraction is thought to be entirely by hydrostatic pressure and retraction is by strong muscles inserting by

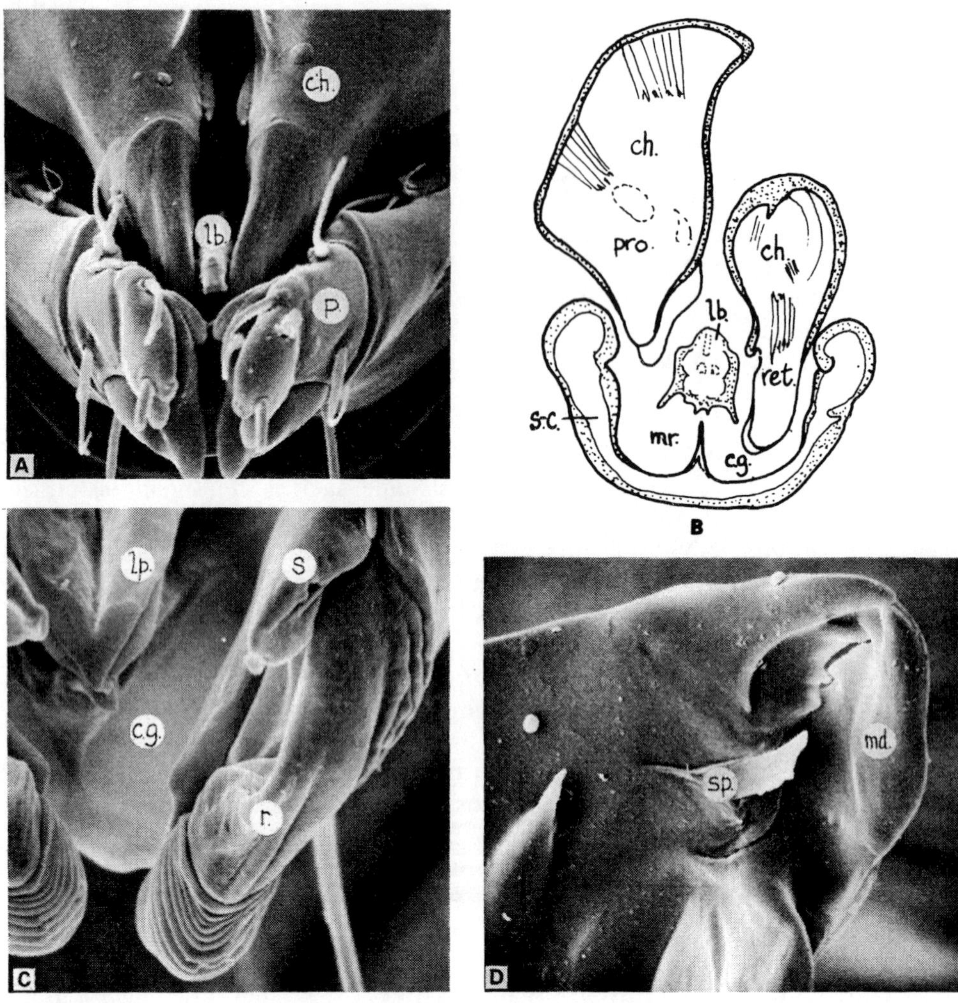

Fig. 6.15. **A**, anterior region of the gnathosoma of *Acarus siro* with the position of the chelicerae in the retracted state; **B**, transverse section of the gnathosoma of *Sancassania* sp. showing the relative position of the chelicerae during their alternate protraction (*pro*) and retraction (*ret*); **C**, anterior region of the subcapitulum of *Lepidoglyphus destructor* illustrating the form of the left cheliceral channel; **D**, paraxial face of the chelicera of *Suidasia medanensis* showing the configuration of the digits and associated structures. *c. g.* = cheliceral channel or groove; *ch.* = chelicera; *lb.* = labrum; *md.* = movable digit; *mr.* = median ridge separating cheliceral channels; *p.* = pedipalp; *r.* = rutellum; *s-c.* = lateral wall of subcapitulum; *sp.* = paraxial chelicral spine.

tendons to the dorsal and ventral posterior margin of the body of each chelicera. In the absence of direct observation, the interpretation of movements of the chelicerae relative to the labrum and sclerotized structures of the anterior region of the subcapitulum is of necessity speculative. During protraction, the digits of the paired chelicerae of *Ac-*

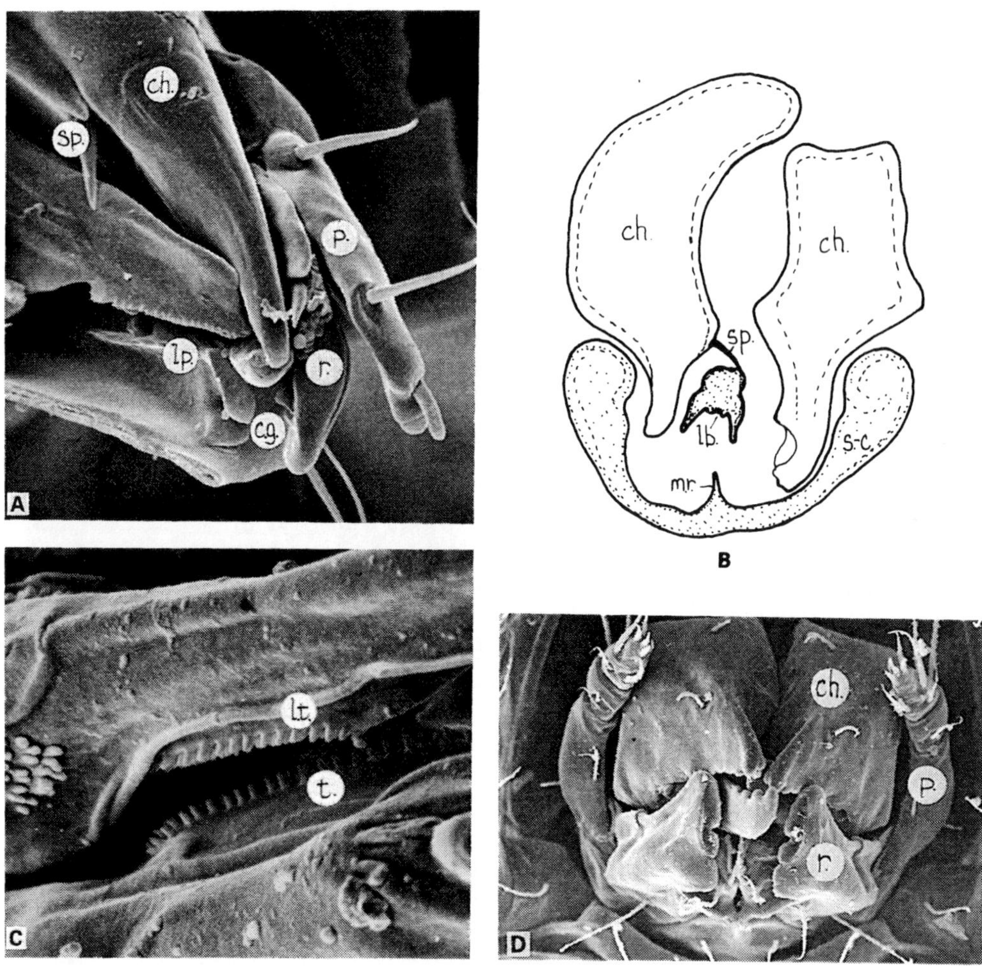

Fig. 6.16. **A**, anterior region of the subcapitulum of *Rhizoglyphus* (Astigmata) showing relationship between the chelicera, paraxial cheliceral spine, labrum, paraxial spine and lateral lips; **B**, transverse section of the gnathosoma of *Sancassania* sp. illustrating the paraxial cheliceral spine and its relationship to the labrum; *C*, teeth on the labrum and floor of the pre-oral cavity in *Carpoglyphus lactis*; **D**, frontal view of the gnathosoma of a phthiracarid mite (Oribatida). *lp.* = lateral lips; *lt.* = labral teeth; *t.* = teeth on floor of pre-oral cavity. Other abbrevations as in Fig. 6.15

arus siro do not appear to move parallel to the longitudinal axis of the body but are directed internally so that the planes of movement of the digits of the two chelicerae cross one another. When the food-bearing chelicera retracts, the movable digit is drawn through the anterior invagination of the hypostome and is thus guided into the cheliceral groove (Figs 6.15C & 6.16A). Food held by the chela meets the rutella and associated teeth-like projections, the prostomal teeth or strigilis, and is either 'torn into smaller pieces' (Akimov, 1979) or reduced in bulk and possibly compacted by the removal of food protruding from the lateral faces of the digit by the shearing action of the rutella. In some species, particularly among the saprophagous forms, the digits are excavated paraxially to form a distinct depression, possibly for the retention of food, and the movable digit appears to be shaped for passing through the hypostomal invagination into the cheliceral groove (Fig. 6.15D). The method by which the food reaches the opening of the pharynx has not been established. Akimov (1979) states that it 'drops' from the chela on to the 'hypopharynx', the floor of the pre-oral cavity, and is swallowed when the labrum is depressed. The relationship between the paraxial spine on the body of the chelicera and the dorsal surface of the labrum in the micrograph of the anterior region of the subcapitulum in *Rhizoglyphus* suggests that the spine may be involved in directing the tip of the elevated labrum against the interdigital region of the protracting chela (Fig. 6.16A). The displacement of the labrum and the close association between the labrum, chela and paraxial spine is also evident in the transverse section of the gnathosoma of *Sancassania* sp. (Fig. 6.16B). It appears as if the lateral displacement of the labrum is brought about by the protracting chelicera and that this displacement might facilitate the deposition of food into the pre-oral cavity by the retracting chelicera. One might also speculate that food deposited in the cheliceral channel is moved towards the pharyngeal opening by the sweeping movements of the digits during successive retractions of the chelicerae. The function of the denticules or pectinate processes situated ventrally on the labrum and the floor of the pre-oral cavity has not been ascertained (Fig. 6.16C). It has been suggested that they form a 'filtration apparatus' which removes excess moisture from the food (Akimov, 1979).

The adaptations of subcapitular structures for the ingestion of different types of food are extensive and some indication of their variety is given by Akimov (1985). Examination of the gnathosoma using scanning electron microscopy reveals a number of microstructural adaptations which are not adequately resolved by light microscopy, especially in their spatial relationship to adjacent structures. This is evident in the micrograph of the anterior region of the subcapitulum of *Lepidoglyphus destructor* (Fig. 6.15C, cf. Fig. 12 in Akimov, 1985). This species is predominantly fungivorous and the rutellum with its grooved pad-like ventral region is probably adapted for dealing with fungal hyphae and spores held in the retracting chelicera. Rutellar adaptations for specialized feeding also occur in parasitic forms and in *Psoroptes* take the form of large labella-like lobes.

The mechanism of food ingestion by rutellate mites is obviously little known. The complex movements of the chelicerae during protraction and retraction and the relationship of the cheliceral digits to subcapitular structures is unlikely to be elucidated until a more functional approach is adopted by those currently working on the descriptive morphology of the

Astigmata and Oribatida. It is surprising that within these taxa, whose exoskeletal morphology has been the subject of such detailed study, our knowledge of the function of components of the gnathosoma in the feeding process is still largely a matter of conjecture.

Oribatida

Oribatid mites attain their maximum population density and species diversity in organic soils and by their feeding activities are considered to influence soil structure by the transmutation of organic matter (Wallwork, 1967; Harding & Stuttard, 1974). Most of the mites inhabit the litter layers or the topmost surface of the mineral soil where the decaying vegetation and the microflora (bacteria, fungi, etc.) associated with its decay provide an abundant source of food. Some species are arboreal and appear to feed on algae and lichens (Aoki, 1973) while some other species, members of the Hydrozetoidea and Ameronothroidea, are secondarily aquatic in littoral or freshwater habitats and probably feed on algae, mosses and higher plants. Much of the information on the food of oribatids is based on laboratory studies or on the examination of gut contents of specimens collected in the field.

The feeding habits of soil-inhabiting species, following the work of Schuster (1956) and Luxton (1972), are considered to comprise six groups:

1. *macrophytophages*, consuming tissues of higher plants;
2. *microphytophages*, feeding on microflora;
3. *panphytophages*, unspecialized plant feeders;
4. *zoophages*, feeding on living animal tissue;
5. *coprophages*, feeding on faeces;
6. *necrophages*, feeding on dead animal material.

With some exceptions, for example, *Orthogalumna terebrantis* feeding on leaf parenchyma of living water hyacinth (Wallwork, 1965) and *Perlohmannia dissimilis* attacking root tissues of potato, strawberry and tulip (Evans *et al.*, 1961), the large majority of terrestrial oribatids classified as macrophytophages ingest only decaying tissues of the plant and are more appropriately classified as saprophages or detritivores. Many species consume the decaying leaf lamina of the plant (phyllophages) and often reduce the leaf to a skeleton while some may ingest woody tissue (xylophages) during some stage of their life cycle and actually burrow into twigs. Oribatids may also show a preference for various kinds of material; for example, *Steganacarus magnus* prefers ash to birch leaves in culture (Murphy, 1955) and the wood of yellow birch to that of hemlock under field conditions (Wallwork, 1967). In the absence of evidence of saprophagy in early derivative groups, Norton (1985) is of the opinion that the saprophagic forms evolved from mycophages which fed on the rich microflora that probably existed in primordial organic soils.

They are often heavily sclerotized and have strongly developed rutella (Fig. 6.16D).

Luxton (1972) considers the Damaeoidea, Oppioidea and possibly the Eremaeoidea to be wholly microphytophagous and to ingest a wide range of fungi, yeasts, bacteria and algae. Some species include pollen as a major part of their diet, for example *Trichoribates setosa* living in flowers and *Saxicolestes auratus, Litholestes altitudinis* and *Zetochestes flabrarius* living on exposed rock faces (Grandjean, 1951c) while lichen is a 'very favourite resort' of such species as *Ommatocepheus ocellatus* and *Mycobates parmeliae* on 'rocks by the seaside' (Michael, 1884). The majority of oribatids inhabiting organic soils are general feeders, the so-called panphytophages, and include microflora and fragments of higher plants in their diet. The wide range of material ingested by these forms is evident from the analysis of food pellets of oribatids living in an acid peat bog by Behan-Pelletier and Hill (1983). The pellets of 15 of the 16 species collected by them contained bacteria and other Monera, fungi, fragments of bryophytes and higher plants, pollen and remains of animals (enchytraeids, collembolans and mites). Only *Chamobates schutzi* appeared to have a relatively restricted diet comprising fungal hyphae and conifer pollen. It was not determined to what extent the arthropod fragments were the result of the ingestion of living prey or the remains of dead animals or exuviae ingested as contaminants of the primary food source. The ingestion of eggs or living tissues of animals is known, however, for a number of oribatid species. Some species ingest 'eggs' from gravid proglottids of anoplocephalid cestodes present in faeces deposited on pastures by the final host and act as intermediate hosts (Segenbusch, 1977). For example, species of *Scutovertex* and *Scheloribates* have been found to be naturally infected with the larvae of *Moniezia expansa* and *Cittotaenia ctenoides* (Stunkard, 1937, 1941). Nematophagy has been reported in a number of Euoribatida (Muraoka and Ishibashi, 1976; Rockett, 1980) and a species of *Pilogalumna* has been observed to feed on injured arthropods and anhydrobiotic nematodes and to be capable of capturing small, slow-moving collembolans (D.E. Walter, personal communication). Cannibalism may occur under conditions of poor nutrition (Stefaniak & Seniczak, 1981). To what extent zoophagy is practised by oribatids under natural conditions has not been ascertained.

The chelicera of the Oribatida is chelate–dentate (more rarely edentate) except in *Gustavia microcephala* which uses the long stylet armed distally with minute teeth and probably representing the movable digit to feed on bacteria and fungal spores (Luxton, 1972). In the chelate forms, the body of the chelicera is either relatively short and compact with the length of the cheliceral body rarely exceeding four times the length of the movable digit or markedly attenuated distally so that the digits are relatively small in comparison with the rest of the chelicera. The latter type occurs, for example, in the Phenopelopidae in which the digits are dentate and in the Suctobelbidae in which they are edentate and scissor-like. Attempts have been made to relate the form of the chelicerae to the feeding behaviour of the mite but the results are somewhat inconclusive. Schuster (1956), for example, used the cheliceral length expressed as a percentage of body length, and the ratio of the length of the cheliceral body to the length of the movable digit as his main criteria. There is a tendency for the length (and to some extent also the breadth) of the chelicera relative to the length of the body to be greater in the macrophytophages than in the microphytophages and the panphytophages, possibly to accommodate the more extensive levator muscles inserting on the well-sclerotized movable digit that produce a stronger bite.

The function of the chelicerae is to cut, tear or crush the food and convey it between the digits to the anterior region of the subcapitulum. It is conjectured, as in the rutellate Astigmata, that the rutella play an important role in the mechanical processing of the food before it enters the pre-oral region of the subcapitulum (Grandjean, 1957c; Dinsdale, 1974a). The atelebasic and pantelebasic rutella, which are present in the macrophytophagous and panphytophagous forms, are flexible structures that diverge as the digits of the protracting chelicera pass between them and converge before the retraction of the food-laden chelicera begins. This flexibility of the rutella is conferred, in part, by the stenarthric or diarthric structure of the subcapitulum. As the chelicera retracts, it is postulated that the digits pass between the rutella and the food protruding on either side of the digits is scraped off so that the amount of food eventually dropping into the pre-oral cavity is of manageable quantity. It is possible that contraction of the cheliceral retractor muscles inserted on the ventral posterior region of the cheliceral body tilts the anterior region of the chelicerae during retraction and forces the food-laden digits between the rutella. The part played by the labrum and Tragardh's organ during the feeding process has not been established. It is interesting to speculate that the latter might function in displacing the labrum during the movements of the chelicerae.

7 Alimentary Canal, Prosomatic Glands and Digestion

Alimentary Canal

The post-oral alimentary canal as in other arthropods is divisible into three primary regions on the basis of embryonic origin. The fore-gut and hind-gut are derived, respectively, from anterior (stomodaeal) and posterior (proctodaeal) ectodermal invaginations and are connected by a mid-gut (mesenteron) of endodermal origin. The differences in embryonic origin are reflected in the different histological structure of the mid-gut compared to the fore-gut and hind-gut. The latter, being invaginations of the body wall, are lined with cuticle and the epithelium which secretes the cuticle resembles that of the integument and is subtended by a thin basement membrane (or basal lamina). Unlike the epithelial cells of the mid-gut, those of the fore- and hind-gut show no signs of secretory or transport functions. The terminology for the components of the alimentary canal has not been standardized and there is considerable confusion in the use of such terms as intestine, rectum and hind-gut, as is evident from a selection of current terminologies presented in Table 7.1. The fore-gut comprises the pharynx and oesophagus, and the mid-gut consists of the ventriculus and a postventricular region which may be clearly divided into an anterior colon and a posterior post-colon (Fig. 7.1). When the postventricular region of the mid-gut is undivided, then it is referred to herein as the postventricular mid-gut. The use of the terms 'hind-gut' and 'excretory organ' for the entire postventricular region of the alimentary canal in the Prostigmata is confusing and should be avoided. Paired excretory or Malpighian tubules, when present, open into the postventricular region of the mid-gut and are not of proctodaeal (ectodermal) origin as in the Myriapoda/Insecta. The proctodaeum as in other Arachnida is not deep and consequently the hind-gut is relatively short and forms only the anal atrium. The ventriculus is usually provided with one or more pairs of caeca or diverticula whose functions are to increase its capacity for the reception of food and its surface area for intracellular digestion. The alimentary tract is supported anteriorly by the pharyngeal muscles

Table 7.1. A comparison of the terminology used by various authors for the components of the alimentary tract.

This text	Mesostigmata[1]	Ixodida[2]	Astigmata[3]	Oribatida[4]	Prostigmata[5]
Pharynx	Pharynx	Pharynx	Pharynx	Pharynx	Pharynx
Oesophagus	Oesophagus	Oesophagus	Oesophagus	Oesophagus	Oesophagus
Ventriculus	Ventriculus	Mid-gut	Stomach	Mesenteron	Ventriculus
Colon*	Colon	Small intestine	Colon	Colon }	Hind-gut or excretory organ
Postcolon*	Rectum	Rectal sac	Postcolon	Rectum }	
Anal atrium	Anal atrium	Rectum	Rectal sac	Anal atrium	Rectum

[1]Woodring & Galbraith (1976); [2]Balashov (1968); [3]Hughes (1959); [4]Woodring & Cook (1962a); [5]Alberti & Crooker (1985).
*When the colon and postcolon are not distinguishable, the region between the ventriculus and anal atrium is referred to as the postventricular mid-gut.

and the synganglion, and posteriorly, in some taxa, by muscles supporting the anal atrium.

The pharynx, oesophagus, ventriculus and anal atrium are stable components of the alimentary canal and the main modifications affect the ventricular caeca and the postventricular mid-gut. In the Tetranychidae, for example, the latter forms an excretory organ which has a narrow connection with the ventriculus but a broad connection with the anal atrium (Blauvelt, 1945), while in the Parasitengona the connection between the ventriculus and the postventricular mid-gut (excretory organ) is lacking (Hughes, 1959; Mitchell, 1964, 1970). The components of the alimentary canal in some of the major acarine taxa are shown diagrammatically in Fig. 7.2A–E.

Fore-gut

The mouth, overhung by the labrum, leads into the pharynx, which is a strongly muscular region of the fore-gut that functions as an effective sucking and/or swallowing organ. The outline in cross-section of the constricted pharyngeal lumen in the Anactinotrichida resembles a Y (Fig. 7.3A) or two Ys connected at their base or, as in the argasoids and some species of *Ixodes*, a triple Y formed by the subdivision terminally of each arm of the Y (Fig. 7.3C). Dilator and constrictor muscles alternate along the length of the pharynx. The dilator muscles comprise:

1. The lateral dilators originating on the lateral wall of the gnathosomatic base and inserting on the lateral walls of the pharynx.
2. The dorsal dilators originating on the subcheliceral plate or on the lateral walls of the gnathosomatic base and inserting on the dorsal arms of the pharynx.

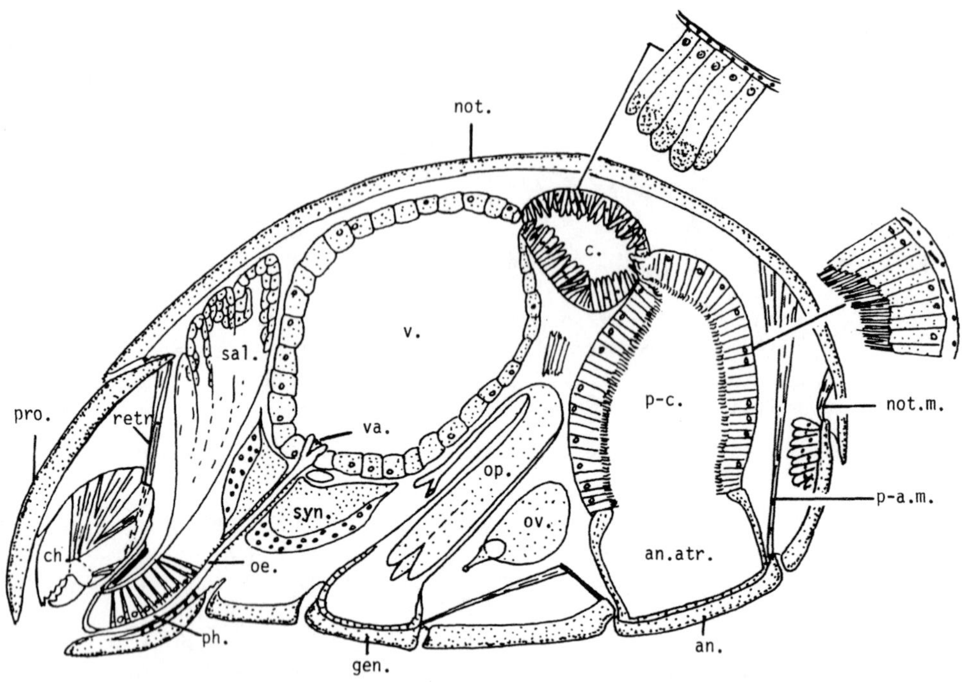

Fig. 7.1. A schematic longitudinal section of a euoribatid mite (*Euzetes*) showing the form of the alimentary tract (based on Hoebel-Mävers, 1967). *an.* = anal plate; *an. atr.* = atrium, *c.* = colon; *ch.* = chelicera; *gen.* = genital plate; *not.* = notogaster; *not. m.* = notogastral muscle; *oe.* = oesophagus; *op.* = ovipositor; *ov.* = oviduct; *p-a. m.* = posterior anal muscle; *p-c.* = postcolon; *ph.* = pharynx; *pro.* = prodorsum; *retr.* = retractor muscle of the chelicera; *sal.* = salivary gland; *v.* = ventriculus; *va.* = valve.

3. The ventral dilators, present in the double-Y condition, originating on the ventral or ventrolateral wall of the gnathosomatic base and inserting on the ventral pharyngeal wall.

Constrictor muscles are arranged so that they connect the arms of the pharynx. Dilation of the pharynx results in an inverted triangular or rectangular-shaped lumen (Fig. 7.3B). Grandjean (1936b) considered the wall of the pharyngeal lumen to describe a number of lips whose points of contact, the commissures, are at the extremities of the arms. Thus the Y-shaped lumen has three lips and the double-Y type four lips. The dorsal lip extends beyond the mouth as the labrum (=epipharynx). This type of pharynx forms an efficient sucking–swallowing organ and is characteristic of predators and parasites ingesting fluid or finely particulate food and of microphages consuming fungal hyphae, spores or pollen grains.

In contrast to the relatively narrow pharyngeal lumen of the Anactinotrichida, that in the Astigmata and Oribatida is extremely wide and capable of considerable dilation. Michael (1901) aptly describes the floor and roof of the pharynx in the Astigmata as 'two more-or-less chitinised

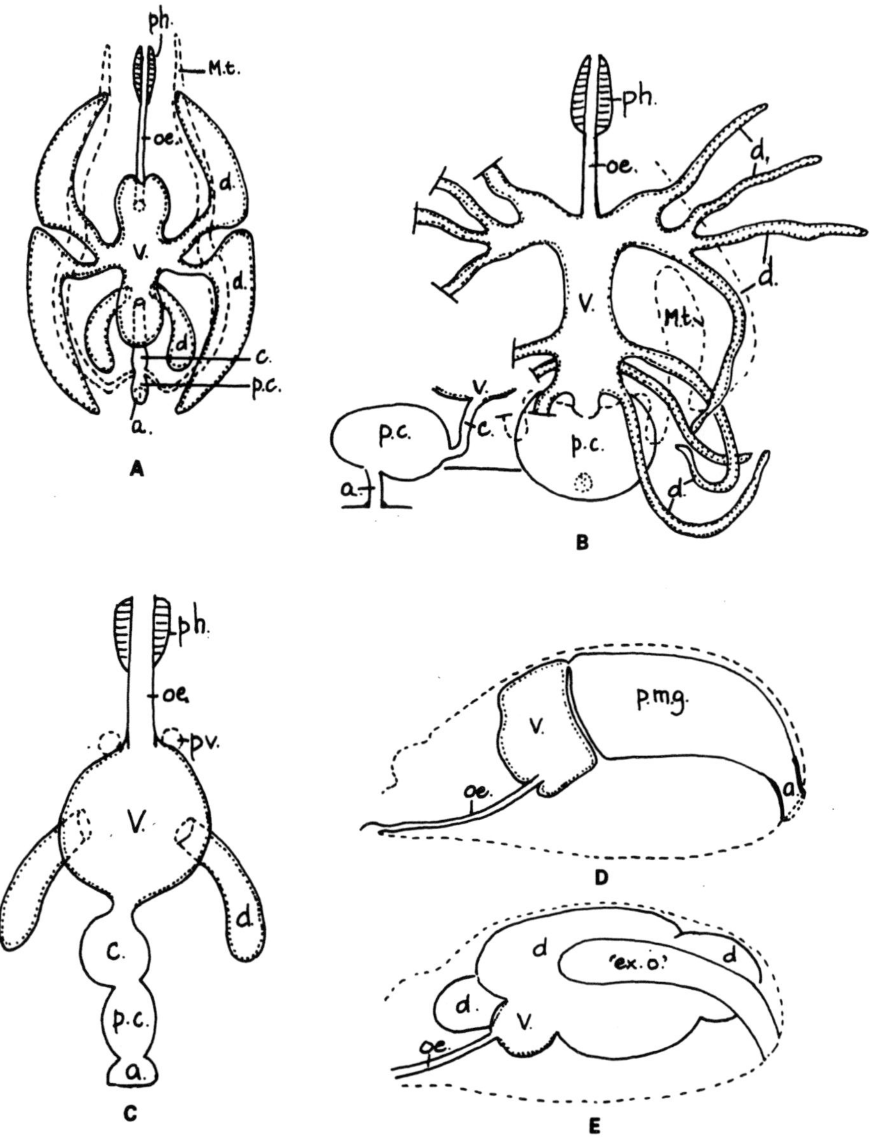

Fig. 7.2. Diagrammatic representation of the major types of alimentary tract in the Acari. **A**, dermanyssid type (Mesostigmata); **B**, ixodid-type (Ixodida); **C**, oribatid-type (Oribatida); **D**, tetranychid-type (Prostigmata); **E**, trombiculid-type (Prostigmata). *a.* = anal atrium; *c.* = colon; *d.* = diverticulum (caecum); *'ex. o.'* = excretory organ; *M. t.* = Malpighian tubule; *oe.* = oesophagus; *p. c.* = postcolon; *ph.* = pharynx; *p. m. g.* = posterior mid-gut; *pv.* = preventricular gland; *v.* = ventriculus.

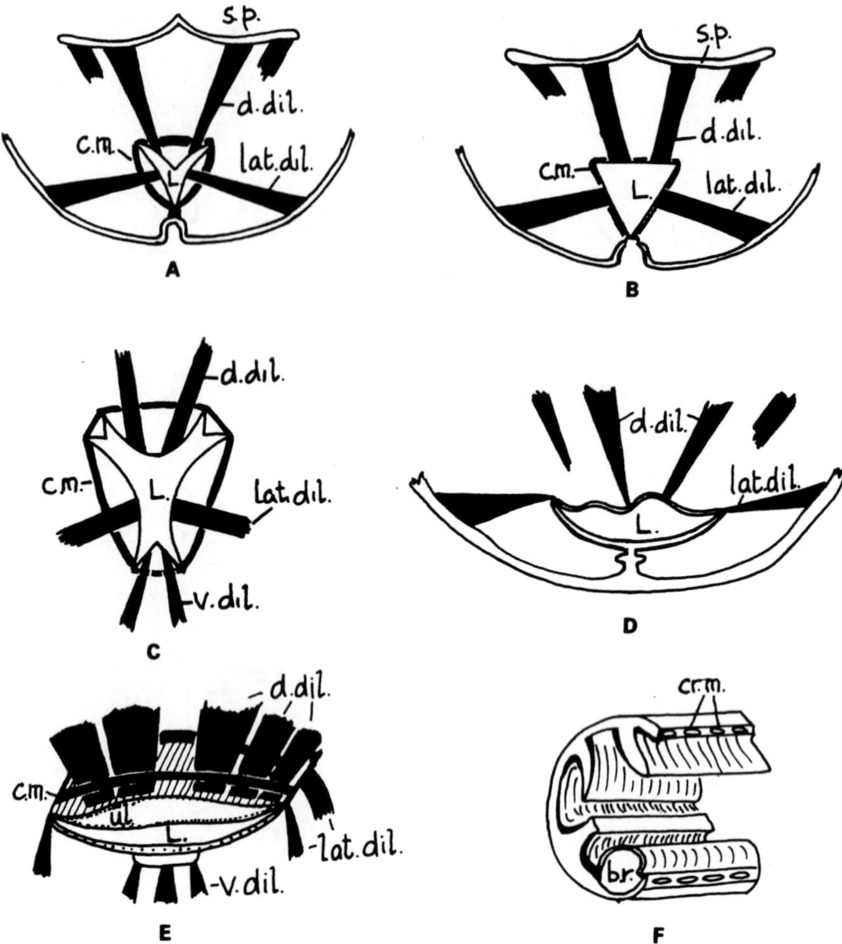

Fig. 7.3. **A–E**, diagrammatic representation of the pharynx and its musculature: **A, B**, in a dermanyssid mite with the pharynx in the constricted (**A**) and dilated (**B**) state; **C**, ixodid tick; **D**, *Caloglyphus* sp. (Astigmata); **E**, euoribatid mite (based on Hoebel-Mävers, 1967). **F**, section of the oesophagus of an oribatid (based on Hoebel-Mävers, 1967). *b. r.* = basal ridge; *cm.* = constrictor muscles; *cr. m.* = circular muscles; *d. dil.* = dorsal dilator muscles; *lat. dil.* = lateral dilator muscles; *L.* = lumen of pharynx; *s. p.* subcheliceral plate; *ul.* = upper lip of pharynx; *v. dil.* = ventral dilator muscles.

half-tubes ... laid one inside the other; both having their convexity downward' (Fig. 7.3D). The lower half-tube is the more heavily sclerotized and either fixed along part of its length with the ventral wall of the subcapitulum or attached to very short ventral muscles. This gives stability to the pharynx. Dilation of the lumen is by the action of muscles inserted on the flexible upper half-tube and originating on the subcheliceral plate. Constriction is achieved by short, transverse constrictor (=occlusor) muscles passing across and above the pharynx and attaching at each end to the edge where the two half-tubes meet. In *Sancassania* muscles originating on the lateral walls of the subcapitulum and inserting on the edges of the flexible roof of the pharynx may also play a role in occluding the pharynx (Fig. 7.3E). A similar type of pharynx also occurs in the Oribatida (Woodring & Cook, 1962a; Hoebel-Mävers, 1967) but in most cases its floor appears to be thinner than the roof and the organ is considered to be kept in place by a ventral connective tissue block attached by short, thick muscles to the floor of the subcapitulum. The pharyngeal lumen of the Prostigmata is also in the form of a relatively broad transverse slit (Henking, 1882; Michael, 1896; Reuter, 1909; Witte, 1978). The downward convexity of the dorsal and ventral walls is often more pronounced than in the Sarcoptiformes. The ventral wall of the pharynx may be connected to the floor of the subcapitulum by a sclerotized bridge, as in the adult of *Erythraeus*. Alternating dilator and constrictor muscles are typically present although constrictors are stated to be absent in the erythraeid *Abrolophus* (Witte, 1978). Specialization of the pharynx for strong pumping action occurs in the phytophagous Tetranychoidea (Blauvelt, 1945; Akimov & Yastrebtsov, 1981). In *Tetranychus*, the dorsal wall of the pharynx is provided with a strong cuticular plunger on which dilator muscles originating on the sigmoid piece and the wall of the infracapitulum are inserted. No transverse constrictor muscles are present but it is considered that elasticity of the cuticular structures of the pharynx may work antagonistically to the dilators (Alberti & Crooker, 1985).

The pharynx merges smoothly with the oesophagus, which is usually subcircular in cross-section. It passes through the synganglion and in the Tetranychidae also through the salivary glands (*Bryobia*) or the silk glands (*Tetranychus*). The oesophagus may be expanded posteriorly in some Oribatida, for example *Damaeus onustus*, to form a distinct crop or ingluvies (Michael, 1884). The thin cuticular lining is often extensively folded longitudinally to form ridges. These are particularly evident in the Euoribatida and in *Euzetes* seven such folds or ridges are present (Fig. 7.3F). The ridges may play a role in mincing or dividing the ingested mass of particulate food. In ticks, the thin cuticular lining consists of fibrillar procuticle and a thin layer of epicuticle (Raikhel, 1979a). Bands of circular muscles separated from the epithelial cells by the basement membrane usually run the length of the oesophagus and food passes along it by peristalsis. The posterior end of the oesophagus, often distinctly widened, extends a short distance into the ventricular lumen and forms a ring-like fold which may be provided with constrictor muscles. This structure, called the proventricular valve in ticks, is considered to prevent the regurgitation of food from the ventriculus into the oesophagus. The oesophagus enters the anterior end of the ventriculus ventrally or centrally. The oesophagus of the Bdellidae is very different from other Acari in that a large stalked diverticulum arises from its dorsal wall immediately behind the pharynx. The diverticulum is provided with circular and longitudinal muscles and is capable of considerable expansion. Michael (1896) named this structure the receptaculum cibi and considered it to be a food reservoir but the structure, subsequently called the supra-oesophageal organ, is now known to produce a stringy exudate used in capturing prey and building moulting nests (Alberti, 1973).

Mid-gut

The ventriculus in the Notostigmata, Mesostigmata and Ixodida is provided with one or more pairs of caeca (Fig. 7.2A,B). In the Notostigmata the sac-like ventriculus is considerably larger than its weakly developed pair of caeca whereas in the other two groups the narrow ventriculus is usually relatively much smaller than the caeca (Jakeman, 1961; Balashov, 1968). Seven pairs of ventricular caeca are considered to be the basic complement in the Ixodida and have been shown to be present in the embryo of *Ornithodoros moubata* by Aeschlimann (1958). Deviations from this number are probably due to secondary variations. The caeca in the Argasoidea are usually shorter than in the Ixodoidea and have a tendency for secondary divisions. Those of the ixodoids are more slender, less often divided secondarily and tend to curve ventrally and form loops. Among the Mesostigmata, three pairs of caeca are typically present in the Dermanyssina and Parasitina, comprising two dorsal pairs, one directed anteriorly and the other posteriorly, and a smaller ventral pair. The latter pair may be absent in males as, for example, in *Haemogamasus hirsutus* (Michael, 1892). Three pairs of caeca are stated to be present in the Uropodina and consist of two anterior pairs having a common origin on each side of the ventriculus, and a posterior pair (Winkler, 1888; Michael, 1889; Woodring & Galbraith, 1976). The two anterior pairs may be formed by subdivision of the anterior pair of caeca, in which case only two of the three pairs of caeca present in the Dermanyssina are represented, the ventral pair being absent. The ventriculus in the *Leiodinychus* is exceptional in that it is considerably larger than the short, broad caeca. Those species which ingest large quantities of their host's tissue fluids, such as some parasitic members of the Dermanyssoidea and the Ixodida, have caeca actually extending into the leg cavities. In *Spinturnix myoti*, in which the body is compressed dorsoventrally, the large anterior and posterior caeca extend into the cavities of legs I and IV respectively, while in argasoids intrusion of the caeca into larval and nymphal appendages is not uncommon.

A pair of globose 'digestive glands', each made up of four to six giant cells, open by way of short ducts into the posterior-ventral surface of the posterior caeca in *Uroactinia agitans* (Woodring & Galbraith, 1976). The function of the glands is unknown but a digestive function is thought to be likely. Other structures associated with ventricular-caecal epithelium have also been described in the Mesostigmata. Hughes (1952) referred to modified cells in the dorsal wall of the ventriculus of *Ornithonyssus bacoti* which contain what appeared to be microorganisms similar to those found in the mycetomes of lice. Mycetome-like bodies have also been found in *Haemogamasus ambulans* (Young, 1968a). The function of these is unknown. Single or doublet 'giant cells' embedded in discrete areas in the mid-gut. have been described in the Macrochelidae and Parasitidae. Bowman (1984) called these 'Coons cells'. In *Pergamasus* they are packed with acidophilic bodies resembling small red blood corpuscles. Their function is unknown but Bowman suggests that they may be microbial in nature.

Two caeca are present in the Oribatida and Astigmata (Fig. 7.2C) and they are usually small in relation to the large sac-like ventriculus which in some Oribatida occupies about two-thirds of the length of the hysterosoma near the dorsal surface (Michael, 1884, 1901). The number and form of ventricular caeca in the Prostigmata is considerably more variable than in other Actinotrichida. This is particularly the case in the Parasitengona where the extensive caeca are often divided into lobes or pouches by the dorsoventral muscles of the hysterosoma (Bader, 1954, 1969; Mitchell, 1964). In the Tetranychidae, two large caeca extend posteriorly from the ventriculus and are separated along the mid-line by the postventricular gut. These caeca are also indented by dorsoventral muscles to form pouches (Alberti & Crooker, 1985).

The ventriculus and its caeca are the main sites of digestion and absorption. Their walls are formed by a single layer of epithelial cells resting on a thin basement membrane. Light and electron microscope studies have shown that the epithelium in the Ixodida and Mesostigmata consists basically of three cell types: reserve (undifferentiated) cells, secretory cells and digestive cells (Hughes, 1954b; Raikhel, 1979a). The characteristics and function of the cell types are discussed below (p. 233). In the Mesostigmata and Ixodida, the epithelial cells and basal lamina of the ventriculus and caeca are enmeshed by a discontinuous network of circular and longitudinal muscles. A muscular layer is also present in the Tetranychidae and Bdellidae (Alberti, 1973), the Oribatida (Michael, 1884; Hoebel-Mävers, 1967) and the Astigmata (Prasse, 1967a).

A pair of preventricular glands located on the ventriculus near the point of entry of the oesophagus occur in the oribatids (Fig. 7.2C). They appear black when seen through the transparent cuticle of nymphs but when dissected out are found to vary in colour from deep yellow to red and dark brown. Michael (1884) considered that they opened by ducts into the ventriculus. Woodring and Cook (1962a) state that these glands in *Ceratozetes cisalpinus* are little more than a few enlarged cells of the ventriculus which lack nuclei and are full of greenish hyaline granules. The function of the preventricular glands is not known.

The movement of the contents of the ventriculus into the remainder of the mid-gut is controlled by a valve (pyloric sphincter). The post-ventricular region of the mid-gut in the Mesostigmata, Ixodida, Astigmata and Oribatida comprises two distinct regions, an anterior colon and a posterior postcolon. *Ornithodoros moubata* appears to be exceptional among the Ixodida in that the colon terminates blindly and its contents cannot be voided in the form of faeces (Burgdorfer, 1951). Sphincter muscles acting as a valve occur between the colon and postcolon and may also be present between the postcolon and anal atrium as in the Mesostigmata. The epithelium of the colon and postcolon is subtended by a basal lamina and surrounded by a plexus of longitudinal and circular muscles. The epithelial cells are columnar or cuboidal depending on the degree of distension of the colon or postcolon. Those of the postcolon in the Astigmata and Oribatida have a deeply striated border. Exceptions occur in *Listrophorus leuckarti* and *Dermatophagoides farinae* in which the 'postcolon' is lined with thin cuticle and may be considered to be of proctodaeal origin (Hughes, 1954a; Brody *et al.*, 1972). In argasoids the postcolon divides posteriorly into two lobes which are conspicuous when the postcolon is distended.

A pair of excretory tubules, the Malpighian tubules, involved in the excretion of the final products of nitrogen metabolism, open into the postventricular region of the mid-gut in the Anactinotrichida and some Astigmata (Fig. 7.2A,B).

Many of the Prostigmata are exceptional in that the postventricular region of the alimentary canal has lost its connection with the ventriculus (Reuter, 1909). This condition does not apply to all prostigmatic mites. A continuous mid-gut has been established in the Eriophyoidea (Nalepa, 1887), Tetranychidae (Blauvelt, 1945), Bdellidae, Erythraeidae and Rhagidiidae (Alberti, 1973; Ehrnsberger, 1981) and Nicoletiellidae (Vistorin, 1980). In the Trombiculidae and Trombidiidae, on the other hand, the ventriculus ends blindly and with the severance of the connection between it and the postventricular mid-gut, the latter has lost the function of transporting food residues but retained its excretory function. It is on this basis that many authors refer to the entire postventricular region of the alimentary tract in this condition as the 'excretory organ'. Mitchell (1970), however, does not consider the 'excretory organ' in the Trombiculidae to represent the combined postventricular mid-gut and hind-gut. He postulates that in the course of evolution of trombiculid mites the 'hind-gut' (including the postventricular mid-gut) was lost and only the excretory tubule retained. Mitchell's postulate seems to be an unnecessary complication when the widespread dual function of food-residue transportation and excretion by the mid-gut is well established in the Actinotrichida. Excretion by the mid-gut predominates even in those species of Astigmata in which the excretory tubules are present but appear to have largely lost their excretory function (Hughes, 1959). It seems more likely that an excretory function of the mid-gut, in the absence of excretory tubules, was established in the Actinotrichida before the severance of the mid-gut and that the postventricular mid-gut required little modification to carry out its present function in those species without a continuous alimentary canal.

Hind-gut

The short cuticle-lined hind-gut forms the anal atrium which connects the postcolon to the anal opening. The latter is usually in the form of a longitudinal slit. The walls of the anal atrium are often plicate. In the Uropodina, the lateral plications or folds extend through the anal orifice and become thickened at their tips to form a pair of anal valevs, while in *Tetranychus* the folds interdigitate distally and thereby close the anus (Woodring & Galbraith, 1976; Alberti & Crooker, 1985). The anal opening in the ticks and mesostigmatic mites is surrounded by an area of sclerotized cuticle which forms an anal ring. Muscles originating on the dorsal shield and inserting on the atrial wall and on the anal valves are present in the Mesostigmata. Those inserting on the atrial wall are considered to act as supporting muscles to prevent the collapse and extrusion of the thin-walled atrium. A pair of muscle bundles extend from the atrial wall to the anal plate in the Ixodida and on their contraction the anal slit opens and the lower part of the atrium slightly everts to facilitate defecation. An atrial muscularis is lacking in these two orders and egestion is probably accomplished by the contraction of the postcolon muscularis which forces the food residue through the atrium. However, muscles underlie the atrial epithelium in the Oribatida and in *Tetranychus* and probably push the excrement through the anus. Forms ingesting solid food and producing boluses in the mesenteron, such as the Oribatida and free-living Astigmata, have large anal orifices. In adult Oribatida, the musculature of the atrial region consists of a depressor system of muscles, comprising the muscularis (circular and longitudinal muscles) of the anal atrium and dorsoventral muscles originating on the dorsal body wall and inserting on the ventral plate, which squeezes out the faecal pellet, and a levator system of muscles which raise and close the pair of large anal plates (Hoebel-Mävers, 1967). Immature oribatid stases lack discrete anal plates and the anal slit is opened by muscles but closed by cuticular tension (Woodring & Cook, 1962a).

The opening of the hind-gut in those Prostigmata in which the ventriculus ends blindly is often referred to as the uropore to distinguish it from the anus of forms with a continuous mid-gut. The structures are homologous and in this work the term anus is used for the opening of the hind-gut irrespective of whether or not the mid-gut is continuous. The postventricular mid-gut in free-living forms, particularly in the weakly sclerotized Prostigmata, is often conspicuous as a white unpaired median structure of the opisthosoma; the colour is due to nitrogenous waste products, chiefly guanine, in its lumen. The posterior opening of the alimentary canal is usually situated posteroventrally, but it may be located terminally as in the Notostigmata and the males of the Tetranychidae, or dorsally on a tubercle as in *Penthaleus*. According to Desch (1988), a hind-gut is lacking in *Demodex*. A short cuticle-lined pouch opening ventrally or subterminally on the opisthosoma of female demodicids was considered, provisionally, to be a proctodaeal structure by Desch *et al.* (1971). Its function is not known.

Prosomatic Glands

The glands located in the prosomatic region of the body of the Acari show considerable variation in number and location. Except in the Ixodida and some Sarcoptiformes, little is known of their function. The majority of the glands discharge their contents, by way of ducts, either into the buccal/subcapitular region of the gnathosoma or on or near the coxae of the first pair of legs. The terminology for individual, groups or the entire complex of glands is various and often confusing. For example, the term salivary glands is restricted by some authors to describe only those glands opening into the buccal region whereas others include the entire complex under this name. In this work, the term salivary glands is used in the restricted sense.

Anactinotrichida

Ixodida

The paired salivary glands are the most constantly occurring exocrine glands in the Anactinotrichida. The multicellular acini of each gland in the Ixodida open by short efferent ducts into all parts of the duct system (Till, 1961; Chinery, 1965; Balashov, 1968; Coons & Roshdy, 1973; Binnington, 1978). In the Argasoidea, the glands are compact and situated above the coxae of the first three pairs of legs and their size is similar in unfed and recently engorged forms. The glands are more extensive in the Ixodoidea, especially the Amblyommidae in which they may extend posterior to coxae IV, and tend to be divided into separate grape-like clusters (Fig. 7.4A). Their size appears to be determined by the physiological condition of the tick and they are larger in feeding than in unfed instars. The cells of the salivary glands comprise non-granular cells which form type-I acini (pyramidal alveoli), granular cells (secretory alveoli) and interstitial epithelial cells. The boundaries between the individual cells comprising type-I acini are hardly distinguishable by light microscopy but morphologically distinct cell types have been defined using transmission electron microscopy (Krolak *et al.*, 1982). The acini are pyramidal, pyriform or irregularly polygonal in shape (Fig. 7.4B). The basal region of the cells in type-I acini in the male of *Dermacentor variabilis* contains complex membrane infoldings extending from the outer cell membrane and enclosing numerous mitochondria and vacuoles (Coons & Roshdy, 1973). Infoldings are fewer in the intermediate region but the vacuoles are more numerous and free ribosomes, mitochondria and some microtubules are present. The apical region of the cells has few or no infoldings and contains some mitochondria, many microtubules, scattered vacuoles and electron-dense bodies. In general, type I acini are found in the anterior two-thirds of the gland and in *A. persicus*, for example, the acini form a compact mass along the dorso-mesial side of each gland and this

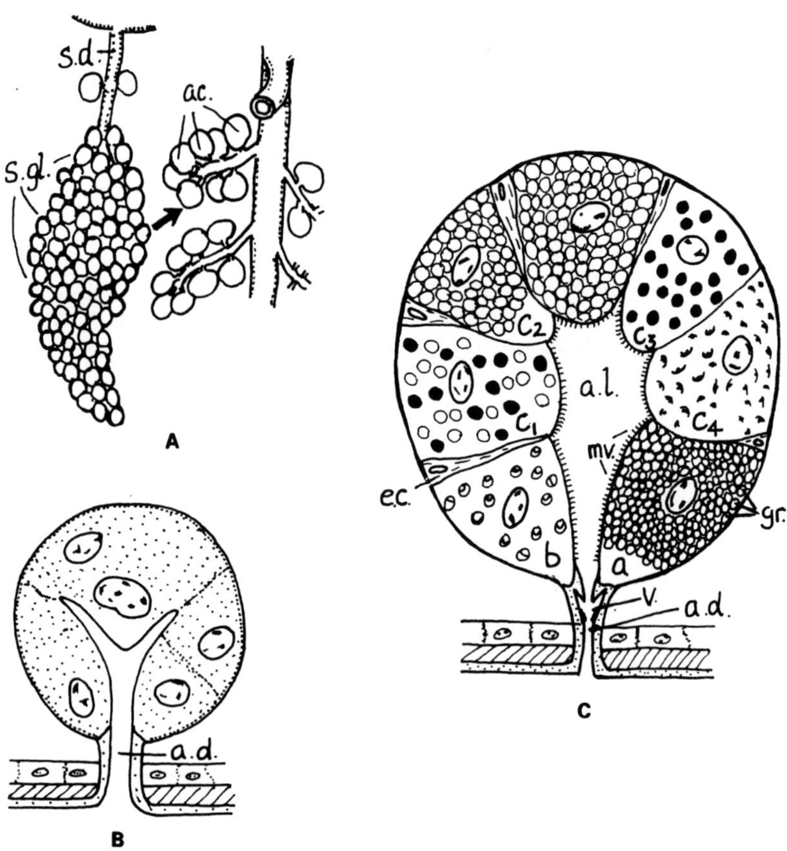

Fig. 7.4. The structure of the salivary glands in ticks. **A**, schematic representation of the form of a salivary gland in ixodoid ticks; **B, C**, acini type I (**B**) and type II (**C**) in the newly attached amblyommid tick (based on Binnington, 1978). *a.* = cell type a; *ac.* = acini; *a. d.* = acinar duct; *a. l.* = acinar lumen; *b.* = cell type b; c_1–c_4 = cell types c; *gr.* = granules; *e. c.* = epithelial cell; *mv.* = microvilli; *s. d.* = salivary duct; *s. gl.* = salivary gland; *v.* = valve.

mass is clearly distinct from the remainder of the gland comprising type II acini.

The granular cells in the Argasoidea form type II acini and consist of three cell types in *Argas persicus* (Roshdy, 1972), *A. arboreus* (Guirgis, 1971), *Ornithodoros tholozani* (Balashov, 1968) and *O. kelleyi* (Sonenshine & Gregson, 1970). In *A. persicus*, small cells (type **a**) situated near the duct of the acinus stain positively for basic proteins but negatively for carbohydrates, whereas type **b** cells are weakly and type **c** cells strongly positive for carbohydrates. Some authors consider that the three cell types are only different developmental stages of one cell type. In the Ixodoidea, with the exception of *Ixodes ricinus* which according to Balashov (1968) has only one type, two types of granular acini (types II and III) occur in both sexes and a third type (type IV acinus) is present in some males. As in the argasoids, different cell types have been found in the granular acini (Fig. 7.4C). Type I acinus of the unfed male of *Dermacentor variabilis*, according to Coons and Roshdy (1973), has type **a** cell at the base and the cell is packed with spherical granules and contains abundant rough endoplasmic reticulum (RER), free ribosomes and a few mitochrondria. The large single **b** cell contains loosely packed granules, compact RER, dense free ribosomes and aggregated mitochondria basally and between the granules. The type **c** cells are divided into three subtypes on the basis of the size and electron density of their granules. These authors also recognize three granular cell types (**d**, **e**, **f**) in the type III acinus although only two types of cells were described for this type of acinus in *Haemaphysalis* and *Hyalomma* by Chinery (1965) and Balashov (1968), respectively. A cuticular valve occurs between the narrow canal leading from the lumen of the acinus to the acinar duct in acini types II and III. A valve is absent in this position in the granular acini of the argasoids. Binnington (1978) distinguished nine different granular cell types in the female (including an additional **c** cell, $\mathbf{c}_4$) and ten in the male of *Boophilus microplus*, and this seems to indicate that the number of cell types may have been underestimated. On the other hand, Kemp *et al.* (1982) point out that some of the histological differences between the granular cell types of ixodoid salivary glands seem to be minor, and that structural and functional similarities are emerging as the result of more detailed studies.

Binnington (1978), in his investigation of the salivary glands of *B. microplus*, considered that in the female cell type **a** of acinus II and cell types **d** and **e** of type III acinus (all situated close to the acinar duct) contained attachment cement precursors. These cells reacted strongly for phenol oxidase, which suggested that a tanning process similar to that in insect cuticle may occur in the cement. Reactions to acetylcholine esterase and a type-B carboxyl esterase were found in cell type $\mathbf{c}_1$ of acinus type II, and also to type-B carboxyl esterase at a lower intensity in acinus type III. Other enzymes present in the granules and/or the cytoplasm of the acini included a protease, leucine aminopeptidase, monoamine oxidase and phosphatases. In the Amblyommidae, type I acini are located along the anterior region of the gland, type II in the anterior two-thirds and type III along the entire gland.

Changes occur in the morphology of the granular cells in the course of attachment and feeding. Attachment to the host appears to initiate the production of granules in some of the cells, as in the case of $\mathbf{c}_2$ and **f** in *Boophilus microplus* (Binnington, 1978). Other cells, especially the putative cement precursor cells, are filled with secretory granules prior to attachment of each instar but have completed their function (secreted their granules) within 72 hours of attachment.

The epithelial cells of acini II and III are located between secretory cells and act as interstitial cells (Fig. 7.4C). Their low-density cytoplasm has few mitochondria and other organelles and no secretory granules.

The apical plasma membrane has numerous microvilli but the basal membrane is not invaginated. The cells join adjacent secretory cells by septate junctions. During feeding in ixodoid ticks, the epithelial cells of type III acini undergo considerable change. They enlarge and a canal system and fold appear in the transparent cytoplasm. The system of canals, formed by basal plasma membrane invagination, perforates the cell cytoplasm up to the apical surfaces. Similar but less pronounced changes also occur in type II acini. Meredith and Kaufman (1973) suggested that in *Dermacentor andersoni* these cells, referred to as 'water cells', are candidates for fluid secretion from the haemolymph to the saliva during feeding. Interstitial epithelial cells are also present in *Ornithodoros moubata* but show little change during feeding (Kaufman & Sauer, 1982). This finding accords well with the virtual absence of salivary gland secretion during feeding in this species.

The cuticle-lined acinar ducts lead to ducteoles or in some cases directly to the main ducts of the salivary duct system, which is of ectodermal origin and formed by invaginations of the body wall. The ducteoles of each gland lead to a main salivary duct which emerges from the gland. The paired salivary ducts pass through the gnathosomatic foramen and extend ventral to the subcheliceral plate before opening into the salivarium. Ridges in the roof of the salivarium at the point of entry of the ducts are considered to impinge on its floor and to stop the flow of secretions until the floor is lowered (Kemp *et al.*, 1982). The cuticle-lined salivary ducts and ducteoles are provided with taenidia which keep them distended. They resemble tracheae in general appearance and have often been confused with them.

Mesostigmata

The paired oral glands in the Mesostigmata are not so extensive nor do they seem to be as structurally complex as in the Ixodida although Bowman (1984) refers, without detail, to five different types of cells in the salivary glands of *Pergamasus longicornis* (Hughes, 1949; Jakeman, 1961). Among the Dermanyssina, the glands appear to be larger in parasitic forms ingesting large quantities of tissue fluids than in predatory forms. Each gland comprises clusters of acini, often resembling a bunch of grapes, which discharge their secretion by way of efferent ducts into the main duct. The cells in *Ornithonyssus* (Hughes, 1949) and *Spinturnix* are vacuolated and have a granular appearance. In the female of *Varroa jacobsoni*, each of the paired glands (adenomere) resembles a truncated pyramid with the narrow end directed towards the gnathosoma and comprises five uniform spherical acini (Akimov *et al.*, 1988). The acini, identical in structure and dimensions, are connected successively along their length by a branched cuticular duct. The lumen of each acinus is surrounded by six conical cells which converge radially and is drained by a short, narrow ducteole in a cuticular duct. The cytoplasm of the cells is vacuolated but there is an absence of granular secretion. A section

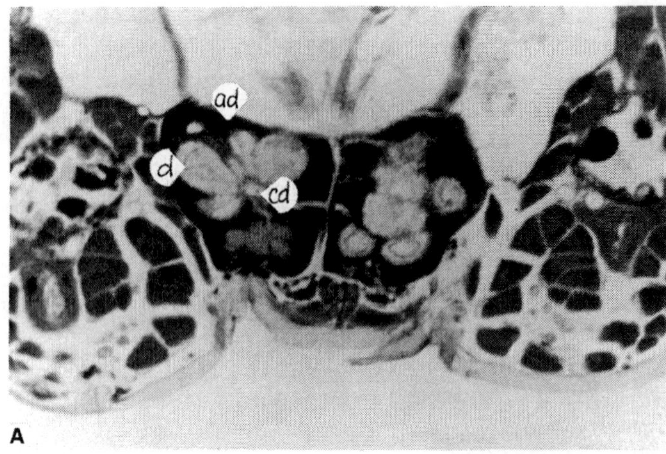

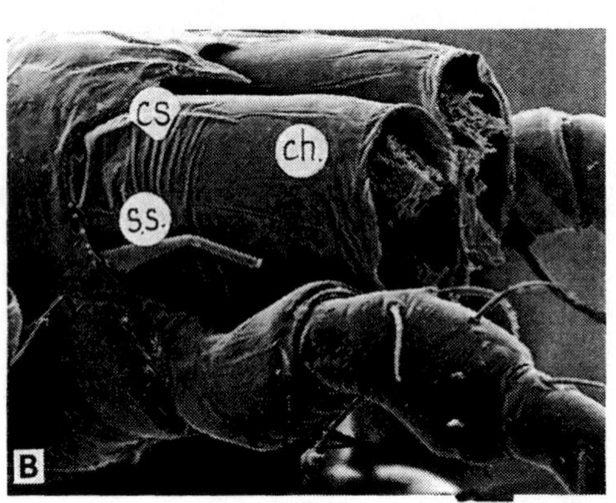

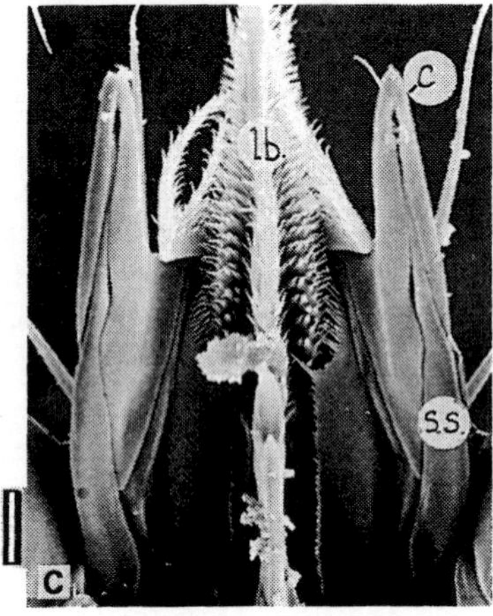

Fig. 7.5. A, transverse section of the salivary glands of *Spinturnix myoti* (Mesostigmata); **B**, lateral view of the gnathosoma of a species of *Paragamasus* showing a salivary stylus, gnathotectum and cheliceral sheath (chelicerae amputated at junction between first and second article); **C**, dorsal view of the hypostomatic region of the gnathosoma of *Hypoaspis aculeifer* (chelicerae removed) with salivary styli lying in dorsal grooves of the corniculi. *ad.* = adenomere; *c.* = corniculus; *cd.* = cuticular duct; *cs.* = corrugated area of cheliceral sheath; *d.* = ducteole; *lb.* = labrum, *s. s.* = salivary stylus.

through the glands of a female *Spinturnix myoti* is shown in Fig. 7.5A The acini in the uropodid Uroactinia agitans are elongate discharge individually into a tubular collecting duct (Woodring & Galbraith; 1976). Their cytoplasm is filled with clear vacuoles and the nuclei are large. Groups of small inactive cells lying in close proximity to the collecting duct are considered to be capable of enlarging and developing into functional secretory cells.

Each of the paired salivary ducts in the Parasitina (with the exception of *Cornigamasus*) runs lateral to the cheliceral frame and extends to the exterior within a sclerotized sheath or stylus, the salivary stylus. Each stylus flanks the protracted chelicera and its anterior end is often bent ventrally so that the secretions are directed towards the hypostomatic region (Fig. 7.5B). In the Dermanyssina and *Cornigamasus*, each duct runs lateroventral to the chelicerae and the secretory stylus extends along the lateral margin of the pre-oral trough and comes to lie in a dorsal groove of the corniculus (Fig. 7.4C). The styli are often enlarged and contribute to the lateral walls of the pre-oral trough, particularly in haematophagous ectoparasites (Evans & Till, 1965; Evans & Loots, 1975). The styli are not usually closely associated with the corniculi in the Uropodoidea. Salivary styli are stated to be absent in the Antennophorina and Cercomegistina (Camin & Gorirossi, 1955).

Notostigmata and Holothyrida

A pair of glands opening by ducts onto the surface of the subcheliceral plate has been described in the Notostigmata and Holothyrida. They are called antennal glands by With (1904) in the former and cheliceral glands by Thon (1905b) in the latter. Those in the Holothyrida are racemose glands comprising about ten acini. A second pair of glands, the maxillary glands, have a syncytial structure and discharge by way of intracytoplasmic ducts into a main duct which unites with the duct of the racemose gland on each side of the body.

Actinotrichida

The Actinotrichida have a number of glands situated in the anterior region of the idiosoma that discharge their secretions into the gnathosomatic region. The most widely occurring and characteristic of these are the paired components of the podocephalic gland-complex and the infracapitular glands. The terminology for these glands follows Alberti (1973) but some alternative terms are listed in Table 7.2. Additionally in the Prostigmata, an unpaired tracheal gland is present in some taxa and paired glands situated entirely within the gnathosomatic region have been described in the Parasitengona.

The podocephalic gland-complex consists of a maximum of four pairs of glands consisting of three pairs of acinose glands (podocephalic glands 1–3) and a pair of coxal or tubular glands (podocephalic glands 4). The coxal glands are considered to have an osmoregulatory function and are described in the section on Excretion (see Chapter 8). The arrangement of the glands and their ducts in the Bdellidae, Trombidiidae and Hermanniidae is shown in Fig. 7.6A–C. The ducts of the podocephalic glands on each side of the body discharge into a common external canal, the podocephalic canal, or into a common internal main duct, which has been variously termed the main common duct (Michael, 1896), the podocephalic duct ('le tube podocéphalique' of Grandjean, 1938b) and the common salivary duct (Moss, 1962). Grandjean (1938b) considered

Table 7.2. A comparison of selected terminologies for the major prosomatic glands of the Actinotrichida.

Michael (1896): Bdellidae	Moss (1962): Trombidiidae	Alberti (1973): Bdellidae	Woodring (1973): Oribatida
	Median salivary gland	Podocephalic gland 1	Rostral gland
Anterior salivary gland	Dorsolateral salivary gland	Podocephalic gland 2	Medial gland
Reniform salivary gland	Lateral salivary gland	Podocephalic gland 3	Lateral gland
Tubular salivary gland	Tubular salivary gland	Podocephalic gland 4 (coxal gland)	Coxal gland
Pericibal gland	Ventral salivary gland	Infracapitular gland	
Azygous salivary gland		Tracheal gland	

that in the Bdellidae the podocephalic duct had evolved from the primitive open gutter-like podocephalic canal through stages which involved the partial closure of the canal, its incorporation within the cuticle as a tube and finally its separation from the cuticle as an internal duct parallel to the surface of the cuticle. He cites the condition in *Cyta latirostris* as the typical open canal, the three-quarters closed canal of *Bdella* as an intermediate form and the internal duct of *Odontoscirus* representing a final stage. According to Alberti (personal communication), the canal in *Cyta* is not an open groove but is similar to that in *Bdella* and cannot be considered to represent the most primitive state (Fig. 7.7A). However, he has found in *Bdella septentrionalis* (Fig. 7.7B) and *Bdellodes longirostris* (Fig. 7.7C) further evidence to support the sinking/closing process in the development of an internal duct. Taenidia-like structures, absent in *Cyta*, appear to be present in the superficial podocephalic canal of *Bdella* and in the internal duct of *Bdellodes* (Fig. 7.7D).

The ducts of the acinose and tubular glands probably developed as ectodermal invaginations in the usual way and originally opened separately on the surface of the body of the mite. At a later evolutionary stage they became connected by the establishment of a cuticular canal whose phylogenetic origin is considered by G. Alberti (personal communication) to be connected with the tubular (coxal) gland. Thus, the podocephalic canal is not a glandular duct but a modified part of the body wall. It is a constant and characteristic feature of the morphology of the Actinotrichida.

The podocephalic glands in the Prostigmata have been studied in varying detail in the Bdellidae (Michael, 1896; Alberti, 1973; Alberti & Storch, 1973), Tetranychidae (Blauvelt, 1945; Alberti & Storch, 1977) and Parasitengona (Thor, 1904; Bader, 1938; Brown, 1952; Moss, 1962; Mitchell, 1964; Vistorin-Theis, 1978; Witte, 1978). Three pairs of acinose glands and a pair of tubular coxal glands are present in the Bdellidae (Fig. 7.6A), Trombidiidae (Fig. 7.6B) and Trombiculidae. The glands on each side of the body in the Bdellidae discharge either into partially open podocephalic canals extending along the lateral walls of the propodosoma from above trochanters I to meet medially on the subcapitulum near the bases of the chelicerae (*Bdella, Cyta*), or into a pair of internal podocephalic ducts (common main ducts) opening in a similar position

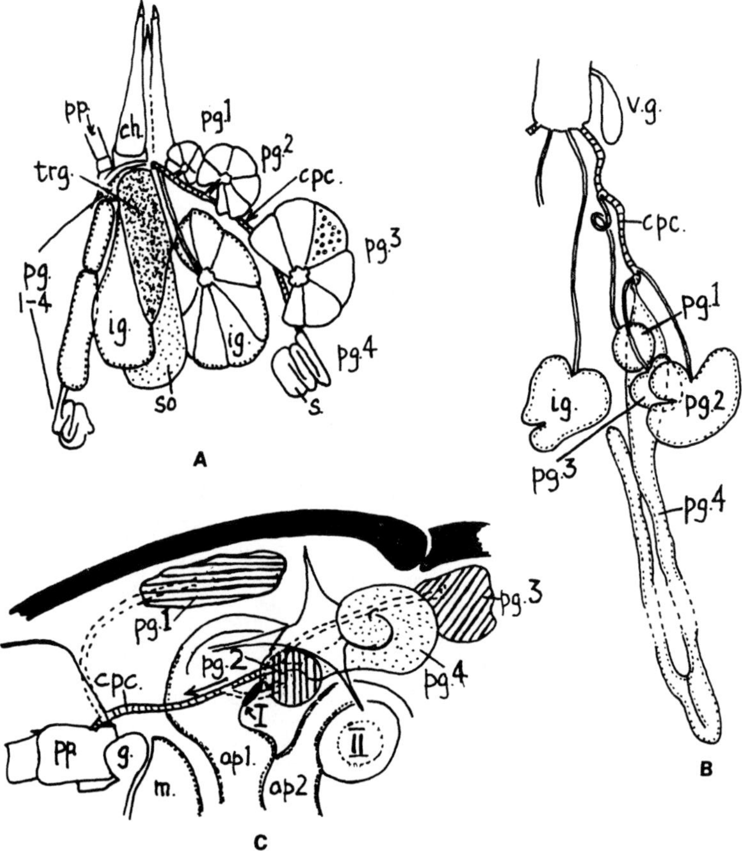

Fig. 7.6. Podocephalic gland complex in the Bdellidae (**A**) (after Alberti & Storch, 1973); Trombidiidae (**B**) (after Moss, 1962) and Hermanniidae (**C**) (after Woodring, 1973). *ap1. ap2* = apodemes; *ch.* = chelicera; *cpc.* = podocephalic canal; *g.* = gena; *i. g.* = infracapitular glands; *pp.* = pedipalp; *s.* = sacculus; *s. o.* = supra-oesophageal organ; *tr.* = tracheal gland; *pg1–pg4* = podocephalic glands 1–4; *v. g.* = putative venom gland; I, II = acetabula I and II.

on the subcapitulum (*Bdellodes, Odontoscirus*). The four pairs of podocephalic glands in the Parasitengona discharge into internal ducts. Only two pairs of acinose glands, in addition to the tubular glands, are present in *Tetranychus urticae*. In the Bdellidae (Alberti & Storch, 1973) and Tetranychidae (Alberti & Crooker, 1985), the cells of the acinose glands contain secretory granules, probably proteinaceous, and organelles characteristic of protein-secreting cells (large nuclei with prominent nucleoli and numerous ribosomes, some attached to ER cisternae). The acinose glands are considered to be the main 'salivary glands' in the

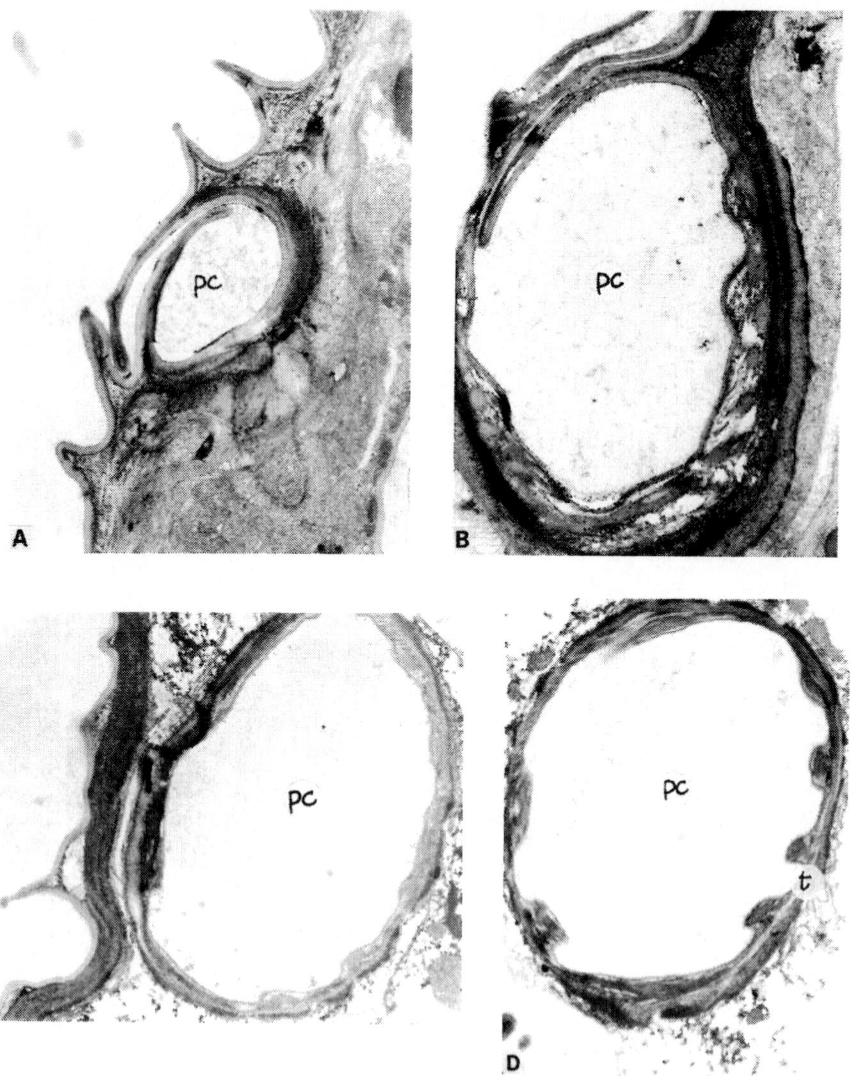

Fig. 7.7. Transverse sections of the podocephalic canal of the Bdellidae (courtesy of G. Alberti). **A**, *Cyta latirostris*; **B**, *Bdella septentrionalis*; **C, D**, *Bdellodes longirostris*. *pc.* = podocephalic canal; *t.* = taenidium.

tetranychids (Mothes & Seitz, 1981b) whereas in the Bdellidae glands 3 are thought to be involved in the secretion of silk (Alberti, 1973). According to Michael (1896), the ringed ducts of the acinose glands, but not of the tubular glands, in the Bdellidae are each provided with a valve-like structure at their commencement within the gland. The valve is formed by the thickening of the walls of the duct into two lips which are in close contact.

The comparative morphology of the podocephalic gland-complex in the Oribatida has been studied by Woodring (1973). The complex consists of a pair of coxal glands (glands 4) and from one to three pairs of acinose glands (glands 1–3) discharging into a podocephalic canal (Fig. 7.6C). The coxal glands, consisting of a thin-walled sacculus and a coiled labyrinth, are present in all active stases. Podocephalic glands 1 and 2 (rostral and medial glands, respectively) appear to be present only in the Archoribatida and each consists of a small number (four to ten) of granular cells. Gland 3 (the lateral gland), however, is present in all active stases of all oribatids. It is usually more or less pear-shaped and appears glandular, having clear vacuoles and numerous cytoplasmic granules. The function of the acinose glands is unknown. Woodring suggests that glands 3 may have an endocrine function associated with the moulting cycle and that glands 1 might function as salivary glands. The podocephalic canal starts as an internal tube at the junction of podocephalic gland 3 duct with the coxal gland duct. From this point, the canal penetrates the cotyloid wall of acetabulum I and continues anteriorly as an open gutter along the lateral body wall to the laterodorsal region of the subcapitulum. In the Archoribatida, the duct of gland 2 penetrates the cotyloid wall of acetabulum I and debouches into the podocephalic canal, while the duct of gland 1 opens through the cuticle into the canal a short distance before it passes from the lateral body wall to the subcapitulum.

The factors controlling the discharge of the secretions of the podocephalic glands and whether the acinose and tubular glands discharge into the podocephalic canal/duct together or individually are not known.

The podocephalic canals in the Astigmata receive the secretions of a single pair of glands, the supracoxal glands, which have an osmoregulatory function (Prasse, 1968b; Brody *et al.*, 1976; Wharton *et al.*, 1979). The duct of each gland opens at the posterior end of a depression or fossa in a crescentic sclerite, the supracoxal sclerite, situated on the lateral body wall above trochanter I (Fig. 7.8A). The slit-like orifice of each duct is surrounded by a sclerotized ring and the sclerite is produced anteriorly into a variously shaped structure called Grandjean's organ. The gutter-like podocephalic canal runs along the ventral border of the sclerite and extends to the basal region of the subcapitulum. A supracoxal seta, often elaborately branched, is usually located a little distance posterior and ventral to the opening of each gland duct and probably protects the orifice from blockage by debris as well as functioning as a sensory organ.

A pair of acinose glands opening by ducts on the labrum or near its base dorsally on the subcapitulum has been described in a number of actinotrichid taxa and are attributed a salivary gland function. In the Bdellidae, the glands lie one on each side of the supra-oesophageal organ and were termed pericibal salivary glands by Michael (1896) and infracapitular glands by Alberti and Storch (1973). The glands resemble podocephalic glands 3 in structure and Alberti (1973) considers them to be involved in the secretion of silk. The cells of each gland radiate from the centre and have large nuclei and nucleoli. The cell body is largely filled by secretory granules. There is a small valve at the commencement of the duct which is long and fine and provided with a taenidial structure. The paired infracapitular or ventral glands in the Trombidiidae, Trombiculidae and Erythraeidae appear to have the same structure as the pericibal glands. In *Trombicula alfreddugesi* (Brown, 1952) and *Allothrombium fuliginosum* (Moss, 1962), the glands are situated ventral to the synganglion but in the Erythraeidae (Witte, 1978) they are located anterior to the synganglion and their ducts open on the posterior part of the subcheliceral plate. The ducts of the ventral glands in the Calyptostomatidae join the podocephalic ducts before they fuse and open in the posterior region of the gnathosoma (Vistorin-Theis, 1978). The 'paired podocephalic glands' of the Eriophyoidea described by Nuzzaci (1979b) may also be considered to be infracapitular glands. Their ducts open into two deep, longitudinal furrows on the subcheliceral plate that terminate at the labrum.

In the Oribatida, the glands lie in the mid-line either dorsal or anterior to the synganglion of adults, but are absent in all immature stases. They are derived from epidermal cells during the tritonymphal pre-moult period (Woodring & Cook, 1962a). Van der Hammen (1986b) referred to these as infracapitular glands and Woodring (1973) as lingular (labral)

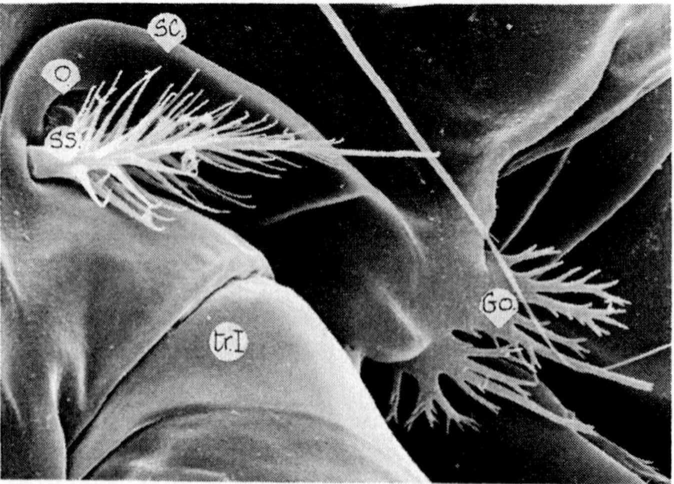

Fig. 7.8. SE micrograph of the supracoxal sclerite (*sc*), supracoxal seta (*ss*) and Grandjean's organ (*Go*) in an acarid mite (Astigmata). *o.* = orifice of supracoxal gland; *trI* = trochanter of leg I.

salivary glands. The glands in the Archoribatida are much smaller in size than the synganglion and lie anterior to it near the base of the labrum whereas those of the Euoribatida are usually larger than the synganglion and lie dorsal or anterior-dorsal to it. The gland ducts open on either side of the labrum, slightly anterior to the middle.

A group of eight or nine cells extending postero-dorsally on each side of the cheliceral ganglion between the cheliceral retractor muscles is considered by Prasse (1968b) to constitute salivary gland cells in the astigmatic genus *Sancassania.* They do not form a compact body but are arranged more or less in a double row, one behind the other. The round nucleus and the acidophilic nucleolus are large and the cytoplasm of the cell is vacuolated. Arising from the anterior end of the group of cells is a string of protoplasm which may represent a duct. This passes under the cheliceral ganglion into the subcapitulum.

An unpaired gland, the tracheal (azygous gland, intercheliceral gland) has been described in the Bdellidae, Erythraeidae, Halacaridae, Hygrobatidae, Tetranychidae, Trombiculidae, Trombidiidae and Rhagidiidae but is absent in the Calyptostomatidae (Vistorin-Theis, 1978). In the Bdellidae, the gland contains smooth ER and unlike other oral glands, electron-transparent granules (Alberti & Storch, 1973). The duct opens between the cheliceral bases. Witte (1978) suggests that the oily secretion of the gland in larval Erythraeidae seals the space between the subcapitulum and the chelicerae and contributes to the protection of the dorsal surface of the subcapitulum along which travel the secretions of the podocephalic and infracapitular glands. Mothes and Seitz (1981b), on the other hand, consider that the gland in *Tetranychus* produces a secretion that facilitates the movements of the cheliceral stylets in the subcheliceral channels. The unpaired gland in the Eriophyoidea, situated anterior to the synganglion, is unusual in that, in addition to its single duct debouching onto the subcheliceral plate, three fine ducts also issue from the gland, pass into each chelicera and continue distally within the stylets. Nuzzaci (1979b) is of the opinion that the gland is the result of the fusion of several glands.

A number of glands situated within the gnathosoma have been described in the Parasitengona. A cheliceral gland lying within each cheliceral shaft of active stases occurs in terrestrial and aquatic species (Schmidt, 1935; Moss, 1962; Witte, 1978). Although these glands are present in the larva of the Erythraeidae, which has hook-like chelicerae, they do not occur within the stylet-like chelicerae of the deuteronymph and adult. The location of the orifice of the gland has not been established with certainty. Witte considers it to be in the arthrodial membrane at the base of the movable digit of the chelicera in the larva of *Abrolophus rubipes.* The function of the cheliceral glands is unknown. Moss suggests that they may produce toxins and function as venom glands. Three pairs of glands, the labial, buccal and pharyngeal, are present in the anterior region of the subcapitulum of the Erythraeidae (Witte, 1978). The openings of the labial and buccal glands have not been established but the pharyngeal gland orifice is situated ventrally in the anterior quarter of the pharyngeal pump. The labial glands occur only in the deuteronymph and adult but the buccal and pharyngeal glands are present in all active stases. The cytoplasm of the labial and buccal gland cells is vacuolated and the glands are stated to produce lipid secretions.

Digestion

Although there have been a number of detailed studies of the histology and histochemistry of the salivary glands and mid-gut epithelium in the Acari, our knowledge of the biochemistry and physiology of the digestive processes is fragmentary. External digestion involving salivary gland secretions and internal digestion, both extracellular and intracellular, have been reported in the Acari but the extent to which each contributes to the digestive process, particularly in mites, is often a matter of conjecture.

External digestion

The secretion of enzymes into the food before its ingestion is a feature of feeding in many carnivorous arachnids. The tissues of the prey are predigested and rendered into a liquid state before entering the alimentary tract. The 'saliva' involved in the predigestive process in the majority of spiders is a mid-gut regurgitation and there are no 'true salivary glands' (Legendre, 1978). A similar process also occurs in scorpions. Regurgitation of mid-gut contents during feeding in the Acari is prevented by the proventricular valve and that of the pharyngeal contents by the downward movement of the labrum into the pre-oral channel and the peristaltic closing of the pharynx. Thus, it would appear that any enzymes produced by the acarine for external digestion of its food originate from salivary glands that debouch their secretions into the buccal or prebuccal region. The functions attributed to the oral secretions of mites are largely conjectural. The role of secretions of the labral salivary glands of the Oribatida was considered by Woodring and Cook (1962a) to be 'lubrication and to some degree digestion' while Hughes (1950c) was of the opinion that in *Acarus siro* the oral secretions contain a mucus which serves as a lubricant for food passage through the pharynx and oesophagus. Evans *et al.* (1961) suggest that 'the massive salivary styli of predacious mesostigmatid mites are suggestive of external digestion' while Mothes and Seitz (1981c) assume that oral secretions destroy the outer membrane of chloroplasts during feeding in *Tetranychus urticae*. It is only in the ticks that the role of salivary gland secretions in feeding has been studied in any detail.

The secretions of the salivary glands of ticks are known to contain a number of agents which may be important in the initial stages of feeding.

Reference has already been made to the production of cement by some cells of acini II and III in ixodoid ticks and its role in the attachment of the parasite to its host. In addition, active agents in the saliva are thought to be responsible for the prevention of blood coagulation, tissue destruction, capillary dilation and extensive haemorrhaging in the feeding lesion (Kemp *et al.*, 1982). However, some caution must be exercised in attributing these effects entirely to particular chemicals in the saliva. The tick's secretions can promote host reactions which destroy the host's own tissues and may simulate many of the postulated functions of the parasite's oral secretions. Balashov (1968) suggested that the anticoagulants originated from the granular cells and were glycoproteins, mucoproteins or some protein–carbohydrate complex. They have been found in the salivary glands of some argasoid and ixodoid ticks. Dilation of blood vessels adjacent to the attachment site of the tick is of common occurrence. Pharmacological agents causing an increase in the permeability of bovine skin capillaries have been shown to be present in the saliva (and glands) of partially fed *Boophilus microplus* by Tatchell and Binnington (1973). The active component is a prostaglandin (or prostaglandins). A prostaglandin from the oral glands in the same tick species was identified as prostaglandin E_2 by Higgs *et al.* (1976). Cytolytic activity in oral gland secretions has been reported in *Argas persicus* by Moorhouse (1975) and tissue destruction occurred before infiltrating leucocytes could have caused autolysis. Howell *et al.* (1975) have indicated the presence of a proteolytic enzyme in the saliva of *Ornithodoros savignyi*. Saliva in some ixodoid species transfers paralysing toxins to the host. In *Ixodes holocyclus*, which has been most extensively studied, the toxin is thought to be produced by cell type **b** in acinus II and peak toxin production in the salivary glands occurs on the 5th–6th day after the beginning of feeding (Kemp *et al.*, 1982). No enzymes which might participate in the digestion of blood have been shown to occur in the saliva of ticks (Akov, 1982).

Internal digestion

The most detailed information on digestion in the Acari refers to the Ixodida. The first step in digestion in ticks is the concentration of the blood meal by the removal of excess water and sodium ions. These pass through the epithelium of the mid-gut into the haemolymph from which they are removed by the salivary glands in ixodoid ticks and by coxal glands in the argasoids. The removal of water and ions in the ixodoids takes place while the tick is attached to its host, but in argasoid ticks excretion from the coxal glands begins towards the end or immediately after engorgement.

In the slow-feeding Ixodoidea, which remain attached to their vertebrate host for long periods, most of the ingested food is blood although females and immature instars take in quantities of non-blood

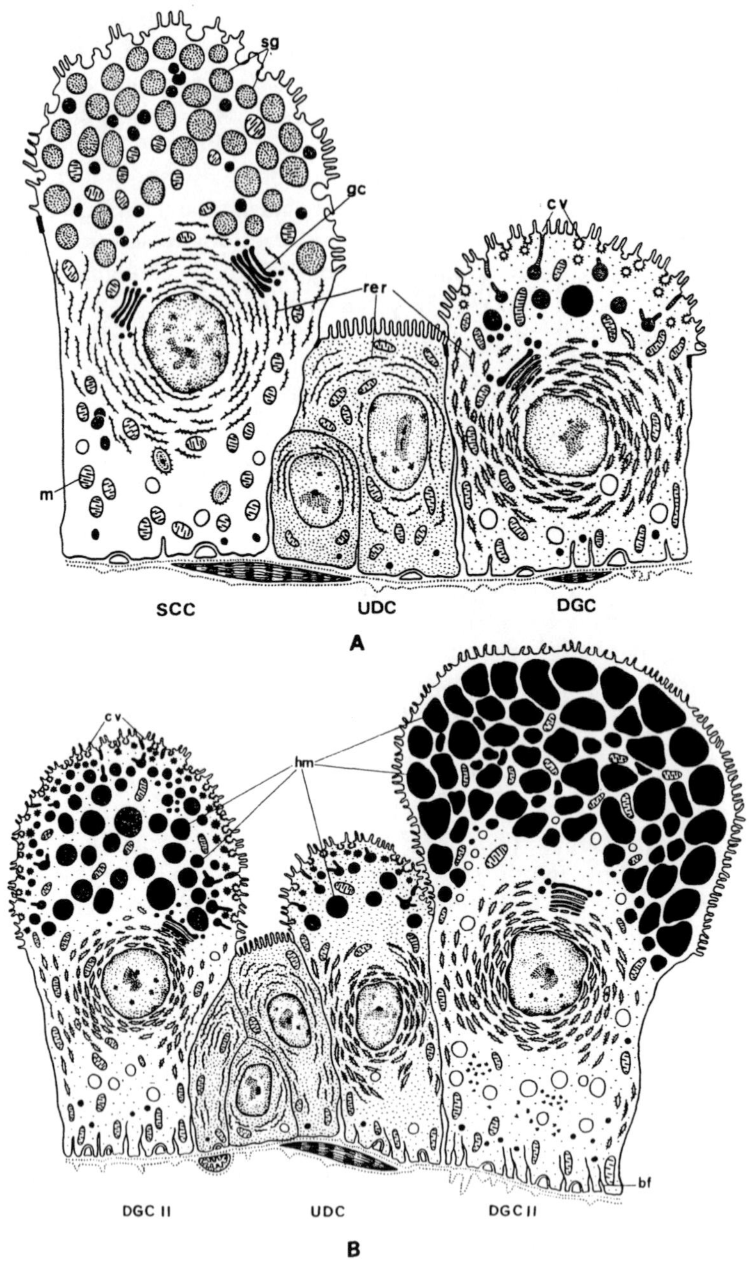

Fig. 7.9. Schematic representation of the ultrastructure of the mid-gut epithelium of an ixodoid tick (after Raikhel, 1979a). **A**, condition of secretory, digestive and undifferentiated cells at the commencement of feeding; **B**, condition of cells during the late phase of feeding when a large volume of blood is ingested. *cv.* = coated vesicle; *bf.* = basal infolding; *gc.* = Golgi complex; *hm.* = haematin granules; *m.* = mitochondria; *rer.* = rough endoplasmic reticulum; *sg.* = secretion granule; *hm.* = haematin granules; *DGC.* = digestive cell; *DGC* II = type II digestive cell; *SCC.* = secretory cell; *UDC.* = undifferentiated cell.

tissue fluids during the initial stage of feeding. Three types of cells (Fig. 7.9A) have been distinguished in the walls of the ventriculus and caeca and are referred to as reserve (or undifferentiated), digestive and secretory cells (Roesler, 1934; Hughes, 1954b). Balashov (1968) postulated that in the female of *Hyalomma asiaticum* all these cells have a common origin with the reserve cell being the initial type from which the other cells have differentiated. Further, secretory and digestive cells are irreversibly specialized and degenerate at the end of their functional activity. Only one exception was noted, when secretory cells transformed into digestive cells. Subsequent histological and histochemical studies on the mid-gut cells of *Boophilus microplus* by Agbede and Kemp (1985) have confirmed Balashov's opinion. Reserve cells (also referred to as stem cells by some authors) have the typical fine structure of undifferentiated cells in that microvilli and basal infoldings are almost or completely absent (Raikhel, 1979a). The cytoplasm contains chiefly free ribosomes and other organelles (Fig. 7.9A, UDC). Differentiation of reserve cells is accompanied by growth, the development of microvilli and basal infoldings, and the appearance of numerous RER cisternae, Golgi complexes and many elongated mitochondria. The secretory cells are columnar or clavate in shape and their dense cytoplasm consists of abundant RER cisternae, roundish mitochondria and Golgi complexes. These organelles occupy the centre of the cell. The apical region of the cell is filled with numerous round or ovoid secretion granules and the cell surface has dense microvilli (Fig. 7.9A, SCC).

Agbede and Kemp (1985) recognized two types of secretory cells in *Boophilus microplus*, (s1) and (s2); the former probably contain glycoprotein granules and may secrete a haemolysin and the latter, found only in females, are highly vacuolated and their colloid-like secretory material forms loosely aggregated spheres in the cytoplasm. Secretory cells (s1) are present at two stages in the feeding cycle: at the time of attachment and just before the phase of rapid engorgement. They apparently release secretory granules into the ventricular and caecal lumina by a process of exocytosis. The secretory material of cell (s2), which may be an acid mucopolysaccharide, is released by the pinching off of the apical part of the cell into the lumen. Cells (s2) have all completed secretion before the beginning of the phase of rapid engorgement. The secretions of both types of secretory cells mix with the gut contents. The secretion of (s2) cells is possibly an anticoagulant while a haemolysin in the secretion of (s1) cells is probably involved in the haemolysis of ingested red blood corpuscles, which occurs rapidly in the first phases of digestion and after engorgement. Hughes (1954b), working with *Ixodes ricinus*, noted that the process was not a simple one but involved the complete digestion of the red cell envelope. This also happens in *B. microplus*. In the absence of biochemical data, it is not possible to evaluate, at present, the importance of the extracellular stage in tick digestion. According to Agbede and Kemp (1987), the basal portions of the s2 cells, after secreting their apical cytoplasm into the gut lumen, are transformed into basophilic cells which appear to be involved in active water transport across the gut wall.

Two types of digestive cells have been described in ixodoid ticks and can be distinguished on the basis of their structure. During the early phase of feeding (2–3 days after attachment) in the female of *Hyalomma asiaticum*, type I digestive cells (DGCI) predominate and are characterized by the uptake of blood components by phagocytosis and pinocytosis (Raikhel, 1979a). Host leucocytes and other fragments of host tissue are engulfed by cytoplasmic projections of the surface of the cell (phagocytosis) and digestive vacuoles are formed (Fig. 7.9B). The cortical cytoplasm of these cells contains structures specialized for pinocytosis such as coated pits of the plasma membrane where proteins are bound before internalization. The majority of the digestive vacuoles show a positive reaction for the lysosomal enzyme acid phosphatase and they are referred to as heterolysosomes. Lipid inclusions and glycogen accumulate in the cells. Type II cells (DGCII), on the other hand, become abundant during the final stages of engorgement and are involved in the uptake of haemoglobin and other proteins of haemolysed blood by pinocytosis (Fig. 7.9B). Pinocytotic vesicles later coalesce to form larger vacuoles. Intracellular digestion is more gradual in these cells than in type I. Haemoglobin splits into haem and globin in the heterolysosomes and only the globin, the protein part of haemoglobin molecules, is further broken down for assimilation. The haem prosthetic group is not needed by the tick and remains in the cytoplasm as haematin granules. Cells II are considered by Balashov (1968) to act as storage sites for reserve proteins, fats and carbohydrates much in the same way as the fat body in insects. Digestive cells that have completed the digestion of their food vacuoles (spent cells) have their cytoplasm full of haematin granules and some lipid inclusions. Many of these detach from the gut wall into the lumen where they rupture or are egested whole.

Male and nymphal digestion is similar to that in the female except in the aphagic male of the Ixodidae where the cycle of anterior mid-gut changes is associated with the utilization of food reserves accumulated in the nymph. Gut cells in the unfed larva are small and flat while in the engorged state the very thin gut wall consists of a few stretched cells. No food inclusions are seen during feeding but, after detachment, mitoses are observed among the gut cells and some of the newly formed cells elongate into the lumen and fill with food inclusions while others later form the gut epithelium of the nymph (Balashov, 1968).

Argasoid ticks are rapid feeders, except for the slow-feeding or non-feeding (some *Ornithodoros*) larvae, and complete engorgement occurs within 0.5 to 2 hours. Unlike the ixodoids, virgin females take the same amount of blood as mated females. Aspects of digestion in the Argasoidea have been studied by Tatchell (1964) in *Argas persicus*, by Balashov (1968) and Balashov and Raikhel (1977, 1978) in *Ornithodoros papillipes* and by Grandjean and Aeschlimann (1973) in *O. moubata*. Stretching of the gut epithelium in the newly engorged female results in the virtual disappearance of the cell boundaries and their transformation into a membrane-like structure with slight thickenings at the sites of the nuclei. In *A. persicus* the blood remains unlysed for 2–3 days while new epithelium develops containing cells which secrete into the gut lumen a saliva-fast PAS-positive colloid that causes haemolysis. The period from 3 to 14 days is one of rapid digestion when the gut cells become involved in the absorption of blood proteins from the lumen and in intracellular digestion. The digestive cells contain acid phosphatase as in the ixodoids. Haematin granules begin to appear in the cytoplasm of the cells, which become club-shaped. Some of these haematin-laden cells detach from the basal lamina and float freely in the mid-gut lumen while in the remainder only the apical portion of the cell detaches and the nucleus remains within the basal portion that is still attached to the lamina. Absorption of haemoglobin and lipids from the erythrocyte stroma occurs chiefly by pinocytosis while particles from the blood meal enter the cell by phagocytosis. Part of the blood meal remains unchanged in *A. persicus* and acts as a food reserve. There appears to be greater morphological variability of the anterior mid-gut epithelial cells in argasoids than in ixodoids. In addition to cells distinguishable as secretory or digestive, some cells contain both secretory granules and digestive vacuoles. Tatchell (1964) considered that in *A. persicus* secretory cells have the potential to develop into digestive cells, if required. He further suggested that the development of one cell type from another is from the proximal part of a cell that remains after the pinching off of its distal part. Some support for this mechanism is given by Agbede and Kemp (1985) working with females of *Boophilus microplus*. They refer to ultrastructural evidence that the basophilic cells of the anterior mid-gut are not derived from reserve cells but from basal remnants of secretory cells.

The histology, and in some cases the histochemistry, of the ventricular-caecal epithelium and the process of digestion have been studied in a number of mite taxa. Food is ingested in a liquid or semi-liquid state in most of the Mesostigmata and Prostigmata and in a solid state in the majority of the Astigmata and Oribatida. The most detailed studies on the histology of the mid-gut in the Mesostigmata have been carried out by Hughes (1952), Akimov and Starovir (1974, 1976, 1977) and Starovir (1982) in the Dermanyssina and by Woodring and Galbraith (1976) in the Uropodina. In the ventriculus and caeca of the unfed blood-feeding *Ornithonyssus bacoti*, Hughes (1952) observed two types of cells: large cells with distal nuclei and one or more large vacuoles and smaller cells with basal nuclei and numerous small vacuoles which appeared empty. The collapsed anterior mid-gut resulting from fasting had the large cells more or less filling the lumen and the basement membrane quite thick. The anterior mid-gut became very distended in the gorged state and the stretching of its wall resulted in the thinning of the epithelium and basement membrane. At this stage a number of large ovoid or spherical cells were evident in the gut contents and were considered to be the large cells of the gut wall which had become detached and shed their secretions into the food mass. Hughes (1952) suggested that the admixture of cells charged with enzymes increased the rate of digestion of the blood meal. About 24 hours after ingestion, the cells floating in the anterior mid-gut contents began to accumulate small refringent granules ('probably derived from haematin') appearing in the gut lumen. As the accumulation increased, these cells became distended and opaque until finally they formed black rugose masses some 3–4 days after feeding. Due to the extreme thinness of the ventricular and caecal walls in the gorged mite, it was suggested that digested food diffused directly to other organs lying in close proximity to the distended ventriculus and caeca. Such a gastrovascular function of the ventricular caeca of the Dermanyssina was suggested by Mégnin as long ago as 1876. Undifferentiated, secretory and digestive cells are present in the ventricular and caecal epithelium of the haematophagous bat mite *Spinturnix myoti* (Fig. 7.10A,B). The digestive cells accumulate haematin *in situ* and some eventually break free into the lumen in much the same way as in ticks (Evans, 1969). The three types of cells are present also in the anterior mid-gut epithelium of *Uroactinia agitans* (Woodring & Galbraith, 1976). The secretory cells are characterized by having numerous clear vacuoles in a basophilic cytoplasm and are relatively uncommon in comparison with the densely packed digestive cells. The latter take up food components by phagocytosis and eventually break free into the lumen. According to Bowman (1984), it is not possible to categorize the gut epithelial cells of *Pergamasus longicornis* into digestive and secretory types.

Gut histology and digestion in the Prostigmata have been studied by, among others, Henking (1882), Thor (1904) and Bader (1938) in the Parasitengona, Wright and Newell (1964) in the Anystidae, Alberti (1973) in the Bdellidae, and Ehara (1960), Akimov and Barabanova

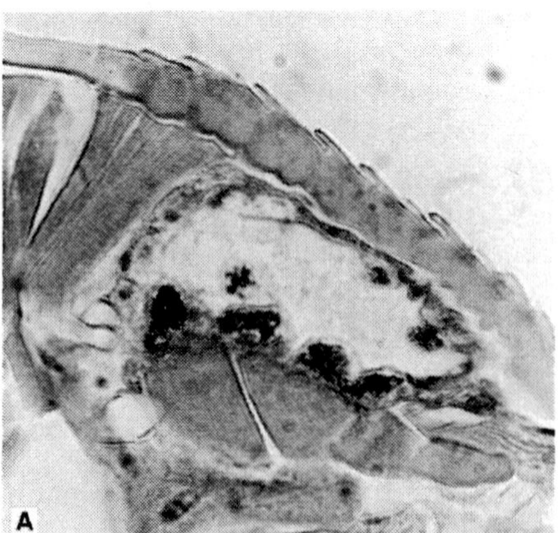

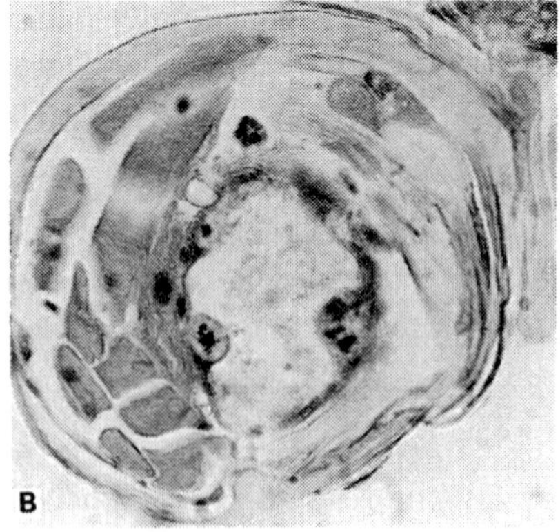

Fig. 7.10. Transverse section of a ventricular caecum or diverticulum in the body (**A**) and cavity of the first leg (**B**) of *Spinturnix myoti* (Mesostigmata) showing secretory and digestive cells.

(1977), Wiesmann (1968), Mothes and Seitz (1981c) and Alberti and Crooker (1985) in the Tetranychidae. In the Bdellidae, reserve, vacuolated secretory and digestive cells are present in the anterior mid-gut epithelium. Food vacuoles in the digestive cells of newly fed forms contain protein, fat and polysaccharide. These cells ultimately become filled with a fine blackish crystalline material and the apical portion of the club-like cells is pinched off into the lumen. Thor (1904) noted a high lipid content of the mid-gut cells in the majority of the parasitengonid mites he studied and this was also observed by Wright and Newell (1964) in *Anystis* and *Dinothrombium*. The mid-gut contents in these species contain yellow to orange granules which the latter authors found to possess the characteristics of chromolipids (inert oxidation products of phospholipids or polymerization products of unsaturated fatty acids). They suggested that lipid taken into the epithelial cells of the mid-gut is deposited in the form of discrete granules, possibly in the modified form of chromolipids, during the degeneration phase of the cells following their detachment from the gut wall. This process was explained by the necessity for lipid detoxification and/or conservation of space owing to the lack of a continuous alimentary tract and hence the inability to eliminate mid-gut contents through the anus.

Only one type of anterior mid-gut epithelial cell is considered to occur in the tetranychids which feed on plant juices and this ranges in shape from a flat cell with large amounts of rough RER, large nucleus, crista-type mitochondria and some microvilli, to a bulbous cell containing heterolysosomes and protruding into the ventricular and caecal lumina. The development of the bulbous cell from the flat form is thought to be due to the resorption of gut contents by phagocytosis and pinocytosis. No secretory cells of the type described in other acarines have been observed. The entire bulbous cell or parts of it may be pinched off from the gut wall and become a free cell ('floating cell', 'food ball') in the lumen. This cell has a large central vacuole surrounded by a narrow cytoplasmic border containing electron-dense inclusions. The vacuole becomes filled with excretory products and the cytoplasmic components appear reduced as the cell nears the posterior mid-gut. Its excretory products may be released into the ventricular lumen. Mothes and Seitz (1981c) suggest that during the phagocytic stage there is a differentiation of cells into those that ingest thylakoid granules and those that function in starch resorption. A number of enzymes capable of hydrolysing carbohydrates have been found in homogenates of whole *Tetranychus urticae* but cellulase and pectinase appear to be absent (Ehrhardt & Voss, 1961).

The published accounts of the histology of the epithelium of the ventriculus and its caeca and the treatment of ingested fragments of food during digestion in the Astigmata and Oribatida are largely descriptive and, as in other mites, little is known of the biochemistry and physiology of digestion. An interesting feature of the epithelium of the anterior mid-gut wall in both groups is the zonation of its different cell types (Brody *et al.*, 1972; Hoebel-Mävers, 1967), a feature which does not appear to occur in mites ingesting liquid or semi-liquid food. Further, a single or compound food pellet (or bolus) is formed in the mid-gut at some stage in the digestive process and may be enclosed by a thin peritrophic membrane. The most detailed study of the histology of the anterior mid-gut wall in the Astigmata is that by Brody *et al.* (1972) on the pyroglyphid *Dermatophagoides farinae*. The cuboidal cells in the dorsal wall of the most anterior region of the ventriculus are flattened, have few microvilli and mitochondria are seldom present whereas, ventrally, there are two rows of active cells extending to the junction of the ventriculus with the colon. These cells have a single nucleus, short microvilli, RER and large vacuoles. Cells floating freely in the ventricular lumen originate from the ventral cell rows and appear as round or elliptical bodies with short microvilli and large vacuoles, filled or empty. The cuboidal cells of the dorsal and lateral epithelium in the more posterior region of the ventriculus have short microvilli, extensive RER, mitochondria, a few lysosomes and multivesicular bodies. Similar cells are typical of the caeca but these also have numerous electron-dense inclusions which Brody and his colleagues consider to be lysosomes. Digestion is stated to be intracellular in the ventriculus–caeca with possibly some extracellular digestion in the colon, which has cells containing actively secreting cytoplasm. In *Sancassania*, Prasse (1967a) recognizes only two distinct cell types in the anterior mid-gut epithelium using light microscopy, namely digestive cells in the outer halves of the caeca, and secretory cells in the ventriculus and inner halves of the caeca. The digestive cells absorb nutrients and gradually swell until club- or flask-shaped, when their apical portion may be pinched off into the gut lumen. Typically, the secretory cells have a large basal nucleus and numerous cytoplasmic vacuoles. Hughes (1950c) suggested that similar cells in *Acarus siro* secreted a mucus which probably contained a digestive enzyme. Both Prasse (1967a) and Akimov (1973) refer, respectively, to the possibility of extracellular digestion in the colon of *Sancassania* and *Rhizoglyphus*.

Assays for amylolytic, chitinolytic and cellulolytic activity have been carried out on a number of astigmatic species. Bowman and Childs (1982), working with species of the genera *Acarus, Glycyphagus, Rhizoglyphus* and *Tyrophagus*, found amylase and chitinase to be present in all the species examined whereas cellulase was detected only in species of *Rhizoglyphus* and *Tyrophagus*. The assays were carried out on homogenates of laboratory cultured mites fed on yeast and ground wheatgerm in a 3:1 ratio. It is suggested that the cellulolytic activity could have originated from gut symbionts rather than from the mites themselves.

The alimentary tract in the elattostasic deuteronymphs of the Astigmata, which ingest no food through the mouth, exhibits varying degrees of regression. According to Boczek *et al.* (1969), this stase in *Acarus farris* is characterized by reduced but not discontinued metabolism. The pharynx, oesophagus and ventricular caeca appear closed and the histology of the ventriculus resembles that of the fore-gut. The post-ventricular region of the gut, on the other hand, does not show such profound changes from that of the feeding protonymph and appears to be functional in that secretions are present in it and the anal atrium appears to be open. Metabolism is maintained through the utilization of materials stored by the protonymph in mesenchymal tissue.

The wall of the ventriculus in the Oribatida has a sparse external musculature and is considerably folded (Dinsdale, 1974b). Certain areas of the wall are smooth whereas others, particularly in the anterior (cardiac) region, are covered by a brush border. According to Hoebel-Mävers (1967), the cells form two zones: an anterior zone of large cuboidal cells and a posterior (pyloric) zone of smaller flat cells. The boundary between the two zones is not sharply defined since one type of cell grades into the other. The caecal wall is almost devoid of a musculature and a brush border is only present in some areas, mainly in the distal region of the caeca. Certain of the cells which lack a brush border become greatly enlarged and vacuolated, and extend into the lumen. Woodring and Cook (1962a), working with *Ceratozetes cisalpinus*, state that these cells finally burst and liberate their contents into the lumen. The secretions of the preventricular glands and the caeca, according to Hoebel-Mävers (1967), mix in the lumen of the ventriculus and form a 'secretion pellet' in which the food particles are embedded. This pellet rapidly increases in size to a diameter of 100–200 μm within the posterior part of the ventriculus. Small bodies found within the ventricular and caecal lumina are considered by Hoebel-Mävers to be 'clots' of merocrine secretions ('Sekretpfropfen'). These became incorporated in the food pellet and gradually dissolve as they travel towards its centre. The food pellet is enclosed in a diffuse peritrophic layer, probably of a mucopolysaccharide nature. Exceptionally, two food pellets come together to form a double-pellet. The more or less compacted food content of each pellet is separated in the area of contact by the peritrophic layer so that their individual form is recognizable within the compound pellet.

Dinsdale (1974b) found the pH in the mesenteron of a phthiracarid mite to range from 5.4 in the caeca to 6.6 in the colon and postcolon. Microchemical tests indicated the presence of protease, lipase and considerable carbohydrase activity in the mesenteron. Lysosomal activity of the gut wall appeared to be involved in the breakdown of material of both external and internal origin. The phagocytosis of larger particles probably occurs in the ventriculus while pinocytosis of finer material takes place in both the ventriculus and caeca. Dinsdale postulated an intracellular mechanism for the digestion of protein but suggested that polysaccharides are broken down at the brush border and in the lumen of the gut.

Investigation of carbohydrase enzymes in whole-body homogenates of saprophagous oribatid mites, particularly members of the Phthiracaridae, by Luxton (1972), demonstrated the presence of enzymes capable of hydrolysing structural carbohydrates of higher plants, such as cellulase and pectinase. The disadvantage of studying whole-body homogenates is the impossibility of determining the origin of the enzymes. Are they produced by the animal or associated with the ingested food or gut symbionts? Subsequent work by Zinkler (1971) and Dinsdale (1974b) failed to find cellulase in phthiracarids and there is increasing evidence that carbohydrase enzymes, including cellulase and chitinase, are produced by gut microflora (Stefaniak & Seniczak, 1976, 1981; Seniczak & Stefaniak, 1978). Norton (1985) speculates that all enzymes acting on structural polysaccharides in oribatid mid-guts may be of microbial and not of endogenous origin.

Posterior mid-gut and egestion

A food pellet is not produced in the ventriculus of liquid-ingesting forms and the waste products of digestion, comprising mainly parts of spent digestive cells and their contents, pass through the posterior ventricular sphincter into the colon or the postventricular mid-gut. The epithelium of the colon in the Ixodida is made up of flattened or cuboidal cells, some of which may bulge into the lumen. Their cytoplasm lacks nutritive inclusions but minute haematin granules are sometimes present and suggest that some intracellular digestion may occur (Balashov, 1968). During and immediately after feeding, the faeces are liquid but in unfed ticks they become dry and powdery. This is probably due to the re-absorption of water from the faeces while in the postcolon. The faeces may be white or black. White faeces consist of the excretory product guanine (and probably other purines) from Malpighian tubules opening into the postcolon; black faeces comprise haematin or a mixture of haematin and guanine. In common with most other acarines, the faeces are not held within the anal atrium and usually pass rapidly through it to the exterior by the action of the muscularis of the postcolon and the muscles controlling the opening of the anal valves. An analysis of the faeces produced by *Hyalomma dromedarii* during feeding showed that 12–35% of the total dry weight consisted of guanine, 25–31% protein and 0.3% haematin (Hamdy, 1973). After feeding, relatively little faecal material was egested and comprised, in total dry weight of faeces, 50–80% guanine and 13–22% unidentified purine but no protein. Most of the metabolic wastes of ixodoid females is never egested but remains in the body of the tick. According to Hamdy (1973), this amounts to 82% of the total guanine and 95% of the total haematin produced by *H. dromedarii*. Guanine, xanthine, hypoxanthine and, in some cases, uric acid have been identified in the white spherules of the excreta of *Argas* and *Ornithodoros* by Dusbábek *et al.* (1991).

Defecation in the Ixodida follows a rhythmical cycle which is less regular in the repeatedly feeding argasoids than in the ixodoids (Enigk & Grittner, 1952). In the first three days after moulting from the nymph, the unfed female of *Haemaphysalis leachi* excretes only guanine, but from 5–10 days the egested matter comprises a mixture of disintegrating cells and granules of excrement. No defecation occurs after this period until feeding commences. From the second day of feeding until three-quarters engorged, the faeces consist only of material from the gut but from the three-quarters engorged state to the first days of oviposition only guanine is produced. Defecation then ceases. In engorged larvae and nymphs, defecation ceases at the onset of apolysis.

Guanine and other purines present in the nitrogenous excreta of ticks act as arrestants, inducing individuals to become akinetic when encountering other ticks or surfaces visited by them (Otieno *et al.*, 1985; Sonenshine, 1985; Dusbábek *et al.*, 1991). The function of these non-specific assembly pheromones is to hold populations at sites where the recruitment of sexual partners is enhanced or where encounters with hosts are most likely (Sonenshine *et al.*, 1982). In argasoids, xanthine appears to be an important component of the assembly pheromone and was found to significantly increase the assembly efficiency of commercial guanine when mixed with it in the ratio of 1:25.

In the Mesostigmata there is no evidence of intracellular digestion in the postcolon. White or a mixture of black and white faecal material is produced by blood-feeding dermanyssoids as in ixodoid ticks (Hughes, 1950b). According to Young (1968b), the anal atrium of *Haemogamasus ambulans* contains faecal material before defecation. Muscles originating on the dorsal shield and inserting on the anal shield contract and draw in the edges of the anal shield thereby exerting pressure on the anal atrium. This results in the opening of the anal valves and the forcing out of the faeces. This is contrary to the process of defecation in *Uroactinia agitans* where the anal atrium does not contain faecal material and the movement of the faeces from the postcolon to the exterior is by the contraction of the postcolon muscularis and of the atrial support muscles which prevent the extrusion of the thin-walled atrium (Woodring & Galbraith, 1976). In the free-living species the fluid faeces are usually milky-white. During defecation, the body is usually tipped so that the anus almost touches the substrate, and in this position a globule of faeces is deposited (Lee, 1974; Bowman, 1984). In members of the family Ameroseiidae (*Ameroseius* and *Kleemannia*) which consume fungal hyphae and spores, fragments of undigested material are often present in the postcolon and are voided through the enlarged anal opening. The fragments do not form a faecal pellet.

A peritrophic membrane or layer is lacking in the Mesostigmata but Rudzinska *et al.* (1982) have described such a structure in the immatures and adults of *Ixodes dammini*. The secretion forming the membrane is produced by ventricular epithelial cells and is released through their microvilli. It appears in the larva (fed on hamsters) about 27 hours after attachment and somewhat earlier in the nymph but is never found in unfed ticks. The peritrophic membrane which contains no fibrils or other organized structures has a spongy appearance and divides the ventricular lumen into two concentric compartments: an inner endoperitrophic space containing haemolysed erythrocytes and reticulocytes, and an outer ectoperitrophic space, about 0.3–0.7 μm in depth, located between the membrane and the epithelium of the ventriculus. The larval membrane increases in thickness from 100 nm to 200–300 nm in the period after repletion. The membrane is considered to act as a mechanical barrier by keeping ingested blood cells from contact with the microvilli of the gut cells and also as a filter which allows only some macromolecules to penetrate it. Rudzinska *et al.* (1982) describe an arrowhead organelle developed by the piroplasm *Babesia microti* that enables it to cross the peritrophic membrane and enter the epithelial cells of the anterior mid-gut.

The posterior mid-gut in the Prostigmata may or may not be connected to the anterior mid-gut. In those forms with a continuous gut, the most detailed studies of the histology and function of the posterior mid-gut (or excretory organ) have been conducted on the Bdellidae (Michael, 1896; Ehara, 1960; Alberti, 1973) and the Tetranychidae (reviewed by Alberti & Crooker, 1985, and Van der Geest, 1985). In the Bdellidae, the cuboidal epithelial cells of the postventricular mid-gut have a basal nucleus and are provided with a very fine brush border. The chyme in the ventriculus passes into the postventricular mid-gut where it is surrounded by thin membrane secreted by the epithelial cells. Re-absorption of water from the faecal mass renders it more compact and it forms a 'faecal ball' ('Kotballen'). Small crystalline excretory material, probably guanine, also occurs in the postventricular mid-gut and often forms a solid rod filling the organ. As a rule, the faecal and excretory products do not mix and are egested separately. The cells of the lateral walls of the postventricular mid-gut in *Tetranychus urticae* are cuboidal and have irregularly shaped processes projecting into the lumen. Their apical portions contain numerous mitochondria and large amounts of glycogen and their basal portions are deeply infolded. They are thought to excrete guanine and also to be involved in osmoregulation. The vacuolated cells of the dorsal wall, however, have few mitochondria, no processes but large amounts of RER. As in the Bdellidae, the faecal and excretory materials are ejected separately. Black faecal pellets, each comprising numerous small faecal balls glued together by a secretion, probably a glycoprotein, come from the ventriculus and caeca whereas the smaller white excrement is formed in the postventricular mid-gut and consists chiefly of guanine. The surface of the excretory pellet is also covered by a thin secretion, probably of the same origin as that of the faecal pellet, and is often anchored to the leaf surface by silken threads.

In the Parasitengona which have lost the connection between the ventriculus and the postventricular mid-gut, the faeces, comprising budded off digestive cells containing waste material, appear to be stored in the caeca. Mitchell and Nadchatram (1969) have shown that in the trombiculid *Blankaartia ascoscutellaris* the waste material in the ventriculus is moved by muscular action into a pair of isolated postero-dorsal caeca which act as faecal pouches. Mites with full caeca appear to have the postero-dorsal region of the opisthosoma under considerable tension and the body wall in an area just posterior to the attachment of a specific muscle group (dorsalia V) may split open. The initial tear is lengthened and opened by muscular action and the faecal lobe is slowly forced through the split (Fig. 7.11A). After extrusion, the lobe is pinched off as the muscles relax and the edges of the split come together. This unique method of eliminating faeces in the Acari is called schizeckenosis and the scar resulting from the process leaves a record of the event. A similar mechanism of defecation was observed in a number of adults of other trombiculid species in laboratory cultures and typical scars resulting from schizeckenosis were found in over 75% of adults of *B. ascoscutellaris* collected from a swamp habitat. Mitchell (1970) postulated that schizeckenosis developed as the result of changes in the diet or habits of adult trombiculid mites which increased the rate of accumulation of indigestible residues in the anterior mid-gut. Not all the additional waste material could be stored in the blind gut and the mites evolved an entirely new means of eliminating faecal material.

The majority of the Astigmata and Oribatida ingest solid food and the residues of digestion are egested in the form of distinct pellets surrounded by a peritrophic membrane. Three pellets are often present in the mid-gut at any one time and comprise the ventricular, colonic and

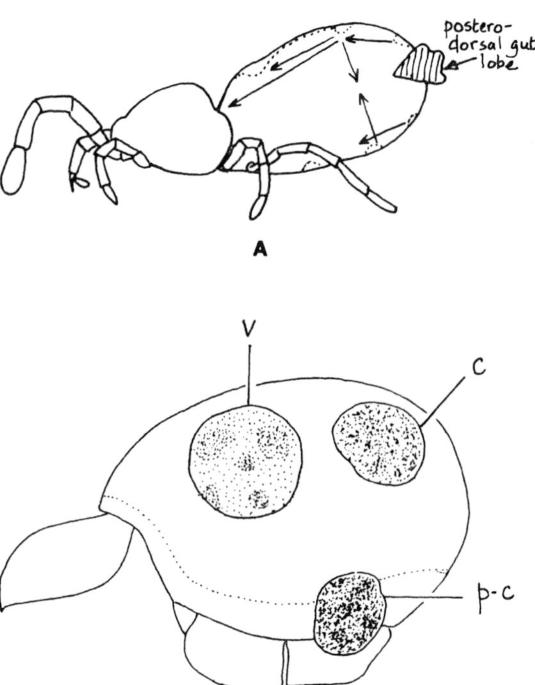

Fig. 7.11. **A**, diagrammatic representation of schizeckenosis in a trombiculid mite (arrows represent direction of contraction of idiosomatic muscles during extrusion of postero-dorsal gut lobe); **B**, food pellets in the alimentary tract of *Steganacarus magnus* (based on Hoebel-Mävers, 1967). *c.* = colonic pellet; *p-c.* = postcolonic pellet; *v.* = ventricular pellet.

postcolonic pellets located, respectively, in the ventriculus, colon and postcolon (Fig. 7.11B). During defecation in the three pellet condition, the extrusion of the postcolonic pellet through the anal orifice is followed by the movement of the colonic pellet into the postcolon and the ventricular pellet into the colon. The pellet becomes more compacted as it travels through the colon and postcolon, probably through the absorption of water by the posterior mid-gut cells (Hughes, 1950c; Hoebel-Mävers, 1967). It is usually extruded as a dry, solid faecal pellet which remains intact in air but breaks down in water.

The peritrophic membrane is considered to be secreted by cells of the wall of the colon in the Astigmata and forms a thin, colourless membrane around the food pellet. It protects the cells of the gut wall from abrasion during passage of the food pellet through the posterior mid-gut. The cells of the spherical colon are flattish, somewhat of the nature of pavement epithelium (Michael, 1901) while those of the postcolon in the Acaridae and Analgesoidea have a brush border of very long microvilli (Hughes, 1950c; Lönnfors, 1930). In the pyroglyphid *Dermatophagoides farinae*, on the other hand, the cells of the 'anterior region of the hindgut' (?colon) have long microvilli and their extensive RER and other organelles are typical of an actively secreting cytoplasm (Brody *et al.*, 1972). Extracellular digestion may occur in this region of the gut. A dense band of electron-dense material between the faecal pellet and the microvilli is thought to contribute to the formation of the peritrophic membrane. The wall of the 'posterior region of the hindgut' is lined with cuticle and covered by a homogeneous electron-dense substance that gives the appearance of a mucus-like lining. Endoplasmic reticulum and putative contractile vacuoles were observed in the epidermis underlying the cuticle. The faecal pellet is made up of three to five food balls held together by the mucus-like material of the 'posterior region of the hindgut'. The cuticular lining of the so-called posterior hind-gut in this species points to a proctodaeal origin. Such an extensive proctodaeum is exceptional for the Acari. A modification of the postventricular gut also occurs in *Listrophorus leuckarti* in that there is no clearly differentiated postcolon and the colon passes into the 'rectum' which is lined by thin cuticle (Hughes, 1954a). Neither the colon nor any region of the 'rectum' has cells with long microvilli. Both species ingest keratin-rich fragments of mammalian origin – dead scale-like cells (corneum) by *D. farinae* and scrapings of hairs by *L. leuckarti*, and the modifications of the posterior region of the alimentary tract may be associated with this specialized type of feeding.

A peritrophic membrane is also present in the Oribatida (Woodring & Cook, 1962a; Hoebel-Mävers, 1967; Dinsdale, 1975). According to Dinsdale, a diffuse peritrophic layer is usually associated with the food pellet or bolus in the ventriculus of *Phthiracarus*. It becomes denser in the colon and postcolon, where it measures 25 nm in thickness, while on a faecal pellet occurring in the postcolon it measured 250–500 nm. The peritrophic membrane has no discernible structure and is probably a mucopolysaccharide. The cells of the colon and postcolon in *Phthiracarus* and *Ceratozetes* have a brush border of microvilli but this was not observed on the colonic cells of a number of species, including *Steganacarus* and *Nothrus*, studied by Hoebel-Mävers. The pellet is supported in the postcolon by compact masses of microvilli.

8 Excretion and Osmoregulation

Excretion involves the elimination from the body of non-gaseous by-products of cellular metabolism. The term is generally restricted to the removal of nitrogenous waste resulting from the metabolism of proteins and nucleic acids although the process is bound up with the regulation of the flux of water and certain electrolytes between the organism and its external environment. Excretion, in the strict sense, is performed by the Malpighian tubules arising from the postventricular region of the alimentary tract or, in their absence or decreased function, by a region of the mid-gut. A number of organs have been found to be involved in ionic and water balance, such as coxal glands, salivary glands, genital papillae and Claparède organs.

Excretion

The Malpighian tubules of the Arachnida develop from the end of the embryonic mesenteron and are considered to be of different origin from those of the Insecta, which arise from the proctodaeum (Millot, 1949). Two pairs of tubules occur in the Holothyrida and a single pair is present in the Notostigmata, Ixodida and Mesostigmata. The tubules are absent in the Prostigmata and Oribatida but are present, albeit weakly developed, in some families of the Astigmata. As in other arachnids, the chief nitrogenous catabolite in the Acari is guanine (2-amino-6-oxypurine) although uric acid (2-6-8-oxypurine) appears to be the main excretory product in some species, for example in the larva of *Eutrombicula splendens* (McLaughlin, 1968). Guanine is the most insoluble of the various purine compounds and, in the same way as uric acid, can be stored in the Malpighian tubules in solid form. The excretion of insoluble catabolites is an important adaptation by arthropods to the terrestrial environment. Their formation reduces the toxicity of nitrogenous excretory products and enables their temporary storage in the organism. Further, this adaptation allows the extensive reabsorption of water during the excretion process. The apparent absence of excretory crystals in the midgut of the Oribatida led Woodring and Galbraith (1976) to postulate that the excretory product in this group is soluble and non-crystalline, possibly urea or ammonia.

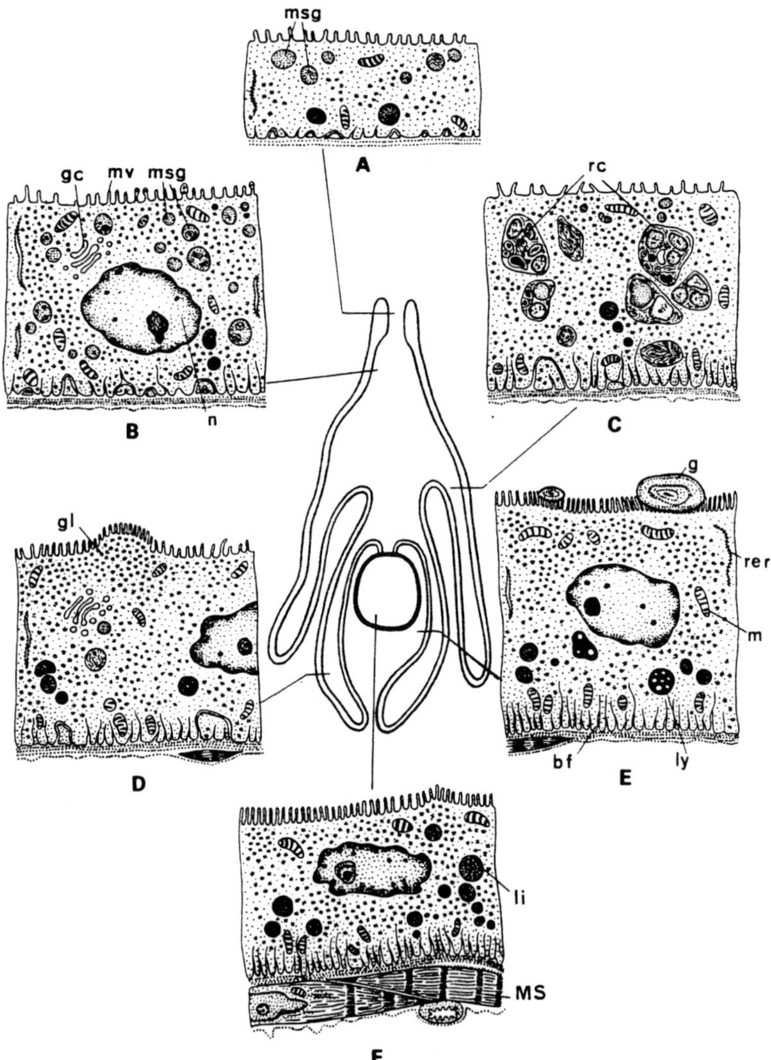

Fig. 8.1. Ultrastructure of the Malpighian tubules in *Hyalomma asiaticum* (Ixodida) (after Raikhel, 1979b); **A**, cell of apical ampulla; **B**, cell of distal region with mucopolysaccaride inclusions; **C**, cell of distal region with rickettsiae; **D**, cell of main region of tubule; **E**, cell of proximal region; **F**, cell of postcolon (rectal sac). *bf.* = basal infoldings; *g.* = guanine spherule; *gc.* = Golgi complex; *gl.* = glycogen; *li.* = lipid droplet; *ly.* = lysosome; *m.* = mitochondrion; *MS.* = muscle; *msg.* = mucopolysaccharide vacuole (secretion granule); *mv.* = microvillus; *n.* = nucleus; *rc.* = colony of rickettsiae; *rer.* = rough endoplasmic reticulum.

Anactinotrichida

The narrow tubules in the Ixodida are two to five times the body length and in females form several loops among the internal organs. Their blind tips extend anteriorly to the level of the synganglion. They open into the postcolon. The wall of the tubule is composed of a single layer of epithelial cells bordered basally by a basement membrane which is covered externally by a muscle layer. The cells are almost flat to cuboidal or columnar in form and have large nuclei (Fig. 8.1). The apical surface of the cells, i.e. the surface facing the lumen of the tubule, has microvilli and the basal infoldings of the plasma membrane are characteristic of transporting epithelia. The cells of the distal region of the tubule of the Amblyommidae (Fig. 8.1A–C) differ from those of the remainder of the tubule in having weakly developed microvilli and basal infoldings of the plasma membrane and in containing the symbiotic rickettsiae *Wolbachia* (Till, 1961; Raikhel, 1979b). The symbionts in *Argas* and *Ixodes*, on the other hand, are found in all parts of the tubule except the part nearest to the postcolon. The function of the symbionts is not known. Cells in the main region of the tubule are identical structurally and numerous glycogen granules are present in the cytoplasm (Fig. 8.1D). The tubule lumen contains numerous spheroid crystalloids, chiefly of guanine. The most extensive guanine crystal formation occurs in the proximal region of the tubule and the microvilli are considered to participate in the process (Fig. 8.1E). The major reabsorption of Na^+ or K^+ ions probably also occurs in this region. Guanine spheroids of various sizes accumulate in the lumen of the tubule, the larger ones showing concentric lamellation. The spheroids are white and strongly birefringent in polarized light. They appear to be concentrated at first in the proximal region of the tubule but later spread into the distal portion until the tubule becomes completely filled and milky-white in colour. Excreta is passed into the postcolon by peristalsis. In the Argasoidea, guanine is the only purine that has been detected and accounts for 93–98% of the dry weight of the white faeces produced by the tick but, in the Ixodoidea, a second unidentified purine has also been found and forms 4–15% of the total purine fraction as opposed to 85–92% (dry weight) of guanine (Kaufman & Sauer, 1982). Only negligible fluid losses occur by way of the Malpighian tubules in ticks.

The histology and form of the Malpighian tubules in the Mesostigmata are essentially similar to those in the ticks. The tubules arise from the lateral or ventrolateral wall of the postcolon, twist dorsally and then run anteriorly to terminate in the anterior region of the idiosoma in free-living species or extend with the anterior ventricular caeca into the cavities of the basal segments of the first pair of legs in blood-feeding parasites. The tubules often lie adjacent to parts of the anterior and posterior caeca (see Fig. 7.2A, p. 211). The cells of the tubule are large with a central nucleus and often possess protuberances directed into the lumen. They all appear to be of the same type. The number of cells in a cross-section of the tubule is usually two or three but never more than five. Guanine crystals are found in the lumen of the tubules and are considered by Hughes (1952) to be precipitated from a solution of nitrogenous material secreted by the cells. Movement of metabolic wastes from the tubules into the postcolon is by peristalsis and in *Echinolaelaps* (Jakeman, 1961) and *Haemogamasus* (Young, 1968a) it is regulated by a sphincter muscle at the proximal end of each tubule. Such a sphincter is lacking in *Uroactinia agitans* (Woodring & Galbraith, 1976) and back-flow of faecal material into the tubules is thought to be prevented by unidirectional peristalsis. Young (1968a) observed peristalsis of the excretory tubules of unhatched larvae of *Haemogamasus liponyssoides* and guanine in the tubules of larval *H. ambulans* at birth.

Actinotrichida

Although a pair of weakly developed Malpighian tubules arising at the junction between the colon and postcolon is present in Acaridae, their role in excretion is considerably less important than in the Parasitiformes (Hughes, 1950c; Prasse, 1967b). In *Acarus* and *Tyrophagus*, the tubules are found to be either empty or filled with a homogeneous mass whereas in *Sancassania* they usually contain a variable amount of unidentified granular material. Guanine is found densely packed in the parenchyma and also appears in the lumen of the postventricular mid-gut. Hughes (1950c) suggested that since guanine was present in the membrane-covered faecal pellet before its entry into the postcolon, its excretion had already taken place in the colon or possibly in the ventriculus. The epithelium of the tubules is similar to that of the mid-gut and has a striated border.

The postventricular mid-gut is considered to be the site of excretion of nitrogenous catabolites in the Prostigmata although little concrete information is available on excretory processes in this group. The thin-walled postventricular mid-gut in the Parasitengona is not connected with the ventriculus and extends nearly the entire length of the body. Most of this so-called excretory organ is in intimate contact with the anterior mid-gut wall (Mitchell, 1964). The excretory product in its lumen is assumed to be guanine. However, in the larva of *Eutrombicula splendens*, in which the postventricular mid-gut is in the form of a narrow tube with a posterior expansion forming a chamber, the semi-fluid granular excretory product (hardening into a solid white pellet when expelled from the body) does not contain guanine or free amino acids. The excreta, on the basis of chemical tests, appeared to contain about 25% uric acid as well as homogeneous, non-granular masses of sialomucin or weakly acidic sulphated mucosubstances probably secreted by the cells of the gut wall (McLaughlin, 1968). In the Tetranychidae, the postventricular gut is connected to the ventriculus and transports both food residues and the excretory product guanine (McEnroe, 1961). The cells of lateral walls of the postventricular mid-gut may be responsible for excreting the guanine and the modified cells of the dorsal wall may produce the matrix in which the crystals of guanine are embedded (Alberti & Crooker, 1985). The report of Malpighian tubules in females of *Tetranychus urticae* by Mothes and Seitz (1980) appears to be erroneous, the structures probably being part of the reproductive tract and having a secretory and not an excretory role (Van der Geest, 1985).

Neither guanine nor uric acid has been detected in the parenchyma or in the cells and lumen of the alimentary tract in the Oribatida. In a species of *Phthiracarus,* Dinsdale (1975) observed large darkly staining bodies up to 2 μm in diameter in the cells of the wall of the ventriculus, but not of the caeca, in glutaraldehyde-fixed tissue in buffered lead nitrate solution. The staining was also localized along the haemocoelic surface of the ventricular wall, particularly within the numerous invaginations. Several of the bodies were found in the membrane-covered faecal pellet and many smaller bodies, showing the same staining reaction but without a limiting membrane, were associated with the brush border of the colon and postcolon. None was detected free in the gut lumen although this may have been due to the solubility of the bodies in the solutions used in the preparation of the tissue for electron microscopy. The plumbophilic excretory material is postulated to accumulate along the outer wall of the ventriculus and to be engulfed in large invaginations of the cell walls. The inclusions resulting from this process pass through the cytoplasm of the cells and appear in the faecal pellet. The composition of the excretory product is not known but is unlikely to contain 'appreciable amounts of uric acid or guanine'.

Osmoregulation

The French physiologist Claude Bernard stated that the constancy of the internal environment is the necessary condition of free life. The maintenance of a proper 'milieu intérieur' depends on a number of equilibrating devices among which, in mites and ticks, those that maintain a constancy in the amount of water (the all-important biochemical reagent) and ions in the tissue fluids are pre-eminent. Acarines like other terrestrial arthropods lose water by diffusion through the body surfaces and respiratory openings as well as in secretions of glands and in faeces. To minimize this, various devices for protection against water loss have evolved including the water-conserving waxy cuticle covering the body and appendages, the mechanical closure of the stigmata, and the retention or reabsorption of water which would otherwise be lost in the excreta. To these may be added behavioural mechanisms which result in the acarine moving to microhabitats where water uptake can occur (Lees, 1948; Camin & Drenner, 1978). For example, unfed, questing *Ixodes ricinus* exposed on the tips of vegetation to dehydrating conditions below its critical equilibrium humidity rapidly loses water at about the rate of 10% per day. When the water balance is normal, the tick avoids the higher humidities of the basal mat of vegetation but this avoiding response (taxis) disappears after desiccation and is replaced by a kinesis, the tick becoming very active in dry air but coming to rest in moist air. This behavioural change results in the tick coming to rest in the vegetation mat where it can replenish its water deficit by absorption of vapour from the atmosphere (Knülle & Rudolph, 1982). A response to light, as well as to humidity, is involved in the control of the vertical movement of larval *Haemaphysalis leporisplaustris*. The unfed water-saturated larva is photopositive but responds negatively to moisture and so moves from the moist basal mat to the tips of the vegetation. The desiccated tick becomes insensitive to light but responds positively to moisture and follows the humidity gradient to the mat.

Actively feeding acarines receive their water by ingesting hydrated food or imbibing free water. Those species which ingest large quantities of fluid of animal origin, often considerably greater than the volume of their body, compensate for the tendency towards dilution of their tissues by excreting hyposmotic fluid. Thus, in ticks, water balance in the feeding phase involves the elimination of water while in the non-parasitic phase it is maintained by conservation and water uptake. According to McEnroe (1963), tetranychids which feed on cell sap use an alternative strategy for fluid excretion in which particulate material, such as chloroplasts, is shunted into the ventricular caeca while the fluid component flows directly into the postventricular mid-gut and is eliminated as 'urine'. However, this mechanism would give no opportunity for the mid-gut to digest much of the ingested food and it is more likely, as Alberti and Crooker (1985) point out, that direct transport of water occurs by way of the anterior mid-gut epithelium to the 'excretory organ' and the coxal glands. Many organ systems are considered to be involved in water and ionic regulation in the Acari although the function ascribed to a particular structure is often inferred only from its morphology, particularly in the mites. Coxal glands, salivary glands, genital papillae and Claparède organs, as stated above, are the main organs involved in ionic and water balance.

Coxal glands

The paired coxal glands (=coxal organs) of the Acari, as in other Arachnida, are probably derived from coelomoducts and each comprises a thin-walled sacculus specialized for the ultrafiltration of the haemolymph and a convoluted tubule or labyrinth whose microvillar walls and high mitochondrial complement suggest the active reabsorption of substances, including ions, from its lumen. Although originally segmentally sequential structures, the coxal glands of extant Arachnida show considerable reduction in number and typically comprise a single pair (two pairs in some Amblypygi and some Araneae) located in the prosomatic region.

Coxal glands, not all showing the typical sacculus/tubule form, have been described from the seven orders of the Acari but it is only in the Argasoidea and Astigmata that the function of the glands has been studied in any detail.

Anactinotrichida

Paired coxal glands having the typical form occur in the Holothyrida, Notostigmata and Argasoidea. The glands in the Holothyrida consist of a membraneous sacculus and looped tubule opening on coxae I (Thon, 1905b) whereas those of the Notostigmata, called tubular glands by With (1904), each open by a duct into a canal or gutter which extends from the posterior margin of coxa I to the base of the tritosternum. The walls of the gland consist of a single layer of cells which merge into each other and have their nuclei situated near the lumen. Other arachnids, such as the Uropygi and Amblypygi, in which a tritosternum (sternapophysis) is present, also have coxal glands discharging into a gutter that connects the gland orifice and the base of the tritosternum. It is probable that the primary role of the tritosternum and subcapitular groove (or analogous structure) in the Arachnida is to transport coxal gland products to the oral region and that their participation in fluid feeding in the Mesostigmata is a secondary development for dealing with copious liquid food.

In the Argasoidea, water and some ions have to be eliminated in order to concentrate the nutrient component of the blood meal and to maintain osmotic and ionic regulation. This is achieved by a pair of coxal glands opening in the articulating cuticle at the base of coxae I (Kaufman & Sauer, 1982). Each gland usually comprises a thin-walled sacculus and a convoluted tubule (Fig. 8.2A) although in *Ornithodoros delanoei* Lees (1946) considers the tubulus to be lacking and parts of the sacculus to probably function as tubular cells. The sacculus is in the form of an ultrafiltration membrane consisting of a layer of podocytes lying on a basement membrane. Their pedicelles in unfed females are spaced about 28–50 nm apart and form pore slits covered by a very fine membrane. Two types of cells are present in the tubule: cuboidal cells with microvilli and large nuclei in the proximal part, and columnar cells with short microvilli and small nuclei in the distal region. Both types have their nuclei located apically. Cytoplasmic vesicles are rare and Golgi bodies are not well developed. They have rich infolding of the basal plasma membrane and numerous mitochondria – features of actively transporting epithelia. The functional difference between the cuboidal and columnar cells has not been elucidated. The coxal fluid of many argasoids is less concentrated than the haemolymph, at least with respect to chloride (Lees, 1946). Ultrafiltration appears to be non-selective in regard to small molecules and it is possible that the tubule is concerned with reabsorption of metabolically active substances, including ions. However, the reabsorption of water does not occur in the tubule. A small gland, the accessory gland, discharges by a short duct into the tubule near the coxal

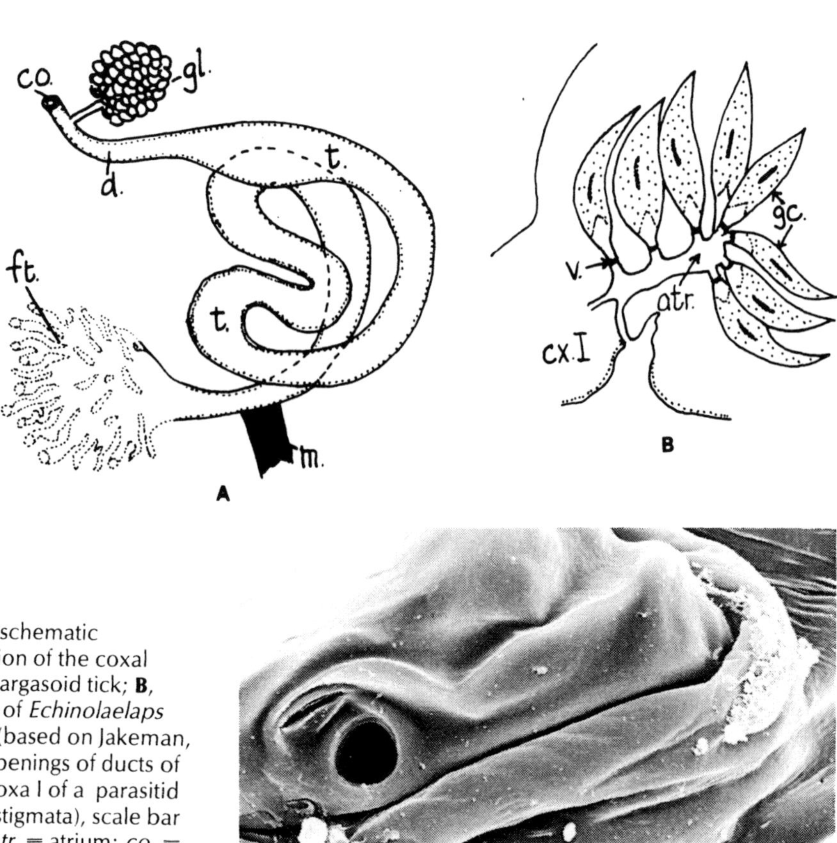

Fig. 8.2. A, schematic representation of the coxal gland of an argasoid tick; **B**, coxal gland of *Echinolaelaps echidninus* (based on Jakeman, 1961); **C**, openings of ducts of glands on coxa I of a parasitid mite (Mesostigmata), scale bar = 10 μm. *atr.* = atrium; *co.* = orifice of coxal gland; *cx I* = coxa I; *d.* = duct; *ft.* = filtration membrane; *gc.* = gland cell; *m.* = muscle; *t.* = tubulus, *v.* = valve.

orifice. Its secretion in *Ornithodoros* is believed to be the source of a sex pheromone present in coxal gland fluid discharged from the body while the tick is feeding or immediately afterwards (Schlein & Gunders, 1981). In *Ornithodoros papillipes*, about 40% of the weight of the blood ingested by the female is eliminated as coxal fluid (Balashov, 1968). Thus, the coxal glands in the argasoids have four main functions: the regulation of the ionic composition of the haemolymph; the regulation of the volume of the haemolymph; the elimination of metabolic waste molecules that are too large to diffuse across the gut membrane, Malpighian tubules or integument; and the dispersal of sex pheromones via the coxal gland fluid.

The paired 'coxal glands' described in the Dermanyssina by Jakeman (1961) and Akimov *et al.* (1988), and in the Uropodina by Woodring and Galbraith (1976) do not have the sacculus/tubule form. In *Echinolaelaps echidninus*, the gland comprises nine or more elongated cells each with a small ventral lumen emptying through a sclerotized valve into a short duct. This duct together with those from the other cells opens into a tubular atrium which has its orifice situated proximally on the ventral surface of coxa I (Fig. 8.2B). Only one pair of glands was considered by Jakeman to be present in the Mesostigmata. However, scanning electron microscopy reveals that three distinct openings are present on the ventral surface of coxa I in the Dermanyssina and Parasitina (Fig. 8.2C), one situated internally and two, usually close together in a depression, externally near the proximal margin (Evans, 1984). This suggests that the ducts of three glands debouch on the ventral surface of coxa I in species of these suborders. The sclerotized valves and ducts of the coxal gland are clearly visible after the specimen has been macerated in lactic acid. The glands are considered to have an osmoregulatory function but this is speculation. Bowman (1984) described the appearance of clear fluid droplets in the region of coxae I during feeding on large prey by *Pergamasus longicornis*. Occasionally, a large and extensive droplet was formed which reached up to and over the peritreme and this was manoeuvred by the mite and deposited on the substrate during a pause in feeding. More commonly, feeding stopped temporarily and the droplet disappeared within 1 or 2 s. Accompanying movements of the labrum or pharynx suggested that the liquid was being imbibed. Whether the droplet originated from an orifice on coxae I or was part of the food stream associated with the gnathosomatic fluid transport mechanism was not established.

The paired coxal glands in *Uroactinia agitans* appear to have the same basic structure as those in the Dermanyssina. They are composed of numerous tubes of varying lengths, each probably representing a single cell. The tubes have a narrow lumen and each opens by way of a separate pore into the wall of a cuticle-lined atrium whose orifice lies near the proximal margin of coxa I (Woodring & Galbraith, 1976).

According to Akimov *et al.* (1988), the latero-dorsal glands in *Varroa jacobsoni* are coxal glands that carry out water–salt exchange. They form a pair of S-shaped tubular structures situated symmetrically on the sides of the prosoma above legs I–IV. They are surrounded by parenchymatous tissue. The glands have straight unbranched intracytoplasmic ducts but do not have ducts that are isolated from glandular tissue. The narrow distal section of each gland bends ventrally and opens by way of a pore located on the ventral surface of coxa I.

A number of glands associated with the coxae of the legs, in addition to the coxal glands, have also been described in the Holothyrida and Notostigmata. Paired acinose glands, called pedal glands by Thon (1905b), open ventrally at the base of coxae I in *Thonius braueri*. Thon considered them to produce a secretion which causes symptoms of poisoning when ingested by warm-blooded animals but this requires confirmation. In addition, two glands (Cruraldrüsen) of unknown function are associated with each of coxae II–IV (Vitzthum, 1940–43). One gland is situated on the inner and the other on the outer side of the coxa. Each appears sac-like in form and the wall of the lumen is formed of cuboidal cells. A pair of elongate glands in the form of compressed horseshoes, which appeared to have their origin in the 'hollow of the first pair of coxae', was described by With (1904) in *Opilioacarus segmantatus*. Coineau and Legendre (1975) have suggested that they are probably the first pair of legs in the process of regeneration after autotomy and not glandular bodies.

Actinotrichida

The paired coxal glands in this taxon discharge into podocephalic canals and contribute to the podocephalic gland complex (see Chapter 7). In the Prostigmata, a study of species from 11 superfamilies showed that typical coxal glands, often referred to as tubular glands, are present in all except *Halacarus basteri*, representing the Halacaroidea (Alberti & Storch, 1977). The gland comprises a sacculus and a looped tubule in the majority of the taxa but a sacculus is absent in the Tetranychidae and the Parasitengona. Three types of sacculi were found: the first with a wide lumen as in *Labidostomma* and considered to be the primitive form; the second with a tubular lumen possessing lateral pouches (*Cheyletus*); and the third, occurring in the majority of species, with a slit-like lumen narrowed by bulging podocytes. The podocytes exhibit, other than their pedicelles, numerous pinocytotic vesicles, lysosome-like inclusions, vacuoles and microtubules. The tubule is divisible ultrastructurally into a maximum of five sections. The main part of the tubule, comprising the two proximal sections, may have deep infoldings of the cellular apex as in the Eupodidae, Tetranychidae, *Cheyletus* and *Cunaxa*, or extensive brush borders apically as in the Parasitengona, Bdellidae, *Labidostomma* and *Anystis*. The tubule opens by means of the ectodermal terminal section into an excretory duct, the ectodermal portion being considered to provide a closing mechanism. It is assumed that the coxal glands are involved in water and ionic regulation. Mills (1973) is of the opinion that the coxal gland in the female of *Tetranychus urticae* may produce a sex-attractant pheromone which is added to the silk threads of the webbing when they are still liquid.

The most extensive study of the morphology of the coxal glands in the Oribatida has been made by Woodring (1973) and Alberti and Storch (1977). Each of the paired organs comprises a sacculus and tubule. The coxal gland duct receives the duct of podocephalic gland 3 before penetrating the cuticle of the body wall and discharging into the posterior end of the podocephalic canal (see Fig. 7.6C, p. 225). Coxal glands are present in all active stases. The sacculus is thin-walled and the podocytes project into its lumen. Their ultrastructure in *Hypochthonius* appears to be similar to that described above for the Prostigmata. The sacculus has no rigidity and is prevented from collapse by three to five saccular muscles. The tubule has either one (type B) or three (type A)

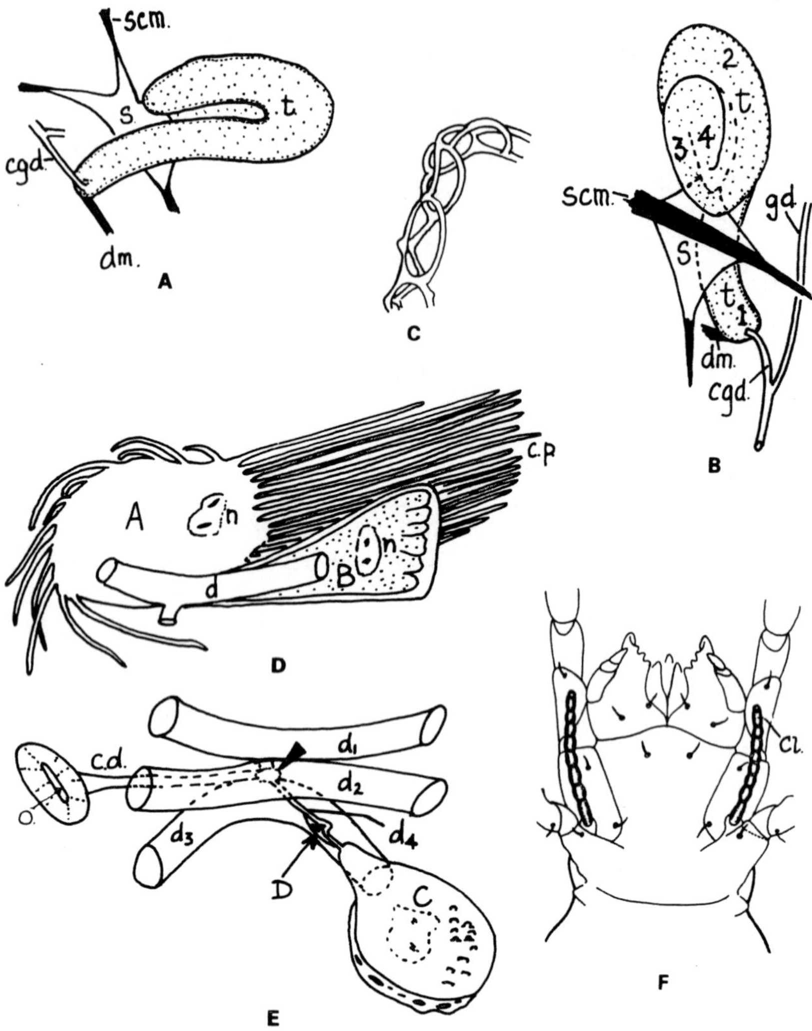

Fig. 8.3. **A**, sacculus and tubule of the coxal gland of an archoribatid mite; **B**, sacculus and tubule of a euoribatid mite; **C**, chitinous support of a tubule showing spiralling strands; **D**, type A and B cells in the supracoxal gland of *Dermatophagoides farinae* (Astigmata); **E**, type C and D cells and duct system of the same; **F**, Claparède organs of a larval epilohmanniid mite (Oribatida). (A–C based on Woodring, 1973; D and E redrawn after Brody *et al.*, 1976; F, after Evans *et al.*, 1961.) A = cell type A; B = cell type B; C = cell type C; *cp.* = filamentous processes of the cell; *d.* d_1–d_4 = ducts; D = cell type D; *c. d.* = common duct; *Cl.* = Claparède organ; *cgd.* = coxal gland duct; *dm.* = duct supporting muscle; *gd.* = lateral gland duct; *n.* = nucleus; *o.* = opening of supracoxal gland duct; *s.* = sacculus, *scm.* = sacculus supporting muscle; *t.* = tubule; 1–4= parts of tubule.

180° curves along its length so that both ends of the tube-like structure are directed anteriorly (Fig. 8.3A,B). Type A tubule occurs in the Archoribatida and type B in the Euoribatida. The tubule has a single layer of cells which lack brush borders and pinocytotic vesicles. The lumen of the tubule of all species forms a spiral which is imposed, in part, by an internal skeletal support. This chitinous support is continuous with the cuticular lining of the coxal gland duct. It is formed of three parallel spiralling strands which never cross one another, the connections forming a Y and not an X configuration (Fig. 8.3C). The length of the tubulus appears to be related to body size and environment. Freshwater species have a greater length of tubulus and littoral species a shorter length than terrestrial species of comparable body size. Oribatids living in fresh water can be expected to have a large influx of water into the body which must be counterbalanced by water elimination. The saccular filtrate in these forms is probably isosmotic to the haemolymph and the reabsorption of essential ions is necessary if the mite is not to become fatally diluted. If, as assumed, the tubule is the site for reabsorption, its greater length in the freshwater species compared with terrestrial forms could be explained by the requirement of a greater reabsorption function by the tubulus and one way of achieving this is to increase its length. The reabsorption of ions is not desirable in those species living in salt water, hence the reduction in the length of the tubule. Woodring (1973) suggests that soluble nitrogenous waste products may also be excreted by the coxal glands.

The so-called supracoxal glands of the Astigmata lack the typical sacculus/tubule structure but have an osmoregulatory function. The glands open by a slit-like pore located above trochanter I into the podocephalic canal running along the ventral border of the supracoxal sclerite (see Fig. 7.8A, p. 228). Each gland in *Dermatophagoides farinae* comprises eight cells that function in pairs as four units, each with a cuticle-lined duct (Brody *et al.*, 1976). Three of the units are similar in size and structure but the smaller fourth unit has a distinctly different morphology. In the similar units, the larger cell of the pair (type A) is characterized by having long mitochondria-rich filamentous processes extending into the haemocoele (Fig. 8.3D). The cell cytoplasm is granular and has few organelles other than the nucleus and a few microtubules. Type A cell envelops part of the other cell of the unit (type B), which contains a T- or Y-shaped cuticle-lined duct. The duct is contained within cell type B except for the end closest to cell type A. The duct-ends appear as porous caps which communicate with the cytoplasm of the cells. The nucleated cell B has numerous microtubules. The free ends of the T- or Y-shaped ducts of the three units coalesce to form a common duct opening to the exterior near the supracoxal seta above trochanter I (see Fig. 7.8B, p. 228). The fourth unit comprises a small type C cell whose components comprise extensive rough endoplasmic reticulum (RER), lysosomes, mitochondria and a nucleus (Fig. 8.3E). This cell is connected to the common duct by a cuticle-lined duct produced by the second cell of the unit (cell D). Cell type A is thought to function in the concentration of materials from the haemolymph, while the duct-producing cell type B probably also has a secretory function. Cell type C with its RER may synthesize a proteinaceous product, possibly an enzyme. It is postulated that the coxal gland produces a hygroscopic secretion rich in salts which is exposed to ambient air on the supracoxal sclerite (Wharton *et al.*, 1979). This vapour-enriched fluid runs by way of the podocephalic canal to the pre-oral cavity from whence it is drawn into the ventriculus by the pumping action of the pharynx. Mites that are water-deficient shrink and show characteristic depressions of the dorsal surface of the idiosoma but as they take up water vapour from the air the body gradually assumes its normal appearance.

Salivary glands

Water and ionic balance in ixodoid ticks are performed primarily by the salivary glands. During feeding, a significant volume of water ingested in the blood meal is eliminated by the glands and returns via the mouthparts to the host (Gregson, 1967; Tatchell, 1967). Kirkland (1971), working with *Haemaphysalis leporispalustris*, suggested that cells of acini type I may be concerned with fluid excretion during feeding, but more recent investigations have indicated that the 'water cells' (=interstitial epithelial cells) of type III acini, and to a lesser extent type II, are better candidates for this role (Kaufman & Sauer, 1982). The 'water cells' are inconspicuous in the unfed tick but increase in size and develop prominent infoldings and mitochondria during feeding. These changes in the morphology of the 'water cells' are accompanied by an increase in the secretory function of the glands. The ionic regulatory function of the oral glands has been examined in *Boophilus microplus* by Tatchell (1969b). He showed that the concentration of Na^+, K^+, Ca^{2+}, Mg^{2+} and Cl^- in the tick was equilibrated to stable values during the rapid phase of engorgement.

In addition to water elimination, the salivary glands in the Ixodoidea also play an important role in the uptake of water vapour in partially dehydrated ticks (Rudolph & Knülle, 1974; McMullen *et al.*, 1976; Knülle, 1984; Needham & Teel, 1986). Compensation of water deficits in non-feeding ticks is possible so long as the RH exceeds the critical equilibrium humidity (CEH) [alternatively expressed as critical equilibrium activity (CEA) in which activity = relative humidity/100] but below this threshold the active transport mechanism ceases to work. The critical equilibrium humidities of different species of hard ticks are relatively high and in many are around 85% RH with a range of 75–94% RH within the superfamily. Non-feeding ixodoid ticks actively absorb water vapour from air by way of the mouth. There is considerable circumstantial evidence that this is mediated by a concentrated hygroscopic secretion produced by agranular type I acini of the salivary glands and swallowed following absorption of atmospheric water. The cells of the acini show ultrastructural features which are characteristic of fluid-transporting epithelia (extensive basal infoldings of plasma membranes with close association of numerous mitochondria and limited luminal spaces) and Balashov (1968) likens them to those present in the nasal salt glands of birds and reptiles. The highly hygroscopic saliva secreted onto the external mouthparts during prolonged exposure to dry air contains sodium, potassium and chlorine in quite different proportions from those found in the copious salivary gland secretions of feeding ticks. Needham and Teel (1986) speculate that glycerol might be a significant component of rehydration saliva in ixodoids.

Kahl and Knülle (1986) have shown that uptake of water vapour from unsaturated air also occurs in fully engorged and detached larvae and nymphs of *Ixodes ricinus*, *I. dammini* and *Haemaphysalis punctata*. They found that the total net gain of water during a single uptake phase in some instances surpassed 15% of body weight. In *I. ricinus*, this ability to actively absorb water vapour persists up to a few days of apolysis and, on the first day after ecdysis, teneral nymphs and adults are able to resume net uptake of water vapour.

The mouth is also the site of water uptake in the argasoids and type 1 acini are similar in ultrastructure to those of ixodoids. It is probable that the salivary glands of haematophagous Dermanyssoidea will also be found to play a role in water balance and osmoregulation. Young (1959) observed droplets of liquid on the mouthparts of *Haemogamasus ambulans* after recently feeding to repletion on blood and suggested that the liquid originated from the salivary glands.

Genital papillae and Claparède organs

In the Actinotrichida, genital papillae (nymphophan organs, genital suckers, genital discs, genital acetabula, etc.) located inside the progenital lips of nymphs and adults of terrestrial mites or on the body of nymphs and adults of some freshwater mites, and a pair of Claparède organs (urpores, urstigmata, bruststiele) situated ventrally between legs I and II on the prosoma of prelarvae and larvae, are considered to play an important role in the uptake of water and ionic regulation (Bartsch, 1973; Alberti, 1979; Fashing, 1988). 'Genital papillae' in the form of eversible sac-like structures located anterior to the genital orifice and protected by the genital verrucae have been described in the Opilioacariformes but neither papillae nor Claparède organs occur in the Parasitiformes. The gential papillae do not function as suckers (Claparède, 1869), sense organs (Michael, 1901; Grandjean, 1946a) or respiratory organs (Vercammen-Grandjean, 1976).

The appearance of the genital papillae during ontogeny follows a sequence which, primitively, comprises one pair in the protonymph, two pairs in the deuteronymph and three pairs in the tritonymph and adult.

Genital papillae never occur in the larva. This series is present in the Oribatida and some Prostigmata. The number of papillae is often reduced in terrestrial Prostigmata and Astigmata and increased in freshwater Parasitengona; for example, known adult Astigmata have a maximum of two pairs while in some freshwater mites more than 100 unprotected papillae are present, usually in the genital area. The papillae located within the genital atrium are extruded by hydrostatic pressure and retracted by muscles attached to the progenital lips (see Fig. 9.13C, p. 283). The papillae appear to be segmental structures and Grandjean (1946a) suggested that they represent the appendages of three opisthosomatic segments, thereby implying a segmental organization of the genital region of the Actinotrichida. He compared them to the spinnerets of spiders but, unlike the Araneae, no trace of opisthosomatic appendages has been observed during embryological development of the Acari. Claparède organs are only present in the prelarvae and larvae whose nymphs and adults have genital papillae (Fig. 8.3F). Grandjean (1946a) considered them to be 'podites' of the second, or possibly the first, pair of legs and postulated their function to be similar to that of the papillae. Alberti (1979) found that ultrastructurally the organs and papillae in species of *Neomolgus*, *Limnesia*, *Piona* and *Arrenurus* have essentially the same structure and concluded that they both have the same function. In *Neomolgus littoralis*, they comprise typical transporting cells characterized by the prevalence of mitochondria in close contact with infoldings of the basal cell membrane. The cells terminate in numerous microvilli which project against the cuticle, which is thinner and lacks the proximal region of ordinary cuticle (Fig. 8.4). In the water mite *Limnesia maculata*, the cuticle of each of the three pairs of genital papillae, which are not covered by the progenital lips, is even more modified in having numerous pores leading into deep pits often containing a dark substance. *Hydrovolzia placophora* is unusual in that the papillae are only found on the epimeral plates (Alberti & Bader, 1990). This location is interesting in that Claparède organs, which are considered to be homonomous structures, are associated with the epimeral regions of prelarvae and larvae. In the astigmatic *Naiadacarus arboricola*, the Claparède organs and genital papillae are unicellular and the covering of the distal dome consists of dense granular material that appears quite different from that of the cuticle. Further, the organs and papillae have differences in their ultrastructure (Fashing, 1988).

All genital papillae and Claparède organs examined to date have numerous mitochondria associated with plasma membrane plications – characteristic of cells with active transport functions. According to Alberti (1979), the cells of the papillae and Claparède organs in mites living in hypotonic fresh water precipitate AgCl after treatment with fixatives containing silver ions. They are considered to be chloride cells with an osmoregulatory function involving the active uptake of ions from the external medium. In terrestrial Oribatida, Astigmata and Prostigmata, cells with a very similar fine structure occur but chloride precipitation is

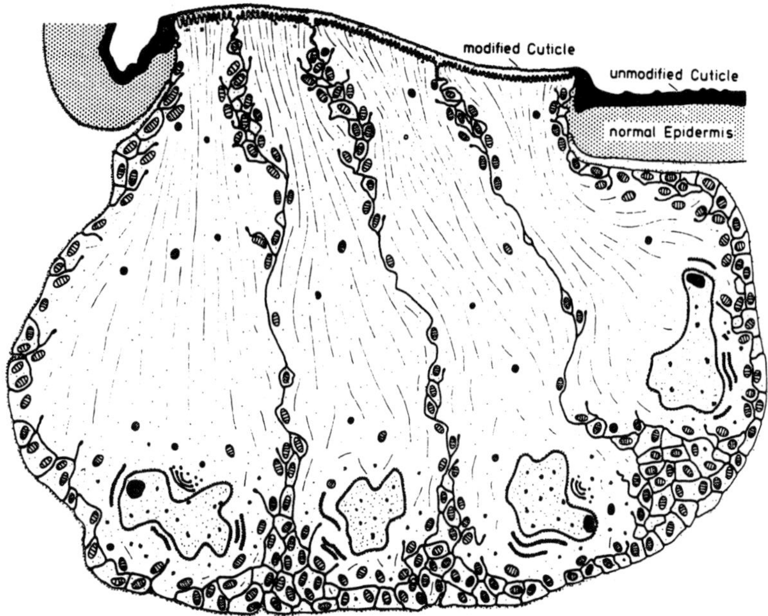

Fig. 8.4. Ultrastructure of a genital papilla of *Neomolgus littoralis* (Prostigmata) (after Alberti, 1979).

absent and it is assumed that the papillae and Claparède organs are mainly responsible for water uptake.

Adults of *Algophagus pennsylvanicus* [Astigmata] living in water-filled tree holes have a convex porous plate on each side of the prosoma between legs I and II. Fashing (1984) referred to these structures as axillary organs. The cuticle of the porous plate differs from surrounding cuticle in being thinner and non-lamellate. The plate covers specialized epidermal cells which are similar in fine structure to chloride cells. Fashing suggests that they have an osmoregulatory function. Earlier, Fain and Johnston (1975) had observed similar structures in other Algophagidae but considered them to act as air chambers and to have a role in flotation and respiration. Axillary organs are present in the larvae, protonymphs, tritonymphs and adults of *Algophagus*. In the larva, each organ is relatively small and situated dorsolaterally above trochanter I (in the region of the supracoxal sclerite) but in succeeding instars it may considerably increase in size and extend ventrally between legs I and II as in *A. pennsylvanicus* or be restricted to a dorsolateral position as in *A. semicollaris*. Claparède organs are lacking and their osmoregulatory function appears to have been taken over by the axillary organs. Whether the larval axillary organs are modified Claparède organs or new structures which have evolved in association with the colonization of semi-aquatic (wading in liquid) to aquatic habitats has not been established. The retention of Claparède organs in the presence of genital papillae (albeit of relatively small size) would be unusual but not unique for the Actinotrichida since H.M. André (personal communication) has observed the presence of putative Claparède organs throughout ontogenetic development in the Tydeoidea and these coexist with genital papillae in the nymphs and adults.

Genital papillae, Claparède organs and axillary organs have a similar fine structure and assumed function to the coxal or eversible organs of such hexapod groups as the Diplura and Thysanura. Eversible vesicles have also been described on the opisthosoma of the Palpigradi (Remane *et al.*, 1976) and structures comparable to Claparède organs possibly occur in larval Solifugae and Pedipalpi.

Sperm access system

Michael (1892) noted that the sacculus and rami of the sperm access system in *Haemogamasus* were capable of endosmosis and exosmosis. This characteristic was also observed by Young (1959) who found that females recently fed to repletion on heparinized blood excreted a few droplets of liquid from the solenostomes (sperm induction pores). Further, a blue dye injected into the haemocoele was taken up by the sacculus and rami and stained the liquid excreted from the solenostome. He suggested that this excretory function of the sperm access system provided a unique method for allowing females that ingest large quantities of fluid to digest the meal quickly without disrupting ionic and water balance.

9 Reproductive Systems

The Acari are dioecious but the form and structure of the reproductive organs show a wide range of variation. The gonads comprise a pair of mesodermal structures that have probably evolved from coelomic sacs whose coelomoducts form the gonoducts. The mesodermal ducts, oviducts in the female and vasa deferentia in the male, no longer open directly to the outside of the body but are connected with a median structure, either the vagina or the ejaculatory duct, formed by the invagination of the ventral body wall. Although the mesodermal components of the reproductive system in both sexes are primitively paired their fusion to form unpaired structures is common. The gonopore is situated on the ventral surface of the body except in males of certain Cheyletoidea, such as the Demodicidae and Psoregatidae in which it is located dorsally. Accessory glands opening into the genital tract are usually present and are particularly well developed in the male.

The Male Reproductive Organs and Spermatogenesis

The male organs consist essentially of paired or unpaired testes, variably situated in relation to the alimentary tract, and leading by way of paired or unpaired vasa deferentia directly or through a median seminal vesicle, to an ejaculatory duct. The terminal section of the latter may be developed into an intromittent or spermatophoric organ, the so-called penis or aedeagus. Accessory glands concerned with the production of seminal fluid and spermatophores open into the seminal vesicle or the ejaculatory duct. Some indication of the range of variation in the form of the male reproductive organs is given in Fig. 9.1A–E.

Testes

Among the Parasitiformes the testes are paired in the Holothyrida (Vitzthum, 1940–43) and the Uropodina (Michael, 1890) but are fused in the mesostigmatic Parasitina, Dermanyssina and Sejina (Winkler, 1888; Michael, 1892). In the Ixodida, they are fused posteriorly in the Argasidae, partially fused along their median surface in the Amblyommidae and connected, at the most, by a thin filamentous strand of tissue in the Ixodidae (Oliver, 1982). Paired testes also occur in the Astigmata and some Prostigmata, for example, Tetranychidae and Erythraeidae (Witte, 1975a), but in most of the Prostigmata they form an unpaired structure. The testes of the Oribatida are exceptional in that they appear as single or bilobed outgrowths of the ventrolateral margins of a large, lobed glandular organ which is often referred to as a seminal vesicle (Michael, 1884; Woodring & Cook, 1962a). It seems likely that this organ is derived by the fusion of the posterior regions of the vasa deferentia which have become glandular (Hughes, 1959).

The testes are considered to be primitively tubular in form and this has been retained by the Ixodida and some Mesostigmata but in other acarines they form regular or irregular globular masses. In the Parasitiformes and Oribatida, the germinal cells of each testis are arranged in cysts or packets. These spermatogonal cysts or spermatocysts are bound within a layer of flattened somatic cells. Spermatogonia (primordial germ cells) in ticks undergo mitotic division to produce several generations of daughter cells, the last generation giving primary spermatocytes which divide meiotically to produce secondary spermatocytes (Fig. 9.2A). Each of these subsequently divides into two spermatids. Meiosis is synchronous in a given cyst. The newly formed spermatids (Fig. 9.2B) are rounded with the nucleus located centrally and the periphery surrounded by cup-shaped subplasmalemma cisternae (Breucker & Horstmann, 1972; Oliver, 1982).

In the course of spermatogenesis in ticks, a large vacuole forms in the sperm cell through the fusion of Golgi-derived vesicles (Breucker & Horstmann, 1972; Oliver & Brinton, 1973; Wuest *et al.*, 1978; Feldman-Muhsam & Filshie, 1979). The wall of the vacuole has numerous cytoplasmic processes of specific substructure (Fig. 9.2C). A cytoplasmic column protrudes into the vacuole from the posterior pole of the sperm cell and an extensive acrosomal vacuole spreads as a flat cisterna under the plasmalemma (Fig. 9.2D). During this time the nucleus comes to lie on the side of the cell in the peripheral cytoplasm and it becomes fusiform in shape. The last phase of spermatogenesis occurs in the female genital tract and is referred to as capacitation (spermateleosis). Certain secretions of the accessory gland are considered to initiate this process, which involves the opening of the operculum of the cell at its anterior pole and the eversion of the cytoplasmic column through it. By this means the entire sperm cell turns inside out so that the wall of the vacuole with its processes forms the outer wall of the capacitated

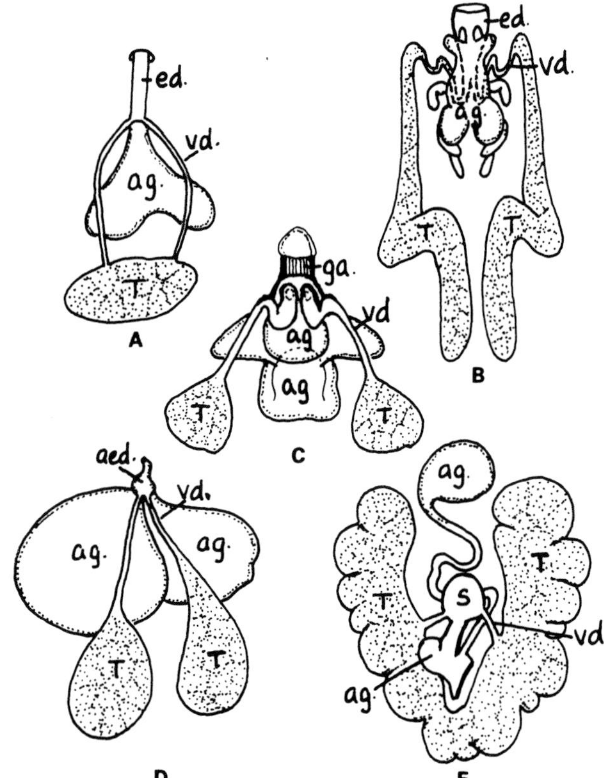

Fig. 9.1. Schematic representation of a range of male reproductive organs in the Acari. **A**, Mesostigmata (Parasitidae); **B**, Ixodida (Amblyommidae); **C**, Mesostigmata (Uropodoidea); **D**, Astigmata (Acaridae), **E**, Prostigmata (Parasitengona). *aed.* = aedeagus; *ag.* = accessory gland; *ed.* ejaculatory duct; *s.* = seminal vesicle; **T.** = testes; *vd.* = vas deferens.

spermatozoon (Fig. 9.2E). There is a doubling of the length of the cell during capacitation and the resultant vacuolated spermatozoon is capable of gliding movements. Germinal cells in the posterior (proximal) part of the tick testis are the first to mature and maturation proceeds towards the anterior (distal) end in a wave-like manner. Tick sperm range from 150 to 1000 μm in length with those of argasoids being longer than those of the ixodoids. All known spermatozoa in the Acari are of the aflagellate type and motility has been acquired secondarily by means of new organelles (Baccetti, 1979).

Spermatogenesis and differentiation of the accessory gland tissue are completed in the nymphal instar of prostriate ixodoid ticks so that recently moulted males have vacuolated sperm cells (prosperms) in their genital ducts (Balashov, 1968). Blood feeding is not required to complete spermatogenesis in these forms, hence the widespread occurrence of aphagia in males of *Ixodes* s.lat. In metastriate forms, however, unfed males are infertile and prosperms only appear in the vasa deferentia 3–5 days after the male commences to feed. Males of some argasoids, for example *Argas persicus* and *Ornithodoros papillipes*, are sterile for several days or weeks after moulting and, when they attempt to copulate with females, fertilization is not accomplished because of the lack of

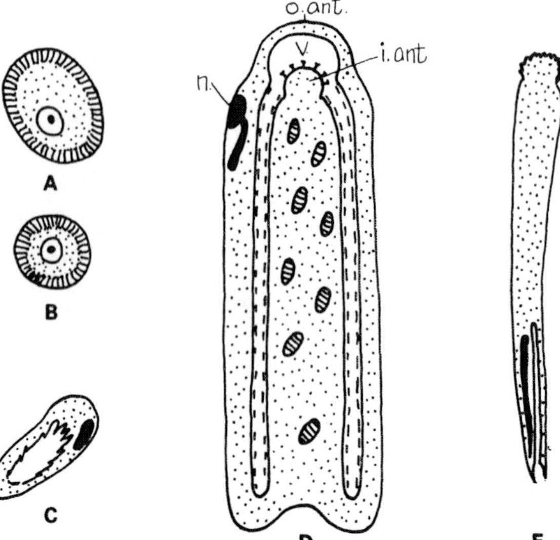

Fig. 9.2. **A**, primary and **B**, secondary spermatocysts of Ixodidae; **C**, spermatid; **D**, uncapacitated sperm (prosperm); **E**, spermatozoon. (A–C, E after Filippova, 1985; D based on Alberti, 1991.) *i. ant.* = inner anterior end; *o. ant.* = outer anterior end; *n.* = *nucleus; v.* = vacuole.

prosperms in the spermatophores. However, feeding can greatly accelerate maturation in these species.

The vacuolated-type spermatozoon of the ticks also occurs in the Notostigmata (Fig. 9.3A) and in some groups of the Mesostigmata, such as the Uropodina (Fig. 9.3B), Epicriina (Fig. 9.3C) and Zerconina (Alberti, 1980a). Characteristic features of the spermatozoa of the Parasitina and Dermanyssina, on the other hand, are the absence of a large vacuole and the presence of longitudinally directed ribbon-like structures or stiff bands which are derived from sac-like invaginations (flat chambers) of the surface of the cell (Witalinski, 1975, 1979b; Alberti, 1980a). Alberti refers to them as 'ribbon spermatozoa' (Fig. 9.3D,E). The flat chambers are assumed to be homogolous to the large vacuole or its smaller precursors of the vacuolated spermatozoa (Alberti & Hänel, 1986). The sperm cell is not turned inside-out during spermatogenesis as in the vacuolated form. The ribbons and cell inclusions in the Parasitina are more complicated than in the Dermanyssina. In both groups, the

Fig. 9.3. **A–C**, three types of vacuolated sperm in the Anactinotrichida, **A**, *Opilioacarus texanus* (Notostigmata); **B**, *Cilliba cassidea* (Mesostigmata); **C**, *Epicrius mollis* (Mesostigmata). **D, E**, ribbon-type sperm in *Parasitus berlesei* (Parasitina); **D**, schematic transverse sections; **E**, transverse section showing peripheral chambers (flat chambers) which constitute longitudinal 'ribbons'. (After Alberti, 1980a.) *Af.* = acrosomal filament; *Ak.* = acrosomal canal; *Apl.* = acrosomal plate; *Av.* = acrosomal vacuole; *äV.* = outer anterior end; *CM.* = cytoplasmic mantle; *CS.* = cytoplasmic column; *Ek.* = inclusion body; *iV.* = inner anterior end; *Lb.* = longitudinal band derived from flat chambers (peripheral chambers); *Ll.* = longitudinal ridge; *KR.* = nuclear region; *MR.* = mitochondrial region; *RE.* = region of inclusion bodies; *Sp.* = sperm apex; *V* = vacuole; *Zk.* = nucleus.

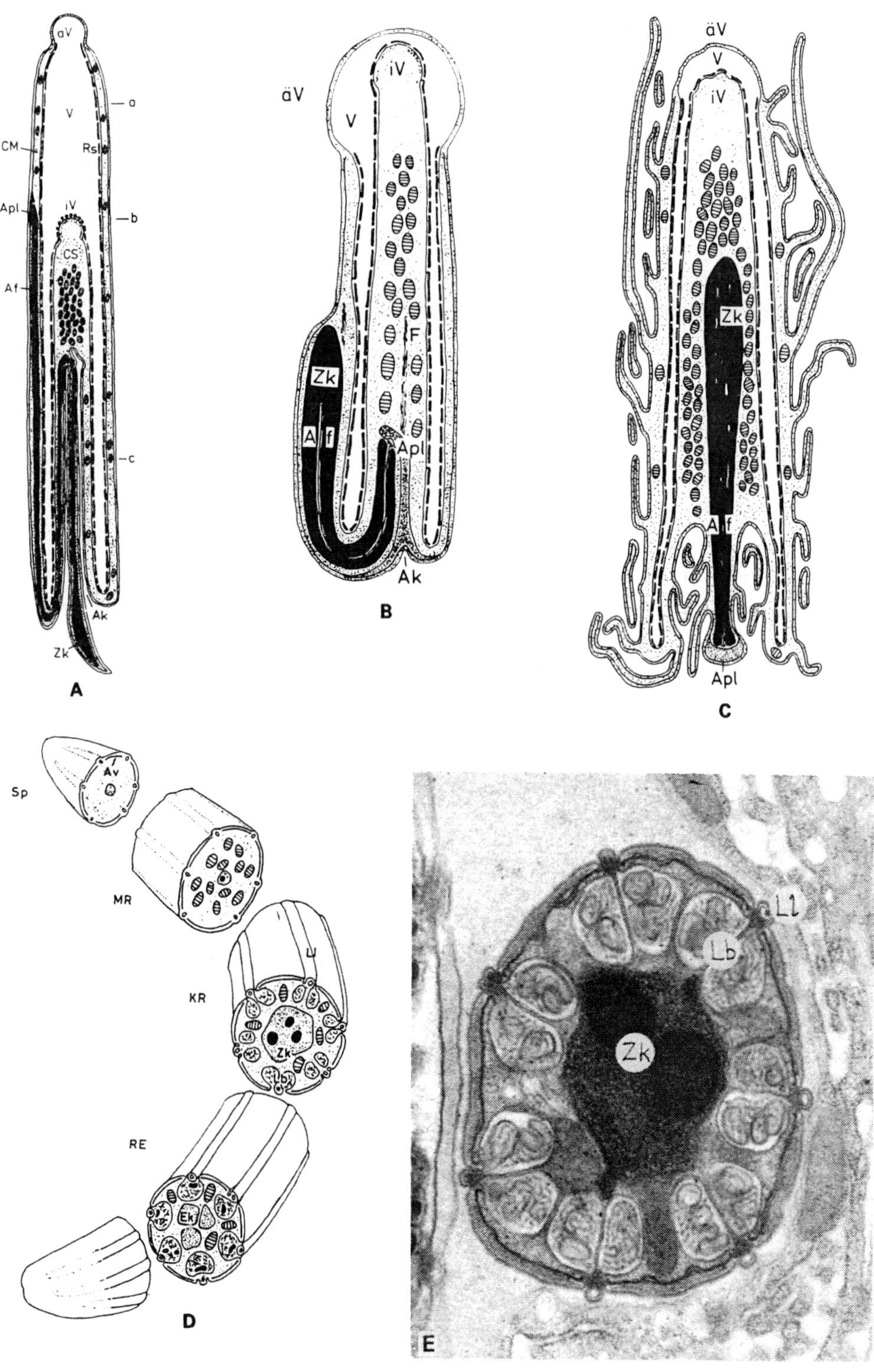
aV
V
a
CM
Rsl
Apl
iV
b
CS
Af
c
Ak
Zk
A
äV
iV
V
F
Zk
A f
Apl
Ak
B
äV
V
iV
Zk
A f
Apl
C
Sp
Av
MR
KR
Zk
RE
Ek
D
L1
Lb
Zk
E

ribbons may be arranged in pairs and they are undulating in those spermatozoa with irregular cell bodies.

The filiform sperm of the Parasitina and the irregularly shaped sperm of the Dermanyssina modify their shape during passage in the female, the latter becoming elongate and the former 'even more filamentous' (Michael, 1892). This requires major changes in the structure of the sperm in the Dermanyssina, in particular, and these appear to be initiated in the rami of the sperm-access system in the female. Thus, during capacitation in the female of the bee parasite *Varroa jacobsoni,* the nucleus elongates and the chromatin changes structure whilst in the cytoplasm fibrils and peripheral filamentous structures become visible (Alberti & Hänel, 1986). The latter are considered to enable the spermatozoa to make slight movements and to arrange themselves in one direction within the receptaculum of the sperm-access system. At present, ribbon spermatozoa have been found only in those Mesostigmata in which the male genital orifice is presternal in position and the chelicerae are modified as gonopods.

The stages of spermatogenesis in the bee parasite *Varroa jacobsoni* are discussed in detail by Alberti and Hänel (1986). They recognize six distinct stages in the male and two within the sperm-access system of the female. In the final stage within the spermatheca of the female, the spermatozoa have completely changed into elongate, fusiform bodies and there are no ribbons of flat chambers. The latter are thought to be transferred to the cell surface and integrated.

Spermatogenesis in cysts probably also occurs in the testes of the Actinotrichida although it is not obvious in the Acaridae where only a few and aberrant sperm cells are produced (Witalinski *et al.*, 1990). In the Oribatida and Prostigmata, late spermatids are separated from each other by extension of somatic cells which give the impression that each spermatid is situated in a vacuole of the large somatic cell and is extended like a secretory vesicle (G. Alberti, personal communication). In *Allothrombium,* the posteriorly joined horseshoe-shaped testes bear pouch-like outgrowths termed testicular follicles and these are arranged irregularly over the tunica propria (see Fig. 9.1E above). The germ cells in each follicle show all stages of development. Spermatids are liberated into the hollow chamber of the follicle (Mathur & LeRoux, 1970). Testes of those species producing spermatophores display a prominent secretory epithelium in addition to the germinal part (Witte, 1975a; Alberti, 1980b). In the Erythraeidae, the glandular part of the testes produces three types of secretion: one forming the thin envelope around the sperm cells; a second giving rise to the spermatophore stalk; and a third, in species of *Abrolophus,* secreting a crown covering the top of the sperm drop (Witte, 1975a).

The spermatozoa of the Actinotrichida tend to be round or oval in shape and there is a greater diversity of types than in the Anactinotrichida (Alberti, 1984). The acrosomal complex in, for example, the terrestrial Parasitengona (*Erythraeus, Dolichothrombium*), Raphignathoidea (*Eust-*

igmaeus), Anystidae and Halacaridae (*Halacarellus*) is complete with both the acrosomal vacuole and acrosomal filament present (Fig. 9.4). A common feature of these taxa is the partial disappearance of the nuclear envelope during spermatogenesis. In water mites (*Limnochares*, *Limnesia*) and the Tarsonemina (*Siteroptes*), the acrosomal filament is lacking whilst in the Oribatida, Astigmata and the Tetranychidae (*Tetranychus*) the complete acrosomal complex is absent.

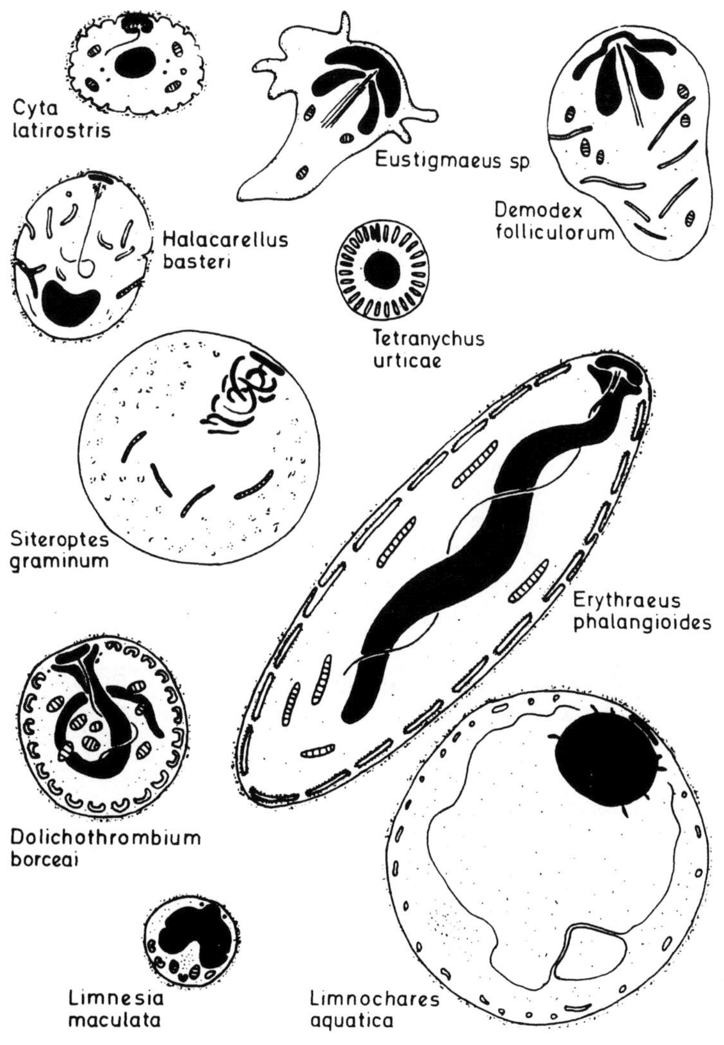

Fig. 9.4. A range of sperm types in the Actinotrichida (after Alberti, 1980b, 1991).

Genital ducts

The vasa deferentia are typically paired ducts connecting the testes with the seminal vesicle or the ejaculatory duct (see Fig. 9.1 above). They are more rarely fused to form a single median duct, as in the Tarsonemina, Eriophyoidea and Halacaridae (Vitzthum, 1940–43). In the Tetranychidae, each vas deferens is connected to its testis by a so-called 'seminal vesicle' which is considered to be a glandular part of the testis by Blauvelt (1945) and Alberti (1980b), and a vas deferens by Ehara (1960). The ducts are lined externally by a thin connective tissue layer and internally by an epithelial layer of columnar or cuboidal cells. Thin muscle fibres may be present. A globular sperm pump with a strong muscular sheath connects the short vasa deferentia to the ejaculatory duct (Alberti & Storch, 1976). The pump is filled with spermatozoa and secretory material of the same composition as that found in the 'seminal vesicles'. The vasa deferentia or the seminal vesicle, when present, lead to the genital atrium, which acts as an ejaculatory duct. In the Oribatida, a sphincter muscle appears to control the passage of sperm cells from the median glandular organ into the vas deferens. The vasa deferentia may fuse anteriorly to form a median common vas deferens or seminal vesicle which is similar to the vas deferens in structure.

The ejaculatory duct is lined with thin cuticle except where secretory cells reach the surface. In ticks, the anterior half of the duct is lined dorsally with hard cuticle and ventrally with soft cuticle while in the posterior half the entire cuticular lining remains soft (Till, 1961). A few bands of circular muscles may be present on the outer surface of the epithelium of the posterior region. The end of the duct is often protrusible and in the Macrochelidae, for example, forms a telescoping ejaculatory organ which deposits a sperm droplet in the hypostomatic region of the male gnathosoma (Krantz & Wernz, 1979). In the Astigmata and some groups of Prostigmata, such as the Tetranychoidea, Raphignathoidea, Cheyletoidea and Tarsonemina, the terminal section of the ejaculatory duct is enclosed in a strongly sclerotized intromittent organ, the aedeagus or penis. It is situated between the bases of the legs in the Astigmata and usually terminally or subterminally in prostigmatic taxa. It is directed posteriorly when extruded. Exceptions occur in some families of parasitic Cheyletoidea, for example Demodicidae, Myobiidae, Psorergatidae and in the Podapolipidae where the aedeagus is usually located dorsally and directed anteriorly. Variously shaped sclerites for muscle attachment are often associated with the aedeagus, especially in the free-living Astigmata. Prasse (1970) states that the protraction of the aedeagus in *Sancassania* is accomplished by hydrostatic pressure and retraction by muscles attached to sclerites in the progenital chamber (genital atrium). The ejaculatory duct passes through a hole in the supporting sclerite of the aedeagus and on its anterior side widens out to form a distensible chamber which leads into the sclerotized aedeagus or penis. The stages in the extrusion of the aedeagus in *S. berlesei* are shown diagrammatically in

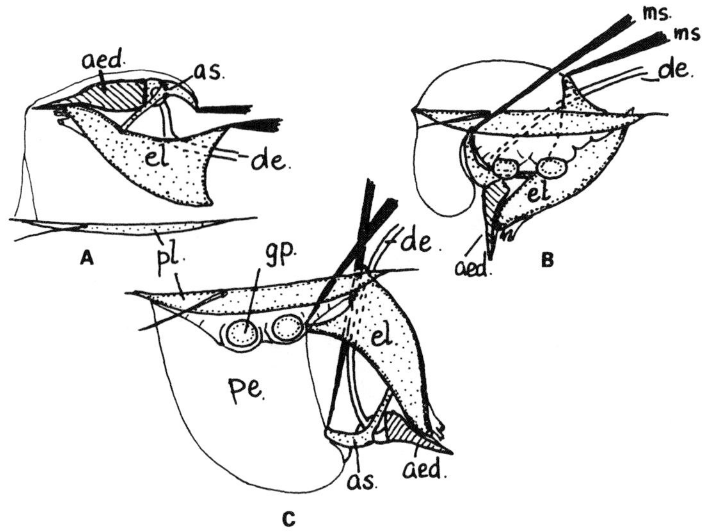

Fig. 9.5. **A–C**, stages in the extrusion of the aedeagus complex in *Sancassania berlesei* (Astigmata) by hydrostatic pressure (based on Prasse, 1970).
aed. = aedeagus; *as.* = postero-dorsal sclerite; *de.* = ejaculatory duct; *el.* = median (epigynal) lip; *gp.* = genital papilla; *ms.* = retractor muscles of aedeagus complex inserted on postero-dorsal sclerite and median lip; *pe.* = bladder-like distension of the genital region; *pl.* = antero-lateral progenital lip.

Fig. 9.5A–C. There is considerable diversity in the size and shape of the aedeagus and its associated sclerites which are used as taxonomic criteria, for example in the genus *Tyrophagus* (Hughes, 1976).

The so-called penis or ejaculatory complex in the Oribatida and in those Prostigmata practising indirect sperm transfer is a spermatophoric and not an intromittent organ. The complex in the Parasitengona is a syringe-like organ for receiving, compacting and expelling the sperm mass and has been the subject of detailed study in the trombiculid *Blankaartia ascscutellaris* by Mitchell (1964) and in the water mite *Hydrodroma* sp. by Barr (1972). A complex series of sclerites and associated musculature is involved in the movements of the ejaculatory complex and Barr considers the structure of the complex to be of value in the classification of water mites. The oribatid penis is essentially a conical double-walled structure, consisting of inner and outer cups, which projects ventrally into the progenital chamber (Woodring, 1970). The penile orifice is situated at the ventral tip of the cone which is provided with a number of pairs of setae (Fig. 9.6A) – for example, seven pairs in *Damaeus onustus* and six pairs in *Podacarus auberti* (Grandjean, 1956a). The dorsal end of the penis receives the vasa deferentia posteriorly and the accessory gland anteriorly. Their common lumina join to form the ejaculatory duct which is the inner cup of the penis (Fig. 9.6B). A chitinous process or tongue arises at the anterior end of

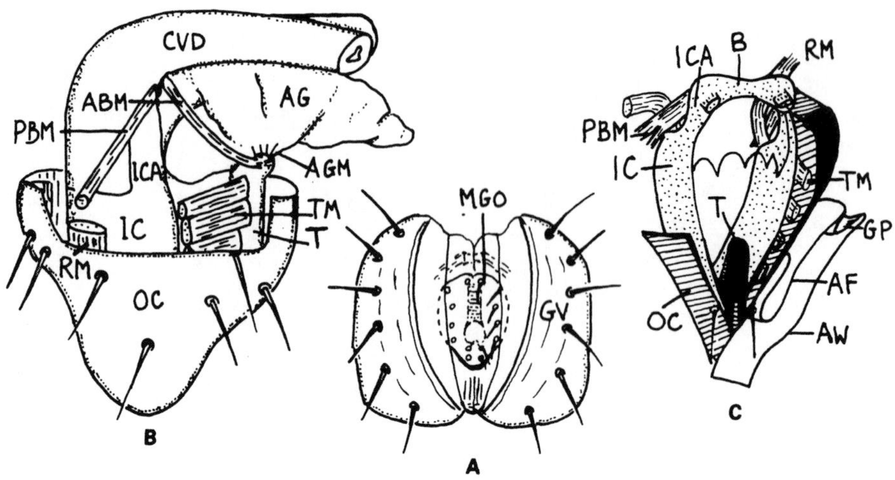

Fig. 9.6. **A**, male genital organ (MGO) of *Damaeus onustus* (Oribatida) exposed by parting the genital plates (GV) (based on Grandjean, 1956a); **B**, the ejaculatory complex in the male of a generalized oribatid mite (based on Woodring, 1970): **C**, diagrammatic view of an isolated strong-type penis with a section of the outer cup (OC) removed to show the tongue (T) and inner cup (IC). *ABM* = anterior bridge muscle; *AF* = atrial folds; *AG* = accessory gland; *AGM* = accessory gland muscle; *AW* = atrial wall; *CVD* = common vas deferens; *GP* = genital papilla; *GV* = genital plate; *ICA* = inner cup arm; *PBM* = posterior bridge muscle; *RM* = retractor muscle; TM = tongue muscles.

the penis at the juncture of the inner and outer cups (Fig. 9.6C). In the majority of the species examined by Woodring, a solid bar of cuticle joins the dorsal ends of the inner cup arms to form a bridge and he referred to this form of penis as the 'strong type' as opposed to the less common 'weak type' which lacks a bridge and has a poorly developed tongue. Strong muscles originating on the lateral surfaces of the tongue insert on the inner lateral walls of the outer cup while those from the dorsal and postero-lateral surfaces insert on the bridge. These tongue muscles may be involved in opening the penile orifice. The penis is extruded by hydrostatic pressure and retracted by the penile retractor muscles originating on the ventrolateral body wall and inserting on the posterior half of the inner cup. Woodring (1970) discusses the comparative morphology of the penis and ovipositor in oribatids and concludes that they are 'completely homologous' structures.

The progenital chamber is usually closed by one or two plates which are opened by internal pressure. In males of the Uropodina, the chamber is covered by a subcircular anterior plate and a smaller sickle-shaped posterior one; these are hinged, respectively, to the anterior and posterior walls of the orifice. A pair of apodemes arise from the antero-lateral edges of the posterior platelet and form the site of attachment of the genital plate retractor muscles which originate on the holodorsal shield (Woodring & Galbraith, 1976). A slight 'tongue and groove' configuration at the interface of the two plates is considered to keep the genital plates firmly closed when the muscles are contracted.

The longitudinal or less commonly the transverse opening of the progenital chamber in the Actinotrichida is considered to be a secondary genital orifice and is closed by a pair of lips or valves, variously termed progenital lips, paragynal lips, genital plates, etc. The chamber contains, in the adults, the eugenital opening and the genital papillae. The progenital chamber in the males of the Acaridae is closed by a pair of anterolateral progenital lips which only partially cover an unpaired median structure which Van der Hammen (1989) considers to be the posterior eugenital lip but is termed the epigynal lip by Grandjean (1938d) and Prasse (1970) – see condition in the female (Fig. 9.13, below). The genital papillae and their retractor muscles are located beneath the progenital lips (Knülle, 1959; Prasse, 1970). In the Oribatida, the progenital chamber is entirely closed by a pair of relatively large setae-bearing plates opening transversely and outward. There is often a ridge or projection of the ventral plate which prevents the plates being withdrawn into the body (Michael, 1884). The plates are opened by hydrostatic pressure and closed by muscles originating on a pre-anal apodeme and inserting on the genital plates (see Fig. 7.1, p. 210).

Accesssory glands

Glands containing various types of secretory cells are connected with the genital tract of the male. In ticks and Mesostigmata, a single multilobed median accessory gland is present and opens into the seminal vesicle near its junction with the vasa deferentia in the former and into the ejaculatory duct in the latter (see Fig. 9.1A,B above). In *Uroactinia agitans*, the opening of the dorsal lobe of the accessory gland into the atrium is separate from the combined opening of the ventral, posterior and lateral lobes (Woodring & Galbraith, 1976). Columnar glandular cells predominate within the gland and it is considered that cells which appear morphologically similar may represent several different secretory cell types. Woodring and Galbraith (1976) state that the secretion (seminal fluid) in *U. agitans* comprises a mixture of at least four different types and is mixed with the densely packed sperm in the ejaculatory duct. The gland cells in ticks secrete mucoprotein and mucopolysaccharide as well as other compounds and the secretions are thought to function in the production of the spermatophore and to be responsible for the initiation of capacitation of the spermatids (Oliver, 1982).

One or two pairs of accessory glands occur in the Astigmata and open into the ejaculatory duct (Nalepa, 1884; Michael, 1901; Prasse, 1970). The single gland in *Sancassania berlesei* and *Acarus siro* opens into the seminal vesicle. In *Talpacarus platygaster* two large glands are present which, according to Michael (1901), differ in form, size and structure. The larger 'receptacular accessory gland' is thin-walled and its hollow interior is full of homogeneous or finely granular material, while the thicker walled 'chambered accessory gland' is so-called because it appears chambered in transverse section owing to the projection of its greatly distended glandular cells into the lumen. The functions of these glands appears to be unknown. One, two or more accessory glands occur in the Prostigmata, except in eriophyid mites (Michael, 1896; Alberti, 1974; Witte, 1975b; Nuzzaci & Solinas, 1984). The oily secretion of the anterior accessory gland in *Abrolophus* seals the genital atrium and prevents the hardening of the spermatophore as the result of air entering the ejaculatory duct during the deposition of the spermatophore stalk (Witte, 1975). Accessory glands are wanting in the Oribatida.

Female Reproductive Organs and Oogenesis

The reproductive organs of the female comprise single or paired ovaries, single or paired oviducts, a median uterus, a vagina, seminal receptacle, accessory glands and progenital chamber (vestibule, genital atrium) (Fig. 9.7A–F). A sperm access system unconnected with the genital orifice and functioning for the reception (and in some cases also the maturation and storage) of sperm is present in the Astigmata and some Mesostigmata (Dermanyssina, Heterozerconina) and Prostigmata (Tetranychidae).

Ovary and oogenesis

The Parasitiformes have a single ovary. In *Argas (Carios) vespertilionis*, the structure appears paired with each ovary connected to each other by a thin strand of tissue devoid of ova (Roshdy, 1961). This condition also occurs in old females of *Argas persicus* that have passed through several gonadotrophic cycles and may be due to the exhaustion of the oocyte supply in the centre of the ovary during previous ovipositions (Balashov, 1968). The lumen of the tubular ovary in ticks is surrounded by a relatively thin wall formed of epithelial cells, oogonia and primary oocytes. The ovarian tissue is separated from the haemolymph by a basement lamina or tunica propria, to which muscle fibres may be connected externally. Oogonia are present in the larval ovarian primordia but primary oocytes first appear in fed nymphs. In adult metastriate species, the oogonia and primary ooctyes are located in a fold or groove extending along the dorsal or antero-dorsal surface of the ovary (Fig. 9.8: 1–3). More advanced oocytes occur outside the groove and as their growth continues the oocytes protrude into the haemocoel giving the surface of the ovary the appearance of a cluster of grapes (Diehl *et al.*, 1982). At this stage, the oocytes are connected to the ovarian wall by a stalk or funicle formed by elongated epithelial cells. They are not surrounded by follicular cells. Stalked oocytes also occur in the ovary of

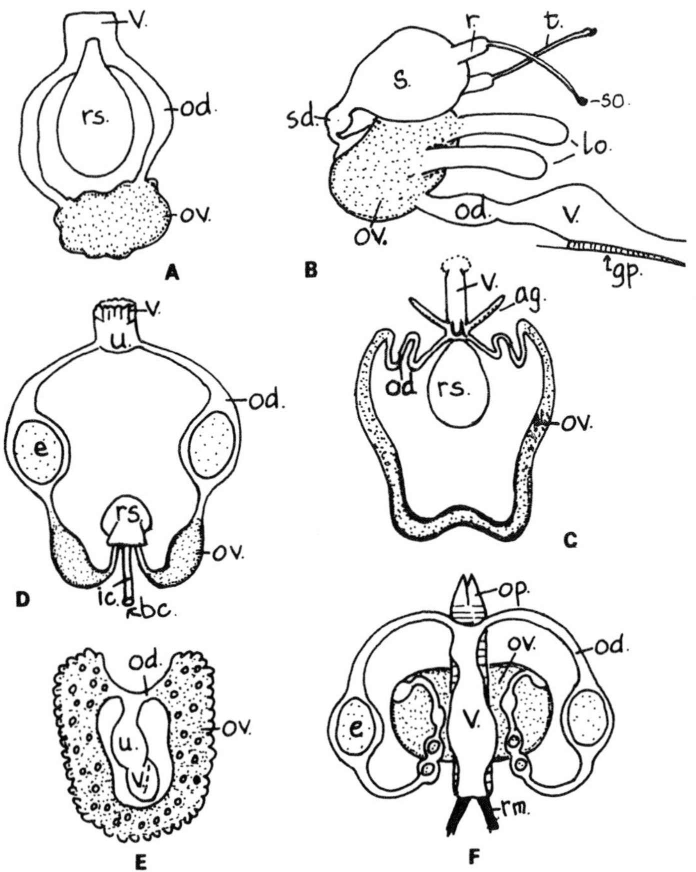

Fig. 9.7. Diagrammatic representation of a range of female reproductive systems in the Acari; **A**, Sejina (Mesostigmata); **B**, Dermanyssina (Mesostigmata); **C**, Ixodida; **D**, Acaridae (Astigmata); **E**, Parasitengona (Prostigmata); **F**, Oribatida. *ag.* = accessory glands; *bc.* = bursa copulatrix; *e.* = egg; *gp.* = genital plate; *ic.* = inseminatory canal; *lo.* = lyrate organ; *od.* = oviduct; *op.* = ovipositor; *ov.* = ovary; *r,* = ramus; *rm.* = retractor muscles of the ovipositor; *rs.* = receptaculum seminis; *s.* = sacculus of sperm access system; *sd.* = sperm duct; *so.* = solenostome; *t.* = tubulus; *u.* = uterus; *v.* = vagina.

the Uropodina and Sejina (Michael, 1892; Woodring & Galbraith, 1976) but oocyte development in the Parasitina and Dermanyssina is internal and the oocytes lack funicles (Michael, 1892; Young, 1968b; Woodring & Galbraith, 1976).

During pre-vitellogenesis in ticks, the oocyte enters a phase of considerable cytoplasmic growth, which usually coincides with the beginning of blood feeding by the female. The nucleus, nucleolus and cytoplasm greatly enlarge and the oocyte protrudes from the wall of the ovary (Fig. 9.8: 4–7). This stage ends a few days after the commencement of feeding with the appearance of yolk granules in the cytoplasm of the

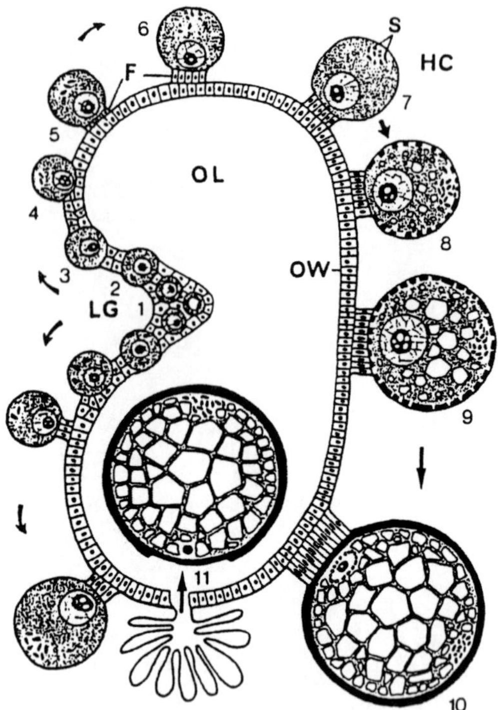

Fig. 9.8. Diagrammatic presentation of oogenesis (in ovarian cross-section) in the Ixodoidea (after Diehl *et al.*, 1982). 1–3, younger oocytes restricted to the longitudinal groove (LG); 4–7, oocytes in the stage of great cytoplasmic growth and protruding from the ovarian wall (OW); 8–10, period of vitellogenesis; 11, ovulation. *F.* = funicle; *HC* = haemocoel; *OL* = lumen of ovary; *S.* = symbionts.

oocyte (Balashov, 1968; Brinton & Oliver, 1971). At this stage the oocyte contains all the organelles necessary for the accumulation of yolk and the deposition of the egg shell (Fig. 9.8: 8–10). Its nucleus is usually situated near the funicle and rickettsia-like symbionts (*Wolbachia*) whose physiological role is unknown, may be present at the opposite pole to the nucleus. Yolk appears to originate from intracellular and extracellular (haemolymphatic) sources. Extracellular yolk proteins are derived from the fat body in *Ornithodoros moubata* and also from the mid-gut in *Rhipicephalus sanguineus* (Coons *et al.*, 1982; Diehl *et al.*, 1982). Those coming by way of the haemolymph enter by micropinocytosis (Brinton & Oliver, 1971). Vesicles of intra- and extracellular origin fuse to form large yolk granules. Oocyte development is not synchronous in the ovary during vitellogenesis. Pressure from the enlargement of the oocyte stretches the basement lamina and forces the mature oocyte through the funicle and into the lumen of the ovary (Fig. 9.8: 11).

The ovary in the Dermanyssina and Parasitina is subglobular in shape and complicated in structure when compared with that of the Uropodina and Sejina (Michael, 1892; Young, 1968b; Akimov & Yastrebtsov, 1984). It is differentiated into a region with a trophic function, the lyrate organ, and a part where the oocytes mature, the ovary in the strict sense (Alberti & Hänel, 1986; Alberti & Zeck-Kapp, 1986). This type of differentiation of the gonad is not known in other arachnids. A conspicuous feature of the ovary are two flattened arms which arise from a central dome-shaped elevation on the upper side of the ovary (Fig. 9.7B). Following Michael (1892), the arms and their 'root' are referred to as the lyrate organ and the elevation as the camera spermatis. There appears to be continuity between the latter and the arms of the lyrate organ, the region where the oocytes develop, the receptaculum seminis of the sperm access system (spermatheca of Alberti & Hänel, 1986) and probably also the oviduct. In *Varroa jacobsoni*, one part of the camera spermatis is made up of soma cells and traversed by nutritive cords originating in a syncytial nutritive tissue in the lyrate organ (Fig. 9.9). The cords connect with oocytes which are at first surrounded by soma cells but later bulge through their sheath into the haemolymphatic space. Ribosomes are one of the most important components transported by way of the nutritive cords to the oocytes. The increase in the number of mitochondria in the middle-stage (stage 2) oocytes may also be due to their transport from the nutritive tissue to the oocytes. A second part of the camera is connected to the receptaculum and is composed of cuboidal cells provided with large nuclei, elongated mitochondria, ribosomes and lysosomal inclusions (Alberti & Hänel, 1986). A restricted area of these cells is not underlain by basement membrane and is thus exposed to the haemolymphatic space. In the third section of the camera, situated ventral to the median portion ('root') of the lyrate organ, the periphery is provided with muscle cells which are in intimate contact with the soma cells, unlike the occasional muscle cells of other parts of the camera. This region of the camera has an indistinct lumen and appears to continue into the oviduct.

The ovary in the Prostigmata and Oribatida is an unpaired structure but in the Astigmata it is paired and each ovary is connected by a narrow duct to the median sac-like seminal receptacle of the sperm access system. Studies by Witte (1975b) on oogenesis in the Erythraeidae have shown marked differences from the same process in the Parasitiformes. Pre-vitellogenesis takes place in the lumen of the ovary and the oocytes are enveloped by a follicle of abortive germ cells (nutritive cells). The oocytes migrate before vitellogenesis to peripheral pouches in the ovary, formed by the basement lamina. This gives the ovary a bunch-of-grapes appearance. Growth of the oocytes is sustained by the abortive germ cells, which have a canal- or stalk-like connection with the oocyte (Fig. 9.10). The role of the abortive germ cells in producing yolk has not been ascertained. It is assumed that yolk precursors are taken out of the haemolymph by the oocytes. Extra-ovarian nutritive cells are not present in the Erythraeidae but such cells do occur in other Prostigmata, for example in the Rhagidiidae, Bdellidae (Alberti, 1974) and Halacaridae (Thomae, 1926). These cells lie beneath the epidermis or around the gut and apparently play a role in the synthesis of yolk proteins.

This nutritive type of oogenesis appears to be widespread in the Prostigmata. In the Tetranychidae, the posterior part of the ovary is encircled by a one-cell-thick layer of slightly flattened epithelial cells which travels into the anterior part of the ovary to provide supporting

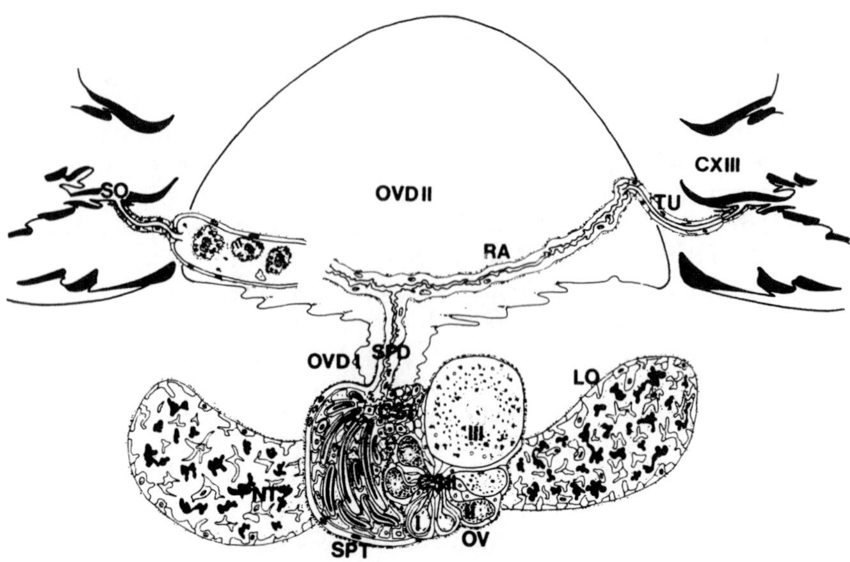

Fig. 9.9. Diagrammatic illustration of the female reproductive system of *Varroa jacobsoni* (Mesostigmata) in dorsal view (after Alberti & Hänel, 1986). *CS II* = part of the camera spermatis traversed by nutritive cords; *CX III* = coxa of leg III; *LO* = lyrate organ; *NT* = syncytial tissue; *OV* = ovary; *OVD I* = oviduct I; *OD II* = vagina; *RA* = ramus of sperm access system; *SO* = solenostome; *SPD* = sperm duct; *SPT* = spermatheca (receptaculum); *TU* = tubulus; I–III = oocytes in different stages of development.

tissue for the developing eggs. The nutritive or nurse cells have three nuclei and develop as sister cells of oogonia (Langenscheidt, 1973). They are connected to the oocytes by nutritive stalks. Nutrients are also supplied to the oocyte from the adjacent mid-gut (Weyda, 1980). As the result of the maturation and differentiation of the germ cells, the ovary becomes divided more or less into three regions comprising an anterior germarium, a median pre-vitellogenic portion and posterior vitellogenic section (Alberti & Crooker, 1985). A nutritive function of the ovary is also suspected by Mathur & LeRoux (1970) in *Allothrombium*. Abortive germ cells form a 'nutritive syncytium' in some Acaridae (Prasse, 1968a). It is assumed that the evolution of nutrimentary oogenesis results in speeding up embryogenesis.

The single ovary of the adult Oribatida forms a large central sac which underlies the ventriculus (Michael, 1884). In *Ceratozetes cisalpinus*, the ova develop on a central stalk or medulla which is not attached to the basement lamina (Woodring & Cook, 1962a). According to these authors development appears to be panoistic, nutritive cells being absent in the ovary. Ultrastructural studies of the ovary of *Hafenrefferia gilvipes* have shown that oogonia and pre-vitellogenic oocytes are provided with microtubule-rich protrusions by which they are interconnected in the

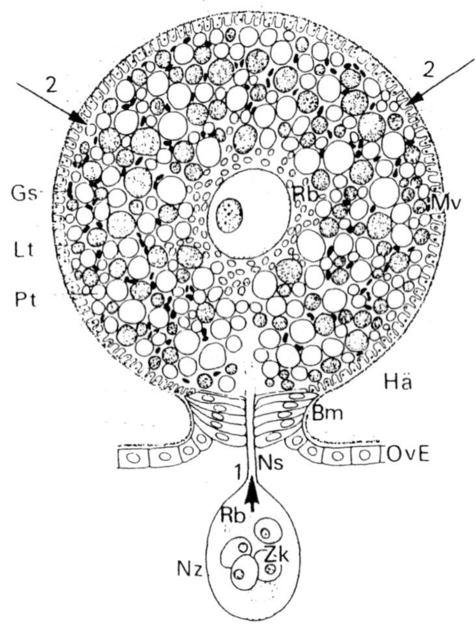

Fig. 9.10. Schematic representation of an oocyte during vitellogenesis in the Erythraeidae (Prostigmata) (after Witte, 1975b). *Bm.* = basement membrane; *Gs.* = glycogen particle; *Hä* = haemolymph; *Lt.* = lipid droplet; *Mv.* = microvilli; *Ns.* = nutritive canal or cord; *Nz.* = nutritive cell; *OvE.* = ovarian epithelium; *Pt.* = protein droplet; *Rb.* = RNA-rich cytoplasmic region; *Zk.* = nuclei; 1 = supply for cytoplasmic components; 2 = supply for yolk components.

centre of each rosette-like group of cells (Witalinski, 1987). Real stalks are absent and the connections, due to the absence of nutritive cells, play a supportive and/or synchronizing role. The oocytes, in the course of growth, move centrifugally from the rosette and this leads to a very early rupture of the protrusions.

In the thelytokous species *Platynothrus peltifer* and *Trhypochthonius tectorum*, the larval and protonymphal ovaries are represented by small groups of cells with rounded nuclei while in the deuteronymph the ovary forms a larger compact structure with the nuclei more strongly chromatic (Taberly, 1987a). Oogonial divisions are frequent at the deuteronymphal stase while premeiotic stages are achieved during the tritonymphal stase and maturation divisions occur in the adult. The oviducts of the tritonymph although formed have a narrow lumen and connect the gonad to the genital bud. A flattened, narrow genital atrium is present but does not open to the exterior. According to Woodring & Cook (1962a), the sex of the 'later tritonymphal stage' may be determined in *Ceratozetes cisalpinus* by the form of the gonad, which is bilobed posteriorly in the male but not in the female.

Genital ducts

The oviducts are paired in all the major taxa except for certain members of the Prostigmata, for example Anystidae, Bdellidae, Eriophyoidea, Halacaridae and Tarsonemina, and mesostigmatic mites of the Parasitina and Dermanyssina in which there is a single median oviduct. A regional enlargement of each oviduct to form an ampulla occurs in many mites and ticks. The epithelial cells of the resting oviduct are plate-like or cuboidal with well-developed circular and often longitudinal muscles on the outside of the basement lamina. The cells are secretory. It is considered by Woodring & Galbraith (1976) that in *Uroactinia agitans* yolk is deposited in the posterior half of the oviduct and the eggshell ('chorion') in the anterior half. Secretions of the oviduct are also stated to be responsible for the formation of the eggshell in, for example, the Oribatida (Michael, 1884), Astigmata (Michael, 1901) and Bdellidae (Alberti, 1974). The oviduct is capable of considerable expansion for the passage of the eggs and in ticks it is often developed into a number of loops. The oviducts lead directly or by way of the uterus (common oviduct) to the vagina and the progenital chamber.

The uterus is well developed in ticks, Astigmata, the majority of the Prostigmata and in the Uropodina and Parasitina but is not defined in the Dermanyssina and Oribatida. It forms a large triangular sac in argasid ticks, storing the endospermatophores and uncapacitated sperm. The uterine epithelium is surrounded by muscles that are more strongly developed than those of the oviduct.

The vagina is of ectodermal origin. Its epithelium comprises flattened cells which secrete the thin, pliable cuticle lining the lumen. The outer wall may be provided with circular muscle fibres. In ticks, the vagina is divided into an anterior vestibular (genital atrium) and a posterior cervical section (Balashov, 1968). Histologically the two sections differ in that the vestibular epicuticle is lined with a thin endocuticular layer which is absent in the cervical section, and the epithelium is less strongly folded (Arthur, 1962). The cervical region serves as a seminal receptacle in prostriate ixodids. In the Acaroidea, the vagina (meatus ovi, pre-oviporal duct) is a large organ with flexible walls that lack a musculature and is capable of considerable expansion to accommodate the large egg during its passage to the exterior (Prasse, 1970).

The anterior region of the vagina in many mites and ticks forms an egg-laying organ. In its simplest form, seen in ticks, the vestibular region of the vagina is everted with the egg by muscular action and forms a short pouch-like ovipositor. Similarly, a short tube-like ovipositor is formed by eversion of the anterior region of the genital tract in the Notostigmata (With, 1904; Juvara-Bals & Baltac, 1977), Tetranychidae (Beament, 1951) and Acaroidea (Prasse, 1970). Three stages in the protraction of the organ in *Sancassania berlesei* are shown diagrammatically in Fig. 9.11A–C. It is withdrawn in this species by retractor muscles inserted on a sclerite associated with the plicated zone of the vagina. The ovipositor in the Oribatida is a more conspicuous structure and forms a relatively long tube or cylinder when evaginated.

Michael (1884) considers the ovipositor in the Oribatida to be 'an extremely beautiful organ' and describes it as 'a long straight tube, which, when not in use, is invaginated like the finger of a tight glove when drawn off; in this way the distal half is withdrawn within the proximal'. The

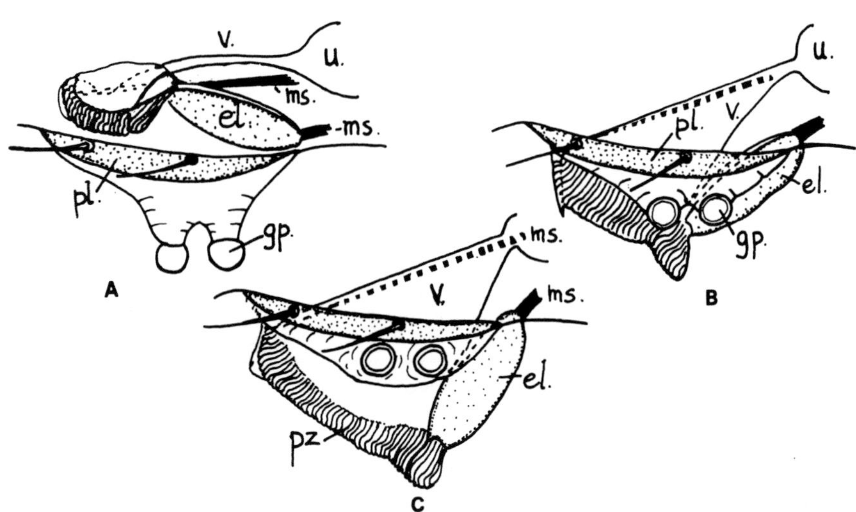

Fig. 9.11. **A–C**, three stages in the extrusion of the vestibular region of the vagina of *Sancassania berlesei* (Astigmata) (based on Prasse, 1970). *el.* = median (epigynal) lip; *gp.* = genital papilla; *ms.* = retractor muscles; *pl.* = antero-lateral progenital lip; *pz.* = plicate zone; *u.* = uterus; *v.* = vagina.

organ is finely plicated by longitudinal folds which allow for its distension during oviposition. The plications are so fine, that according to Michael, 'when viewed with an amplification of about fifty diameters the ovipositor seems iridescent like a diatom seen with a power not sufficient to resolve it'. The evaginated ovipositor is divided into two parts by a circular constriction, the distal part being shorter and slightly narrower than the proximal (Fig. 9.12A). It terminates in three eugenital lobes or cusps which surround the egg-laying orifice, the eugenital orifice (Grandjean, 1956b). The latter, at rest, is in the form of an inverted Y and this trifid configuration is a primitive feature throughout the Actinotrichida. Each lobe is partially or entirely sclerotized along its surface of contact with the other lobes. One of the lobes is always ventral and unpaired while the other two form a latero-dorsal pair (Fig. 9.12B). This arrangement of the lobes is a constant feature but there is considerable diversity in their size and shape. Michael (1884) considers that the function of the lobes is 'to seize the egg as it passes from the ovipositor and to hold it in position during deposition'. Each lobe bears four setae and in the Desmonomata and Euoribatida six setae are situated around the constriction. The setae of the ovipositor are frequently hollow and resemble eupathidia. Their form suggests that they may have a mechanochemosensory function. The ovipositor is everted by hydrostatic and possibly egg pressure and is retracted by powerful muscles originating on the notogaster and inserting at the junction of the anterior and posterior parts of the ovipositor (Woodring & Cook, 1962a).

Three eugenital lobes also occur on the ovipositor of some

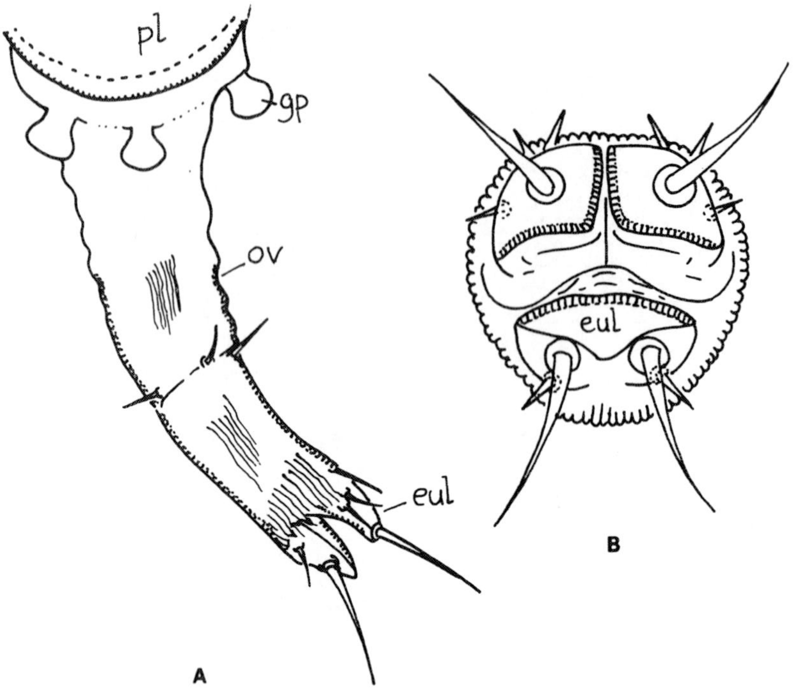

Fig. 9.12. **A**, everted ovipositor of *Heminothrus targionii* (Oribatida); **B**, frontal view of the region of the three eugenital lobes. (Based on Grandjean, 1956b.) *eul.* = eugenital lobe; *gp.* = genital papilla; *ov.* = ovipositor; *pl.* = genital plate.

Endeostigmata *sensu* Grandjean (1939b), such as *Speleorchestes* and *Terpnacarus*. The ovipositor in the former is long, as in the Euoribatida, and contrasts with the short form in *Terpnacarus*. The description by Hirst (1917) of four eugenital lobes on the ovipositor of *Speleorchestes poduroides* is erroneous; only three lobes are present. A well-developed ovipositor also occurs in the Bdellidae and some Eupodoidea and it may be divided by a circular constriction into two sections, as in *Cyta* and *Thoribdella* (Alberti, 1974).

The progenital orifice in the Actinotrichida is in the form of a longitudinal, more rarely transverse, slit and is closed by a pair of progenital lips. Genital papillae, when present, are situated in the progenital chamber. There is only one progenital lip or plate in the Tetranychidae while in the Acaridae three lips are discernible comprising a pair of antero-lateral progenital (=paragynal) lips and a medial (=epigynal) lip that is only partially covered by the paired lips (Fig. 9.13A–C). The medial lip is probably the posterior eugenital lip. The two pairs of genital papillae in the Acaridae are situated, in their retracted state, within cavities of the progenital lips. The cavities containing the papillae open through diachilous slits (Grandjean, 1938d; Van der Hammen, 1989). The

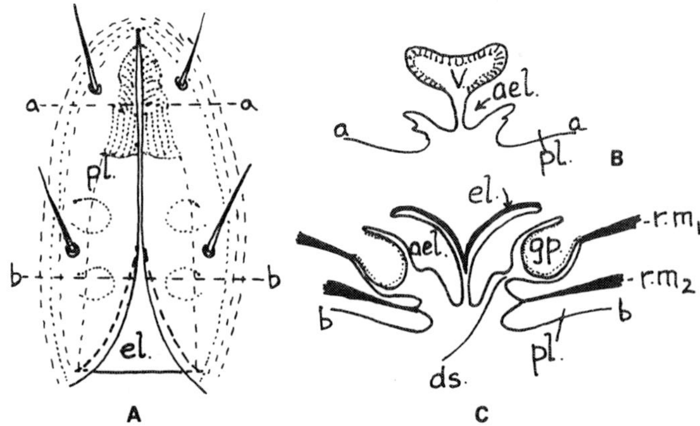

Fig. 9.13. **A**, progenital and epigynal lips in *Sancassania berlesei* (Astigmata); **B**, cross-section at a–a in **A**; **C**, cross-section at b–b in **A**. (After Prasse, 1970.) *ael.* = eugenital (inner paragynal) lip; *ds.* = diachilous slit; *el.* = median (epigynal) lip; *gp.* = genital papilla; *pl.* = antero-lateral progenital lip (external paragynal fold of Witalinski *et al.*, 1990); rm_1 = retractor muscle of the genital papilla; rm_2 = retractor muscle of the progenital lip; *v.* = vagina.

form of the genital valves protecting the progenital orifice in the Oribatida is essentially the same as in the male.

The genital orifice in the Anactinotrichida is in the form of a transverse slit. In the Mesostigmata one, three or four sclerotized shields or plates are associated with the genital atrium and four shields cover the orifice in the Holothyrida. Lying beneath the genital shield in many groups of Mesostigmata, such as the Parasitina, Uropodina and Zerconina, are one or more sclerotized structures which are variously referred to as the endogynum, perigynum or genital sclerites. They are present only in those species in which insemination is tocospermic, i.e. by way of a spermatophore introduced through the genital orifice. The function of these structures has not been satisfactorily elucidated. Michael (1890) considers the shoe-shaped, spiniferous perigynum in '*Uropoda ovalis*' to be involved in a mechanism which controls the entry of spermatozoa from the seminal receptacle into the vagina. It is possible that some of the structures may be involved in holding and preventing the escape of the spermatophore after its implantation beneath the genital shield (Winkler, 1888).

Accessory glands

The occurrence of accessory glands in the female is less common than in the male and they are absent in the Oribatida and in most of the Prostigmata and Astigmata. Paired tubular glands opening into the vagina occur in all ticks and they produce a protein-rich secretion during the period of oviposition which coats the egg as it passes through the vagina (Oliver, 1982). In ixodoid ticks, the epithelium of the anterior (vestibular) region of the vagina detaches from the cuticle and becomes glandular to form the lobular accessory gland. This gland produces a lipid-rich secretion that is discharged onto the surface of the egg. Paired accessory glands opening into the vagina also occur in the Parasitina and Uropodina but are absent in the Dermanyssina (Winkler, 1888; Michael, 1890, 1892; Woodring & Galbraith, 1976).

Well-developed accessory glands are present in the Erythraeidae (Witte, 1975b). An unpaired dorsomedian gland discharges into the uterus and produces a protein-rich secretion as in the case of the tubular glands of the ticks. A further resemblance is seen in the adaptation of the vaginal epithelium to produce the so-called 'Lipiddruse' whose secretion forms a thin lipid layer around the egg.

Seminal receptacles and sperm access systems

A receptacle for the storage of sperm material opens into the vagina in metastriate ticks, Uropodina and Sejina. Paired receptacles occur in the non-parthenogenetic species of the erythraeid genus *Abrolophus* and in the Trombiculidae (Witte, 1975b; Mitchell 1964). The Bdellidae are unusual in that the receptacle and its duct, when present, are located near the anterior end of the ovipositor (Alberti, 1974). In all these cases, sperm material enters the reproductive tract by way of the genital orifice. However, in the Dermanyssina (Michael, 1892) and probably the Heterozerconina, and in the Astigmata (Michael, 1901) and Tetranychidae (Crooker & Cone, 1979) sperm is not introduced through the genital orifice but through a copulatory pore (bursa copulatrix, solenostome, sperm induction pore) leading to a sperm access system (spermatheca of some authors) which does not join the oviduct or the vagina.

The copulatory pore in the Tetranychidae, absent in parthenogenetic females, is located ventrally on a slight elevation between the genital opening and the anus. It leads by way of a cuticle-lined tube to an ovoid sac-like receptacle which consists of a simple columnar epithelium (Alberti & Storch, 1976). Sperm are deposited directly into the lumen of the receptacle by the aedeagus and later enter the columnar epithelial cells. Here they slightly enlarge and become arranged in packets of between three and ten sperm (Feiertag-Koppen & Pijnacker, 1985). About one day after insemination, sperm are also found dorsal to the oviduct and near the ovary. Later, sperm enter the ovarian cavity and spread in the ovary. Towards the end of vitellogenesis one sperm enters the growing oocyte, probably by way of the nutritive cord. Apparently, sperm do not reach the early developing oocytes in time for fertilization so that these remain male-determined. The method by which the aflagellate spermatozoa reach the oocytes has not been ascertained.

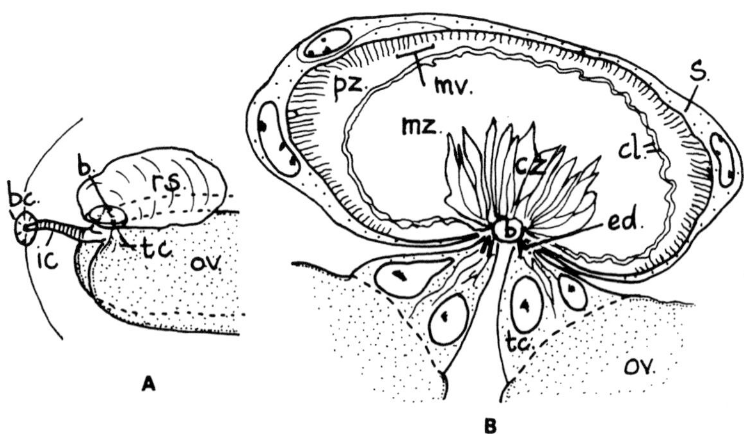

Fig. 9.14. **A**, schematic representation of the sperm access system in *Acarus siro* (Astigmata); **B**, schematic representation of the ovaries (*ov*), transitory cone (*tc*); and receptaculum seminis (*rs*). (After Witalinski *et al.*, 1990.) *b.* = basal part of receptaculum seminis; *bc.* = bursa copulatrix; *cl.* = outer double cuticular lamina; *cz.* = central zone; *ed.* = efferent duct; *ic.* = inseminatory canal; *mv.* = microvilli; *mz.* = middle zone; *pz.* = peripheral zone; *s.* = receptacular sac.

Sperm storage and maturation in the Astigmata take place in a sperm access system opening by a single pore, the bursa copulatrix, situated at or near the posterior region of the opisthosoma. A simple cuticle-lined inseminatory canal (sperm access tube, canalis copulator) extends from the bursa copulatrix to a cuticular cup-shaped basal part of a saccular organ of globular or elliptical form called the sacculus or receptaculum seminis (Fig. 9.14A). Its wall comprises an epithelial layer with the cells having large nuclei and long microvilli (Prasse, 1970). A pair of short efferent sperm ducts (ducti conjunctivi) leave the basal part of the sacculus and connect with the ovaries. In *Acarus siro* (Fig. 9.14B), numerous cuticular lamellae protrude from the basal part into the lumen of the sacculus and divide it into three regions or zones: central, middle and peripheral (Witalinksi *et al.*, 1990). The spermatozoa appear to remain in close contact with the lamellae. The efferent ducts in this species enter the ovaries by way of their conical terminal parts (transitory cones). The form of the organ is a useful taxonomic character in some Acaridae (Fain, 1982).

The sperm access system of the Dermanyssina, so elegantly described by Michael (1892), opens to the exterior by paired pores or solenostomes (Fig. 9.15). Typically, each solenostome is situated towards coxa IV in the arthrodial membrane connecting the coxa of leg III to its acetabulum. However, in the Rhodacaroidea in particular, there is considerable variation in the location of the solenostomes and they may be found opening into or near the acetabula of legs IV, on the metapodal shields posterior to coxae IV or on the trochanters and femora of legs III (Athias-

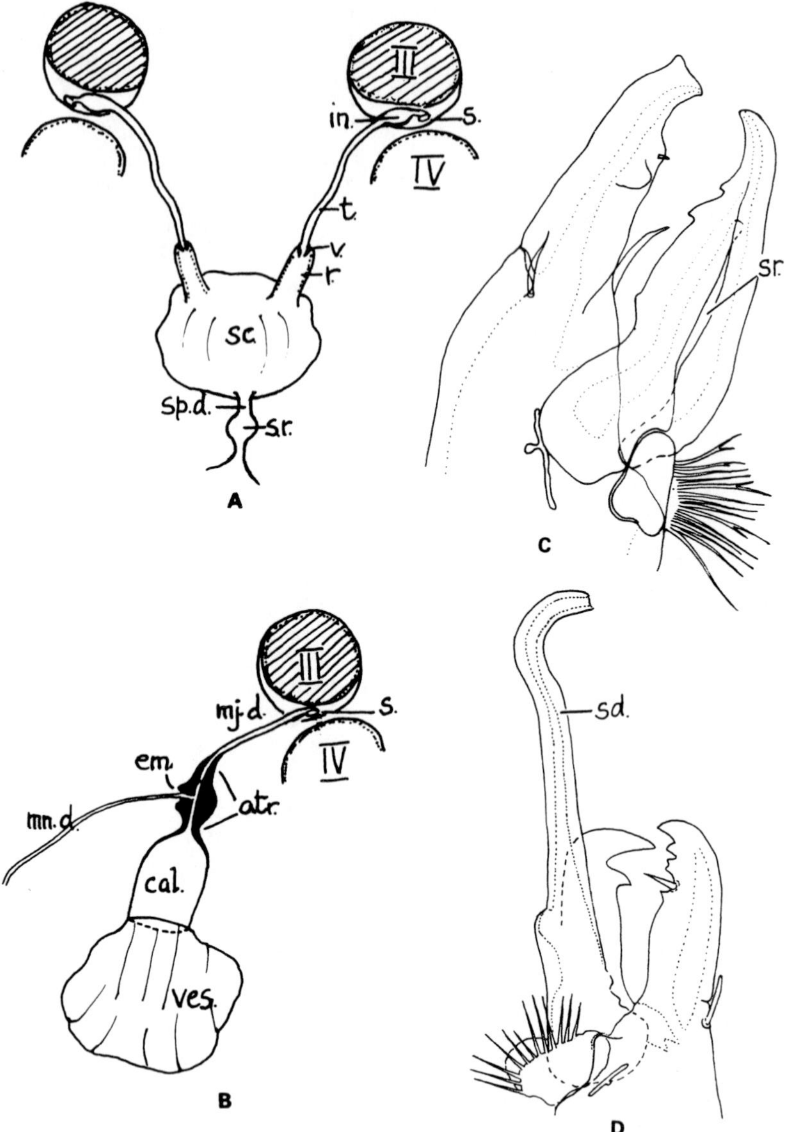

Fig. 9.15. **A**, **B**, two major types of sperm access system in the Dermanyssina (Mesostigmata); **A**, laelapid-type; **B**, phytoseiid-type. **C, D**, male chelicera of a parasitid (**C**) mite showing the form of the spermatotreme (*sr*) and of a laelapid mite (**D**) illustrating a spermatodactyl (*sd*). *atr.* = atrium; *cal.* = calyx; *em.* = embolus; *in.* = infundibulum; *mj. d.* = major duct; *mn. d.* = minor duct; *r.* = ramus; *s.* = solenostome; *sc.* = sacculus; *sp. d.* = sperm duct; *sr.* = sperm reservoir (receptaculum); *t.* = tubulus, *v.* = valve; *ves.* = vesicle; III, IV = coxae of legs III and IV.

Henriot, 1968; Lee, 1974; Evans & Till, 1979). In view of the connection of the copulatory pore with the appendages, this type of insemination is termed podospermy, as opposed to tocospermy where the sperm is introduced by way of the genital orifice (Athias-Henriot, 1968). Two main forms of the system are present: the laelapid and phytoseiid types. In the laelapid-type (Fig. 9.15A), each solenostome leads by way of a flat chamber or infundibulum into a cuticle-lined duct, often with taenidia, termed the tubulus (tubulus annulatus, sperm access duct, canal adducteur). The tubulus leads into a wider tubular structure, the ramus, and the two rami join and usually enlarge to form a bladder-like sacculus, referred to as the sacculus foemineus by Michael (1892) and 'poche de maturation' by Fain (1963a). According to Alberti and Hänel (1986), a sacculus is not developed in *Varroa jacobsoni*. These authors were unable to detect cell boundaries in the epithelium of the ramus, and its wall is either composed of very large cells or represents a syncytium. A cuticular valve projects into the lumen at the junction of the tubulus and ramus. Leading from the sacculus or fused rami is an unpaired median duct termed the sperm duct (spermathecal egress duct) which, histologically, is a continuation of the rami/sacculus. The sperm duct, which may be enlarged regionally to form a sperm reservoir, is connected to a large sac-like organ, referred to as the spermatheca by Alberti & Hänel (1986), whose wall consists of a thin syncytium but has no cuticular lining. Its thin basement lamina is continuous with that of the sperm duct. The use of the term spermatheca for this structure is confusing since it is also used by some authors to describe the sperm access system as a whole. Akimov and Yastrebtsov (1984) refer to the structure as a receptaculum seminis, which appears appropriate and obviates the confusion which is unavoidable when the same name is applied to an organ and to one of its components. The combined sperm duct and receptaculum seminis was referred to as the cornu sacculi by Michael. The cornu sacculi may or may not have a connection with the camera spermatis of the ovary. Akimov and Yastrebtsov (1984) consider that spermatozoa in *V. jacobsoni* reach the ovary by the penetration of the wall between the receptaculum and the ovary. On the other hand, Alberti and Hänel (1986) suggest that contact between the sperm duct and the camera spermatis is maintained by free (mesodermal) cells which by penetration of the sperm duct from its base could form an area of the receptaculum filled with 'inner cells'. The latter are a characteristic feature of the contents of the receptaculum of *V. jacobsoni*. A similar invagination probably occurs into the camera spermatis and this would enable insemination by way of active penetration by capacitated spermatozoa. They did not observe a connection between the ovary and the uterus as reported by Akimov and Yastrebtsov (1984).

The phytoseiid-type of sperm access system, occurring in the Phytoseiidae, Otopheidomenidae and some Ascidae, comprises a discrete structure leading from each solenostome (Fig. 9.15B). The histology of the paired structures has not been investigated in detail but their importance as a taxonomic character at species level in the Phytoseiidae, an economically important group of predatory mites, has led to considerable interest in the shape and size of those components which resist maceration during preparation for microscopical study. It is not possible on a functional or structural level, at present, to homologize the various components of the laelapid and phytoseiid types of sperm access system and the terminology used herein for the components of the latter is largely based on Athias-Henriot (1971) and Karg (1982) and is given in Fig. 9.15B. Most North American acarologists refer to the calyx as the cervix and care should be taken not to confuse this term with neck (cervix) as used by European acarologists for the connection between the calyx and atrium. It has not been established whether there is a connection between the sperm access system and the ovary or the genital ducts. Although the term sperm duct is sometimes used as an alternative name for the minor duct, it seems unlikely that a tubule of such narrow diameter would allow the passage of spermatozoa along its length unless it is capable of considerable distension. It is possible that fertilization takes place in the ovary but how the sperm reach that organ from the vesicle of the sperm access system is unknown. There have been no detailed studies of the sperm access system in the Heterozerconidae.

Young (1968b) postulated that the rami and sacculus, on the basis of their involvement in water balance, are derived from paired coelomoducts or nephridia and may represent coxal glands. According to Alberti and Hänel (1986) neither comparative morphology nor ultrastructure supports this hypothesis. These authors suggest that podospermy has evolved from primary tocospermy through a stage when the spermatozoa penetrated the oviducal wall into the haemolymphatic space. This was followed by the transfer of the spermatophore by the chelicerae not to the gonopore but to a region between coxae III and IV where penetration of the integument occurred and the spermatozoa were liberated into the haemolymphatic space. Spermatozoa reached the ovary by active migration. An analogous type of insemination has been described by Storch and Ruhberg (1977) in the onychophoran *Opisthopatus cinctipes.* It is possible, as suggested by Athias-Henriot (1969), that penetration of the integument was facilitated by using the pores of cuticular glands. The final stages in the development of a sperm access system would be the formation of access ducts (tubuli, rami and sperm duct), either by deepening the copulatory pore or by adapting existing cuticular gland ducts, and a connection between the sperm duct and the ovary with the assistance of free (mesosdermal) cells.

In all males of podospermic species, the genital orifice is in a presternal position and the movable digit of each chelicera is provided with a sperm injection appendage, the spermatodactyl (spermatophoral process, spermatostyle), whereas in tocospermic mites, except the Parasitina, the genital orifice is usually located within the sternogenital shield, typically in the region of coxae II to III, and the movable digit of the chelicera is not modified for sperm transfer. *Liroaspis togatus* (Sejina),

Celaenopsis badius (Antennophorina) and *Fuscuropoda billi* (Uropodina) are exceptions in having the opening in the presternal position. The anterior position of the genital orifice facilitates the pick-up of the spermatophore by the chelicerae and this is also evident in the Parasitina which although tocospermic in respect to the site of introduction of the spermatophore nevertheless show a resemblance to the Dermanyssina in the position of the genital opening and in the modification of the movable digit of the chelicera for carrying the spermatophore. A further resemblance is seen in their possessing ribbon and not vacuolated spermatozoa as in the tocospermic Uropodina, Epicriina, Sejina and Zerconina. Thus, the Parasitina exhibit characteristics which are intermediate between tocospermic and podospermic forms.

The spermatophore carrier, or spermatotreme, on the movable digit of the chelicerae of the Parasitina is in the form of a foramen midway along the digit through which the neck of the spermatophore passes during its transfer to the female (Fig. 9.15C). The spermatodactyl in the Dermanyssina, on the other hand, is a free appendage arising from the basal region of the digit and its shape is invariably species specific (Fig. 9.15D). In some of the parasitic dermanyssoid mites, the spermatodactyl may be fused with the greater part of the digit but remains free distally. It is usually provided with a sperm transfer tube (Young, 1959). During insemination the tip of the spermatodactyl is introduced through the solenostome and into the flat chamber (infundibulum) leading to the tubulus (Young, 1959; Amano & Chant, 1978). The method by which the contents of the spermatophore are introduced into the tubulus has not been fully determined. Young, working with *Haemogamasus ambulans*, found the sperm transfer tube of the spermatodactyl, when inserted in the solenostome, to contain sperm and/or sperm cysts. The latter were greatly compressed and elongated in the tube and became ruptured as they were forced into the female. Pressure created by the downward movement of the chelicerae onto the spermatophore was thought to force the cysts into the sperm access system.

Alberti and Hänel (1986) consider that podospermy is a more efficient sperm transfer method than tocospermy and as a consequence the number of spermatozoa introduced into the female (30–40 in *Varroa jacobsoni*) is much reduced in comparison with the numerous spermatozoa in the spermatophores of tocospermic forms such as the Uropodidae (Faasch, 1967). They further suggest that the specialized genital system of the Dermanyssina with its nutrimentary egg development and podospermy may provide a key to the understanding of the success of this species-rich group of free-living and parasitic mites.

Chromosome Types, Numbers and Sex Determination

Metacentric, acrocentric and holocentric chromosomes have been reported in the Acari. The holocentric type, in which the centromere is actively diffused along the length of the chromosome, appears to predominate in the Prostigmata (Tarsonemina, Tetranychoidea, Raphignathoidea and Cheyletoidea). Acrocentric chromosomes occur in the majority of the genera of ixodoid ticks and in some Mesostigmata (e.g. Phytoseiidae) while metacentric chromosomes have been found in *Haemaphysalis* (Kahn, 1964) and occur in some species of the Amblyseiinae (Blommers-Schlösser & Blommers, 1975; Hoy, 1985).

The karyotypes of a range of acarine species are given by Oliver (1977) and Helle and Wysoki (1983). Diploid (2n) numbers in the Ixodida range from 12 to 34. Triploidy has been proved in the parthenogenetic race of *Haemaphysalis longicornis* and is suspected in some species of *Ornithodoros* s.lat. and in *Haemaphysalis leporispalustris* and *Ixodes tasmani* (Oliver, 1982). Diploid numbers of 6 to 18 pairs of chromosomes occur in the Mesostigmata but parthenogenesis is widespread in this order; for example, the Phytoseiidae contains arrhenotokous and thelytokous species in which the basic chromosome number is $n = 4$, except in a few aberrant species of Amblyseiinae with $n = 3$ (Wysoki, 1985). A diploid number of 18 and a haploid number of 9 appears to be characteristic of the Oribatida although some forms have been reported with a haploid number of 8 (Taberly, 1988). *Nothrus silvestris* is exceptional in having 36 chromosomes but it has not been established whether this species is tetraploid. Among the Astigmata diploid numbers range from 4 to 18 pairs and in the Prostigmata from 4 to 22.

The main sex-determining mechanisms comprise XX-XY and XX-XO chromosome systems, arrhenotoky and thelytoky. XX females and XY males is the sex chromosome system in argasoid ticks and in the majority of ixodoid species of the genus *Ixodes* s.lat. Metastriate ticks, with few exceptions, have XO males in which the sex chromosome is often much larger than the autosomes. XO males also occur in the Astigmata. Multiple sex chromosomes ($X_1X_1X_2X_2$: X_1X_2Y) have been demonstrated in two species of *Amblyomma*.

Sexual Dimorphism

The extent to which the sexes can be recognized on the basis of external morphological characteristics varies considerably throughout the Acari. In general, sexual dimorphism is weak in argasoid ticks, Notostigmata and Oribatida. Separation of the sexes in the Notostigmata is usually based on the numbers of setae in the genital and pregenital regions and the ornamentation of the pregenital area (Van der Hammen, 1966, 1977b). The presence or absence of an ovipositor in macerated specimens is often the only way the sexes can be distinguished in the Oribatida when the female is not gravid, although some females also differ from males by their larger size, including the dimensions of the genital valves. Secondary sexual characters are relatively uncommon and not always striking when

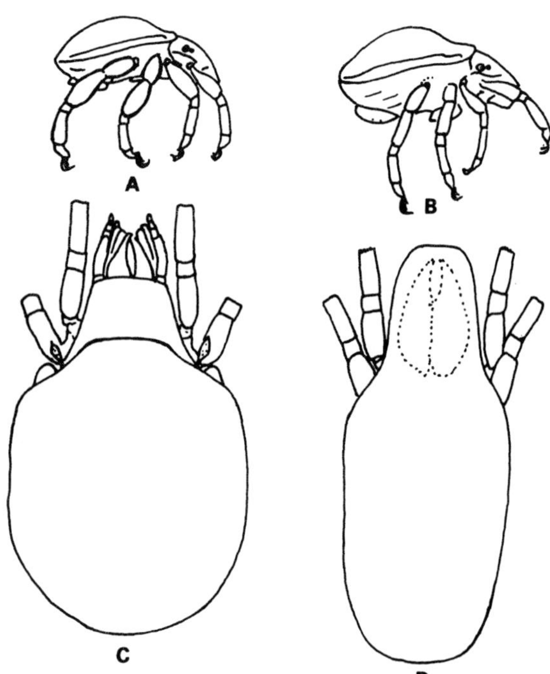

Fig. 9.16. Sexual dimorphism in the oribatids. *Podacarus auberti* (**A**, male; **B**, female) and *Pirnodus detectidens* (**C**, female; **D**, male).

based on setal differences. However, they are very marked in *Podacarus auberti* (Fig. 9.16A), where the legs of the smaller male, particularly the fourth pair, are much stouter than in the female, and in *Pirnodus detectidens* (Fig. 9.16B), where the differences in the shape and structure of various areas of the body are so extreme that on cursory examination the two sexes would appear to belong to different species (Grandjean, 1955; Travé, 1959).

Males of ixodoid ticks differ from females (and immature instars) in the increased sclerotization of the dorsum of the idiosoma; the dorsal shield or scutum covers almost all the dorsum in the male but is restricted to the anterior region of the idiosoma in the female. In metastriate forms such as *Hyalomma* and *Rhipicephalus* the males have distinct adanal shields which are absent in females (Arthur, 1962). The sexes in the Holothyrida and Mesostigmata are readily recognized by the relative shape and in some cases also the position of the shield(s) covering the genital orifice. The male aperture is subcircular and covered by one or two shields and is located intercoxally (see Fig. 12.7, p. 402). The larger genital orifice of the female is covered by four shields in the Holothyrida (see Fig. 12.3A, p. 390) and by one, three or four in the Mesostigmata (see Fig. 12.8, p. 403). Secondary sexual characters occur in the Mesostigmata, and in the Parasitina and Dermanyssina are represented by the spermatotreme or spermatodactyl on the movable digit of the male chelicera and by the hypertrophy of certain setae of legs II and sometimes IV to form spurs for holding the female during mating. The shape and setation of the dorsal shield may also be affected as in the ameroseiid genus *Neocypholaelaps* and the uropodid *Uroseius acuminatus* (Evans, 1963a; Hughes, 1976). In some Dermanyssoidea, for example *Hemipteroseius dysderci*, male and female deuteronymphs may be distinguished by the chaetotaxy of the opisthogaster (Evans, 1963b).

Sexual dimorphism is marked in the Astigmata. In addition to the presence of a sclerotized aedeagus or penis, the male of some species may also be distinguished from the female by the presence of copulatory suckers, one pair flanking the anus and a pair (rarely a single sucker) on each of tarsi IV (see Fig. 12.23A,E, p. 448). The suckers are modified setae. In some parasitic forms such as the Analgidae and Psoroptidae, the posterior extremity of the body of the male may be lobate (see Fig. 12.25C, p. 455). Andropolymorphism occurs in some free-living and parasitic taxa (see Fig. 9.17 below).

The extent of sexual dimorphism varies considerably in the Prostigmata. It is most apparent in the Tarsonemina and Tetranychidae. In the former, the males are smaller and usually of different body shape from the female and the fourth pair of legs may be enlarged and modified for grasping rather than walking (Fig. 12.21F). Physogastry occurs in the gravid females of some species of Pygmephoridae and Pyemotidae and the opisthosoma becomes enormously distended (see Fig. 12.21B, p. 443). The posterior region of the opisthosoma in males of the Tetranychidae is attenuated and the sclerotized aedeagus is a conspicuous feature in macerated specimens. Sexual dimorphism is weakly expressed in the Eriophydoidea, Tydeidae, Eupodoidea and Trombidioidea and the sexes are mainly distinguished by the chaetotaxy and/or configuration of the genital region. Andropolymorphism is found among the Cheyletidae and affects, in particular, the size and shape of the pedipalps, those of the heteromorphic male being considerably longer (Volgin, 1969). In the genus *Cheyletus* forms intermediate between the heteromorphic and normal males are common.

Gynandromorphism

Species with morphological abnormalities sometimes appear in populations of mites and ticks. They have been well documented in the metastriate ticks and are apparent in the asymmetry of the scutum, duplication of the anus, absence of one stigma and enlarged eyes (Vitzthum, 1940–43). Abnormalities affecting the legs are not uncommon among polymorphic astigmatic mites; for example, of 200 heteromorphic males of *Sancassania dampfi* examined by Oudemans (1928), two specimens had a normal leg III on one side and a homomorphic form of the leg on the other side.

The most striking abnormalities are seen in the gynandromorphs or sex mosaics which are tetralogical forms in which some parts of the body exhibit female characteristics while the remaining parts are male. The phenomenon is most frequently seen in metastriate ticks, such as members of the genera *Amblyomma*, *Dermacentor*, *Hyalomma* and *Rhipicephalus*. Gynandromorphy is usually bilateral, i.e one side of the body exhibits male features and the other side female characteristics, but antero-posterior gynandromorphs and forms with an irregular distribution of sexual characters also occur. Most of the literature simply gives descriptions of the abnormalities but Oliver & Delfin (1967) have studied the feeding, sexual behaviour, fecundity and the development of the F_1 progeny in two bilateral gynandromorphs of *Dermacentor occidentalis* produced by crossing males, treated with apholate-acetone solution, with untreated females. The gynandromorphs possessed feminine characters on the left side of the body and masculine plus intersex characters on the right side. Both had female-type genital regions. During feeding, the female side of the body became engorged while the male side remained small. One mated gynandromorph died without ovipositing but the other laid nearly 800 eggs. The incubation time and percentage hatch for the egg batch were normal as were the feeding and moulting periods of larvae and nymphs. The adults also appeared to be normal.

The method of development of the two gynandromorphs of *D. occidentalis* is unknown but Oliver & Delfin (1967) think it unlikely that they are the result of double fertilization or somatic crossing-over between chromosomes. The gynandromorphs were clearly predominantly female and probably initially female (XX). They suggest that 'male cells' in the adult may have resulted from non-disjunction or elimination of the paternal X-chromosome in early cleavage. This would result in only one sex chromosome – the male condition (XO) – in one cell line which would function genetically as male tissue. An alternative hypothesis considers that the chemical treatment changed the female determining properties of genes on the paternal X-chromosome without affecting its division and segregation kinetics.

Andropolymorphism

Andropolymorphism, in the sense of the existence of more than one form of male in the life cycle of an animal, is not uncommon in the Actinotrichida. It has a genetic basis although this seems to be weakly expressed in the mites where it is often strongly influenced by environmental conditions. Male polymorphism is most evident in the external appearance of the morphs but differences may also occur in ecology, physiology and behaviour.

The phenomenon is found among the grain mites and feather mites (Astigmata) and in the prostigmatic families Anystidae and Cheyletidae. Four types of male are distinguished by Türk and Türk (1957) in the Rhizoglyphinae (Acaridae) and these exhibit two basic body forms (Woodring, 1969). The homotype is similar in shape and in the length of its dorsal setae to the ungravid female while the bimotype differs from the

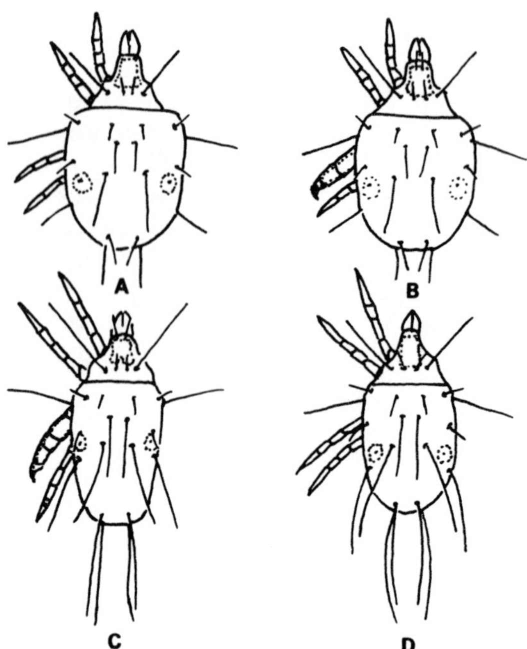

Fig. 9.17. Andropolymorphism in the Rhizoglyphinae (Astigmata). **A**, homomorphic male; **B**, heteromorphic male; **C**, pleomorphic male; **D**, bimorphic male.

ungravid female in the shape of the body and in having longer dorsal setae (Fig. 9.17A–D). The homotype is represented by the homomorphic and heteromorphic males and bimotype by bimorphic and pleomorphic males. In the heteromorphic and pleomorphic males the third pair of legs are modified, being usually enlarged and terminating in a stout claw-like structure.

Woodring (1969) has observed two different patterns of male polymorphism in the Rhizoglyphinae. In the primary form, any one of the homotypes and one of the bimotypes occur in a regular ratio under given environmental conditions. For example, in *Sancassania berlesei* the ratio is between pleomorphic and homomorphic males. The more common secondary andropolymorphism is characterized by the predominance of one male form within a basic body form, either the homotype or bimotype. In *Rhizoglyphus echinopus*, for example, the homomorphic male predominates and the heteromorphic male is uncommon under normal conditions while in *Sancassania boharti*, the bimorphic male is far commoner than the pleomorphic type. The suppression in culture of the heteromorphic and pleomorphic males of *R. echinopus* and *S. boharti*, respectively, appears to be due to the production of a non-specific volatile pheromone, possibly a metabolic waste product, by actively feeding adults and immatures. In the absence of actively feeding forms, developing larvae of *S. boharti* produced as many pleomorphic as bimorphic males.

Both primary and secondary andropolymorphism can occur in the same species. All four types have been observed in *Sancassania anomalus* with pleomorphic and homomorphic males occurring in a regular ratio and heteromorphic and bimorphic forms appearing only rarely. Pleomorphic males, using their enlarged pair of legs as weapons, kill and feed on homomorphic males, thereby elevating the pleomorphic/homomorphic ratio, and even fight among themselves. In normal populations, fighting decreases to an equilibrium level if no new adults appear but soon resumes if new males emerge or strange males are introduced.

Temperature and type of food can affect the male ratios in forms showing primary andropolymorphism but do not appear to influence male ratios in the secondary type. The pleomorphic/homomorphic male ratio in *S. anomalus* is, for example, highest at 20°C but decreases above and below this temperature. Immatures receiving a diet of animal tissues produce a higher ratio than those feeding on wheatgerm and yeast. In *S. berlesei*, it has been shown that the ratio of male types produced by breeding the male morphs with virgin females is not dependent on the male type of the parent but on environmental conditions under which they are reared (Timms *et al.*, 1982).

The selective advantage of pleomorphs in a population has been studied by Timms *et al.* (1980). Working with *S. berlesei*, they found that females mated with pleomorphs produced more offspring earlier than those mated with bimotypes. This enabled a more rapid build-up of the population numbers which could exploit favourable environmental conditions. They also observed the greater longevity of pleomorphs in spring and considered that this, in conjunction with the ability of these males to produce more young earlier, could enhance survival.

Parthenogenesis

Parthenogenesis is a common method of reproduction in the Acari and has been reviewed by Oliver (1971). It is facultative when it coexists with bisexual reproduction or obligatory when males are absent or rare. The phenomenon is known to occur in the Anactinotrichida and Actinotrichida. Arrhenotoky or male-producing parthenogenesis in which the males are haploid and the females diploid is the most widespread form and occurs in the Mesostigmata, Prostigmata and Astigmata. Thelytoky (female-producing parthenogenesis) is less common although there are records of its occurrence in the three orders referred to above as well as in the Oribatida and Ixodida. A third type of parthenogenesis, deuterotoky, in which both males and females are produced from unfertilized eggs, is rare but has been reported in the Anoetidae (Heineman & Hughes, 1969), in ticks (Davis, 1951) and in a laboratory population of *Tetranychus urticae* by Jesiotr and Suski (1970).

Arrhenotoky appears to be the predominant form of parthenogenesis in the mesostigmatic families Ascidae, Macrochelidae, Macronyssidae and Laelapidae; for example, of 21 species of Macrochelidae examined by

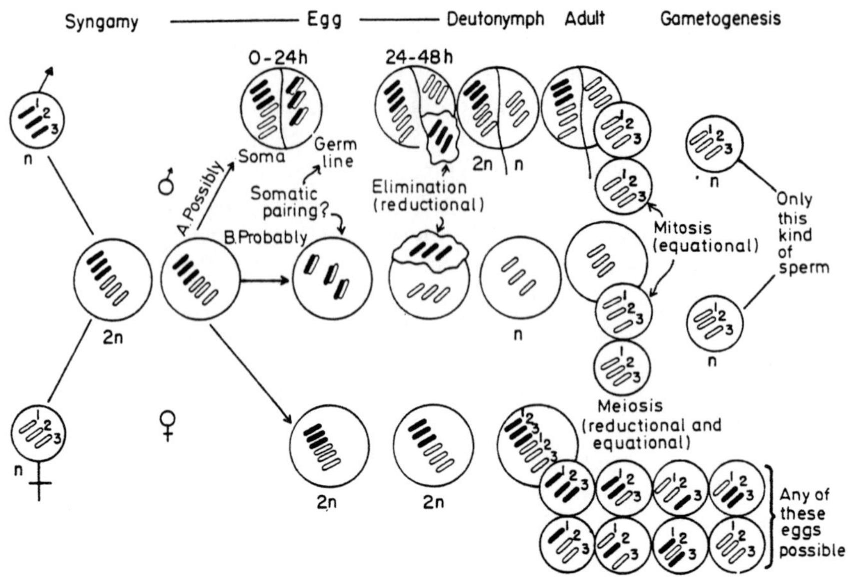

Fig. 9.18. A model showing possible and probable mechanisms for pseudo-arrhenotoky in *Metaseiulus occidentalis* (Mesostigmata) (from Nelson-Rees *et al.*, 1980).

Filipponi (1964) 19 were arrhenotokous and two were thelytokous. In some Phytoseiidae and Dermanyssidae, however, haploid males are produced only after the females have mated. Oliver (1971) attributed this to pseudogamy and referred to the phenomenon as arrhenotokous gynogenesis. Radiation experiments and cytological studies by Helle *et al.* (1978) and Nelson-Rees *et al.* (1980), respectively, demonstrated that the phytoseiid males are of biparental origin and not the product of parthenogenesis. The mode of reproduction in which haploid males develop from fertilized eggs by the elimination of the paternal set of chromosomes was attributed to parahaploidy by Nelson-Rees *et al.* (1980). These authors suggested two mechanisms for parahaploidy in *Metaseiulus occidentalis*, which are shown diagrammatically in Fig. 9.18. In approximately day-old embryos half of the chromosomes are considered to be eliminated from some (possible line A) or all (probable line B) cells. Somatic pairing and a reductional division separate two sets of chromosomes and the paternal set is eliminated from the cell. Subsequently, the germ line of the male is haploid whereas female embryos, nymphs and adults are diploid. The role of the paternal set of chromosomes before its elimination is not known but suggestions have been made that its presence is required to induce embryonic development or that it functions in sex determination or in determining the fertility of the sons (Schulten, 1985). The term parahaploidy is used by Hartl and Brown (1970) to cover all cases in which genetic haploidy is attained by heterochromatinization, suppression or elimination of one chromosome set. It is considered by some authors to be confusing when applied to the peculiar haplo-diploidy in phytoseiid mites and they prefer the term pseudo-arrhenotoky, proposed by De Jong *et al.* (1981) for all cases in which male haploidy is brought about by the complete elimination of a set of chromosomes (Helle & Pijnacker, 1985). This term denotes the important fact that the haploid–diploid condition in the Phytoseiidae is not achieved by arrhenotoky.

In the Prostigmata, arrhenotoky is found in the Anystoidea, Cheyletoidea, Eriophyoidea, Pterygosomatidae, Raphignathoidea, Tarsonemina (Pyemotidae and Tarsonemidae) and Tetranychoidea. Among the Astigmata it appears to occur only in the family Anoetidae (Heineman & Hughes, 1969). Helle *et al.* (1984) suggest that arrhenotoky also occurs in some species of Oribatida, for example *Humerobates rostrolamellatus* and *Orthogalumna terebrantis*, on the basis of chromosome numbers in egg squashes, some of the eggs having a diploid number and others a haploid number of chromosomes.

Thelytoky in the Mesostigmata has been reported in the families Macrochelidae (Filipponi, 1964), Phytoseiidae (Wysoki, 1985), Ascidae, Rhodacaridae and Laelapidae (Walter & Oliver, 1989). Males occur rarely in *Rhodacarus denticulatus*, *Protogamasellopsis corticalis* and *Gamasellodes vermivorax* but in *Protogamasellus massula* they form about 5% of reared populations although they do not attempt to mate with females (D.E. Walter, personal communication). In ticks thelytoky is rare but appears in both the Argasidae (*Ornithodoros*) and Ixodidae, usually in a very small percentage of the siblings. *Haemophysalis longicornis* is exceptional in that bisexual and parthenogenetic races occur in different geographical areas. Amongst the Prostigmata, thelytoky occurs only rarely in the Tarsonemidae, Cheyletidae and the Tetranychoidea except for the Bryobinae. The latter is considered to be the most primitive group of the Tetranychidae and thelytoky is relatively common (Helle & Pijnacker, 1985). Diploid thelytoky is assumed to occur in the Tetranychidae while haploid thelytoky has only been observed in the Tenuipalpidae. Males occur very rarely in some thelytokous species of Tenuipalpidae and Bryobinae. As in the case of arrhenotoky, the only reference to thelytoky in the Astigmata is for the Anoetidae.

Luxton (1981) listed 35 species of Oribatida in which males were absent while in another 11 species their occurrence was exceptional. This list was largely compiled from Grandjean's seminal study (Grandjean, 1941b) of sex ratios and parthenogenesis in oribatids. Thelytokous parthenogenesis was inferred mainly from sex ratios of populations sampled in the field. Grandjean (1941b) recognized three categories or groups of sex ratios: group P (parthenogenetic forms) in which males were absent or only present exceptionally; group S (normal bisexual forms) in which the ratio of males to females approximated to unity; and group SP in which there was a tendency towards parthenogenesis in some genera, e.g. *Hydrozetes* and *Scutovertex*. He considered the males, which consistently comprised only a small proportion of the adult population of a species, to be relics of bisexual ancestors and referred to them as atavistic males. The production of such males (also referred to as 'spandaric') appears to be common among obligate thelytokous animals. The males of the thelytokous species *Platynothrus peltifer* and *Trhypochthonius tectorum* occur in low proportions (1–5%) in all populations and are capable of producing spermatophores. However, spermatophores deposited by them appear to be ignored by young females. Spermatogenesis in these males is similar to that of males of bisexual species but apparently stops at a late stage in spermiogenesis so that the sperm do not mature (Taberly, 1988). The genital organs in two of 43 males of *T. tectorum* showed hermaphroditic structures. The males in both species have the same complement of chromosomes (n = 9) as the females. A study of oogenesis in these two species by Taberly (1987b) has shown that diploidy is restored by a joining of the anaphase plates of the second meiotic division, a mechanism which he referred to as 'mixocinèse'. At present, experimental rearing of female progeny from virgin females appears to have been accomplished only in *Platynothrus* and *Trhypochthonius.*

According to Norton and Palmer (1991), there are only about 26 oribatid species in which thelytoky has been proven and of these only one, *Oppiella nova*, is a member of the Euoribatida (= Brachypylina). These authors suggest that ancestors of large, completely parthenogenetic macropyline families, for example Brachychthoniidae, Camisiidae, Malaconothridae and Nanhermanniidae, were parthogenetic and that speciation and evolutionary radiation occurred in the absence of biparental reproduction.

Paedogenesis

Reproduction in a pre-adult stage is rare in the Acari. According to Baker (1979) paedogenetic eggs occur in both nymphal instars of a species of *Brevipalpus* and fully grown larvae were observed in some nymphs. The larvae probably escape from the body of the nymphs, which lack a genital orifice, by the rupture of the integument of the opisthosoma. Volkonsky (1940) considers that a paedogenetic egg is produced by the larva of *Podapolipus grassi* (Tarsonemina).

10 Methods of Sperm Transfer, Mating Behaviour and Oviposition

The Acari practise internal fertilization and achieve sperm transfer in a variety of ways. There are two main methods of sperm transfer: indirect transfer involving the deposition by the male of a stalked spermatophore on a substrate to be taken up later by the female; and direct transfer in which sperm is introduced by the male, usually by way of a spermatophore, into the female genital orifice or into the copulatory pore of a sperm access (spermathecal) system. Arachnida probably possessed spermatophores before their emergence as a terrestrial group and this enabled a more direct and efficient method of insemination than is achieved by liberating the sperm cells into water. The majority of terrestrial arthropods practising indirect transfer live in humid habitats where presumably the risk of the sperm drying out is reduced. Schaller (1971) considers this to indicate that many of these forms are incompletely adapted to existence in air and that only those species in which the mating behaviour allows close contact between the sexes and the rapid transfer of the sperm droplet are able to live in dry conditions.

Indirect sperm transfer is the primitive method of reproduction in the Arachnida and may or may not involve the pairing of the sexes (Schaller, 1954; Alexander, 1964; Weygoldt, 1966). From an evolutionary standpoint, Alexander (1964) considers that association between the sexes during sperm transfer preceded 'dissociation' (spermatophore deposition irrespective of the presence of females) and supports his hypothesis with reference to the pairing behaviour shown by primitive arthropods such as *Limulus* (Merostomata) and *Peripatus* (Onychophora). According to Alexander, dissociation as well as direct sperm transfer are derived from pair formation. This is accepted by Thomas and Zeh (1984) who suggest that dissociation evolved 'soon after the arthropods colonized terrestrial habitats' and postulated that low desiccation stress and high mortality risks from interactions with mates are among the conditions necessary for males to switch from pairing to dissociation. In contrast, Schaller (1979) is of the opinion that 'sperm transfer without pair formation represents the more primitive mode' and presents supporting evidence from the pseudoscorpions, the morphologically and ecologically primitive families of this group showing the primitive form of sexual behaviour.

In terrestrial and secondarily aquatic Acari practising indirect sperm transfer, there are trends towards the elaboration of the structure of the spermatophore in those species showing dissociation and of the behavioural events associated with spermatophore transference in 'pairing' forms. Insemination in those species with direct sperm transfer is by way of the female gonopore, or through a single or paired copulatory pore leading to the receptaculum of a sperm access system. In the latter, the female gonopore, the oviporus, is used only for egg-laying. The most evolved method of insemination occurs in those actinotrichid species which practise copulation by way of a copulatory pore and in which the sperm is inserted by an aedegus without the formation of a spermatophore. The diversity in methods of sperm transfer in the Acari is astonishing and to quote Schaller (1979) 'just about all modes that can be considered for arthropods are found among them'. Ecological factors have probably played an important part in the differentiation of the methods of reproduction.

Chemical communication involving sex pheromones is an important facet of the mating behaviour of both mites and ticks. Pheromones affect the behaviour of individuals of the same species and function as attractants, arrestants and mating stimulants (Sonenshine, 1984). The part sex pheromones play in the mating behaviour of the Acari is discussed, where appropriate, in the accounts of indirect and direct sperm transfer.

Indirect Sperm Transfer

Indirect sperm transfer occurs in the Oribatida (Pauly, 1952) and in a number of groups of the Prostigmata, such as Adamystidae (Coineau, 1976), Anystidae (Schuster & Schuster, 1966), Bdellidae (Alberti, 1974), Eriophyoidea (Oldfield *et al.*, 1970; Sternlicht & Griffiths, 1974), Labidostommatidae (Vistorin, 1978), the terrestrial and a majority of the aquatic Parasitengona (Witte, 1984), and some Tydeidae–Tydeinae (Schuster & Schuster, 1970). Indirect transfer is not known for the Anactinotrichida although the method of insemination has not been established in the Notostigmata, which show weak sexual dimorphism.

The method is reminiscent of the form of insemination in some other groups of Arachnida such as scorpions, pseudoscorpions and amblypygids in which the male also deposits a stalked spermatophore on a substrate (Alexander, 1957, 1962; Weygoldt, 1966). The structure carrying the sperm droplet typically comprises a stalk, attached to the substrate, and an apical sperm droplet support of varying complexity (Fig. 10.1A–D). The entire structure supporting the sperm mass is produced from secre-

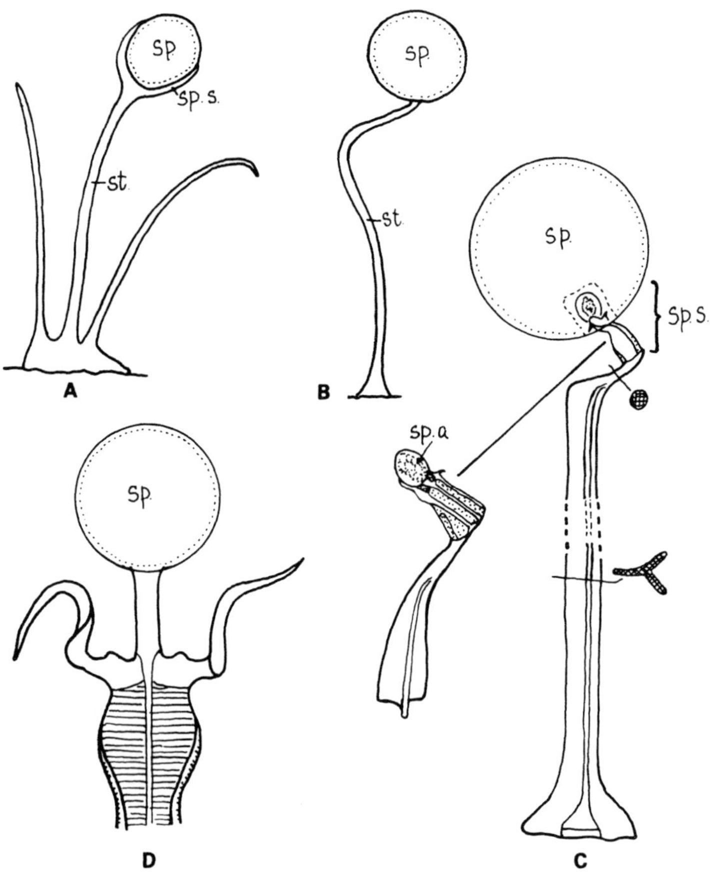

Fig. 10.1. **A**, branched spermatophore of a trombiculid mite; **B, C**, spermatophores of oribatid mites showing structure of the stalk and sperm support in *Damaeus quadrihastatus* (**C**) (after Cancela da Fonseca, 1969); **D**, head of the spermatophore in *Nicoletiella jaquemarti* (Prostigmata) (based on Vistorin, 1978). *sp.* = sperm droplet; *sp. a.* = spermatophoric ampoule; *sp. s.* = sperm droplet support; *st.* = stalk.

tions of the male genital tract and in some species it is formed into its specific shape by the 'penis', a spermatophoric and not an intromittent organ in forms with indirect sperm transfer. Witte (1975a) considers that in the Erythraeidae the stalk and apical region of the spermatophore and the thin envelope surrounding the sperm cells are each formed from a different type of protein secretion produced by the glandular part of the testes. The secretions forming the stalk and apical region of the spermatophore are hardened by oxidation. Premature hardening caused by the entry of air into the ejaculatory duct during stalk deposition is prevented by lipid secretions of the anterior accessory gland which seal the progenital chamber. Lipid secretions from the posterior lateral gland ('hintere Lateraldrusen') act as a glue to attach the stalk to the substrate and to connect the different components of the spermatophore. In the eriophyoids *Trisetacus juniperinus* and *Phytoptus avellanae*, on the other hand, four different substances produced by secretory cells of the 'seminal vesicle' (the region of the genital tract between the short vas deferens and the ejaculatory duct) are involved in the formation of the spermatophore (Nuzzaci & Solinas, 1984). Two of these form the main part of the 'sac' (*sensu* Oldfield *et al.*, 1970), including the external protective layer and the final sperm embedding and storing substance. The other two components are considered to play a part in forming the stalk and the apical region (head) of the spermatophore. Sperm sacs occur in the spermathecae of both sexually immature (without mature eggs) and ovigerous females of these two species. In the former, the spherical sac is complete, albeit with a reduced external protective layer, but in the latter only the sperm embedding and storage substance and the spermatozoa are present. Seminal fluid and stalk substance in the Oribatida are produced by the 'seminal vesicle' (probably a fused portion of the vasa deferentia) and the mixture passes by way of the vasa deferentia to the ejaculatory duct where it is somehow separated into its two components with the stalk substance nearest the penile orifice (Woodring, 1970). Coincidental with penile extrusion, a portion of stalk substance is extruded, attached to a substrate and as the body of the mite is raised the stalk is drawn out. A ball of semen adhering to the tip of the stalk is drawn through the penile orifice.

The extent of mating behaviour as a preliminary to insemination within groups practising this type of sperm transference varies considerably. In the Oribatida, pre-mating contact between the sexes very rarely occurs and numerous spermatophores are usually deposited by the male in the absence of females. The females appear to locate the spermatophores without obvious signals although it is probable that some form of chemical cue is involved. Schaller (1971) considers that the greater the population density of a species, the less intimate is the contact between the pairs and cites the Oribatida as an extreme example. Grandjean (1956c) observed probable mating behaviour in a South American species of Galumnidae (*Erogalumna zeucta*) in which the smaller male walked behind the female with his prodorsum beneath her opisthosoma and his first pair of legs over the posterior part of her body. Spermatophore deposition was not observed.

Cancela da Fonseca (1969) divides the process of spermatophore deposition in the male of *Damaeus quadrihastatus* into four distinct phases:

1. The male having found a suitable deposition site lowers its body so that the genital plates are almost in contact with the substrate and begins a series of body contractions.
2. The contractions cease and the base of the spermatophore is deposited on the substrate followed by the slow and steady elevation of the body during which time the stalk of the spermatophore is produced.

3. The male becomes immobile for a few seconds before suddenly raising the body further by means of the posterior legs and producing the sperm support.
4. Finally, a fraction of a second later, the sperm droplet is deposited on the support and is held to it by the 'ampoule spermatophorique', a component of the support which has the property of adhering to surfaces with which it comes into contact (Fig. 10.1C).

Details of the structure of the spermatophore in the Oribatida are given by Pauly (1952), Taberly (1957) and Cancela da Fonseca (1969).

In the Parasitengona and Anystidae, contact between the sexes is widespread and when it occurs is a prerequisite to spermatophore production (Witte, 1984). This partner-related form of spermatophore deposition and sperm transfer occurs, for example, in some species of the terrestrial genera *Allothrombium*, *Anystis*, *Trombidium* and *Leptus*, and of the aquatic genus *Eylais*. The pairing behaviour involves an encircling dance in which the partners make tapping contacts, usually with their first pair of legs. During the dance both partners may move at the same pace while encircling each other (*Eylais extendens*, *Erythraeus phalangioides*) or the male may be more mobile than the more or less stationary female and runs in circles and loops around her (*Anystis*, Fig. 10.2A). In *Allothrombium fuliginosum*, the female, walking slowly in loops or circles, keeps tapping contact with her partner making similar movements within her ambit (Fig. 10.2B). The encircling dance is reduced to simple tapping contacts in such species as *Abrolophus quisquiliarum*, *Charletonia cardinalis*, *Hydrachna cruenta* and *Limnesia maculata* (Witte, 1991). Some males of the families Johnstonianidae, Calyptostomatidae, Erythraeidae, Trombiculidae, Hydrachnidae and Limnocharidae, on the other hand, deposit spermatophores in isolation or without contact with the female. Witte (1991) refers to this as habitat-related deposition. Pre-mating contacts were possibly part of the reproductive behaviour of the stem species of the Parasitengona and reduction in contact between the sexes, which appears in many families of the group, may be the product of convergent evolution.

The presence of females has also been found to be necessary for the deposition of spermatophores in *Nicoletiella denticulata* (Nicoletiellidae) by Vistorin (1978) and to increase the rate of spermatophore production in the Nanorchestidae (Schuster & Schuster, 1977). In several other species of Nicoletiellidae, the presence of a female is not a requirement for spermatophore deposition.

Females may be alerted to the presence of or guided to the spermatophores by elastic threads, secretion tracks or pheromones produced by the male (Witte, 1984). The threads and tracks are secreted by way of the genital orifice while the pheromones appear to emanate from fields of spermatophores. A male of *Allothrombium fuliginosum* may deposit signalling threads in a circular area in the absence of a female. The thread appears to consist of the same substance as the spermatophore stalk and

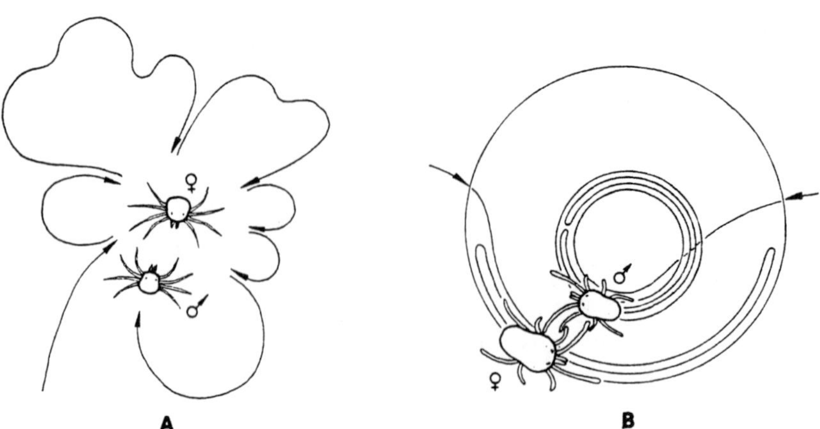

Fig. 10.2. Diagrammatic illustration of the pairing dances in two species of Prostigmata. **A**, movements of a male *Anystis* sp. around a stationary female; **B**, the movements of the male and female of *Allothrombium fuliginosum*, each of which makes tapping contacts during the course of the dance. (After Witte, 1991.)

runs just above the ground. This so-called primary signalling field, as well as subsidiary fields formed in the surrounding area, may be repeated on a new site at daily intervals. Spermatophores are deposited in the primary field when a female enters the territory and the pairing dance takes place. The area is only recognized by both sexes on the day the threads are laid. The primary signal field in this species has evolved into a mating territory which is defended vigorously against intruding males by the occupying male (Moss, 1960). In the Anystidae and most Erythraeidae, a large area surrounding the primary signalling site is sparsely covered with threads which assist the female in her search for spermatophores that are deposited only in the primary site.

The zig-zag secretion track provides a more elaborate signal and is laid down by males of the terrestrial genera *Johnstoniana*, *Trombidium* and *Camerothrombidium* as well as the aquatic genera *Hydrachna* and *Limnochares* (Fig. 10.3A). The tract leads to the spermatophore and in *Johnstoniana* the angles of the tract are marked by signal stalks each with an apical droplet (Fig. 10.3B). These stalks resemble spermatophores except for their smaller size and absence of sperm cells. *Camerothrombidium errans*, on the other hand, circles the spermatophore while following a zig-zag path and touches the ground at intervals with its genital orifice (Fig. 10.3C).

New strategies for ensuring insemination have evolved in actively swimming water mites. Two main strategies are evident in these forms and involve either the attraction of the female by pheromone gradients spreading from spermatophores deposited on the substratum, or a pre-acquisition behaviour involving closer association between pairs. The former strategy occurs in species of *Limnesia* and *Hydrachna* in which

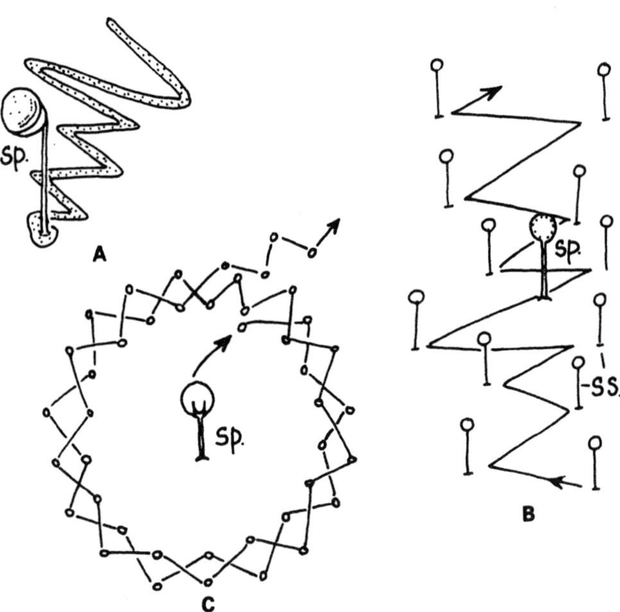

Fig. 10.3. Examples of zig-zag secretion tracks associated with spermatophores in *Limnochares aquatica* (**A**), *Johnstoniana errans* (**B**) and *Camerothrombidium rasum* (**C**) (after Witte, 1991). *sp.* = spermatophore; *ss.* = signalling stalks.

males deposit fields of several hundred spermatophores, while *Arrenurus globator* has adopted the latter strategy. The male of *A. globator* uses the hind legs to position the female so that the postero-dorsum of his opisthosoma abuts her opisthogaster. An adhesive substance secreted by the male is applied to the female venter and attaches her to him in that position. The female is then carried into a position over the spermatophore which is deposited after their attachment.

The male of the terrestrial species *Saxidromus delamarei* also carries the female to the spermatophore (Coineau, 1976). He moves under the female from the front, grasps the basal region of her posterior legs on each side with his first pair of legs and lifts her from the ground (Fig. 10.4A). In this position the opisthosoma of the female overhangs the anterior region of the male (Fig 10.4B). Still carrying the female on his back, the male deposits a spermatophore (Fig. 10.4C,D) and then rotates his body so that the female genital orifice is brought over the spermatophore (Fig. 10.4E). The large head of the spermatophore enters the female genital tract as she is lowered over it (Fig. 10.4F). At this stage the male pushes the female backwards which results in the breaking of the stalk. The male releases his hold on the female and the partners separate.

In the least specialized (?primitive) form of spermatophore in the Parasitengona, for example *Allothrombium, Trombidium* and *Eylais*, the sperm droplet lacks a sheath and the matrix secretion of the sperm forms

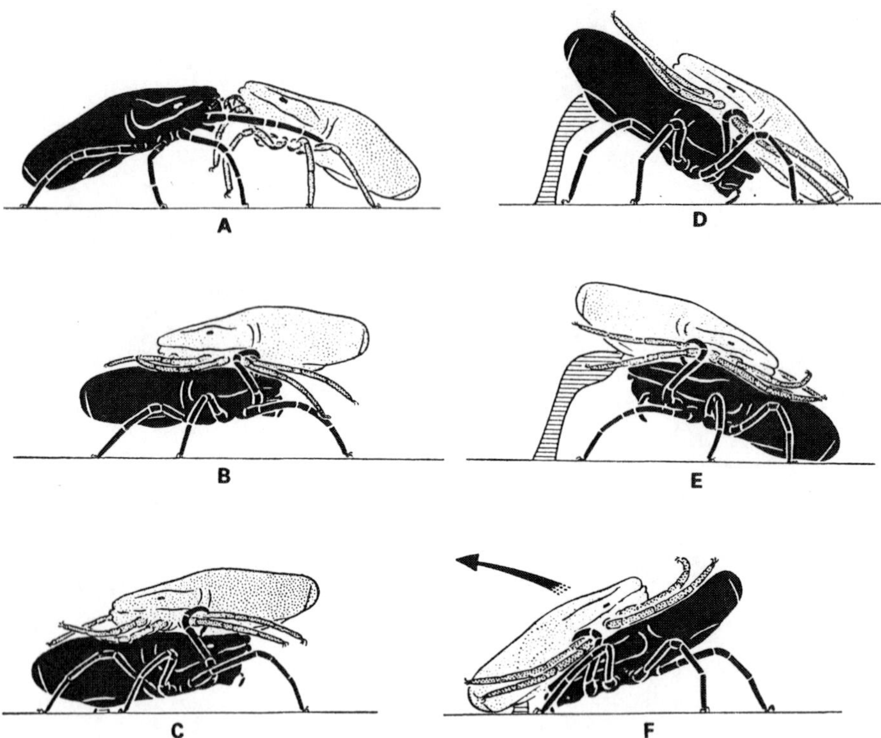

Fig. 10.4. Some stages in the behaviour associated with sperm transfer in *Saxidromus delamarei* (Prostigmata) (after Coineau, 1976). **A, B**, male (in black) lifting the female (dotted) from the ground; **C, D**, male depositing a spermatophore; **E**, after rotation, the male lowers the genital opening of the female over the head of the spermatophore; **F**, the male pushes the female backwards which results in the breaking of the spermatophore stalk.

the periphery of the sperm mass (Witte, 1984). Elaboration of the spermatophore is apparent in the development of sperm droplet sheaths and in the specialization of the sperm support. For example, the sheath covering the sperm droplet in *Erythraeus* is composed of large electron-dense proteinaceous droplets, secreted by the glandular part of the testis, and covers the whole droplet (Fig. 10.5A), while in *Abrolophus* a homogeneous secretion, enriched by lipids, covers the sperm and the protein droplets are restricted to the crown of the spermatophore (Fig. 10.5B). Intrafamily differences also occur in the form of the spermatophore in the Bdellidae. The sperm mass is protected by a secretion layer of similar material to that forming the stalk in the subfamilies Bdellinae, Odontoscirinae and Spinibdellidae but is uncovered in the Cytinae (Alberti, 1974). A penis is extruded during spermatophore deposition in the Bdellinae and Odontoscirinae, and in these two sub-

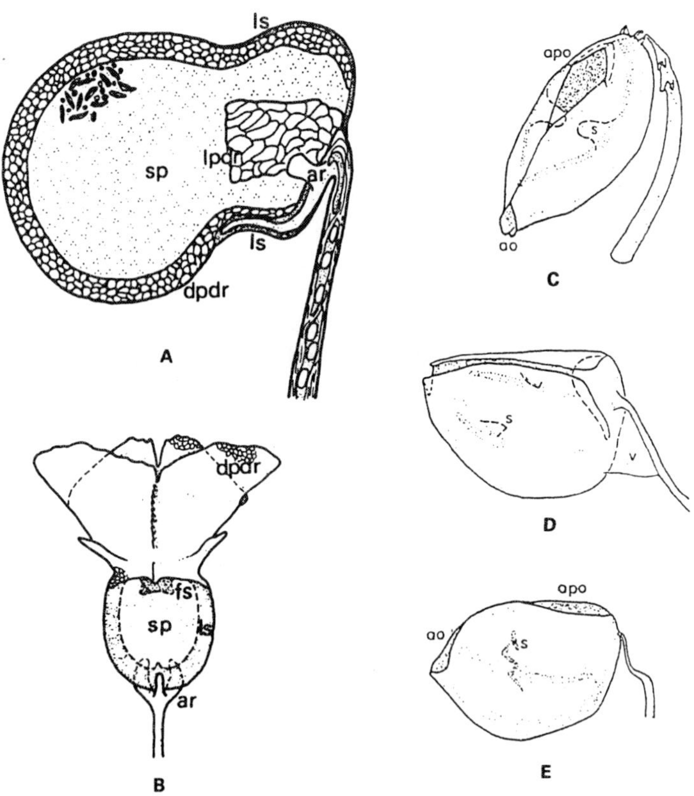

Fig. 10.5. A, B, sheath structures and location of secretory materials in the sperm droplet of *Erthyraeus phalangioides* (**A**) and *Abrolophus rubipes* (**B**). **C–E**, sperm pouches of the spermatophores of three species of water mites: *Thyas barbigera* (**C**), *Sperchon setiger* (**D**) and *Limnesia maculata* (**E**) (after Witte, 1991). *ao.* = anterior opening; *apo.* = apical opening; *ar.* = apical rami; *dpdr.* = electron-dense proteinaceous droplets; *fs.* = fibrillar secretion; *lpdr.* = electron-light proteinaceous droplets; *ls.* = lipid-rich secretion; *s.* = lateral spine; *sp.* = sperm; *v.* = velum.

families and in the Spinibdellinae the ovipositor is provided with a seminal receptacle.

The development of the apical region of the spermatophore into an intricate container or pouch for the sperm droplet is characteristic of those water mites in which the males deposit fields of spermatophores. The sperm droplet support has the form of a container or pouch with two openings (Fig. 10.5C–E). A pair of spines may project into its cavity to assist in the retention of the sperm droplet. The female sucks the sperm into her vagina from the posterior opening while water flows into the pouch by way of the apical aperture. This device ensures the retention of the sperm droplet on the stalk during the disturbance of the surrounding water by the male in the course of depositing adjacent spermatophores.

The success of indirect sperm transfer on land requires the survival of the sperm droplet for a period in air where it may be subject to low and fluctuating humidities. Not all the mites practising this type of insemination live in permanently humid habitats where the risk of the sperm drying out is reduced. Witte (1984) suggests that certain structural and physiological features of the spermatophore have enabled mites, particularly Parasitengona and Anystidae, to adapt to dry as well as humid conditions. The uptake of water vapour from unsaturated air by the sperm droplet appears to be a primitive feature of the spermatophores of terrestrial Parasitengona and enables the maintenance of a stable water balance. The mechanism is probably a passive one depending on the hygroscopic properties of the matrix secretion of the sperm (Witte, 1991). In *Johnstoniana*, for example, water is taken up reversibly from humid air and the sperm droplets and droplets of the signalling stalks can be observed to shrink and swell repeatedly, whereas in *Allothrombium* and *Trombidium* droplets in contact with water quickly burst.

A further interesting aspect of indirect sperm transfer refers to methods which ensure that only spermatophores with viable sperm cells are picked up by the female. Males of several taxa in the Parasitengona, Anystidae, Tydeidae and Bdellidae press down 'old' spermatophores before new ones are added while in the Oribatida it is the female that destroys them. The apical part of the stalk of old spermatophores of *Allothrombium fuliginosum* bends to the ground, probably due to desiccation, and cannot be taken up by the female.

A type of behaviour not previously known in mites practising indirect sperm transfer has been described by Michalska and Boczek (1990) in eriophyid mites of the genera *Acalitus* and *Vasates*. During the migration of females of *Vasates robiniae* and *V. allotrichus* from leaves to overwintering sites on the branches of *Robinia pseudoacacia*, the males were observed to remain on the leaves and to form aggregations around quiescent female deuteronymphs to which they were attracted during the whole period of quiescence. The males guarded the deuteronymphs and deposited spermatophores around them. Aggressive behaviour between males occurred within colonies when there were more than ten males to one quiescent female. This type of behaviour, in some of its aspects is reminiscent of that associated with pre-copulation in certain actinotrichid mites practising direct sperm transfer.

Direct Sperm Transference

This method of insemination involves the transfer by the male of a spermatophore or liquid sperm directly into the reproductive system of the female. The sperm material may be introduced into the genital orifice, or into a copulatory (sperm induction) pore which leads by way of a duct to a sperm receptacle (spermatheca, seminal receptacle, sacculus foemineus). Athias-Henriot (1970a) refers to these two types of insemination in the Mesostigmata as tocospermic and podospermic, respectively.

The latter is confusing when applied generally to non-tocospermic insemination in the mites. Although the term podospermic describes the location of the paired copulatory pores in most mesostigmatic mites, i.e. on or near to the base of the posterior two pairs of legs, it does not apply to the condition obtaining in the Actinotrichida where the single pore is situated in the posterior half of the opisthosoma either ventrally or terminally. Porospermy, implying insemination by way of a pore, is a more appropriate descriptive term and the use of podospermic should be restricted to the Mesostigmata. Both tocospermy and porospermy are broad categories, each including a variety of sperm delivery methods which have evolved independently.

When present, the spermatophore in those Acari with direct sperm transfer comprises an outer sac-like structure containing the seminal fluid and sperm. The outer case is not taken into the female reproductive system but usually remains adhering to the genital orifice or, in podospermic forms, to the chelicerae. As Schaller (1979) has pointed out the term spermatophore is used uncritically for any form of sperm carrier irrespective of its ultrastructure, chemical composition or mode of formation. Feldman-Muhsam (1967), dealing with the complex spermatophore of ticks, retained the term spermatophore for the structure as a whole and proposed the term ectosphermatophore for the outer bulb or sac of the fully formed spermatophore, and endospermatophore for the structure found initially in the neck of the ectospermatophore and after copulation in the vagina. This distinction in terminology between the two components of the spermatophore has merit in practical as well as in structural and functional terms. Amano and Chant (1978) have already adopted this terminology for the spermatophore of the Phytoseiidae. As a practical measure and without implying homologies, the term endospermatophore is broadened herein to include any sperm packet or sac formed *within* the sperm access system during porospermic insemination. Ectospermatophore is used generally for the outer sac of the spermatophore which is not taken into the female genital tract.

The events surrounding direct sperm transfer result in close contact between the partners, usually a venter-to-venter position, and the transfer of the spermatophore or sperm is accomplished by the mouthparts and/or legs of the male, or by an intromittent organ (aedeagus, penis). Direct sperm transfer is the only method of insemination in the Mesostigmata, Ixodida and Astigmata, and has also been established among the Prostigmata in such taxa as the Tetranychoidea, Cheyletoidea, Tydeidae (Pronematinae), Tarsonemina and some water mites (e.g. Pionidae and Mideidae).

Mesostigmata

Tocospermic and podospermic insemination occur in this suborder. The former is characteristic of the Parasitina, Epicriina, Zerconina, Uropodina,

Sejina and probably also the Antennophorina, while the latter is found in the Dermanyssina and Heterozerconina (Athias-Henriot, 1969). In tocospermic forms other than the Parasitina, the male chelicerae are not modified for sperm transference and the male genital orifice is usually located within the sternal region (typically at a level with coxae II to III). Podospermic Dermanyssina have a free appendage on each movable digit of the chelicera for sperm transfer and the genital orifice lies in a presternal position. According to Alberti (1988), tocospermy in the Parasitina is different from that found in other tocospermic Mesostigmata and is more comparable to the insemination route of the podospermic Dermanyssina. He suggests that the similarities between the Parasitina and Dermanyssina could be expressed by the term 'neospermy' to contrast with the 'archispermy' of the Ixodida, Uropodina, Epicriina and Sejina.

The mode of delivery of the spermatophore to the genital region of the female orifice in the tocospermic forms is by the gnathosomatic appendages, assisted in some cases by the first pair of legs (appendicular transfer). In the Uropodina and Sejina, the sac-like spermatophore may be manipulated by the chelicerae only, the chelicerae and palps, or the palps and the first pair of legs but none of the appendages is specially modified for this purpose. Spermatophore production in the Uropodina is preceded by the partners adopting a venter-to-venter position with both facing in the same direction. The attainment of this mating position is achieved in a variety of ways. The most spectacular behaviour occurs in *Leiodinychus krameri* during which the male mounts the back of the female and by a series of rocking movements rolls the female on her back before mounting her for spermatophore transference (Radinovsky, 1965). The superior position of the male in this species contrasts with the inferior position adopted in species of *Uropoda*, *Fuscuropoda* and *Caminella* (Faasch, 1967; Compton & Krantz, 1978). Mating behaviour in the semi-aquatic polyaspidid, *Caminella peraphora*, is accompanied by the female producing a plastic secretion which is manipulated by the last pair of legs of the male to form gas-filled pockets on the back of the female; these probably act as a float.

The male inserts only the neck of the spermatophore into the female genital orifice and the endogynum present in many uropodids may assist in holding the spermatophore. According to Faasch (1967) the males of *Uropoda orbicularis* and *Fuscuropoda marginata* use their palps to press the sperm out of the spermatophore into the vagina. After the passage of the sperm material into the vagina the ectospermatophore may be seen hanging from the genital shield.

The Parasitina differ from other tocospermic Mesostigmata in having the male chelicerae specially modified for transporting the spermatophore. The movable digit is provided with a distinct foramen, the spermatotreme, which receives the neck of the spermatophore. Michael (1892) suggests that the spermatophore is 'blown like a bubble right through the hole'. During mating, the male slips under the female and the pair adopt a venter-to-venter position with both facing in the same direction. Legs II to IV of the male, particularly the second pair which are usually provided with spurs (hypertrophied setae) and are often crassate, hold the female. The second pair of legs grasps the female's fourth pair, and his second and third pair are applied to the sides of her opisthosoma (Zukowski, 1964). While in this position the spermatophore is produced and carried by one of the chelicerae to the female genital opening. The genital shield is lifted by the chelicerae and the spermatophore introduced into the opening. According to Michael, the anterior portion of the spermatophore is broken off during the withdrawal of the chelicerae.

Podospermic insemination is practised by the Dermanyssina. The sperm material enters the female by way of the copulatory pores (solenostomes, sperm-induction copulatory pores) located in the acetabula of coxae III, less commonly on the trochanter or femur of legs III (in some Rhodacaroidea), and on or near the base of coxae IV (e.g. Veigaiidae and Pachylaelapidae) and more rarely lateral to the genital shield between coxae IV or posterior to coxae IV in association with the metapodal shields (some Rhodacaroidea). The structure of the different forms of sperm access system is discussed in the section on reproductive organs (see Chapter 9). Males of this suborder exhibiting podospermic insemination have each movable digit of the chelicerae provided with a sperm transfer appendage, the spermatodactyl.

There is considerable variety in the form of the spermatodactyl in the Dermanyssina, both in structure and in size relative to the length of the movable digit. It is usually provided with a canal which extends along its length. In many of the parasitic Dermanyssoidea, the elongate spermatodactyl is entirely fused with the reduced movable digit and in *Alphalaelaps saplodontia* the long spermatodactyl is about five or six times the length of the cheliceral shaft (Evans & Till, 1966). The longest described spermatodactyl relative to the size of the male appears to occur in *Pargamasellevans michaeli*, a small euedaphic species from S. Africa. The spermatodactyl is accommodated in a tubular pouch opening dorsal to the genital orifice and extending posteriorly to about the level of the posterior margin of the sternal shield (Loots & Ryke, 1968, and personal observation). The tips of the paired spermatodactyls, which are bent back on themselves, protrude from the mouth of the pouch. The spermatodactyls appear to break off after insemination of the female and have been found in the tubules of the sperm access system, which opens on the integument on either side of the female genital shield.

The mating behaviours of representatives of a number of families have been observed in detail, for example the Ologamasidae (Lee, 1974) and Phytoseiidae (Amano & Chant, 1978). In the species of ologamasids studied by Lee, the males mount the dorsum of the female before adopting one of two basic mating positions: ventral (Fig. 10.6A) or lateral (Fig. 10.6C). In the former, the male comes to lie between the legs of the female with their venters in apposition and uses his spurred second pair of legs to firmly hold the fourth pair of legs of the female. His initial

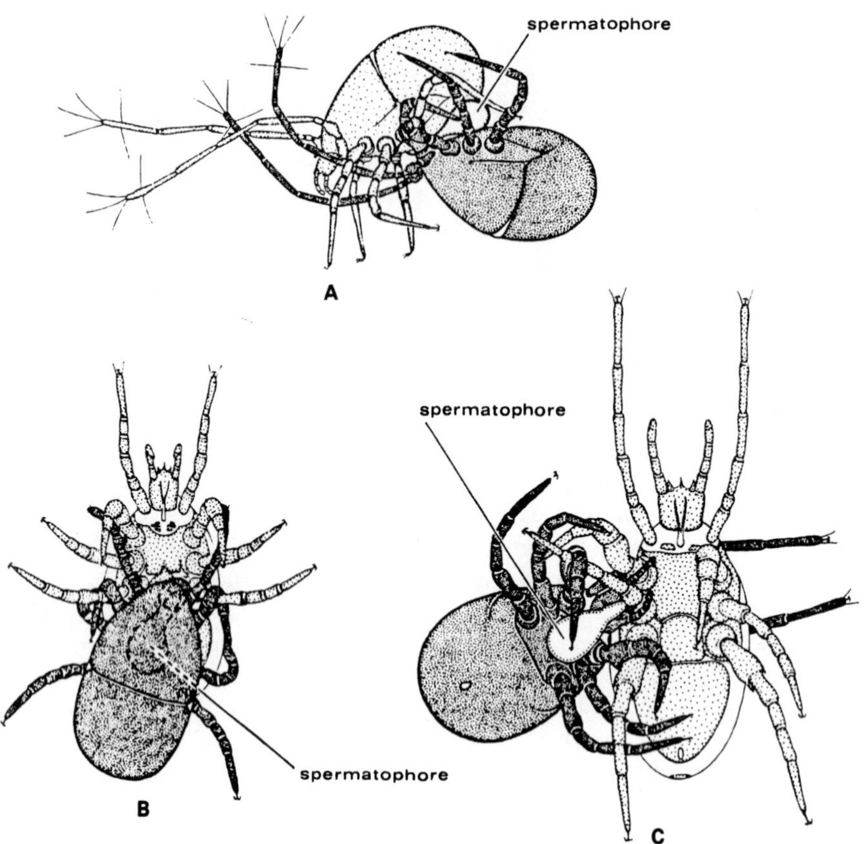

Fig. 10.6. Positions adopted by the male (darkly shaded) and female during mating in the Ologamasidae (Mesostigmata) (after Lee, 1974). *Euepicrius filamentosus* (**A**); *Gamasellus tragardhi* (**B**); *Athiasella dentata* (**C**).

position relative to the female is parallel so that their longitudinal axes coincide but changes to a skewed ventral position for sperm transfer (Fig. 10.6B). In species having a lateral position (*Athiasella dentata*), the male's idiosoma lies outside the sternogenital region of the female with his venter in apposition to her side while gripping her legs III and IV on one side of the body with his second pair of legs (Fig. 10.6C). When the mating position has been achieved sperm transfer commences. A spermatophore issues from the genital orifice with the aid of a protrusible ejaculatory organ and gradually distends to form a flask-shaped capsule attached by its short neck to the orifice. The gnathosoma is then flexed and the protracted chelicerae are passed one on each side of the spermatophore which is held paraxially between the digits of one of the chelicerae. In this position, the spermatophore is carried to the solenostome of the sperm access system. Lee observed in one species both spermatodactyls lying on either side of the spermatophore with their tips apparently inserted in the solenostome. After insemination, the palps are used to remove the ectospermatophore from the chelicera.

A more detailed description of sperm transfer is given by Krantz and Wernz (1979) for the macrochelid *Glyptholaspis*. The male deposits a sperm droplet in the hypostomatic region of the gnathosoma by means of a telescoping ejaculatory organ. This droplet is transferred to the region of the arthrodial membrane at the base of the movable digit of one of the chelicerae and held in the angle between its open movable digit and the cheliceral shaft during transfer to the solenostome at the base of coxa III. The spermatodactyl of the other chelicera, which is closed and uncharged, probes the area of the solenostome and its tip is inserted into the opening followed by rapid pulsations indicating the initiation of sperm transfer. Apparently seminal fluid is transferred from the sperm droplet on the open chelicera to the spermatodactyl of the closed chelicera by way of a fluid bridge between the droplet and a paraxial lobe of the arthrodial membrane. Seminal fluid passes to the intercondylar region of the closed chelicera, and it is suggested that the fluid flows into a 'reservoir' on the movable digit from which issues a canal extending to the tip of the spermatodactyl. Males transfer sperm with either chelicera and often alternate between the paired solenostomes in consecutive matings.

The venter-to-venter insemination position in the Phytoseiidae (with the male in the inferior station) is achieved in various ways. In the *Phytoseiulus*-type, the pair makes contact 'face to face' with their palps and first pair of legs touching (Amano & Chant, 1978). The male moves to the underside of the female from a frontal, lateral or posterior position without mounting her. Initial face-to-face contact may or may not occur in the *Amblyseius-Typhlodromus*-type before the male mounts the female from a frontal, lateral or posterior direction. The male wanders over the female's back momentarily during which time their mouthparts and anterior legs may touch. This is followed by the male crawling under the female posteriorly. The omission of the mounting phase in the mating behavioural sequence in *Phytoseiulus* (Prasad, 1967), and apparently also in *Phytoseius*, appears to be atypical for the podospermic Dermanyssina in which mounting has been noted in the Laelapidae (Jakeman, 1961), Macrochelidae (Oliver & Krantz, 1963), Haemogamasidae (Young, 1968b) as well as in the Ologamasidae.

According to Dosse (1959), the spermatophore is transported to the female by the chelicerae and although mites have been observed with a spermatodactyl inserted in the major duct of the sperm access system, the method by which the contents of the ectospermatophore are actually transferred to the female has not been determined. A sperm packet or endospermatophore is visible in the receptaculum of the sperm access system after copulation. It resists maceration in lactic acid and is therefore conspicuous in preparations for microscopic study but in live individuals it appears to disappear quickly from the receptaculum.

Sex pheromones have been shown to be involved in the mating

behaviour of the Phytoseiidae. Hoy and Smilanick (1979), working with *Metaseiulus occidentalis*, demonstrated the production of a pheromone(s) by deuteronymphs, by virgin, mated and gravid females and by some protonymphs. The males respond to the pheromone by contact and display a 'hovering' behaviour with deuteronymphs and some protonymphs which involves holding their palps and anterior pairs of legs in contact with the dorsum of the mite. Although several males may hover over a single deuteronymph no aggressive behaviour was observed. The pheromone appears to function as an arrestant, and in contrast to the condition described by Rock *et al.* (1976) in *Neoseiulus fallacis*, it does not appear to act as an attractant. The source and chemical nature of the pheromones are unknown.

Competition between males for the chance to inseminate females occurs in some species groups of *Macrocheles* (Costa, 1967). Mating takes place only once during the life of the female and is only possible when she is newly emerged, presumably before the cuticle of the entrance into the sperm access system hardens. Costa observed, in culture, older males of *Macrocheles parapisentii* frequently killing newly moulted males. Older well-sclerotized males when kept together did not attempt to kill one another. Thomas and Zeh (1984) consider that this competitive behaviour 'imposes strong selection for rapid development'.

There is no detailed account of the mating behaviour in the Heterozerconina. The modification of the male chelicerae for sperm transfer in the Heterozerconidae apparently affects the fixed digits which have a helicoidal canal (Vitzthum, 1940–43). The Discozerconidae, on the other hand, appear to have a small spermatodactyl arising from the fixed digit.

Ixodida

In the course of tocospermic insemination in ticks, the spermatophore is introduced into the genital orifice of the female by the mouthparts. Argasid ticks copulate off the host, metastriates (Amblyommidae) copulate on the host and prostriates (Ixodidae) copulate on or off the host. Initially, the male climbs on the back of the female and crawls under her by way of the posterior end of her body so that their ventral surfaces are in apposition and his gnathosoma reaches her genital orifice. Spermatophore production is preceded in the argasoids by a short period when the male gnathosoma (chelicerae and palps) is introduced into the female aperture and withdrawn and reinserted several times (Feldman-Muhsam, 1969, 1986). In *Ornithodoros savignyi* three stages in spermatophore formation are evident. A white, transparent ectospermatophore is extruded from the male gonopore followed a few seconds later by the ejaculation of uncapacitated sperm, proteinaceous granules and seminal fluid into this empty sac which becomes opaque. Finally, the male extrudes sperm symbiotic yeast-like fungi (*Adlerocystis*) into the ectospermatophore followed by a bilobed endospermatophore which is

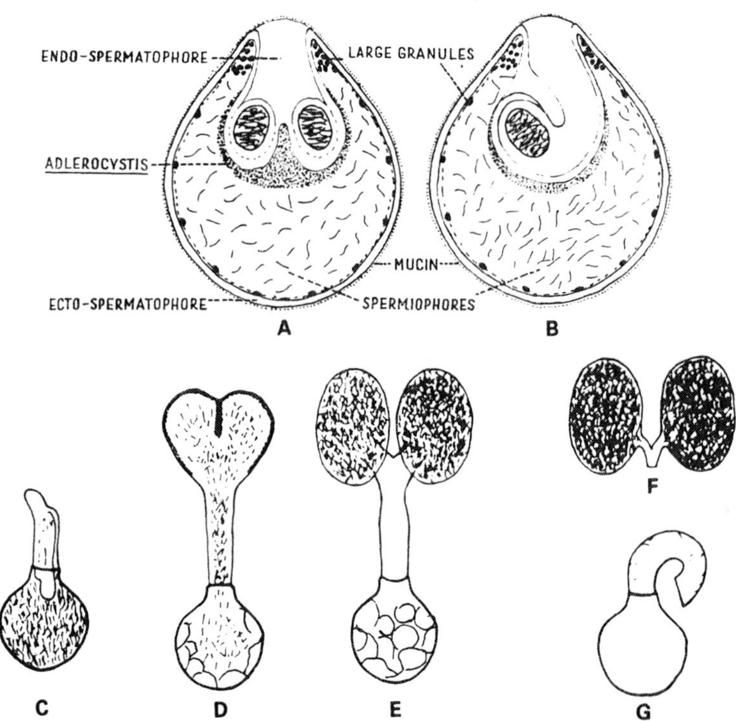

Fig. 10.7. **A**, **B**, schematic representations of the fully formed spermatophore in *Ornithodoros* (Ixodida), in longitudinal section (**A**) and in a longitudinal section perpendicular to section **A** (**B**). **C–G**, schematic illustrations of the evagination of the endospermatophore of *Ornithodoros savignyi*: **C**, commencement of evagination; **D**, eversion of the two lobes; **E**, lobes transformed into two capsules; **F**, capsules as found in female genital system after copulation; **G**, empty ectospermatophore found outside female genital opening. (After Feldman-Muhsam, 1969.)

connected to the rim of the ectospermatophore and closes it like a stopper (Fig. 10.7A,B). The adlerocysts are believed to play a part in preserving the viability of the sperm until the female has taken a blood meal. Each lobe of the endospermatophore contains a proteinaceous sac containing a high concentration of spermine phosphate. The whole process takes about 30 s. The wall of the pear-shaped ectospermatophore comprises three layers: an outer mucin layer, an intermediate proteinic layer and an inner mucopolysaccharidic layer.

Soon after the spermatophore is completed, the male begins to secrete saliva and, holding the neck of the spermatophore with the palps, uses the cheliceral digits to implant the tip of the spermatophore into the female aperture. Saliva continues to be secreted and spreads over the spermatophore thereby keeping it moist and preventing it from adhering to either of the partners. From *in vitro* observations, Feldman-Muhsam postulates that once in the genital orifice a finger-like projection

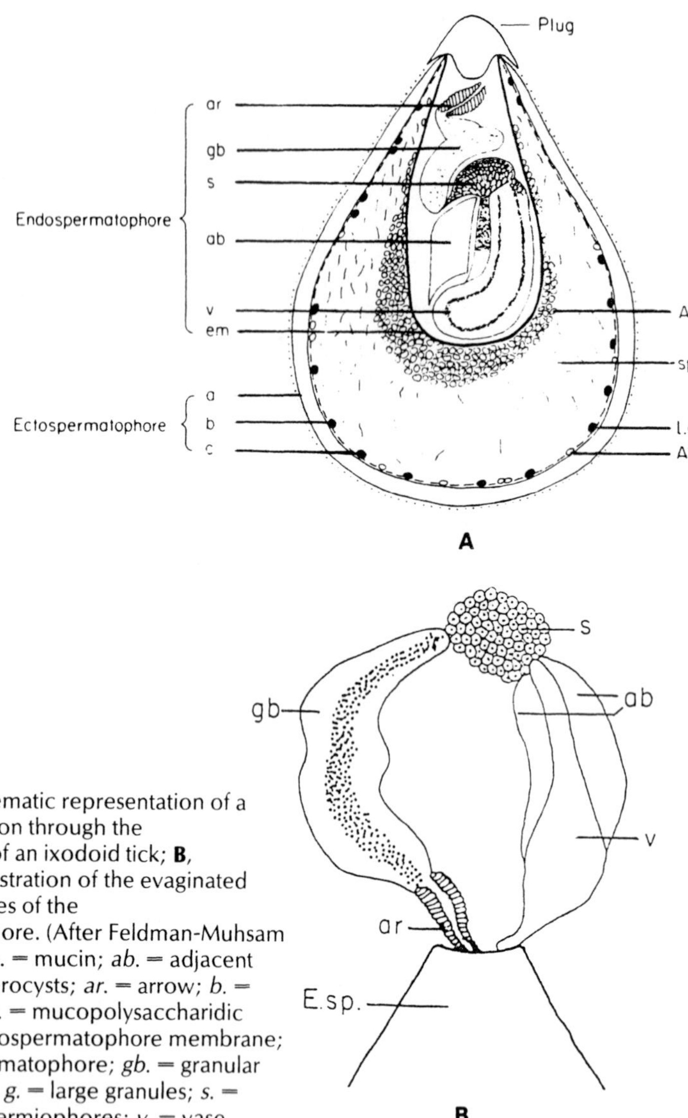

Fig. 10.8. A, schematic representation of a longitudinal section through the spermatophore of an ixodoid tick; **B**, diagrammatic illustration of the evaginated chain of organelles of the endospermatophore. (After Feldman-Muhsam & Borut, 1984.) *a.* = mucin; *ab.* = adjacent body; *Ad.* = adlerocysts; *ar.* = arrow; *b.* = proteinic layer; *c.* = mucopolysaccharidic layer; *em.* = endospermatophore membrane; *E. sp.* = ectospermatophore; *gb.* = granular body = tubule; *l. g.* = large granules; *s.* = sponge; *sp.* = spermiophores; *v.* = vase.

protrudes from the spermatophore (Fig. 10.7C) to be followed by the evagination of the two lobes of the endospermatophore (Fig. 10.7D). Finally, the mass of sperm and the adlerocysts stream from the ectospermatophore into the everted lobes by pressure created from bubbles of carbon dioxide produced in the ectospermatophore. The lobes separate to form two capsules connected to each other and to the ectospermatophore (Fig. 10.7E). The connection between the capsules and ectospermatophore breaks near the bifurcation leading to the capsules and the ectospermatophore, with the attached neck of the endospermatophore, hangs from the genital aperture to shrivel and fall off (Fig. 10.7F,G). The length of time the sperm capsules remain attached to each other depends on the number of subsequent copulations and the pressure resulting from the arrival of additional capsules. The connecting tube between the capsules eventually breaks and the sperm leave each capsule by a short tube-like appendix.

In some species of *Ornithodoros*, for example *O. canestrinii* and *O. delanoei*, the everted endospermatophore does not form two distinct capsules but appears as an elliptical, short-necked sac with a trace of a division into two capsules distally. Only one capsule is formed in *Argas persicus* but two distinct capsules are produced in *Argas reflexus*. The adlerocysts attach externally to the sperms or are present in the acrosomal canal of the sperm cells in species of *Ornithodoros* but it seems that they do not do so in other Ixodida.

The structure and contents of the ectospermatophore in ixodoid ticks is similar to that in the argasoids except that the proteinic layer of its wall is extremely firm and rigid in the ixodoids but has some elasticity in the argasoids. However, it is in the structure of the endospermatophore that major differences between the taxa occur (Feldman-Muhsam & Borut, 1984). In ixodoids it forms a single lobed sac containing five organelles, all of which are absent in argasoids. The organelles are attached in a chain and are enveloped by the endospermatophoric membrane, which separates them from the contents of the ectospermatophore (Fig. 10.8A). The endospermatophore which closes the ectospermatophore is itself closed by a separate plug. After the spermatophore is implanted in the female aperture, the chain of organelles evaginates from the ectospermatophore in the form of a loop since those organelles at either end of the chain are attached to the ectosphermatophore (Fig. 10.8B). The everted membrane of the endospermatophore appears as a small bleb at the opening of the ectospermatophore and sperm flows into it. This bleb becomes the sperm capsule but in contrast to the argasoids the capsule is provided with two tube-like appendices.

Unlike argasoids, the palps of the male ixodoid are not introduced into the female genital aperture during any stage of the implantation of the spermatophore. The hypostome and chelicerae enter the aperture in prostriate ticks but only the tips of the chelicerae appear to do so in metastriates. This difference is related to the form of the genital aperture in the two taxa; it is narrow but capable of considerable extension in the former but very narrow and bordered with sclerotized flaps that prevent its stretching in the latter (Feldman-Muhsam & Borut, 1971). Further, the mouthparts are not involved in transporting the spermatophore from the male orifice to the female because of the anterior position of the gnathosoma which even in the flexed state is unable to reach the male orifice to pick up the spermatophore. In *Hyalomma* and *Rhipicephalus*, the male presses his anal plates against the body of the female and creates a space between their venters for the extrusion of the spermatophore. The spermatophore appears to be pushed by the body movement of the male to the region of the female gonopore. At this stage the male withdraws the chelicerae from the female gonopore and uses them to implant the spermatophore while supporting it with the fourth article of his palps.

Pheromones play an important role at all stages of the mating behaviour or courtship of ticks (Sonenshine, 1984, 1986). The existence of a volatile sex attractant pheromone was first observed in female ticks of the genera *Amblyomma* and *Dermacentor* by Berger *et al.* (1971), and the pheromone was identified as 2,6-dichlorophenol (DCP). Subsequently, sex attractants have been recorded in a number of other species of ixodoid ticks and all are phenolic in nature (Graf, 1978). They have the effect of exciting and attracting the male. Males detect the pheromone by means of chemosensory sensilli on the tarsi of the first pair of legs. In metastriate ticks the secretion is produced by the paired foveal glands whose ducts open through the pores of the foveae dorsales, while in prostriate females glands in the genital tract are considered to be involved (Sonenshine *et al.*, 1981).

Once the female has been mounted, the male palpates her body surface before moving to her venter to probe for the gonopore. Gonopore location and, in some cases, species recognition are controlled by the contact genital sex pheromone which is located in the vulva and detected by the pit sensilli of the cheliceral digits of the male. In *Dermacentor andersoni* and *D. variabilis*, for example, the males are unable to distinguish conspecific females until they probe the gonopore and if they do not receive the specific stimulus they terminate the courtship and depart (Sonenshine *et al.*, 1982). This delay in specific recognition until the final stage of courtship contrasts with the situation in other ixodoids such as the camel ticks *Hyalomma dromedarii* and *H. anatolicum excavatum* in which differences in the relative amounts of the sex attractant DCP emitted by their females enable the males to identify conspecific females. The genital sex pheromone is considered to comprise two active components, one of which may be 20-hydroxyecdysone (or a closely related steroid). The lobate accessory glands of the distal region of the vagina are suggested as a possible source of the pheromone.

In the case of the argasoids *Ornithodoros tholozani* and *O. erraticus*, the males are attracted to ambulating females which they mount. They are believed to detect compounds in the coxal fluid secretion of the female which act as a mating stimulant (Schlein & Gunders, 1981). The stimulant appears in the coxal fluid at about the fourth day after a blood meal. Males do not react to inactive females but will mount and attempt to copulate with females and nymphs of other species coated with female coxal fluid. The composition of this interspecific contact sex pheromone is not known.

Females of three species of ixodoid ticks (*Aponomma hydrosauri*, *Amblyomma albolimbatum* and *Am. limbatum*) parasitizing the sleepy lizard (*Trachydosaurus rugosus*) produce a species-specific pheromone which induces males to detach from beneath the host scales and search for females (Bull & Andrews, 1984; Bull, 1986). The signal produced by this excitant is considered to be different from that of the non-specific sex attractant (?DCP) which allows searching males to orientate towards females over short distances, since it works only after the male has fed and the searching response it induces is non-directional. The attractant, on the other hand, works equally well on fed or unfed males and the searching response of the male is directional.

Prostigmata

Direct sperm transfer in this group is either tocospermic or porospermic (opisthospermic). The former occurs in water mites of the family Pionidae in which spermatophores are carried from the male gonopore to that of the female by means of the opposed tarsal claws of the third pair of legs of the male (Mitchell, 1957b). The mating position in this family is one in which the venters of the partners come together and the male faces in the opposite direction from the female. This is accomplished in the Pioninae by the male locking his modified fourth pair of legs around the bases of a pair of the female's legs, while in the Tiphysinae the male is relatively inactive and it is the female that positions the male by holding him in a basket formed from her flexed legs. In the latter subfamily the male becomes firmly attached to the female venter, possibly by an adhesive secreted by her. The accomplishment of insemination necessitates the carriage of a spermatophore by the male before pairing begins.

Lundblad (1929) describes a somewhat different method of insemination in *Midea orbiculata* (Mideidae) which does not involve the appendages in transferring a spermatophore. The partners lie venter to venter with their genital regions touching and both facing in the same direction. The male lying beneath the female holds her by means of his legs and occasionally the palps, and produces a 'sperm ball' which glues the partners together. It is considered that the sperm enters the female gonopore directly from the sperm ball.

In those members of the Prostigmata in which males have a protrusible aedeagus, it is difficult to establish from the literature whether insemination is tocospermic or porospermic. It appears that only in the Tetranychidae has porospermy been established unequivocally although it is also considered to occur in the Tarsonemina. Much of the information on the structure of the reproductive system in other taxa which practise direct sperm transfer refers to cuticle-lined structures which resist maceration during preparation for microscopy and conclusions reached on their function are speculative and often confusing. Other Actinotrichida (Astigmata) which have an intromittent penis or aedeagus are porospermic and it is probable that this will also prove to be the case in the Prostigmata. It is certainly an aspect of sperm transfer that requires further study.

Female deuteronymphs of *Tetranychus urticae* spin a web of silk strands before entering a quiescent phase prior to moulting. The web aids males in finding a pharate female, which becomes increasingly attractive to them. This attraction is due to the production of a sex pheromone by the pharate female which acts as an attractant–arrestant for males (Cone, 1979, 1985). The attractant property of this substance terminates after initial male contact, and subsequently acts as an arrestant with a holding range of about two body lengths. Initially there is a 'hovering' response during which the male is positioned over the pharate female with legs I to III twitching and its chelicera and palps probing the area where the old cuticle will split before the emergence of the female. This is followed by a 'guarding' phase as one or more males adopt resting positions around the pharate female. Male webbing laid over the pharate form soon after contact probably allows the guarding male to detect any movements made by the female during emergence or to be alerted to the arrival of a new male (Penman & Cone, 1974b). Combat between males over moulting deuteronymphs frequently occurs and ends in the retreat, injury or even death of one of the combatants. The source of the sex pheromone has not been ascertained. Males attempt to mate with fresh deuteronymphal exuviae and this suggests that moulting fluid may play a part in male attraction. Regev and Cone (1975) consider farnesol to be a component of the sex attractant–arrestant. The adaptive advantage of the male commencing his mating activities with an inactive pharate female, termed precopulation, is that it 'increases the probability of mating and might enhance the reproductive potential under natural conditions' (Shih *et al.*, 1976). According to Everson and Addicott (1982), spider-mite males are able to distinguish the time individual pharate females will take to reach maturation. By selecting a more mature female to guard, the male will be able to decrease the length of guarding time and thus increase the number of females inseminated by him during his lifespan.

Mating takes place as soon as the female emerges from the exuvium (*Tetranychus tumidus*) or even before she is entirely free (*T. urticae*). The position adopted by the male is unusual. He slips under the posterior end of the female, dorsum uppermost, and clasps her legs III and IV from behind. While in this position he flexes his opisthosoma upwards almost at a right-angle to the remainder of the body in order to bring his extruded aedeagus into contact with the female copulatory pore. The latter is situated on a small elevation between the genital orifice and the anus. Multiple mating occurs in this genus but second matings following a 'complete' first mating are totally ineffective. The mechanism by which the first complete mating precludes fertilization by a later insemination is not known (Cone, 1985).

Precopulation occurs widely in the Tarsonemina and has also been observed in the Pronematinae (Tydeidae) and Cheyletidae (Summers & Witt, 1973). In the Tarsonemina, it has been recorded in the Dolichocybidae (Rack, 1967), Pyemotidae (Herfs, 1926), Pygemephoridae (Eickwort, 1979), Scutacaridae (Ebermann, 1982) and Tarsonemidae (Nucifora, 1963; Suski, 1972). The male uses his genital capsule as a suction organ to attach to the pharate female ('pupa'), enclosed in the cuticles of the calyptostatic nymph and larva, and with the aid of his modified fourth pair of legs carries her on his back. An exception occurs in the pygmephoroids where only the fourth pair of legs are used since the genital capsule lacks a membraneous adhesive structure (Lindquist, 1986). The male is able to distinguish between pharate males and females, and Ebermann (1982) suggests that a sex pheromone may be involved. The pharate female is carried in various positions by the male but the attachment is generally over her opisthosomatic region. Ecdysis normally occurs on a substrate after release by the male and copulation takes place immediately with the emerged female. More rarely, copulation takes place with older females and, according to Karl (1965), this allows unmated females to be inseminated by their sons. During copulation the male and female are pointing in opposite directions and attached caudally so that their bodies do not overlap. The male uses legs III and IV to hold down the female's posterior two pairs of legs and attaches the genital capsule to the postero-ventral extremity of the female so that the aedeagus is in a position to enter her copulatory pore. Although the male may hold the pharate female for up to 24 hours before she moults, copulation is relatively rapid and in *Tarsonemus* lasts from a matter of seconds to about five minutes. This mating position adopted by the pair is called retroconjugate by Regenfuss (1973) and contrasts with the proconjugate positon in the Podapolipidae, a highly specialized group of permanent parasites of insects, in which the male copulates with the larval female and has dispensed with precopulation. In the proconjugate position, both sexes face in the same direction with the anterior region of the male under the female opisthosoma so that his dorsally placed genital capsule is in contact with the postero-ventral region of the larval female. Insemination is probably porospermic but this has to be confirmed.

The mating behaviour of the tydeid *Homeopronematus aconi* is described in detail by Knop (1985). During the precopulation phase males commonly guard pharate females, the male standing on top or by the side of the female. As in *Tetranychus* antagonistic behaviour between guarding and intruding males occurs although the pushing movements by the forelegs and jabbing with the chelicerae do not appear to cause damage to the combatants, one of whom eventually moves away. Pheromones are probably involved in this behaviour. Contact between the first pair of legs of two adults appears to be important at the precopulatory and the copulatory stages of mating behaviour. When it occurs between two males it results in antagonistic behaviour but between male and female it usually initiates pairing leading to copulation. During copulation the female's body, supported on the substrate only by her fourth pair of legs, is situated at an acute angle over the back of the male with her third pair of legs resting or tapping the area over his metapodosoma. A hook on the femur and a 'latch' on the genu of each of legs IV of the male attach to the pretarsi of the female's second pair of legs and this attachment is considered to play an important part in attaining the mating position.

Males mate more than once and appear to be sexually active during the whole of their adult life.

Mating in *Cheyletus malaccensis* also involves precopulatory and antagonistic behaviour by males. During the latter, the two males adopt a face-to-face position and grasp each other with their raised pedipalps so that they resemble the partners in a 'promenade à deux'. They engage in a pushing match in the vicinity of the pharate female but neither appears to suffer injury (Summers & Witt, 1973).

Astigmata

Porospermic insemination by way of a sperm access system in this order probably evolved from a form of dermal copulation. This probably involved the penetration of the female integument by the male aedeagus and the liberation of sperm into the haemocoel. The copulatory pore or bursa copulatrix in the Acaridae and Lardoglyphidae is situated dorsolaterally on the posterior margin of the opisthosoma (or retro-anal according to Michael, 1901) and is surrounded by a saucer- or funnel-shaped sclerotized disc, while in the Glycyphagidae it is located terminally on a sclerotized projection. The pore leads by way of a relatively narrow inseminatory canal to a sclerotized cup- or bell-shaped neck (calyx) or a distensible sac-like seminal receptacle, which is connected to each ovary by a short efferent duct (see Fig. 9.14, p. 285). In *Acarus*, *Lardoglyphus* and *Austroglycyphagus*, Griffiths and Boczek (1977) consider the sperm to be transferred as a spermatophore which attains its final form in the sperm access system. Since the aedeagus is introduced into the canal connecting the bursa copulatrix to the seminal receptacle during copulation, it is unlikely that an ectospermatophore is produced. Michael (1901) refers to the possible hardening of the periphery of the sperm mass 'near to the entrance of the tube from the bursa' to form an almost round structure within the receptacle and this is supportd by Griffiths and Boczek's suggestion that the fluid sperm mass solidifies to form a 'spermatophore' in the anterior region of the receptaculum. The 'spermatophores' illustrated by these two authors in the seminal receptacle of mated females of *Lardoglyphus konoi* after maceration in Heinze mountant are reminiscent of the sperm packets or endospermatophores in the sperm access system of macerated females of the Phytoseiidae (see Fig. 1 in Dosse, 1959). Further work is necessary to elucidate this very interesting aspect of the insemination process in the Acaroidea. Species of the genera *Tyrophagus* and *Glycyphagus* do not produce endospermatophores.

During mating in the Acaridae, the male mounts the back of the female so that the posterior regions of the opisthosoma of both are in contact while facing in opposite directions (retroconjugative form). Thus, the male aedeagus is in position for insertion into the female bursa copulatrix. The male secures the female in the mating position by means of a pair of anal suckers and by tarsal suckers on the fourth pair of legs. In the Glycyphagidae, in which the male lacks such attachment organs, the pair face in the same direction during copulation (OConnor, 1982b). Copulation in *Acarus siro* lasts for more than 30 min (Griffiths & Boczek, 1977) and in *Tyrophagus putrescentiae* for an average of 1.5–2.0 h (Boczek, 1974). Boczek and Griffiths (1979) found that in an isolated virgin pair of *A. siro* the female is often the first to become aware of its partner and she appears to follow his trail around the container. The male does not appear to become active until the female's opisthosoma is close to him. This suggests the release of an attractant stimulus by the female.

Observations made by Griffiths and Boczek (1977) on virgin pairs of *A. siro* and *L. konoi* showed that one complete mating produced only one endospermatophore. Virgin pairs of the latter kept in isolation produced from eight to 25 endospermatophores over a period of 15 days. In *Acarus*, the body of the endospermatophore breaks down completely, except for a short tail, soon after entry into the posterior region of the receptaculum, whereas in *Lardoglyphus* the endospermatophore retains its shape and the sperm seem to be released when its anterior margin breaks down.

Precopulatory behaviour occurs among the Psoroptidae and Chirodiscidae. In the former, males use their adanal suckers to attach to a pair of posterior copulatory tubercles on the tritonymphs (pubescent females) and mate with the females soon after they emerge from the tritonymphal cuticle. Pairs may remain attached for 48 hours (Evans *et al.*, 1961). Copulatory lobes or discs occur on the posterior extremity of the larva, protonymph and tritonymph of the female line of some Chirodiscidae, such as *Schizocarpus*. The males, showing a variety of attachment organs in the form of suckers and/or sclerotized cylindro-conical tubes, attach to immatures of the female line (Fain *et al.*, 1984).

Oribatida

A form of direct sperm transfer possibly occurs in *Collohmannia gigantea* which shows distinct sexual dimorphism (Schuster, 1962). The male, considerably smaller than the female, places his forelegs on the female's flanks and he follows her for about an hour. During this time, he occasionally places the anterior end of his prosoma under her opisthosoma and pushes upwards with such force as to cause her to fall over. When she comes to rest, he releases her and moves to a frontal position where he stands with the posterior part of his body erect. He extrudes his penis which unlike that in all other oribatids is very long and, at first sight, resembles an ovipositor. He deposits on his modified posterior legs a drop of fluid and rubs the legs, one against the other, before advancing towards the female with these legs extended. She appears to 'lick' his fluid-smeared legs many times before moving away. The nature and function of the fluid has not been ascertained but a spermatophore is not deposited as in other Oribatida. It is possible that the function of the fluid 'food' produced by the male is to entice the female for mating and that transference of sperm material occurs during direct contact of their genital orifices.

Syngamy

In common with other Arachnida, the micropylar orifice is lacking or is a transitory structure in the eggs of the Acari (Warren, 1941; Witalinski, 1986). The penetration of the oocyte by the sperm occurs within the genital tract of the female but the actual site is often unknown and a matter of conjecture. Brinton *et al.* (1974), working with *Dermacentor andersoni*, suggest that fertilization takes place within the ovary and that the tip of the sperm attaches to a micropyle-like area before releasing its nucleus into the egg. There is a tendency in this species for the spermatozoa to penetrate the wall of the oviduct during migration to the ovary. On the other hand, Goroshchenko (1965) considers fertilization in argasoid ticks to occur in the ampullate parts of the oviducts with the sperm attaching to cells of the oviduct by their anterior part. As the oocyte descends into the lower oviduct the acrosomal region of the sperm dissolves an area of the egg shell and releases its nucleus into the oocyte. Ixodoids lack oviducal ampullae and, in *Hyalomma asiaticum*, Balashov (1968) states that the oocytes are fertilized when they pass into the anterior region of the oviduct.

According to Mothes and Seitz (1981a) syngamy in *Tetranychus urticae* appears to take place in the vitellogenic region of the ovary before the closure of the 'egg shell pores'. Feiertag-Koppen and Pijnacker (1985), however, suggest that the sperm enters the oocyte by way of the nutritive cord. The ovary is also stated to be the site of syngamy in *Sancassania* (Prasse, 1968a) and this is probably also the case in the podospermic Dermanyssina (Alberti & Hänel, 1986). In the tocospermic Uropodina, Woodring and Galbraith (1976) consider that sperm passes from the seminal receptacle onto the egg as it travels through the vagina. Alberti (1988) demonstrated that spermatozoa reach the ovary through a haemocoelic route in the tocospermic Parasitina and probably invade the ovary from behind. He considers that the connection of developing oocytes to a nutritive tissue in the Parasitina makes the ovary a solid structure and spermatozoa are unable to move up the oviducts to the ovary as in other tocospermic mites and ticks in which the lyrate organ is lacking. Fertilization is most likely to occur through the region where the nutritive cord originates from the oocyte.

Eggs and Oviposition

The egg is of the centrolecithal type, i.e. the yolk is concentrated in the centre so that the cytoplasm is peripheral. (An exception occurs in the alecithal eggs of the Spinturnicidae.) It is usually large in relation to the size of the body of the gravid female. In the egg of the argasoid *Ornithodoros moubata* (Fig. 10.9A,B), the yolk spheres are held in a cytoplasmic reticulum and the cytoplasm is bounded by a thin vitelline envelope or membrane which underlies a two-layered shell produced by the egg itself and not by ovarian follicles as in the case of the chorion of insect eggs (Aeschlimann, 1958). The smooth inner-shell layer lacks pore canals and is probably composed of protein while the possible function of the incomplete outer granular layer is to form a substrate for the coating of lipid-rich secretion from Gené's organ during oviposition (Lees & Beament, 1948).

Kolata *et al.* (1973) describe four layers external to the vitelline envelope in the pre-oviposition egg of *Fuscuropoda marginata* comprising an innermost dense layer, which they refer to as chorion, followed by a layer of degenerating follicle cells (the theca), a wax layer and a surface vesicular layer of tissue associated with the follicle wall. Immediately following oviposition, the surface layer is lacking and the wax layer is more closely adpressed to the surface of the theca which shows evidence of cellular degeneration. It was not possible to determine whether the egg coverings were formed by follicle cells or by the oocyte alone. According to Witalinski (1986), the oocytes are not covered by follicle cells in *Euryparasitus emarginatus* (Mesostigmata), *Erythraeus phalangoides* (Prostigmata) and *Hafenrefferia gilvipes* (Oribatida). In his detailed account of egg-shell formation in *Pergamasus barbarus*, Witalinksi (1987) refers to the inner layer of the shell, a single layer resulting from the transformation of the vitelline envelope during passage of the egg through the reproductive tract, as the endochorion and the outer layer, produced by secretions from accessory glands associated with the tract, as exochorion. The same two layers are also recognized in the egg of the water mite *Limnochares aquatica* (Witalinski, 1988). The egg of *Tyrophagus putrescentiae* has its external surface provided with irregular raised mounds and locular chambers which Callani and Mazzini (1984) consider to be produced by follicle cells during oogenesis and not from secretions of glandular cells in the genital tract. In the absence of aeropyles, they suggest that the locular chambers, which are open basally, may be involved in gaseous exchange.

The deposition on the surface of the egg of a layer of lipid-rich secretion prior to oviposition is characteristic of mites. The layer is produced entirely from secretions of accessory glands opening into the reproductive tract and usually covers the entire surface of the egg. It functions in preventing an excessive loss of water and in protecting the embryo from mechanical injury and from attacks by predators. The surface of the egg is often developed into a lattice or basket weave-like

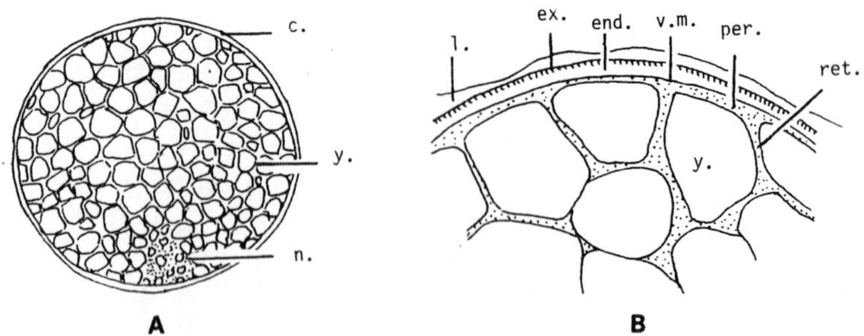

Fig. 10.9. A, schematic representation of structure of the egg of *Ornithodoros moubata* (Ixodida); **B**, detail of the outer layers of the egg. (After Aeschlimann, 1958). *c.* = cuticle; *ex.* = exocuticle; *end.* = endocuticle; *l.* = wax (lipoid); layer; *n.* = nuclear region; *per.* = periplasm; *ret.* = reticulum; *v.* = vitelline membrane; *y.* = yolk spherules.

structure as in *Rhysotritia* or produced into spiny or filamentous processes as in some Bdellidae and Ologamasidae. In the latter, a surface pile (or filaments) is considered to be a protective device and occurs in those eggs which are not buried (Lee, 1974). Morphological differences in the structure of the egg shell between non-diapausing and diapausing eggs occur in the Tetranychidae (Lees, 1961; Veerman, 1985). For example, in *Petrobia latens* the summer eggs are subspherical, slightly ribbed and surmounted by a tapering projection whereas the winter eggs are enclosed in a waterproof wax case closed by a heavily sculptured lid. The winter egg was aptly described as 'une petite marmite' by Hammer (1804). The thick outer wax layer in *Panonychus ulmi* is attached to the egg shell by a cement layer of oil and protein and additional water-proofing is achieved by wax secreted inside the shell by the embryo (Beament, 1951). The internal layer is laid down soon after oviposition in summer eggs but is complete before oviposition in winter eggs, which are retained by the female until a later stage of embryogenesis. Dittrich and Streibert (1969) suggest that the internal layer may arise from clear yolk and not from the embryo.

The newly formed outer layer ('exochorion') of the egg of *Limnochares aquatica* comprises numerous tightly folded lamellae separated by an electron-dense material. Within a few hours of deposition in water, the layer becomes swollen and assumes a foamy appearance. This is facilitated by the separation of the lamellae, which are considered to be prevented from sticking together by the surfactant action of the electron-dense material (Witalinski, 1988). The foamy layer is thought to be a protective layer much in the same way as the gelatinous envelope of amphibian eggs. According to Sokolov (1977), this outer layer is lacking in the eggs of a species of *Hydrachna* which are deposited in the stems and leaves of water plants where they are protected from predators and mechanical damage.

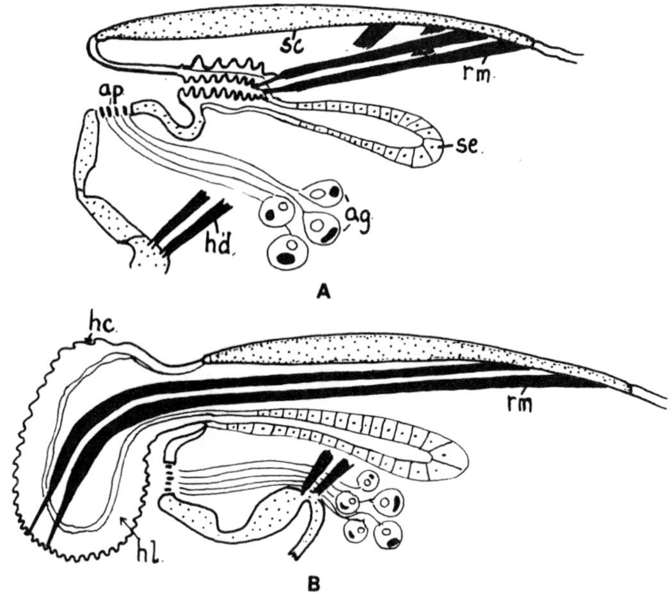

Fig. 10.10. **A**, **B**, schematic longitudinal sections of Gené's organ in the retracted (**A**) and everted (**B**) state in *Boophilus microplus* (Ixodida) (based on Booth *et al.*, 1984). *ag.* = acinar accessory glands; *ap.* = area porosae; *hc.* = horn cuticle; *hd.* = hypostomatic depressor muscles; *hl.* = lumen of horn; *rm.* = retractor muscles of Gené's organ; *sc.* = scutum; *se.* = secretory epithelium.

In ticks, waterproofing of the egg is performed by Gené's organ of the female after leaving the genital tract. It is absent in immatures and males and only differentiates in the female during feeding. Argasoid eggs receive their wax entirely from this organ but ixodoid eggs are also provided with a wax covering from secretions of the lobate accessory gland of the genital tract. Gené's organ (cephalic gland) is an eversible, sac-like structure situated at the anterior end of the idiosoma above the gnathosoma (Fig. 10.10A,B). It consists of an area of glandular epithelium of the body wall, which forms the tubular glands, and a cuticular protrusible sac with paired lobes or horns. Four lobes are present in the Ixodidae and two in the Argasoidea and Amblyommidae (Arthur, 1962; Balashov, 1968). The tubular glands always remain in the tick but they are continuous with the lumen of the horns (Booth *et al.*, 1984). The material in the lumen of the glands is amorphous and less granular than that in the horns. During oviposition the tick's hypostome is adpressed to the ventral body wall and Gené's organ is fully everted by hydrostatic pressure through a slit in the folded cuticle connecting the anterior margin of the scutum to the posterior dorsal margin of the gnathosomatic base. The horns of the everted organ manipulate the egg as it extrudes from the short ovipositor and smear it with wax from pore canals opening

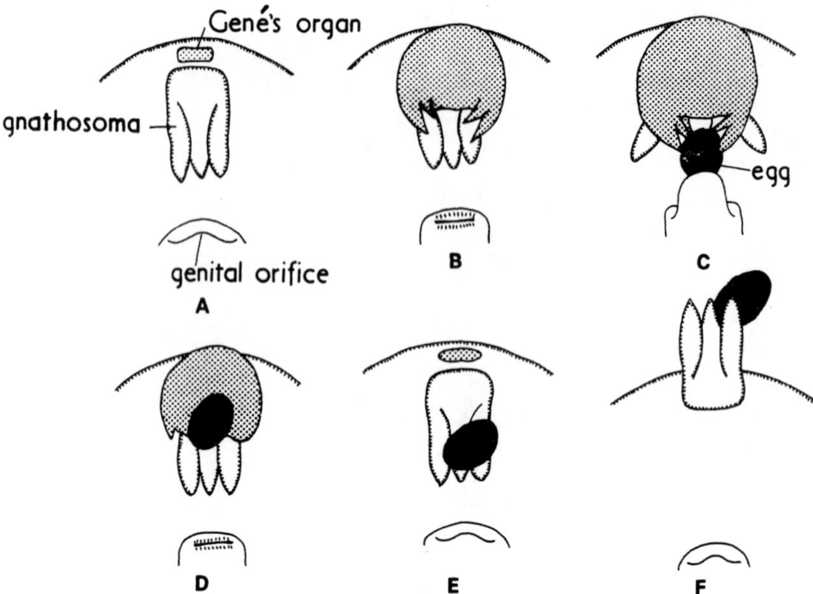

Fig. 10.11. A–F, stages in oviposition and waxing of the egg in an ixodoid tick (after Evans *et al.*, 1961).

on its surface (Fig. 10.11). Minute cuticular projections on the parts of the surface of the horns coming into contact with the egg may play a part in gripping the egg as it leaves the vagina. Retraction of the organ by muscles originating at the posterior end of the scutum and inserting into the horns, results in the egg being left adhering to the dorsal surface of the hypostome and when the latter returns to its normal position the egg is carried with it. As oviposition proceeds, the eggs of ixodoid ticks, in which the waxes are sticky, tend to accumulate on the dorsal surface of the female and form an adherent mass. The wax is complex in composition and in *Boophilus microplus* consists of branched and non-branched chain alkanes, steroids, fatty acids and alcohol, and corresponding wax and cholesteryl esters (Booth *et al.*, 1984).

The secretion from the acinar accessory glands opening on the areae porosae, situated on the dorsal surface of the gnathosomatic base of female ixodoid ticks only, is incorporated into the egg wax during oviposition. The function of the secretion is not known. Feldman-Muhsam and Havivi (1960) suggested that it acted as a lubricant for the evagination and invagination of Gené's organ but this has been rejected by Atkinson and Binnington (1973) who considered its function in *Boophilus microplus* to be that of inhibiting the autoxidation of unstable Δ2,4,6-triene steroids in the wax layer covering the egg. However, chemical studies by Vermeulen *et al.* (1986) on the 'outer chorion surface' of the eggs of *Rhipicephalus evertsi* have shown that the porose area secretion is probably not involved in preventing autoxidation of Δ2,4,6-triene steroids and consider it unlikely to be involved in the chemical protection of the eggs when they are exposed to environmental conditions.

The majority of mites are oviparous, i.e. they lay eggs in which embryos have as yet developed little, if at all. The length of time the eggs are retained in the female reproductive tract varies considerably, and Filiponni and Francaviglia (1964) have shown that in some species of *Macrocheles* this is influenced by availability of food; plentiful food favours oviparity whereas a shortage of food produces larviparity through the longer retention of the egg in the female genital tract. Ovoviviparity, in the sense of eggs being laid containing embryos in an advanced state of development with eclosion occurring soon after they are laid, is not uncommon. This phenomenon occurs, for example, in many species of nest-inhabiting and ectoparasitic Dermanyssoidea and in many parasitic Astigmata. In some species of mites, the female produces living young instead of eggs. The majority of the forms practising viviparity in the Mesostigmata are parasitic in habit, such as females of endoparasitic Rhinonyssidae, which give birth to larvae, and ectoparasitic Spinturnicidae, which produce protonymphs. In the case of the latter, embryonization of the early stages of ontogenesis takes place resulting in the suppression of a hexapod larval phase (Akimov & Yastrebtsov, 1990). This reduction in the number of active instars is an adaptation for a parasitic mode of life and it is possible that 'exotrophic nutrition' of the embryo takes place at the expense of the female. Unlike the condition of larviparity induced by shortage of food in *Macrocheles*, viviparity in *Spinturnix vespertilionis* occurs in the presence of plentiful food (blood of chiropteran host) which is probably essential for the maintenance of a high nutritional status in the female during the extended period of embyronic development when an 'embryo' increases in size from about 50 to more than 600 microns.

Prelarviparity and larviparity have been recorded in the Oribatida. The former is common in the Phthiracaroidea and Camisiidae while larviposition is widespread in the aquatic and semi-aquatic members of the genus *Ameronothrus* (Strenzke, 1946; Luxton, 1964). Larviparity also occurs in the Astigmata and is characteristic of species living in water-filled treeholes (Fashing, 1977). *Sancassania moniezi* is of particular interest since it produces two female morphs; one is of normal size (1050–1250 μm) and lays eggs, whereas the other is, in comparison, enormous (2000–2262 μm) and viviparous, retaining larvae, protonymphs and even completely developed deuteronymphs in the genital tract (Zachvatkin, 1941). In the physogastric females of the Pyemotidae development to the adult takes place within the body of the female and adults emerge through her genital orifice.

Development of eggs in some Euoribatida is unusual in that the eggs are not laid but remain within the body of the female after her death. The eggs hatch later within the 'shell' that remains of the parent's body and the larvae escape through the genital or anal apertures from which the protecting plates have dropped off (Michael, 1884). This form of egg retention is referred to as 'aparity'. Willmann (1931) suggested that 'aparity' is restricted to the last batch of eggs and that earlier batches must have been laid normally if populations of the species are to be maintained, especially in those with low fecundity. It is unlikely that 'aparity' is a normal reproductive process and probably occurs fortuitously as the result of developing eggs not being deposited before the death of the mother but continuing to develop and hatch within her body. The suggestion by Woodring and Cook (1962b) that most of the records of the occurrence of immature forms in dead adults may be due to the immatures gaining entry through the empty camerostome to feed saprophytically on the dead mite seems less plausible and is unlikely. 'Aparity' may also occur in the Uropodina where egg-shell remains and one or more fully developed larvae have been found in the body cavity of dead females of two species of *Trichuropoda* (Kielczewski & Wisniewski, 1977) and about 30 larvae in the body cavity of two species of *Macrodinychus* whose genital orifices seem to be too small to allow viviparous birth (Hirschmann, 1975).

The egg passes into the vagina by the peristaltic movements of the oviduct and uterus. Although many eggs may be present simultaneously in the oviduct only a single egg is found in the vestibule of the vagina. The egg passes through the genital orifice as the result of the contraction of the muscles of the vagina and/or the dorsoventral muscles of the opisthosoma. The egg may be delivered to the exterior by means of an ovipositor, which can take the form of a long telescopic tube-like structure as in the Oribatida and some Prostigmata, or a short tube formed by the eversion of the vagina through the genital orifice as in ticks and tetranychid mites. Alternatively, egg delivery may be via a shute-like structure formed by the anterior extension of the genital shield as in some Dermanyssina.

Eggs are deposited in situations adapted for the requirements of the offspring. Forms with elongate ovipositors such as the Oribatida select an oviposition site and the egg is placed in crevices in wood, moss or lichens upon which the larvae feed. Parasitic mites lay their eggs on the body or in the nest of the host. Those which are permanent parasites deposit their eggs, or their offspring in the case of viviparous forms, on the host. For example, the myobiid *Acanthophthirius polonicus*, parasitic on *Myotis dasycneme*, glues its eggs to the bat's hair (Pater de Kroot *et al.*, 1979) and the itch mite *Sarcoptes scabiei* deposits its eggs in burrows in the cornified epithelium of the skin of the host.

Some mites use their chelicerae to prepare an oviposition site. The eggs of the water mite *Hydrachna processifera* occur in longitudinal galleries ranging in length from 0.5 to 3.00 mm just below the epidermis of the stems of the water plant *Alisma*. The mite uses its awl-like chelicerae to bore a hole in the stem and then presses its hypostome through the aperture to form a short gallery. When this is completed in one direction, the hypostome is withdrawn and reinserted in the opposite direction resulting in the lengthening of the gallery (Soar & Williamson, 1925). During egg-laying, the genital orifice is closely applied to the opening of the gallery and the egg after deposition is probably pushed into the gallery by the hypostome. In the moth ear mite *Dicrocheles phalaenodectes*, the gravid female uses her chelicerae to prepare an egg-laying site by puncturing and scarifying a small area of the soft conjunctiva in the outer ear or tympanic recess of its host. The egg extrudes from the genital orifice between the coxae of the fourth pair of legs and as it passes forward from beneath the gnathosoma it is grasped by the palps and pressed against the area of scarified cuticle (Treat, 1975). Many species of 'box mites' tunnel into decaying leaf and woody tissue and females deposit their eggs at random within the tunnels. *Phthiracarus anonymum* is of particular interest in that the prelarvae are lodged in small hemispherical cavities along the tunnel wall (Grandjean, 1940b). These are thought to be excavated by the chelicerae of the female.

Many prostigmatic mites protect their eggs by covering them with silk, as in the case of *Spinibdella cronini* whose thin-shelled egg has a spun covering (Alberti, 1974). In *Cheyletus eruditis*, the tacky eggs are loosely bound together by strands of silk and the female remains within the 'nest' or very close to the egg batch and attacks intruders (Summers & Witt, 1972). Spider mites (Tetranychidae), as their common name implies, utilize silk for a variety of purposes including the construction of egg covers and guy ropes (Saito, 1985). Woven egg covers are observed in *Tetranychus viennensis* and *Schizotetranychus schizopus* and guy ropes extending from the terminal spike of the egg to the substrate and contributing to the anchorage of the egg are seen in *Panonychus ulmi*. Many spider mites produce silk threads during walking and these accumulate on the leaf surface to form a loose web under which the mites feed and oviposit (Saito, 1985).

There is considerable variety in the shape of the egg. The majority are ovoid, globular or elliptical but some are flattened and others are curved longitudinally so as to have a slight crescentic shape. In the Notostigmata, the eggs (*in utero*) are most unusual in having a distinct process at each pole, one of which is short and blunt and the other narrower and more elongate (Van der Hammen, 1966). Eggs are generally membraneous in texture and white or milky at first but with a tendency to become yellower and light brown with age. The elasticity of the egg membranes permits oviposition through small orifices. Notable exceptions are seen in the eggs of some Oribatida in which the shell is hard, brittle and dark-brown to black in colour and the surface may be perforated by minute pores (Michael, 1884). In *Damaeus onustus*, for example, the shell of the newly deposited egg is uniformly brown but as development proceeds a pale expandable strip of cuticle appears around the periphery. Eggs of some Prostigmata are coloured and this is particularly the case in the Tetranychidae in which various shades of green, yellow and red are encountered. A colour difference may occur between summer (non-

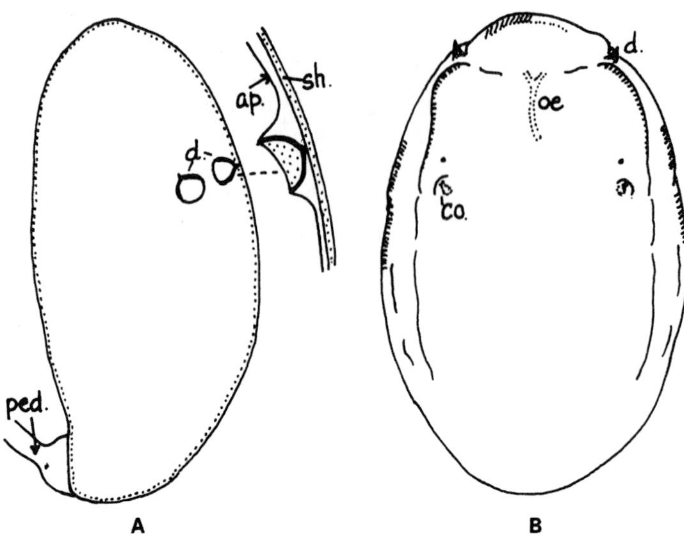

Fig. 10.12. A, egg of *Myialges macdonaldi* (Astigmata) in dorsolateral view (after Evans *et al.*, 1963); **B**, prelarva of *Pilogalumna allifera* (Oribatida) from below (after Grandjean, 1962). *ap.* = apoderma; *co.* = organ of Claparède; *d.* = putative egg bursters; *oe.* = oesophagus; *ped.* = pedicel; *sh.* shell of egg.

diapausing) and winter eggs as in the cherry-red coloration of the summer egg and the glistening white winter egg of *Petrobia latens.*

The highest fecundity occurs in ticks, particularly among the Ixodoidea. The argasoids lay their eggs in batches after successive blood meals and a typical series produced after seven meals in *Argas persicus*, for example, is 131, 159, 133, 110, 97, 95 and 47 eggs (Arthur, 1962). The high numbers in the earlier ovipositing periods are a common feature of egg-laying in this group but it has not been determined whether the decline in numbers is due to inadequate insemination or to some physiological factor inherent in the female. The eggs of ixodoids are laid in one batch and the number produced are among the highest of any haematophagous arthropod. Females of large species, such as those in the genera *Hyalomma* and *Amblyomma*, produce from 15 000 to 20 000 eggs; those of the medium size members of the genera *Dermacentor*, *Rhipicephalus* and *Boophilus* produce 3000 to 6000 eggs, and the smaller burrow-inhabiting *Ixodes* and *Haemaphysalis* only about 1000 eggs (Balashov, 1968). Egg productivity in this family is determined by the nutritional status of the female. Oviposition is only possible when the female has attained a minimum body weight. For example, this minimum weight for incompletely fed females of *Ixodes ricinus* is 20–70 mg, as opposed to 200–300 mg for females in the replete state. At the minimum weight only three eggs per milligram body weight are produced whereas nine eggs per milligram body weight are deposited by replete females.

Egg production in other Acari does not approach the high levels of the ixodoids although *Acarus siro*, under optimum conditions (15°C, 90% RH) in the laboratory, averaged an output per female of 435 eggs with a maximum of 858, but the females needed to mate repeatedly to attain maximum production (Cunnington, 1985). Among the Oribatida, 250 eggs in a single year is stated by Grandjean (1950) to be produced by a female of *Platynothrus peltifer* but this is probably exceptional. The most comprehensive numerical data on fecundity in the mites refer to species of economic importance in the families Tetranychidae and Phytoseiidae (Helle & Sabelis, 1985). Mites may lay their eggs singly or in batches and in the former condition only one large mature egg is present at any one time in the reproductive tract. The majority of the Mesostigmata produce their eggs singly but exceptions do occur as, for example, in the uropodid *Uroseius acuminatus* and the ichthyostomatogastrid *Asternolaelaps fecundus* in which many eggs are visible in the oviducts of the gravid female. Deposition of egg batches occurs among the Oribatida (Woodring & Cook, 1962b) and Prostigmata, particularly in the water mites.

A pair of sclerotized bosses occurring on the prelarval cuticle (apoderma) of the epidermoptid *Myialges macdonaldi* was considered by Evans *et al.* (1963) to function as egg bursters (Fig. 10.12A). The structures lie immediately beneath the egg shell. Subsequently, similar egg bursters have been found to be common in free-living Astigmata (Fain & Herin, 1979). In *Dermatophagoides pteronyssinus*, the bursters, in the form of bosses, are located beneath the shell above the anterodorsal region of the body of the larva, near the base of the gnathosoma, and after emergence of the larva they are seen to lie on the same side of the cleavage line that encircles the egg along its longitudinal axis (Spieksma, 1967). The egg cleavage line passes between the two bosses in *D. farinae* (Furumizo, 1973). According to Grandjean (1954a, 1962), a pair of cuticular projections (dent *k*) situated anteriorly or anterolaterally on the prelarvae of the Euoribatida (Fig. 10.12B) and *Oribotritia* may function as egg bursters ('dents d'eclosion'). They are not present in all euoribatid prelarvae; for example, they are absent in *Liodes* and probably also in *Hermanniella*.

11 Development and Dispersal

Embryonic Development

Detailed information on the embryology of the Acari is meagre and our present knowledge, largely based on ticks, is summarized by Hughes (1959), Anderson (1973), Ivanova-Kazas (1979) and Aeschlimann (1984). Development appears to take the same general course in all species which have been examined although there are differences between major taxa.

The type of cleavage is influenced by the considerable amount of yolk present in the egg. In the majority of the Acari, cleavage is partial and superficial, the nuclei dividing within the cytoplasm (intralecithal) and migrating to the surface prior to cleavage. Further divisions of the nuclei produce a uniform monocellular blastoderm layer surrounding the yolk. The embryo is now at the stage of a periblastula with the blastocoele obliterated by the central yolk mass (Aeschlimann, 1984). The small eggs of some actinotrichid mites, such as those of the Tarsonemina and some Astigmata, contain less yolk and a phase of total cleavage precedes blastoderm formation (Brucker, 1900). The specialized sequence of total cleavage, which Anderson (1973) considers to indicate a secondary condition for the Acari, has been described in *Sancassania* by Prasse (1968a). According to Akimov and Yastrebtsov (1990), cleavage is total and asynchronous in the alecithal eggs of the Spinturnicidae and leads to the formation of a stereoblastula (Fig. 11.1A).

The first external sign of embryonic development in eggs having partial and superficial cleavage is the formation of a mid-ventral germ disc with a temporary gastral groove. The appearance of the germ disc may be considered as gastrulation. The germ disc, which represents the future telson, migrates in a posterior direction over the surface of the yolk and a broad germ band is formed by the proliferation of blastoderm cells in front of the disc (Fig. 11.2A). Proliferation of cells at the side of the gastral groove results in a compact group of large cells below the germ disc, some of which sink the yolk mass where they divide and form vitellophages while others remain and form posterior mid-gut rudiments

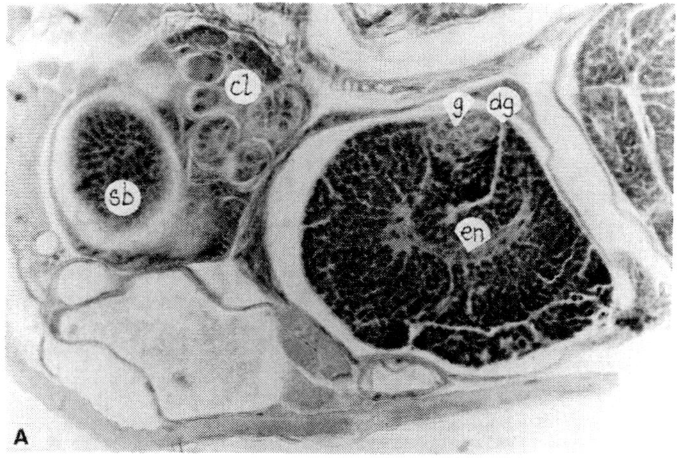

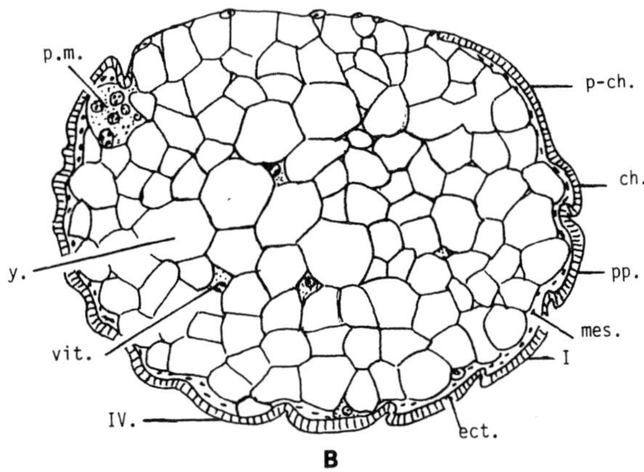

Fig. 11.1. **A**, transverse section of part of the idiosoma of a gravid female of *Spinturnix myoti* (Mesostigmata) showing early stages of embryogenesis; **B**, sagittal section of the embryo of *Ornithodoros* (Ixodida) at the segmented germ band stage (after Aeschlimann, 1958). *ch.* = cheliceral segment; *cl.* = oocytes showing cleavage; *dg.* = dorsal groove or fissure; *ect.* = ectoderm; *en.* = entoderm; *g.* = genital germ; *mes.* = mesoderm; *p-ch.* = pre-cheliceral lobe; *p. m.* = posterior mid-gut rudiment; *pp.* = pedipalpal segment; *sb.* = stereoblastula stage; *vit.* = vitellophage; *y.* = yolk mass; *I. IV* = segments of ambulatory appendages I and IV.

(Fig. 11.1B). Mesoderm arises along the length of the germ band and in ticks forms paired bands which break up into somite rudiments.

The formation and metamerization of the germ band has been studied in detail in ticks but information for mites is incomplete. In the argasoid *Ornithodoros moubata* (Fig. 11.1B), the germ band increases in length without showing metamerization until it is U-shaped with both ends on the dorsal surface of the yolk mass (Aeschlimann, 1958). When

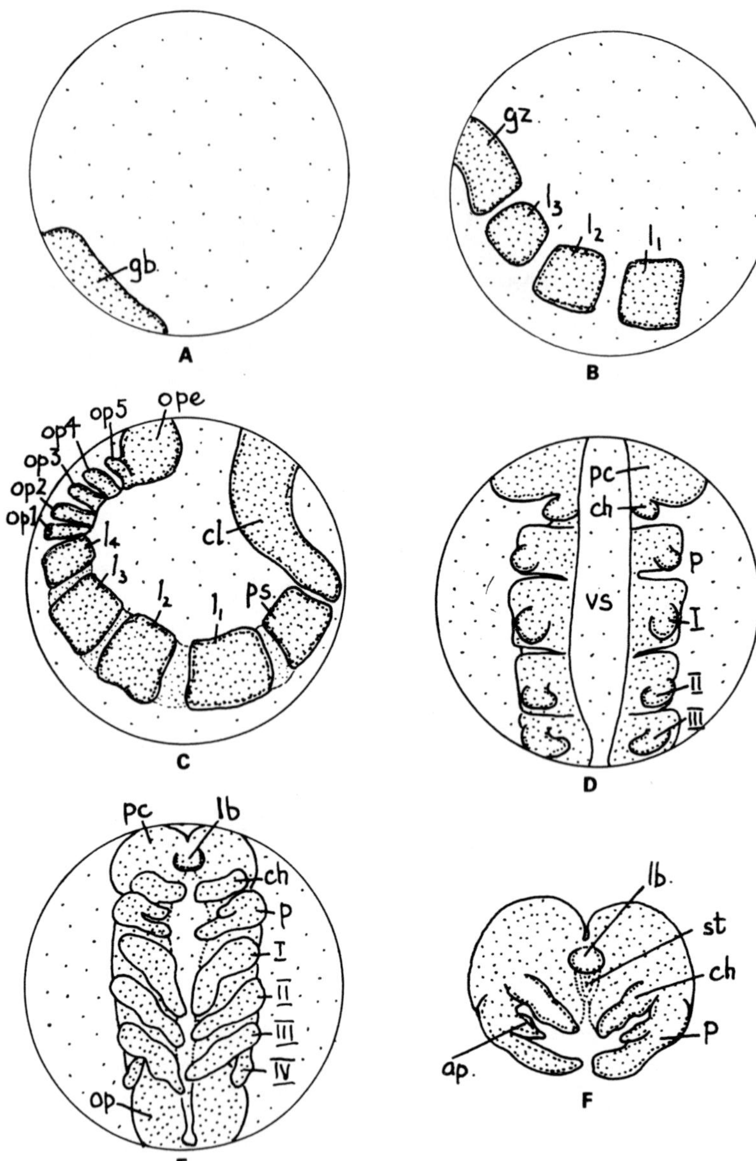

Fig. 11.2. Germ band formation and segmentation in *Hyalomma dromedarii* (Ixodida) (after Anderson, 1973). **A**, early growth of germ band; **B**, beginning of formation of the prosoma; **C**, further formation of the prosoma and segmentation of the opisthosoma; **D**, formation of limb buds and form of the ventral sulcus; **E**, closure of the ventral sulcus, elongation of the prosomal appendages and obliteration of opisthosomatic segmentation; **F**, anterior extremity of the prosoma showing chelicerae becoming pre-oral and the apophyses on the pedipalps. *ap.* = apophysis on pedipalpal base; *ch.* = chelicera; *cl.* = cephalic lobe; *gb.* = germ band; *gz.* = growth zone; *lb.* = labrum; l_1–l_4 = first to fourth ambulatory segment; op_1–op_5 = opisthosomatic segments 1–5; *op.* = opisthosoma; *ope.* = posterior end of opisthosoma; *p.* = pedipalp; *pc.* = precheliceral lobe; *ps.* = pedipalpal segment; *st.* = stomodaeum; *vs.* = ventral sulcus; I–IV = ambulatory appendages I–IV.

metamerization begins the two halves of the germ band separate to form a broad ventral sulcus. Three ambulatory segments are delineated at first, followed by the pedipalpal, the fourth ambulatory and five opisthosomatic segments. The formation of the cheliceral metamere from the precheliceral (cephalic) lobes is delayed. Later in development, external evidence of primitive metamerism is lost. As in Opiliones, there is no inversion of the germ band.

The development of the germ band in the ixodoid *Hyalomma* (Anderson, 1973) is more specialized in that it separates precociously into two divergent halves and gives rise to the rudiments of the 3rd and 4th leg segments and five opisthosomatic segments and a telson while the germ band occupies the postero-ventral surface of the egg mass (Fig. 11.2B,C). Further, the anterior part of the germ band arises *in situ* as bilateral aggregates of cells forming in succession the second and first ambulatory, and the pedipalpal segments. Formation of the cheliceral segment is delayed and occurs at the posterior end of the cephalic lobe (Fig. 11.2D). The resulting U-shaped band is similar to that in *Ornithodoros*. Limb buds develop on the prosomatic segments when the bilateral segmented germ band is formed, those of the ambulatory segments appearing before the pedipalpal and cheliceral pairs (Fig. 11.2D,E). No limb buds occur on the opisthosomatic segments.

A similar bilateral germ band to that in ticks, in which the ends lie on or near the dorsal surface of the yolk mass, occurs in the limited number of Actinotrichida which have been studied. The anterior end of the contracted germ band remains at the anterior end of the yolk mass but its posterior end withdraws onto the ventral surface of the yolk. In the Mesostigmata, on the other hand, the anterior end of the germ band does not extend further than the anterior pole and the posterior end remains on the dorsal surface as the anterior end withdraws along the ventral surface (Camin, 1953a, Zukowski, 1964). This results in the contracted germ band being curved around the posterior pole. However, pulsations shift the germ band forward to the normal position on the antero-ventral surface of the yolk at a later stage. The elongated germ band of the Dermanyssina divides into seven unequal parts of which the posterior element represents the integrated opisthosoma and telson. The remaining six parts comprise the buds of prosomatic segments I–VI. The opisthosoma never shows traces of segmentation. During embryogenesis in *Spinturnix*, a temporary dorsal groove, resulting from the invagination of ectoderm, has been observed (Fig. 11.1A) and the germ band initially divides into four parts followed by the division into two segments of each of the three anterior parts (Akimov & Yastrebtsov, 1990). The terminal part forms the unsegmented opisthosoma and telson as in other Dermanyssina. In the Tarsonemina, two transverse grooves are stated to divide the germ band into a short posterior region, which will become the opisthosoma, a longer anterior region, the future gnathosoma, and between them a region which will later form the podosoma. The first three ambulatory appendages are delineated first followed by legs IV, the pedipalps and finally the chelicerae. The opisthosoma is considered to show traces of six segments, probably including the telson (Brucker, 1900; Reuter, 1909).

Longitudinal contraction of the entire segmented germ band (blastokinesis) in ticks occurs as soon as the limb buds begin to develop and allows the future larva to take its definitive form according to already fixed polarity and symmetrical axes. The ventral sulcus is eliminated as the two halves of the germ band meet in the mid-line (Fig. 11.2E). It is as a result of blastokinesis that there is a disappearance of external primitive metamerism and the prosoma and opisthosoma fuse to form an idiosoma. At the end of blastokinesis in the Ixodida and Mesostigmata, a median unpaired labral bud arises from the acron anterior to the stomodaeal invagination and a median apophysis or lobe (the basis corniculi of Zukowski, 1964) appears on each pedipalp coxa (Fig. 11.2F). The chelicerae later migrate from a post-oral to a pre-oral position and come to lie in front of the stomodaeum. In *Ornithodoros* the chelicerae undergo a 90-degree rotation during migration. The pedipalps after migration lie lateral to the stomodaeum. The median lobes of the pedipalps are considered by many authors to form the hypostome but the discovery of a pair of post-stomodaeal lips in the embryo of *O. moubata* suggests that they may also take part in the morphogenesis of the hypostome (Aeschlimann, 1984).

Temporary regression of the rudiments of the fourth pair of legs occurs later in embryonic development so that the larva is hexapod (Warren, 1941; Zukowski, 1966). According to A.V. Yastrebtsov (personal communication) the regressed legs in the majority of the Mesostigmata are concealed under the larval cuticle but their structural elements (muscles, nerves, etc.) only develop to the stage found in prelarval morphogenesis. An exception occurs in the Spinturnicidae in which there is considerable embryonization of the early stages of ontogenesis as an adaptation to parasitism. All four pairs of legs develop simultaneously and the temporary regression of the fourth pair of legs does not occur (Akimov & Yastrebtsov, 1990). This results in the lack of a hexapod larval stage. All other species which have been investigated have a reduced fourth ambulatory segment (segment VI) in the embryo and external rudiments of its appendages occur only in the Anactinotrichida and Tarsonemina. The reports in earlier literature of the presence of a fourth pair of limb buds in the Astigmata are not supported by the work of Hughes (1950a) on *Acarus siro*. The larva of a species of Notostigmata has been described with vestiges of the fourth pair of legs (Coineau & Van der Hammen, 1979) but nothing is known of the embryology of this group or of that of the Holothyrida.

Development of the mesenteron has been studied in some detail in ticks but to a lesser extent in mites. In ticks, the mesenteron is of entodermal origin and appears to arise largely from cells in the posterior mid-gut rudiment which retains its position beneath the germ disc during the formation of the germ band and later, in the contracted germ band, comes to lie beneath the posterior region of the opisthosoma. Aeschlimann (1958) has shown that in *O. moubata* the rudiment produces a pair of lateral processes which detach as excretory tubules. Their proximal ends enlarge and unite to form a median 'rectal sac' (postcolon) which establishes a connection, after the first moult, with the proctodaeum but not with the mid-gut sac, which is probably formed from the remaining cells of the rudiment. Vitellophages do not appear to be a major source of mid-gut epithelial cells. The proctodaeum meets the postcolon of the mesenteron, except in argasid ticks, and forms the short anal atrium. The stomodaeum gives rise to the epithelium of the pharynx and oesophagus. In the Mesostigmata, the mid-gut and Malpighian tubules are also of entodermal origin and the connectin between the mid-gut and the stomodaeum and proctodaeum occur during larval morphogenesis.

The mesenteron in *Acarus siro* arises from three distinct sources according to Hughes (1950a). The anterior part arises from cells in the anterior polar cap and these grow back over the yolk and send a proliferation forwards as the oesophagus. The posterior part (colon and postcolon) arises from a posterior proliferation of cells arising from the posterior cap of cells and these indent the yolk mass in the median line. The dorsal part of the cell mass becomes the colon and the ventral region forms the postcolon which becomes united to the proctodaeum. The walls of the ventricular caeca are formed from vitellophage cells and the Malpighian tubules arise at the upper end of the postcolon.

The embryonic development of the gonads is poorly understood. According the Bonnet (1907) those of *Hyalomma aegyptium* appear late in the embryo as groups of cells, probably mesoderm of the second opisthosomatic segment, flanking the postcolon. These paired rudiments lengthen and curve upwards to meet above the postcolon. A gonoduct is produced at the posterior end of each gonad and later these connect with a mid-ventral ectodermal exit duct.

The nervous system arises in argasoid ticks after segmentation of the germ band but before the growth of the prosomatic limbs, as a pair of longitudinal thickenings of the mid-ventral ectoderm of the germ band. Simultaneously, ganglia rudiments are formed from ectoderm of the cephalic lobe. At a later stage in development, it is possible to distinguish a chain of paired ganglia connected to longitudinal and transverse commissures running the length of the germ band. At this stage in *O. moubata*, 11 paired ganglia comprising one precheliceral, six prosomatic and four opisthosomatic are discernible. However, at the time of the contraction of the germ band, the ganglia fuse to form a compact synganglion surrounding the oesophagus. The cheliceral ganglia move forward on either side of the mouth and fuse with the precheliceral ganglia to form the pre-oesophageal or supra-oesophageal mass, while those of the legs and the opisthosoma form the post-oesophageal or sub-oesophageal mass of the synganglion. The pedipalpal ganglia are situated laterally and appear to unite the two masses. In the mites *Acarus siro* and

Spinturnix vespertilionis, the structure of the central nervous system develops directly into the definitive form without the formation, at any stage, of a long ventral chain of segmental ganglia (Hughes, 1950a; Akimov & Yastrebtsov, 1990).

Anderson (1973) considers embryonic development in the Acari to more closely resemble that of the Opiliones than any other group of Arachnida for which embyrological data are available. The resemblances are evident in the absence of inversion of the germ band, in the mode of formation and bilateral separation of the prosomatic section of the germ band and in the germ disc being the opisthosomatic rudiment and not a full embryonic primordium. The embryonic development of the Ricinulei, which are considered by Weygoldt and Paulus (1979b) to be the most closely related group to the Acari, is unknown.

Ontogeny

The terminology applied to the ontogeny of the Acari is largely derived from the work of Grandjean (1938c, 1940b, 1951b, 1957a, 1970a). At the basis of this terminology is the concept of the stase. A stase is one of the successive forms through which the acarine passes in the course of ontogenetic development. Stases are idionymic, i.e. they differ from one another by discontinuous external (surface) characters which enable them to be homologized with corresponding stases in other species of the same group.* Differences in dimensions of existing structures between successive forms and allometric changes are rejected as discriminating characters. The requirement of a fundamental difference in character between successive forms in the life cycle and the nature of the difference which rests on binary data (on presence or absence data) characterizes the stase within the broader concept of the instar in which the change of skin (the moulting process) and not the change in character is usually emphasized (André, 1988). Although *a stase is always an instar*, the converse, according to Grandjean (1970a), is not necessarily the case and he cites as examples successive forms arising from 'repetition-moults' that are identical, even in size, or arising from 'growing-moults' and presenting only differences in dimensions. The occurrence of 'growing moults' has not been established in the Acari and supernumerary argasid nymphs which are cited as examples of forms displaying 'repetition moults' do present differences in their chaetotaxy (Siuda, 1982). Developmental forms which differ from one another by discontinuous characters but which cannot be homologized with corresponding forms in other

*Johnston and Wacker (1967) have used the Greek word *stasis*, from which the French form *stase* is derived, to denote 'a period of development delimited by molts'. In more recent literature the French form, stase, has been retained. Grandjean's original terminology for normal and regressive stases is followed in this work.

species of the same group have been termed *stasoids* by Van der Hammen (1975). The term instar when used in this work includes both stases and stasoids.

In addition to the normal stase (one which feeds, moves and grows), one or more stases of the life cycle may be subject to regressive development and lose their mouthparts so that they are unable to feed, or lose their mouthparts and ambulatory appendages so that they can neither feed nor walk. Grandjean called a morphologically regressive stase of the former type an *elattostase* and of the latter type a *calyptostase*. The degree of morphological regression exhibited by calyptostases is extensive and ranges from the extreme form in which the stase is represented by an apoderma (the cuticle of a calyptostase lacking appendages and body setae) to one in which simplified appendages and some body setae are retained. In the majority of cases in the Acari, the calyptostase is an endostase, i.e. it does not emerge from the egg or the exuviae of the preceding stase, whereas the elattostase is an ectostase and does emerge from the egg or from the cuticle of the preceding instar. The stase terminology was originally applied to the ontogenetic development of the Actinotrichida but is applicable to the Anactinotrichida. The occurrence of a special stase at the beginning or at a later stage in ontogeny has been termed, respectively, protelattosis and metelattosis by Van der Hammen (1975, 1978).

A maximum of six stases occur in the life cycles of the Acari, comprising prelarva (deutovum, prolarva), larva (tritovum), protonymph, deuteronymph (hypopus), tritonymph and adult. Each represents a developmental level in the ontogeny of the Acari. At the first level is the prelarva, at the second the larva, at the third, fourth and fifth, respectively, are the protonymph, deuteronymph and tritonymph, and at the sixth level is the adult. Although in the early phylogenetic history of the Acari, the prelarva is considered to have been an active hexapod instar, in extant species it is a calyptostase, with the rare exceptions of the prelarvae of *Saxidromus* (Adamystidae) and *Speleorchestes* (Nanorchestidae) which are elattostases. The normal larval stase is also hexapod and the normal nymphal stases are octopod but in both phases the legs may be completely lacking as in some calyptostases or reduced in number as in the Eriophyoidea where only two pairs of ambulatory appendages are present in all active stases. The egg is not considered a stase. When a sequence of stases in a life cycle show a similarity in appearance they are referred to as homostasis while a stase which differs markedly from the preceding one is called a heterostase (Grandjean, 1954b). The similar nymphs in most acarine life cycles are examples of homostases while the adult stase of some Oribatida which differs markedly in appearance from the nymphal stases is an example of a heterostase. One or more stases which resemble and succeed each other constitute a phase. Thus, in the Acari it is possible to distinguish a maximum of four phases: prelarval, larval, nymphal and adult. The acarine does not moult after attaining the adult phase except in some Prostigmata (Michener, 1946; Immamura,

1952) and possibly in the Notostigmata (Coineau & Legendre, 1975).

The concept and terminology of a stase are also applicable to the postembryonic development of other Arthropoda. Calyptostases, particularly endocalyptostases, occur commonly at the beginning of the ontogenetic development of arthropods and may be followed by elattostases as in spiders and Thysanoptera (André, 1988).

Anactinotrichida

The full complement of six stases appears to occur, with certainty, only in the Notostigmata; the complete life cycle of *Phalangiacarus brosseti* has been described by Coineau and Van der Hammen (1979). The prelarva is intermediate in form between an elattostase and a calyptostase in having only a 'slightly regressive' gnathosoma and three pairs of long, jointed legs without setae. The larva is hexapod with vestiges of the fourth pair of legs.

In the other three orders of the Anactinotrichida, with the possible exception of the Holothyrida for which the life cycle is incompletely known but which is considered to have three nymphal stases, there is a reduction in the number of stases in the developmental cycle. If the occurrence of six postembryonic stases is accepted to be the 'normal' condition obtaining in the Acari then the apparent reduced number of stases in the Mesostigmata and Ixodida could be due to the regression of one or more stases to calyptostases or to the embryonization of the early stages of ontogeny. There is some evidence in these taxa that a moult (or moults) occurs within the egg before the eclosion of the larva, such as the report by Winkler (1888) for the Parasitina and Bonnet (1907) and Aeschlimann (1984) for the Ixodida. Sitnikova (1978) considers this to indicate the existence of a prelarval phase. Akimov and Yastrebtsov (1990) point out that the substantial transformation in the internal organization and histolysis of internal organs which usually characterizes the transition between phases does not occur between the so-called larval and protonymphal instars in the Mesostigmata. In fact, larval and protonymphal morphogenesis complement each other to such a degree that some of the processes characteristic of larval morphogenesis can be transferred to the protonymphal instar. On this basis they consider that it would be more correct to consider that the Mesostigmata have a trinymphal non-larval ontogenesis with substantial embryonization of the early phases of ontogeny. The terminology used in this work for the stases of the nymphal phase in these two groups follows current usage.

Mesostigmata

Four stases comprising larva, protonymph, deuteronymph and adult are typical for this taxon. An interesting modification of the normal life cycle for dispersal from transient habitats occurs in some Uropodina and

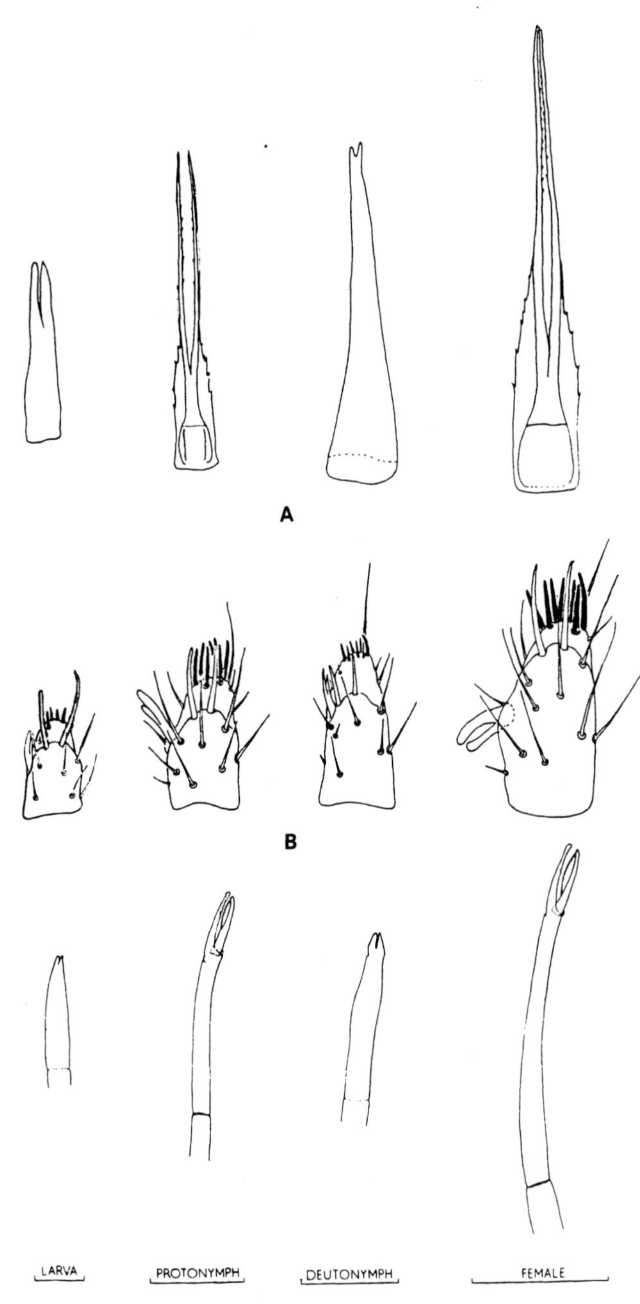

Fig. 11.3. Form of the tritosternum (**A**), pedipalp tarsal sensilli (**B**) and chelicerae (**C**) in the non-feeding larva and deuteronymph, and the feeding protonymph and female in *Ornithonyssus bacoti* (Mesostigmata) (after Evans & Till, 1965).

Microsejina where a phoretic deuteronymph (Wandernymph) may be produced as an alternative to the normal form (Dauernymph) under certain environmental conditions (Faasch, 1967; Evans & Till, 1979). The phoretic deuteronymph attaches to its host, usually another arthropod, by an anal pedicel or stalk.

In many predatory species of the Dermanyssina, the larva has non-functional chelicerae and does not feed, and may be considered an elattostase. Morphological regression in one or more stases, however, is more commonly encountered as an adaptation to a parasitic mode of life, especially in obligatory parasitic forms, and the regressed stases are usually of shorter duration than normal stases. This leads to an acceleration in the duration of the life cycle. The larva and deuteronymph in the Macronyssidae are ellatostases in which the structures associated with feeding, such as the chelicerae, palptarsal sensilli and tritosternum, are reduced or non-functional (Fig. 11.3), while the protonymph and adult are normal feeding forms (Evans & Till, 1965). In *Spinturnix* there is considerable embryonization of the early stages of ontogenesis resulting in the absence of a hexapod larval phase and the birth of an octopod protonymph (Akimov & Yastrebtsov, 1990). The larvae and adults of the Halarachnidae inhabiting the respiratory tract of Pinnipedia, for example *Orthohalarachne*, are normal instars but both nymphal instars are non-feeding fragile forms with reduced idiosomatic setation and mouthparts (Furman, 1977). This also applies to the halarachnid *Pneumonyssus simicola* living in the lungs of monkeys, and Furman *et al.* (1974) found that, *in vitro*, the protonymph and deuteronymph may each moult completely before the emergence of the adult, or the protonymphal cuticle may be retained by the deuteronymph so that the adult emerges from the protonymphal and deuteronymphal cuticles simultaneously. In the halarachnids, the larva is the inter-host transfer stage. The adaptive significance of the various types of postembryonic development in the Mesostigmata is discussed by Athias-Binche (1987).

Parasitic members of the Mesostigmata tend to show a reduction in idiosomatic sclerotization (to allow for the expansion of the idiosoma during feeding) and in the setal complement of the appendages in comparison with free-living forms (Evans & Till, 1965). The reduced number of setae on palpal and leg segments of the adult is often due to localized neoteny whereby the larval complement of setae is retained by the adult (Evans, 1964). Hypertrichy of the idiosoma occurs in many nest-inhabiting forms, which may be ectoparasitic (some *Haemogamasus*) or free-living (*Pneumolaelaps*).

Ixodida

The ticks are obligate haematophagous parasites and are considered to have only three instars: larva, nymph and adult. Aeschlimann (1984) states that the larva of *Ornithodoros moubata* moults under its egg shell at least once, if not twice, before hatching. The significance of these

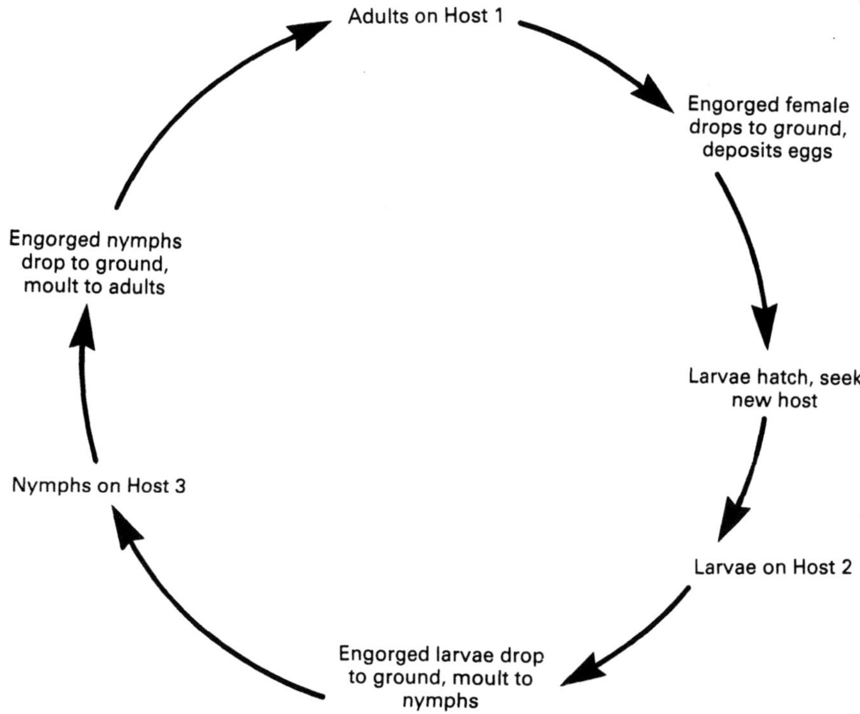

Fig. 11.4. Life cycle of a three-host ixodoid tick.

'intrachorionic moults' has not been determined although they may indicate the presence of regressed stases, as referred to above. In the Argasoidea, the life cycle is complicated by the appearance of additional nymphal instars; for example, in *O. moubata* the male usually appears after the fourth or fifth nymphal moult and the female after the fifth of sixth. Each nymphal moult is preceded by a blood meal. The additional instars have been considered to be isophena (forms resultIng from 'repetition or growing moults') by Van der Hammen (1975) who referred to the phenomenon as plethomorphosis. However, Siuda (1982) has shown that the three nymphal instars in *Argas (A.) polonicus* may be differentiated on the basis of the setal numbers of the palps and of the tarsi of legs I and II, although there is some intra-instar variation and inter-instar overlap in some of the numerical data. Thus, it appears that the successive 'supernumerary' instars are not only different in size but also show differences in surface structures and as such cannot be considered to be isophena. Grandjean (1970a) suggested that if there should be a definite change in character at each ecdysis in argasoid nymphs, then the instars are an example of neostasy.

Ixodoid ticks feed only once during each active instar except for the males of some species of Ixodidae which are aphagic. The majority of nymphal and adult argasoids, with some exceptions, take a number of short blood meals. Exceptions are seen in the non-feeding and immobile larvae of *O. moubata* and *O. savignyi*, and the non-feeding females of *Otobius* and *Antricola* (Balashov, 1968). A multi-host life cycle is characteristic of most species of argasoids except the one-host species of the genus *Otobius* and the two-host *Alveonasus lahorensis*.

In the Ixodoidea, there is only one nymphal instar and since the larva, nymph and adult each feeds only on one host, the maximum number of host changes in the life cycle is three. A three-host life cycle occurs in all the Ixodidae and in the majority of the Amblyommidae and is shown diagrammatically in Fig. 11.4. The total period of parasitism is extremely short in relation to the time spent off the host. For example, in *Ixodes ricinus* the larva, nymph and adult female spend a total of only about 25 days parasitizing their hosts in a life cycle of up to three years' duration (Evans *et al.*, 1961). The immatures and adults of species living in nests and burrows such as *Ixodes plumbeus* parasitic on sand martins and *I. trianguliceps* on small mammals tend to feed on the same host since their chances of encountering alternative hosts are restricted. Field species, on the other hand, are not restricted in this way and the larvae and nymphs usually feed on birds and/or small mammals and the adults on larger animals.

Some species have a two-host cycle during which the larva remains attached to the host after engorgement and moults to the nymph which feeds before detaching and falling to the ground to moult. The adult parasitizes a second host, which may be the same or a closely related species as in *Rhipicephalus bursa* and *Hyalomma detritum* feeding on ungulate species, or a very different host species as in *H. plumbeum* whose larvae and nymphs feed on birds and rodents, and adults on ungulates (Balashov, 1968). In the genus *Boophilus* the larva and nymph moult on the one host and only the engorged female disengages. This one-host life cycle is considered to be the best adapted to nomadic host animals.

The type of life cycle in some species is not rigid and may be changed by such factors as host and temperature. For example, *H. anatolicum* has a three-host cycle on cattle but only a two-host cycle on rabbits; larvae of this species fed on rabbits have a predominantly two-host cycle at 20–22°C but 40% develop into the three-host type at 30°C. Modifications of the one-host life cycle are less common although Ammah-Attoh (1966) has reported some *Boophilus decoloratus* passing through a two-host cycle on cattle.

It is considered by Pomerantsev (1948) that the transition from a three-host to a two- and one-host developmental cycle is one of the most important adaptations of ixodoid ticks for their successful parasitism of large nomadic animals. The potential for finding a host, which is essential for the survival of the tick, is greatly reduced in ungulates feeding in open steppe and desert habitats. A reduction in the number of host encounters required to complete the life cycle increases the chances of the survival of the parasite and this has been achieved through one or both immature instars remaining on the host to moult after engorgement.

The activity of ixodoid ticks is usually synchronized, through diapause, with cyclic climatic changes and has been the subject of extensive research, which is summarized by Arthur (1962), Balashov (1968) and Gray (1991).

Actinotrichida

There is greater diversity in the ontogenetic development of this group than in the Anactinotrichida, particularly in relation to the incidence of morphologically regressed stases (Table 11.1). However, the occurrence of six stases in the developmental cycle is remarkably constant and it is only in a small number of taxa, for example the Tetranychidae and the Tarsonemina, that one or two nymphal stases appear to be completely lacking. The prelarval phase is often markedly regressed and represented only by an apoderma while calyptostasic or elattostasic forms may occur during the nymphal phase. The most drastic regression occurs in some parasitic species of the Astigmata and Prostigmata.

Prostigmata

Five normal stases succeed the calyptostasic or, more rarely, the elattostasic prelarva in many families, for example the Adamystidae, Anystidae, Caeculidae, Bdellidae, Eupodidae, Iolinidae, Nanorchestidae, Speleorchestidae, Terpnacaridae and Tydeidae. The degree of regression of the prelarva shows considerable variation within these families. In *Saxidromus* (Coineau, 1979) and *Speleorchestes* (Schuster & Pötsch, 1988), the prelarva emerges from the egg as an ellatostase, being mobile but having non-functional mouthparts, while in the Anystidae, Caeculidae and Bdellidae it is calyptostasic but retains simplified appendages. It is very

Table 11.1. A selection of ontogenies from the Actinotrichida showing the incidence of regressive stases (in italic type).

Oribatida	Astigmata	Prostigmata		
Egg	Egg	Egg[1]	Egg[2]	Egg[3]
Prelarva	*Prelarva*	*Prelarva*	*Prelarva*	
Larva	Larva	Larva	Larva	Larva
Protonymph	Protonymph	*Protonymph*	*Protonymph*	
Deuteronymph	*Deuteronymph*	Deuteronymph	*Deuteronymph*	*Nymph*
Tritonymph	Tritonymph	*Tritonymph*	*Tritonymph*	
Adult	Adult	Adult	Adult	Adult

[1]As in *Allothrombium* (Trombidiidae).
[2]As in *Boydaia* (Ereynetidae).
[3]As in *Tarsonemus* (Tarsonemidae).

regressed in the Tydeidae and Iolinidae and represented by an apoderma. A regression of the larva to an elattostase is a feature of the ontogenetic development of the Nicoletiellidae and some Rhagidiidae but the three nymphal stases and the adult are normal and predatory (Grandjean, 1945).

The ontogeny of the Parasitengona has been the subject of considerable study since the classic work of Henking (1882) on the development of the trombidiid *Allothrombium fuliginosum*. He observed the presence of two distinct developmental phases, between the end of the active larval instar and the eclosion of the nymph, and between the end of the active nymphal instar and the eclosion of the adult. The first phase, represented by the nymphochrysalis which succeeds the active larva and by the teleiochrysalis which follows the active nymph, terminates with the appearance of an internal membrane, the apoderma. The second phase extends from the appearance of the apoderma to eclosion of the nymph or the adult. Henking named the mite at this stage the nymphophanstadium or the teleiophanstadium. An apoderma also appears at the egg stage and the stadium extending from its appearance to the eclosion of the larva was termed the schadonophanstadium. The adult was referred to as the prosopon. A similar developmental cycle occurs in the water mites and Walter (1920) referred to the three apoderma as schadonoderma, nymphoderma and teleioderma according to the stadium at which each appeared. This author observed on the nymphoderma and teleioderma the presence of small discs, nymphophan and teleiophan organs (=genital papillae), similar in position and form to those in the genital region of the active nymph and adult. On this basis, Walter concluded that there originally existed three nymphs between the larva and adult of which only the second remained an active instar, the first and third being represented by the nymphoderma and teleioderma, respectively. The schadonoderma, nymphoderma and teleioderma represent, respectively, cuticles of the calyptostasic prelarva, protonymph and tritonymph. The terms nymphochrysalis and teleiochrysalis refer, respectively, to the 'resting' larva and deuteronymph (the *pupe* of French authors, see Cassagne-Mejéan, 1969). The schadonophanstadium is equivalent to the calyptostasic prelarval phase and the nymphophanstadium and teleiophanstadium represent, respectively, the calyptostasic protonymphal and tritonymphal stases of the nymphal phase. The terms nymphochrysalis and teleiochrysalis are often used incorrectly to include the nymphophan and teleiophan stadia. Grandjean (1938c, 1940b) rejected the terminology of Henking and Walter for the ontogenetic development of the Parasitengona and proposed that the terms prelarva, larva, protonymph, deutonymph (=deuteronymph), tritonymph and adult should be used for the six postembryonic developmental stases of the Acari irrespective of whether they are calyptostases, ellatostases or normal forms.

In the more primitive (least regressed morphologically) prelarva of the terrestrial Parasitengona, such as the trombidiid *Campylothrombium barbarum*, the chelicerae, palps and three pairs of legs are conspicuous

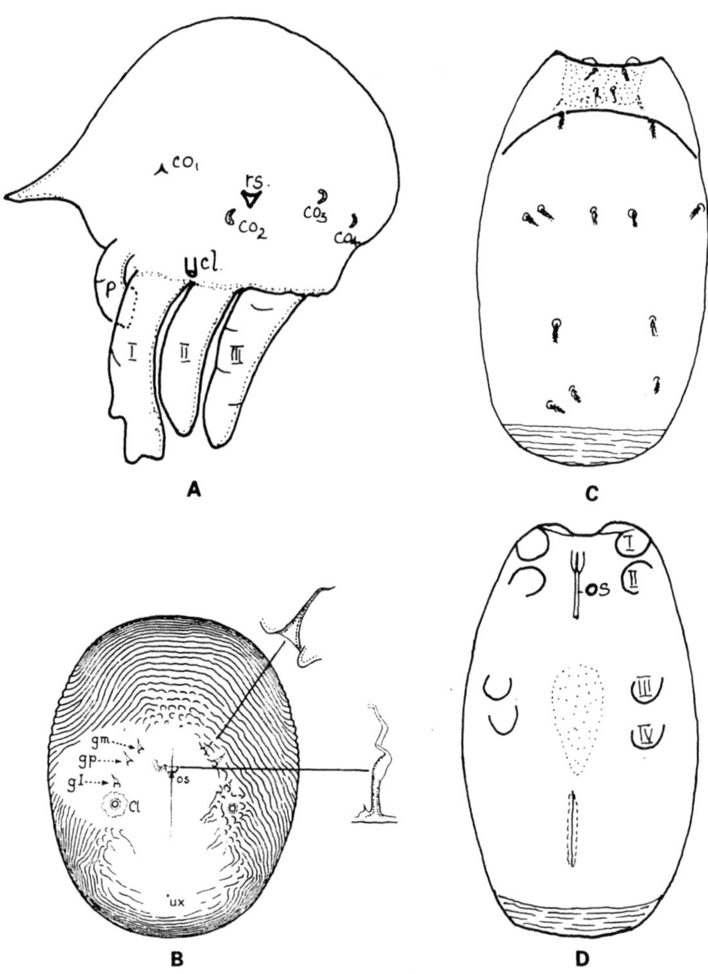

Fig. 11.5. **A**, prelarva of *Campylothrombium barbarum* (Prostigmata) (based on Robaux, 1974); **B**, ventral surface of the prelarva of *Balaustium florale* (after Grandjean, 1957b); **C, D,** dorsum and venter of the calyptostasic protonymph of *C. barbarum* (after Robaux, 1974). *cl.* = organ of Claparède; co_1–co_4 = cuticular organs; *gl. gm. gp* = excretory canals of putative coxal organs; *os.* = vestige of mouth, pharynx and oesophagus; *p.* = palp; *rs.* = residual setae; *ux.* = vestige of anus.

with the palps and legs showing some evidence of division into articles (Fig. 11.5A). Vestiges of the mouth, pharynx and the Claparède organs (situated dorsal and slightly posterior to the bases of the first pair of legs) are also discernible (Robaux, 1974). The remaining structures on the body comprise a pair of 'residual setae' situated above the second pair of legs and four pairs of 'cuticular organs' which Robaux suggested could be the openings of coxal glands although no traces of the excretory ducts of the glands are evident. At the other extreme, as in the erythraeid *Balaustium florale*, the prelarval stase (Fig. 11.5B) is very regressive and

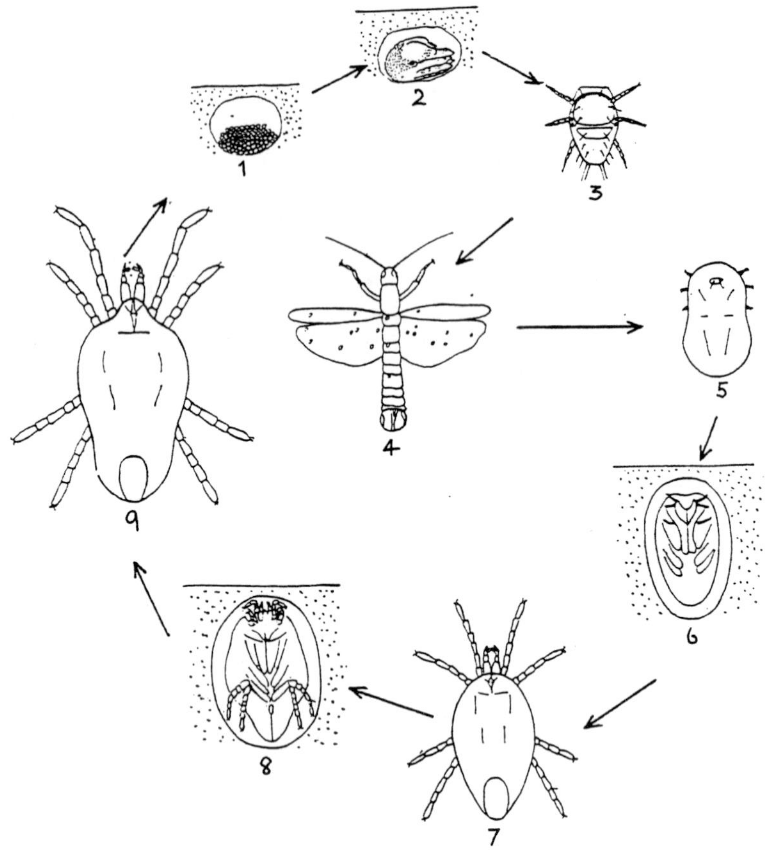

Fig. 11.6. Diagrammatic representation of the life cycle of *Eutrombidium rostratus* (Prostigmata) (based on Evans *et al.*, 1961). **1** = egg-mass in soil; **2** = prelarva; **3** = unfed larva; **4** = orthopteran host; **5** = engorged larva; **6** = deuteronymph developing within the cuticle of the calyptostasic protonymph in the soil; **7** = normal deuteronymph; **8** = adult developing within the cuticle of the calyptostasic tritonymph in the soil; **9** = adult female.

has lost its legs and other appendages although retaining vestiges of the mouth, pharynx, Claparède organs and the putative excretory ducts of three pairs of coxal glands (Grandjean, 1957b). The cuticle of the calyptostasic prelarva of water mites may be smooth, as in *Arrenurus*, *Eylais* and *Lebertia*, or furnished with spines, as in *Hydrovolzia*, *Sperchon* and *Thyas*. In the latter genus, the spines are considered to exert pressure on the egg shell, which splits as a consequence and the portion of the prelarval cuticle which is exposed is protected by a broad band of short pointed papillae.

The form of the calyptostasic protonymph and tritonymph varies considerably throughout the terrestrial Parasitengona to the extent of the occurrence of vestiges of the pharynx, genital papillae (= nymphophan and teleiophan organs), legs and body setae. In *Trombidium mediterraneum* for example, vestiges of the pharynx and legs, the latter in the form of small protuberances, as well as some body setae are present (Fig. 11.5C,D) whereas in *Allothrombium* there are no body setae (Robaux, 1974). On the other hand in *Balaustium* a vestige of the pharynx and well-developed body setation are evident but genital papillae and appendages are lacking.

In both terrestrial and aquatic Parasitengona, the normal larva is a heterostase adapted, with few exceptions, for a parasitic mode of life on invertebrate or vertebrate hosts and as such is an important agent in the dispersal of the species. There is a variety of postembryonic developmental cycles in the Parasitengona. The common type in terrestrial and aquatic species comprises three calyptostases (prelarva, protonymph and tritonymph) and three active feeding stases (larva, deuteronymph and adult). The heterostasic larva is parasitic and the deuteronymph and adult free-living and usually predatory or, as in the case of *Balaustium florale*, the larva, deuteronymph and adult are pollen feeders (Grandjean, 1946b). This type of life cycle is shown diagrammatically for *Eutrombidium rostratus* in Fig. 11.6. The nymphal calyptostases may occur entirely within the cuticle of the preceding stase as in *Eutrombicula splendens* (Johnston & Wacker, 1967) or the larval and the deuteronymphal cuticles may split transversely into two parts between legs II and III with the anterior section remaining in position but the posterior being cast off from behind as in *Balaustium forale* (Grandjean, 1946b). A number of modifications of this type of cycle occur in water mites by the extension of the larval association with the host or by the elimination of larval parasitism. The former condition is apparent in the Eylaidae and Hydrachnidae where the larva remains attached to the host until the deuteronymph is fully formed and emerges from the larval and calyptostasic protonymphal skins (Mitchell, 1957c). These two families are unrelated and have acquired this specialized type of life cycle independently. Larval parasitism is eliminated in the life cycles of *Unionicola* and *Najadicola* (Unionicolidae) which parasitize molluscs and sponges. The larvae probably never feed but through their swimming are largely responsible for the dispersal of the species. The deuteronymph and adult have secondarily acquired a parasitic mode of life as ectoparasites of their unusual hosts. An even more specialized development occurs in *Piona* and many other still-water forms where there is no active larval stase and the deuteronymph emerges from the egg (Lundblad, 1927). Thus, this developmental pattern has four calyptostases and only two normal stases. Among the terrestrial species, the most dramatic development is probably seen in the trombiculid *Vatacarus*, in which the active deuteronymph is suppressed and the larva and adult are the only normal stases, the prelarval and nymphal phases being calyptostases (Audy *et al.*, 1972).

The regression of the prelarva, protonymph and tritonymph to calyptostases occurs in *Pimeliaphilus podapolipophagus* (Pterygosomatidae) and is described in detail by Cunliffe (1952). In *Boydaia nigra*, an ereynetid mite living in the nostrils of birds, the three nymphal stases are calyptostasic (Fain, 1972a). The vestige of the pharynx of each nymph is visible through the larval skin.

Only two active nymphal stases are present in the Tetranychidae and both are normal stases like the larva and adult. The 'missing nymph' is considered by Van Impe (1985), on leg chaetotactic criteria, to be most likely the protonymph although there does not appear to be an apoderma present between the larval and the first nymphal cuticles which could represent an inhibited stase. The related Tenuipalpidae apparently have only one nymphal stase and adults of the genera *Phytoptipalpus* and *Larvacarus* are exceptional in being hexapod and retaining nearly all the characters of the larva (Baker & Pritchard, 1952). All the three active ontogenetic stases of the Eriophyoidea have only two pairs of legs but the status of the two immature stases has not been determined.

The ontogeny of the Tarsonemina is characterized by the absence of a normal nymphal phase. In the Tarsonemidae, the hexapod larva is usually an active stase. However, in *Acarapis* the second and third pair of legs are reduced and the larva is capable of only limited movement, and in the Acarphenacidae, Pyemotidae and Dolichocybidae the entire larval phase is spent within the body of the physogastric female. Larvae enter an inactive swollen state before the emergence of the adult. Reuter (1909) observed a regressed nymphal stase represented by an apoderma beneath the larval cuticle of the pyemotid *Siteroptes graminum* and the presence of such a calyptostasic nymph (quiescent larva, apodous nymph, apodous pupal stage, pharate nymph) has also been observed in the Tarsonemidae (Ewing, 1922; Cameron, 1925; Fain, 1970a; Lindquist, 1986). The adult develops within the skins of the nymph and larva. There does not appear to be a record of the presence of a prelarval phase in this group.

Astigmata

Of the six postembryonic stases encountered in the free-living members of this order, the prelarva and deuteronymph, when present, are regressive stases. The prelarva is represented by an apoderma within the egg and occurs commonly in the Acaridia but is absent in the majority of the Psoroptidia (Fain & Herin, 1979). The deuteronymph is a facultative or an obligatory instar which is either an elattostase provided with a posteroventral clasping or sucker device for attaching to a host for dispersal (Fig. 11.7) or a calyptostase. The larva, protonymph, tritonymph and adult are normal stases. Additional regressive stases occur in the specialized life cycles of species in which the deuteronymph lives under the skin or in the cellular tissues of birds.

The specialized deuteronymphal stase of astigmatic mites is commonly called the hypopus (*pl.* hypopodes) and the status of this heterostase has been the subject of considerable confusion in the literature of the last century when it was variously considered a representative of a separate family of adult mites, an immature mesostigmatic mite

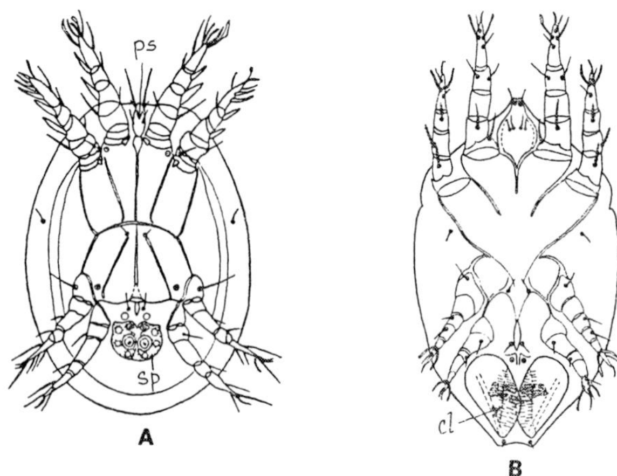

Fig. 11.7. **A**, venter of a hypopus (elattostasic deuteronymph) of an acarid mite (Astigmata) showing palposoma (*ps*) and sucker plate (*sp*); **B**, venter of the hypopus of *Labidophorus* sp. showing clasping organs (*cl*) (after Evans *et al.*, 1961).

and the adult of both sexes of some species of *Tyroglyphus* (Michael, 1888b). The hypopus is more resistant to desiccation and other adverse environmental conditions than the normal stases. It lacks functional mouthparts and the gnathosoma, if present, is markedly reduced. The mouth and chelicerae are completely lacking, and the palpal components in their least reduced form comprise a subrectangular basal part from which arise two anteriorly directed arms each bearing one or two setae, a lyrifissure and a long terminal solenidion. In the more evolved hypopodes only the solenidia and the setae may remain. Fain (1968b) refers to the entire structure as the palposoma. Associated with the loss of functional mouthparts is a regression of the alimentary tract, and Oboussier (1939) recognizes three types:

1. forms in which the tract is open from anus to ventriculus (e.g. *Carpoglyphus*);
2. open only from anus to postcolon (e.g. *Sancassania*);
3. with no lumen or anal opening (e.g. *Glycyphagus*).

A deuteronymphal stase appears to be absent in obligatory parasitic species, such as skin mites of the families Sarcoptidae and Psoroptidae, and feather mites, but is usually present in those species in which the other stases, except the prelarva, are free-living (with the notable exception of the Pyroglyphidae). The deuteronymph is very often a facultative instar and its appearance in the developmental cycle is induced in part by adverse environmental conditions which frequently prevail in the transient habitats that many species occupy, for example nests, stored food products and accumulations of organic debris. Deterioration of food quality often associated with high population density has been found to affect deuteronymphal development. For example, Griffiths (1966) working with species of *Acarus* found that deuteronymphal formation appeared to be favoured by malnutrition in the presence of a certain minimum quantity of vitamin B and that the period of response included the larval and protonymphal stases. Although the response seemed to be reversible during the larval life, it became irreversible in the course of the protonymphal stase. This suggests that the period for developmental option is limited to the larval and early protonymphal stases when the quality of the food activates a switch mechanism initiating alternative developmental programmes, either with or without the deuteronymph (Stratil & Knülle, 1985). Protonymphs which moult to the deuteronymphal stase have a considerably longer developmental period than those developing directly into tritonymphs. The extension in the duration of the protonymphal stase was considered by Griffiths to be associated with physiological and developmental changes and not due to starvation. Dietary experiments with *Acarus immobilis* suggested that ergosterol may play a role in suppressing deuteronymphal development and that ribonucleic acid is important to individuals which by-pass the deuteronymph (Griffiths, 1969).

Knülle (1987) found that the induction of the deuteronymphal stase in the developmental cycle of *Lepidoglyphus destructor*, a species living in stored food products and a number of field habitats, is regulated by a genotype–environment interaction. The deuteronymphal-producing genetic units are considered to behave in an additive manner so that if present in individuals at sub-threshold concentrations development is continuous (without deuteronymph) regardless of the quality of the food, whereas at higher accumulations they exceed an induction threshold which is low with inferior and high with superior diets. Physiologically, this may be seen as a dosage phenomenon with hormone concentrations corresponding with additive gene action and controlling the sensitivity of a switch mechanism activated by factors in the food.

The hypopus or deuteronymph is the dispersal phase in the life cycle of free-living Astigmata. In calyptostasic forms, dispersal is largely achieved by air currents and in ellatostatic forms by phoresy (phoresis). Morphological adaptations for phoresy are many and include the dorsoventral flattening of a strongly sclerotized body and the development of a range of organs for attaching to the cuticle of arthropods and the hair of mammals (Fain, 1971; OConnor, 1982b). These are discussed below in the section on dispersal.

Elattostasic deuteronymphs occur facultatively in the life cycles of

Fig. 11.8. Stages in the life cycle of *Hypodectes propus* (Astigmata) (after Fain & Bafort, 1967). **A**, female; **B**, anterior region of body of a heteromorphic male showing the large chelicerae; **C**, membranes of the egg after removal of the 'shell' and the enclosed deuteronymph, striated cuticle with setae represents the larval phase and the verrucose cuticle is that of the protonymph; **D**, deuteronymph (hypopus) from the subcutaneous tissues of a pigeon; **E**, a ruptured hypopodal cuticle showing the presence of a verrucose cuticle, representing the tritonymph, surrounding the developing adult.

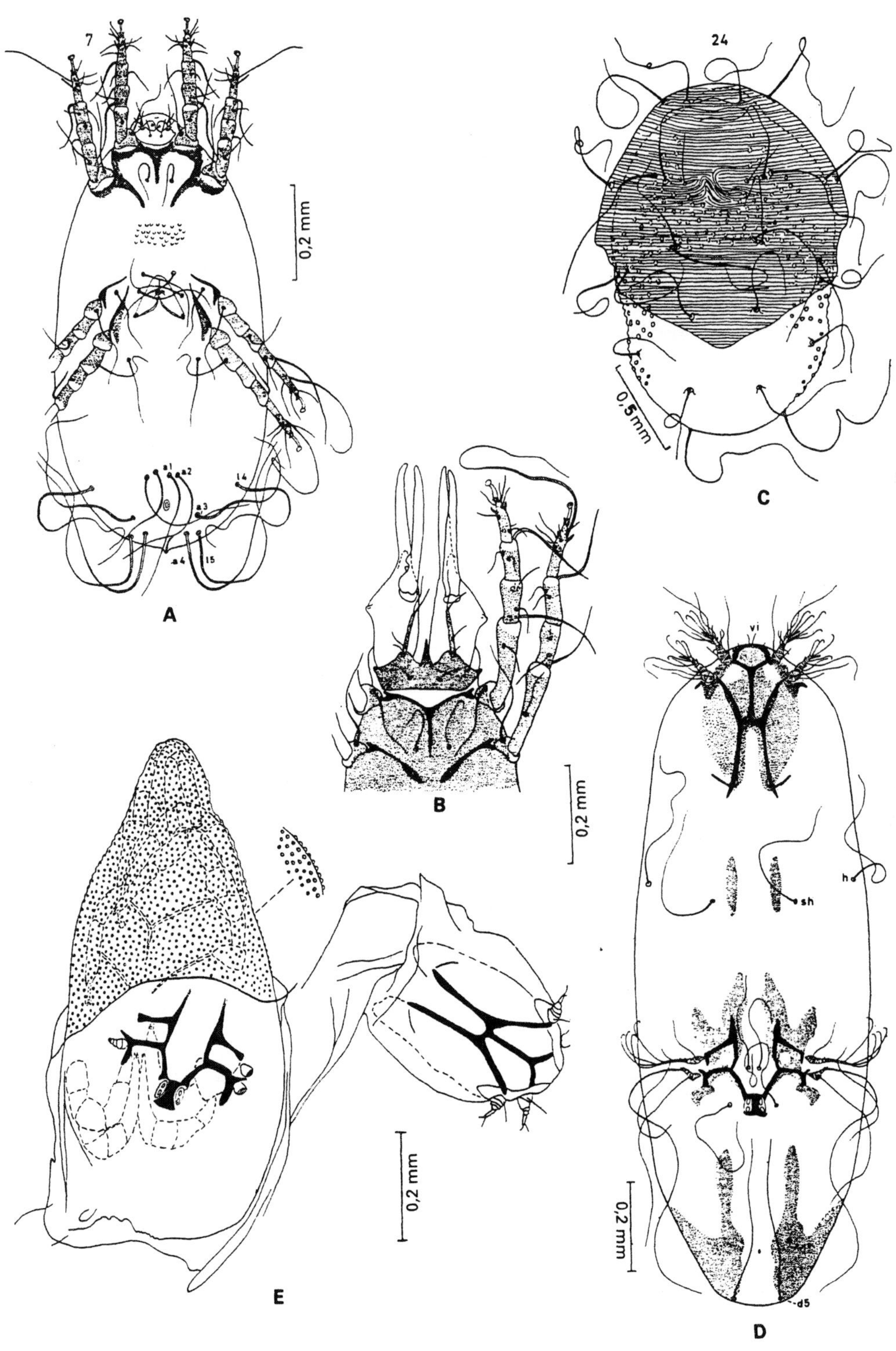
7
0,2 mm
a1
a2
l4
a3
a4
l5
A
B
0,2 mm
24
0,5 mm
C
vi
h
sh
0,2 mm
d5
D
0,2 mm
E

Acaridae and Glycyphagidae but are probably obligatory in the Anoetidae, Chaetodactylidae, Hypoderidae and Labidophoridae. They have not been found in the Canestriniidae (except in *Megacanestrinia*), Hyadesiidae, Linobiidae, Rosensteiniidae and Pyroglyphidae. In the genus *Acarus*, two types of hypopodes are present comprising a typical active elattostatic form in *A. siro* and *A. farris*, and an inert form with short hypotrichous legs that is capable of weak movements but incapable of supporting the body and with a reduced attachment organ in *A. immobilis* and *A. gracilis* (Griffiths, 1964a). Inert deuteronymphs also occur in some species of Glycyphagidae, and in *Glycyphagus domesticus* the calyptostasic hypopus resembles an oval cyst and is enclosed within the cuticle of the protonymph.

The most dramatic specialization of an astigmatic developmental cycle occurs in the hypoderid *Hypodectes propus*, an inhabitant of the nests of columbiform birds (Fain, 1969a). The hypopus of this species lives in the subcutaneous tissues of the pigeon and was once thought, erroneously, to represent the heteromorphic deuteronymph of the feather mite *Falculifer rostratus*. The larval and protonymphal stases are reduced to apoderma and surround the developing deuteronymph within the shell of the egg (Fig. 11.8). On emergence, the small deuteronymphs penetrate the skin of the nestlings, greatly increase in size and remain in the subcutaneous tissues until the host begins to incubate its eggs. At this time, the large hypopodes migrate through the subcutaneous tissues of the abdomen and pass through the skin into the nest. They then undergo rapid development into the adult phase but between the cuticle of the hypopus and the developing adult is the apoderma of the tritonymph. The female is non-feeding and the chelicerae are markedly reduced. Homomorphic and heteromorphic males occur in the nest; the latter being characterized by the massive form of the chelicerae in comparison with the normal type in the homomorph. It has not been established if the male feeds. The female produces numerous eggs which develop into hypopodes in the nest. The method by which the hypopus increases in size and accumulates sufficient food reserves for the laying of numerous eggs by the non-feeding female is not known.

Oribatida

All the oribatids have a calyptostasic prelarva but the larva, the three nymphal stases and the adult are normal (Grandjean, 1957a). The immature stases of the Euoribatida are always very different from the adults, particularly in the lesser extent of body sclerotization and the absence or poor development of idiosomatic tecta. In the Archoribatida (Lower Oribatida), with the exception of the Euptyctima, the immatures resemble the adults. The aptychoid immatures of the Euptyctima differ considerably from the ptychoid adults.

The prelarva of the Euoribatida is very regressive. It is usually ovoid and lacks appendages although vestiges of the mouth, pharynx, Claparède's organ and a pair of small denticles, considered to function as egg bursters, may be present (Fig. 10.12B). Prelarvae in the Lower Oribatida, on the other hand, are not so regressed and show vestiges of the chelicerae, palps and three pairs of legs. *Palaeacarus histricinus*, the least regressive form, has two pairs of long setae and the appendages are well developed although lacking segmentation (Lange, 1960; Travé, 1976). The larva emerges from the ruptured cuticle of the prelarva and from the shell of the egg if this remains closely associated with the prelarval cuticle.

Diapause

Dispause may be defined as a neurohormonally mediated state of low metabolic activity which can manifest as delayed morphogenesis, increased resistance to environmental extremes and altered or reduced behavioural activity (Tauber *et al.*, 1986). Unlike quiescence, which occurs in direct response to adverse physical conditions and is terminated as soon as favourable conditions return, diapause begins long before the commencement of unfavourable conditions and terminates well after the ending of such conditions (Beck, 1980). Biophysical and physiological mechanisms are involved in its initiation, maintenance and termination. Temperature, food, moisture and photoperiod have been implicated in diapause induction but photoperiod, including the light (photophase) and dark (scotophase) phases, is considered to be the most important. In the Acari, diapause is invariably facultative in that not every individual of every generation enters diapause. It may occur in the egg, larva, nymph or female but for each species the suppressed development normally occurs within a particular stage. The phenomenon has been studied in phytophagous tetranychoid and eriophyoid mites and their phytoseiid predators, and in ticks.

Diapause is a widespread phenomenon in the Tetranychidae but only occurs either in the egg or in the adult female (Veerman, 1985). Egg or embryonic diapause is found in the genera *Bryobia, Petrobia, Schizonobia, Aplonobia, Eurytetranychus, Panonychus, Schizotetranychus* and *Oligonychus* (except for a single species which displays adult diapause). Adult diapause occurs in *Eotetranychus, Neotetranychus, Platytetranychus* and *Tetranychus.* Differences between diapause and non-diapause forms in the spider mites often affect morphology, coloration, physiology and behaviour. Morphological differences occur in both egg and adult diapause. The summer eggs of *Petrobia latens* are subspherical in shape with faint ribbing, cherry-red in colour and produced at the top into a slender process, whereas the winter eggs are completely protected by a wax coating formed into a characteristic shape. Non-diapausing females of *Eotetranychus* and *Tetranychus* living in temperate and northern regions have distinct lobes on the integumentary striae of the dorsal surface of the body but winter forms lack such lobes (Pritchard & Baker, 1952).

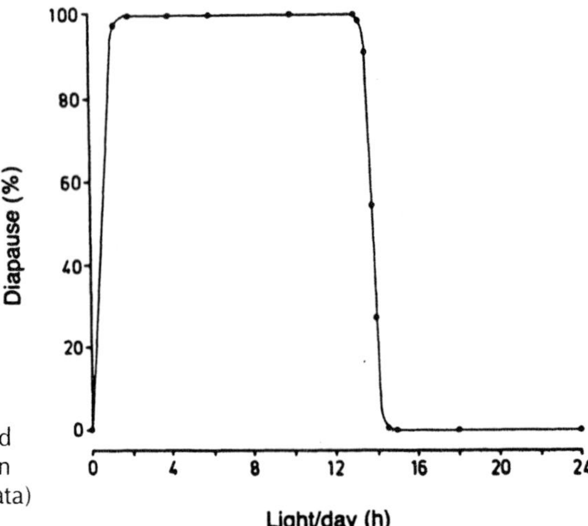

Fig. 11.9. Effect of photoperiod on the incidence of diapause in *Tetranychus urticae* (Prostigmata) (from Veerman, 1991).

Colour changes are a conspicuous feature of diapause in many spider mites and it is particularly evident in *Tetranychus urticae.* Summer females of this species are yellowish-green in colour while in winter the diapausing forms assume a bright orange coloration. The pigments involved are hydroxy-keto carotenoids which do not occur in the leaves of the plant host (Veerman, 1974). Physiological differences are seen in the amount of lipid or sugar present, and in the rate of respiration. Fat deposition in body tissues and decreased oxygen consumption are characteristic of diapausing *T. urticae* while sorbitol accumulation takes place in the diapausing eggs of *Panonychus ulmi.*

Mites induced to diapause or to lay diapause eggs often exhibit differences in behaviour compared with summer forms. In *P. ulmi,* for example, summer females are relatively inactive and often lay all their eggs on one leaf whereas those depositing winter eggs become highly active at the time of egg maturation and walk away from the leaves upon which they have been feeding to lay their eggs on the bark of the branch or in crevices in dormant buds.

Adult diapause in *T. urticae* is manifested in the arrest of feeding and oviposition. The induction response curve is of long-day type, i.e. long photophases favour continuous development whereas short photophases result in the incidence of diapause (Fig. 11.9). Veerman (1977a) gives the point of transition between very high and very low incidences of diapause (critical photophase) as 13 h 50 min under laboratory conditions. No diapause induction occurs in constant darkness but diapause incidence is almost complete at 1 h photophase:23 h scotophase. Veerman and Vaz Nunes (1987) have shown that it is essentially night length that is being measured by the photoperiod clock in this species although light plays an important part in the photoperiodic induction of diapause. These authors suggest that short-night and long-night regimes and critical night length are more appropriate terms than the currently used long-day and short-day regimes and critical day length. Dietary studies have indicated that vitamin A is essential for photoperiodic induction. It is possible that a vitamin A or a derivative serves as a photoreceptor pigment for the photoperiodic clock (Veerman, 1991).

At high temperatures (25°C for a Dutch strain) no diapause occurs at any photophase in *T. urticae* (Helle, 1962). Photoperiod sensitivity is maximal at the end of the first nymphal instar and declines during the succeeding nymphal instar. Termination of diapause depends on a period of chilling before the return of warm weather in spring but there is considerable variation in the period of chilling required to break diapause in individuals from different localities. For example, Lees (1953a) found that an English strain ended diapause after exposure to temperatures below 10°C for 100 days while in Leningrad forms only 55 days at 3–6°C were required (Bondarenko, 1958). Sensitivity to photoperiod remains in females during the first few months of diapause and has a far greater influence than temperature on reactivation (Glinyanaya, 1972; Veerman, 1977b). This sensitivity probably serves to retain the diapausing state through autumn and until the arrival of low temperatures in the winter.

The fruit tree red spider mite, *Panonychus ulmi*, overwinters in the egg stage. According to Lees (1953a,b) only winter females (those laying diapausing eggs) are produced at daily photophases of 6–13 h at a temperature of around 15°C but none is produced at photophases of 15–16 h. However, in times of restricted food supply winter females may be produced when other external conditions favour the production of summer forms. Unlike *T. urticae*, about 60% diapause occurs in continuous darkness. The second nymphal instar is the most sensitive to diapause-inducing factors although egg-laying females can be induced to produce the alternative egg by imposing antagonistic conditions. Egg diapause can be broken by chilling to a temperature of 1–9°C for 120–150 days. Winter eggs are slightly larger and more heavily pigmented than summer eggs, and the internal waterproofing wax layer secreted by the embryo is completed before oviposition whereas in summer eggs the layer is laid down about 6 h after oviposition (Beament, 1951).

Diapause is a feature of the life cycles of those eriophyoid mites living on deciduous plant hosts in the northern hemisphere. Two different types of female occur in the life cycle: the protogyne, which is a non-diapausing female, and the deuterogyne (deutogyne), which is the overwintering form of this sex. Factors inducing diapause do not appear to have been studied in detail. Easterbrook (1984) noted that the production of deuterogynes of *Aculus schlechtendali* and *Epitrimerus piri* seemed to be affected by leaf condition; their production began earlier when the leaves had been damaged by the feeding of large spider mite or rust mite populations early in the season. Diapause appears to be obligatory in some species (Schliesske, 1984). Deuterogynes cannot lay eggs in the same year as they appear but must experience winter chilling before they are able to reproduce. Males occur only with protogynes and resemble them in general appearance. In most species, marked differences are

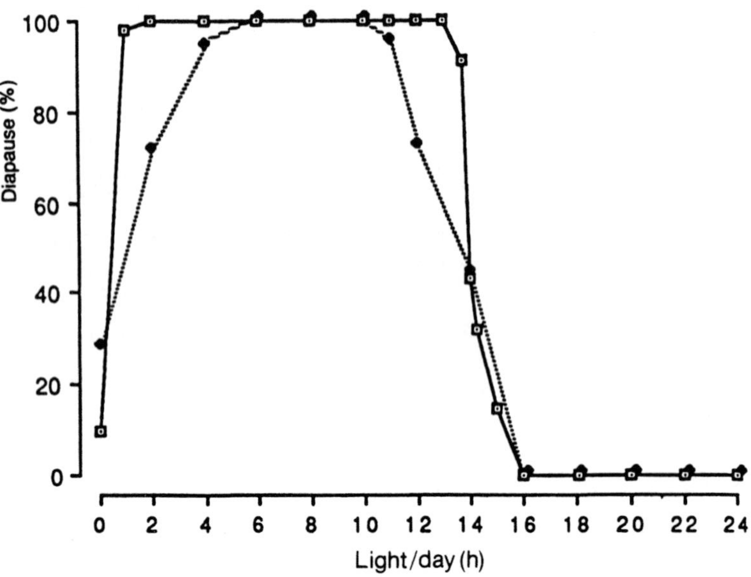

Fig. 11.10. Effects of photoperiod (solid line) and thermoperiod (broken line) on diapause induction in *Amblyseius potentillae* (Mesostigmata). For photoperiod the horizontal axis represents hours of light per day at 19°C. For thermoperiod, comprising a thermophase at 27°C and a cryophase at 15°C in constant darkness, the horizontal axis stands for hours per day at 27°C. After Veerman (1991) from data in Van Houten *et al.* (1988).

apparent in the external characteristics of the two forms of female. Deuterogynes may show reduced microtuberculation of the idiosoma, narrow tergites, and fewer ridges, furrows and projections of the body than conspecific protogynes.

Facultative adult reproductive diapause is photoperiodically induced in phytoseiid mites and is of the long-day type. According to Hoy (1975a,b), the critical photophase in *Metaseiulus occidentalis* is 11.2 h at 19°C and 11.6 h at 16°C. Diapause is entirely averted at 22–30°C and development is continuous under conditions of complete light or dark. Chilling does not appear necessary for ending diapause. The eggs are the most sensitive to diapause-inducing factors. Mites transferred as protonymphs into inductive conditions yield only 10% of adults in diapause while of those transferred as deuteronymphs or newly emerged females none enters diapause. In *Amblyseius potentillae*, the critical photophase in a Dutch strain was 14.5 h and this varied only little between 15.0 and 22.5°C. Sensitivity to photoperiod was found in all postembryonic phases of development but appeared to be maximal at the protonymphal instar (Van Houten, 1989). The mite remains sensitive to diapause-inducing and diapause-averting day lengths during the adult phase and a second diapause can be induced after completing the first. Photoperiodic induction was shown to be temperature dependent and, of particular interest, was the observation that diapause could be induced in continuous darkness by a thermoperiod, i.e. a temperature cycle comprising a warm phase (thermophase) and a cool phase (cryophase). Thus, from the thermoperiodic response curve in Fig. 11.10, based on a 27°C:15°C thermoperiod in darkness, thermoperiods with cryophases of from 14 to 18 h result in 100% diapause whereas in those of 8 h or less no diapause occurred (Van Houten *et al.*, 1988). It was found that photoperiodic and thermoperiodic responses in *A. potentillae* are dependent on the presence of vitamin A in the diet of the mites.

Diapause in ixodoid ticks ensures the synchronization of the life cycle with favourable seasons of the year and enables the species to resist unfavourable conditions such as drought, extremes of temperature and lack of food (Belozerov, 1982). It appears as delayed metamorphosis in engorged larvae and nymphs of a number of species of *Ixodes* and *Haemaphysalis*, as interrupted oogenesis in, for example, *Dermacentor marginatus* females engorging in summer and autumn, and as delayed embryogenesis in the egg stage. Unfed ticks may exhibit a behavioural change manifested by decrease in or suppression of host-questing activity and by a reluctance to feed even when placed on a host. This so-called 'behavioural diapause' occurs, for example, in the larvae, nymphs and females of *Ixodes ricinus*, *Dermacentor variabilis* and *D. pictus*. *Dermacentor* spp., in response to appropriate seasonal cues, do not quest or feed, even when placed on hosts (Sonenshine, 1988). In contrast, the phenomenon in nymphal and adult *Ixodes ricinus* results in reduced readiness to attack which does not prevent feeding on experimental animals, and appears to be readily disturbed by exogenous factors (Gray, 1991). For example, Gray (1987), working in Ireland, showed that behavioural diapause in the female of *I. ricinus* could be terminated by exposure to males.

Photoperiod, and to a lesser extent temperature, appear to be the predominant regulatory mechanisms of diapause in ticks. Working on a Leningrad population of *I. ricinus*, Belozerov (1967a) showed that larvae maintained at 18°C moult, after engorgement, into nymphs in long-day conditions (18–24 h photophase) but enter diapause in short-day conditions (9–14 h photophase). In those maintained at 25°C, however, non-diapause activity continued irrespective of photoperiod. Diapause also seems to be influenced by the age of the larva; the older ones show the greater tendency to diapause. Nymphs have a convertible diapause regulatory mechanism in which diapause in unfed forms is of the short-day type and in fed ones of the long-day type (Belozerov, 1967b). Under field conditions in continental Europe, ticks of this species feeding in spring respond to increasing day length during questing and feeding and thus develop without a morphogenetic diapause. After moulting, however, the majority show a behavioural diapause which prevents them from feeding again that year so that they overwinter as unfed ticks. Those ticks feeding in autumn respond to shortening day lengths which induces morphogenetic diapause. They resume development in the following summer and the next instar feeds again in the autumn. Thus, most individuals only feed once a year and a 3-year life cycle is usual in continental climates. In mild oceanic climates where the grazing season is long and winters are mild, selection pressures on ticks to develop overwintering mechanisms are not as strong as under the more vigorous conditions obtaining in continental Europe and, as in Ireland, behavioural diapause is less strongly developed and many spring-feeding ticks may feed again in the autumn (Gray, 1987).

Khalil and Shanbaky (1976) suggest that facultative reproductive diapause in *Argas arboreus* results from gonadotrophic hormone (GTH) deficiency in the presence of a diapause-inducing factor (DIF), which is formed under the influence of a short photophase in unfed females. DIF is degraded slowly during short photophases but rapidly during long photophases. The topical application of ecdysone can terminate larval diapause in *Dermacentor albipictus* (Wright, 1969).

Dispersal

Dispersal is an important aspect of the life of many Acari, particularly those that live in temporary habitats such as the nests of insects, birds and mammals, and accumulations of organic debris (compost, manure and tidal debris). In phytophagous species, dispersal mechanisms enable mites to spread over large areas of cultivated crops and to colonize widely separated plants as well as to escape from natural enemies. Structural, developmental, physiological and behavioural adaptations for dispersal are common and varied (Binns, 1982). The main methods of dispersal in mites are by air currents (anemohoria) or by utilizing other animals for transportation (zoohoria) although crawling is also a means of dispersal over relatively short distances as within a host plant or between a concentration of host plants.

Anemohoria

Mites have been trapped at altitudes of up to about 3000 m and aerial drift appears to be a common phenomenon among the Tetranychidae (Fleschner *et al.*, 1956). At least two different behavioural mechanisms are involved in aerial dispersal in this family (Kennedy & Smitley, 1985). Many species, for example *Panonychus citri, Oligonychus ununguis* and *Eotetranychus sexmaculatus*, descend from foliage on silk threads produced from a proteinaceous secretion of a unicellular silk gland associated with each pedipalp which opens through a hollow spigot-like seta on the palptarsus (Alberti & Storch, 1974). The threads are attached to a substrate and are caught in air currents which break them and carry the mites aloft. This type of behaviour is often referred to as 'ballooning', a term which is used for a specific type of aerial dispersal mechanism in spiders. In the latter, silk issuing from the spinnerets of the raised opisthosoma is drawn out by a current of air and carries the spider with it. Since the two mechanisms are not the same and to avoid confusion, Kennedy and Smitley (1985) consider that the term ballooning should not be used to describe the behaviour in mites. A more complex type of behaviour occurs in *Tetranychus urticae* and involves the adoption of a distinct dispersal posture by the female, less frequently by the immature instars but rarely by the adult male (Kennedy & Smitley, 1985). The mite, facing the wind, holds the forelegs in an upright position and is carried aloft; one in the normal standing posture is not affected. Desiccation, population density and the condition of the host plant are involved in conditioning the response with desiccation probably being the most significant. The dispersal posture usually happens when the mites are concentrated in the upper parts of the plant and on the leaf apices due to increased activity and a stronger response to light resulting from a low ambient relative humidity caused by depletion and desiccation of the food source (Suski & Naegele, 1963; McEnroe & Dronka, 1971).

Dispersal by air currents also occurs in the Eriophyoidea and the adult, more commonly the female, is the disperser. The mites adopt a dispersal posture by rearing up on their terminal lobes and caudal setae while waving their anterior legs (Jeppson *et al.*, 1975). Occasionally numbers of mites form chains by crawling on each other. Nault and Styer (1969) consider wind, temperature and light to affect dispersal in *Eriophyes tulipae* infesting grasses. For example, an increase in temperature from 12 to 24°C resulted in an eight-fold increase in the numbers trapped.

The adoption of a disperal posture has been observed in the phytoseiid *Amblyseius fallacis* (Johnson & Croft, 1976). Under laboratory conditions, it occurred in pre-ovipositing and ovipositing females and to a lesser extent in males. When the wind speed exceeded about 1 mph (0.45 m s^{-1}), the mites altered their behaviour and this was seen in the change from a random searching movement to a directional movement towards the edge of the experimental arena followed by the termination of all forward movements and orientation towards the air flow. At this stage, the mites stopped waving their forelegs and assumed a stance with the forelegs, second pair of legs and the anterior part of the body off the ground and with the third and fourth pair of legs giving support to the body. The mite was orientated parallel to but facing away from the direction of the air flow. Among the biotic factors involved in the initiation of aerial dispersal in predatory phytoseiids, prey density is very important. Johnson and Croft (1981) found that in apple orchards predator dispersal by air currents and by crawling down tree trunks to the ground occurred when the numbers of prey became a limiting factor to the reproduction of the predator. Hoy *et al.* (1985), working with *Metaseiulus occidentalis* in Californian almond orchards observed most aerial movements to occur between 16.00 and 22.00 hours when wind speed and relative humidity increased and temperature decreased.

According to Sabelis and Dicke (1985), the dispersal posture is not essential for successful take-off in *A. fallacis* and the posture is never exhibited by *Phytoseiulus persimilis.*

Aerial dispersal is also considered to occur in the calyptostasic deuteronymph (inert hypopus) produced by some Astigmata such as *Glycyphagus* and *Lepidoglyphus.* Zachvatkin (1941) suggested that feather, leaf or fan-shaped setae on the body of active immatures and adults of some astigmatic species acted as 'sails' for wind transportation, but this requires confirmation.

Zoohoria

The use of other animals for transportation and disperal commonly occurs in mites, particularly those living in habitats which are discrete and temporary. One type of association between the mite and its carrier is referred to as phoresy and has been defined by Farish and Axtell (1971) as:

> a phenomenon in which one animal actively seeks out and attaches to the outer surface of another animal for a limited time during which the attached animal (termed the phoretic) ceases both feeding and ontogenesis, such attachment presumably resulting in dispersal from areas unsuited for further development, either of the individual or its progeny

Phoresy in the strict sense of these authors is of widespread occurrence in the Mesostigmata, Astigmata and Tarsonemina. Another type of association leading to dispersal is more intimate and the mite obtains sustenance from the host during transit. It is most widespread in the Parasitengona in which the parasitic larva is the dispersal phase and the deuteronymph and adult are free-living and usually predatory. Data on the attachment of larvae of water mites to dragonflies show highly non-random, 'clumped' distributions and Binns (1975a) considers this to indicate that the association with the host is primarily dispersive, as in phoresy, rather than parasitic.

Phoresy

This is the most common method by which mites disperse from and colonize temporary accumulations of organic matter. Monocultural plant stands and ecological climaxes, on the other hand, are not characterized by phoretic associations (Binns, 1982). Early phoretic colonizers of temporary habitats arrive as passengers on various co-inhabitating arthropods or other animals. Although the location of suitable habitats is determined by the behaviour of the host, the final selection of a habitat is probably made by the phoretic in the presence of a stimulus which leads to detachment. Most species practising phoresy are *r*-strategists as evidenced by their rapid development and high rates of reproduction and this enables them to rapidly colonize and exploit temporary habitats.

The extent of structural, life cycle and behavioural adaptations exhibited by phoretic mites varies considerably. The phoretic in most Dermanyssina shows little or no structural adaptation and resembles the equivalent instar in a confamilial species which does not practise phoresy. At the other extreme, the phoretic instar in the Astigmata is highly specialized in structure, behaviour and physiology and is of a very different life form from the non-phoretic instars in its life cycle. The main structural adaptations for phoresy are seen in the development of organs for attaching to the host such as the enlargement of the claws of the first pair of legs as in some Tarsonemina, the modification of certain setae of the opisthogastric region to form suckers or claspers as in some hypopodes of the Astigmata (see Fig. 11.7 above and Fig. 11.13 below) and the secretion of an anal anchoring stalk as in the phoretic deuteronymph of some Uropodina (see Fig. 11.11A below). Life cycle adaptations are most spectacular in those species in which the phoretic is an alternative to a non-phoretic instar, as in the phoretomorphic deuteronymph of some Uropodidae, or is a facultative instar as in the case of the elattostasic deuteronymph (hypopus) of some species of Astigmata. In many species, the life cycle of the mite is synchronized to that of the host species and may involve accelerated or suspended development of the phoretic instar. Pheromones or hormones produced by the host appear to be involved in the synchronization. Facultative parthenogenesis is a relatively common phenomenon in phoretic females, particularly in the Mesostigmata, and enables a rapid colonization of new habitats by a single or few individuals. Adaptations in behaviour are seen in the increased activity of some phoretics as they seek out a host, contrasting with their inactivity when on the host, and in the decreased activity of other phoretics as they assume a questing or perching posture when lying in wait for a host. Cessation of feeding during transit on the carrier, and often for a time before transit, is a characteristic of the majority of phoretic forms.

The associations between the phoretics and their hosts range from the unspecific or euryxenous, in which the mites utilize a variety of carrier species, to the specific or stenoxenous, where the mites use a single host species for dispersal. The latter type of relationship commonly obtains when the mite inhabits the nest or brood chamber of the carrier and occurs in both mite–vertebrate and mite–insect phoretic associations. Examples of stenoxenic associations between mites and vertebrates are found among the endofollicular hypopodes (Ctenoglyphidae and Labidophoridae) of rodents and the subcutaneous hypopodes (Hypoderidae) of birds. Stenoxenic mite–insect associations occur, for example, between species of *Dinogamasus* and xylocopid bees of the genus *Mesotrichia* and between *Poecilochirus* and silphid beetles of the genus *Nicrophorus*.

MESOSTIGMATA

Phoresy is relatively common in the Mesostigmata (Costa, 1969; Lindquist, 1975). It is not unusual to find a number of phoretics of different species on the same host. For example, Hyatt (1959) recorded up to six species occurring on the dung beetle *Geotrupes stercorarius* with one beetle carrying 488 mites comprising 188 deuteronymphs of *Parasitus coleoptratorum*, 147 adults of *Alliphis halleri*, 141 females of *Macrocheles glaber*, five deuteronymphs of *Parasitius intermedius*, four females of *Scamaphis equestris* and three adults of *Scarabaspis inexpectatus*. The *Parasitus* and *Macrocheles* were found on the surface of the body of the beetle and the remainder under the elytra. None of these mites exhibits a specialized body form or conspicuous attachment organs. This is usually, but not invariably, the case of phoretics in the Antennophorina, Parasitina, and Dermanyssina (Eviphidoidea, Ameroseiidae, Ascidae and Digamasellidae). A notable exception occurs in *Antennoseius (Vitzthumia) janus*, which has two morphologically distinct types of female. The non-dispersing female has a granular integument, barbed setae, invaginated dorsal shield and fully developed sternal shield, whereas the dispersing female has a soft opisthonotal integument, smooth setae, an entire dorsal shield and reduced sternal shield (Lindquist & Walter, 1989). Many *Macrocheles* attaching to the body bristles of scarabaeid beetles by their cheliceral digits have a remarkably constant digital dentition including a strong bicuspid tooth on the movable digit which opposes a deep concavity on the fixed digit. The bristle lies in the concavity and is held firmly by the bicuspid tooth (Evans & Hyatt, 1963). More conspicuous adaptations occur in the females of *Hoploseius* (Lindquist, 1975) and the deuteronymph of *Gamasodes* in which, respectively, the fourth and the second pair of legs are enlarged and provided with opposable hypertrophied setae for attaching to the host.

Only the female is phoretic in the Macrochelidae and she is facultatively arrhenotokous or, more rarely, thelytokous. This enables the establishment of the species within the new habitat by a single or few individuals (Filipponi, 1955). In some species, both sexes and deuteronymphs are involved in dispersal, for example, *Alliphis halleri* on dung beetles and *Thinoseius brevisternalis* on sandhoppers (Evans, 1963b). In others, such as the Parasitinae, only deuteronymphs participate in phoresy. For example, in the stenoxenic association between species of *Poecilochirus carabi*-complex and burying beetles of the genus *Nicrophorus* spp., the mites reproduce in the brood chamber or crypt which the beetles build around the buried carcass of a small vertebrate and the majority of the phoretic deuteronymphs are transported by the parent beetles when they leave the crypt in search of other small carcasses in order to reproduce further (Korn, 1982; Schwarz & Müller, 1992). Deuteronymphs on overladen carriers tend to transfer between burying beetles which meet at carcasses.

Although most phoretic Dermanyssina appear to be predominantly nematophagous, some species actually predate the eggs and/or early developmental instars of their host species. *Macrocheles muscaedomesticae* feeds on the eggs and pre-imaginal instars, except puparia, of houseflies in manure (Filipponi, 1955; Jalil & Rodriguez, 1970) and *Arctoseius cetratus* predates eggs and first instar larvae of the sciarid *Lycoriella auripila* in mushroom compost (Binns, 1972). Females of *A. cetratus* attach, probably by their chelicerae, to the pleural folds of the anterior region of the abdomen of the host and a maximum of 26 mites was found on one fly. There is no evidence that the phoretic feeds on the host. This association between the mite and its insect host is considered to be host specific.

The behaviour of *Blattisocius tarsalis* and one of its hosts, the herminiine noctuid *Epizeuxis aemula*, is particularly interesting. The larvae of this North American moth feed on dead leaves on the forest floor but, on occasion, have been reported to be injurious to corn fodder. Over a seven-year period (1967–73), Treat (1975) found that about 24% of the moths collected by him at 'black light' were infested with adults of *B. tarsalis*. In rearing experiments of the mite and its host, it was found that the pupa of the moth was seldom visited by the female mite until some hours before the emergence of the moth. At this time, female mites assembled on or in the silken cocoon surrounding the pupa and came to rest near the head end. The few male mites that assembled moved about and attempted to copulate with the females but copulation did not take place and it was assumed that the females had been inseminated before moving onto the pupa. Eclosion of the adult moth was extremely rapid but as soon as the pupal skin split the mites moved onto the strip of exposed surface of the moth in the cleft lateral to the antennae and were later observed on the body of the fully emerged moth. Treat suggested that the highly evolved behavioural pattern displayed by the mites may be in response to a kairomone produced by pharate adult moths, particularly since they were able to distinguish between cocoons containing pharate adults from those containing younger pupae. The preferred sites of occupancy appeared to be at the bases of the hind wings on the dorsolateral surface of the third thoracic and first abdominal segments but, unlike *M. muscaedomesticae* and *A. cetratus*, the mites feed on the host, as evidenced by feeding scars in the occupied sites. Mites on larvae of the moth were observed ingesting tissue fluids. The chelicerae with their reduced fixed digits and well-developed dentate movable digit are adapted for piercing the cuticle of the host. It is not known whether *B. tarsalis* shows the same behavioural patterns on its common microlepidopteran hosts infesting stored grain.

A close relationship between the mite and its host also occurs in those species which live in abdominal pouches or acarinaria of their host, such as species of *Dinogamasus* on female carpenter bees (Skaife, 1952). The mites leave the acarinarium when the bee begins egg-laying and undergo an inactive, non-feeding phase in the cells of the host's progeny. They only resume activity when the developing bees reach the quiescent

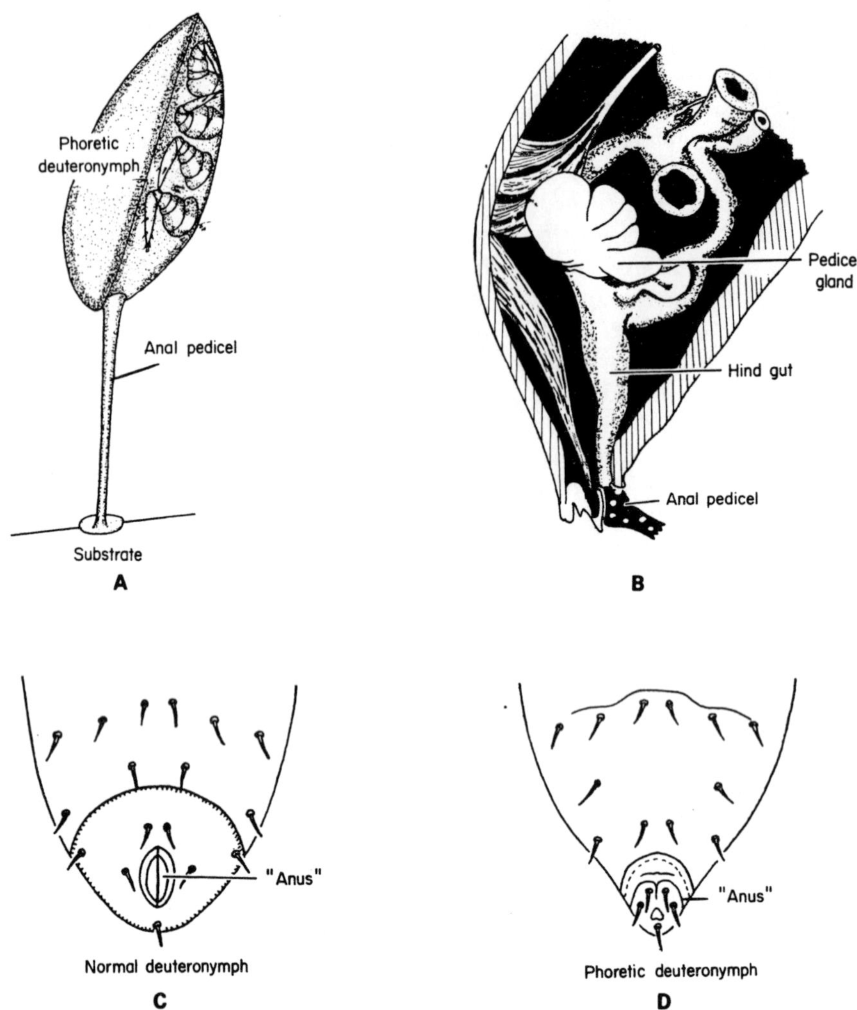

Fig. 11.11. **A**, phoretic uropodine deuteronymph attached to a substrate by its anal pedicel; **B**, sagittal section of the posterior region of the idiosoma of a uropodine mite showing location of the pedicel gland; **C, D**, anal region of a normal (**C**) and phoretic (**D**) deuteronymph of *Microsejus* sp. (after Evans & Till, 1979).

prepupal stage. They run over the surface of the prepupae and apparently feed, without causing harm, on integumentary exudates produced by the prepupae. The mites die if separated from the bees at this time. At pupation the female mite begins to lay eggs and during the period of pupal development the entire life cycle of the mite is completed rapidly so that newly emerged female bees leave the nest with young female *Dinogamasus* in their acarinaria. The mechanism by which the development of the mite attains synchrony with that of its host is unknown.

The phoretic in the life cycle of the Uropodina and Microgyniina is a specialized deuteronymphal instar which is produced as an alternative to the normal deuteronymph under certain environmental conditions (Faasch, 1967; Evans & Till, 1979; Athias-Binche, 1984). In some species all the deuteronymphs are phoretomorphs (*Uropoda orbicularis*) while in others normal and phoretic deuteronymphs are produced (*Fuscuropoda marginata*). The phoretic deuteronymph ('Wandernymph') may be readily distinguished from the normal form by the structure of the anal region, which is associated with the production of an anal pedicel or stalk produced from secretions of a gland opening into the postventricular region of the alimentary tract (Fig. 11.11A,B). In early literature, the stalk was referred to as the uropod, hence the generic name *Uropoda*. The stalk is used to attach the mite to the cuticle of its arthropod host and the gland producing it only occurs in the phoretic nymph. The anus in the phoretic is enlarged and covered by a single disc-like plate bearing setae whereas that of the normal nymph is much smaller and covered by two anal shields which lack setae (Fig. 11.11C,D). Camin (1953b) refers to a putative tritonymphal instar in the life cycle of *Dyscritaspis whartoni* (Polyaspididae) but the so-called 'tritonymph' has the anal characteristics of a phoretic nymph and probably represents the phoretic deuteronymph as in other Uropodina. There is certainly no evidence at present to indicate the occurrence of three *successive* nymphal instars after the larval in the life cycle of the Uropodina, or in other Mesostigmata for that matter, and it is misleading to apply the term tritonymph to a nymphal instar in this group.

Species of Uropodina which produce phoretic deuteronymphs inhabit non-static habitats such as manure, compost, rotting tree stumps and tree holes. In *Fuscuropoda marginata*, studied by Faasch (1967), the phoretic deuteronymph is more resistant to drought and lack of food than the normal form and has a more rounded pulvillus and proximally directed claws on legs II to IV which probably enable it to cling to the smooth surface of its beetle host before secreting the anal pedicel. The phoretomorph shows a preference for species of *Geotrupes* which it seeks out and detects by the sensilli of the tarsi of the first pair of legs. Numerous individuals are often found attached by their anal pedicel to the cuticle of the beetle's body and appendages. Faasch found that the stimulus provided by fresh manure induces the nymphs to detach and after a period of two to three weeks they moult into adult male or female.

PROSTIGMATA

Among the Prostigmata, phoresy is a relatively common phenomenon in the Tarsonemina and the phoretic is usually a female. In the Tarsonemidae it occurs in those species associated with insects, such as members of the genus *Iponemus* on ipine beetles and *Tarsanonychus emblematus* under the elytra of hispine beetles of the genus *Botryonopa* (Lindquist, 1986). There do not appear to be any marked morphological adaptations for phoresy in this group. Dimorphic females, comprising normal and phoretic females, occur in the Pyemotidae (Moser & Cross,

1975), Scutacaridae (Norton, 1977) and Pygmephoridae (Clift & Larsson, 1987). The differences between the normal and phoretic females are sometimes so great that the two morphs have been classified in different genera and even in different families. This applies to species of the pygmephorid subgenus *Pediculaster* of the genus *Siteroptes* in which the phoretomorph differs from the normal female in the shortened tarsus of leg I being fused with the tibia and the claw being sessile and not pedunculate. Differences are also seen in the form and number of the setae of certain podomores (Martin, 1978).

The phoretic association of Scutacaridae with insects has been studied by Cross and Bohart (1969), Norton and Ide (1974) and Binns (1979). The latter, working with *Scutacarus baculitarsus* and the phorid *Megaselia halterata*, found that prior to attachment to its host the female mite adopted an upright perching stance – the 'ambush' position – after a short searching period. In this stance, the anterior legs were held out and often moved in a questing manner. Attachment behaviour was elicited through contact by means of the outstretched legs and the mites clustered round the anterior region of the abdomen of both sexes of the fly so that their bodies overlapped like tiles on a roof. The phoretomorphs appeared to be capable of surviving for considerable periods without feeding. The frequency of attachment of phoretic females in samples of flies taken from a number of different sources around a mushroom farm showed a negative binomial distribution. Binns (1975b) considers this type of distribution to be characteristic for phoretic mites. Clift and Larsson (1987) have found the same type of distribution of the pygmephorid *Brennandania lambi* on mushroom flies in Australia. This species detached from *M. halterata* in the presence of *Agaricus* fungus but remained attached when the fungus was absent. According to Clift and Toffolon (1981) the mite only reproduces in the presence of mycelium of *Agaricus* spp.

Phoresy is relatively uncommon in other Prostigmata. Van Eyndhoven (1964) gives an interesting account of the occurrence of females of *Cheletomorpha lepidopterorum* (Cheyletidae) on a noctuid moth (*Peridroma saucia*). A total of 227 mites was found on the moth and almost all occurred on the wing surfaces. The mites were attached by their anal region and it was thought that they secrete an adhesive substance which sticks them to the scales. Species of the tydeid genus *Pronematus* have also been recorded from noctuids and *P. pyrrohippeus* is found in apparently undamaged tympanic recesses of the host (Treat, 1975). The association between these mites and the moths is considered to be phoretic. The big bud mite, *Cecidophyopsis ribis*, spreads from bush to bush by clinging onto insects and other arthropods and the adult mite has been found on the larvae and adults of coccinellid beetles, hive bees, currant aphids and spiders (Massee, 1928). The mites are capable of leaping short distances to grasp the bodies of passing insects (Warburton & Embleton, 1902).

ASTIGMATA

In the Astigmata, phoresy is common and widespread among free-living species. The phoretic is an elattostatic deuteronymph (hypopus) which may be a facultative or an obligatory instar in the life cycle. It is more resistant to adverse temperature and humidity conditions than other stases in the developmental cycle (Woodring & Carter, 1974). It is probably more specialized morphologically for phoresy than any other form of phoretic mite. The chelicerae are lacking and the remainder of the gnathosoma is much reduced. There is no mouth or apparent functional alimentary tract so that the hypopus is non-feeding. The nervous system, on the other hand, is well developed. The body is usually well sclerotized and dorsoventrally flattened with the short, stout legs having a modified chaetotaxy. A conspicuous feature of most hypopodes is seen in the modification of the paraproctal region to provide organs for attaching to arthropod or vertebrate hosts (see Fig. 11.7 above). The attachment organs in acaroid hypopodes associated with arthropods are usually located on a rigid plate-like structure and have been studied in detail for *Sancassania boharti* by Woodring and Carter (1974). The plate has two pairs of suckers more or less surrounding the anus, two pairs of discs or conoides (one lateral and one paramedial) and five oval or circular dark areas representing the sucker plate apodemes which give rigidity to the plate (Fig. 11.12A). Both the suckers and discs contain birefringent cuticular elements and are considered to be modified setae. The discs may have a sensory function. The four suckers are attached by a flexible cuticular ring to a flange or socket in the plate. Retractor muscles originating on the dorsal body wall insert into the centre of each sucker and the inward bowing of the centre of the sucker resulting from their contraction is considered to create the vacuum necessary for their functioning

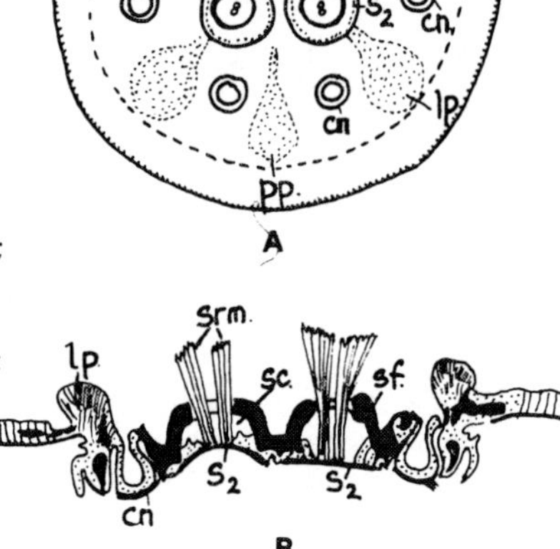

Fig. 11.12. A, anal sucker plate of the deuteronymph of an acarid mite (Astigmata); **B**, transverse section at level of the posterior pair of suckers (s_2) (based on Woodring & Carter, 1974). *an.* = anus; *ap.* = anterior sucker plate apodeme; *cn.* = conoid (discs); *lp.* = lateral sucker plate apodeme; pp. = posterior sucker plate apodeme; s_1, s_2 = suckers; *sc.* = cavity of sucker; *sf.* = sucker support flange; *srm.* = sucker retractor muscles.

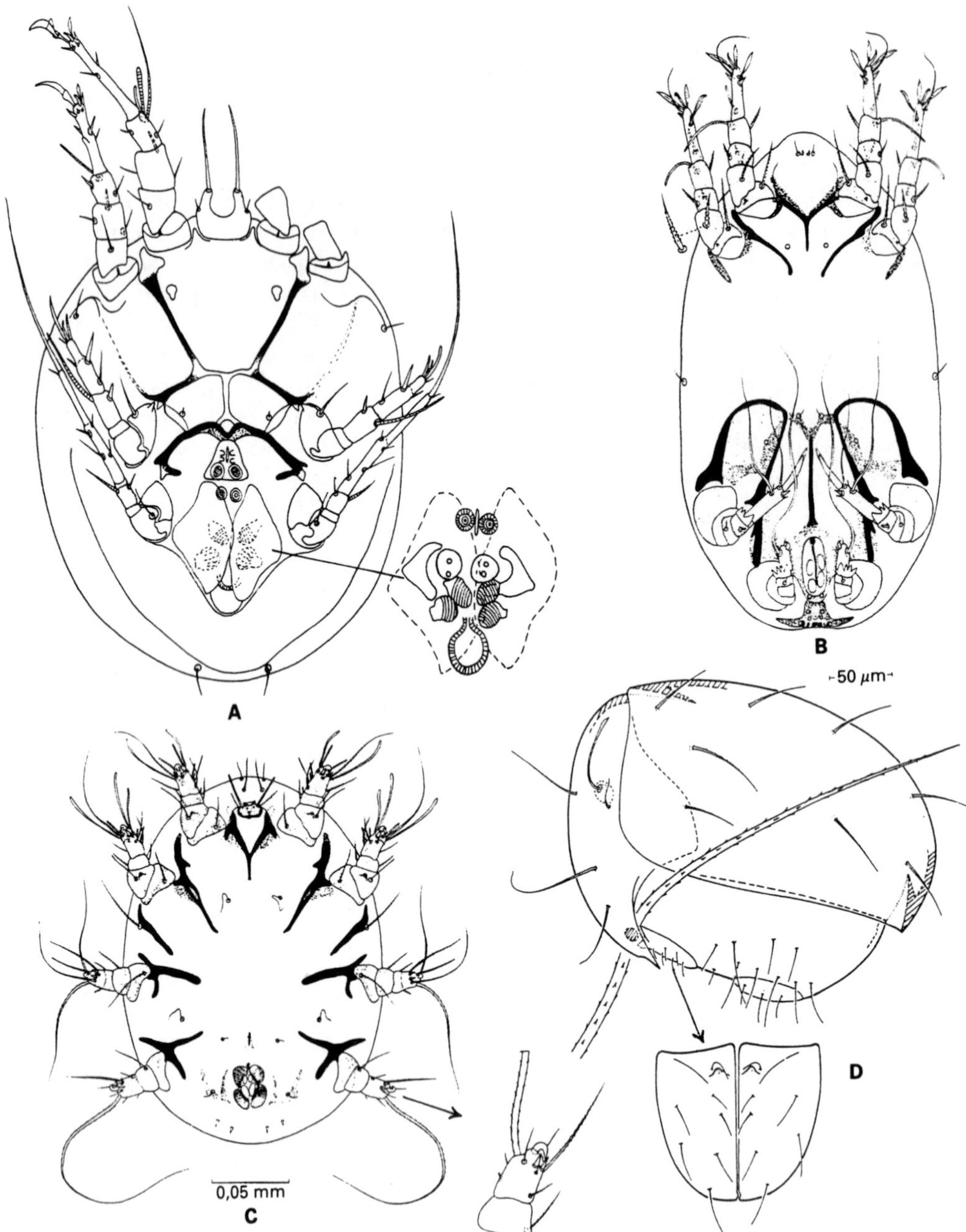

Fig. 11.13. **A**, venter of of the deuteronymph of *Fibulanoetus longitarsis* (Astigamata) showing form of the clasping organ; **B**, venter of the deuteronymph of *Rodentopus claviglis* illustrating the fan-shaped tibial setae on legs III and IV; **C**, venter of the deuteronymph of *Echimyopus brasiliensis* showing the ambulacral claw opposing spines on the tarsus; **D**, method of attachment of *Mesoplophora* sp. (Oribatida) to insect bristle. (A, after Fain *et al.*, 1980; B, C, after Fain, 1969b; D, after Norton, 1980.)

(Fig. 11.12B). It is possible that the very flat external surfaces of the five apodemes contribute to the mechanism of attachment.

On the basis of the method of attachment to and location on the host, the elattostasic hypopodes may be divided into four main groups (Fain, 1969b, 1971): entomophilous, pilicolous, endofollicular and subcutaneous. The organs of attachment of the entomophilous hypopodes are located on a sucker plate of the form described above. This type of structure is adapted for attaching the mite to the smooth surface of the cuticle of insects and other arthropods and is characteristic, for example, of the Acaridae, Anoetidae (=Histiostomatidae), Chaetodactylidae, Hemisarcoptidae and Winterschmidtiidae. The hypopus in the Anoetidae, which appears to be an obligatory instar, is exceptional in having the two posterior pairs of legs directed more or less anteriorly (see Fig. 11.13A). According to Hall (1959) these legs are used for jumping onto the host. Entomophilous hypopods are usually very active and tactile and chemosensory sensilli of the forelegs are used in host location and selection. Some hypopodes are known to be attracted to their hosts by chemical cues (Carpenter & Greenberg, 1960). Host selection has developed to a high degree in some species; for example, in the acarid *Naiadacarus arboricola* inhabiting water-filled tree-holes and phoretic on syrphid flies, the hypopodes attach almost exclusively to female flies which, unlike males, return to tree-holes. In *Hericia hericia* living in sap exuding from trees, the production of hypopodes in May and June coincides with the time that their host flies visit the sap (Robinson, 1953). The synchronization of the developmental cycle of the mite with that of the host also occurs in winterschmidtiid mites associated with wasps (Vespidae) and is seen in the appearance of hypopodes in the nest coinciding with the emergence of adult wasps (Krombein, 1961). The factors which induce detachment from the cuticle of the host do not appear to have been investigated in any depth but probably involve chemical cues as in the case of the uropodid phoretomorph. Host oviposition has been observed or inferred to be a stimulus for detachment in a number of species; for example, rhythmic pulsations of the abdomen of the female wasp during oviposition appear to stimulate detachment of the hypopus of *Vespacarus* so that the mite drops into the nest or into a new cell (Krombein, 1961). In the case of *Histiostoma polypori* whose hypopodes reattach to the same host (an earwig) with successive moults, chemical changes in the host's cuticle are thought to provide the detachment stimulus (Behura, 1956).

The relationship between the entomophilous deuteronymph of *Hemisarcoptes cooremani* and its coccinellid host *Chilocorus cacti* is particularly interesting in that the mite changes in colour from amber to cream and in body shape from flattened to swollen while in contact, subelytrally, with the host (Houck & OConnor, 1991). Radiolabelling experiments showed that the mite acquired materials directly from the beetle and the source of nutrients appeared to be reflexed beetle haemolymph which the insect produces to defend itself against predators.

It is possible that the entry of nutrients into the mite is by way of a vestigial anal opening into the hind gut and that the discoidal 'suckers' of the attachment organ play a role in this process but this requires further study. These authors suggest that the deuteronymph of *H. cooremani* represents a transitional form (a 'semiparasite') from a free-living to a parasitic state.

In the pilicolous hypopodes, the sucker plate is replaced by a complex organ adapted for clasping mammalian hair (Fain, 1969b). It comprises a pair of superficial flaps or valves covering two pairs of modified setae (see Figs 11.7B and 11.13A above). The setae are clubbed and provided with transverse ridges. The host's hair is held between the two pairs of clubbed setae and the contraction of the flaps by muscular action compresses the setae so that the hair is firmly held. Most of the species with pilicolous hypopodes are inhabitants of nests of mammals and belong to the families Glycyphagidae (*Dermacarus, Glycyphagus, Myacarus, Zibethacarus*) and Labidophoridae (*Labidophorus, Xenoryctes*). An exception occurs in the hypopodes of the anoetid genus *Fibulanoetus* in which both a pilicolous clasping organ and a sucker plate are present (Fig. 11.13A). This species attaches to the hairs of scarabaeid beetles by means of the clasping organ, which is similar to that in the Glycyphagidae (Fain *et al.*, 1980). The reduced sucker plate has a pair of small adanal suckers anterior to the flaps of the clasping organ and two deeply situated rounded structures between the suckers and the anterior pair of claspers. Fain *et al.* (1980) consider the retention of adanal suckers by these species to indicate a more primitive condition than that found in the pilicolous glycyphagid hypopodes in which only the clasping organ is present. The similarity in the form of the attachment organ of *Fibulanoetus* and pilicolous Glycyphagidae is probably the result of convergent evolution.

The endofollicular hypopodes, with the exception of *Pedetopus* which has a rudimentary sucker plate, have neither a sucker plate nor a hair clasper situated in the opisthogastric region of the idiosoma. They live in the hair follicles of rodents, principally in those of the region of the tail. In the *Rodentopus*-type occurring in the Ctenoglyphidae and Labidophoridae, the tibiae of legs III and IV have strongly modified setae that are flattened, fan-shaped and have their distal margin provided with spines (Fig. 11.13B) while in the *Echimyopus*-type, the opposition of two spines and the tarsal claw on the legs forms a simple pincer (Fig. 11.13C). These structures are considered to play a role in securing the hypopodes within the hair follicle. The hypopus of *Melesodectes auricularis* is unusual in that it is found in scrapings and wax from the ears of badgers. The strong 'knife-like' claws of the hypopus (typical also of many Hypodectinae) may be used to penetrate the epidermis of the newly born badger (Lukoschus & De Cock, 1971).

Organs of attachment are entirely lacking in subcutaneous hypopodes of the family Hypoderidae which live under the skin of birds (Hypoderinae) or rodents (Muridectinae). The life cycle of *Hypodectes propus*, whose hypopodes occur in the tissues of pigeons, has been described by Fain and Bafort (1967) and is discussed above (see also Fig. 11.8 above). The synchronization of the development of the mite with the reproductive cycle of the host occurs in *H. propus* and it is considered that this is dependent on the hormone prolactin, which is present in the host tissues at the time of brooding. It is probable that this type of synchrony is of widespread occurrence in subcutaneous and endofollicular hypopodes and that it ensures the dispersal of the mite through the infestation of young before they leave the nest. Although lacking feeding organs and a functional alimentary tract, the hypopus of *H. propus* markedly increases in size during the period in the host tissues (from a body length of 150 to 1500 μm). The mechanism by which this is achieved is not known but it is possible that nutrients are absorbed through the cuticle of the mite.

ORIBATIDA

Phoresy in the Oribatida has been reviewed by Norton (1980). Adults of *Mesoplophora, Paraleius, Metaleius, Lincnocepheus, Oppia, Euscheloribates* and *Tectocepheus* have been recorded from Coleoptera, especially passalids. A species of *Metaleius* has also been recorded from Diptera and one of *Mesoplophora* from Dictyoptera. On passalids, *Metaleius* occupies the depressed region between abdominal sternites I and II while the small *Oppia* is restricted to integumentary grooves irrespective of the region of the body on which it occurs. The species of the ptychoid genus *Mesoplophora* cling to the bristles of their host. The attachment to the bristle is achieved in one of two ways, either by clasping the middle of the bristle between the rostrum of the aspis and the anterior portion of the genital plates (Fig. 11.13D) or, on shorter hairs, by clasping the bristle terminally. In the former method a tooth-like tubercle directed antero-ventrally on each genital plate helps to lock the bristle in position and up to nine individuals have been found attached in a linear manner to one bristle. In one instance of terminal attachment, the bristle was swallowed by the mite so that its tip extended into the pharynx.

Dispersal in the Parasitengona

Unlike phoresy in which the phoretic normally does not feed on the host, dispersal in the Parasitengona is achieved by means of a parasitic larva which becomes firmly attached to the host by its mouthparts. The larva is a heterostase and with few exceptions, for example the mussel parasites *Najadicola* and *Unionicola*, it is the only parasitic instar in the life cycle. The larvae (chiggers, red-bugs, harvest mites) parasitic on vertebrates are predominantly members of the family Trombiculidae and although most commonly occurring on mammals also parasitize birds, reptiles and amphibians. The nymph and adult in this family feed on eggs and early larval instars of small arthropods. Larvae of the Erythraeidae and Trombidiidae are frequently found attached by their mouthparts to the intersegmental cuticle of the body and legs of a range of arthropods. One of the earliest records is that by De Geer (1778) who described a larval erythraeid mite (*Leptus*) attached to a harvestman. The insect hosts of larval Trombidiidae include Orthoptera, Hemiptera, Dictyoptera, Lepidoptera, Diptera and Hymenoptera, but usually a particular species is associated with a single order of insects. For example, *Allothrombium fuliginosum* occurs on aphids (Homoptera) and *Eutrombidium trigonum* and *Microtrombidium fasciatum*, respectively, parasitize Orthoptera and Diptera.

One of the most detailed studies of the behavioural biology of larval Parasitengona has been carried out on water mites by Mitchell (1967, 1968). Larvae of *Arrenurus* living in ponds and parasitizing adult Odonata find the host when it is in the terminal naiad stage. The larvae assemble on the naiad without feeding and remain inactive when it leaves the water for ecdysis. During the course of ecdysis, the larvae become active and move from the naiad exuvium to the emerging imago. They become quickly attached to a site by their mouthparts and, after feeding, detach and drop off the host. The larvae must drop into water for the life cycle to continue and during the aquatic phase of their life the deutero-nymph and adult are predatory. In *Arrenurus fissicornis* and *A. reflexus* parasitizing libellulid dragonflies, the majority of the larvae attach to the membraneous areas of the 6th to 9th abdominal segments of the host and this appears to arise from a synchronization of larval activity with host ecdysis rather than specific site selection by the mite. Larval crawling activity on the naiad exuvium appears towards the end of ecdysis when only the last few segments of the host abdomen are in contact with the exuvium. Larvae move onto the abdomen of the dragonfly and attach close to the point of contact. However, if the larvae are transferred to another region of the body or, by manipulation the normal course of emergence of the imago is disturbed, they will attach to the first membrane they contact irrespective of the site on the thorax or abdomen. The numbers of mites arranged linearly on the right and left sides of the host abdomen often show an asymmetrical arrangement – in one instance Mitchell (1967) found an individual of *Leucorrhinia intacta* with 191 mites on the left and only 65 on the right. This arises from the asymmetry of larvae clustered on the neck and prothorax of naiads. As the naiad skin splits along the mid-line the mites are separated into left and right groups and this division is maintained on the body of the host since the larvae rarely cross the mid-line as they move from naiad to imago. Males of *L. intacta* stay around water after maturation and the larvae drop off from one side of a segment at a time. The female, on the other hand, returns to water only for oviposition and there is a synchronous drop from all segments which is considered to be triggered by a complex of stimuli associated with oviposition.

12 Classification of the Acari

Introduction

Extant members of the Arthropoda comprise three major assemblages: Chelicerata (Xiphosura, Arachnida and Pycnogonida), Crustacea and Uniramia (Onychophora, Myriapoda and Hexapoda). Their interrelationships are problematic, Manton (1977) considered them to have evolved independently with 'arthropodization' occurring more than once in the history of the group. On the other hand, a monophyletic origin of the Arthropoda has been proposed by, for example, Snodgrass (1938) and Boudreaux (1979) in which the Crustacea, Myriapoda and Hexapoda are united within the taxon Mandibulata. Similarly, opinions on the interrelationships of the major taxa within the Chelicerata are divided and since 1950 a number of fundamentally different classifications of the assemblage has been proposed. This diversity of opinion is not surprising when one considers the different approaches to classification.

Most classifications have been produced by the so-called orthodox method in which taxa are grouped on the basis of features that appear to reflect their evolution. This method is without a strict theoretical basis and relies a great deal on the experience (and intuition) of the taxonomist. However, since the 1960s two other methods of classification, the phenetic (numerical taxonomy) and the cladistic (phylogenetic systematics), have been introduced which are based on different concepts of relationship. In the phenetic method, relationships between taxa are based strictly on overall similarity with no phylogenetic assumptions; this involves the analysis of a great deal of information (Sneath & Sokal, 1973). The cladistic method, based on the works of Hennig (1950, 1966) and Remane (1952), considers all taxa to be holophyletic (monophyletic) and the establishment of a primitive/ancestral (plesiomorphic) or derived/advanced (apomorphic) state of each character is regarded to be essential for any meaningful consideration of phylogeny (Wiley, 1981). Further, in the cladistic method, only the common possession of derived characters (synapomorphies) can be used to establish relationships since only these are inherited from the most recent ancestor. The determination of whether a character is primitive or derived is usually by the 'outgroup method'. A character state occurring in the nearest relative of the group under consideration, i.e. its sister group, is hypothesized to be primitive within that group whereas a character state restricted to the group under study is considered to be derived. Thus, the cladistic method involves the critical analysis and assessment of characters and a consideration of a broad spectrum of groups in constructing lineages. Both methods have developed their own (although sometimes overlapping) terminology (Holmes, 1980).

Classification of the Chelicerata

The Chelicerata was considered by Bergström (1979) to represent three main lineages: the Agalaspida, with no living representatives, the Merostomata and the Arachnida. His Merostomata included the Xiphosura, Eurypterida (extinct) and Scorpiones. The separation of the Scorpiones from other arachnids and their classification with the Eurypterida agreed with the concepts of Sharov (1966) and also reflected, to some extent, the opinions of Anderson (1973), based on developmental data, that the Xiphosura and Arachnida diverged from a common chelicerate stock with the scorpions diverging early from other arachnids and possibly even evolving independently of them. Weygoldt and Paulus (1979a,b), in their comprehensive study of morphology, taxonomy and phylogeny of the Chelicerata using the cladistic method, include the scorpions within the Arachnida, regarding them as the sister group (Ctenophora or Pectinifera) of all other epectinate arachnids (Lipoctena or Epectinata). They proposed the division of the Chelicerata into two major groups, the Euchelicerata, containing the Xiphosura, Eurypterida and Arachnida, and the Agalaspida (Fig. 12.1A). The following ten taxa are included in the Arachnida:

1. Scorpiones
2. Uropygi (including Thelyphonida and Schizomida)
3. Amblypygi
4. Araneae
5. Palpigradi
6. Solifugae
7. Chelonethi (or Pseudoscorpiones)
8. Opiliones
9. Ricinulei
10. Acari

For comprehensive accounts of the characteristics of the above taxa, excluding the Acari, the reader is recommended to Grassé (1949) and Kaestner (1968).

A more radical classification of the Chelicerata has been developed by Van der Hammen (1989). The subphylum is divided into two major

Fig. 12.1. Diagrammatic representation of the phylogeny of the Chelicerata according to: **A**, Weygoldt & Paulus (1979b); **B**, Van der Hammen (1989).

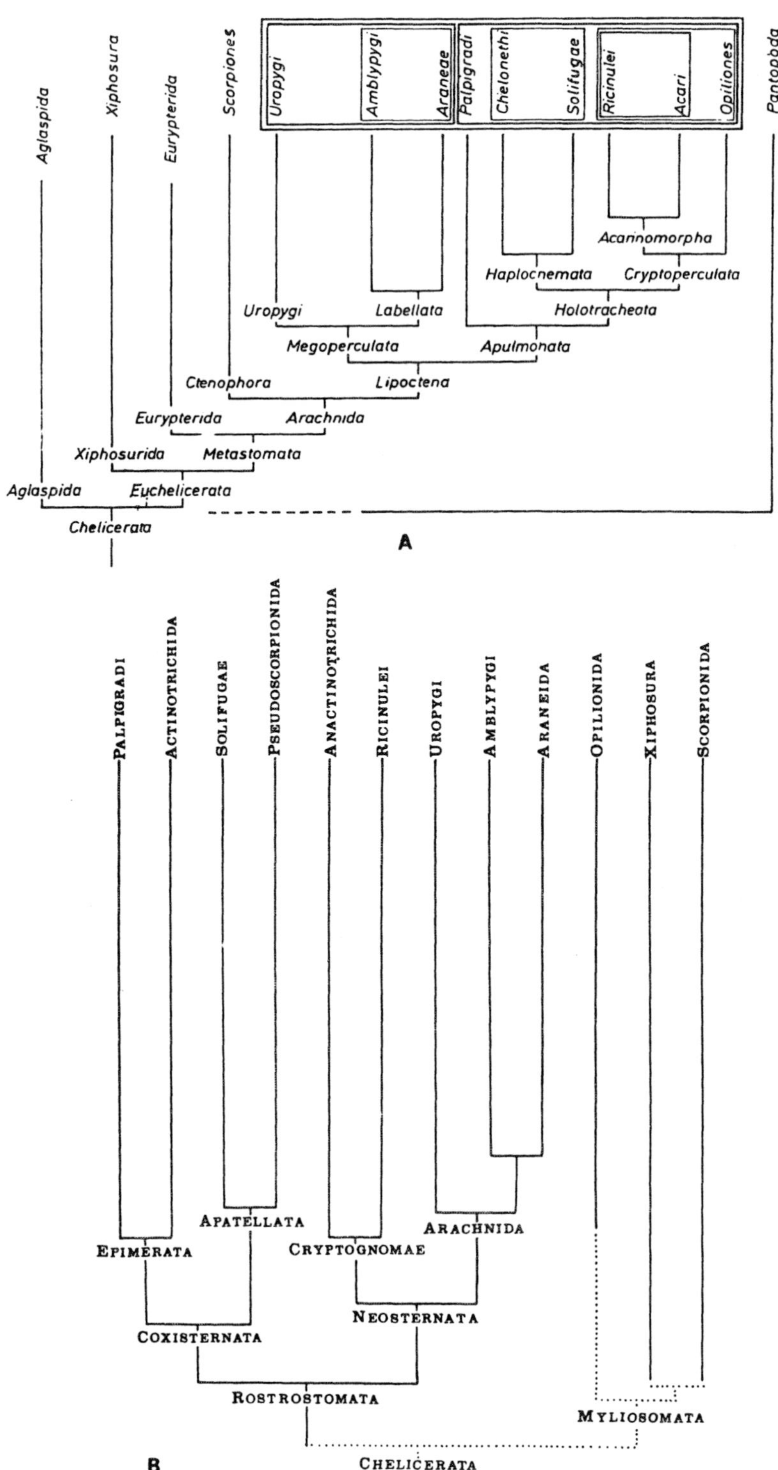
Aglaspida
Xiphosura
Eurypterida
Scorpiones
Uropygi
Amblypygi
Araneae
Palpigradi
Chelonethi
Solifugae
Ricinulei
Acari
Opiliones
Pantopoda
Acarinomorpha
Haplocnemata
Cryptoperculata
Uropygi
Labellata
Holotracheata
Megoperculata
Apulmonata
Ctenophora
Lipoctena
Eurypterida
Arachnida
Xiphosurida
Metastomata
Aglaspida
Euchelicerata
Chelicerata
A
PALPIGRADI
ACTINOTRICHIDA
SOLIFUGAE
PSEUDOSCORPIONIDA
ANACTINOTRICHIDA
RICINULEI
UROPYGI
AMBLYPYGI
ARANEIDA
OPILIONIDA
XIPHOSURA
SCORPIONIDA
APATELLATA
ARACHNIDA
EPIMERATA
CRYPTOGNOMAE
NEOSTERNATA
COXISTERNATA
ROSTROSTOMATA
MYLIOSOMATA
B
CHELICERATA

groups on the basis of their different feeding methods. Thus, the Myliosomata comprising the Opiliones, Xiphosura and Scorpiones practise coxisternal feeding, i.e. the apophyses of some of the leg coxae are involved in the feeding process, whereas the Rostrostomata, containing all other arachnids, are characterized by rostral feeding in which the leg coxae are not involved (Fig. 12.1B). The character state of coxisternal feeding was not definitely stated by Van der Hammen but is generally considered to be a primitive (plesiomorphic) condition in the chelicerates (Bergström, 1979; Weygoldt & Paulus, 1979b). If this is the case, then, according to the cladistic method, the grouping of the three taxa within the Myliosomata has no relevance phylogenetically. The Rostrostomata is divided into two classes, the Coxisternata containing the Palpigradi, Actinotrichida, Solifugae and Pseudoscorpiones, and the Neosternata comprising the Anactinotrichida, Ricinulei, Uropygi, Amblypygi and Araneae. The Coxisternata are characterized, in part, by the presence of epimera and apodemes and the legs ancestrally having two femora whereas the Neosternata have a neosternum and coxae and the legs only a single femur ancestrally.

Relationships of the Acari to other Arachnida

The ongoing debate on the relationship of the Acari to other arachnids centres on whether the Acari is a monophyletic or a diphyletic group. The first classification of the Chelicerata which postulated a diphyletic origin of the Acari was proposed by Zachvatkin (1952). He accepted Grandjean's (1936a) division of the Acari into three major groups, Anactinochitinosi, Notostigmata and Actinochitinosi, but substituted the names Parasitiformes, Opilioacarina and Acariformes, respectively, for the three taxa. Further, he extended Grandjean's use of the birefringence of body and leg setae in distinguishing the Actinochitinosi from the Anactinochitinosi to the Arachnida generally and this criterion appeared among the nine contrasting characters which he used to distinguish the 'phalangoid-complex' (=Actinochaeta) from the 'arachnoid-complex' (=Actinoderma). The Actinoderma contained the taxa Amblypygi, Araneae, Ricinulei, Parasitiformes, Opiliones and Opilioacarina while the Actinochaeta included the Palpigradi, Solifugae, Acariformes, Tartarides (=Schizopeltida) and Chelonethi.

A diphyletic origin of the Acari has also been postulated by Van der Hammen (1977a, 1979, 1989). The Actinotrichida (=Acariformes) are included with the Palpigradi in the taxon Epimerata (=Acoxata) and the Anactinotrichida (Parasitiformes + Opilioacarina) are classified with the Ricinulei within the Cryptognomae (Fig. 12.1B). The criteria used by Zachvatkin and Van der Hammen in the development of their classifications have been discussed and assessed by Lindquist (1984) in his review of the current theories on the evolution of the major groups of the Acari.

In contrast, the monophyly of the Arachnida and the unity of the Acari are stressed in the classifications of Dubinin (1959), Sharov (1966), Yoshikura (1975), Savory (1977), Weygoldt and Paulus (1979b) and Lindquist (1984) although the arguments in support of their conclusions have little in common. Dubinin included the Acariformes (=Actinotrichida), Parasitiformes and Opilioacarina in the class Acaromorpha and considered the Solifugae to be an outgroup of that taxon. This followed the suggestion by Grandjean (1936a, 1954d) that the Actinotrichida and the Solifugae shared a common ancestry on the basis of a number of morphological similarities, for example the presence of a prelarval organ of Claparède and the structure of the chelicerae, pretarsus and prodorsal shield. More recently, Alberti (1980b) has identified synapomorphies in the sperm structure of these two groups. Weygoldt and Paulus (1979b) in their cladistic analysis of the Arachnida concluded that the Acari is a monophyletic group on the basis of the presence of a gnathosoma and that the Acari and Ricinulei are sister groups since they share a similar postembryonic development characterized by the presence of a hexapod larva and three octopod nymphal instars. These two sister groups formed the taxon Acarinomorpha. The Opiliones, whose close relationship with the Acari was emphasized in many earlier classifications of the Arachnida, was recognized as an outgroup of the Acarinomorpha. Lindquist (1984) agreed with the monophyletic status of the Acari and the concept of the two sister groups, Acari and Ricinulei. However, he pointed out that a gnathosoma is not restricted to the Acari but also occurs in the Ricinulei and presented three additional synapomorphies in support of the Acarinomorpha:

1. a movable gnathosoma separated by a circumcapitular suture from the idiosoma;
2. a roughened scaly or denticulate labrum;
3. trochanters of legs III and IV divided into two articulating segments.

Shultz (1989) introduced a number of appendicular characters into the classification of the Arachnida and presented a cladogram showing the phylogenetic relationships of the arachnid orders suggested by the characters revealed in his study. The most important, and controversial, conclusion of this work was that the Opiliones, Scorpiones, Chelonethi and Solifugae appear to form a monophyletic group based on a unique arrangement of the femoropatellar joint and that the scorpions are derived and not 'more primitive' than the other arachnid orders, as is the traditional opinion. But, as the author states, his hypothesis 'requires corroboration from other lines of evidence'. The study supported a monophyletic origin for the Arachnida.

It appears, at present, that there is no convincing evidence for a polyphyletic origin of the Arachnida or a diphyletic origin of the Acari. Certainly, each arachnid order seems to be more closely related to some other arachnid lineage than to any known non-arachnid group (Shultz, 1989). The two major acarine lineages, the Anactinotrichida and Actinotrichida, probably had a remote but common ancestry and no other arachnid order is more closely related to either of them than they are to each other (Grandjean, 1970a; Lindquist, 1984).

Higher classification of the Acari

The foundations of the classification of the Acari at ordinal or subordinal level, according to whether the taxon is given, respectively, subclass (Evans *et al.*, 1961) or ordinal (Baker & Wharton, 1952) rank, were laid by Kramer (1877) and Canestrini (1891). The orders were defined on the basis of characteristics of the respiratory system, as evidenced by the taxa Mesostigmata, Metastigmata, Prostigmata, Heterostigmata, Astigmata and Cryptostigmata. The taxon Notostigmata was added by With (1904) and the Holothyroidea by Reuter (1909). A number of alternative names for these taxa have been introduced subsequently, mainly on the pretext that not all members of a particular order exhibit the characteristic implied by its name, but then this is also true, for example, of the Diptera and the Hemiptera! The preferred names used in this work and listed below, with alternative names in parentheses, reflect to some extent current usage; for example, Tetrastigmata, Metastigmata and Cryptostigmata, respectively, have been largely replaced by Holothyrida, Ixodida and Oribatida:

Notostigmata (=Opilioacarida)
Holothyrida (=Holothyroidea; Tetrastigmata)
Ixodida (=Metastigmata, Ixodoidea)
Mesostigmata (=Gamasida)
Prostigmata (=Actinedida + Tarsonemida)
Astigmata (=Acaridida)
Oribatida (=Cryptostigmata; Oribatei)

The interrelationships of the acarine orders were indicated to some extent in the classification proposed by Reuter (1909); for example, the Mesostigmata, Holothyrida and Ixodida, but not the Notostigmata, formed the Parasitiformes (or Gamasiformes) while the Prostigmata and Heterostigmata constituted the Trombidiformes and the Cryptostigmata (comprising Oribatida, Astigmata and Demodicidae!) was placed in the Sarcoptiformes. The Holothyrida was excluded from the Parasitiformes by Oudemans (1923) who later (Oudemans, 1931b), indicated the natural affinities of the Trombidiformes, Sarcoptiformes and Tetrapodili (=Eriophyoidea) by combining them in a new taxon, the Trombidi-Sarcoptiformes. Vitzthum (1940–43) retained this modified concept of the Parasitiformes and included the Acaridiae (=Astigmata) and the Oribatei as taxa of equal status within Sarcoptiformes.

Grandjean (1936a), on the other hand, considered the Acari to comprise three major groups of equal taxonomic status, namely the Actinochitinosi (equivalent to the Trombidi-Sarcoptiformes of Oudemans, 1931b), Anactinochitinosi and Notostigmata. The Actinochitinosi differed, among other features, from the other two groups in the birefringence of the body and leg setae (except for the isotropic solenidia) which was attributed to the presence of 'actinochitin' in the setal axis. The Notostigmata displayed a number of characteristics found in the other two groups and consequently its members were referred to as 'acariens synthétiques'. Evans *et al.* (1961) combined the Notostigmata and Anactinochitinosi within their Acari-Anactinochaeta and gave it equal status to the Acari-Actinochaeta (=Actinochitinosi). This division of the Acari into two major assemblages was followed by Van der Hammen (1968a) who in 1961 had introduced the new names Actinotrichida and Anactinotrichida chosen by Grandjean (see Grandjean, 1970a) to replace, respectively, the Actinochitinosi and Anactinochitinosi. Van der Hammen (1961) also proposed the name Opilioacarida for Notostigmata. Grandjean (1970a), however, defended and retained his earlier concept of the higher classification of the Acari. He agreed that the Notostigmata are more closely related to the Anactinotrichida than they are to the Actinotrichida but considered them to be too far removed from the Anactinotrichida to visualize them, in the present state of our knowledge, as primitive forms of that group.

Lindquist (1984) in his cladistic analysis of the Acari concluded that they formed a unified group comprising two major lineages, the Anactinotrichida (=Opilioacarida + Parasitiformes) and Actinotrichida (=Acariformes). The Actinotrichida comprised the Notostigmata (=Opilioacarida), Holothyrida, Ixodida and Mesostigmata, and the Actinotrichida contained the Prostigmata, Astigmata and Oribatida. The Notostigmata was recognized as a sister group of the Parasitiformes (comprising Holothyrida, Ixodida and Mesostigmata).

The results of comparative spermatological studies by Alberti (1984) support the position of the Notostigmata within the Anactinotrichida and the 'placement of only two major taxa within the Acari'. He considered on the basis of the synapomorphic vacuolated-sperm type that the Anactinotrichida formed a monophyletic group. The spermatozoa of the Anactinotrichida and the Actinotrichida, however, was found to be very different and no synapomorphic relationship could be recognized. This points to a wide separation of the two groups but, to date, spermatological studies have not been able to provide further insights into the relationships between each of these taxa and other arachnids.

Unlike the Anactinotrichida, the ordinal classification of the Actinotrichida is in a state of flux. OConnor (1984), on the basis of a cladistic analysis of the taxon, supported its division into two assemblages, the Sarcoptiformes and Trombidiformes. His concepts of the relationships within the Actinotrichida differed from those of previous authors. The Sarcoptiformes included the Astigmata, Oribatida and Endeostigmata (excluding the families Sphaerolichidae and Lordalycidae). Members of the Endeostigmata had previously been classified within the Prostigmata (Kethley, 1982). The Trombidiformes comprised the Prostigmata and the two families of the Endeostigmata excluded from the Sarcoptiformes and for which he proposed the taxon Sphaerolichida. OConnor (1984) considered that in a phylogenetic classification, the Oribatida constitutes a paraphyletic group ('excluding some descendants of a common ancestor, the Astigmata') and should be rejected as a formal group. He proposed combining the Astigmata and Oribatida (excluding the

Palaeosomata) to form the 'Higher Sarcoptiformes' with the remainder of the Sarcoptiformes comprising the Palaeosomata and the 'endeostigmatic' groups Alicorhagiida, Bimichaeliida, Oehserchestida and Terpnacarida. Some support for the closer relationship of the rutellate Endeostigmata to the Oribatida-Astigmata lineage than to the Prostigmata appears to come from spermatological studies on *Speleorchestes* (Endeostigmata; Nanorchestidae), which show that 'the very strange sperm cells' in this taxon have some characteristics in common with the Astigmata (Alberti, 1991). Further comparative studies are required to firmly establish these relationships which if confirmed would have far-reaching effects on the classification of the Actinotrichida. As stated elsewhere, the author has retained, for practical reasons, the traditional classification of the 'Endeostigmata' within the Prostigmata.

The classification adopted in this work, and outlined below, reflects the monophyly of the Acari and its subdivision into two major assemblages, the Anactinotrichida and Actinotrichida.

Subclass ACARI

Superorder ANACTINOTRICHIDA	Superorder ACTINOTRICHIDA
Order Notostigmata	Order Prostigmata
Order Holothyrida	Order Astigmata
Order Ixodida	Order Oribatida
Order Mesostigmata	

Further, the Holothyrida, Ixodida and Mesostigmata are collectively referred to as the Parasitiformes, following Lindquist (1984), and their sister group, the Notostigmata, is assigned to the Opilioacariformes. Similarly, the Astigmata and Oribatida comprise the Sarcoptiformes and the Prostigmata form the Trombidiformes, more or less in the sense of Vitzthum (1940–43). There is, at present, no uniform and generally acceptable hierarchic system for the higher taxa of the Acari. The ranks applied to the higher taxa in this work follow Evans *et al.* (1961) but alternative systems in which the superorders and orders are relegated, respectively, to orders and suborders are equally acceptable in the present fluid state of the classification. The rank of cohort and its derivatives have been widely used in the Acari for classificatory levels between subclass and order (Woolley, 1988), between order and suborder (Krantz, 1978) and between suborder and family (Vitzthum, 1940–43), but should be discouraged in favour of group-terms in the Linnaean hierarchy (OConnor, 1984).

A detailed treatment of the classification of the Acari is beyond the scope of this work and the following account is primarily intended as a guide to the systematic position of the taxa referred to in the preceding chapters with some further information on their external morphology and additional observations on their biology. The most recent synopses of the classification of the Acari are given in Parker (1982) and Woolley (1988), the former providing concise diagnoses of the major taxa. Key works of a comprehensive nature, however, are few. Krantz (1978) and Dindal

(1990) are recommended for keys to families, the latter including only free-living forms. On a more regional basis, the publications on the mite fauna of the USSR by Zachvatkin (1941), Gilyarov and Krivolutsky (1975), Gilyarov and Bregetova (1977) and Gilyarov (1978), and of Czechoslovakia by Daniel and Černý (1971) are well illustrated and provide keys to genera and, in some cases, to species. Publications dealing with the external morphology and classification of specific groups will be referred to in the appropriate part of the text.

Our knowledge of structure and function in the Acari has increased at an impressive rate during the last decade but much of this new information has yet to be assessed and introduced into the classification. Taxonomy in many groups is still at a descriptive stage and the classification at suprafamily levels is in a state of flux. The recent application of the cladistic method to the classification of some groups of the Acari is already producing results which often disagree with current concepts of relationships between and within major taxa but their piecemeal introduction into the existing classification, based mainly on more traditional methods, is fraught with difficulty and has not been attempted in this work.

The two major divisions of the Acari, each given superordinal status in this work, may be recognized by the following external morphological characters:

	Anactinotrichida	**Actinotrichida**
1.	Leg coxae free and movable (soluticoxate).	Leg coxae fused with venter of podosoma and often forming coxisterna (fixicoxate).
2.	Body and leg setae isotropic.	Body and leg setae anisotropic.
3.	Palp typically with ambulacrum represented by paired terminal claws or a movable subterminal tined claw-like structure, lost in ticks and some holothyroids.	Palp without vestige of ambulacrum
4.	Idiosoma without dorso-sejugal furrow.	Dorso-sejugal furrow typically present.
5.	Podocephalic canal absent.	Podocephalic canal present.
6.	Chelicera typically with dorsal slit organ.	Chelicera without dorsal slit organ.
7.	Free-living forms with subcapitular gutter or groove present and usually with associated tritosternum.	Subcapitular gutter or groove absent, never with tritosternum.
8.	The tarsi of legs II–IV with peripodomeric fissure associated with slit organs.	Tarsi of legs II–IV without peripodomeric fissure and slit organs.
9.	Apparently no anamorphosis.	Anamorphosis present with maximum increase of three opisthosomatic segments during nymphal phase but reduced or absent in more derivative taxa.

Anactinotrichida (*cont.*)	**Actinotrichida** (*cont.*)
10. Organs of Claparède and genital papillae lacking (analogous genital structures may be present in the Notostigmata).	Organs of Claparède and genital papillae generally present.

Superorder Anactinotrichida

This superorder forms a well-defined monophyletic group comprising four orders whose adults may be distinguished as follows:

1. Tarsus of pedipalp with one or two terminal claws (see Fig. 12.2D below); opisthosoma with four pairs of simple dorsolateral stigmata without peritremes; two or three pairs of prodorsal ocelli; large terminal anus; trochanters III and IV divided by a eudesmatic joint into two podomeres .... NOTOSTIGMATA (p. 387)

– Tarsus of pedipalp without terminal claws, ambulacrum represented, at the most by a tined claw-like structure at the inner basal angle of the palptarsus (see Fig. 5.9A, p. 141), often absent in parasitic taxa; one pair of lateral stigmata in the region of coxae II–IV and with associated peritremes or a pair of complex stigmatic plates; at the most one pair of prodorsal ocelli; anus usually ventral or subterminal; trochanters III and IV undivided 2

2. Subcapitulum with hypostome developed into an elongate holdfast structure armed with rows of retrorse denticles (see Fig. 12.4A below); without elongate peritremes; pedipalp without vestige of ambulacrum; modified cheliceral digits working in a horizontal plane; haematophagous ectoparasites of vertebrates ...... IXODIDA (p. 390)

– Subcapitulum without such a hypostomatic structure; peritremes present, except in some endoparasites; cheliceral digits working in a vertical plane 3

3. Venter of subcapitulum with a maximum of four pairs of setae and a pair of corniculi, tritosternum with base and lacinia(e), rarely absent in parasitic taxa (see Fig. 12.9A below); anal valves usually nude, rarely with one pair of setae; gnathotectal process(es) usually present .... MESOSTIGMATA (p. 399)

– Venter of subcapitulum with six to numerous ventral setae and corniculi (see Fig. 12.3C below); reduced tritosternum rarely present; anal valves with many setae; gnathotectal process lacking .... HOLOTHYRIDA (p. 389)

Order Notostigmata

Mites of this order are large, adults ranging from 1.5 to 2.3 mm in length, and have the general appearance of small harvestmen. The idiosoma has a rather weakly sclerotized leathery cuticle without well-defined sclerotized areas and is more or less constricted behind the podosoma. The legs are long and slender with I and IV usually longer than the idiosoma. A plane of autotomy occurs between the coxa and the trochanter. The body and legs, basically a pale yellowish brown, are ornamented with bluish stripes and bands due to subcuticular pigment granules.

The prodorsum bears two or three pairs of lateral ocelli but no lyrifissures (Fig. 12.2A). The opisthosoma, on the other hand, is richly provided with lyrifissures and sigilla and their orientation and grouping are considered to indicate primitive segmentation (see Fig. 2.2, p. 24). The idiosoma is considered to comprise a total of 17 or 19 segments of which 11 or 13 form the opisthosoma. Numerous short, usually papilliform, setae are present on the prodorsum but are sparse on the opisthosoma and usually restricted to the posterior region except in the genera *Panchaetes* and *Salfacarus* in which the increased setation is associated with a decrease in the number of lyrifissures. The transverse genital orifice in both sexes lies at the bottom of a cuticular fold behind legs IV. A pair of eversible genital papillae, covered by genital verrucae (wart-like valves) is present lateral to the orifice. The female has a short eversible ovipositor. A pair of sternal verrucae, similar in shape to the genital verrucae, occur between coxae II. Immediately behind the gnathosoma is a tritosternum completely divided to the base (Fig. 12.2B). A narrow canal or gutter runs from the opening of a coxal gland posteroventral to each coxa I to the base of the tritosternum. The terminal, retractable anal tubercle consists of two valves which cover the anal opening.

Tritonymphs and adults have four pairs of simple stigmata situated dorsolaterally on the four anterior segments of the opisthosoma but only two and three pairs are present in the protonymphs and deuteronymphs, respectively.

The gnathosoma is very mobile and is not contained in a camerostome. The chelate-dentate chelicerae have three articles with the middle article bearing three to five setae and two lyrifissures and the basal article none, one or two setae in the tritonymph and adult. The ventral surface of the subcapitulum has a capitular groove and bears 12 or more pairs of setae of which the two most anterior pairs are hypertrophied to form rutella (usually atelebasic in form) and With's organs (Fig. 12.2C). Hypertrophy of a third pair of anterior setae to form a small spine (often referred to as a corniculus) may also occur. There are prominent lateral lips and a labrum provided with many denticles on its lower surface. The pedipalp has six free palpomeres including an ambulacrum represented by two terminal claws (Fig. 12.2D). The palptibia-tarsal articulation is oblique. A pair of salivary ducts opens below the chelicerae.

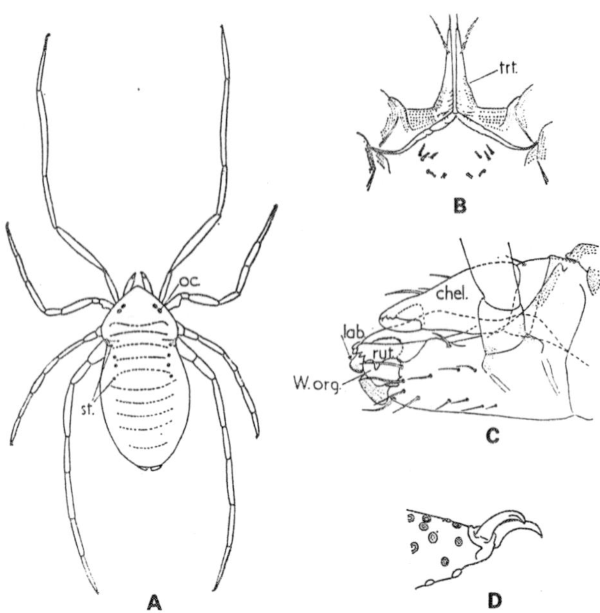

Fig. 12.2. *Opilioacarus segmentatus* (Notostigmata). **A**, dorsal view; **B**, tritosternum and channel; **C**, lateral view of the gnathosoma; **D**, pedipalpal claws (apotele). (After Evans *et al.*, 1961.) *chel.* = chelicera; *lab.* = labrum; *oc.* = ocelli; *rut.* = rutellum; *st.* = stigmata; *trt.* = tritosternum; *W. org.* = With's organ.

Legs I and II have seven fully articulating podomeres while III and IV have eight due to the presence of two trochanters. Tarsi II–IV have an apicotarsus (acrotarsus) in the tritonymph and adult. Laterocoxal setae are present on legs I and II. Ambulacra of all legs have two claws and a pulvillus and the ambulacral stalks (=pretarsi) of legs II–IV have two pairs of setae. A group of chemosensory organules occurs distally on the dorsal surface of tarsus I, and behind the group on the antiaxial face of the podomere is an elongate cavity containing two sensilli, one of which protrudes to the exterior. Ordinary and papilliform setae occur on the legs. The setae are optically isotropic with the exception of a small number distally on tarsi II–IV.

There are prelarval, larval and three nymphal instars. The prelarva in some species is an elattostase and the larva has vestiges of the fourth pair of legs.

Classification and Distribution

Members of this order comprise eight genera belonging to a single family, Opilioacaridae: ***Adenacarus*** (Arabia); ***Opilioacarus*** (southern Europe, North Africa, India and as subgenus *Neocarus* in America); ***Paracarus*** (central Asia); ***Panchaetes*** (West Africa); ***Phalangiacarus*** (Gabon, Ivory Coast); ***Salfacarus*** (Madagascar, South Africa, Tanzania) and ***Simacarus*** and *Vanderhammenacarus* from Thailand. They occur mainly under stones and in litter in rather dry situations. *Simacarus* is cavernicolous.

References
Grandjean (1936a); Juvara-Bals and Baltac (1977); Leclerc (1989); Lehtinen (1980); Naudo (1963); Van der Hammen (1966, 1968c, 1977b, 1989).

Order Holothyrida

The holothyrids are large to very large (adults ranging from 2 to 7 mm), moderately to strongly sclerotized mites. The oval idiosoma is covered dorsally by a single shield, usually arched and densely covered with short setae. Lyrifissures and numerous pores, openings of integumentary glands, are also present. A pair of ocelli with convex lenses occurs on the pronotum of *Australothyrus.*

The venter of both sexes is covered by a holoventral shield which encloses the genital and anal regions. The genital orifice of the male is situated between coxae IV and covered by two shields while the considerably larger genital area of the female has four shields comprising a large square postgenital, a narrow anterior pregenital and a pair of lateral laterogynals (Fig. 12.3A). All the genital shields are provided with setae and pores. The anus is closed by two setae-bearing valves. A characteristic pit, of unknown function, with numerous setiform cuticular projections and termed the peridium is present posterior to each of acetabula IV in the Allothyridae. A tritosternum represented by two simple laciniar elements occurs in *Allolothyrus* but is absent in other taxa. A pair of slit-like lateral stigmata with associated peritremes is present.

The gnathosoma is situated in a camerostome beneath an anterior projection (rostral tectum) of the dorsal shield. The chelate-dentate chelicerae comprise three articles with the basal article, lacking setae, movably articulated with the middle article bearing one or two setae and two lyrifissures, one dorsal and the other antiaxial (Fig. 12.3B). The subcapitulum bears six to numerous setae and a pair of corniculi, each of which may have a ventral denticle as in *Holothyrus.* The hypostome is extended into a pair of rounded or pointed lobes bearing a fringe of elongate papilliform processes (Fig. 12.3C). The large labrum, provided with strong spine-like processes and often referred to as radula-like, is flanked dorsolaterally by a pair of paralabral styli. The pedipalps are six-segmented including the ambulacrum represented by two or three separate or basally fused 'claws' near the inner basal angle of the tarsus (Fig. 12.3D). In the Holothyridae and Neothyridae, a comb of stout setae is situated distally on the paraxial face of the palptibia.

All the legs have seven primary articulating podomeres with the ambulacrum comprising two claws and on legs II–IV also a median pulvillus. A pair of setae is present on ambulacral stalks (=pretarsi) II–IV. Secondary peripodomeric fissures, associated with lyrifissures, are present

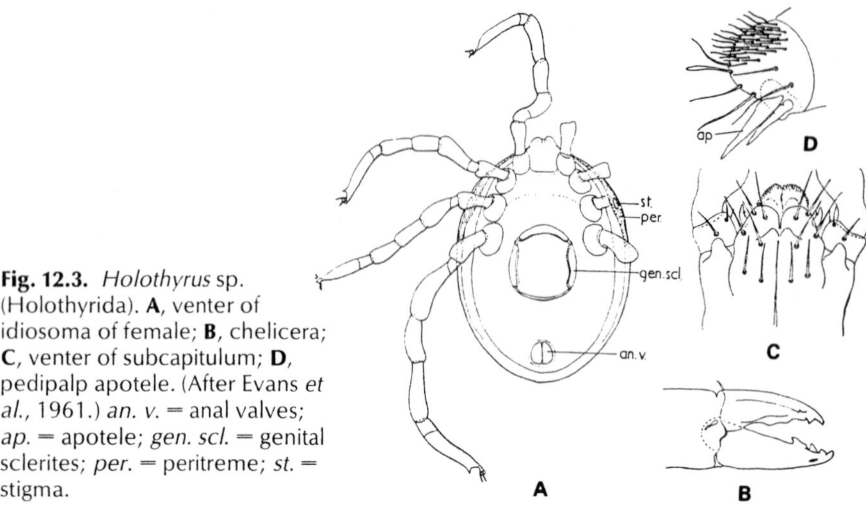

Fig. 12.3. *Holothyrus* sp. (Holothyrida). **A**, venter of idiosoma of female; **B**, chelicera; **C**, venter of subcapitulum; **D**, pedipalp apotele. (After Evans *et al.*, 1961.) *an. v.* = anal valves; *ap.* = apotele; *gen. scl.* = genital sclerites; *per.* = peritreme; *st.* = stigma.

on all femora and on tarsi II–IV. Tarsus I in the Allothyridae has a subterminal suture which cuts off an apicotarsus. Some of the chemosensory organules of tarsus I are located in a cavity which is reminiscent of Haller's organ in ticks.

The ontogeny is incompletely known. Three octopod nymphal instars and possibly a hexapod larva are thought to occur in the life cycle.

Classification and Distribution

The 14 or so described species are classified in five genera and three families. The genera *Holothyrus*, *Hammenius* and *Thonius* (Holothyridae) occur on the islands of the Indian Ocean and New Guinea; *Neothyrus* (Neothyridae) is Neotropical and *Allothyrus* and *Australothyrus* (Allothyridae) are Australasian. Holothyrids inhabit litter and soil under trees, grasses and ferns and are considered to be predacious. *Holothyrus coccinella* is reported to produce a secretion which is toxic to poultry and can cause an inflammatory condition when in contact with mucous membranes of man.

References

Domrow (1955); Hirst (1920, 1922); Lehtinen (1981); Van der Hammen (1961, 1983, 1989).

Order Ixodida

Ticks are considerably larger than mites, adult Ixodoidea ranging in length from 1.7 to 6.1 mm and Argasoidea from 3.6 to 12.7 mm in the unfed state. Engorged individuals of both taxa may attain 20–30 mm. The dorsal surface of the idiosoma may be leathery and mammillated or wrinkled (Argasoidea) or it may be regionally sclerotized to form a

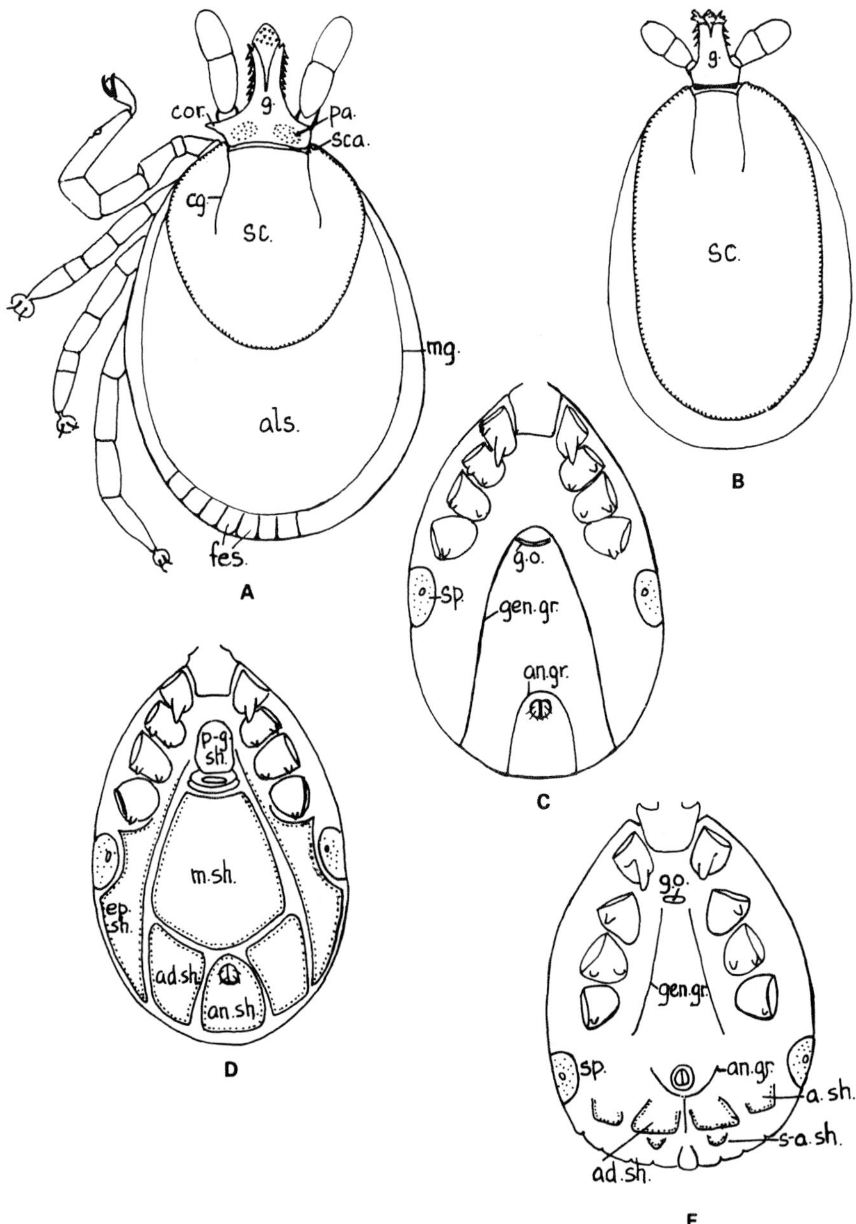

Fig. 12.4. **A**, dorsum of a generalized female ixodoid tick (Ixodida); **B**, form of dorsal sclerotization in a male ixodid tick; **C**, venter of the idiosoma of a female ixodid tick; **D**, venter of the idiosoma of a male ixodid tick; **E**, venter of the idiosoma of a male amblyommid tick. *a. sh.* = accessory shield; *ad. sh.* = adanal shield; *als.* = alloscutum; *an. sh.* = anal shield; *an. gr.* = anal groove; *cg.* = cervical groove; *cor.* = cornua; *ep. sh.* = epimeral shield; *fes.* = festoons; *g.* = gnathosoma; *g. o.* = genital orifice; *gen. gr.* = genital groove; *m. sh.* = median shield; *mg.* = marginal groove; *pa.* = porose area; *p-g. sh.* = pregenital shield; *sc.* = scutum; *sca.* = scapula; *sp.* = spiracular plate; *s-a.sh.* = *subanal shield.*

distinct scutum (Ixodoidea) which covers only the antero-median region in females and immatures but almost the entire dorsum in males (Fig, 12.4A,B). In the Nuttallielloidea, the extremely convoluted integument is differentiated into an anterior pseudoscutum resembling the scutum of female ixodoids in size and position and in the argasoid *Nothoaspis* a more extensive anterior area lacking the ornamentation of the remainder of the idiosoma defines a shield-like structure (Keirans *et al.*, 1976, 1977). A small median dorsal scutum may be present in the larva of some Argasoidea. The region lateral and posterior to the scutum in females and immatures is termed the alloscutum. The posterior region of the idiosoma in some Ixodoidea is divided into a number of festoons (Fig. 12.4A). A characteristic reversible sac-like egg-waxing organ, Gene's organ, is located antero-dorsally on the idiosoma, above the gnathosomatic base, in the female. Ventrally, the idiosoma bears shields (adanal, subanal and accessory shields) only in males of the Ixodoidea (Fig. 12.4D,E). The genital orifice of both sexes is in the form of a relatively short transverse slit located in the intercoxal region. The anus lies mid-ventrally, approximately equidistant from the level of the fourth pair of legs and the posterior margin of the body, or subterminally and the anal valves in all active instars bear one or more pairs of setae. A pair of stigmata is present laterally in the region of coxae IV in the nymphal and adult phases. Ocelli or other photoreceptors may be present. A pair of coxal glands occurs in the Argasoidea and each opens by a slit-like aperture between coxae I and II.

The gnathosoma (=capitulum) is movably articulated to the idiosoma. It lies in a camerostome and is usually concealed from above by the anterior region of the idiosoma in nymphs and adults of the Argasoidea while in the Ixodoidea, Nuttallielloidea and larval Argasoidea it is articulated anteriorly and visible in dorsal aspect. In the Ixodoidea, porose areas are present on the gnathosomatic base in females (Fig. 12.4A) and the scutum in all instars is emarginated anteriorly for articulation with the gnathosoma and has antero-lateral shoulder-like projections called scapulae. The hypostome forms a characteristic anteriorly directed conical structure bearing backwardly projecting denticles. The chelicera consists of a shaft terminating in two 'digits' with outwardly directed teeth. The pedipalp usually has three or four articles which are more or less cylindrical in the Argasoidea but, other than the terminal article, tend to be enlarged and flattened in the Ixodoidea and Nuttallielloidea. The apotele is lacking.

All the legs have seven podomeres with the ambulacrum comprising a pair of claws and a pulvillus, except in the Argasoidea in which pulvilli are rudimentary or absent. Peripodomeric scissures are present proximally on the femora I–IV and on tarsi II–IV. The coxae may have retrorse spurs in the Ixodoidea but are unarmed in the other two superfamilies. Tarsus I has a battery of sensory organules situated dorsally in a capsule and adjacent pit which form Haller's organ.

Three major taxa are recognized within the Ixodida and are herein given superfamily rank following Camicas and Morel (1977). Two of these, the Argasoidea and Ixodoidea, are worldwide in distribution and are represented by numbers of species whereas the Nuttallielloidea is based on a single species collected in South Africa and Tanzania.

Key to superfamilies

1. Dorsum of idiosoma of immatures and adults with a strongly sclerotized scutum; gnathosomatic base with porose areas in female; one or two pairs of post-hypostomatic setae; stigmatic plate relatively large, round or comma-shaped and located posterior to coxae IV; coxae of legs often with backwardly projecting spurs .... Ixodoidea

– Dorsum of idiosoma without strongly sclerotized scutum, at the most a pseudoscutum, differentiated from surrounding integument by its ornamentation, or small isolated sclerotized discs for muscle attachment; no porose areas; with one pair of post-hypostomatic setae; stigmatic plate when present small and located laterally in the region of coxae III–IV; leg coxae lacking spurs 2

2. Gnathosoma ventral or subterminal in nymphs and adults and usually concealed from above, camerostome with distinct cheeks; pedipalps linear and comprising four cylindrical articles; supra- and subcoxal folds present; coxal glands present and opening to the exterior between coxae I and II .... Argasoidea

– Gnathosoma visible from above; pedipalps of three articles with the two basal articles enlarged and quadrate or subtriangular in shape, terminal article arising from deep depression in ventro-internal surface of penultimate article; supra- and subcoxal folds lacking; without coxal glands .... Nuttallielloidea

Superfamily Argasoidea

Members of this superfamily are commonly referred to as 'soft ticks' due to the lack of a well-sclerotized dorsal scutum. The dorsal and ventral surfaces of the idiosoma are usually sharply demarcated from one another by a definite sutural line in the Argasidae (Fig. 12.5A) whereas in the Ornithodoridae the lateral suture is invariably lacking and the body is more or less convex with well-developed lateral surfaces. Small discs of sclerotized cuticle for muscle attachment are usually present on the dorsal and ventral surfaces of the idiosoma. A system of ventral grooves and folds, which is influenced by the degree of feeding is present but better developed in the Ornithodoridae than in the Argasidae. They comprise marginal, paired dorsoventral, longitudinal, supracoxal, longitudinal subcoxal (two pairs), pre-anal, transverse postanal and median anomarginal grooves (Fig. 12.5B,C). The legs often bear dorsal humps which are useful diagnostic features.

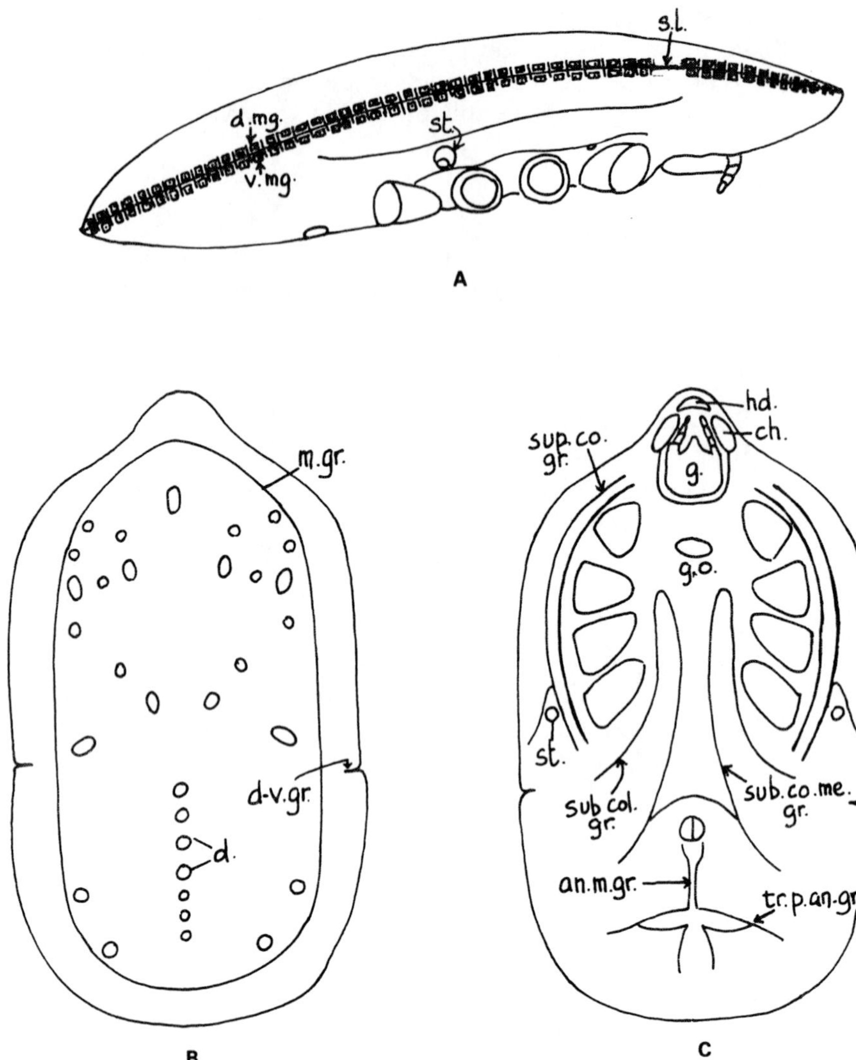

Fig. 12.5. ***A****, lateral view of an argasid tick (Ixodida) showing the lateral line separating the dorsal and ventral regions of the idiosoma;* ***B****,* dorsum of the idiosoma of an ornithodorid tick; **C**, venter of the idiosoma of an ornithodorid tick. **B, C**, based on Balashov (1968). *an. m. gr.* = anomarginal groove; *ch.* = cheek; *d.* = discs for muscle attachment; *d. mg.* = dorsal margin of idiosoma; *d-v. gr.* = dorsoventral groove; *g.* = gnathosoma; *g. o.* = genital orifice; *hd.* = hood; *m. gr.* = marginal groove; *s. l.* = suture line; *st.* = stigmatic plate; *sub. co. me. gr.* = subcoxal median groove; *sub. col. gr.* = subcoxal lateral groove; *sup. co. gr.* = supracoxal groove; *tr. p. an. gr.* = transverse postanal groove; *v. mg.* = ventral margin of idiosoma.

The sides of the camerostome are developed into a pair of folds known as cheeks and a lobe formed at its anterior margin is referred to as the hood. (Fig. 12.5C). The hypostome is rounded or notched at its tip with the fine apical denticles forming the corona. In general, the hypostomatic armature in the slow-feeding nymphs and adults of the Argasoidea is reduced in comparison with the Ixodoidea. Porose areas are absent from the gnathosomatic base. The four articles of the pedipalp are about equal in length and movably articulated with one another. The fourth article terminates in a group of contact chemoreceptors.

The argasoids parasitize a range of terrestrial vertebrates, especially birds and bats. They are essentially multihost, habitat ticks which as nymphs and adults feed rapidly to engorgement on the blood of their host. Their classification at generic and subgeneric levels has been discussed by Pospelova-Shtrom (1946,1969), particularly in relation to the divergent opinions on generic concepts of Russian and North American tick workers, and by Edwards (1975) on the basis of chaetotactic studies on the legs and pedipalps of the larvae. These works together with those of Clifford *et al.* (1964) and Kohls *et al.* (1965) should be consulted for details. The monograph of the 'Argasidae' by Nuttall *et al.* (1908b) and the revision of the 'Argasidae' of the USSR by Filippova (1966) and that of the Sudan by Hoogstraal (1956) give keys to species and are well illustrated.

Superfamily Ixodoidea

The so-called 'hard ticks' are readily recognizable from other Ixodina by the combined presence of a well-sclerotized dorsal scutum and anteriorly directed gnathosoma in immatures and adults. The scutum may have paired cervical grooves, depressions extending posteriorly from the inner angles of the scapulae (see Fig. 12.4A above). Marginal grooves commencing at the posterior border of the scutum and running along the side of the body in females and corresponding lateral grooves arising near the scapulae on the scutum of males are often present. The dorsal alloscutum of some females also bears paired postero-lateral grooves and a single postero-median grove. The scutum is highly ornate in the genera *Amblyomma* and *Dermacentor*. Ocelli are absent in the Ixodidae but a pair may be present on the lateral margins of the scutum in the Amblyommidae. The venter of the idiosoma is provided with genital and anal grooves (see Fig. 12.4C above). The anal groove surrounds the anus in front in the Ixodidae while it contours the anus behind or is faint and indistinct in the Amblyommidae. The form of the anal groove in these two taxa is the basis of the older division of the hard ticks into the Prostriata (=Ixodidae) and Metastriata (=Amblyommidae). A stigmatic plate with aeropyles and situated postero-lateral to each coxa IV is present in the nymphal and adult phases. The denticles of the hypostome are usually more numerous than in the Argasoidea and the hypostome is adapted for long periods of attachment to the host. The tip of the hypostome may be rounded, pointed, spatulate or notched. Two pairs of post-hypostomatic setae are present in the Ixodidae but only one pair occurs in the Amblyommidae. The fourth, terminal, article of the pedipalp is inconspicuous and sunk into a ventral depression of the third article, articles 2 and 3 are usually the longest and widest.

The ixodoids parasitize a wide range of vertebrates and are important as vectors of disease to man and domesticated animals. So far as their classification is concerned, there is general agreement on generic concepts in the Amblyommidae but some difference of opinion exists in the classification of the Ixodidae. The classic monographs of Nuttall and Warburton (1911,1915) and Robinson (1926) laid the foundation of the classification of the 'Ixodidae' and contain good descriptions and figures. Subsequent key works tend to be regional in their treatment, for example Cooley (1938, 1946), Cooley and Kohls (1945) and Gregson (1956) cover North America; Pomerantsev (1950) the USSR; Hoogstraal (1956) the Sudan; Arthur (1963) the British Isles and Feider (1965), Romania. A synonymic list of ixodoid genera is given by Camicas and Morel (1977).

Superfamily Nuttallielloidea

The only representative of this superfamily, *Nuttalliella namaqua*, exhibits a unique combination of characters which excludes its classification in either of the other two superfamilies of the Ixodida (Keirans *et al.*, 1976). The idiosomatic integument of the female is extremely convoluted, forming closely spaced pits surrounded by elevated rosettes. Discs and mammillae, however, are absent. Setae are lacking on the dorsal surface (excluding pseudoscutum) but many of the pits of the venter have short setae. The ornamentation of the pseudoscutum comprises a network of irregular, deep cavities some of which have a single seta. The genital orifice is situated at the level of coxae II and the subterminal anus has two valves, each bearing about 35 setae. An 'organ of unknown function' is located immediately posterior to each coxae IV. A stigma, postero-lateral to each coxa IV, has the form of a relatively small rectangular plate with a fenestrated or latticed area. The gnathosomatic base appears to be divided by a deep suture into a posterior rectangular part with long, subtriangular cornua and an anterior part with a dome-like projection bearing the pedipalps. The latter are composed of three articles of which the relatively small elongate third article is articulated to the ventro-internal face of the second article. The hypostome may be concealed by internal flanges of the palptrochanters. The distal ends of the leg podomeres form a protective collar around the base of the succeeding podomere and are particularly conspicuous between the trochanters and basifemora.

Characters considered to reflect a relationship with the Ixodoidea are the presence of a pseudoscutum (possibly a true scutum), absence of subcoxal and supracoxal folds and the apical position of the gnathosoma while those showing argasoid resemblances are the lack of porose areas, general form of the integument, unarmed coxae and reduced dentition of the hypostome. Uniquely nuttallielloid features are seen in the protective collars of the leg joints, the form of the spiracular plates, the pedipalp of three articles and the 'organs of unknown function'. Females of *N. namaqua* are known from semi-arid areas in Namaqualand, Cape Province, South Africa (host uncertain) and from a small carnivore, a rodent and nests of swallows (*Hirundo abyssinica unitatis*) in Tanzania. Males are unknown.

Ticks as vectors of disease

Ticks are capable of transmitting a greater variety of infectious agents to man and domesticated animals than any other group of blood-feeding arthropods. A number of these agents cause zoonoses, i.e. diseases that are transmitted from animals to man under natural conditions. Disease organisms transmitted by ticks include viruses, spirochaetes, rickettsiae, anaplasmas, bacteria, piroplasmas and filariae (Arthur, 1962; Philip, 1963; Balashov, 1968). Pathogens are transmitted from ticks to their vertebrate hosts through a number of routes including saliva, regurgitation, coxal fluid and faeces.

Arbovirus transmission by ticks has been discussed by Hoogstraal (1966). Examples of tick-borne viral diseases are Russian spring–summer encephalitis (RSSE), occurring in the forest zone of the USSR, and Colorado tick fever (CTF) from the USA, both transmitted to man, and louping ill of sheep occurring in Britain. The virus of RSSE has been isolated from species of *Ixodes*, *Haemaphysalis* and *Dermacentor*, particularly from *I. persulcatus* and *I. ricinus*. Ticks receive the virus during feeding on wild mammals and, more rarely, birds and both trans-stadial and trans-ovarial transmission are known to occur. The major host of the virus of CTF, a febrile condition accompanied by lumbar pains and anorexia, is *Dermacentor andersoni*. Transmission is trans-stadial and the disease reservoirs are rodents.

Ticks, chiefly of the genera *Ornithodoros* and *Ixodes*, are vectors of a number of spirochaetes of the genus *Borrelia*, a group of bacteria pathogenic for man and other vertebrates. African tick-borne relapsing fever, caused by *Borrelia duttoni*, is widespread in East, Central and South Africa where the vector is *Ornithodoros moubata*. The spirochaetes are ingested with infected blood and migrate form the gut lumen into the haemocoele where they multiply before invading the salivary glands and other internal organs. Spirochaetes are discharged with the saliva and coxal fluid during feeding. Trans-ovarial transmission has been established, the spirochaete penetrating immature oocytes before the formation of the 'chorion'. *Borrelia burgdorferi* is the causal agent of Lyme disease or Lyme borreliosis, which only appeared as an important zoonosis in the early 1980s (Burgdorfer *et al.*, 1982). The disease occurs over most of the northern temperate regions and causes skin conditions, arthritis, and heart and neurological problems. The chief vectors are

Ixodes dammini and *I. pacificus* in North America, *I. ricinus* in Europe and *I. persulcatus* in Asia (Piesman, 1989; Kahl, 1991; Kahl & Gray, 1991). The spirochaetes in unfed ticks are restricted to the mid-gut, which has led to suggestions that transmission from the vector to its host is by regurgitation during feeding and by faecal contamination. However, Zung *et al.* (1989) have found that the spirochaetes migrate to the salivary glands of the tick during feeding which indicates probable salivary transmission. Important vertebrate reservoir hosts are rodents; the white-footed mouse (*Peromyscus leucopus*) in North America and probably species of *Apodemus* and *Clethrionomys* in Europe. Trans-ovarial transmission is not thought to be significant.

Tick-borne rickettsiae are the causal agents of a number of human diseases including Rocky Mountain spotted fever (RMSF), Siberian tick typhus and Q-fever. RMSF was the first tick-borne disease to be investigated scientifically and is widespread in the USA and in regions of Canada and Mexico (Hoogstraal, 1967). The chief vectors of the infective agent *Rickettsia (Dermacentroxenus) rickettsi* are *Dermacentor variabilis* and *D. andersoni* and trans-ovarial passage as well as transmission to females by way of infected sperm has been demonstrated. *R. (D.) sibiricus*, the agent of Siberian tick typhus, is biologically closely related to *R. (D.) rickettsi* and causes a similar illness. *Dermacentor nuttalli* is a major vector and rodents act as reservoirs. Q-fever is caused by *Coxiella burneti* and, unlike the effects of other rickettsiae, a rash is not developed in cases of infection. Ixodoids rather than argasoids appear to be most closely associated with the organism in natural foci. Trans-stadial and trans-ovarial transmission have been demonstrated. The rickettsiae are present in the tick faeces and contamination of the skin followed by the passage of the organisms through abraded or unabraded skin is the probable method of infection.

Ticks are the most important vectors of anaplasmosis, a blood parasitic febrile disease. They ingest *Anaplasma* in the course of feeding and the organism is later transmitted trans-stadially or trans-ovarially. The method of transmission between vector and host has not been established. Anaplasmas occur in the red blood cells of ruminants and are widespread in cattle (*Anaplasma marginale*) and sheep and goats (*A. ovis*) in USSR and Africa.

Dermacentor andersoni in North America and *D. marginatus* in northern Asia are reservoirs and transmitters of *Francisella tularensis*, the causal agent of tularaemia in man, rabbits and sheep. Rabbits appear to be the important source of human infection in North America and rodents in northern Asia. The bacilli show a close affinity with the plague-causing bacterium. Transmission in the vector is trans-stadial. Infection of the vertebrate host is either by inoculation via the saliva during the feeding of the tick or by faecal contamination.

Protozoa of the families Babesiidae and Theileriidae (Sporozoa; Piroplasmida) are transmitted by ixodoid ticks to a range of domesticated animals. Members of both families inhabit erythrocytes of the host but the theileriids also invade other blood cells, usually lymphocytes. Infective stages of the parasite occur in the salivary glands of the tick and transmission to the host is by inoculation during feeding. Trans-ovarial and trans-stadial transmission occur in the cases of *Babesia* but trans-ovarial passage is considered to be absent in *Theileria*. Babesiids parasitize reptiles, birds and, mainly, mammals and many produce disease (babesiasis) in cattle, horses and dogs. In cattle and deer, the disease is associated with lysis of the erythrocytes and excretion of the released haemoglobin in the urine (haemoglobinuria), a condition in cattle commonly referred to as redwater. *Babesia bigemina* and *B. bovis* are important agents of redwater in cattle and deer, the former in America, Africa and Australia and the latter in Europe (including the former Soviet republics) and Africa. The vectors of *B. bigemina* are species of *Boophilus*, *Haemaphysalis* and *Rhipicephalus* and of *B. bovis* species of *Ixodes*, *Boophilus* and *Rhipicephalus*. *Babesia canis*, transmitted by a number of amblyommid species to dogs, and *B. caballi*, transmitted by *Dermacentor marginatus* to horses, cause biliary fever. *Theileria* spp. are essentially parasites of ruminants. *T. parva* is the causal agent of East Coast fever (or Rhodesian redwater fever) transmitted by *Rhipicephalus* spp. to cattle and buffalo in southern Africa; *T. annulata*, carried by *Hyalomma*, is responsible for Mediterranean Coast fever.

Ornithodoros species are known to transmit larval filariae of *Macdonaldus oschei* and *Dipetalonema witei* (Nematoda; Filariidae) to pythons and gerbils, respectively (Balashov, 1968). Ticks ingest the prelarvae (microfilariae) while feeding on infected blood and the parasites migrate from the gut to the haemocoele, Malpighian tubules and muscules. The infective stages for the vertebrate host concentrate in the gnathosomatic muscles and the pharynx of the tick. During feeding, the larval filariae pass through the pharyngeal wall of the tick and into the host.

Order Mesostigmata

The mites of this order range in size from 200 to 2500 μm. Many of the smaller forms, particularly euedaphic species, are weakly sclerotized and pale in colour but generally the idiosoma is partially or entirely covered by a number of chestnut-brown shields separated by whitish striated cuticle. For descriptive purposes, the idiosoma is considered to be divided into a prosoma and opisthosoma. The division between the two regions is defined by the posterior limit of an anterior scutal element when present as a discrete shield, or by its chaetotaxy when fused with other dorsal shields. The dorsal region of the prosoma and opisthosoma are termed, respectively, the pronotum and opisthonotum and the venter of the opisthosoma is referred to as the opisthogaster.

The sclerotization of the dorsum of the idiosoma may be considered to comprise three scutal elements: a pronotal shield covering the prosomatic

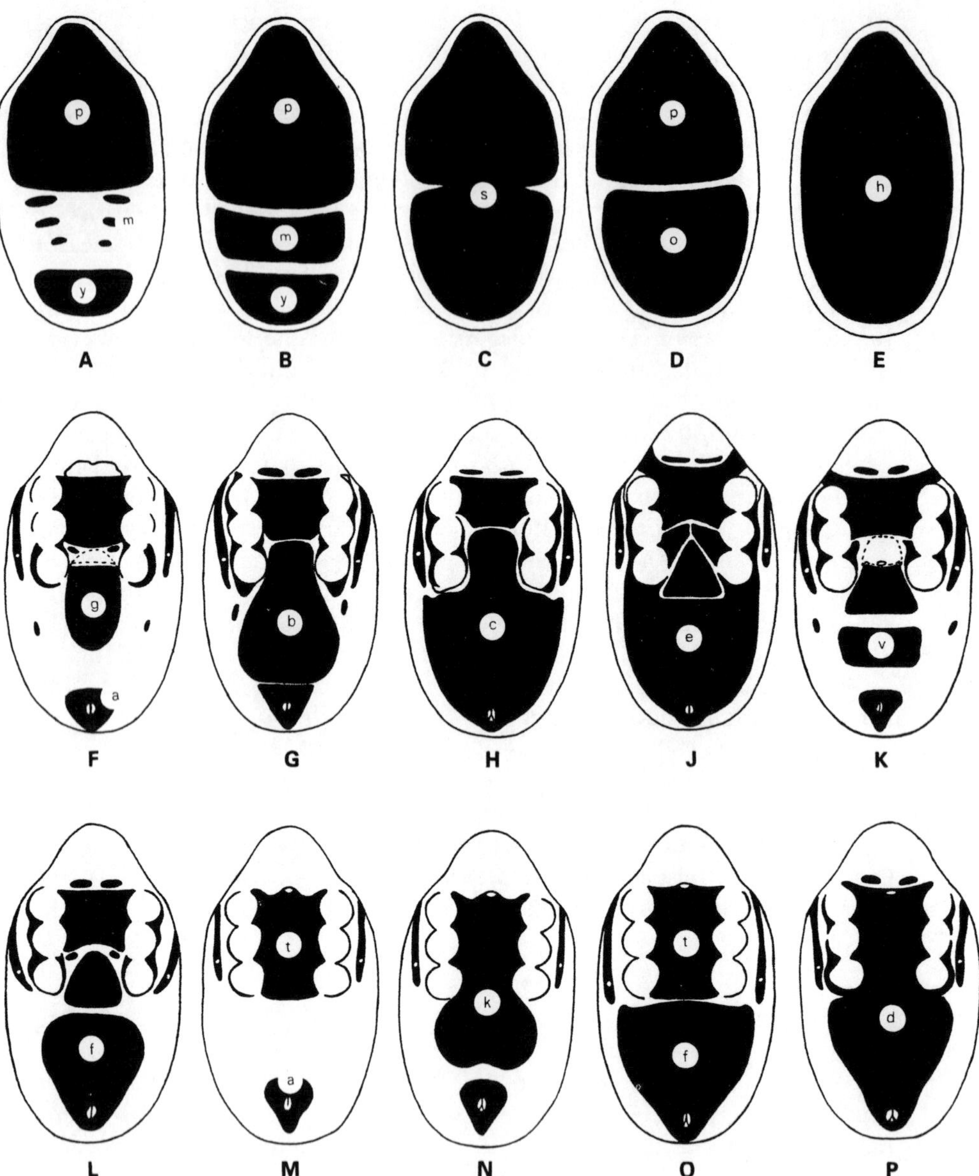

Fig. 12.6. Diagrammatic representation of a range of types of dorsal and ventral idiosomatic sclerotization in the Dermanyssina and Parasitina (Mesostigmata): **A–E**, dorsal sclerotization of females; **F–L**, ventral sclerotization of females; **M–P**, ventral sclerotization of males. (After Evans & Till, 1979.) *a.* = anal shield; *b.* = opisthogenital shield; *c.* = hologastric shield; *d.* holoventral shield; *e.* = opisthogastric shield; *f.* = ventro-anal shield; *g.* = genital shield; *h.* = holodorsal shield; *k.* = sternoventral shield; *m.* = mesonotal shield; *o.* = opisthonotal shield; *p.* = pronotal shield; *s.* = schizodorsal shield; *t.* = sternogenital shield; *v.* = ventral shield; *y.* = pygidial shield.

region, a pygidial shield restricted to the posterior region of the opisthosoma and between these, a single mesonotal shield or two or more pairs of mesonotal scutella (Fig. 12.6A,B). Various combinations of these scutal elements usually occur during ontogeny. For example, fusion of the pygidial and mesonotal shields results in an opisthonotal shield (Fig. 12.6D) while that of the three elements forms a holodorsal shield (Fig. 12.6F) or a schizodorsal shield if the fusion between the pronotal and opisthonotal shields is incomplete laterally, as indicated by the lateral incisions (Fig. 12.6C). A marginal shield partially or entirely surrounding the holodorsal shield occurs in the deuteronymphs and adults of certain Uropodina (see Fig. 12.11C–E below). The dorsal surface of the idiosoma is provided with setae, gland pores and slit-organs and various systems of terminology have been proposed for them, particularly for the Dermanyssina, Parasitina and Uropodina. The most generally used system has been discussed and illustrated (see Fig. 2.3B, p. 25) in Chapter 2 and also by Evans and Till (1979).

The sternal shield(s), associated setae and slit-organs (=lyrifissures), and the genital orifice, with the rare exception of the uropodoid *Metagynella*, are located in the intercoxal region (Figs 12.7A & 12.8). Additional sclerites may occur anterior to the sternal shield and are referred to as presternal shields whilst those carrying the first pair of sternal setae are termed jugular shields. A tritosternum is usually present medially near the posterior margin of the gnathosoma and typically comprises a rectangular base from which arises a single lacinia or a pair of laciniae. In the Uropodoidea, the tritosternum is usually concealed by the enlarged coxae I (see Fig. 12.11F below). The female genital orifice, except in the Cercomegistina and Antennophorina, is protected by a single genital shield which may be extended anteriorly as a hyaline flap. This shield extends posteriorly into the opisthogastric region in the Dermanyssina, Parasitina, Epicriina and Zerconina but is usually restricted to the intercoxal region in the Sejina, Microgyniina, Uropodina and Diarthrophallina. In the Cercomegistina and Antennophorina, three basic genital shields are present comprising a median mesogynal shield and paired latigynal shields (Fig. 12.8C). Fusion of the mesogynal and latigynals with podal elements may occur so that the region posterior to the genital orifice is occupied by an entire mesogynal–latigynal shield (Fig. 12.8D). In some taxa, especially those in which the mesogynal and latigynal shields are reduced in size, a fourth genital cover, the sternogynal shield, is developed and considered to comprise parts of the sternal shield and possibly a sclerotized region of the vagina (Fig. 12.8E,F). The genital orifice of the male is located either at or near the anterior margin of an intercoxal shield, the sternogenital, or within such a shield, usually at the level of coxae II–III (Fig. 12.7B,C). In the latter position, the subcircular orifice is protected by two shields, the anterior of which may bear a pair of setae.

The coxae of legs I are situated with the gnathosoma in a gnathopodal cavity. Those of legs II–IV are inserted in acetabula surrounded by a

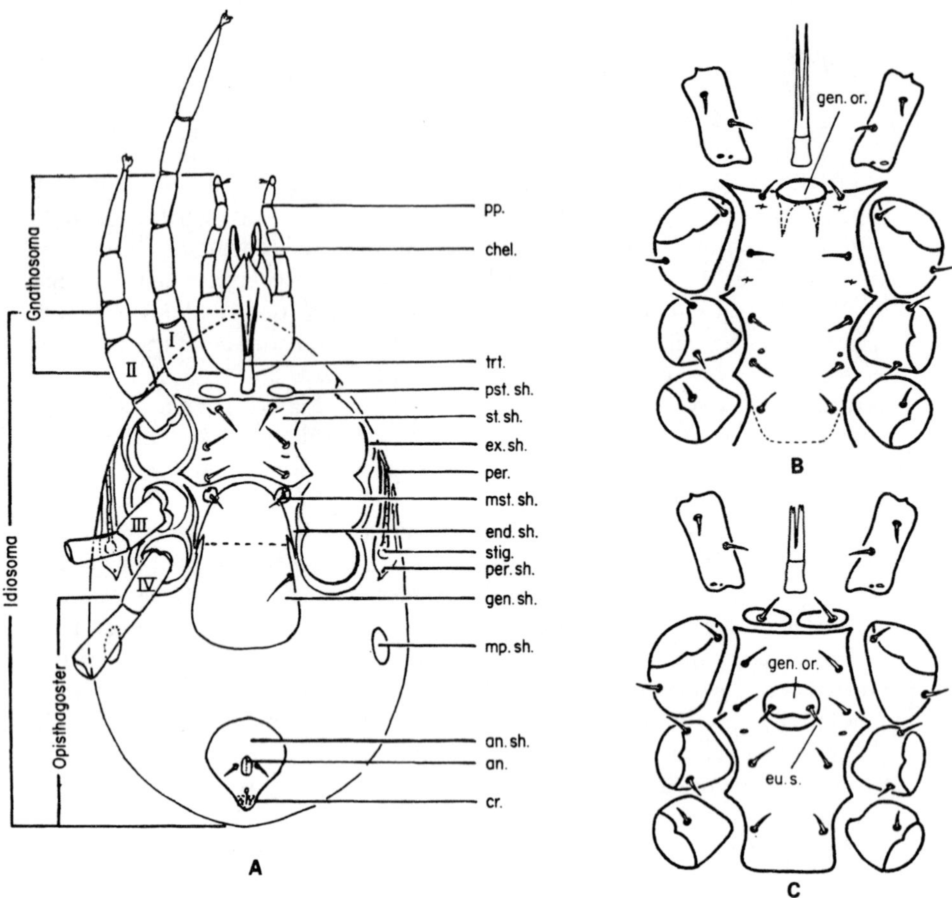

Fig. 12.7. **A**, diagrammatic representation of the venter of a dermanyssoid mite (Mesostigmata) illustrating external morphology; **B**, sternogenital region of a male parasitid mite; **C**, sternogenital region of a male epicriid mite. (After Evans & Till, 1979.) *an.* = anus; *an. sh.* = anal shield; *an. v.* = anal valve; *chel.* = chelicera; *cr.* = cribrum; *end. sh.* = endopodal shield; *ex. sh.* = exopodal shield; *eu. s.* = eugenital seta; *gen. sh.* = genital shield; *gen. or.* = genital orifice of male; *mp. sh.* = metapodal shield; *mst. sh.* = metasternal shield; *per.* = peritreme; *per. sh.* = peritrematic shield; *pp.* = pedipalp; *pst. sh.* = presternal shield; *st. sh.* = sternal shield; *stig.* = stigma; *trt.* = tritosternum; *I–IV.* = legs I–IV.

holopodal shield or by individual exo- and endopodal shields. The endopodal shields may be free or fused with the sternal or sternitogenital shield and the exopodal shields are often fused with the peritrematic shields. The exopodal shields are particularly well developed in the Uropodoidea and Fedrizzioidea and border depressions or pedofossae into which legs II–IV can be withdrawn for protection (see Fig. 12.11F below). Podal shields are absent or weakly developed in larvae and protonymphs. In the majority of the females of the Dermanyssina, the paired openings of the sperm access system are located in the arthrodial cuticle of the acetabula III.

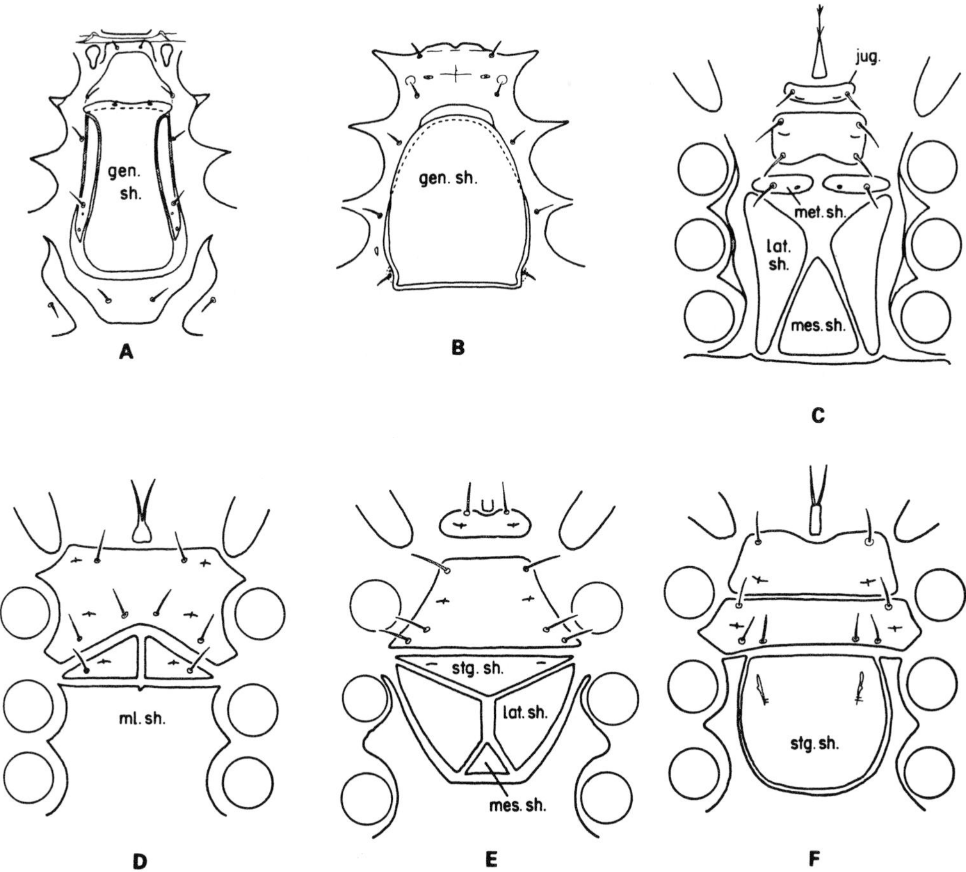

Fig. 12.8. Sclerotization and chaetotaxy of the intercoxal region of females of the Uropodina (**A**, trachytid-type; **B**, polyaspidid-type); Cercomegistina, (**C**, holocercomegistid-type); Antennophorina (**D**, celaenopsid-type; **E**, klinckowstroemiid-type; **F**, fedrizziid-type). (After Evans & Till, 1979.) *gen. sh.* = genital shield; *jug.* = jugularium; *lat. sh.* = latigynal shield; *mes. sh.* = mesogynal shield; *met. sh.* = metasternal shield; *ml. sh.* = fused mesogynal-latigynal shield; *stg. sh.* = sternogynal shield.

The anus is usually subterminal and enclosed in an anal shield bearing three setae, the paired adanals (=paranals) and unpaired postanal, or a more extensive shield, the ventro-anal, incorporating a number of opisthogastric setae. A spiculated area, the cribrum (see Fig. 12.7A above), may be present posterior to the postanal seta and is considered to provide a dispersal platform for sex pheromones produced by the cribral glands, whose ducts open by solenostomes closely flanking the cribrum (Krantz & Redmond, 1988). The anal valves may bear a maximum of one pair of euanal setae, usually in the early instars. In some females, the ventro-anal shield may fuse with the genital shield to form a hologastric shield (see

Fig. 12.6H above) while in males a fusion of the sternitogenital and ventro-anal shields is common and produces a holoventral shield (see Fig. 12.6P above). A variety of types of ventral sclerotization are shown diagrammatically in Fig. 12.6F–P above. One or two pairs of metapodal shields are usually present posterior to coxae IV in the Dermanyssina.

The stigmata in the nymphs and adults are located in the podosomatic region ventrolateral to coxae III–IV or, in the case of the Uropodoidea, coxae II–III (Figs 12.7A & 12.11F). An anteriorly directed canal, the peritreme, is usually associated with each stigma. The peritreme is often strongly reduced or lacking in endoparasitic Dermanyssoidea and in the Diarthrophallina and Epicriina. The stigma and peritreme are typically enclosed by a peritrematic shield which shows varying degrees of fusion with the exopodal shields and the dorsal shield.

The tubular gnathosoma is movably articulated to the idiosoma and lies in a distinct gnathopodal cavity. The gnathotectum is well developed and usually produced anteriorly into one or more gnathotectal processes which overhang the hypostomatic region of the subcapitulum (see Fig. 5.2A, p. 133). The chelicerae comprise three articles and are normally chelate-dentate. Mechanosensory and chemosensory sensilli are usually present and may consist of a dorsal (*id*) and an antiaxial lateral (*iα*) slit-organ and a dorsal seta (*d*) near the basal region of the fixed digit, and a pilus dentilis and one or more putative chemosensory sensilli in the anterior half of the fixed digit (see Fig. 5.12A, p. 150). The arthrodial membrane at the base of the movable digit is often produced into a coronet of setiform processes or one or two brush-like structures. Filamentous, dendritic or brush-like digital excrescences, particularly of the movable digit, are usually present in adults of the Antennophorina and Cercomegistina. The male chelicera is adapted for sperm transfer in the Dermanyssina, Parasitina and Heterozerconina. It takes the form of a spermatodactyl in the Dermanyssina and a spermatotreme in the Parasitina, both structures being associated with the movable digits (see Fig. 5.12D, p. 150, & Fig. 9.15C,D, p. 286). In the Heterozerconidae, however, it is the fixed digit which is modified for sperm transfer and is provided with a helicoidal canal. The chelicerae in parasitic species exhibit considerable modification in the degree of development of the fixed digit, the dentition of the digits and the relative lengths of the first and second articles. The hypostomatic region bears a maximum of three pairs of setae and a pair of corniculi while the gnathosomatic base has one pair of setae (Fig. 12.9A). The subcapitular groove may be provided with rows of anteriorly directed denticles. The pedipalps normally comprise six articles of which the apotele is represented by a movable two-, three- or four-tined claw-like structure at the inner basal angle of the palptarsus. A fusion of the tibia and tarsus to form a tibiotarsus occurs in some Uropodina and Antennophorina. The trochanter lacks setae in the larval phase but otherwise the trochanteral to tarsal palpomeres have a setal complement. The number of setae on the palpomeres and their ontogenetic development provide useful taxonomic criteria (Evans, 1964) and

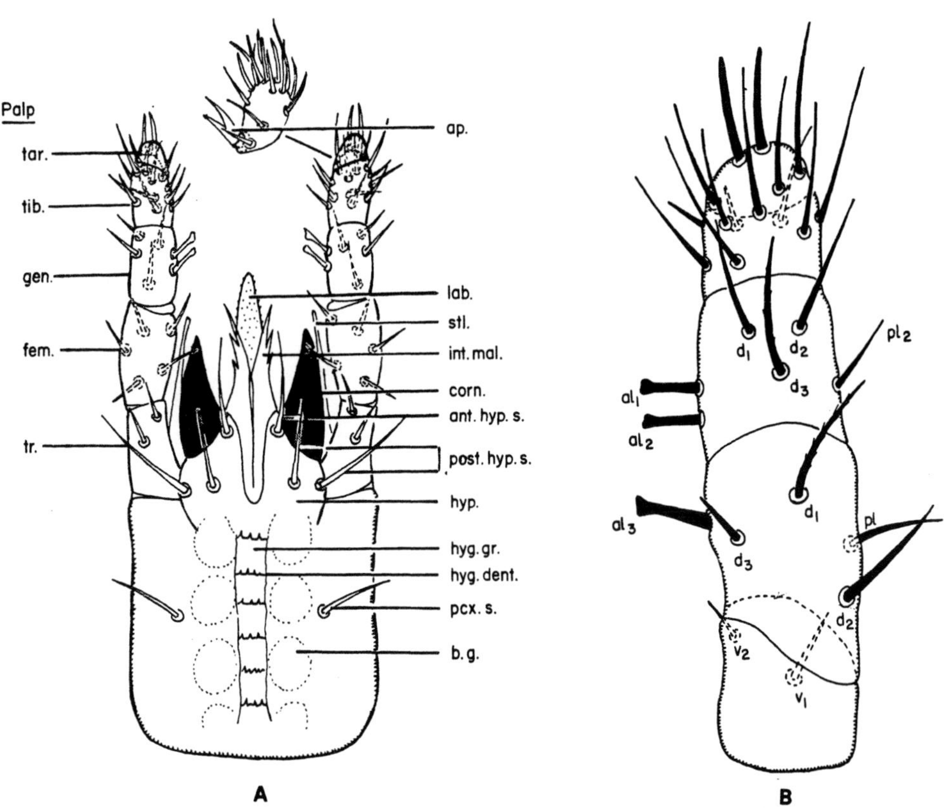

Fig. 12.9. **A**, diagrammatic representation of the venter of the gnathosoma of a laelapoid mite (Mesostigmata) (after Evans & Till, 1979); **B**, pedipalp trochanter to tibia of an ameroseiid mite illustrating the notation for the setae of the trochanter, femur and genu (*al.* = antero-lateral seta; *d.* = dorsal seta; *pl.* = postero-lateral seta; *v.* ventral seta) (after Evans, 1964). *ant. hyp. s.* = anterior hypostomatic seta; *ap.* = apotele; *b. g.* = gnathosomatic base; *corn.* = corniculus; *fem.* = femur; *gen.* = genu; *hyg. dent.* = subcapitular denticles; *hyg. gr.* = subcapitular groove; *hyp.* = hypostome; *int. mal.* internal mala (hypostomatic process); *lab.* = labrum; *pcx. s.* = pedipalp coxal seta; *post. hyp. s.* = posterior hypostomatic setae; *styl.* = salivary stylus; *tar.* = tarsus; *tib.* = tibia; *tr.* = trochanter.

the terminology for the setae of the three basal movable palpomeres, based on the system used for the podomeres, is given in Fig. 12.9B.

The legs have seven podomeres with the femora and tarsi II–IV divided by a peripodomeric suture associated with slit organs. Tarsus I in some phoretic deuteronymphs of the Parasitina has an apicotarsus. The system of terminology for the leg setae is discussed in Chapter 2. The number, distribution and ontogenetic development of the setae on the podomeres provide useful taxonomic characters at all levels of classification. The ambulacrum typically comprises two claws and a pulvillus but may be lacking on the first pair of legs. Hypertrophy of one or two setae on certain podomeres of legs II and, less commonly, legs IV of the male occurs in some Parasitina, Dermanyssina and Uropodina, with formation of spur-like structures (See Fig. 12.15E below).

There are normally a larval and two nymphal instars. A prelarva is not known to occur. The larva is hexapod and lacks setae on the palptrochanter. The octopod nymphs lack an opening to the genital system and, normally, the protonymph has one palptrochanteral seta while the deuteronymph has two. A heteromorphic phoretic deuteronymph may be present in the life cycle of the Uropodina and Microgyniina.

A detailed account of the external morphology of the Mesostigmata is given by Evans and Till (1979).

Classification

The order Mesostigmata is considered to comprise the following 12 suborders: Antennophorina, Arctacarina, Cercomegistina, Dermanyssina, Diarthrophallina, Epicriina, Heterozerconina, Microgyniina, Parasitina, Sejina, Uropodina and Zerconina. Their interrelationships have been the subject of considerable discussion but remain problematical, Camin and Gorirossi (1955) used the number of primary shields associated with the female genital orifice as the major criterion in the division of the order into two cohorts. The Trigynaspida, comprising the Antennophorina and Cercomegistina, was distinguished from the Monogynaspida, comprising all other taxa, in having basically three genital shields as opposed to one in the latter group. Additional characters which support this division are the presence in the Trigynaspida of an unpaired mesonotal shield lying between the pronotal and pygidial in the deuteronymph, an unpaired seta posterior to the vertical setae of the pronotum, and the presence of excrescences on the movable digits of the chelicerae. Hirschmann (1975), on the other hand, considered the presence or absence of setae on the larval pygidial shield to be a basis for the division of the group into two supercohorts, the Trichopygidiina (with pygidial setae) and Atrichopygidiina (without pygidial setae). The Atrichopygidiina contained two cohorts, the Uropodina and Trachyuropodina, while all other cohorts were included in the Trichopygidiina. Although the classification of Camin and Gorirossi (1955) has been widely adopted, particularly by North American acarologists, that of Hirschmann has received little or no support.

In contrast, Athias-Henriot (1975a) introduced both reproductive and morphological criteria into the higher classification of the Mesostigmata. The major division of the order was based on the method of insemination, namely podospermic or tocospermic. Within the podospermic forms, the Dermanyssina was distinguished from the Heterozerconina by the different modifications of the male chelicerae for insemination and by their adenotaxy. The presence or absence of a spermatotreme and excrescences on the movable digit of the chelicerae, and the type of adenotaxy were used to characterize the Parasitina, Zerconina, Epicriina, Uropodina and Antennophorina (s. lat.).

Studies on sperm structure and the evolutionary aspects of methods of insemination in the Mesostigmata by Alberti (1984, 1988, 1991) have necessitated a reappraisal of the broad concept of tocospermy as presented by Athias-Henriot (1975a). On the basis of sperm structure, two types of sperm cells are present: the vacuolated form, which is considered to be plesiomorphic, occurs in the Epicriina, Sejina, Uropodina, Zerconina and probably also in the Antennophorina (including Cercomegistina) while the derived ribbon-type is present in the Dermanyssina and Parasitina. Males with ribbon-type sperm cells have the chelicerae modified as gonopodes and a presternal genital orifice while the females have the ovary differentiated into nutrimentary and germinative parts and an unpaired oviduct. Those taxa with vacuolated sperm cells have unmodified chelicerae and, with rare exceptions a mid-sternal genital orifice in the male and a non-nutrimentary tubular ovary and paired oviducts in the female. Thus, tocospermy in the sense of Athias-Henriot (1975a) covers two basically different systems, one of which is derived and found in the Parasitina and the other plesiomorphic and known or assumed to occur in the taxa with vacuolated sperm cells. The tocospermic condition in the Parasitina, to some extent, occupies an intermediate position between podospermy and the type of tocospermy occurring in the other mesostigmatic taxa. Although our knowledge of methods of insemination and sperm structure is incomplete or lacking in many of the 'tocospermic' taxa, present data do suggest that the plesiomorphic and apomorphic sperm types can be used as basis for the higher classification of the Mesostigmata. Following Alberti (1988), the two major taxa could be referred to as Neospermatida (for the tocospermic Parasitina and podospermic Dermanyssina) and Archispermatida (for the tocospermic Epicriina, Sejina, Uropodina, Zerconina and probably the Antennophorina) and given hierarchical rank between order and suborder, but this action would seem to be premature at present.

The following key to the suborders of the Mesostigmata is based chiefly on adults which exhibit the characteristic morphology of their suborder and does not necessarily accommodate neotenic and highly specialized parasitic forms.

1. Tarsus of leg IV with a minimum of 20 setae, setae av_1 and pv_1 always present (see Fig. 3.8C, p. 74); euneoadenic 2

– Tarsus of leg IV with a maximum of 18 setae, setae av_1 and pv_1 absent; protoadenic or euneoadenic 4

2. Without excrescences on movable digits of chelicerae; female with one primary genital shield (monogynaspid condition) bearing six to numerous setae .... SEJINA (see Fig. 12.10A,B below)

– Movable digits of the chelicerae with filamentous, dendritic or brush-like excrescences; female genital orifice covered by three primary shields (trigynaspid condition) which may be variously coalesced or reduced, when coalesced female without outwardly opening genital shield; salivary styli absent 3

3. With two or more dorsal shields; if one, then a line of fusion apparent between pronotal and opisthonotal elements; vaginal sclerites absent; gnathotectal process with anterior projection or serrations but without ventral keel; tritosternal laciniae fused, rarely separated; palpgenu normally with six setae .... CERCOMEGISTINA (see Fig. 12.8C above)

With typically a single dorsal shield, if divided, vaginal sclerites present; gnathotectum triangular, smooth and normally with ventral keel; tritosternal laciniae rarely fused proximally; palpgenu usually with seven setae .... ANTENNOPHORINA (see Fig. 12.8D–F above)

4. Both sexes usually with a pair of sucker-like organs on the venter of the opisthosoma; some taxa (Heterozerconidae) with six setae on the palpfemur; poorly known mites associated with myriapods and snakes .... HETEROZERCONINA (see Fig. 12.16J below)

– Without such sucker-like organs; palpfemur with not more than five setae 5

5. Opisthonotal shield with two pairs of large, subcircular muscle-insertion scars near posterior margin; idiosoma typically subtriangular in outline; peritreme short, often comma-shaped; male genital orifice mid-sternal; palpal claw two-tined .... ZERCONINA (see Fig. 12.14A,B below)

– Opisthonotal shield, when present, without large muscle scars 6

6. Dorsal shield with a polygonal network of relatively large bi- or trifurcate tubercles; legs I long, without ambulacrum and with tarsus bearing two or more long clubbed setae ventrally; peritreme markedly reduced and enclosed with stigma in small depression or lacking; male genital orifice mid-sternal; palpal claw three-tined .... EPICRIINA (see Fig. 12,14C–F below)

– Dorsal shield without such tubercles; peritreme usually conspicuous except in some parasitic species and commensals of insects 7

7. Female with sternal setae I–II and III–IV on separate shields or all sternal setae on separate shields; dorsum of adults with pronotal, mesonotal (single or paired) and pygidial shields; genital orifice of male mid-sternal; life cycle with facultative pedunculate phoretic deuteronymph .... MICROGYNIINA (see Fig. 12.10C,D below)

– Female with arrangement of sternal setae and shields otherwise; dorsum with one or two shields, if mesonotal and pygidial elements present then male with genital orifice presternal 8

8. Genital shield of female oval, subtriangular or tongue-shaped, partially or entirely enclosed by a sternal–endopodal–ventral shield and typically located within the intercoxal region; male with subcircular genital orifice mid-sternal; subcapitular setae forming more or less a longitudinal row on either side of the subcapitular groove, rarely, posterior hypostomatic setae forming a transverse row; male chelicerae not modified for sperm transfer; dorsal slit-organ of tarsi II–IV usually closely associated with the peripodomeric fissure, if well-separated, dorsum with pygidial shield(s) or paranal setae greatly elongated 9

– Genital shield of female flask-, wedge-shaped or subtriangular usually extending well into the opisthogastric region, if surrounded by fused sternal–endopodal–ventral sclerites, then posterior hypostomatic setae forming a transverse row; male with genital orifice presternal rarely mid-sternal (Arctacarina); male chelicera with or without modification for sperm transfer; dorsal slit-organ of tarsi II–IV widely removed from the peripodomeric fissure 10

9. Dorsum in adults and immatures with a single dorsal shield, never flanked by marginal shield; dorsal chaetotaxy markedly hypotrichous but up to six pairs of elongate marginal setae conspicuous; anal shield with very long paranal setae; specialized parasites of passalid beetles .... DIARTHROPHALLINA (Fig. 12.13)

– Dorsum in immatures with more than one dorsal shield, adults usually with pygidial shield(s) or marginal shield in addition to dorsal element; dorsal chaetotaxy usually hypertrichous; stigmata and peritremes often displaced anteriorly by the development of pedofossae; tritosternum and gnathosoma may be partially or entirely covered by enlarged coxae I; tritosternum with narrow base rarely well removed from coxae I, usually concealed by coxae I or with a broad base abutting coxae I; scabellum often present; paranal setae not conspicuously elongated; heteromorphic phoretic deuteronymph usually present .... UROPODINA (see Fig. 12.12 below)

10. Metasternal shields in the female large, free or fused medially with the sternal shield and flanking a triangular genital shield antero-laterally; endogynum strongly developed and often with sclerotized teeth-like structures; genital orifice of male presternal; male chelicera with spermatotreme .... PARASITINA (see Fig. 12.15C–E below)

– Typically without large metasternal shields in the female, the metasternal setae being situated on small shields, on striated cuticle or fused with the sternal shield; endogynum absent or weakly developed; male chelicera unmodified or with spermatodactyl; male genital orifice presternal or mid-sternal 11

11. Female with podonotal shield and very small mesonotal scutella and a pygidial scutellum; pilus dentilis of female bifurcate; without sperm access system; male genital orifice mid-sternal with anterior genital valve bearing a pair of setae; a single family restricted to boreal regions .... ARCTACARINA (see Fig. 12.15A,B below)

– Free-living females without above combination of dorsal sclerotization; pilus dentilis simple, setiform, inflated or absent; sperm access system unpaired or paired and opening by solenostomes in the region of coxae III and IV or on legs III, rarely on opisthogaster; male genital orifice presternal; movable digit of male chelicera with spermatodactyl; largest suborder with free-living and parasitic representatives . ... DERMANYSSINA (see Fig. 12.16A–H below)

Suborder Sejina

Members of this suborder, at one time referred to as the Liroaspina, are considered to be among the most primitive of the Mesostigmata. They are euneotrichous (exhibiting primary neotrichy) and euneoadenic. The dorsum of the idiosoma of the female has two (pronotal and opisthonotal or pronotal–mesonotal and pygidial) or three to six (comprising pronotal, mesonotal and pygidial elements) shields (Fig. 12.10A) while the male usually has a holodorsal shield or two dorsal shields. The pliable cuticle is often tuberculate. The large female genital shield with six to numerous setae occupies the greater part of the intercoxal region with the result that the area of sternal sclerotization is much reduced and fragmented (Fig. 12.10B). The genital orifice of the male is mid-sternal and subcircular. Both sexes usually have a ventro-anal shield and large anal orifice. The subcapitulum has four pairs of setae and a pair of corniculi, usually strong and dentate and resembling rutella. In both sexes the chelicerae possess dentate digits that lack processes. Tarsus IV has at least 20 pairs of setae (setae av_4 and pv_4 present).

The classification of this suborder requires revision. Johnston (1982) recognizes three families (Sejidae, Ichthyosomatogastridae and Uropodellidae), but Athias-Henriot (1977) considers that all present known genera should be included in a single family, the Sejidae (=Liroaspididae). Its members occur in litter and decaying wood, in nests of seabirds and small mammals. Deuteronymphs of *Archaeopodella scopulifera* have been collected from *Rattus fuscipes* in Australia. The suborder is cosmopolitan. Species with strong dentate corniculi and large anal openings appear to ingest solid food. In *Asternolaelaps*, fragments of pollen, fungal spores and arthropod cuticle have been observed in the alimentary tract (Athias-Henriot, 1972). Hirschmann (1991) has revised the genus *Sejus*.

Suborder Microgyniina

The adults are small, delicate mites bearing a superficial resemblance to the Sejina in having podonotal, mesonotal (pair or single) and pygidial sclerites on the dorsum of the idiosoma (Fig. 12.10C). The chaetotaxy is hypertrichous. The females have the sternal shield divided into two parts, each bearing two pairs of sternal setae, and the genital shield may be entire with a single pair of setae or divided into two small sclerites, each

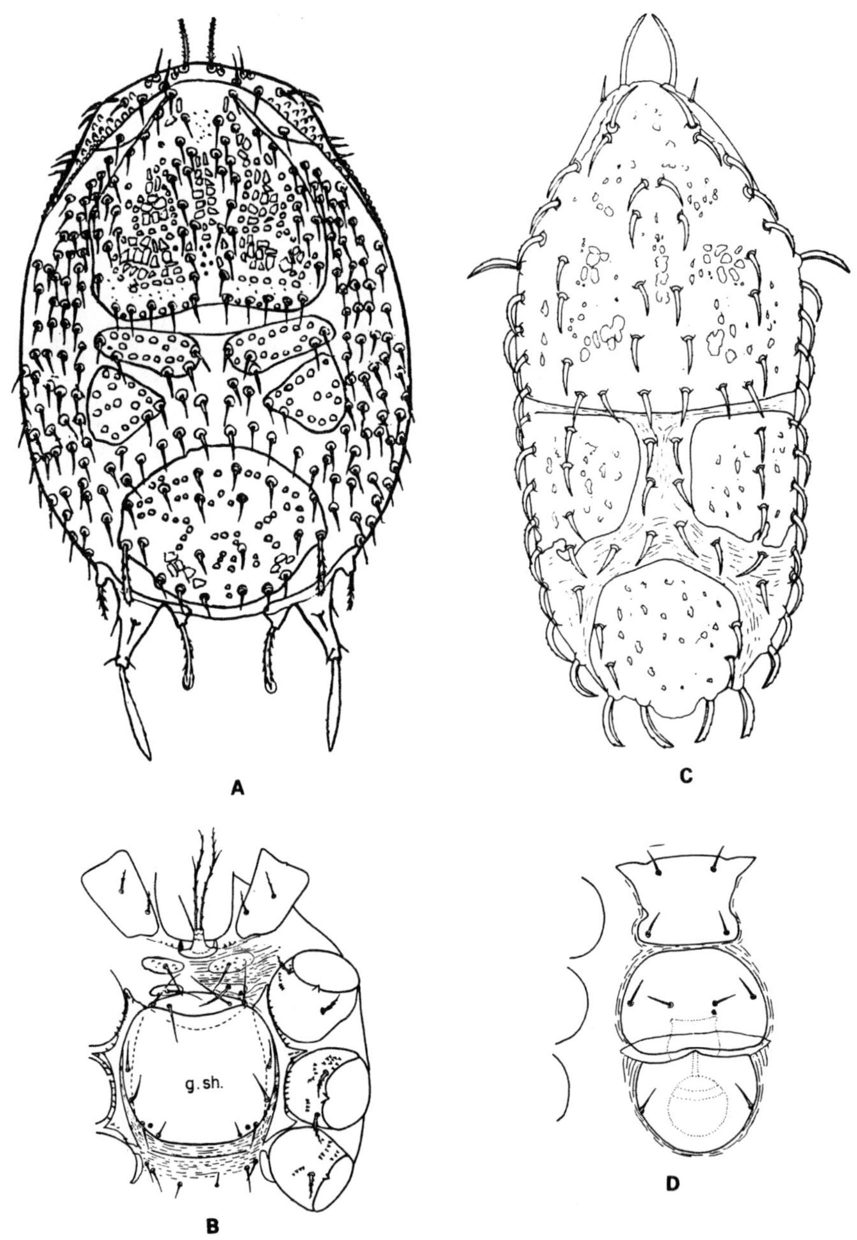

Fig. 12.10. A, B, dorsum (**A**) and intercoxal region (**B**) of *Sejus togatus, female (Sejina); C, D, dorsum (**C**) and intercoxal region (**D**) of the female of Microsejus trucicola* (Microgyniina). After Evans & Till (1979).

with a seta (Fig. 12.10D). A ventro-anal shield is usually present in both sexes. Phoretic deuteronymphs with modified anal region may be present in the life cycle. The male genital orifice is mid-sternal. The cheliceral digits are weakly dentate and those of the male are not modified for sperm transfer. The claws on legs I are sessile; tarsus IV is without setae av_4 and pv_4.

The suborder comprises two genera classified in a single family (Microsejidae). The species are inhabitants of tree stumps and rotting wood in the Holarctic region. Their feeding habits are unknown.

References

Evans and Till (1979); Gilyarov & Bregetova (1977); Tragardh (1942).

Suborder Cercomegistina

The dorsum of the idiosoma of adults has separate pronotal and opisthonotal shields or pronotal, mesonotal and pygidial shields, more rarely a holodorsal shield with transverse suture. The idiosomatic chaetotaxy is usually hypertrichous. Females have sternal, latigynal and mesogynal shields which vary in size and degree of sclerotization but latigynals and mesogynal may be fused (Seiodidae). Vaginal sclerites are absent. Sternal setae I are usually situated on unsclerotized cuticle or on a jugularium (see Fig. 12.8C above). The male orifice is mid-sternal. The gnathotectal process in both sexes is typically serrate or subdivided but without a ventral keel. The chelicerae have the movable digit or both digits with a row of closely-set teeth. The movable digit bears distinct excrescences. The palpgenu has the normal complement of six setae. The tritosternal laciniae are typically fused basally. Tarsus IV has setae av_4 and pv_4.

Four families are recognized of which members of the Asternoseiidae, Davacaridae and Seiodidae are free-living in litter and humus while some Cercomegistidae are associated with other arthropods.

References

Athias-Henriot (1959); Camin and Gorirossi (1955); Evans (1958); Kethley (1977); Kinn (1972).

Suborder Antennophorina

This suborder together with the Cercomegistina formed the division Trigynaspida of Camin and Gorirossi (1955). A holodorsal shield is usually present in the adults. The genital orifice of the male is typically mid-sternal but is presternal in the Celaenopsoidea. A maximum of four sclerites, paired latigynals, mesogynal and sternogynal are associated with the genital orifice and may be free or variously fused (see Fig. 12.8D,E above). Vaginal sclerites are present. The gnathotectum is typically mucronate and has a ventral keel. The movable digit of the chelicera bears filamentous or dendritic excrescences. The palpgenu usually bears seven setae (a unique condition for the Mesostigmata), except in highly specialized forms. The tritosternal laciniae are separate. Tarsus IV has av_4 and pv_4 present.

Johnston (1982) defines six superfamilies. The majority of species are associated with insects, particularly ants (Antennophoroidea and Aenictequoidea) and passalid beetles, but associations with myriapods, lizards and snakes are not uncommon. Little is known of their feeding habits. Donisthorpe (1927) describes the feeding behaviour of *Antennophorus* species on ants of the genus *Lasius*. The ant disgorges a droplet of fluid in response to the mite stroking its mouth and this is imbibed by the mite. Species of *Megisthanus* are considered to be predatory.

References

Butler and Hunter (1968); Camin and Gorirossi (1955); Kethley (1977).

Suborder Uropodina

Members of this suborder show a trend towards the development in the adult of a highly specialized body form associated with adaptations for the protection of the gnathosoma and legs. The ontogenetic development of the dorsal sclerotization of the idiosoma is characteristic. Larvae have a sclerotized pronotal shield with a postero-median extension for the attachment of the cheliceral retractor muscles, mesonotal scutella and a small pygidial shield, usually but not invariably without setae (Fig. 12.11A). The protonymphal sclerotization resembles that of the larval (Fig. 12.11B) except in the Trachyuropodidae in which the pronotal and mesonotal elements are fused to form a single shield. The dorsal shields are usually coalesced in the deuteronymph and adult but the pygidial sclerite may remain free from an antero-dorsal shield (Fig. 12.11D). A marginal shield showing various degrees of development is usually present in the adults and in some deuteronymphs (Fig. 12.11C–E). The dorsal chaetotaxy is frequently hypertrichous (Fig. 12.12D). The sternum is completely fused and typically surrounds a trapdoor-like genital shield in the female and a round or oval orifice protected by two sclerites in the male (Figs 12.8B & 12.12E). The sternum is usually fused with the opisthogastric shield in the adults so that the entire venter is sclerotized (Fig. 12.12E). In the majority of the Polyaspidoidea and Uropodoidea, depressions or pedofossae occur lateral to legs II–IV into which the legs can be withdrawn for protection. The tritosternum has its base longer than wide with two laciniae or wider than long with a single simple or divided lacinia. The former type is either covered by enlarged coxae I or widely separated from them, whereas the latter type usually abuts coxae I (Fig. 12.11F,G). The cheliceral digits are usually short relative to the long shaft and the fixed digit is typically provided with two or more sensory organules subapically. The male chelicera is not modified as a gonopod in these tocospermic mites. In some genera of the Uropodoidea, a conspicuous sclerotized node is developed on the levator tendon near its

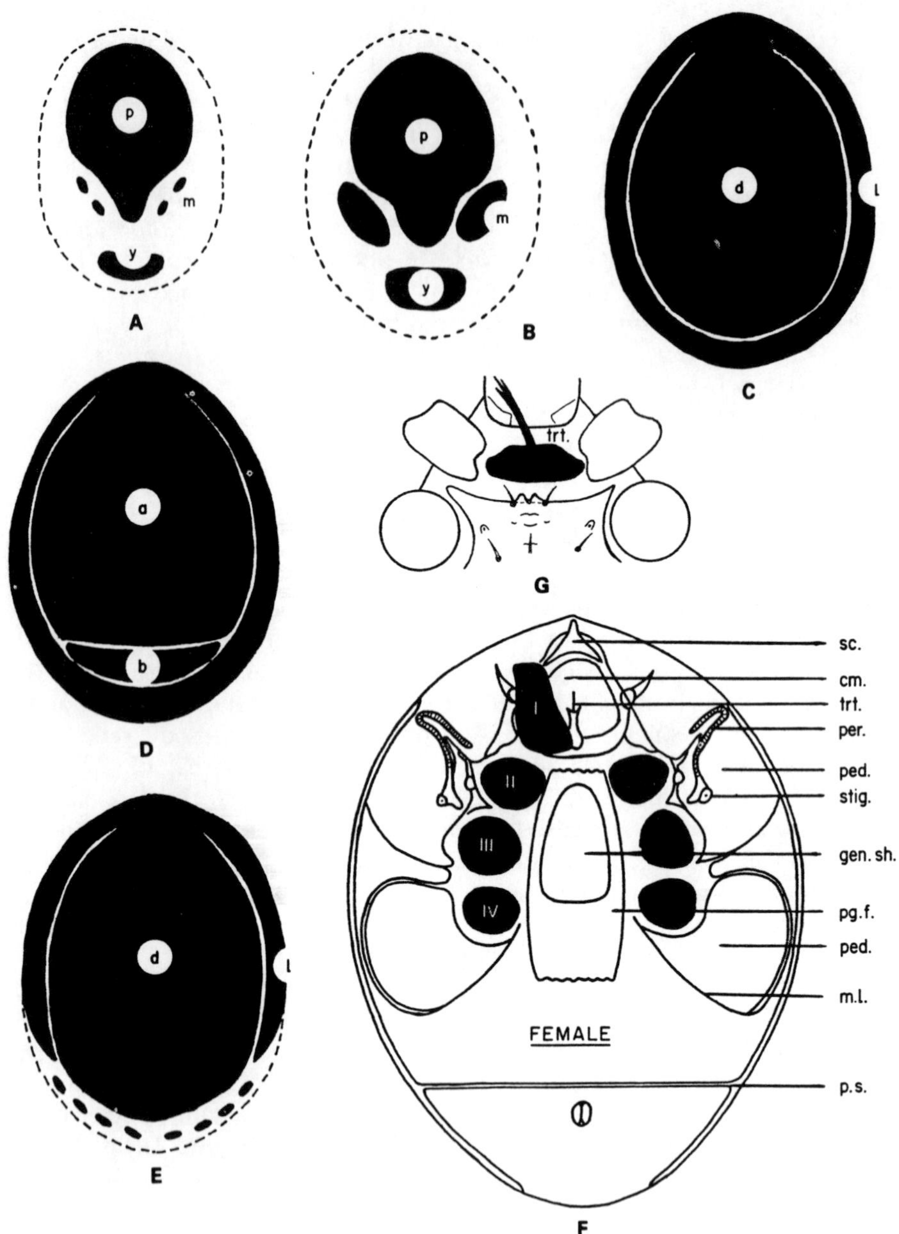

Fig. 12.11. **A–E**, diagrammatic representation of the dorsal sclerotization of the larva (**A**), protonymph (**B**) and adults (**C–E**) of the Uropodoidea (Mesostigmata); **F**, venter of a generalized uropodoid mite showing external morphology; **G**, tritosternum of polyaspidid mite. (After Evans & Till, 1979.) *a.* = antero-dorsal shield; *b.* = postero-dorsal shield; *cm.* = gnathopodal cavity; *d.* = dorsal shield; *gen. sh.* = genital shield; *l.* = marginal shield; *m.* = mesonotal shield; *m. l.* = metapodal line; *n.* = podomesonotal shield; *p.* = pronotal shield; *ped.* = pedofossa; *per.* = peritreme; *pg. f.* = perigenital field; *p. s.* = pre-anal suture; *sc.* = scabellum; *stig.* = stigma; *trt.* = tritosternum; *y* = pygidial shield.

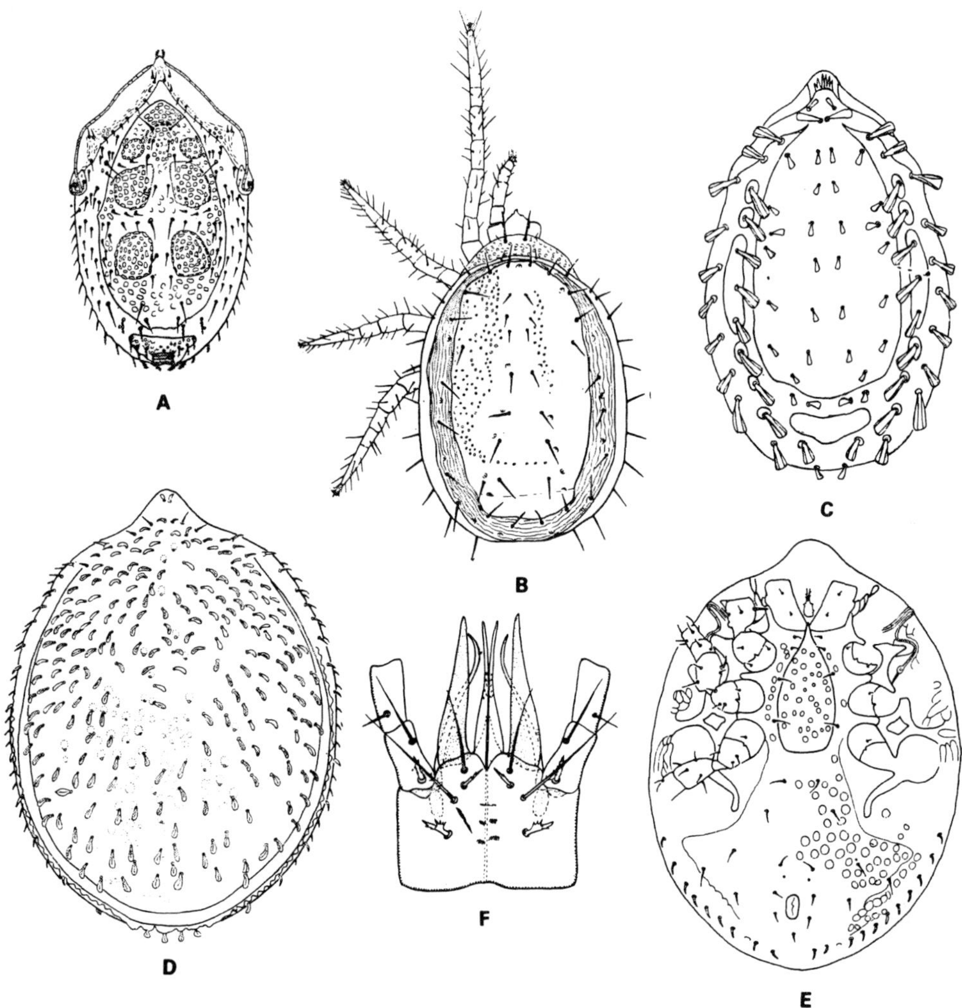

Fig. 12.12. **A**, dorsum of *Thinozercon michaeli* (Thinozerconoidea); **B**, dorsum of *Protodinychus punctatus* (Thinozerconoidea); **C,F**, *Polyaspis patavinus* (Polyaspidoidea), dorsum of idiosoma (**C**) and venter of gnathosoma (**F**); **D, E**, dorsum (**D**) and venter (**E**) of *Oodinychus janeti* (Uropodoidea). After Evans & Till (1979).

attachment to the movable digit (see Fig. 5.12D, p. 150). The hypostomatic processes are often complex but the corniculi are usually short but may be long as in the Polyaspididae (Fig. 12.12F). The four pairs of subcapitular setae are arranged in a longitudinal row on either side of the subcapitular groove. The pedipalp may have the tibia and tarsus fused and the chaetotaxy is often reduced. The legs are usually short and an ambulacrum may be lacking from leg I. The coxae of legs I of those species of the Uropodoidea with pedofossae are markedly enlarged and are capable of concealing the tritosternum and gnathosoma. In such species, a scabellum flanking the gnathopodal cavity is present and provides a platform for legs I when they are withdrawn for protection while flanges on the femora of legs II–IV protect some of the joints of the legs resting in the pedofossae. The podomeres have a reduced chaetotaxy. Pedunculate phoretic deuteronymphs may be present in the life cycle. Legs II of the male rarely have spurs.

The suborder is considered to comprise three superfamilies – Thinozerconoidea (Fig. 12.12A,B), Polyaspidoidea and Uropodoidea. The familial and generic classifications, particularly of the Uropodoidea, require revision. The Uropodina are mainly inhabitants of organic matter, either of a permanent nature, as in forest soils, or of a transient nature as in dung, compost, tidal debris, stored products and nests of birds and mammals. Dispersal from transient habitats is usually by means of a phoretic deuteronymph. Some uropodoids are inhabitants of ants' nests and show a range of associations with these insects (Donisthorpe, 1927). In some myrmecophilous species, the dorsum of the idiosoma extends ventrally to form a groove which fits over a specific article of the antenna or podomeres of the ant host. This type of clasping device occurs for example, in *Coxequesoma umbocauda* and *Antennequesoma rettenmeyeri* which attach, respectively, to the leg coxae of *Eciton hamatum* (Elzinga, 1982a) and the antennae of *Nomamyrmex esenbecki* (Elzinga, 1982b).

References

Ainscough (1981); Athias-Binche (1982); Athias-Binche and Evans (1981); Berlese (1904); Evans (1972); Evans and Till (1979); Hirschmann (1979); Hirschmann and Zirngiebl-Nicol (1964); Karg (1989); Wiśniewski (1979). Extensive contributions to the taxonomy of the Uropodina have been made by W. Hirschmann and his co-workers and are published in a series of papers in *Acarologie, Schriftenreihe für vergleichende Milbenkunde*, Hirschmann-Verlag, Fürth/Nürnberg. A list of papers and new taxa appearing in this publication from 1957 to 1980 is given in Folge 27, pp. 57–120 (1980).

Suborder Diarthrophallina

This suborder consists of oval or elongate highly specialized mites showing morphological adaptations for parasitism. The single dorsal shield in both sexes (and also immature instars) is extremely hypotrichous with some of the marginal setae, up to six pairs, long and whip-like (Fig. 12.13A). The ovoid shield covering the genital orifice of the female is surrounded by the sternal element of a compound sternito-ventral shield in much the same way as in the Uropodoidea (Fig. 12.13B). A weak endogynum is present. In some of the elongate species, the female genital orifice is situated more posteriorly, between coxae IV. The male aperture is mid-sternal and protected by sclerites. The anal shield has a pair of very long paranal setae and the peritremes are reduced or absent. The small,

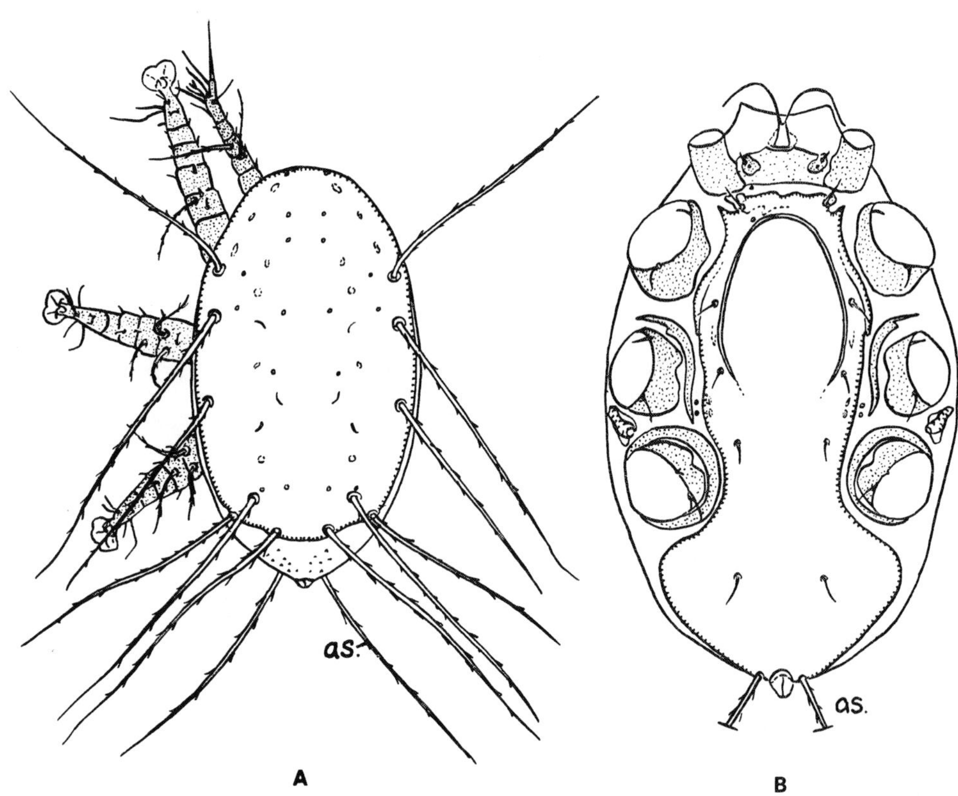

Fig. 12.13. Female representative of the Diarthrophallina in dorsal (**A**) and ventral (**B**) view. *as.* = paranal setae.

chelate chelicerae have a hyaline appendage bearing fimbriae or weak spurs while the gnathotectum is mucronate or produced into simple or fringed processes. The pedipalpal chaetotaxy is much reduced and the palpal-claw is lacking. Legs I are shorter or subequal in length and more slender than legs II and lack ambulacra. Legs II–IV have ambulacra comprising well-developed pulvilli but reduced claws. Coxa I is often fragmented resulting in one of the coxal setae lying on a small sclerite postero-lateral to the tritosternum. In general, the podomeric chaetotaxy of all legs is hypotrichous.

All instars of the Diarthrophallidae (the only family) are associated with beetles of the family Passalidae and considered to be parasitic. They are most commonly found in Neotropical and Australasian regions. The suborder appears to be most closely related to the Uropodina.

References

Schuster & Summers (1978); Womersley (1961a,b).

Suborder Zerconina

Both sexes have two subequal, reticulated dorsal shields with serrated margins. The opisthonotal shield has four conspicuous muscle-attachment scars near its posterior margin (Fig. 12.14A). Sternal setae I are situated on a normal sternal shield. The female genital shield has one pair of setae and is rounded or truncate posteriorly while the male genital orifice is mid-sternal. The inguinal glands are multiple and both sexes have a large ventro-anal shield (Fig. 12.14B). Peritrematic shields are strongly developed and may extend posterior to the level of coxae IV. The peritreme is long in the deuteronymph but in the adults is reduced in length and often sickle-shaped. Euanal setae may be retained by the deuteronymph and adult. The chelicerae are small with few teeth, and the male's chelicerae are not modified for sperm transfer. The third pair of hypostomatic setae forms more or less a transverse row with the second pair of setae. The chaetotaxy of the pedipalps is normal (adults with 2-5-6-14-15). Leg I has an ambulacrum; leg II of the male is not modified for mating.

This tocospermic group contains a single family (Zerconidae) and its members inhabit litter and humus of woodlands and grasslands, mosses and nests of small mammals. The species are considered to be predominantly oligophagous predators.

References

Blaszak (1975, 1979); Gilyarov and Bregetova (1977); Halašková (1970).

Suborder Epicriina

The members of this suborder are readily recognizable by the ornamentation of the holodorsal shield in adults which comprises a number of bi- or trifurcate tubercles forming a polygonal pattern (Fig. 12.14C). The tubercles, according to Alberti *et al.* (1981), are formed of secretory material (cerotegument). The dorsal chaetotaxy is hypotrichous. Sternal setae I are situated on jugularia in both sexes. The flask-shaped or posteriorly truncate genital shield of the female carries one or two pairs of setae; both sexes have ventro-anal shields (Fig. 12.14D). Inguinal glands are typically multiple. The male genital orifice is mid-sternal with the anterior genital valve bearing a pair of setae (12.14E). The peritreme is markedly reduced or absent and enclosed with the stigma in an oval depression. The cheliceral digits are weakly dentate but the dorsal seta is long. The corniculi are slender and digitiform. Tarsi I lack ambulacra but have long subterminal setae and from two to four long clubbed ventral setae (Fig. 12.14F); other tarsi have pulvilli and two claws. Leg II of the male is not modified for mating.

This small group of mites is represented by two genera (*Epicrius* and *Berlesiana*) and the species occur in forest litter, mosses and decaying wood in the northern hemisphere.

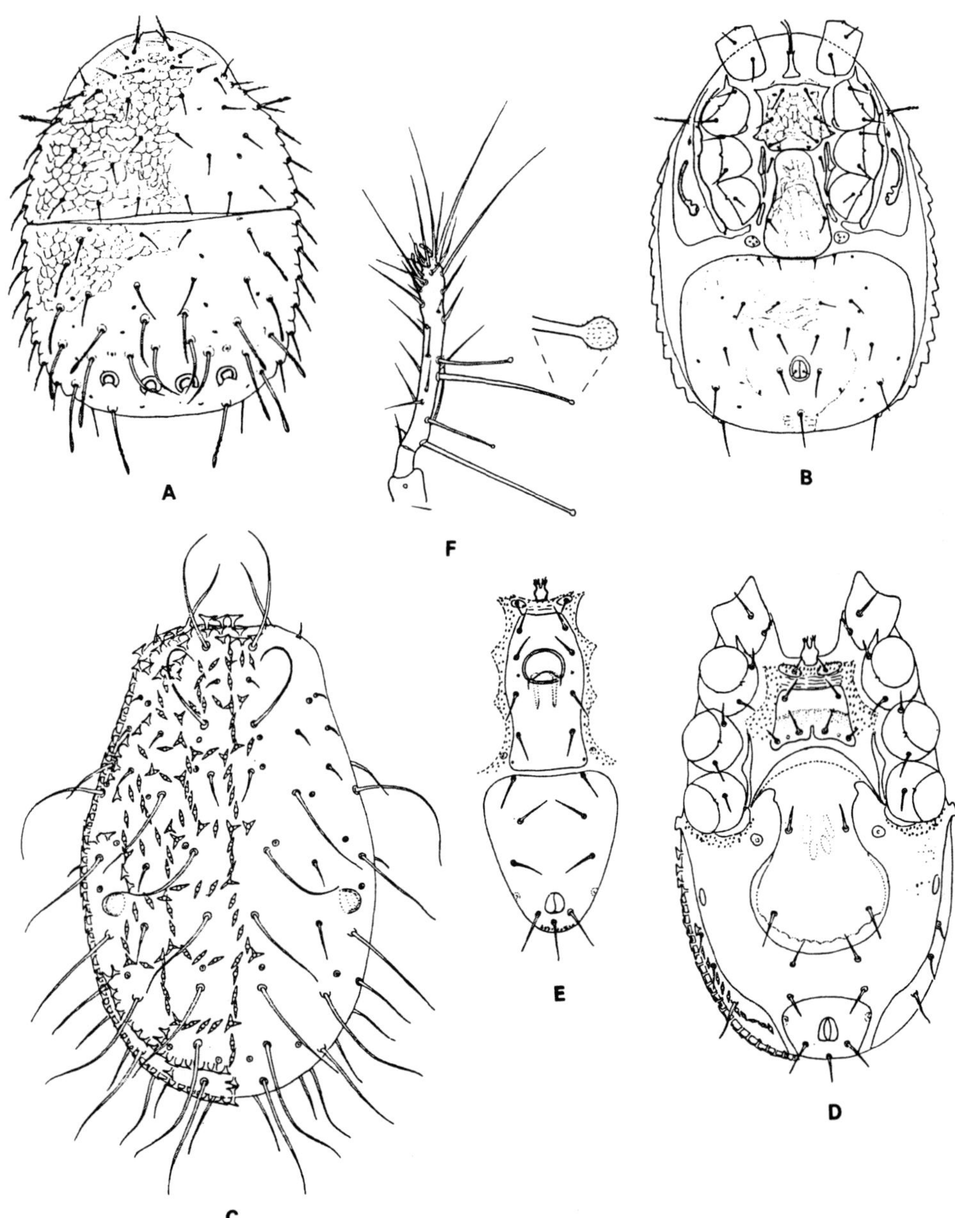

Fig. 12.14. **A**, **B**, dorsum and venter of the idiosoma of the female of *Zercon triangularis* (Zerconina); **C–E**, *Epicrius spinituberculatus* (Epicriina), dorsum of female (**C**) and venter of female (**D**) and male (**E**); **F**, tarsus I of *Epicrius mollis*. (After Evans & Till, 1979.)

References

Evans (1955a); Gilyarov and Bregetova (1977).

Suborder Arctacarina

This small group of free-living mites shows a superfical resemblance to members of the Parasitina in life-form. The male has a holodorsal shield but the female dorsum is partially covered by a large anterior shield, either in the form of a pronotal shield or a more extensive sclerite incorporating some opisthonotal setae, a small pygidial sclerite without setae and, in some taxa, mesonotal scutella (Fig. 12.15A). The sternal shield is well developed in both sexes and bears two or three pairs of setae in the female (Fig. 12.15B). The genital shield of the female is tongue-shaped and lacks setae while the male orifice is mid-sternal with the anterior of the two genital sclerites bearing a pair of setae. An endogynum is absent but a small vaginal sclerite may be present. An anal shield is present in the female and a ventro-anal one in the male. The stigmata and peritremes are normal. The gnathosoma is strongly developed with the gnathotectal process broadly triangular or anteriorly attenuated. The large dentate chelicerae have a bifurcated pilus dentilis. The movable digit is not modified in the male. The subcapitulum has the normal four pairs of setae and the corniculi are strong and horn-like. Pedipalps are normal with a three-tined palpal-claw. All the legs are long and relatively slender with the ambulacrum of legs I represented by a pair of small claws, or lacking altogether. The male has a spur on each of femora II and IV.

The suborder is represented by a single family (Arctacaridae) whose members are restricted to the boreal regions of North America and north-eastern Asia. They occur in litter and humus of soils and the form of the gnathosoma suggests that they are predacious.

References

Evans (1955c); Gilyarov and Bregetova (1977).

Suborder Parasitina

The adults are medium to large, well-sclerotized, free-living mites. The idiosoma of the female is covered by a holodorsal or schizodorsal shield or by separate pronotal and opisthonotal shields while the male has a holodorsal shield with at the most a transverse suture-line indicating the fusion of the two subequal dorsal sclerites which occur in all deutero-nymphs. The pronotal chaetotaxy is usually normal but that of the opisthonotum is commonly hypertrichous (Fig. 12.15C). The sternal shield in the female bears three pairs of setae and the metasternal shields are large and flank the antero-lateral margins of a triangular genital shield (Fig. 12.15D). The latter overlies a pair of sclerites, referred to as 'nymphae' by Berlese (1906); the relationship, if any, of these sclerites to

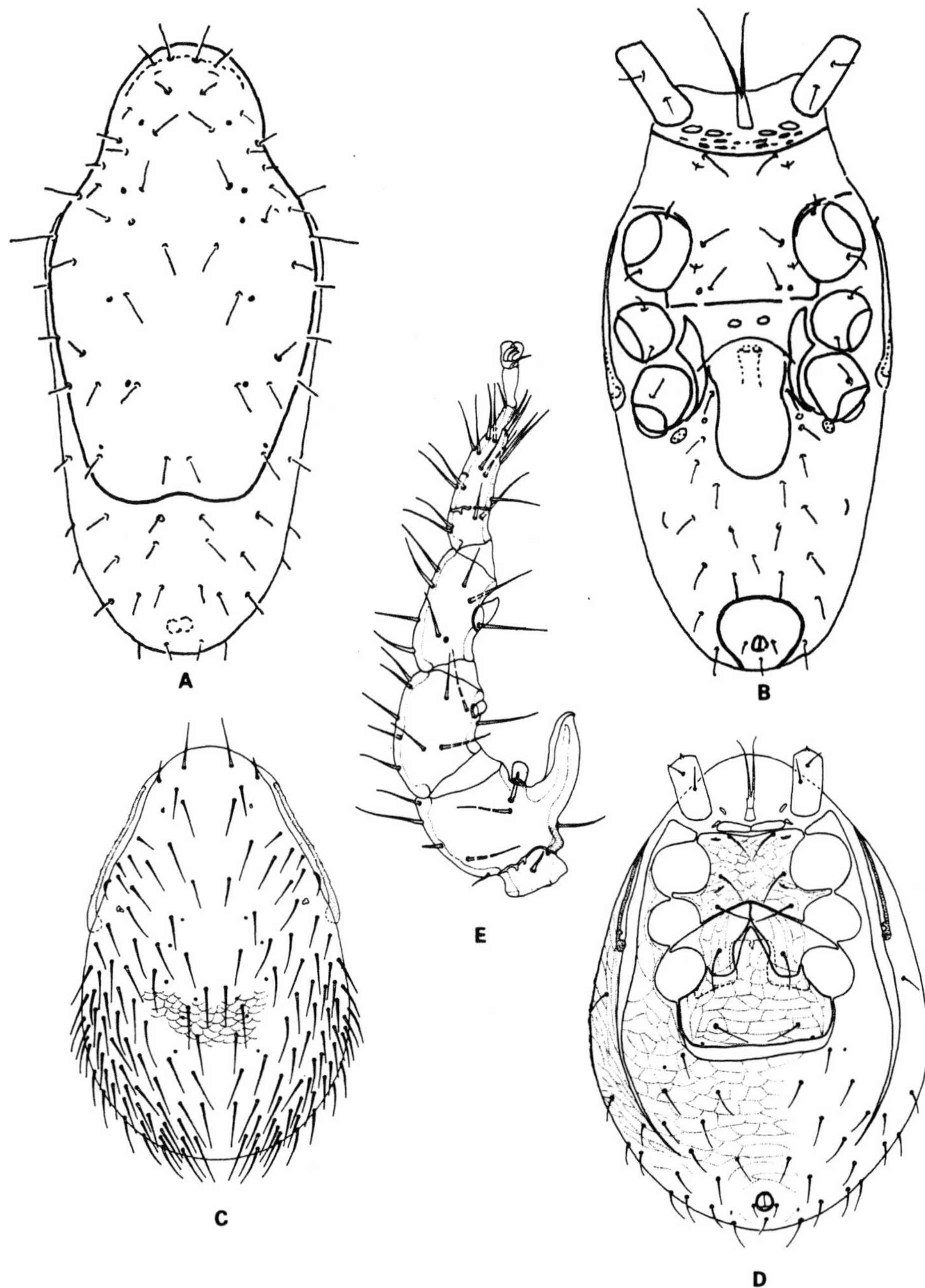

Fig. 12.15. A, B, dorsum (**A**) and venter (**B**) of the female of *Arctacarus rostratus* (Arctacarina); **C**, dorsal shield of the female of *Pergamasus crassipes* (Parasitina); **D**, venter of the female of *Holoparasitus pollicipatus*; **E**, leg II of the male of *Paragamasus* sp.

latigynal shields has not been established. The endogynum is usually large and conspicuous and may be provided with teeth. Presternal sclerites are usually present. The male orifice is presternal and the genital sclerite, on which strong retractor muscles are inserted, overlies the reduced base of a biramous tritosternum and probably functions as an ejaculatory organ. The ventro-anal shield in both sexes is variously fused with the podal, peritrematic and, more rarely, the dorsal shield. The inguinal glands are multiple. The postanal seta in the larva is usually very long but never so in the nymphs and adults. The stigmata and peritremes are normal, the latter rarely enlarged. The chelicerae are strong and dentate with a simple pilus dentilis and the movable digit in the male has a spermatotreme. There are four pairs of subcapitular setae and ten or more transverse rows of subcapitular denticles. The hypostomatic processes are usually large and laterally fimbriate and the corniculi are horn-like. The legs are long with ambulacra, legs I only rarely lacking ambulacra. Leg II of the male and some phoretic deuteronymphs (*Gamasodes*) have spurs (hypertrophied setae) on the femur, genu and tibia (Fig. 12.15E). The tarsus of leg I in the phoretic deuteronymphs of the Parasitinae is divided into an apicotarsus (acrotarsus).

There is only a single family (Parasitidae) with two subfamilies. The members of the Pergamasinae are predominantly soil inhabitants which with the Veigaiidae and Rhodacaridae s. str form the major mesostigmatic predators of forest and grassland soils in the Palaearctic region. Their deuteronymphs are never phoretic. The Parasitinae, on the other hand, occur commonly in temporary accumulations of organic debris, such as compost, manure, tidal debris and nests of bumblebees and small mammals. Colonization of these transient habitats is by phoretic deuteronymphs that are transported mainly by insects. The Parasitinae are cosmopolitan in distribution.

References

Athias-Henriot (1967, 1979b); Berlese (1906); Bhattacharyya (1963); Evans and Till (1979); Holzmann (1969); Hyatt (1980); Karg (1971); Micherdzinski (1969).

Suborder Dermanyssina

This is by far the most species-rich and biologically diverse group of the Mesostigmata and contains free-living as well as a range of parasitic and

Fig. 12.16. A–G, a selection of species of Dermanyssina. **A**, dorsum of *Euryparasitus emarginatus* (Rhodacaroidea), female; **B**, dorsum of *Rhinoceius oti* (Dermanyssoidea), female; **C**, dorsum of *Anthoseius rhenanus* (Phytoseioidea); **D**, venter of female of *Haemogamasus nidi* (Dermanyssoidea); **E**, venter of female of *Eviphis ostrinus* (Eviphidoidea); **F**, venter of female of *Zerconopsis remiger* (Ascoidea); **G**, venter of male of *Laelaps hilaris* (Dermanyssoidea); **H**, position of sperm access system (arrowed) in the Macrochelidae; **J**, venter of the female of *Afroheterozercon spirostreptus*, female showing posterior 'suckers' (Heterozerconina) (after Fain, 1989). A–H after Evans & Till (1979).

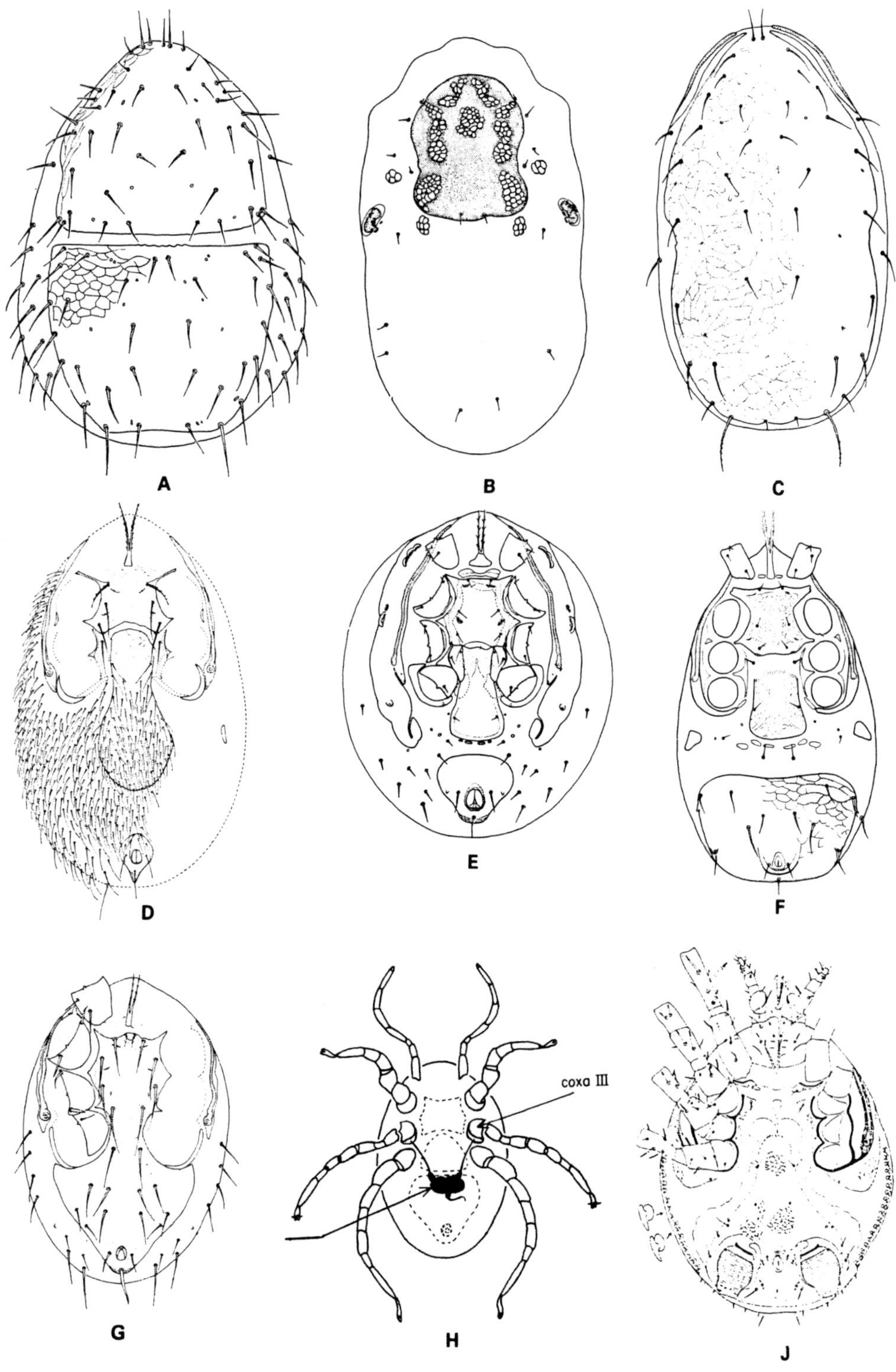
A
B
C
D
E
F
G
coxa III
H
J

paraphagic species occurring on arthropods and vertebrates. The dorsum of the idiosoma in free-living adults is typically covered by a holodorsal or a schizodorsal shield or by subequal pronotal and opisthonotal shields (Fig. 12.16A) although, rarely, pronotal, mesonotal and pygidial sclerites may be present. In haematophagous parasites, a reduction in dorsal sclerotization often occurs to allow for expansion of the body during feeding (Fig. 12.16B). The chaetotaxy is normally holotrichous but hypotrichy (Fig. 12.16C) and hypertrichy (Fig. 12.16D) are not uncommon, especially in endoparasitic species. Ventral sclerotization in the females shows considerable variety based on the extent of the fusion of the four basic ventral sclerites; sternal, genital, ventral and anal. The most common combinations are separate sternal, genital or genito-ventral and anal (Fig. 12.16E) or ventro-anal shields (Fig. 12.16F). In the male, either a holoventral shield or a separate intercoxal shield with a ventro-anal or anal shield is usually present (see Fig. 12.6M–P above). Presternal sclerites may be present in both sexes. The tritosternum is typically biramous but is reduced or absent in some specialized parasitic species. The female genital orifice lies at the level of the anterior margins of coxae IV and the male orifice is presternal and opens through an eversible ejaculatory organ (Fig. 12.16G). The suborder is podospermic and the ducts of a single (laelapid-type, Fig. 12.16H) or paired (phytoseiid-type) sperm access system in the female open by way of solenostomes in the region of legs III or IV, or on legs III. Peritrematic shields are well developed in free-living forms and the stigmata and peritremes are normal except in some parasitic species in which the peritremes are often markedly reduced or absent. The periteme is often enlarged in species living in wet habitats and may function as a plastron. The chelicerae in predatory forms are chelate-dentate but the digits are variously modified in parasitic species (Evans & Till, 1965). The movable digit in the male carries a sperm-transfer organ, the spermatodactyl. The gnathotectum is usually produced into one or more gnathotectal processes. The corniculi are horn-like in predatory forms but may be membraneous in blood-feeding forms. Salivary styli are present and the hypostomatic processes are often fimbriate. The pedipalps are normal with the genu and tibia with never more than five and six setae, respectively. The palp-claw may be two-, three- or four-tined and in some taxa (Veigaiidae) has an associated hyaline flap. The legs are usually relatively long and slender and the chaetotaxy of the podomeres provides stable taxonomic characters (Evans, 1963a). Legs I may lack an ambulacrum but legs II–IV have ambulacra although the claws may be reduced. Legs II and more rarely IV may be armed with spurs.

This is the largest suborder of the Mesostigmata but the superfamilial classification is not settled as is evident from the different concepts of Karg (1971) and Johnston (1982). The latter recognizes five superfamilies. The Veigaioidea comprises a single family of predatory mites commonly found in mosses and the litter and humus of temperate forest soils. Members of one genus (*Cyrthydrolaelaps*) live in rock crevices and under stones on the seashore. The Eviphidoidea are free-living in litter and humus of soils or in compost and manure heaps, or are associates of other arthropods, especially beetles. They feed chiefly on nematodes and other small arthropods. Members of the family Parholaspididae are entirely free living but those of the families Eviphididae, Macrochelidae and Pachylaelapidae contain free-living as well as species associated with Coleoptera, and sometimes Diptera, in their deuteronymphal and/or adult stages. Within the Rhodacaroidea, species of the Ologamasidae and Rhodacaridae are soil-inhabiting predators, those of the latter family being voracious predators of the euedaphon of temperate soils. Members of these two families form the dominant group of mesostigmatic predators in tropical soils. Some Digamasellidae inhabit the tunnels of wood-boring beetles. The Ascoidea contains a diverse group of free-living mites which Karg (1971) considers to represent two superfamilies (Ascoidea and Phytoseioidea). A number of species of Halolaelapidae are intertidal or estuarine and commonly occur in tidal debris, while others occur in compost and manure. The Ameroseiidae appear to be mainly microphytophagous while the Phytoseiidae and Ascidae contain predatory and microphytophagous forms. The Otopheidomenidae are associated with insects as are some species of Antennoseiidae. The superfamily Dermanyssoidea includes predatory soil and nest-inhabiting species as well as species showing a range of associations with other arthropods and vertebrates. The majority of the free-living and nest-inhabiting species are found in the family Laelapidae. Members of the families Dermanyssidae, Macronyssidae, Spinturnicidae and Histrichonyssidae are obligatory blood-feeding ectoparasites, mainly of birds and mammals, while those of the Halarachnidae and Rhinonyssidae are endoparasitic in the respiratory tract of birds and/or mammals. The Varroidae and Iphiopsidae are associated with other arthropods. The adoption of a parasitic mode of life by members of this superfamily has been accompanied by a variety of morphological and life-cycle modifications.

References

Bregetova (1956); Evans (1955b); Evans and Till (1965, 1966, 1979); Gilyarov and Bregetova (1977); Karg (1971); Krantz (1978); Newell (1947b); Radovsky (1967, 1969); Samšiňák and Dusbábek (1971).

Suborder Heterozerconina

The external morphology of members of this suborder has not received detailed study and the following account is based on some species of Heterozerconidae. All species are characterized by the presence of a pair of ventral sucker-like organs on either side of the anus (Fig. 12.16J). Strong retractor muscles originating on an apodeme arising from each of the endopodal shields flanking coxae IV insert in the region of each sucker and their contraction creates a suction within the bell-like suckers when these are applied to a substrate. Both sexes have a hypertrichous holodorsal shield, the hypertrichy being conspicuous marginally. In females the sternal shield is divided so that at least some of the sternal setae and lyrifissures are situated on sternito-endopodal shields which flank the anterior extension of the genito-ventro-anal shield. Males have either similar sternito-endopodal shields, flanking a narrow sternal part of a holoventral shield (*Asioheterozercon*), or a holoventral shield fused with endopodal elements (*Afroheterozercon*). Sternal setae I are often on jugularia. The male genital orifice is presternal with a large genital sclerite (? =ejaculatory organ) covering the basal region of the biramous tritosternum. The suborder is podospermic and females have a sperm access system. The digits of the chelicerae of females are usually long and slender, dentate or edentate and often with an interdigital process bordered with closely set setules. A sperm-transfer process on the male chelicerae either arises near the basal region of the fixed digit and has a helicoidal 'duct' (Heterozerconidae) or is slender and arises from the movable digit (Discozerconidae). The hypostomatic processes, at least in *Asioheterozercon*, are developed into labella-like lobes (see Fig. 5.5, p. 138). The two anterior pairs of hypostomatic setae are widely separated from the third pair. The palptrochanter usually has a large internal process bearing seta v_2, a palpfemur with six setae, and a three-tined palpal-claw. The femur of leg II in the male has one or two strong spurs (hypertrophied setae) and a single spur may also occur on the genu and tibia.

These mites have been collected mainly on myriapods, particularly millipedes, in tropical and warm-temperate regions, but some species of Heterozerconidae have been found on snakes. Little is known of their biology and the two families are in need of revision.

References

Fain (1989); Finnegan (1931).

Superorder Actinotrichida

This superorder is considered, at present, to comprise three orders which may be distinguished on the basis of the following morphological characteristics of the adults:

1. Pedipalps small, comprising one or two articles (rarely three as in *Schizoglyphus*, known from deuteronymph) and closely adpressed to the sides of the subcapitulum (see Fig. 12.23B below); chelicerae chelate; subcapitulum typically with rutella; ambulacra of legs with a single claw and small caruncle (see Fig. 2.17B,C, p. 53), parasites and commensals often with large ambulacral disc and vestigial claws (see Fig. 12.23C below); stigmata and tracheal system absent but 'genital tracheae' may occur rarely; idiosoma usually weakly sclerotized and without trichobothria, never covered by overlapping sclerites or vermiform .... ASTIGMATA

– Pedipalps usually with three to five articles and conspicuous, if small with fewer articles and relatively inconspicuous, then idiosoma either vermiform or with overlapping dorsal sclerites, or with prodorsal trichobothria or chelicerae styliform; chelicerae chelate or movable digits (both digits in Eriophyoidea) modified into piercing stylets or hook-like organs; a respiratory system may be present; prodorsum often with trichobothria 2

2. Subcapitulum with conspicuous rutella (see Fig. 5.7, p. 141); chelicerae chelate-dentate, very rarely styliform; one pair of prodorsal trichobothria comprising a piliform, barbed or clavate sensillus arising from conical bothridium (see Figs 12.27 & 12.31A below); pedipalps simple, palptibia never with a distal claw-like hypertrophied seta; ambulacra of legs I–IV usually with one or three claws, bidactyl condition uncommon, never with pad-like or rayed empodia; tracheal system when present in adults opening to the exterior in the acetabular cavities of legs I and III and in the sejugal furrow; brachytracheae and genital tracheae may be present; usually heavily sclerotized in the adult and with ridge- or wing-like expansions of the idiosoma, which is never vermiform; lateral opisthosomatic glands present except in the more primitive taxa .... ORIBATIDA

– Subcapitulum rarely with rutella; chelicerae may be chelate but usually modified by the loss or reduction of the fixed digit and development of the movable digit into hook-like or styliform structures (see Fig. 6.10, p. 184); pedipalps various, often with tibial-claw and tarsus forming a 'thumb-claw'; ambulacra of at least legs II and III usually with two lateral claws with or without a median empodium which can be pad-like, claw-like, bell-shaped or rayed; tracheal system when present usually opening by stigmata situated between the cheliceral bases or on the anterior region of the prodorsum and often with associated peritremes; usually weakly sclerotized, never with ridge or wing-like expansions of the idiosoma, which may be vermiform or covered dorsally by a number of discrete or overlapping sclerites; lateral opisthosomatic glands absent........... PROSTIGMATA

Order Prostigmata

This is the most heterogeneous of the acarine orders with adults ranging in size from 100 μm to 16 mm. The idiosoma, typically ovoid but possibly elongated or fusiform, is usually weakly sclerotized and sclerites may be present or absent. It is usually divided into two regions, the propodosoma and hysterosoma, by a dorso-sejugal furrow (Fig. 12.17A). The prodorsum bears 3–6 pairs of dorsal setae and 0–2 pairs of trichobothria or is hypotrichous. If only one pair of trichobothria is present, it is not situated postero-laterally. A frontal protuberance (the naso) at the anterior extremity of the prodorsum and overhanging the cheliceral bases occurs in some taxa. An anterior ocellus and one or two pairs of lateral ocelli

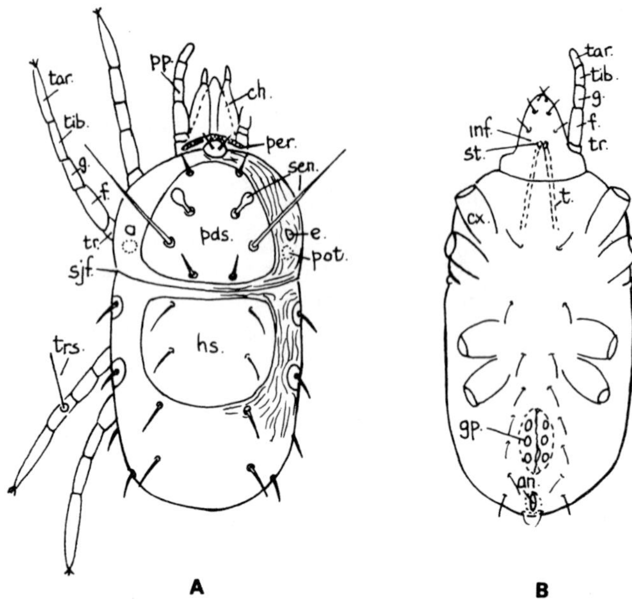

Fig. 12.17. Diagrammatic representation of the dorsum (**A**) and venter (**B**) of a generalized prostigmatic mite showing some basic features of external morphology. *an.* = anal orifice; *ch.* = chelicera; *cx.* = coxa; *e.* = ocellus; *f.* = femur; *g.* = genu; *gp.* = genital papilla; *hs.* = hysterosomatic shield; *inf.* = subcapitulum (infracapitulum); *pds.* = prodorsal shield; *per.* = peritreme; *pot.* = post-ocular tubercle; *pp.* = pedipalp; sen. = sensilli; *sjf.* = sejugal furrow; *st.* = stigma; *t.* = trachea; *tar.* = tarsus; *tib.* = tibia; *tr.* = trochanter; *trs.* = trichobothrium.

may be present on the prodorsum. The hysterosoma may have the holotrichous actinotrichid chaetotaxy of 16 pairs but in some families this complement is reduced while in some other families there is extreme hypertrichy (see Fig. 12.19C below). A maximum of six pairs of lyrifissures (cupules) may be present but more usually there are only four pairs. Some taxa show external evidence of segmentation in the form of transverse integumentary furrows or scissures, or dorsal sclerites. Lateral opisthosomatic glands (latero-abdominal glands) are absent. Anamorphosis is evident in some taxa by the addition of 1–3 post-larval segments (AD, AN and PA) but is absent, with PS remaining paraproctal throughout ontogeny, in the more derivative forms. The leg coxae are plate-like or completely fused with the venter of the idiosoma. The larvae of those species with genital papillae in the nymphs and adults have a pair of Claparède organs situated between coxae I and II. The genital orifice is longitudinal, rarely transverse, generally located at the anterior end of the opisthosoma and closed by a pair of progenital valves (Fig. 12.17B). The genital aperture in the male is dorsal in some specialized parasitic forms (see Fig. 12.20B below). Genital papillae, when present number three pairs, more rarely two pairs, and an ovipositor may be present or absent.

The anus is usually subterminal or terminal, rarely dorsal. Fertilization is either by copulation or spermatophores deposited on a substrate.

The gnathosoma is generally conspicuous except in the Tarsonemina, in which the cheliceral bases are fused with the subcapitulum to form a capsule-like structure (see Fig. 5.15B,C, p. 159) and some parasitic forms such as the Demodicidae (Fig. 12.20F) where the entire structure is much reduced in size. The chelicerae may be chelate-dentate, with one, two or more setae, but usually setae are absent and the fixed digit is reduced while the movable digit is edentate and often styliform (see Fig. 6.10, p. 184). There are no Tragardh's organs or paraxial spines on the cheliceral bases, which may be free or fused together. The fused bases, often forming a stylophore, may be integrated with the subcapitulum. The subcapitulum bears 1–3 pairs of adoral setae and a number of subcapitular setae. Rutella are present in some Endeostigmata but labiogenal sutures are never present. The pedipalps are extremely variable in form and have from one to five free articles. They are often raptorial. An ambulacrum (apotele) is never present but there may be claw-like hypertrophied setae present dorsodistally on the palptibia. A palpal supracoxal spine is usually present.

The legs have the normal podomeric complement except in those species in which there is either an increased number of podomeres resulting from the division of the femur into a basi- and telofemur or a reduced number due to the fusion of certain podomeres. The ambulacrum typically bears two lateral claws with or without a median empodium, which may be claw-like, pad-like, rayed (often in the form of crochet-like 'tenent hairs') or bell-shaped (see Fig. 2.17D–F, p. 53). The lateral claws may also have tenent hairs or simple setiform processes. Trichobothria may be present on the tibiae and tarsi.

Classification

The classification of the Prostigmata and the hierarchic rankings given to the various suprafamilial groups are not settled and the order is in need of fundamental revision. The subordinal taxa adopted in this work, mainly following Kethley (1982), have a practical rather than a phylogenetic basis. Aspects of phylogenetic relationships within the Prostigmata are discussed by Lindquist (1976). The following key to suborders is based on morphological characteristics of the adult phase.

1. Idiosoma fusiform and annulated with either two pairs of legs (plant mites, see Fig. 12.20C below) or four pairs of short stumpy legs with reduced number of podomeres; ambulacrum with two claws (skin parasites, see Fig. 12.20F below), [Eriophyoidea, Demodicidae] .... RAPHIGNATHINA (in part)

– Idiosoma more or less elliptical or round; if fusiform, then with four pairs of normal legs and ambulacrum with two claws and a setulate empodium 2

2. Prodorsum of female with a pair of stigmata and associated tracheae, and a pair of capitate sensilli arising from bothridia, both generally present in females but typically absent in males (except Tarsocheyloidea and Heterocheyloidea); hysterosoma with four or five dorsal sclerites (tergites) which may be reduced in number by fusion in the male (see Fig. 12.21A,C below); cheliceral bases forming a stylophore which may be integrated with the subcapitulum to form a gnathosomatic capsule (see Fig. 5.15B,C, p. 159) .... TARSONEMINA

– Prodorsum of adults without stigmata which when present are located between or on the cheliceral bases; capitate prodorsal sensilli may be present; hysterosoma without such tergites 3

3. Terrestrial mites (free-living, parasites of animals and plants)........... 4

– Aquatic mites 14

4. Subcapitulum with rutella. [Alicorhagioidea, Bimichaelioidea, Oehserchestoidea, Terpnacaroidea) .... ENDEOSTIGMATA (in part)

– Rutella not developed 5

5. Idiosoma and legs densely covered with setae; idiosoma with one or two pairs of eyes, sessile or stalked, and usually one or two pairs of sensilli located within areae sensilligerae with an associated crista metopica (Fig. 12.19B,C below); palptibia with distal claw usually opposing the tarsus (see Fig. 5.9F, p. 144); femora of legs divided by a eudesmatic joint into a basifemur and telofemur. [Trombidioidea, Erythraeoidea, Calyptostomatoidea] .... PARASITENGONA (in part)

– Idiosoma not densely covered with setae, usually arranged in transverse rows; eyes, if present, never stalked; prodorsal sensilli when present without associated areae sensilligerae and crista metopica; femora of legs entire or divided 6

6. Very characteristic yellow or orange mites with the idiosoma completely armoured, the dorsum covered by a single ornamented shield; prodorsal region with an anterior median and a pair of lateral eyes and two pairs of branched sensilli (see Fig. 12.18G,H below); chelicerae independent, large, chelate (see Fig. 6.10A, p. 184); pedipalps linear with four articles; leg femora divided .... LABIDOSTOMMATINA

– Without the above combination of characters 7

7. Two pairs of prodorsal trichobothria present; subcapitulum anteriorly elongated and snout-like; pedipalps raptorial terminating in a claw-like structure or one or two long setae, never with thumb-claw structure (see Fig. 12.18E,F below); chelicerae separate, chelate or movable digit hook-like, fixed digit reduced; peritremes absent. [Bdelloidea] .... EUPODINA (in part)

– With none, one or two pairs of prodorsal trichobothria; if with two pairs then either pedipalps non-raptorial or conspicuous peritremes present 8

8. Basal parts of the chelicerae fused with each other and with the subcapitulum *or* entire gnathosoma represented by a pair of dagger-like sclerites; movable digits, when present, styliform; idiosomatic trichobothria absent; palptibia with conspicuous claw in free-living species but often lacking and number of palpal articles reduced in parasites of birds and mammals in which the idiosoma may be elongate or subcircular and the first pair of legs may be adapted for grasping hairs (see Fig. 12.20G). [Cheyletoidea] .... RAPHIGNATHINA (in part)

– Cheliceral bases independent, partially fused or coalesced to form a stylophore but always distinct from subcapitulum 9

9. Bases of chelicerae fused into a stylophore, movable digits very long, styliform and strongly recurved basally (see Fig. 6,10F,G, p. 184); peritremes run over antero-dorsal region of the idiosoma (see Fig. 4.8B,E,F, p. 114); prodorsal trichobothria and genital papillae absent; with or without palptibial claw. [Tetranychoidea] .... RAPHIGNATHINA (in part)

– Cheliceral bases separate, adnate or fused; if forming a stylophore then stylets shorter and not recurved basally; genital papillae present or absent 10

10. Empodium characteristic, consisting of one to five pairs of divergent tenent hairs, sessile or arising from a median process; palptibia usually with distal claw; bases of chelicerae independent, fused or coalesced to form a stylophore; peritremes present or absent; without prodorsal trichobothria and genital papillae. [Raphignathoidea] .... RAPHIGNATHINA (in part)

– Empodium when present not of this form; palptibia with or without a distal claw; chelicerae independent or fused; peritremes present or absent; prodorsal trichobothria and genital papillae present or absent 11

11. Tarsus I, at least, with one or more recumbent solenidia (see Fig. 3.6G, p. 71), typically inserted in recessed area of the cuticle (rhagidial organ); cheliceral bases independent; two or three pairs of genital papillae; stigmata present at base of chelicerae but peritremes absent [Eupodoidea] .... EUPODINA (in part)

– Rhagidial organs absent; genital papillae present or absent; stigmata and peritremes present or absent; cheliceral bases independent or fused 12

12. Palptibia usually with one to three claw-like setae; if simple, then cheliceral bases bulbous and naso with pair of trichobothria or body elongate and three pairs of *ad* setae on anal valves; peritremes usually present on anterior margin of prodorsum or at base of chelicerae; some taxa with long radiating legs densely covered with setae (Anystidae, see Fig. 12.19A below) or strong internal setae on podomeres of rake-like legs I (Caeculidae); parasitic forms without empodia on legs I–IV .... ANYSTINA

– Pedipalps linear and without claw-like setae on tibia; stigmata usually absent, never with emergent peritremes; legs never radiating or rake-like 13

13. With at the most one pair of prodorsal trichobothria (see Fig. 12.18C below); cheliceral bases fused or adnate, fixed digit reduced or absent; movable digit linear and needle-like; empodium in the form of a setulate pad; genital papillae present or absent; paired stigmata and neostigmata may be present between cheliceral bases; a pair of opisthosomatic trichobothria and an ampullaceous sensillus on tibia I may be present (Ereynetidae) [Tydeoidea] .... EUPODINA (in part)

– With two pairs of prodorsal trichobothria; cheliceral bases independent, fixed digit with a group of comb-like teeth or blunt-ended; respiratory system absent; empodium rayed or large, claw-like; two or three pairs of genital papillae. [Sphaerolichoidea] .... ENDEOSTIGMATA (in part)

14. Mites occurring in fresh water, some in brackish water; pedipalps normally with five articles but may be reduced by fusion of femur-genu or femur-genu-tibia, linear, uncate or weakly chelate; idiosoma elongate or globular, weakly to heavily sclerotized, dorsum with single or many sclerites; majority of species capable of swimming and may have swimming setae on the legs; larvae heteromorphic in respect of nymph and adult, usually parasitic on other arthropods or molluscs; protonymph and tritonymph calyptostasic. [Arrenuroidea, Eylaioidea, Hydrachnoidea, Hygrobatoidea, Hydrophantoidea, Hydrovolzioidea, Lebertioidea, Stygothrombioidea] .... PARASITENGONA (in part)

– Mites occurring principally in marine habitats from interstitial to abyssal; pedipalps linear or raptorial with three or four articles; idiosoma typically flattened and with four dorsal and four ventral plates which may show subdivision or fusion; non-swimmers with claws of legs strong, often divided and may have comb-like projections; larvae homomorphic, protonymph and deuteronymph normal stases. [Halacaroidea] .... EUPODINA (in part)

Suborder Endeostigmata

This is considered to comprise the most primitive members of the

Actinotrichida and the species show a resemblance in their general body plan and chaetotaxy to certain Enarthronota (Oribatida). Reference has been made elsewhere to the uncertain systematic position of this taxon; some authors consider that the rutellate forms belong to the Astigmata–Oribatida lineage and not to the Prostigmata. Idiosomatic sclerotization is usually weak and a dorso-sejugal furrow is generally present. Segments AD and in some cases also AN and PA are added during ontogeny. The prodorsum, bearing 1–2 trichobothria, is developed into a naso and may have ocelli. The podocephalic canal is superficial. The opisthosoma is holotrichous or hypertrichous (Bimichaeliidae and Nanorchestidae) and may show evidence of primary segmentation (Fig. 12.18A). There are 2–3 pairs of genital papillae and eugenital setae are usually present in the adults (Fig. 12.18B). The chelicerae are chelate with 1–2 setae and the subcapitulum has a pair of rutella except in the Lordalychidae and Sphaerolichidae. The pedipalps are linear, never raptorial and without a tibial claw. Stigmata and peritremes rarely occur (Alicorhagiidae; *Stigmalychus*). The femora of the legs are usually divided and the ambulacrum may have both paired claws and an empodium, or claws only or an empodium only. The empodium may be claw-like or occasionally pad-like or rayed. Sperm transfer in the suborder is indirect by spermatophores. A prelarval, homomorphic larval and three (occasionally two) nymphal stases occur in the life cycle.

The suborder comprises predatory and microphytophagous forms (Walter, 1988). The species are found in mosses, forest litter and grassland and appear to be very common in subtropical soils of South Africa (Olivier & Ryke, 1965; Loots & Ryke, 1966). Some species are saltorial.

References

Grandjean (1939b, 1941c, 1943b); Kethley (1982); Theron (1976); Theron and Ryke (1969, 1975a,b); Van der Hammen (1989).

Suborder Eupodina

The idiosoma is weakly sclerotized or has prodorsal and hysterosomatic sclerites which are most extensive in the Halacaroidea. The body is markedly elongate (vermiform) in the Nematalycoidea. The prodorsum is produced into a naso in the Eupodoidea. A dorso-sejugal furrow is present (Fig. 12.18C) or absent. There is usually no anamorphosis but segment AD may be added in some Bdellidae and Eupodidae (Baker, 1987). The genital orifice is generally longitudinal but transverse in the Iolinidae (Tydeoidea). Eugenital setae are present or absent and there are 2–3 pairs of genital papillae except in some Halacaroidea and the majority of the Tydeoidea. The chelicerae are chelate or the fixed digit is reduced or absent and the movable digit usually elongate and styliform (see Fig. 6.10B–E, p. 184). The cheliceral bases may be separate or fused

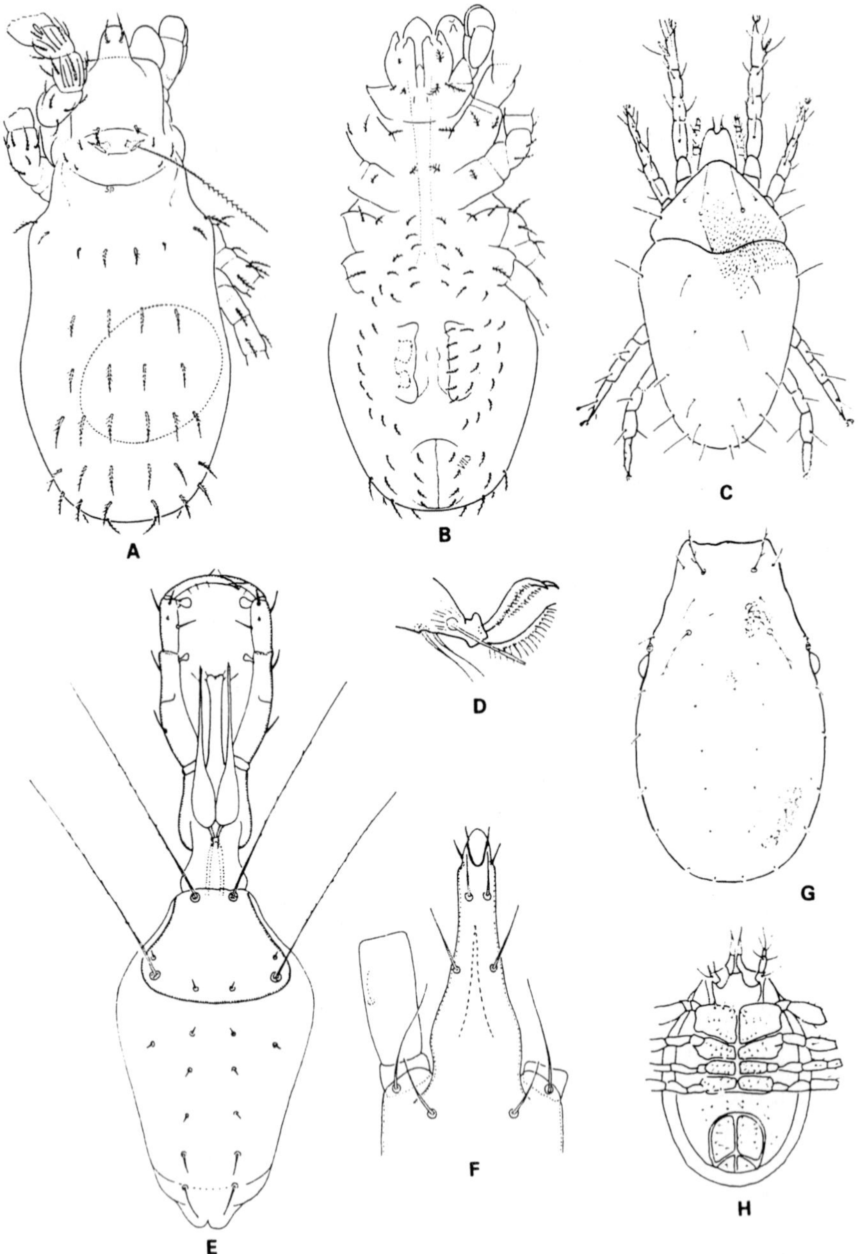

Fig. 12.18. **A**, **B**, dorsum (**A**) and venter (**B**) of female of *Alicorhagia fragilis* (Endeostigmata) (after Grandjean, 1939b); **C**, **D**, dorsum of a representative of the Tydeidae (**C**) with detail of the ambulacrum (**D**) (after Baker *et al.*, 1958); **E**, **F**, dorsum of the idiosoma (**E**) and venter of the subcapitulum (**F**) of *Dactyloscirus inermis* (Cunaxidae); **G**, **H**, dorsum (**G**) and venter (**H**) of a member of the Labidostommatina.

as in some Tydeoidea. The chelicerae have 0–2 setae although the number may be increased in some Bdellidae. Rutella are lacking but the subcapitulum may be produced into a snout-like structure (Fig. 12.18F). The pedipalps are linear or raptorial and usually have four or five articles (Fig. 12.18E) but a reduction to three articles occurs in some Halacaroidea and to two or three in the Iolinidae. The tracheal system opens to the exterior by neostigmata situated at the bases of the chelicerae but is absent in the Halacaroidea. Peritremes are absent. Genital tracheae are present in some Bdellidae. The leg femora are often subdivided and the ambulacrum usually has lateral claws and an empodium (Fig. 12.18D). The latter, when present, may be pad-like, claw-like or rayed. Sperm transfer is either indirect by spermatophores or, less commonly, direct by way of an aedeagus as in the Cunaxidae, Iolinidae and Halacaroidea. A prelarval, larval and three nymphal instars are usually present in the life cycle but a regression of the nymphal stases to calyptostases occurs in some parasitic members of the Ereynetidae.

The majority of the free-living members of the suborder inhabit the organic strata of soils, algae, lichens, mosses, trees and shrubs and are considered to be predominantly predacious but with microphytophagous and omnivorous representatives. The Nematalychidae live in the deeper layers of grassland soils or coastal sands while the Halacaroidea are mainly predatory or algivorous marine forms. Some members of the Ereynetidae are obligate parasites of terrestrial slugs, subelytral parasites of aquatic beetles or inhabitants of the nasal tracts of amphibians, birds and mammals.

References

André (1979, 1980); Atyeo (1960, 1963); Baker (1965); Bartsch (1972); Den Heyer (1981); Fain (1962, 1963b); Gilyarov (1978); Newell (1947a); Strandtmann (1971); Strenzke (1954); Zacharda (1980).

Suborder Labidostommatina

The mites of this small suborder, comprising the single family Nicoletiellidae (=Labidostommatidae), are medium to large in size and heavily sclerotized. A single, strongly ornamented shield covers the entire dorsum of the mite and there is no dorso-sejugal furrow (Fig. 12.18G). The prodorsum is provided with two pairs of trichobothria and three ocelli (two lateral and an antero-median). The dorsal chaetotaxy of the hysterosoma is normal and the group does not appear to be anamorphic. Eugenital setae are absent but there are two pairs of genital papillae. The form of the ano-genital region shows sexual dimorphism, genital and anal regions of the male being distinctly separated by sclerotized cuticle and those of the female contiguous (Fig. 12.18H). The separate chelate chelicerae have enlarged digits and two dorsal setae (see Fig. 6.10A, p. 184). Rutella are absent and the linear palps have four articles. Simple intercheliceral stigmata each with a tracheal trunk are present. The coxal fields are extensive and contiguous and the femora of the legs are subdivided into three articles (basi-, meso- and telo-femora). The ambulacra of legs I are bidactyl or lacking while those of legs II–IV comprise a pair of true claws and a claw-like empodium. Sperm transfer is indirect. The larva is an elattostase but the three nymphal stases are active forms.

The nicoletiellids are found in mosses, lichens and the organic layers of soils and are predators of other arthropods. The family is more or less worldwide in distribution.

References
Feider and Vasiliu (1969); Greenberg (1953); Robaux (1977).

Suborder Anystina

The body sclerotization is usually weak but a prodorsal shield is present in some taxa. Prodorsal trichobothria are lacking in the Pomerantzioidea and Pterygosomatoidea but one or two pairs are usually present in the other superfamilies (Fig. 12.19A). A naso may be present. The dorsal chaetotaxy is often hypertrichous and segment AD (and also AN in Adamystidae) may be added during ontogeny. Genital papillae are usually present but never occur in the Pterygosomatidae. An eversible aedeagus is present only in the Pterygosomatidae. The cheliceral bases are separate with the movable digit typically hook-like and the fixed digit reduced or absent. The Adamystidae are the exception in having the cheliceral bases bulbous, the dentate movable digit weakly hooked and the fixed digit normally developed or reduced. Rutella are never present. The pedipalps are usually raptorial with four or five articles and with one or more claw-like distal setae on the palptibia, but may be linear and non-raptorial. The stigmata at the bases of the chelicerae generally have prominent and often emergent peritremes which may extend onto the anterior region of the prodorsum. The coxal fields are arranged radially in the rapidly moving species of the Anystidae which are commonly referred to as 'whirligig mites' but are linear in slow-moving forms. The leg femora are usually divided but are entire in the parasitic species. The ambulacra have either two claws, with (Pterygosomatidae) or without (Pomerantziidae) tenent hairs, and no empodium or two claws and a membraneous, cup-like or claw-like empodium. Sperm transfer is indirect except in the Pterygosomatidae in which it is probably direct. Prelarval, larval and three active nymphal stases are usually present but the prelarva, protonymph and tritonymph are calyptostases in the Pterygosomatidae.

The suborder as defined in this work comprises a broad assemblage of taxa whose interrelationships have not been satisfactorily established. For example, the systematic positions of the Adamystidae, Paratydeidae, Pomerantziidae and Pterygosomatidae are problematic and their inclusion in the Anystina is debatable. Some authors classify the Adamystidae in the Endeostigmata, the Pomerantziidae in the Raphignathina and the Pterygosomatidae within the Parasitengona. The concepts adopted in this work follow Kethley (1982).

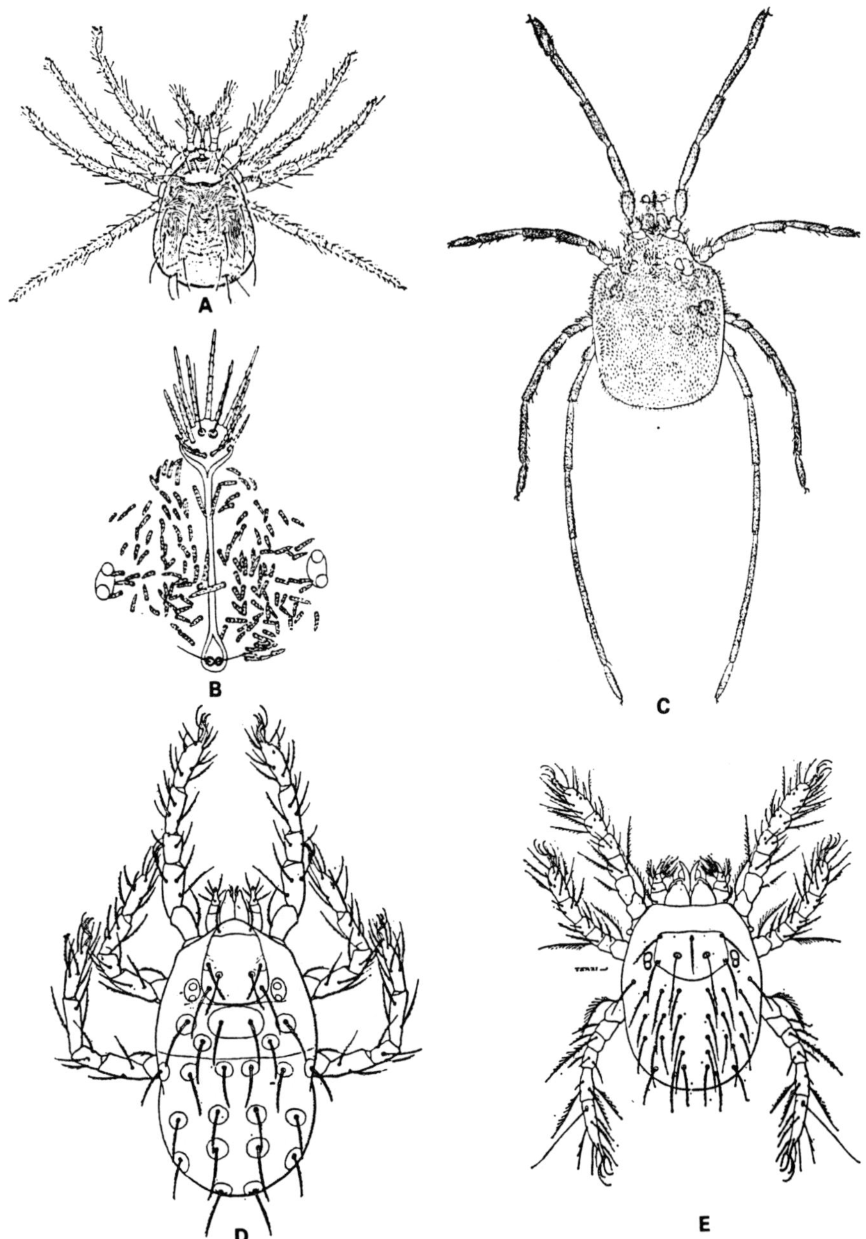

Fig. 12.19. **A**, *Anystis* sp. (Anystina); **B**, crista metopica and ocelli in a trombidiid mite (Trombidioidea) **C**, *Erythraeus phalangioides* (Erythraeidae); **D**, **E**, larval phase; **D**, *Allothrombium fuliginosum* (Trombidiidae), and **E**, *Neotrombicula autumnalis* (Trombiculidae).

Members of this suborder, other than the Pterygosomatidae, are free-living predators living in the litter and humus of soils and in the case of the Anystidae also on low-growing vegetation and on trees. Some species of the Anystidae, Teneriffiidae and Caeculidae occur on the seashore. The Pterygosomatidae are all parasitic either on other arthropods (*Pimeliaphilus*) or on lizards.

References

Coineau (1974, 1979); Davidson (1958); Grandjean (1943a); Jack (1962); McDaniel *et al.* (1976); Oudemans (1936b); Smith Meyer and Ueckermann (1987).

Suborder Parasitengona

The members of this suborder comprise highly specialized terrestrial and secondarily aquatic species. They have a characteristic life cycle in which the larva is parasitic and heteromorphic in respect of the nymphal stases (except in the Calyptostomatoidea). Further, at least the prelarva, protonymph and tritonymph are calyptostasic. The aquatic species have been referred to in the literature as the Hydracarina, Hydrachnellae or Hydrachnida.

The terrestrial representatives, comprising the superfamilies Calyptostomatoidea, Erythraeoidea and Trombidioidea, are relatively weakly sclerotized but nymphs and adults have a linear prodorsal sclerite, the crista metopica, provided with one or two pairs of trichobothria (Fig. 12.19B). This sclerite provides attachment sites for the cheliceral and certain subcapitular muscles. One or two pairs of prodorsal ocelli, sometimes borne on stalks, may be present. The idiosoma is generally very hypertrichous (Fig. 12.19C). In the aquatic families, the sclerotization of the idiosoma is variable and ranges from forms with only a single prodorsal shield to those with two to several sclerites covering the entire dorsum. The idiosoma is rarely hypertrichous in this group. Anamorphosis is considered to be absent. Sexual dimorphism is weak in the terrestrial moiety and restricted to the genital region but is often more pronounced in the water-mites and may involve the legs and palps. Eugenital setae are lacking but genital papillae are usually present. These comprise three pairs restricted to the genital atrium in the terrestrial species but in aquatic forms the papillae may be numerous and superficial and may occur outside the genital region.

The cheliceral bases are separate and never chelate, the fixed digit being markedly reduced or absent and the movable digit sickle-shaped (see Fig. 5.15A, p. 159), or the entire chelicera may be styliform as in the Erythraeidae and Calyptostommatidae. The pedipalps generally have five articles but this number may be reduced in some aquatic mites by the fusion of the femur and genu, or of the femur, genu and tibia as, for example, in the Stygothrombiidae and Limnocharidae. Among the terrestrial forms, the palps are linear and non-raptorial in the Calyptostomatoidea but are raptorial with a pronounced tibial claw in the other taxa (see Fig. 5.9F, p. 144). In the aquatic species, the palps show a variety of forms. They may be linear and simple as in the Eylaioidea, have the tibia produced dorsodistally so as to oppose the tarsus and form a chelate condition as in some Hydryphantoidea, or have a ventrodistal prolongation of the tibia so that the tarsus is able to fold against it (uncate state) as in some Arrenuroidea. Larval water-mites, unlike their nymphs and adults, usually have a claw-like seta on the palptibia. The entire gnathosoma is retractile in the Calyptostomatoidea and in some Erythraeoidea. Stigmata at the base of the chelicerae and short peritremes are present in the nymphs and adults of the terrestrial families but functional prostigmata do not appear to occur in the water-mites.

The leg femora are divided by a eudesmatic joint into a basi- and telofemur in the terrestrial forms but may be entire in certain aquatic species. The podomeres are highly hypertrichous in terrestrial species but this condition only occurs in the actively swimming forms among the water-mites. The ambulacra of legs II–IV usually have a pair of claws but lack an empodium, except in some larvae. Legs I generally have two claws but only one claw is present in some water-mites. Sperm transfer is indirect by stalked spermatophores.

The larvae are usually parasitic on arthropods (Fig. 12.19D) or vertebrates (Fig. 12.19E). Nymphs and adults of the terrestrial groups inhabit mosses, leaf litter and humus of soils. Little is known of their feeding habits but the majority are considered to be predatory while some species feed on pollen. Adults of the large red velvet mites of the genus *Dinothrombium* (Trombidiidae) live deep in the sandy soils of semi-arid regions and after heavy rain emerge for a few hours to feed on termites swarming at the same time, and to mate and oviposit (Newell & Tevis, 1960). The remainder of the year is spent in vertical burrows in the soil. Water-mites occur in ponds, lakes, streams, rivers and also in marshes and swamps. Adults of the Unionicolidae parasitize freshwater molluscs but this is exceptional and nymphal and adult water-mites are thought to be predominately predacious on other arthropods.

References

Terrestrial Parasitengona; Berlese (1912); Feider (1955); Newell (1957); Robaux (1974); Southcott (1961, 1963, 1986); Thor and Willmann (1947); Vistorin-Theis (1976); Wharton and Fuller (1952).

Aquatic Parasitengona; Cook (1966, 1967, 1974, 1980); Lundblad (1957); Prasad and Cook (1972); Smith (1976); Viets (1925).

Suborder Raphignathina

These are small to medium-sized mites with the body generally oval in shape (Tetranychidae, Fig. 12.20A) although it may be circular (Psorergatidae, Fig. 12.20B), elongated (Syringophilidae, Fig. 12.20E) or fusiform (Eriophyoidea, Fig. 12.20C; Demodicidae, Fig. 12.20F). The idiosoma is

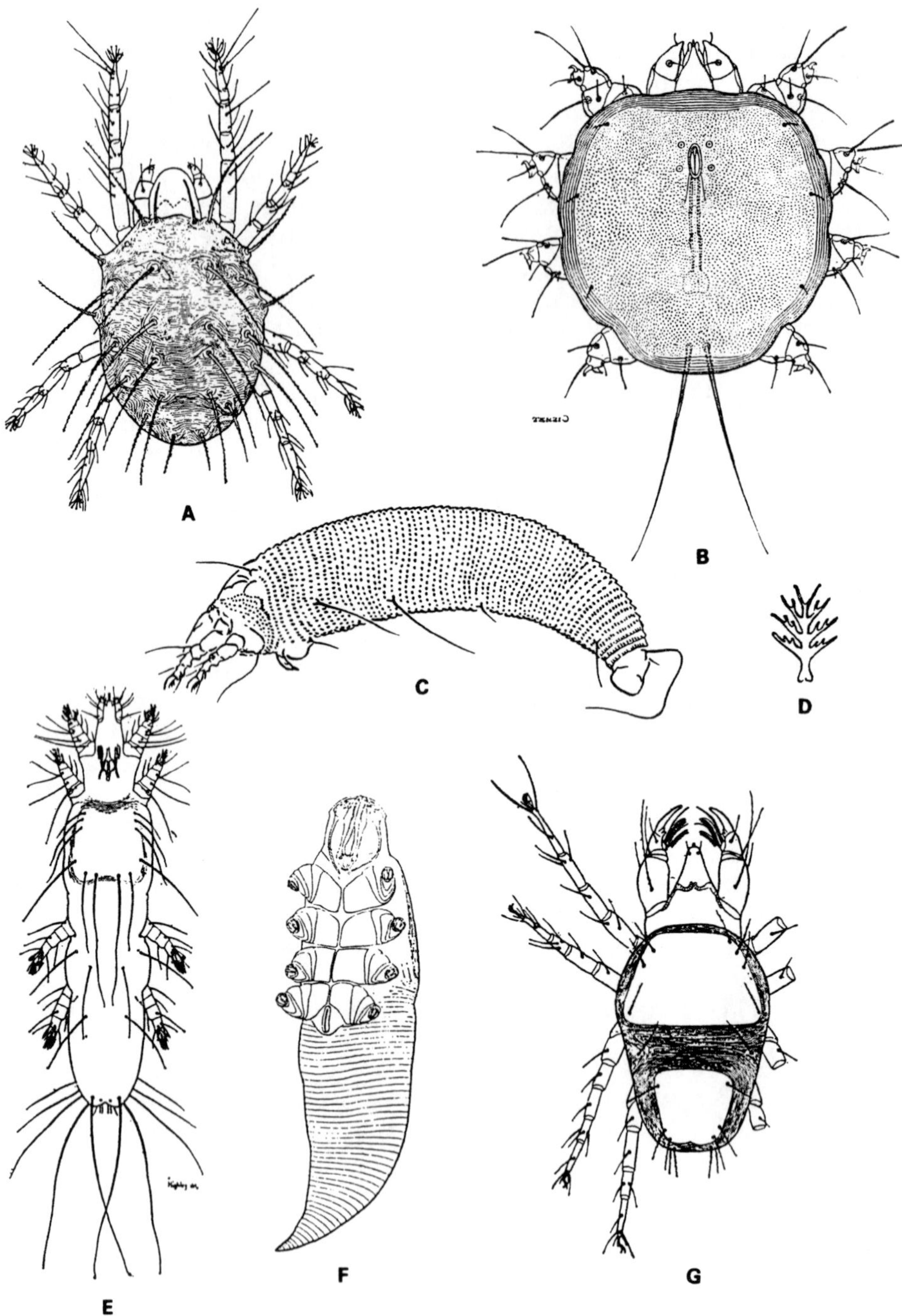

Fig. 12.20. A, female of *Panonychus ulmi* (Tetranychidae); **B**, male of *Psorergates musculinus* (Psorergatidae); **C D**, *Eriophyes insidiosus* (Eriophyidae) (**C**) and empodial claw (**D**) (after Keifer & Wilson, 1956); **E**, *Syringophilus bipectinatus* (Syringophilidae); **F**, *Demodex phylloides* (Demodicidae); **G**, *Cheyletus eruditus* (Cheyletidae).

typically weakly sclerotized but may have one to many dorsal sclerites, particularly in the Rhaphignathoidea and Cheyletidae. Prodorsal trichobothria and a naso are never present. The maximum of 14 pairs of 'dorsal' setae is often reduced and hypertrichy occurs in some taxa of the Raphignathoidea, Tetranychoidea and Cheyletoidea. There is no evidence of anamorphosis. Genital papillae and eugenital setae are absent. The genital and anal openings may be weakly separated, contiguous or coalesced. The male genital orifice is subventral, terminal or dorsal and an aedeagus is present except in the Eriophyoidea. The cheliceral bases may be separate but are often fused with each other and sometimes with the subcapitulum. The fixed digit, except in the Eriophyoidea in which it is styliform, is markedly reduced and the movable digit is styliform. There are no cheliceral setae. The pedipalps generally have four or five articles and a tibial claw-like seta is present in many families of the Raphignathoidea, Tetranychoidea and Cheyletoidea (Fig. 12.20G) but is absent in the Tenuipalpidae and Eriophyoidea and the majority of the obligatory parasitic members of the Cheyletoidea. A reduction in the number of palpomeres is characteristic of parasitic species, which may only have one, two or three articles as, for example, in the Psorergatidae, Ophioptidae, Myobiidae and Demodicidae. Stigmata, when present, may be situated between or on the cheliceral bases and the peritremes are either restricted to the dorsal surface of the gnathosoma or extend to the membraneous anterior region of the prodorsum. The podocephalic canal is superficial.

Only two pairs of legs are present in all active instars of the Eriophyoidea and the typical octopod condition of the nymphs and adults in the other major taxa of this suborder may be reduced in some gall-forming Tenuipalpidae to three pairs while the posterior two pairs of legs are reduced or absent in some Harpyrhynchidae. The leg femora are not divided but families of parasitic species often exhibit a reduction in the number of podomeres and/or modifications for grasping the hairs of their host. The ambulacra are various and may comprise two claws with or without an empodium or lack true claws and have a rayed empodium (Eriophyoidea, Fig. 12.20D). The claws and the pad- or claw-like empodium may have tenent hairs. Sperm transfer is direct except in the Eriophyoidea in which it is indirect by way of stalked spermatophores. There may be a prelarval, larval and two (never three) nymphal instars but a reduction in the number of instars may occur in the derivative taxa. The Eriophyoidea have only two active immature instars which are termed nymphs.

The Raphignathoidea mainly comprise predatory species but some Stigmaeidae are phytophagous. They occur in mosses, the litter and humus of soils, on low-growing vegetation and under the bark and on foliage of trees. Some species of Xenocaligonellidae and Stigmaeidae are phoretic on beetles and sandflies, respectively. Members of the Tetranychoidea and Eriophyoidea are obligate plant-feeders and many of the former, in particular, are of economic importance as pests of protected and field crops. They are worldwide in distribution. The Cheyletidae consist of predatory species, more or less restricted to the family Cheyletidae, and a range of parasitic species living on the surface or in the skin of birds (Harpyrhynchidae, Cheyletiellidae) and mammals (Psorergatidae, Cheyletiellidae), in the fur of mammals (Myobiidae), in the quills of birds (Syringophilidae) and underneath the posterior margin of the body scales of snakes (Ophioptidae).

References

Baker (1949); Baker and Pritchard (1953); Fain (1972b); Farkas (1965); Gilyarov (1978); Helle and Sabelis (1985); Hirst (1919); Jeppson *et al.* (1975); Smith Meyer (1974); Pritchard and Baker (1955, 1958); Smith Meyer (1979); Summers (1962); Summers and Price (1970); Summers and Schlinger (1955); Volgin (1969).

Suborder Tarsonemina

This suborder, as defined in this work, is equivalent to the Heterostigmata of Lindquist (1976). He considered that the Heterostigmata together with its sister group the Raphignathina warranted a higher ranking, the Eleutherengona. This concept is followed by Kethley (1982). Van der Hammen (1972) created a separate order for the Tarsonemina which he considered to be more closely related to the Astigmata than to the Prostigmata but this has been rejected by Lindquist (1976).

The Tarsonemina are small to medium-sized mites of variable sclerotization in which the idiosoma is elongate or subcircular and covered by sclerites (Fig. 12.21A). A region of the opisthosoma is greatly distended during engorgement and prior to ovigenesis in females of some taxa, and this is termed physogastry. The condition in *Pyemotes* in which the dorsal area of distensible cuticle occurs between tergites EF and H is shown in Fig. 12.21B. The larvae and adults of the free-living taxa retain the tergal, setal and lyrifissure elements of the six opisthosomatic segments (or five segments and the anal block) characteristic of the larvae of the Actinotrichida but the tergites of segments E and F are fused so that setae *e* and *f* and the cupules *im* and *ip* are borne on a common sclerite thereby giving the impression of only four or five segments (Lindquist, 1986). The first and second opisthosomatic tergites are also fused in the Heterocheyloidea (Fig. 12.21C). Various degrees of fusion of segments H and PS, associated with the housing of the genitalia, occur in the male. A pair of bothridia with capitate or ampulliform trichobothria is present in all females but only in males of the Tarsocheyloidea. In the Scutacaridae, the prodorsum is completely covered by the large anterior sclerite of the hysterosoma (Fig. 12.21D). A pair of prodorsal stigmata with associated tracheae is present in all females and also in some males of the Tarsocheylidae. The podocephalic canal is superficial. Genital and anal openings are separate and each is covered by a pair of setae-bearing valves in the Tarsocheyloidea and Heterocheyloidea while the genital valves are flanked or surrounded by aggenital plates. In the other Tarsonemina, the anus is terminal and a sperm access system appears to open ventro-subterminally. The egg-laying aperture, at least in the Tarsonemidae, is located above the base of a medioventral plate, immediately posterior to legs IV. Paired aedeagal shafts occur in males other than those of the Tarsocheyloidea and Heterocheyloidea.

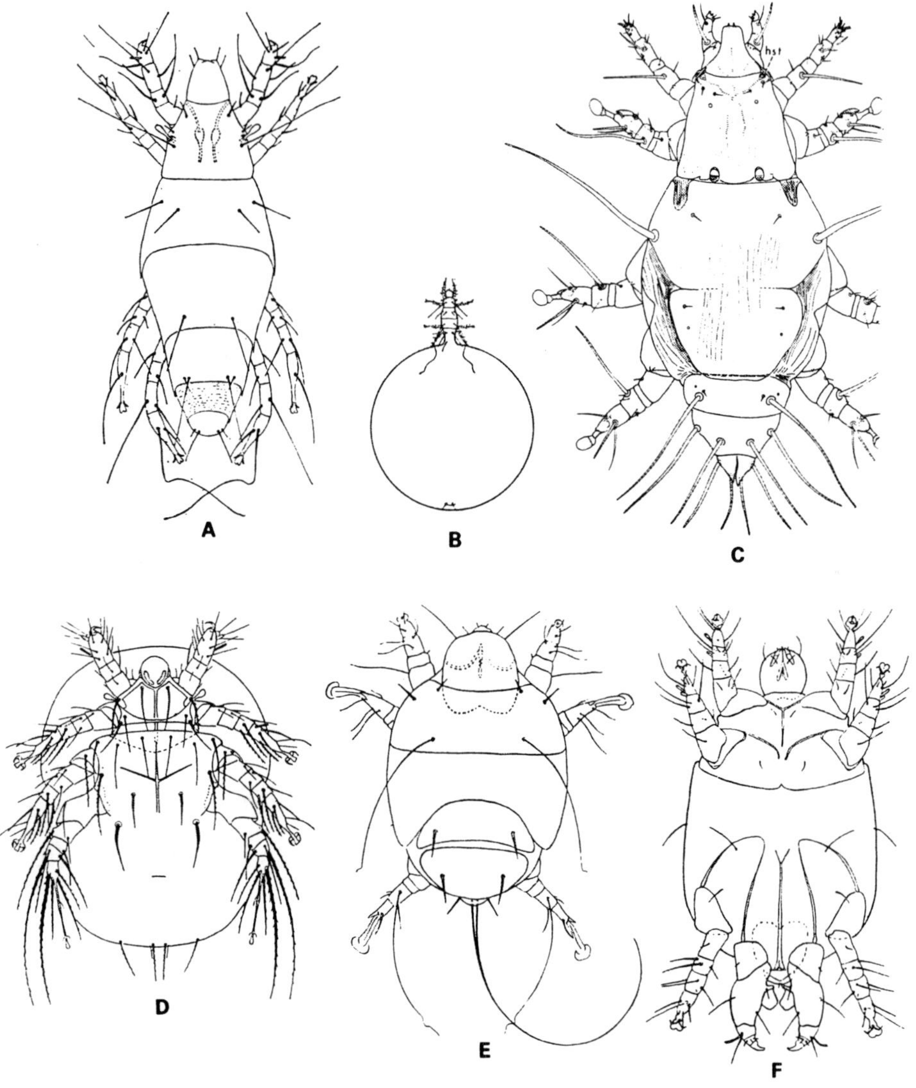

Fig. 12.21. **A**, *Pyemotes* sp. (Pyemotidae), female; **B**, female of *Pyemotes* in physogastric state; **C**, *Heterocheylus* sp. female (after Lindquist & Kethley, 1975); **D**, venter of *Imparipes intermissus* (Scutacaridae) (after Karafiat, 1959); **E**, female of *Podapolipus grassi* (Podapolipidae); **F**, venter of *Steneotarsonemus spinosus*, male (Tarsonemidae) (after Schaarschmidt, 1959).

The cheliceral bases are fused into a stylophore which is not integrated with the subcapitulum in the Tarsocheyloidea and Heterocheyloidea. Integration of the stylophore with the subcapitulum in the other taxa of the suborder results in a capsule-like gnathosoma (see Fig. 5.15B, p. 159). The movable digits are curved and blade-like and partly retractable in the Tarsocheyloidea and Heterocheyloidea while in the other superfamilies (Pyemotoidea, Pygmephoroidea and Tarsonemoidea) they are styliform and completely retractable. The fixed digit is lacking. The conspicuous linear pedipalps extend beyond the apex of the gnathosoma in the Tarsocheyloidea and Heterocheyloidea and comprise three palpomeres (trochanter, femorogenu and tibiotarsus with terminal claw) in the former but with two or three palpomeres in the latter. The palps in the remaining taxa are inconspicuous and consist of one or two ill-defined articles. The gnathosomatic capsule is reduced and non-functional in the males of the Scutacaridae and Pygmephoridae.

There are four pairs of legs in the post-larval instars except in the Podapolipidae (Fig. 12.21E) in which males have 3–4 and the females 1–4 pairs of legs. The Tarsocheyloidea and Heterocheyloidea have coxal plates III and IV separated medially and legs II–IV similar to one another in the number of podomeres and the form of the ambulacra, whereas in the remainder of the Tarsonemina these coxal plates are united and legs II–IV are rarely similar in shape. Usually leg IV of the male is enlarged and modified for copulation (Fig. 12.21F) and leg IV of the female may be very different in shape and number of podomeres from II and III. The ambulacra of legs I–IV in the Tarsocheyloidea and Heterocheyloidea have a stalked membraneous empodium. Other Tarsonemina have membraneous empodia and paired claws on legs II–III but legs IV may lack ambulacra and legs I, in which the tibia and tarsi are often fused, are without empodia but may have a single or paired claws which are sometimes sessile. The method of sperm transfer is not known in the Tarsocheyloidea and Heterocheyloidea but is direct in the other taxa. There are four post-embryonic instars (larva, two nymphs and adult) in the Heterocheyloidea (and this probably also applies to the Tarsocheyloidea) but active nymphal stages have been eliminated from the life cycle of other Tarsonemina.

The tarsocheyloids are widely distributed in rotting wood, litter in treeholes and decaying leaf litter whereas the heterocheyloids are mainly subelytral symbionts of passalid beetles. The majority of the Pyemotoidea are parasites of insects while the Pygmephoroidea and Tarsonemoidea contain free-living mites, which are predominantly phytophagous, as well as symbionts of a variety of insects. Females of the parasitic species are usually capable of physogastry with the areas of integumental expansion in *Pyemotes*, for example, occurring in the region between the two posterior tergal elements (Fig. 12.21A,B).

References

Beer (1954); Cross (1965); Husband (1980, 1984); Karafiat (1959);

Krczal (1959); Lindquist (1976, 1986); Lindquist and Kethley (1975); Mahunka (1965); Martin (1978); Schaarschmidt (1959).

Order Astigmata

The majority of the mites of this order are pearly white to yellow-brown in colour with the cuticle smooth, striated or covered with microtrichia. The body is usually weakly sclerotized except in many phoretic deutero-nymphs (hypopodes) and certain commensals and parasites, such as members of the Analgoidea and Psoroptoidea, in which the idiosoma is more or less entirely sclerotized or provided with sclerites. The sejugal furrow is well developed in some free-living forms, for example the Acaridae and Hemisarcoptidae. The prodorsum is generally provided with a prodorsal sclerite or shield and bears a maximum of four pairs of setae – excluding a pair of supracoxal setae on legs I (*eI*) – arranged in two transverse rows. The anterior row comprises the vertical internal setae (*vi*), usually overhanging the gnathosoma, and the vertical externals (*ve*) situated lateral or postero-lateral to the vertical internals (Fig. 12.22A). The posterior row consists of the scapular internals (*si*) and scapular externals (*se*). In epizooic forms, one or more pairs of prodorsal setae may be lost. The maximum number of setae in the hysterosomatic region is 18 pairs (Griffiths *et al.*, 1990) and includes four pairs on segment C (c_1–c_3, cp), two pairs each on segments D (d_1,d_2) and E (e_1,e_2), one pair (f_2) on segment F and three pairs on each of segments H (h_1–h_3), PS (ps_1–ps_3) and AD (ad_{1-3}). Other systems of nomenclature that have been applied to the setae of the hysterosoma of the Astigmata are discussed by Griffiths *et al.* (1990) and the one used by Hughes (1976), based on Zachvatkin (1941), is shown in Fig. 12.22A. Cupules (putative lyrifissures) *ia, im, ip* and *ih* are normally present but *iad* and *ian* are lacking, i.e. no cupules are added to the larval complement during ontogeny. A peranal segment is never developed. Latero-opisthosomatic glands ('oil glands') are generally present (*gl.* in Fig. 12.22). Details of openings of the supracoxal glands and associated structures are given in Chapter 7 (see Fig. 7.8, p. 228).

The coxisternal region usually has conspicuous apodemes but the coxal fields are generally not sclerotized (Fig. 12.22B). The apodemes of leg I may be united to form a V- or Y-shaped 'sternum' and in some feather mites, for example the Freyanidae, the coxal fields are all closed. The coxisternal setae are denoted by the letter *a* and the coxisterna (1–4) to which they belong. Griffiths *et al.* (1990) consider a maximum of four pairs (*1a, 3a, 3b, 4a*) to be present. The setae designed *4a* have previously been considered to be the anterior genital or the aggenital setae. The inverted Y-shaped progenital opening (=oviporus) in the female is normally bordered by progenital lips (=genital folds or valves) and usually lies between the bases of legs III and IV although in some species it occupies a more anterior position. In parasitic and some free-living species, there are no separate progenital lips, which are considered to be fused with the ventral body wall, and the oviporus is more or less transverse or inverted U- or V-shaped and may have associated genital apodemes (see Fig. 12.25B below). A maximum of one pair of genital (or progenital) setae (g) is present. A sclerite situated anterior to the oviporus and termed the pregenital sclerite or epigynum, may be present and in some species is united with the coxisternal apodemes. Two pairs of genital papillae, usually situated in the progenital chamber, are present in both sexes except in the deuteronymph of *Schizoglyphus* in which there are three pairs. An exception occurs in the Anoetoidea in which the two pairs are enlarged, superficial and variable in position in relation to the genital region of the adults (see Fig. 12.24D below). The bursa copulatrix, the opening of the sperm access system, is located subter-

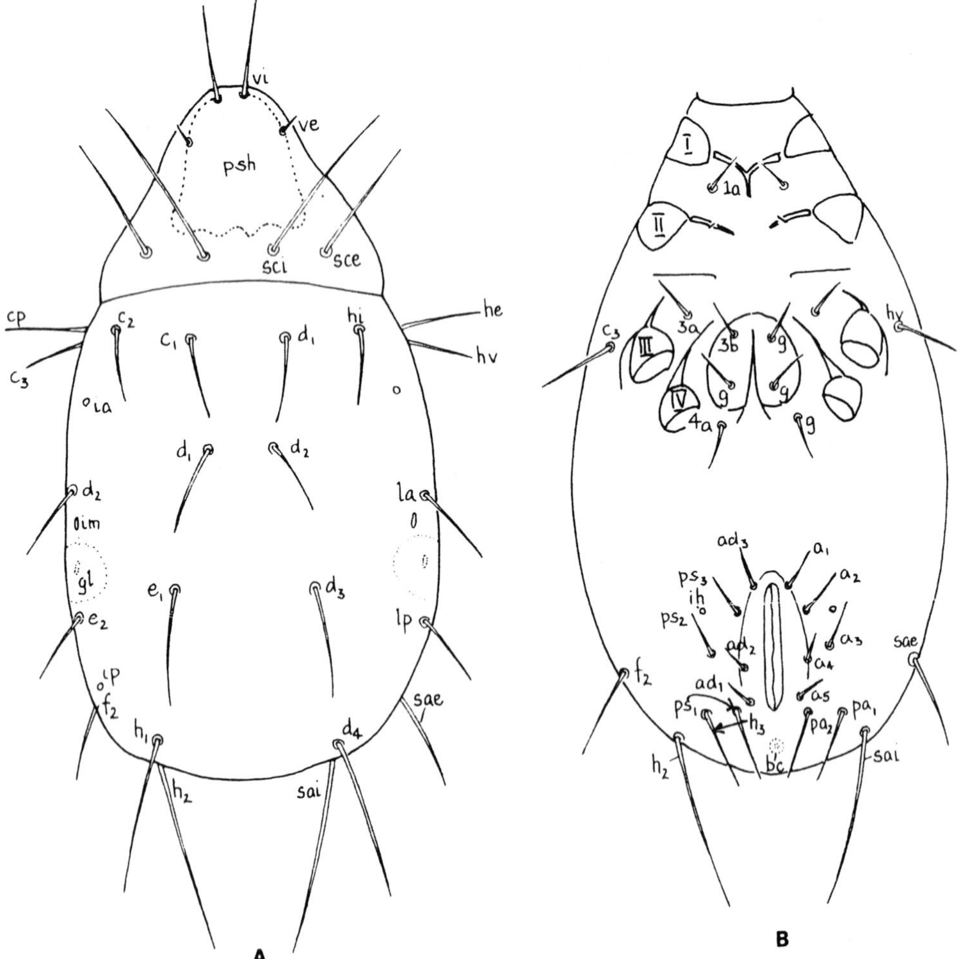

Fig. 12.22. Dorsum (**A**) and venter (**B**) of an acarid mite with a comparison of the notations proposed for the dorsal and ventral idiosomatic setae by Griffiths *et al.* (1990), on the left of the mid-line, and by Hughes (1976), based on Zachvatkin (1941), on the right. See text for details.

minally, terminally or postero-dorsally (Fig. 12.22B). The slit-like anal opening is relatively large in the free-living forms for the expulsion of the faecal pellet. The penis or aedeagus of the male usually consists of a strongly sclerotized tube-like structure with supporting sclerites which is protracted by hydrostatic pressure and retracted by muscles. A copulatory sucker often called the adanal or paranal sucker, may occur on either side of the anal opening of the male as in, for example, the Acaridae, Chortoglyphidae and Pyroglyphidae (Fig. 12.23A). The posterior of the idiosoma in males may be bilobed as, for example, in the Analgidae (see Fig. 12.25C below) and Psoroptidae.

The gnathosoma (Fig. 12.23B) is movably articulated to the idiosoma and each chelicera, with its compact body, is typically chelate-dentate except in the Anoetoidea (=Histiosomatoidea) in which the digits are flattened laterally and bear a number of small closely set teeth (see Fig. 12.24C below). A paraxial spine and one or more protuberances may be present. Rarely, as in *Hypodectes* (Hypoderidae), the palps and chelicerae are vestigial in the female but enlarged in the heteromorphic male. The malapophyses constitute the paired anterior region of the subcapitulum and appear to form the so-called rutella which, in free-living forms, are usually provided with one or more tooth-like processes (prostomal teeth or strigilis) latero-dorsally on their internal faces. Adoral setae are absent and there is only one pair of subcapitular setae. Supracoxal setae *e* are often relatively large in free-living species. The pedipalps are reduced and usually comprise only two articles although three articles have been described in the Schizoglyphidae deuteronymph (the only known instar).

The legs have six movable podomeres including the apotele. The podomeres have a well-defined chaetotaxy and the terminology for the setae has been discussed in Chapter 2 (see Fig. 2.19B, p. 57). The ambulacrum normally has an empodial claw, bifurcate in some adult Lardoglyphidae, and a membraneous pad or caruncle (see Fig. 2.17B,C, p. 53) but may be variously modified in commensals and parasites (Atyeo, 1979). An enlargement of the claw and reduction in the size of the caruncle is not uncommon while in some feather mites, for example the Analgidae and Pterolichidae, there is a large ambulacral disc provided with secondary sclerites (Fig. 12.23C). In the Analgoidea, the ambulacrum can be withdrawn into the ambulacral stalk (Fig. 12.23F). The legs in some parasitic forms may be adapted for attaching to the fur of the host; for example in the Myocoptidae, the genu, tibia and tarsus of legs III and IV in the female and legs III in the male curve inward towards the enlarged femur to form a clasping organ. Heteromorphic and pleomorphic males in certain Acaridae have a modified third pair of legs which are not used for walking but are held away from the ground and used as a weapon in fighting young homomorphic males. The modified leg is enlarged and usually terminates in a single claw. An enlargement of legs III and/or IV occurs in the males of some Psoroptida, and in *Dinalloptes* (Alloptidae) marked asymmetry of legs II and IV is evident with leg II on one side and leg IV on the other hypertrophied. In males of

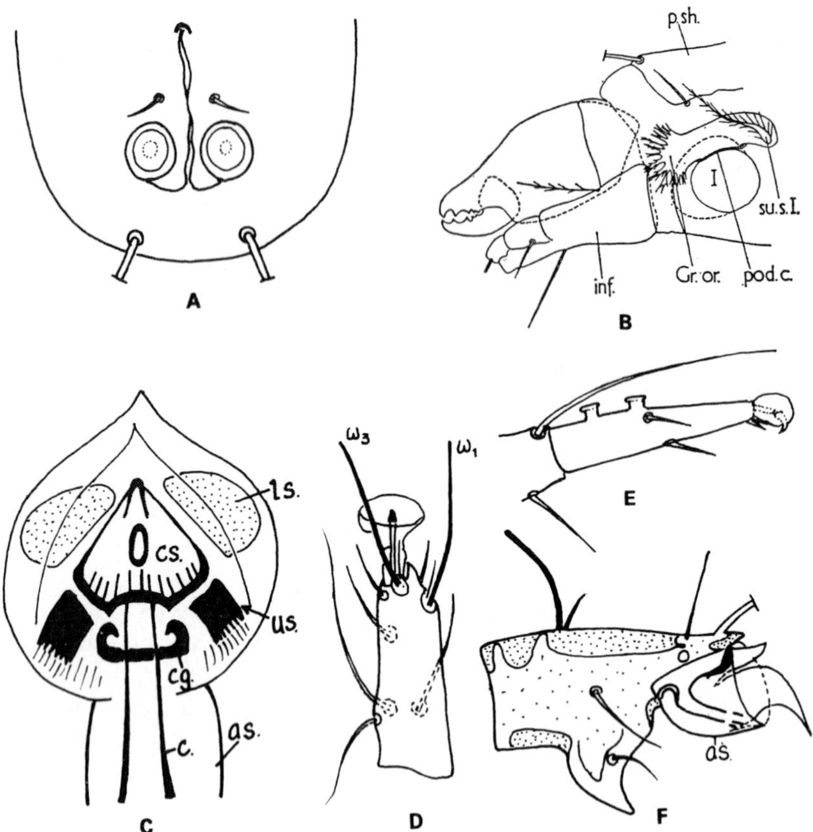

Fig. 12.23. **A**, diagrammatic representation of the adanal suckers of discs of an acarid male; **B**, lateral view of the gnathosoma of an acarid mite; **C**, sclerites of the enlarged caruncle of a species of Alloptidae (Analgoidea); **D**, tarsus I of the male of *Dermatophagoides pteronyssinus* (Pyroglyphidae) showing distal position of the solenidia; **E**, tarsus IV of a male acarid showing the tarsal 'suckers'; **F**, retracted ambulacrum of an analgid mite. C and F based on Atyeo (1979). *as.* = ambulacral stalk; *c.* = condylophore; *c.g.* = condylophore guide; *cs.* = fused basilar piece and claw; *Gr. or.* = Grandjean's organ; *inf.* = subcapitulum; *ls.* = lateral sclerite; *p.sh.* = prodorsal shield; *pod.c.* = podocephalic canal; *su.s. I* = supracoxal seta; *us.* = unguiform sclerite.

the Acaroidea, two sucker-like setae usually occur dorsally on the tarsus of leg IV (Fig. 12.23E).

Sperm transfer is direct by way of the aedeagus into the solenostome of the sperm access system. A prelarva represented by an apoderma is present in some species. The larva, protonymph and tritonymph are usually normal active instars but the deuteronymph is a facultative or obligatory ellatostasic, more rarely calyptostasic, instar which is adapted for phoresy and/or for resisting adverse environmental conditions (see sections on Ontogeny and Dispersal in Chapter 11). It is commonly referred to as the hypopus (*pl.* hypopodes). The deuteronymphal stase is generally lost in the obligatory parasitic Astigmata.

Classification

Although the Astigmata form a well-defined natural group, the higher classification is in a state of flux. Yunker (1955) recognized three major taxa (cohorts) within the Acaridiae (=Astigmata), namely the Acaridia, Ewingidia and Psoroptidia. The Ewingidia, comprising highly specialized species adapted to clinging to the gills of pagurid crabs, is no longer recognized as a major taxon but is either relegated to superfamily status (Atyeo, 1979) or its constituent genera are accommodated in the family Acaridae (OConnor, 1982a). The Acaridia and Psoroptidia represent two ecomorphological groups, the former comprising mainly species that are free-living or associated with other arthropods and the latter, with few exceptions, consisting of commensals or parasites of birds and mammals throughout their life cycle. This convenient division of the Astigmata into two groups is often retained in current classifications and is followed in this work.

DIVISION ACARIDIA

The superfamilial classification of the non-psoroptid astigmatic mites presented by OConnor (1982a), and adopted in this work, has a strong ecological basis and, in some respects, differs in its concepts from the more traditional classifications, as for example those of Yunker (1955) and Krantz (1978). This is most apparent in the treatment of the hemisarcoptid–glycyphagid–acarid complex.

Superfamily Schizoglyphoidea

This taxon is based on the morphological characteristics of the phoretic deuteronymph of *Schizoglyphus biroi* collected from a tenebrionid beetle in New Guinea (Mahunka, 1978). The normal stases of the genus are unknown. Unlike other astigmatic deuteronymphs, the pedipalp has three articles, the subcapitulum is relatively well-developed and bears both supracoxal and subcapitular setae and there are three pairs of genital papillae. It is considered by O'Connor (1982a) to represent the earliest known derivative lineage in the Astigmata.

Superfamily Anoetoidea

The characteristic feature of the normal stases in this group is the adaptation of chelicerae and palps for filter-feeding (Fig. 12.24A–D). The body is usually weakly sclerotized in the Anoetidae, except for the male, but is completely sclerotized in both sexes in the Guanolichidae whose members inhabit bat guano. The oviporus in the Anoetidae is in the form of a transverse slit which is considered to be formed by the open anterior margins of the fused progenital valves. The two pairs of 'genital papillae' are large, round or oval in shape and superficial (Fig. 12.24B). The solenostome of the sperm access system is dorsal or, less commonly, terminal in position. The setae of the legs tend to be spine-like. The

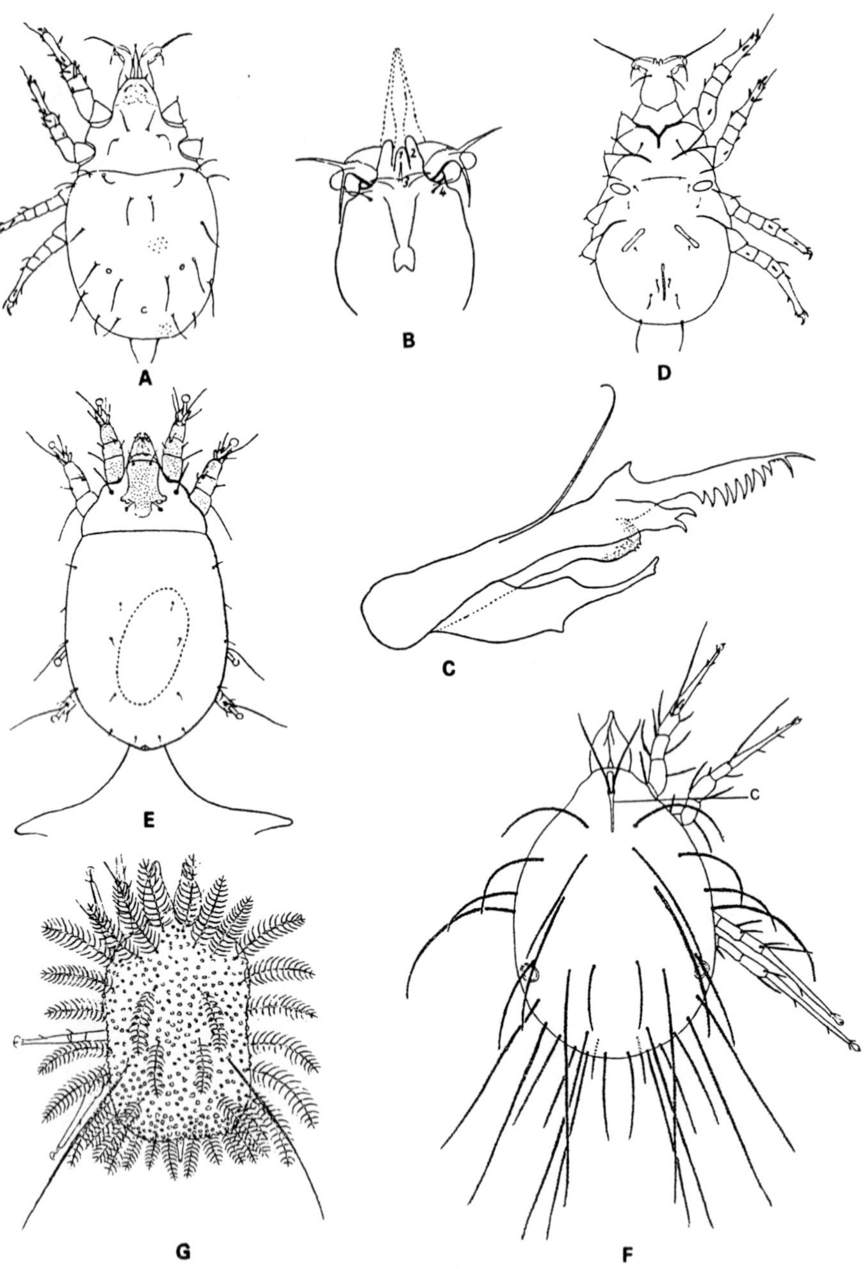

Fig. 12.24. **A**, dorsum, and **B**, venter of *Histiostoma myrmicarum* (Anoetidae) (after Scheucher, 1957); **C**, venter of subcapitulum, and **D**, chelicera of *Histiostoma feroniarum*; **E**, *Hemisarcoptes malus*, female (Hemisarcoptidae); **F**, *Glycyphagus domesticus*, female (Glycyphagidae) showing form of 'crista' (c); **G**, *Ctenoglyphus canestrinii*, female (Glycyphagidae) (after Hughes, 1976).

phoretic deuteronymphs of the Anoetidae (not known in the Guanolichidae) lack vertical setae and have the femora and genua of the anteriorly directed legs III and IV fused.

The Anoetoidea on the whole live in wet habitats although some species of *Anoetus* are associated with halicitid bees, feeding on microorganisms growing on stored pollen and nectar in the bee cells, and members of the genus *Spinanoetus* live in vertebrate carrion. The phoretic deuteronymphs mainly occur on insects but those of species inhabiting mammalian nests attach to the hairs of mammals.

References

Hughes and Jackson (1958); Scheucher (1957).

Superfamily Canestrinioidea

These mites are commensals or parasites of adult Coleoptera during their entire life cycle. The adults are large, usually weakly sclerotized with a striate or reticulate cuticle in the Canestriniidae but dorsally well sclerotized in the Heterocoptidae. The sejugal furrow is present or absent and a complete dorsal idiosomatic chaetotaxy may be present or reduced by the absence of *ve* and c_3 (Heterocoptidae). The oviporus is in the form of an inverted V and there are two pairs of genital papillae in the adult, those of the Heterocoptidae having a segmented appearance. The ambulacrum of the legs is usually developed into a disc-like structure with secondary sclerotization from which the apex of the empodial claw, when present, projects. The posterior legs of males of the Canestriniidae are specialized for attaching to immature females and the posterior region of the idiosoma may bear large processes. Deuteronymphs appear to be known only in the genus *Megacanestrinia*.

The majority of the normal stases live under the elytra of a range of beetle taxa and are thought to feed on exudates. Those occurring on passalids, however, tend to be found externally and some feed from the host mouthparts. The superfamily has a wide geographical distribution but is not represented in the Nearctic.

References

OConnor (1979); Samšiňák (1971).

Superfamily Hemisarcoptoidea

According to OConnor (1982a), most of the species of this superfamily are associated with 'wood-related habitats' although members of the family Chaetodactylidae occur only in the nests of solitary bees and those of the algophagous Hyadesidae inhabit intertidal zones. The Algophagidae are aquatic or semi-aquatic living in sap exudates or water-filled tree-holes while the Hemisarcoptidae and Winterschmidtiidae (=Saproglyphidae) occur in a variety of habitats including arboreal situations. Some species of Carpoglyphidae are pests of stored foods with a high sugar content. The mites, with the exception of some well-sclerotized aquatic species, are soft-bodied with the sejugal furrow present or absent (Fig. 12.24E). The dorsal idiosomatic chaetotaxy is complete. Ocelli are often present on the prodorsum. Males typically lack adanal suckers but leg II may be markedly enlarged in the Hyadesiidae. The ambulacrum comprises a well-developed caruncle and an empodial claw which is commonly enlarged in aquatic species but lacking in the Hemisarcoptidae. The condylophores are separate or fused basally to give a 'V'-configuration.

References

Fain (1972c); Fain and Johnston (1975); Luxton (1989); Türk and Türk (1957); Zachvatkin (1941).

Superfamily Glycyphagoidea

This superfamily comprises weakly to strongly sclerotized mites in which the cuticle is smooth, striated or densely microtrichous. A sejugal furrow is normally lacking. The prodorsal sclerite, when present, may be subrectangular or linear. In the latter form, it is often referred to in the older literature as the 'crista metopica' (Fig. 12.24F). The dorsal idiosomatic chaetotaxy is complete and the setae are generally long and often barbed or, more rarely, short and simple (e.g. Chortoglyphidae). Leaf-like and strongly bipectinate setae occur in some Glycyphagidae (Fig. 12.24G). The males usually lack adanal suckers and sucker-like setae on tarsus IV except in the Chortoglyphidae. The legs are often slender and elongate and the ambulacrum has a caruncle with or without a small empodial claw. The caruncle is usually relatively small and membraneous but in the Rosensteiniidae it may be in the form of an elaborate disc showing secondary sclerotization. The condylophores are slender and often elongate.

Members of this superfamily occur in a variety of habitats including stored food products but the majority are associated with mammals, birds and insects. Those associated with mammals, for example some species of Glycyphagidae and Echimyopodidae, occur in the nests and the deuteronymphs are either external phoretics which attach to mammalian hair by means of a postero-ventral clasping organ or are follicular or subcutaneous 'parasites'. The Euglycyphagidae, represented by a single species, live in the nests of birds of prey while the deuteronymphs are phoretic on trogid beetles co-inhabiting the nest. Some species of Aeroglyphidae and Rosensteiniidae inhabit bat roosts and feed in bat guano. Many species of Glycyphagidae have successfully invaded stored food and assumed pest status although the majority feed on fungi associated with the stored product. Two species, *Glycyphagus domesticus* and *Lepidoglyphus destructor*, are commonly found on furnishings, clothing and walls of houses in which damp conditions encourage the growth of fungi.

References

Fain (1969b); Gilyarov and Krivolutsky (1975); Hughes (1976); Samšiňák (1977); Zachvatkin (1941).

Superfamily Acaroidea

The majority of specialists working on the Astigmata include the Hemisarcoptoidea, Glycyphagoidea and Hypoderoidea within this superfamily. The idiosoma is generally weakly sclerotized with the cuticle smooth, striated or, less commonly, scaly or covered with microtrichia. A sejugal furrow is usually present (see Fig. 12.22A above). The prodorsal sclerite is shield-like, in the form of linear longitudinal sclerites, or absent. The dorsal setae are often long and pilose but may be short and simple. Progenital lips are conspicuous and the oviporus and the male opening are typically situated between legs III and IV (Fig. 12.22B). Adanal suckers are present in the males of most taxa as are sucker-like setae on tarsi IV (see Fig. 12.23A,B above). Legs II of the male are enlarged in *Acarus* and each is provided with a femoral apophysis bearing seta *vF.* Enlargement of legs III occurs in the male of the Glycacaridae, in the heteromorphic male of the Lardoglyphidae and the heteromorphic and pleomorphic males of certain Acaridae. The ambulacrum is broadly connected to the tarsus and has a caruncle and well-developed empodial claw. The condylophores are usually short and stout.

The Acaridae comprise saprophagous, fungivorous, phytophagous and graminivorous mites which commonly occur in stored food products, especially species of *Acarus* and *Tyrophagus.* Members of Lardoglyphidae and Glycacaridae inhabit nests of birds and/or mammals but some species of *Lardoglyphus* are important pests of stored meat (*L. zacheri*) and dried fish (*L. konoi*). The family Suidasiidae contains species which are associated with insects, nests of birds and stored food products. The deuteronymphs are typically phoretic on arthropods and have a well-developed attachment organ in the form of a sucker-plate.

References

Gilyarov and Krivolutsky (1975); Griffiths (1977); Hughes (1976); Zachvatkin (1941).

Superfamily Hypoderoidea

This superfamily comprises only the family Hypoderidae of which 14 of the 16 known genera are represented only by the deuteronymph (OConnor, 1982a). In the described adults, the body cuticle is smooth or scaly and the sejugal furrow is reduced or absent. The progenital lips are well developed and the oviporus is bordered anteriorly by a large inverted-U-shaped epigynum. The male idiosoma is bilobed posteriorly and the adanal suckers may be large, as in *Neottialges,* or small, as in *Hypodectes.* The gnathosoma is normally developed in *Neottialges* but strongly modified in *Hypodectes* in which the chelicerae and palps are vestigial in the female but greatly enlarged in the male (see Fig. 11.8, p. 355). Claws and condylophores are lacking on all ambulacra. The deuteronymphs are modified for life as subcutaneous or, rarely, visceral-tissue parasites of mammalian (Muridectinae) or avian (Hypodectinae) hosts. They lack attachment organs and coxisternal setae and the palposoma is very reduced or absent. The deuteronymph markedly increases in size in the host tissues notwithstanding the absence of a mouth and alimentary tract. The hypoderids are cosmopolitan in distribution. Within the Acaridia, the Hypoderidae is considered to be the most closely related group to the Psoroptidia. This is seen, for example, in the form of the oviporus and in the loss of setae *ve* and Grandjean's organ.

References

Fain (1967a); Fain and Bafort (1967); Fain and Beaucournu (1972).

DIVISION PSOROPTIDIA

Members of the Psoroptidia typically lack setae of the AD segment, have the progenital lips fused to the ventral body wall and show a reduction in their leg chaetotaxy (loss of pl″ from tarsus I and the tibial seta v″ from legs I and II). The deuteronymphal stase does not occur in the life cycle. OConnor (1982a) includes the following four superfamilies within this division.

Superfamily Pterolichoidea

The majority of the species of this superfamily are commensals or external parasites on the wing and tail feathers of birds (Fig. 12.25A,B). The Syringobiidae are an exception in living inside the quills of flight feathers of their hosts. Few morphological features are consistently present in the pterolichoids. The round or oval ambulacral disc is usually large and provided with secondary sclerites. The condylophores may be elongate or short and thick with tapering ends. There is no condylophore guide and the disc cannot be retracted into the ambulacral stalk as in the Analgoidea. The majority of males have adanal suckers but those on tarsus IV are often reduced. In the Gabuciniidae, the genital papillae are not associated with the genital opening as in members of the other families but remain between legs IV whereas the oviporus lies between legs II–III and the male orifice is situated posterior to legs IV. The oviporus is usually in the form of an inverted V but may be longitudinal as in the Rectijanuidae. An epigynum may be present or absent. The posterior region of the male idiosoma is usually weakly to strongly bilobed.

OConnor (1982a) recognizes 13 families within this superfamily with the majority of the species distributed between the Pterolichidae, Gabuciniidae, Eustathiidae, Syringobiidae and Freyanidae. The various types of feather provide a variety of niches which have been fully exploited by both the Pterolichoidea and Analgoidea and there is a tendency for a given species to be restricted to one type of microhabitat (Dubinin, 1951, 1956; Evans *et al.*, 1961). The Pterolichoidea are rarely found on passeriform birds.

References

Dubinin (1951, 1956); Gaud (1966); Gaud and Atyeo (1975,1978).

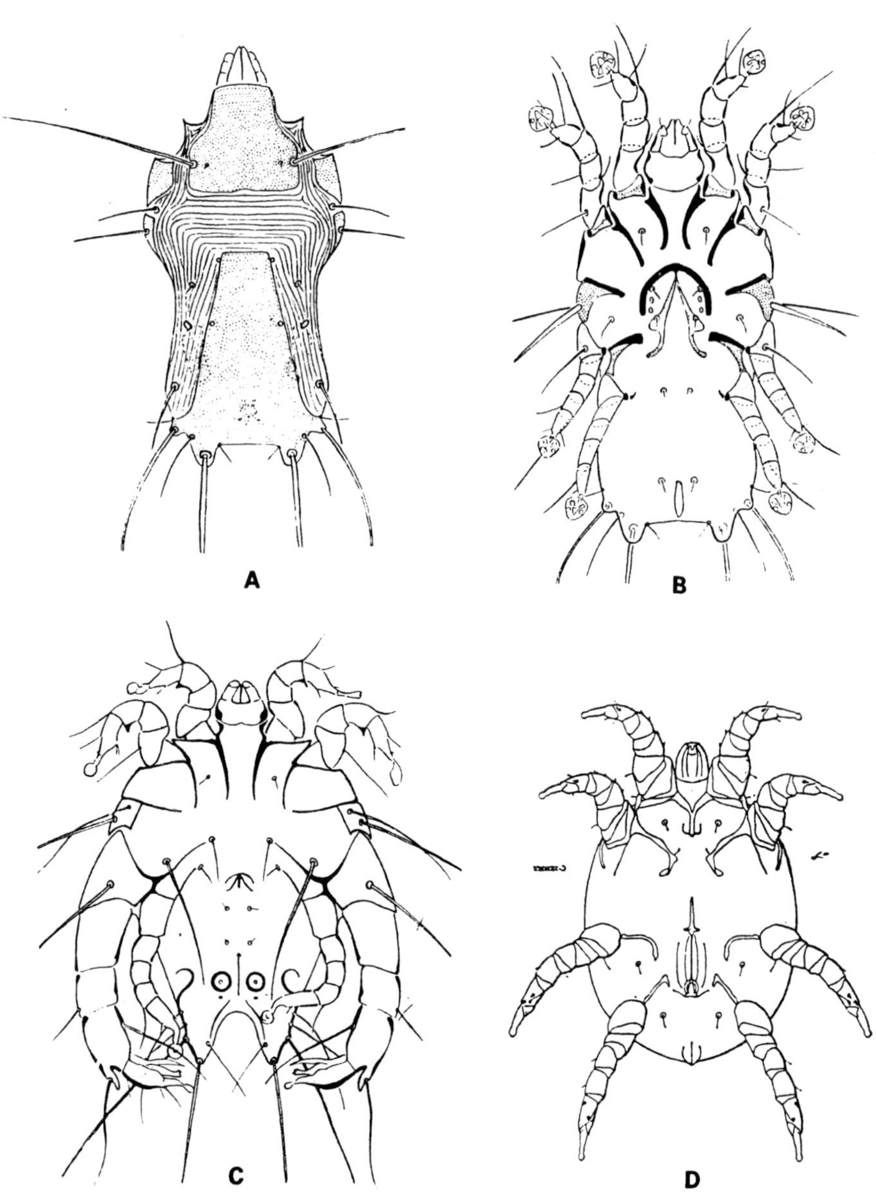

Fig. 12.25. **A**, dorsum and, **B**, venter of *Mouchetia viduata* (Pterolichidae) (after Gaud, 1962); **C**, venter of male of *Megninia carreti* (Analgidae) (after Gaud & Mouchet, 1959); **D**, venter of the female of *Cytodites nudus* (Cytoditidae).

Superfamily Analgoidea

As in the case of the Pterolichoidea, the species of this superfamily are all associated with birds. They not only inhabit the feathers but some species are found in feather follicles, in the skin or within the respiratory tract of their hosts. The main characteristic of this group is seen in the form of the ambulacra in which the folded ambulacral disc is capable of retraction into the ambulacral stalk (see Fig. 12.23F above). A condylophore guide may be present (see Fig. 12.23C above). Adanal suckers and tarsal IV suckers are typically present in the male but are lost in the Apionacaridae, Dermoglyphidae, Laminosioptidae and Cytoditidae. Legs III and/or IV are enlarged in the males of some taxa (Fig. 12.25C).

The Alloptidae, Analgidae, Avenzoariidae, Proctophyllodidae, Trouessartiidae and Xolalgidae are found externally on the feathers of the body and wings of a wide range of birds including the Passeriformes. The Epidermoptidae, Knemidokoptidae and Laminosioptidae are parasites on the skin or in the feather follicles. In the epidermoptid genera *Myialges* and *Microlichus*, the females are obligate or facultative hyperparasites of Hippoboscidae or Mallophaga whereas the males and immature stases live in the skin of their avian host. The Apionacaridae and Dermoglyphidae live within the quills of the flight and contour feathers while the Cytoditidae (Fig. 12.25D) are found in the respiratory tract and air sacs.

References

Dubinin (1951, 1956); Evans *et al.* (1963); Fain (1965b); Fain and Bafort (1964); Gaud and Atyeo (1967); Krantz (1978).

Superfamily Pyroglyphoidea

Although the majority of pyroglyphoid species are commensals or parasites of birds and occasionally mammals, some are free-living and occur in the nests of birds and in human habitations. They share with the Analgoidea and Psoroptoidea the presence, in some taxa, of a condylophore guide at the base of the ambulacral disc and the lack of proral and unguinal setae on all leg tarsi. A characteristic feature of the Pyroglyphoidea is the apical position of solenidion ω_1, as well as the famulus and ω_2, on tarsus I and the presence of cuticular hooks at the tips of, at least, tarsi I and II (see Fig. 12.23D above). The body is often weakly sclerotized with the cuticle striate but may be completely sclerotized in some Pyroglyphidae. A prodorsal sclerite is usually present and some taxa have an extensive dorsal hysterosomatic shield. A sejugal furrow may be present. The genital region, between legs III and IV in both sexes, is large and the oviporus is in the form of an inverted V or U (Fig. 12.26A). An epigynum is typically present. Adanal suckers and sucker-like setae of tarsi IV are generally present in the male whose body is bilobed in the majority of the parasitic forms. In many males, legs III are conspicuously larger than legs IV (Fig. 12.26B).

The family Pyroglyphidae contains forms which live on the feathers, and less commonly, in the quills of birds as well as in their nests and in

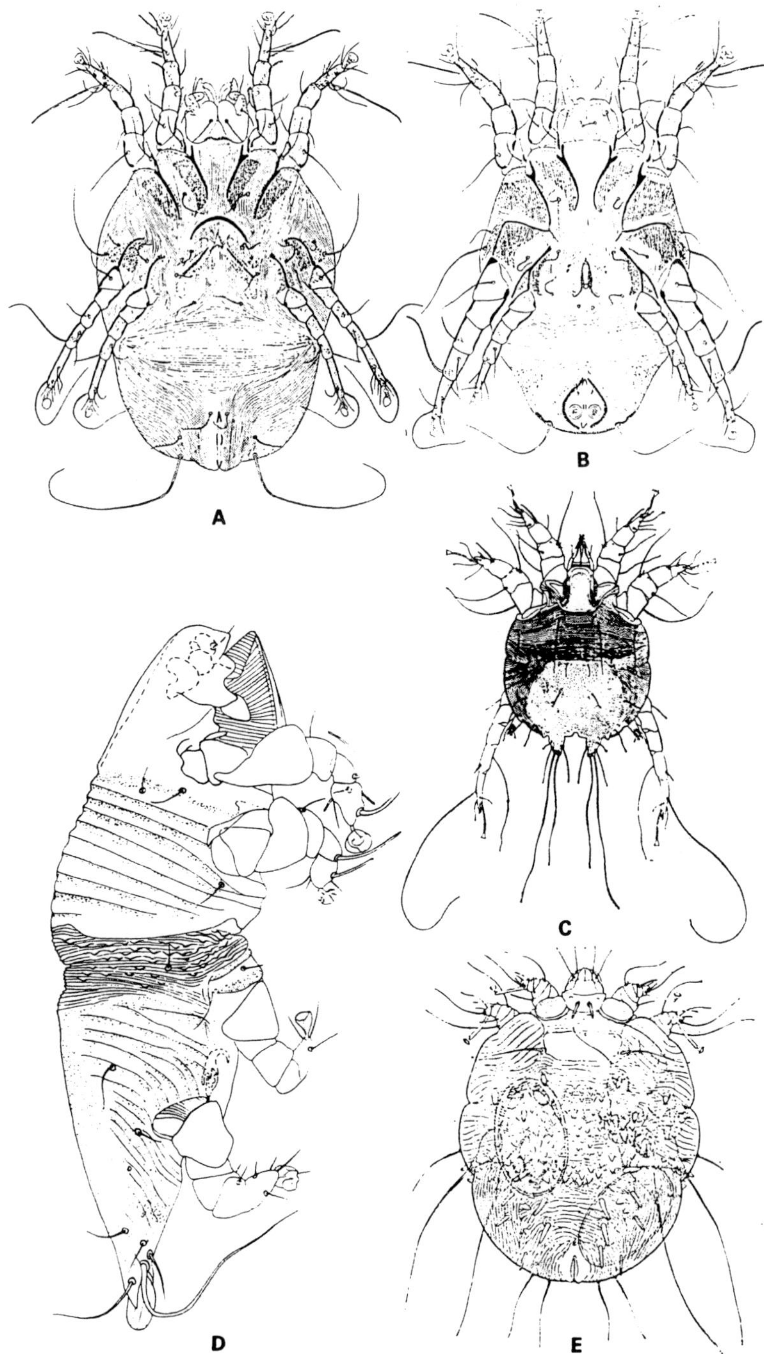

Fig. 12.26. **A**, **B**, venter of the female (**A**) and male (**B**) of *Dermatophagoides pteronyssinus* (Pyroglyphidae) (after Fain, 1966); **C**, male of *Psoroptes ovis* (Psoroptidae); **D**, *Aeromychirus aeromys*, female (Listrophoridae) (after Fain, 1979); **E**, *Sarcoptes scabiei*, female (Sarcoptidae).

the dust of houses. The species living in human habitations, chiefly those of *Dermatophagoides* and *Euroglyphus*, are the source of allergens causing bronchial asthma and rhinitis in atopic individuals. The Turbinoptidae live in the nasal passages of their avian hosts and the Ptyssalgidae are found within the quills of hummingbirds.

REFERENCES

Atyeo and Gaud (1979); Fain (1965c, 1967b, 1977); Hughes (1976).

Superfamily Psoroptoidea

The mites of this superfamily are mainly skin and fur parasites of mammals but some species live in the respiratory passages of their host. Morphologically, they show considerable resemblance to some members of the Analgoidea and Pyroglyphoidea although typically the ambulacral disc cannot be retracted into the ambulacral stalk as in the former and ω_1 is usually basal and not apical in position as in the latter. Prodorsal and hysterosomatic sclerites may be present and the posterior end of the body of the male is often bilobed (Fig. 12.26C). Adanal suckers are usually lacking in endoparasitic forms, for example Gastronyssidae, Lemurnyssidae and Pneumocoptidae. Legs III and/or IV of the male are often enlarged but leg IV in the Psoroptidae may be reduced in size in relation to the other legs. Fur mites have evolved various means of attaching to the hairs of their host. In the Listrophoridae this is achieved by means of two striated membraneous flaps, arising from coxal fields of legs I, acting as a clasper (Fig. 12.26D) while in the Myocoptidae legs III and IV of the female and legs III of the male are modified to form powerful clasping organs. The Chirodiscidae, on the other hand, attach to hairs by an expanded membraneous structure originating from tarsi I and II. The legs of mites of the family Sarcoptidae which live in skin and hair follicles are short and the cuticle of the dorsal surface of the body is often scaly and provided with spine-like setae (Fig. 12.26E). These integumental structures help to maintain the mite in the shallow channels it excavates in the skin of the host.

REFERENCES

Fain (1965d, 1968d, 1970b); Fain and Hyland (1974); OConnor (1982a).

Order Oribatida

Mites of this order range in size from about 200 to 1400 μm in length and the body is usually well sclerotized in the adult. A complete articulation between the proterosoma and hysterosoma occurs in the Palaeosomata, Enarthronota, Parhyposomata, Mixonomata and Euptyctima. In the dichoid condition as seen in the Mixonomata, the sejugal zone is entirely formed of soft cuticle so that the prodorsum can be retracted under the anterior edge of the notogaster whereas in the ptychoid condition occurring in the Euptyctima and some families of the Enarthronota articulation of the sejugal zone and of the coxisternum allows the legs to be concealed as the prosoma is folded down on the opisthosoma (see Fig. 2.7, p. 31). An unusual secondary division of the hysterosoma in the Elliptochthoniidae results in a podo-opisthosomatic articulation as well as the protero-hysterosomatic one. Norton (1975)

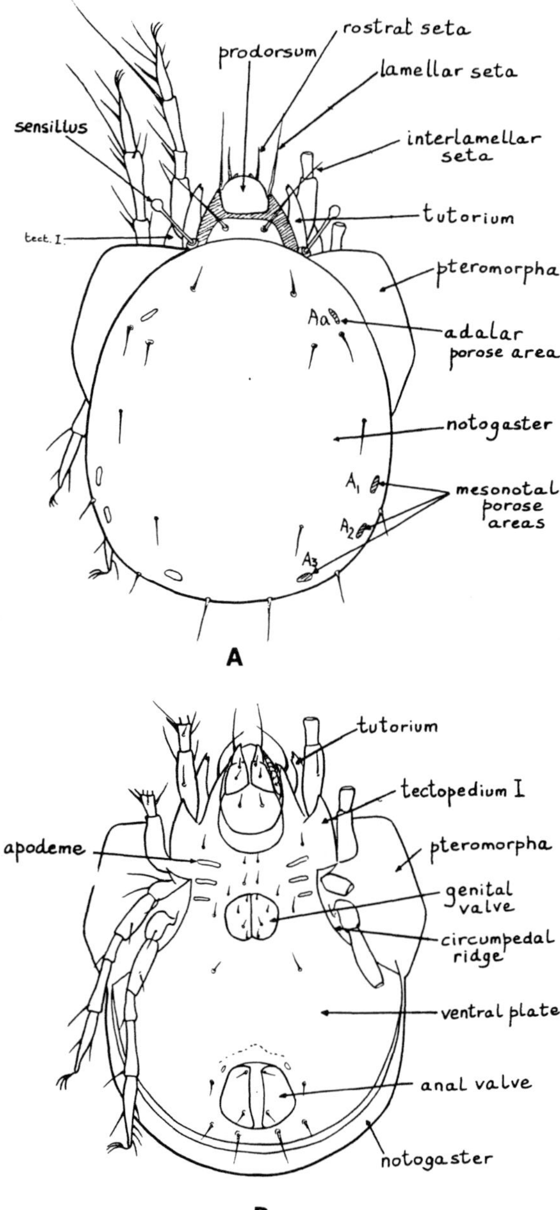

Fig. 12.27. Diagrammatic representation of the external morphology of a euoribatid (brachypylinid) mite; (**A**) dorsal, (**B**) ventral.

refers to this condition as trichoidy. In holoid forms (Desmonomata (=Nothronota) and Euoribatida), the cuticle of the sejugal zone is partially or entirely sclerotized so that movement between the proterosoma and hysterosoma does not occur in this region. Grandjean (1969) proposed the descriptive terms Dichosomata for those oribatids which are dichoid or ptychoid and Holosomata for holoid forms. A thin, broad lobe of the prodorsum, the rostral tectum, extends anteriorly over the gnathosoma and its undersurface forms the roof and sides of the camerostome. The prodorsum has a maximum of six pairs of setae. The most characteristic are the sensilli (=pseudostigmatic organs), absent only in a small number of mainly aquatic genera, situated at the basal corners of the prodorsum and each arising from a deep bothridium (=pseudostigma). The interlamellar setae lie between the pseudostigmatic organs and a pair of lamellar setae is situated more anteriorly near the level of the posterior margin of the rostral tectum (Fig. 12.27A). The remaining three pairs of setae comprise a pair of rostral setae positioned near the anterior margin of the prodorsum and two pairs of exobothridial (=exosensillary) setae, often lacking in the Euoribatida, lateral or ventral to the bothridia. A number of blade-like tecta may be present on the prodorsum of the Euoribatida. The most conspicuous are the lamellae, which typically extend from just internal to the bothridia to the bases of the lamellar setae, and the translamella joining the lamellae distally. In some families, low ridge-like thickenings of the prodorsal cuticle termed costulae are present in place of the lamellae. A prolamella may pass on either side from the distal end of a lamella to the prodorsal margin. A sublamella, when present, extends from a point external to the bothridia to the tip of the lamella. A tutorium may be developed lateral and ventral to each lamella and immediately above the basal podomeres of the first leg.

The hysterosomatic plate covering the dorsum and sides of the hysterosoma is termed the notogaster and in the majority of adults is entire (Fig. 12.27A). In the Palaeosomata, the notogaster is not recognizable as such and a region of unsclerotized cuticle, the asthenic zone, may be present posterior to the prodorsum. In a number of Desmonomata, a pair of longitudinal bands of soft cuticle divide the notogaster into a dorsal unpaired notaspis flanked on either side by a lateral pleuraspis while in the Enarthronota (excluding the Mesoplophoroidea) the notaspis is itself subdivided into two to four sclerites by one to three transverse scissures and a subdivision of the pleuraspis may also occur (see Fig. 12.29C below). Small platelets lying between the notaspis and pleuraspis are referred to as superpleurals. The transverse bands of soft cuticle dividing the notaspis may be protected by the overhanging posterior margin of the anterior plate or by a narrow intercalary sclerite (Hypochthoniidae and Cosmochthoniidae). In some Euoribatida, a pair of tecta, termed pteromorphae, arise from the antero-lateral corners of the notogaster. They vary in size from small strap-like structures to larger subtrianguar or auriculate wings and may be delimited by a line of desclerotized cuticle. The number of setae on the adult notogaster varies between four and 16 pairs. The full complement, the holotrichous condition, is considered to be archetypal and comprises six transverse rows, *c,d,e,f,h,ps*, with, respectively, *4,2,2,2,3,3* pairs of setae (see Fig. 2.5D, p. 28). The loss of one or more pairs of setae of the holotrichous condition is common: unideficient (15 pairs) through quadrideficient (12 pairs) to multideficient (less than 12 pairs) states are common. A maximum of seven pairs of lyrifissures or cupules (*ia, im, ip, ih, ips, iad, ian*) may occur laterally on the notogaster and, in some Euoribatida, a series of areae porosae, usually four pairs, are present and comprise an anterior adalar pair and three postero-lateral pairs comprising the mesonotal series (Fig. 12.27A). One or two pairs of the porose areas may be lost or their number increased by subdivision while in certain genera the areas may be replaced by sacculi. A lenticulus is sometimes present near the anterior margin of the notogaster in some Euoribatida. A pair of large latero-opisthosomatic glands (=latero-abdominal glands) opening by circular or slit-like pores on the surface of the notogaster may be present. The openings of the glands are borne on tubular outgrowths in the Hermanniellidae.

The ventral region of the podosoma associated with each leg is a coxisternite or epimeron as described in Chapter 2 and in the Archoribatida and some Euoribatida, a pair of minute supracoxal spines (*eI*) is situated on coxisternum I slightly anterior to the trochanteral articulations. The arrangements of the ventral plates of the opisthosoma in the majority of the oribatids are shown in Fig. 12.28A–E. The genital and anal plates, flanked, respectively, by the aggenital and adanal plates, are usually conspicuous and in the majority of the Archoribatida they together occupy almost the whole length of the opisthosoma in the mid-ventral line, the so-called macropyline condition (Fig. 12.28A). Various fusions of these four plates may occur; for example, the aggenitals and adanals on each side fuse to form long aggenito-adanal plates (Fig. 12.28B) or there may be fusion of the genital with the aggenital and the anal with the adanal plates. In the Euoribatida, the relatively smaller genital and anal plates are usually well separated and are surrounded by a large ventral plate, the so-called brachypyline condition (Fig. 12.27B). A small pre-anal plate may also be present in some archoribatid families and in the Brachychthoniidae a pair of setae-bearing plates, the peranals, occurs between the anal plates. In the Eulohmanniidae, the ventral margins of the notogaster are acutely pointed and, although not meeting in the mid-ventral line, separate the genito-aggenital region from the ano-adanal region (Fig. 12.28C). This condition is called pseudodiagastry. The notogaster in the Nanhermanniidae, on the other hand, is continuous across the venter of the opisthosoma between the genital and anal regions and the condition is termed diagastry (Fig. 12.28D). The separation of the genito-aggenital region from the ano-adanal region by a transverse scissure in the Epilohmanniidae is termed schizogastry (Fig. 12.28E). The plates covering the genital and anal openings bear setae and their number and arrangement are used as taxonomic characters. One, two and three pairs of genital papillae are present in the protonymph, deuteronymph and tritonymph, respectively, and may be used to distinguish these nymphal stases. The adults retain the tritonymphal complement of papillae.

The gnathosoma lies in a camerostome and is articulated to the idiosoma by means of condyles located at the posterior corners of the

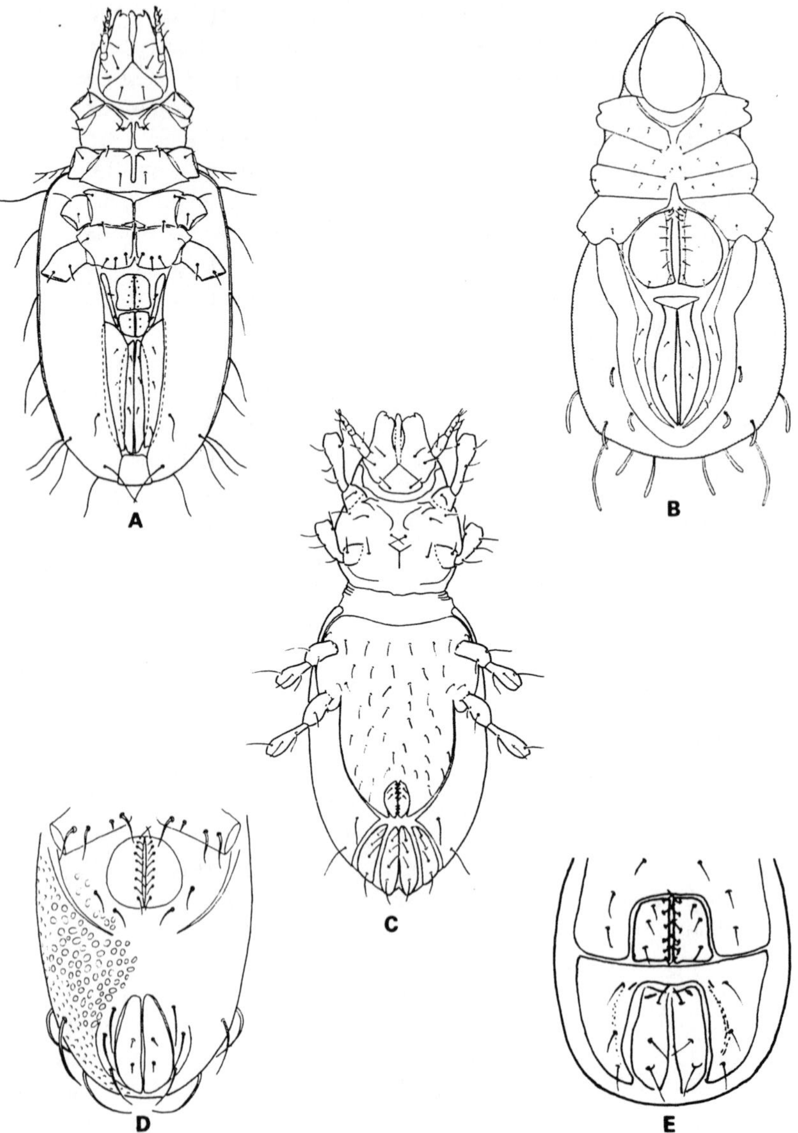

Fig. 12.28. Types of ano-genital regions in the Archoribatida (Macropylina). **A**, *Perlohmannia*; **B**, *Nothrus*; **C**, *Eulohmannia*; **D**, *Nanhermannia*; **E**, *Epilohmannia*.

camerostome. The rostral tectum may be very reduced so that the gnathosoma is partially visible from above, the astegasime condition (see Fig. 12.29A below), or be more extensive so as to conceal the gnathosoma from above, the stegasime condition. The chelicerae are compact and chelate except in the Gustaviidae in which the single digit resembles a long stylet with minute distal teeth (see Fig. 12.31C below). The cheliceral body or shaft, elongated in the Suctobelbidae and Phenopelopidae (see Fig. 12.31D below), typically bears two setae (*cha, chb*) and a paraxial Tragardh's organ; sometimes a lower Tragardh's organ may be present. The digits are generally dentate. The subcapitulum is rutellate and a labiogenal articulation, absent in the most primitive families, separates a posterior unpaired hysterostome or mentum from a pair of anterior genae and may be anarthric, stenarthric or diarthric in form (see Fig. 5.7, p. 141). Two or three pairs of adoral setae, a pair of supracoxal spines (near the bases of the pedipalps) and up to five further pairs of setae comprising anterior subcapitulars, median subcapitulars and hysterostomals occur on the subcapitulum. the pedipalps generally have five palpomeres with the predominant setal formula (0-2-1-3-9). In some families, for example the Epilohmanniidae and Euphthiracaridae, the number of palpomeres is reduced. The pedipalps are never raptorial.

The legs have six podomeres except in the Palaeosomata in which the femur is divided into a basifemur and telofemur. The apotele usually comprises one (monodactyl) or three (tridactyl) claws and a basilar piece; two claws (bidactyl) are uncommon. In the triheterodactyl condition, the lateral claws are thinner than the median claw and/or differently shaped. The chaetotaxy of the legs has been discussed in Chapter 2 (see Fig. 2.19A, p. 57). The tibial solenidia are long and whip-like as in the Astigimata.

Six stases are represented in the postembryonic development of the Oribatida, namely prelarva, larva, three nymphs and adult. Sperm transfer is indirect by stalked spermatophores except in *Collohmannia*. This large order is cosmopolitan in distribution with the greatest species diversity occurring in the tropics. Its members usually form the numerically dominant acarine component of the soil fauna and are chiefly saprophagous or microphytophagous. Some species spend part or all of their life cycle on low-growing vegetation or on trees.

Our present knowledge of the external morphology of the Oribatida, and much of the descriptive terminology, is based on the studies of the late F. Grandjean whose publications (reprinted in seven volumes between 1972 and 1976 under the title *Ouvres Acarologiques Complètes*) should be consulted for detailed information.

Classification

The classification of the Oribatida adopted in this work follows, in the main, that of Johnston (1982) which is largely based on Grandjean's (1969) classification in which six major groups of apparent equal status were recognized but not given hierarchic rank (Grandjean, 1969). These groups have subsequently been given the rank of cohorts or suborders, the latter being adopted in this work. The major division of the Oribatida into two groups, the Archoribatida or Macropylina and the Euoribatida or Brachypylina is retained. An alternative classification of the Oribatida has been introduced by Lee (1984) in which the order is divided into two suborders, the Dismalida and Comalida, distinguished chiefly on the form of the rutella (=external malae). The morphological features used in the following key relate mainly to the adult stase.

1. Leg genua similar in shape and subequal in length to corresponding tibiae (see Fig. 12.29B below); propodo-hysterosomatic articulation present or absent, some ptychoid; stegasime or astegasime; genital and anal regions in adults usually touching or narrowly separated, together extending almost the entire length of the opisthosoma and with narrow aggenital and adanal plates on either side either free or fused *or* diagastric, pseudodiagastric or with one or two ventral plates (see Fig. 12.28 above); without long tracheae opening into acetabular cavities of legs or sejugal furrow, at the most brachytracheae; notogaster with at least 15 pairs of setae (except Mesoplophoridae with eight pairs) and may be divided by transverse scissures (see Fig. 12.29C below); dehiscence never circumgastric. [Division ARCHORIBATIDA (=Lower Oribatida; Macropylina)] 2

– Leg genua usually very differently shaped from and shorter than corresponding tibiae (see Fig. 12.31E below); no propodo-hysterosomatic articulation; stegasime; opisthosoma venter typically covered by a ventral plate surrounding well-separated genital and anal plates (Fig. 12.27B); notogaster with four to 15 pairs of setae and never divided by transverse scissures or furrows; adults often with long tracheae opening into some of the leg acetabula and sejugal furrow; dehiscence circumgastric. [Division EUORIBATIDA (=Higher Oribatida; Brachypylina)] .... suborder CIRCUMDEHISCENTIAE

2. Femora of legs divided into basi- and telofemora (see Fig. 12.29B below); stegasime (see Fig. 12.29A below); latero-opisthosomatic glands absent; adults weakly sclerotized .. suborder PALAEOSOMATA

– Femora of legs entire; stegasime or astegasime; latero-opisthosomatic glands present or absent; adults usually well-sclerotized 3

3. Notogaster subdivided by one or more transverse intersegmental scissures or furrows (see Fig. 12.29C,E below); peranal segment may be present 4

– Notogaster without transverse intersegmental scissures or furrows; peranal segment lacking 5

4. Latero-opisthosomatic glands absent; lyrifissures *iad* and *ian* absent; notogaster with one to three transverse scissures; ptychoidy in some taxa (see Fig. 12.29D below) .... suborder ENARTHRONOTA

– Latero-opisthosomatic glands present; lyrifissures *iad* and *ian* present; notogaster with single transverse scissure; usually small, weakly sclerotized and astegasime forms .... suborder PARHYPOSOMATA

5. Body dichoid, i.e. sejugal zone of flexible cuticle (see Fig. 12.30A below) .... suborder MIXONOMATA

– Body holoid or ptychoid (in adult only) 6

6. Body more or less compressed laterally and usually ptychoid in the adult, i.e. with dorsal hinge between prosoma and opisthosoma allowing the former to be folded down and the legs to be concealed (see Fig. 12.30B below) .... suborder EUPTYCTIMA

– Body holoid, articulation between propodosoma and hysterosoma lost (see Figs 12.30D,E, & 12.31B) .... suborder DESMONOMATA

The most comprehensive key works on the Oribatida have been produced by J. Balogh and his co-workers. A key to the oribatid genera of the world appears in Balogh (1972) while the oribatids of the Neotropical region and the Archoribatida of the Palaearctic region are dealt with, respectively, by Balogh and Balogh (1990) and Balogh and Mahunka (1985). The works of Sellnick (1929, revised 1960) and Willmann (1931) deserve mention for their service to soil ecologists especially in the period 1945–1960.

DIVISION ARCHORIBATIDA OR MACROPYLINA

Suborder Palaeosomata

The mites of this suborder are considered to be the most primitive of the Oribatida. They are weakly sclerotized and astegasime (Fig. 12.29A). Six pairs of setae, including the simple or sickle-like sensilli arising from deep bothridia, are present on the prodorsal plate. There is no distinct notogaster and the number of setae on the dorsal, lateral and posterior regions of the hysterosoma in nymphs and adults always exceeds 32. The setae may be very long and heavily pigmented as in the Palaeacaridae and may be situated on locally sclerotized areas. The unpigmented setae are only birefringent at their base while the brown-coloured setae are isotropic. Lyrifissures *ian* and *iad*, the latero-opisthosomatic glands and the peranal segment are never present. Seven pairs of setae occur on the subcapitulum of which three pairs are adoral. The rutella are simple. The legs in the larva and protonymph have only a single femur but in the deuteronymph the femur of leg I is divided, while in the tritonymph and adult all legs have a basifemur and telofemur. The genua of the legs are similar in length and form as the tibiae (Fig. 12.29B). The ambulacra in the adult are tridactyl, with the central claw well developed or much reduced in size, or, less commonly, bidactyl. Unlike other Oribatida, tridactyl and bidactyl forms occur in the immatures. The number of stases in the life cycle is normal and the immatures are not heteromorphic in respect of the adult.

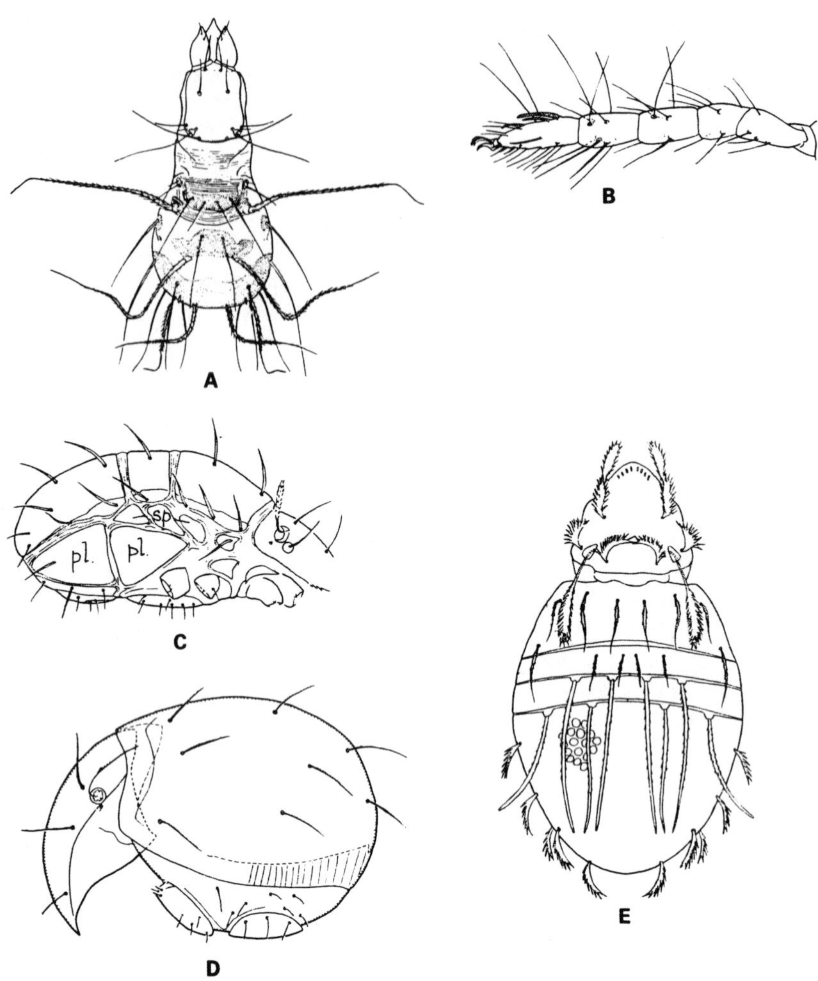

Fig. 12.29. **A**, dorsum and **B**, leg I of *Palaeacarus histricinus* (Palaeosomata) (based on Grandjean, 1954d); **C**, *Eobrachychthonius* sp. (Enarthronota); **D**, *Mesoplophora* sp. (Enarthronota); **E**, *Cosmochthonius lanulatus* (Enarthronota). *pl.* = pleural platelets; *sp.* = superpleural platelets.

Grandjean (1954d) recognized three families: Palaeacaridae; Archeonothridae (with two subfamilies, Archeonothrinae, Acaronychinae) and Ctenacaridae (with three subfamilies, Ctenacarinae, Adelphacarinae and Aphelacarinae). Subsequently, the subfamilies have been given family rank (Gilyarov & Krivolutsky, 1975; Johnston, 1982). The order is probably cosmopolitan in distribution but the majority of the species appear to favour warm climates. A marked exception is *Adelphacarus sellnicki* which has been found in Swedish Lapland. The majority of the known species have been collected from soil habitats but *Aphelacarus acarinus* also occurs in dust in houses.

Suborder Enarthronota

This diverse group of oribatids is generally recognized by the division of the notogaster into discrete sclerites by one to three transverse scissures (Fig. 12.29C). The Mesoplophoridae are exceptional in lacking these scissures (Fig. 12.29D). In some taxa such as Brachychthoniidae, the scissures probably reflect segmentation but in those having an intercalary sclerite (Fig. 12.29E) and erectile setae (type-S scissure) the scissure tends to divide the primitive segments (Norton, 1984). The mites of this suborder are small to medium-sized and are stegasime or astegasime. Setae f_1 and f_2 are present but lyrifissures *iad* and *ian* and latero-opisthosomatic glands are lacking. Ptychoidy occurs throughout ontogeny in the families Mesoplophoridae and Protoplophoridae which constitute the 'Arthroptyctima' of Grandjean (1967). The maximum number of notogastric setae is 32. The notogaster of the Mesoplophoridae is exceptional in that it bears only eight pairs of setae while setae of the *f*, *h*, and *ps* series are carried on the ventral plate. A peranal segment is present in the Brachychthoniidae, Pterochthoniidae and Phyllochthoniidae. Ocelli occur in *Heterochthonius*. The femora of the legs are usually undivided. Adults and immatures are similar.

The suborder comprises 13 families divided among seven superfamilies. Members of the Brachychthonoidea and Hypochthonoidea are represented by a number of species and are commonly found in mosses, grassland and forest soils, particularly in the temperate regions. Members of the other superfamilies are relatively rare.

References

Gilyarov and Krivolutsky (1975); Grandjean (1946c, 1949b, 1969); Moritz (1976); Niedbala (1976); Norton (1984).

Suborder Parhyposomata (=Parhypochthonata)

The body of the mites of this suborder either lacks sclerotization, as in the Parhypochthoniidae and Gehypochthoniidae, or is moderately sclerotized as in the Elliptochthoniidae. A transverse scissure is present between segments D and E (see Fig. 2.6A, p. 29). The notogastric chaetotaxy is normal (16 pairs) and there is a complete set of lyrifissures. The latero-opisthosomatic glands are present and a peranal segment occurs in the Parhypochthoniidae. There is only one pair of aggenital setae. Members of the Elliptochthoniidae are stegasime but the other two families are astegasime. The ambulacrum is tridactylous with a regressive median claw in the Parhypochthoniidae and Elliptochthoniidae but the median claw is lacking in the Gehypochthoniidae. The famulus has one or several bracts.

Each of the three families contains only one genus. The species are relatively uncommon and have been collected either from tree-holes or soil. The suborder is represented in Europe, North America and Asia.

References

Grandjean (1934c, 1969); Norton (1975); Strenzke (1963).

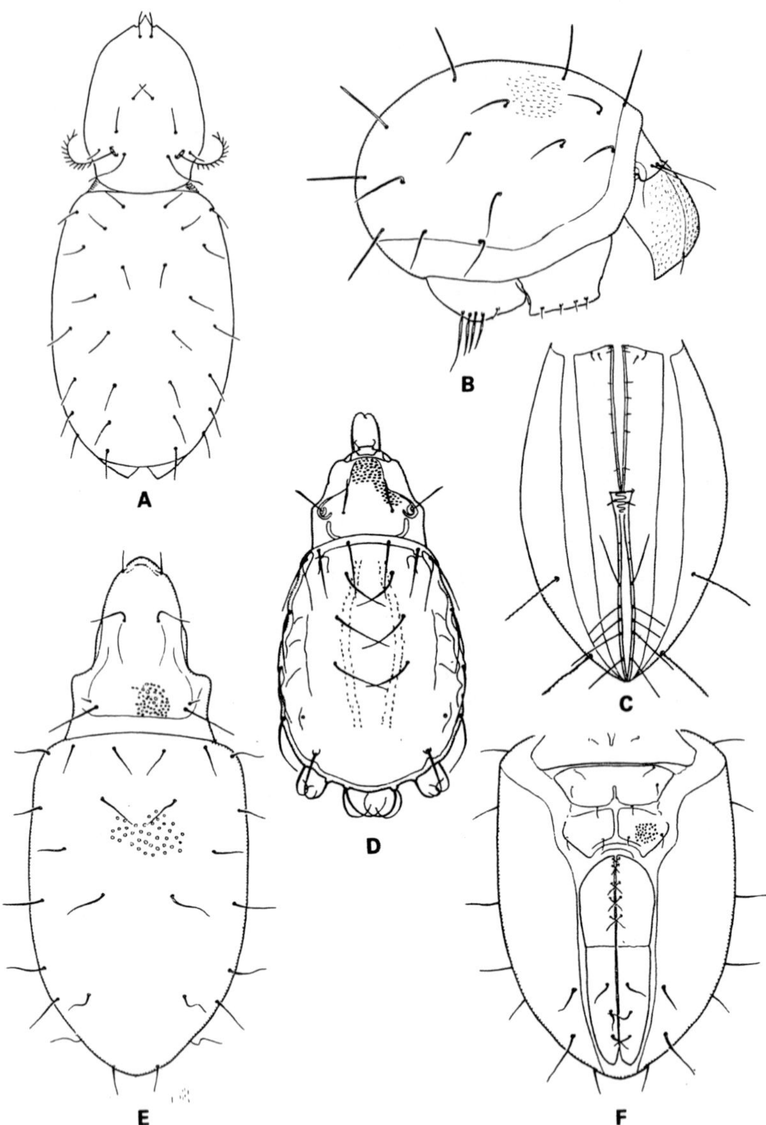

Fig. 12.30. **A**, *Eulohmannia* sp. (Mixonomata); **B**, *Steganacarus* sp. (Euptyctima; Phthiracaridae); **C**, ano-genital region of *Rhysotritia* sp. (Euptyctima; Euphthiracaridae); **D**, *Platynothrus* sp. (Desmonomata); **E**, **F**, dorsum (**E**) and venter (**F**) of *Malaconothrus* sp. (Desmonomata).

Suborder Mixonomata

These well-sclerotized mites are dichoid and stegasime (Fig. 12.30A). The notogastric chaetotaxy is holotrichous or neotrichous and a peranal segment is not developed. The Eulohmanniidae are pseudogastric and the Epilohmanniidae are schizogastric. The genital plates in the Perlohmanniidae and some Lohmanniidae have a transverse suture (see Fig. 12.28A above). Discrete aggenital plates are present in the Epilohmanniidae and Perlohmanniidae. The latero-opisthosomatic glands are present or absent. The pedipalps have two to five palpomeres.

Grandjean (1969) included the ptychoid superfamilies Euphthiracaroidea and Phthiracaroidea, which formed his Euptyctima, and the Collohmanniidae within the Mixonomata. These three families have subsequently been removed by some authors, for example Johnston (1982), from the Mixonomata and are given, collectively, their own subordinal (or cohort) status, the Euptyctima. The Mixonomata comprises the four families mentioned above and the Nehypochthoniidae. They are predominantly soil-inhabiting forms.

References

Gilyarov and Krivolutsky (1975); Grandjean (1969); Johnston (1982).

Suborder Euptyctima

These are medium to large-sized mites in which the articulation of the sejugal zone and the coxisternum may result in the development of ptychoidy in the adult but not in the immatures (Fig. 12.30B). The opisthosoma is large in relation to the prosoma and is laterally compressed. The notogaster bears 14 or more pairs of setae and latero-opisthosomatic glands are present or absent. The elongate ano-genital region occupies almost the entire venter of the opisthosoma and is narrow in the Collohmanniidae and Euphthiracaridae (Fig. 12.30C) but is broad in the Phthiracaridae. The aggenitals and adanals are free or fused, respectively, with the genital and anal plates. The anal plates in the Euphthiracaridae articulate anteriorly in a triangular zone of interdigitating processes. The pedipalp has two to five articles. Sperm transfer in *Collohmannia gigantea* appears to be unique among the Oribatida in not involving stalked spermatophores attached to a substrate.

The suborder, containing the three families referred to above, is cosmopolitan. The euphthiracarids and phthiracarids are most common in forest litter where some species burrow into twigs and pine needles.

References

Gilyarov and Krivolutsky (1975); Jacot (1930); Märkel (1964); Van der Hammen (1963b).

Suborder Desmonomata (=Nothronota)

These are medium-sized to large, well-sclerotized mites without an articulation in the sejugal zone (Fig. 12.30D). The cuticle has an ornamental birefringent cerotegument in the Malaconothridae. An unpaired notaspis and on either side a lateral pleuraspis are present in some families. There are 16 or fewer pairs of notogastric setae. Latero-opisthosomatic glands and lyrifissures *ian* (except in *Hermannia*) are present. Sensilli and bothridia are lost in Malaconothridae (Fig. 12.30E,F). The notogaster is continuous mid-ventrally in front of the anal region (diagastric condition) in the Nanhermanniidae (see Fig. 12.28D above). The genital and anal regions are situated in a broad ventral plate in the Hermanniidae but there is no separate ventral plate in members of the other families. Aggenitals and adanals are free or fused. A well-developed coxisternum is present and the coxisternal setation may be neotrichous, as in the Nothridae. The line of dehiscence is in the form of an inverted T. Thelytoky is widespread in this group.

Seven families are represented in this suborder, which was referred to as Nothronota by Johnston (1982); they comprise the Crotoniidae, Camisiidae, Trhypochthoniidae and the four families referred to above. The systematic position of the Hermanniidae is problematical and its inclusion in this suborder is provisional. The family has features in common with the Circumdehiscentiae and this relationship is supported by recent work on the structure of oribatid spermatophores by Alberti *et al.* (1991) which indicates an affinity of the Hermanniidae with the Belbidae. The majority of the species of the Desmonomata are found in woodland soils but members of the Malacanothridae and Trhypochthoniidae are common in semi-aquatic habitats especially in mosses of moorlands and bogs.

References

Gilyarov and Krivolutsky (1975); Knülle (1957); Luxton (1982); Sellnick and Forsslund (1955); Van der Hammen (1952).

DIVISION EUORIBATIDA OR BRACHYPYLINA

Suborder Circumdehiscentiae

This is the largest suborder of the Oribatida and the name of the taxon reflects a major characteristic of these mites, namely the occurrence of a circumgastric line of dehiscence in the immature stases. The adults are well sclerotized and stegasime. The prodorsum may be provided with variously shaped lamellae and other tecta (Fig. 12.31A,B). Sensilli and bothridia may be reduced or lacking in the adults of some aquatic species (Hydrozetidae and Ameronothridae). One or both pairs of exobothridial setae are lost. The notogaster has a maximum of 15 pairs of setae and may be provided with pteromorphae. The Grandjean systems of notogastric setal notation for holotrichous, unideficient (lacking f_1, Fig. 12.32B) and multideficient taxa (such as *Dometorina*) are given below. The alternative notation for the notogastric chaetotaxy proposed by Lee (1981, 1987) and based on three longitudinal series J, Z, S (see p. 26) is given in brackets.

Holotrichous	Unideficient	Multideficient (*Dometorina*)
c1 (J1)	c1 (J1)	
c2 (Z1)	c2 (Z1)	ta (Z1)
c3 (S1)	c3 (S1)	
cp (S2)	la (Z2)	te (Z2)
d1 (J2)	da (J2)	
d2 (Z2)	dm (J3)	
e1 (J3)	dp (J4)	
e2 (Z3)	lm (Z3)	ti (J3)
f1 (J4)		
f2 (Z4)	lp (Z4)	ms (Z4)
h1 (J5)	h1 (J5)	r1 (J5)
h2 (Z5)	h2 (Z5)	r2 (Z5)
h3 (S5)	h3 (S5)	r3 (S5)
ps1 (J6)	ps1 (J6)	p1 (J6)
ps2 (Z6)	ps2 (Z6)	p2 (Z6)
ps3 (S6)	ps3 (S6)	p3 (S6)

Latero-opisthosomatic glands are present. An octotaxic system of porose areas (Aa, A_1, A_2, A_3) or sacculi (Sa, S_1, S_2, S_3) may be present or absent. A lenticulus is present in some families (e.g. Phenopelopidae, Hydrozetidae). The coxae form a coxisternum and there is a well-defined ventral plate (Fig. 12.27B). The genital area in both sexes is relatively small and well separated from the anal area. The cheliceral body is normal or attenuate (Fig. 12.31D) and a Tragardh's organ is usually present. The Gustaviidae are exceptional in that the single digit is styliform and provided with denticles in its anterior third (Fig. 12.31C). The subcapitulum is typically diarthric. The genua of the legs are reduced in size (Fig. 12.31E). A tracheal system opening at the bases of the legs and ventrally in the sejugal zone is usually present in the adults. The immature stages are very different in appearance from the adults.

The classification of the suborder and its status in relation to the Desmonomata is in need of further study, particularly in the re-evaluation of the characters used in its higher classification. At present, the Circumdehiscentiae is usually divided into two groups on the basis of the presence or absence of the octotaxic system of areae porosae or sacculi on the notogaster and are referred to, respectively, as the Poronoticae and Pycnonoticae. Grandjean (1954e), in his classification of the suborder, used mainly the types of dehiscence, characteristics of the immature stages and the chaetotaxy of the notogaster and genital plates in the subdivision of these two major taxa. This system is difficult to incorporate into the traditional hierarchic classification. Opsiopheredermal (Hermanniellidae), Eupheredermal (a large group including Belbidae, Cepheidae, Eremaeidae) and Apheredermal sections are recognized in the Pycnonoticae. The apherederms are further divided into two groups on the basis of whether the nymphs are dorsodeficient (Metrioppiidae [=Ceratoppiidae]

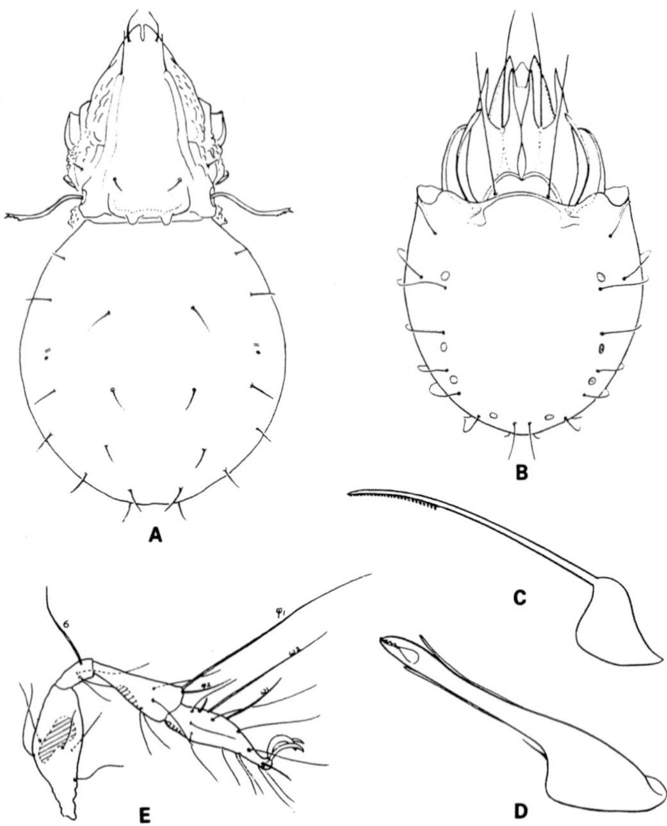

Fig. 12.31. **A**, *Conchogneta* sp. (Circumdehiscentiae: Autognetidae); **B**, *Oribatella* sp. (Circumdehiscentiae: Oribatellidae); **C**, chelicera of *Gustavia* sp. (Gustaviidae); **D**, chelicera of *Eupelops* sp. (Phenopelopidae); **E**, leg I of a euoribatid mite showing the unequal lengths of the genu and tibia.

and Liacaridae) or unideficient and sometimes bideficient (Oppiidae, Suctobelbidae). The dorsodeficient taxa are at least quadrideficient. Three main types of apheredermal nymphs occur in the Poronoticae:

1. those with a 'wrinkled' (or plicate) cuticle as in the Achipteriidae and Phenopelopsidae (Fig. 12.32A);

2. species with a dorsal macrosclerite in the form of a large hysterosomatic shield as in, for example, the Ceratozetidae, Euzetidae and Galumnidae (Fig. 21.32B);

3. forms with dorsal microsclerites in the form of porose platelets eccentrically placed around the bases of certain notogastric setae ('Excentrosclerotosae') as in the Scheloribatidae and Oribatulidae.

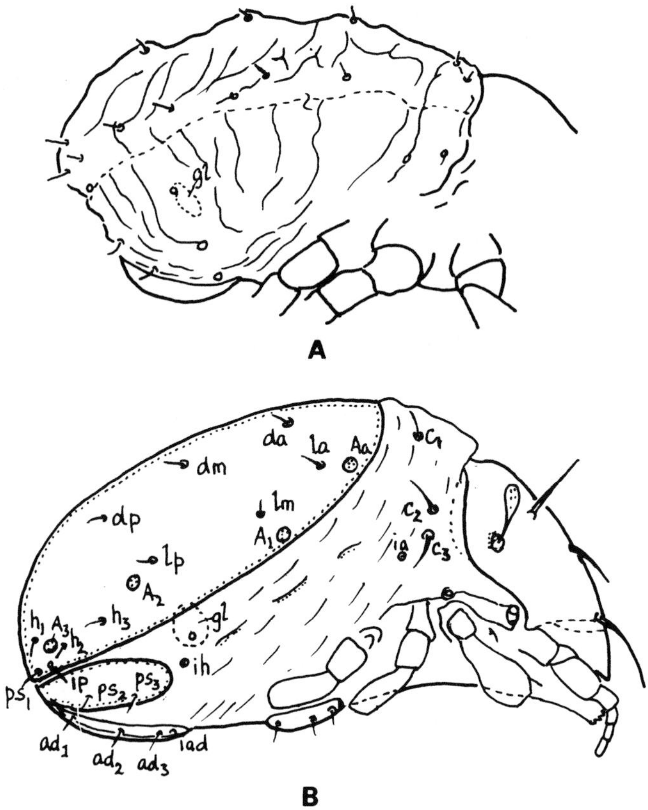

Fig. 12.32. **A**, protonymph of *Tectocepheus* sp. showing 'wrinkled' cuticle; **B**, deuteronymph of *Galumna tarsipennata* illustrating the presence of a large hysterosomatic sclerite (macrosclerite). See text for details of setal notation (after Travé, 1970).

According to Johnston (1982), the Pycnonoticae contains 75 families and the Poronoticae 29 families. A superfamilial classification of the Euoribatida is given by Balogh (1972).

These mites are cosmopolitan in distribution and form a significant component of the mite population of forest soils. Many species are arboricolous (e.g. in the Liacaridae, Carabodidae, Galumnidae) while others live in vegetation in fresh water (Hydrozetidae) or are littoral (Ameronothridae).

References

Balogh (1972); Gilyarov and Krivolutsky (1975); Grandjean (1954e); Norton (1990); Van der Hammen (1963a).

References

Aeschlimann, A. (1958) Développement embryonaire d'*Ornithodoros moubata* (Murray) et transmission transovarienne de *Borrelia duttoni*. *Acta Tropica* 15, 15–64.

Aeschlimann, A. (1984) What is our current knowledge of acarine embryology? In: Griffiths, D.A. & Bowman, C.E. (eds), *Acarology VI*, Vol. 1. Ellis Horwood, Chichester, pp. 90–99.

Agbede, R.I.S. & Kemp, D.H. (1985) Digestion in the cattle-tick *Boophilus microplus* (Canestrini). Light microscope study of the gut cells in nymphs and females. *International Journal of Parasitology* 15, 147–157.

Agbede, R.I.S. & Kemp, D.H. (1987) *Boophilus microplus* (ixodid tick): fine structure of the gut basophilic cell in relation to water and ion transport. *Experimental and Applied Acarology* 3, 233–242.

Ainscough, B.D. (1981) Uropodine studies. 1. Suprageneric classification in the cohort Uropodina Kramer, 1882 (Acari: Mesostigmata). *International Journal of Acarology* 7, 47–56.

Akimov, I.A. (1973) On the morphological and physiological characteristics of the alimentary canal of the bulb mite *Rhizoglyphus echinopus* (Fumouze & Robin). In: Daniel, M. & Rosický B. (eds), *Proceedings of the 3rd International Congress on Acarology*, Academia, Prague, pp. 703–706.

Akimov, I.A. (1979) Morphological and functional characteristics of the mouthparts of the Acaridae mites (Acaridae Ewing et Nesbitt, 1942). In: Piffl, E. (ed.), *Proceedings of the 4th International Congress of Acarology*. Akadémiai Kiadó, Budapest, pp. 569–574.

Akimov, I.A. (1985) *Biological Foundations of Harmfulness in Acaroid Mites*. Naukova Dumka, Kiev, 160pp [in Russian].

Akimov, I.A. & Barabanova, V.V. (1977) Morphological and functional peculiarities of the digestive system of tetranychid mites (Trombidiformes, Tetranychoidea). *Entomologischeskoe obozrenie*. 1977 (4), 912–922 [in Russian].

Akimov, I.A. & Starovir, I.S. (1974) Morpho-functional peculiarities of the digestive system in *Phytoseiulus persimilis* Athias-Henriot (Gamasoidea, Phytoseiidae). *Vestnik Zoologii* 1974, 60–64 [in Russian].

Akimov, I.A. & Starovir, I.S. (1976) Structure of the digestive system in *Amblyseius andersoni* and *A. reductus*. *Vestnik Zoologii* 1976 (4), 7–13 [in Russian].

Akimov, I.A. & Starovir, I.S. (1977) Morpho-functional peculiarities of the digestive system of *Amblyseius andersoni* (Gamasoidea, Phytoseiidae). *Vestnik Zoologii*, 1977 (3), 13–18 [in Russian].

Akimov, I.A. & Yastrebtsov, A.V. (1981) Structure and function of the mouth apparatus and leg muscles in *Tetranychus urticae* C.L. Koch (Trombidiformes, Tetranychoidea). *Vestnik Zoologii* 1981 (3), 54–59 [in Russian].

Akimov, I.A. & Yastrebtsov, A.V. (1984) Reproductive system of *Varroa jacobsoni* 1. Female reproductive system and oogenesis. *Vestnik Zoologii* 1984 (6), 61–68 [in Russian].

Akimov, I.A. & Yastrebtsov, A.V. (1987) Muscular system and elements of the skeleton of the parasitic mite *Spinturnix vespertiliones* (Gamasina, Spinturnicidae). *Parasitologiya* 21, 445–449 [in Russian].

Akimov, I.A. & Yastrebtsov, A.V. (1988) Skeletal-muscular system of gamasid mites (Mesostigmata: Gamasina). *Zoologische Jahrbuecher. Abteilung fuer Anatomie und Ontogenie der Tiere* 117, 397–439.

Akimov, I.A. & Yastrebtsov, A.V. (1989) Muscular system and skeletal elements of the mite *Nothrus palustris* (Acariformes, Oribatida). *Zoologicheskii Zhurnal* 68, 57–67 [in Russian].

Akimov, I.A. Yastrebtsov, A.V. (1990) Embryonic development of the mite *Spinturnix vespertiliones* (Parasitiformes: Spinturnicidae). *Acarologia* 31, 3–12.

Akimov, I.A. Hänel, H., Yastrebtsov, A.V. & Romanovskii, I.A. (1986) The nervous system of the mite *Varroa jacobsoni* (Parasitiformes, Varroidae), a honey bee parasite. General morphology of the synganglion and its development in ontogenesis. *Vestnik Zoologii* 1986 (2), 45–50 [in Russian].

Akimov, I.A., Starovir, I.S., Yastrebtsov, A.V. & Gorgol, V.T. (1988) *The Varroa Mite – the Causative Agent of Varroatosis of Bees. A Morphological Outline.* Nukova Dumka, Kiev, 118pp [in Russian].

Akov, S. (1982) Blood digestion in ticks. *Current Themes in Tropical Science* 1, 197–211.

Alberti, G. (1973) Ernährungsbiologie und Spinnvermögen der Schnabelmilben (Bdellidae, Trombidiformes). *Zeitschrift für Morphologie und Ökologie der Tiere* 76, 285–338.

Alberti, G. (1974) Fortpflanzungsverhalten und Fortpflanzungsorgane der Schnabelmilben (Acarina: Bdellidae, Trombidiformes). *Zeitschrift für Morphologie und Ökologie der Tiere* 78, 111–157.

Alberti, G. (1975) Prälarven und Phylogenie der Schnabelmilben (Bdellidae, Trombidiformes). *Zeitschrift fuer Zoologische Systematik und Evolutionforschung* 13, 44–62.

Alberti, G. (1979) Fine structure and probable function of the genital papillae and Claparède organs of Actinotrichida. In: Rodriguez, J.G. (ed.), *Recent Advances in Acarology*, Vol. 2. Academic Press, New York, San Francisco & London, pp. 501–507.

Alberti, G. (1980a) Zur Feinstruktur der Spermien und Spermiocytogenese der Milben (Acari). 1. Anactinotrichida. *Zoologische Jahrbuecher. Abteilung fuer*

Anatomie und Ontogenie der Tiere 104, 77–138.

Alberti, G. (1980b) Zur Feinstruktur der Spermien und Spermiocytogenese der Milben (Acari). II. Actinotrichida. *Zoologische Jahrbuecher. Abteilung fuer Anatomie und Ontogenie der Tiere* 104, 144–203.

Alberti, G. (1984) The contribution of comparative spermatology to the problems of acarine systematics. In: Griffiths, D.A. & Bowman, C.E. (eds), *Acarology VI*, Vol. 1. Ellis Horwood, Chichester, pp. 479–490.

Alberti, G. (1988) Genital system of Gamasida and its bearing on phylogeny. In: Channa Basavanna, G.P. & Viraktamath, C.A. (eds), *Progress in Acarology*, Vol. 1. Oxford & IBH Publishing, New Delhi, pp. 197–204.

Alberti, G. (1991) Spermatology in the Acari: systematic and functional implications. In: Schuster, R. & Murphy, P.W. (eds), *The Acari. Reproduction, Development and Life-History Strategies.* Chapman & Hall, London, pp. 77–105.

Alberti, G. & Bader, C. (1990) Fine structure of the external 'genital' papillae in the freshwater mite *Hydrovolzia placophora* (Hydrovolziidae, Actinedida, Actinotrichida, Acari). *Experimental and Applied Acarology* 8, 115–124.

Alberti, G & Crooker, A.R. (1985) Internal anatomy. In: Helle, W & Sabelis, M.W. (eds), *Spider Mites, their Biology, Natural Enemies and Control*, Vol. 1A. Elsevier, Amsterdam, pp. 29–62.

Alberti, G. & Ehrnsberger, R. (1977) Rasterelektronemikroskopische Untersuchungen zum Spinnvermögen der Bdellidae und Cunaxidae (Acari: Prostigmata). *Acarologia* 19, 55–61.

Alberti, G. & Fernandez, N.A. (1988) Fine structure of a secondarily developed eye in the freshwater moss-mite, *Hydrozetes lemnae* (Coggi, 1899) (Acari; Oribatida). *Protoplasma* 146, 106–117.

Alberti, G & Fernandez, N.A. (1990) Aspects concerning the structure and function of the lenticulus and clear spot of certain oribatids (Acari Oribatida). *Acarologia* 31, 65–72.

Alberti, G. & Hänel, H. (1986) Fine structure of the genital system in the bee parasite. *Varroa jacobsoni* (Gamasida: Dermanyssina) with remarks on spermiogenesis, spermatozoa and capacitation. *Experimental and Applied Acarology* 2, 63–104.

Alberti, G. & Storch, V. (1973) Zur Feinstruktur der "Munddrüsen" von Schnabelmilben (Bdellidae, Trombidiformes). *Zeitschrift für Wissenschaftliche Zoologie, Leipzig* 186, 149–160.

Alberti, G. & Storch, V. (1974) Über Bau und Funktion der Prosoma-Drüsen von Spinnmilben (Tetranychidae, Trombidiformes). *Zeitschrift für Morphologie und Okologie der Tiere* 79, 133–153.

Alberti, G. & Storch, V. (1976) Ultrastruktur-Untersuchungen am männlichen Genitaltrakt und an Spermien von *Tetranychus urticae* (Tetranychidae, Acari). *Zoomorphologie* 83, 283–296.

Alberti, G. & Storch, V. (1977) Zur Ultrastruktur der Coxaldrüsen actinotricher Milben (Acari, Actinotrichida). *Zoologische Jahrbuecher. Abteilung fuer Anatomie und Ontogenie der Tiere* 98, 394–425.

Alberti, G. & Zeck-Kapp, G. (1986) The nutrimentary egg development of the mite, *Varroa jacobsoni* (Acari, Arachnida), an ectoparasite of honey bees. *Acta Zoologica* 67, 11–25.

Alberti, G., Storch, V. & Renner, H. (1981) Über den feinstrukturellen Aufbau der Milbencuticula (Acari: Arachnida), *Zoologische Jahrbuecher. Abteilung fuer Anatomie und Ontogenie der Tiere* 105, 183–236.

Alberti, G., Fernandez, N.A. & Kümmel, G. (1991) Spermatophores and spermatozoa of oribatid mites (Acari: Oribatida). Part II: Functional and systematical considerations. *Acarologia* 32, 435–449.

Alexander A.J. (1957) Courtship and mating in the scorpion, *Opisthophthalmus latimanus. Proceedings of the Zoological Society of London* 128, 529–544.

Alexander, A.J. (1962) Biology and behaviour of *Damon variegatus* Perty of South Africa and *Admetus barbadensis* Pocock of Trinidad, W.I. (Arachnida, Pedipalpi). *Zoologica* 47, 25–37.

Alexander, R.D. (1964) The evolution of mating behaviour in arthropods. In: Highman, K.C. (ed.), *Insect Reproduction. Royal Entomological Society of London Symposium No. 2*, pp. 78–94.

Alexander, R.M. (1968) *Animal Mechanics.* Sedgwick & Jackson, London, 346pp.

Altner, H. (1977) Insect sensillum specificity and structure: an approach to a new typology. In: Le Magen, J. & Macleod, P. (eds), *Olfaction and Taste VI.* IRL Press, London, pp. 295–303.

Altner, H. & Prillinger, L. (1980) Ultrastructure of invertebrate chemo-, thermo- and hygroreceptors and its functional significance. *International Review of Cytology* 67, 69–139.

Amano, H. & Chant, D.A. (1978) Mating behaviour and reproductive mechanisms of two species of predaceous mites, *Phytoseiulus persimilis* Athias-Henriot and *Amblyseius andersoni* (Chant) (Acarina: Phytoseiidae). *Acarologia* 20, 196–213.

Ammah-Attoh, V. (1966) Development of *Boophilus decoloratus* (Koch, 1844) on the ox. *Nature* [*Lond.*] 209, 629–630.

Amosova, L.I. (1979) The integument. In: Balashov, Yu. S. (ed.), *An Atlas of Tick Ultrastructure.* Nauka, Leningrad [in Russian]. Translation in: *Miscellaneous Publications of the Entomological Society of America*, 1983, pp. 23–34.

Anderson, D.T. (1973) *Embryology and Phylogeny in Annelids and Arthropods*, International Series of Monographs on Pure and Applied Biology: Zoology Division, Volume 50. Pergamon Press, Oxford, 495pp.

André, H.M. (1977) Note sur le genre *Mediolata* (Actinedida: Stigmaeidae) et description d'une nouvelle espèce corticole. *Acarologia* 18, 462–474.

André H.M. (1979) A generic revision of the family Tydeidae (Acari: Actinedida). Introduction, paradigms and general classification. *Annales de ls Société Royale Zoologique de Belgique, Bruxelles* 108, 103–168.

André, H.M. (1980) A generic revision of the family Tydeidae (Acari: Actinedida). 4. Generic descriptions, keys and conclusions. *Bulletin et Annales de la Société Royale Entomologique de Belgique, Bruxelles* 116, 103–130 and 139–168.

André H.M. (1988) Age-dependent evolution: from theory to practice. In: Humphries, C.J. (ed), *Ontogeny and Systematics.* Columbia University Press, New York, pp. 137–187.

André, H.M. & Remacle, C. (1984) Comparative and functional morphology of the gnathosoma of *Tetranychus urticae* (Acari: Tetranychidae). *Acarologia* 25, 179–190.

Aoki, J.-I. (1973) Soil mites (oribatids) climbing trees. In: Daniel, M. & Rosický, B. (eds), *Proceedings of the 3rd International Congress of Acarology*. Academia, Prague, pp. 59–65.

Arthur, D.R. (1951) The capitulum and feeding mechanisms of *Ixodes hexagonus* Leach. *Parasitology* 41, 66–81.

Arthur, D.R. (1956) The morphology of the British Prostriata, with particular reference to *Ixodes hexagonus* Leach. *Parasitology* 46, 261–307.

Arthur, D.R. (1957) The capitulum and feeding mechanism of *Dermacentor parumapertus* Neumann. *Parasitology* 33, 229–258.

Arthur, D.R. (1960) *Ticks. A monograph of the Ixodoidea. Part V. The genera Dermacentor, Anocentor, Cosmiomma, Boophilus, Margaropus.* Cambridge University Press, 251pp.

Arthur, D.R. (1962) *Ticks and Disease.* Pergamon Press, Oxford, 445pp.

Arthur, D.R. (1963) *British Ticks.* Butterworths, London, 213pp.

Athias-Binche, F. (1982) A redescription of *Thinozercon michaeli* Halbert, 1915 (Uropodina: Thinozerconoidea). *Proceedings of the Royal Irish Academy, Dublin* 82B, 261–276.

Athias-Binche, F. (1984) Phoresy in the Uropodina (Anactinotrichida): occurrence, demographic involvement and ecological significance. In: Griffiths, D.A. & Bowman, C.E. (eds), *Acarology VI*, Vol. 1. Ellis Horwood, Chichester, pp. 276–285.

Athias-Binche, F. (1987) Significance adaptive des differents types de développements postembryonnaires chez les gamasides (acariens: anactinotriches). *Canadian Journal of Zoology* 65, 1299–1310.

Athias-Binche, F. & Evans, G.O. (1981) Observations on the genus *Protodinychus* Evans, 1957 (Acari: Mesostigmata) with descriptions of the male and phoretic deuteronymph. *Proceedings of the Royal Irish Academy, Dublin* 81B, 25–36.

Athias-Henriot, C. (1959) Rédescription du stade adulte de *Seiodes ursinus* Berlese, 1887 (Parasitiformes, Antennophorina). *Zoologischer Anzeiger, Leipzig* 163, 11–25.

Athias-Henriot, C. (1967) Observations sur les *Pergamasus.* 1. Sous-genre *Paragamasus* Hull, 1918 (Acariens. Anactinotriches, Parasitidae). *Mémoires du Musém National d'Histoire Naturelle, Paris* 49, 1–198.

Athias-Henriot, C. (1968) L'appareil d'insemination laelapoide (Acariens anactinotriches, Laelapoidea. Premières observations. Possibilité d'emploi á des fins taxonomiques. *Bulletin Scientifique de Bourgogne.* 25, 175–228.

Athias-Henriot, C. (1969) Les organes cuticulaires sensoriels et glandulaires des Gamasides, poroidotaxie et adénotaxie. *Bulletin de la Société Zoologique de France, Paris* 94, 485–492.

Athias-Henriot, C. (1970a) Notes sur la morphologie externe des Gamasides (Acariens, Anactinotriches). *Acarologia* 11 (1969), 609–629.

Athias-Henriot, C. (1970b) Un progrès dans la connaissance de la composition métamérique des gamasides: leur sigillotaxie idiosomale (Arachnida). *Bulletin de la Société Zoologique de France, Paris* 96, 73–85.

Athias-Henriot, C. (1971) Nouvelles notes sur les Amblyseiini (Gamasides Podospermiques, Phytoseiidae). 1. La dépilation des génuaux et tibias des pattes. *Acarologia* 13, 4–15.

Athias-Henriot, C. (1972) Gamasides Chiliens (Arachnides) II. Revision de la famille Ichthyostomatogasteridae Sellnick 1953 (=Uropodellidae Camin, 1955). *Arquivos de Zoologia Sao Paulo* 22, 113–191.

Athias-Henriot, C. (1975a) The idiosomatic euneotaxy and epineotaxy in gamasids (Arachnida, Parasitiformes). *Zeitschrift fuer Zoologische Systematik und Evolutionforschung* 13, 97–109.

Athias-Henriot, C. (1975b) Nouvelles notes sur les Amblyseiini, 2. Le relevé organotaxique de la face dorsal adulte (Gamasides, Protoadiniques, Phytoseiidae). *Acarologia* 17, 20–29.

Athias-Henriot, C. (1977) A new Australian mite, *Archaeopodella scopulifera* gen. et sp. n. (Gamasida: Liroaspidae). *Journal of the Australian Entomological Society* 16, 225–235.

Athias-Henriot C. (1979a) Un problème arachnologique: l'interprétation phylogénétique des deux états idiosomaux des gamasides – protadénie et eunéoadénie (Parasitiformes). In: Piffl, E. (ed.), *Proceedings of the 4th International Congress of Acarology*. Akadémiai Kiadó, Budapest, pp. 443–445.

Athias-Henriot, C. (1979b) Sur le genre *Eugamasus* Berlese, 1892 (Parasitiformes, Parasitidae). I. Redéfinition descriptions des petites espèces. *Acarologia* 20, 3–18.

Athias-Henriot, C. (1982) Analyse d'ouvrage. *Revue Arachnologique* 4, 167–174.

Atkinson, P.W. & Binnington, K.C. (1973) New evidence on the function of the porose areas of ixodid ticks. *Experientia* 29, 799–801.

Atyeo, W.T. (1960) A revision of the family Bdellidae in North and Central America (Acarina: Prostigmata). *University of Kansas Science Bulletin* 40, 345–499.

Atyeo, W.T. (1963) The Bdellidae (Acarina) of the Australian Realm. Part 1. New Zealand, Campbell Island and the Auckland islands. Part 2. Australia and Tasmania. *Bulletin of the University of Nebraska State Museum* 4, 113–210.

Atyeo, W.T. (1979) The pretarsi of astigmatid mites. *Acarologia* 20 (1978), 244-269.

Atyeo, W.T. & Gaud, J. (1979) Ptyssalgidae, a new family of analgoid feather mites. *Journal of Medical Entomology. BP Bishop Museum Honolulu* 16, 306–308.

Audy, J.R. (1968) *Red Mites and Typhus*, University of London, Athlone Press, 191pp.

Audy, J.R., Nadchatram, M. & Vercammen-Grandjean, P.H. (1963). La "néosomie" un phénomène inédit de néoformation en acarologie, alliée à un cas remarquable de tachygénèse. *Bulletin de la Academie Royale de Belgique, Classe de Science, Bruxelles* 49, 1015–1026.

Audy, J.R., Radovsky, F.J. & Vercammen-Grandjean, P.H. (1972). Neosomy: radical intrastadial metamorphosis associated with arthropod symbiosis. *Journal of Medical Entomology. BP Bishop Museum, Honolulu* 9, 487–494.

Avery, D.J. & Briggs, J.B. (1968) The aetiology and development of damage in young fruit trees infested with the fruit tree red spider mite, *Panonychus ulmi. Annals of Applied Biology* 61, 277–288.

Axtell, R.C., Foelix, R.F., Coons, L.B. & Roshdy, M.A. (1973) Sensory receptors in ticks and mites. In: Daniel, M. & Rosický, B. (eds), *Proceedings of the 3rd International Congress of Acarology*, Academia, Prague, pp. 35–39.

Baccetti, B. (1979) Ultrastructure of sperm and its bearing on arthropod phylogeny. In: Gupta, A.P. (ed.), *Arthropod Phylogeny*. Van Nostrand Reinhold, New York, pp. 609–644.

Bader, C. (1938) Beitrag zur Kenntnis der Verdauungsvorgänge bei Hydracarinen. *Revue Suisse de Zoologique, Genève* 45, 721–809.

Bader, C. (1954) Das Darmsystem der Hydracarien. *Revue Suisse de Zoologique, Genéve* 61, 505–549.

Bader, C. (1969) Contributions to the taxonomy of water mites. In: Evans, G.O. (ed.), *Proceedings of the 2nd International Congress of Acarology*. Akadémiai Kiadó, Budapest, pp. 89–92.

Bailey, L. (1982) Isle of Wight disease of honey bees. *Antenna* 6, 289–292.

Baker, A.S. (1987) *Caleupodes*, a new genus of eupodid mite (Acari: Acariformes) showing primary opisthosomal segmentation. *Bulletin of the British Museum (Natural History) (Zool.)* 53, 103–113.

Baker, E.W. (1949) Paratydeidae, a new family of mites. *Proceedings of the US National Museum* 99, 267–320.

Baker, E.W. (1965) A review of the genera of the family Tydeidae (Acarina). In: Naegele, J.A. (ed.), *Advances in Acarology* Vol. 2. Cornell Press, Ithaca, NY, pp. 95–133.

Baker, E.W. (1979) A note on paedogenesis in *Brevipalpus* sp. (Acari: Tenuipalpidae), the first such record for a mite. *International Journal of Acarology* 5, 335–336.

Baker, E.W. & Pritchard, A.E. (1952) *Larvacarus*, a new genus of spider mites. *Proceedings of the Entomological Society of Washington* 54, 130–132.

Baker, E.W. & Pritchard, A.E. (1953) The family categories of tetranychoid mites, with a review of the new families Linotetranidae and Tuckerellidae. *Annals of the Entomological Society of America* 46, 243–258.

Baker, E.W. & Wharton, G.W. (1952) *An Introduction to Acarology*. Macmillan, New York, 465pp.

Baker, E.W., Camin, J.H., Cunliffe, F., Woolley, T.A. & Yunker, C.E. (1958) Guide to the families of mites, with contributions by R.E. Crabill, jr. and G. Nunes. *Institute of Acarology, University of Maryland*, Contrib. No. 3, 1–242.

Baker, R.A. (1991) Development and life-history strategies in mussel mites (Hydrachnellae: Unionicolidae). In: Schuster, R. & Murphy, P.W. (eds), *The Acari. Reproduction, Development and Life-History Strategies*. Chapman & Hall, London, pp. 65–73.

Baker, W.V. (1967) Some observations on predation in an anystid mite. *Entomologist's Monthly Magazine* 103, 58–59.

Balashov, Yu. S. (1960) Growth and expansion of the body integument of ixodid ticks during bloodsucking. *Parazitologicheskii Sbornik, Leningrad* 19, 263–290 [in Russian].

Balashov, Yu. S. (1961) The structure of digestive organs and blood digestion in argasid ticks. *Parazitologicheskii Sbornik, Leningrad* 23, 8–18 [in Russian].

Balashov, Yu. S. (1963) Anatamo-histological peculiarities of moulting of the tick *Hyalomma asiaticum* (Acarina, Ixodoidea). *Zoologicheskii Zhurnal* 42, 345–348 [in Russian].

Balashov, Yu. S. (1965) Structure of the mouthparts and bloodsucking mechanism in ixodid ticks. *Trudy Zoologichesko Instituta Akademiya Nauk SSR Leningrad* 35, 251–271 [in Russian].

Balashov, Yu. S. (1968) *Blood-Sucking Ticks (Ixodoidea) – Vectors of Disease of Man and Animals.* Nauka Publishers, Leningrad, 1967, 319pp [in Russian].

(Translation in: *Miscellaneous Publications of the Entomological Society of America* 8, No. 5, 1972, 161–376.)

Balashov, Yu. S. (1979) Salivary glands. In: Balashov, Yu. S. (ed.), *An Atlas of Ixodid Tick Ultrastructure.* Nauka Publishers, Leningrad [in Russian]. (Translation in: *Miscellaneous Publications of the Entomological Society of America* 1983, 98–128).

Balashov, Yu. S. & Leonovich, S.A. (1976) Morphological characteristics of Haller's organ in ticks of the tribe Amblyommatini (Acarina, Ixodidae). *Entomologischeskoe Obozrenie* 55, 949–952 [in Russian].

Balashov, Yu. S. & Raikhel, A.S. (1977) Ultrafine structure of the midgut of hungry nymphs of *Ornithodoros papillipes* (Acarina, Argasidae). *Parazitologiya* 11, 122–128 [in Russian].

Balashov, Yu. S. & Raikhel, A.S. (1978) Ultrafine structure of the midgut in nymphs of *Ornithodoros papillipes* (Acarina, Argasidae) during assimilation of blood. *Parazitologiya* 12, 21–26 [in Russian].

Balashov, Yu. S., Ivanov, V.P. & Ignatev, A.M. (1976) Fine structure of the palpal receptor in ixodid ticks (Acarina: Ixodoidea). *Zoologicheskii Zhurnal* 55, 1308–1317 [in Russian].

Balogh, J. (1959) Neue Oribatiden aus Ungarn. *Acta Zoologica Hungarica* 5, 24–53.

Balogh, J. (1972) *The Oribatid Genera of the World.* Akadémiai Kiadó, Budapest, 188pp.

Balogh, J. & Balogh, P. (1990) Oribatid mites of the Neotropical region 2. *Soil Mites World* 3, 1–333.

Balogh, J. & Mahunka, S. (1985) *Primitive Oribatids of the Palaearctic Region.* Elsevier, Amsterdam, 372pp.

Barr, D. (1972) The ejaculatory complex in water mites (Acari: Parasitengona): morphology and potential value in systematics. *Contributions Life Science Division Royal Ontario Museum* No. 81, 1–87.

Barrington, E.J.W. (1967) *Invertebrate Structure and Function.* Nelson, London, 549pp.

Bartsch, I. (1972) Ein Beitrag zur Sytematik, Biologie und Ökologie der Halacaridae (Acari) aus dem Litoral der Nord- und Ostsee. 1. Systematik und Biologie. *Abhandlungen und Verhandlungen des Naturwissenshaftlichen Vereins in Hamburg* 16, 155–230.

Bartsch, I. (1973) *Porohalacarellus alpinus* (Thor) (Halacaridae, Acari) ein morphologischer vergleich mit marinen Halacariden nebst Bemerkungen zur Biologie dieser Art. *Entomologisk Tidskrift* 94, 116–123.

Batelli, A. (1891) Note anatomo-fisiolische sugli Ixodini. Communicazione preventiva (2nd part). *Monitore Zoologica, Italia* 2, 98–104.

Beadle, D.J. (1973) Muscle attachment in the tick *Boophilus decoloratus* Koch (Acarina: Ixodidae). *International Journal of Insect Morphology and Embryology* 2, 247–255.

Beament, J.W.L. (1951) The structure and formation of the egg of the fruit tree red spider mite, *Metatetranychus ulmi* Koch, *Annals of Applied Biology* 38, 1–24.

Beck, S.D. (1980) *Insect Photoperiodism,* 2nd edn. Academic Press, New York.

Beer, R.E. (1954) A revision of the Tarsonemidae of the Western hemisphere

(Order Acarina). *Kansas University Science Bulletin* 36, 1091–1387.

Beer, R.E. (1963) Social parasitism in the Tarsonemidae, with description of a new species of tarsonemid mite involved. *Annals of the Entomological Society of America* 56, 153–160.

Behan-Pelletier, V.M. & Hill, S.B. (1983) Feeding habits of sixteen species of Oribatei (Acari) from an acid peat bog, Glenamoy, Ireland. *Revue d'Ecologie et de Biologie du Sol* 20, 221–267.

Behura, B.K. (1956) The relationship of the tyroglyphid mite, *Histiostoma polypori* (Oud.) with the earwig, *Forficula auricularia. Journal of the New York Entomological Society* 64, 85–94.

Belozerov, V.N. (1967a) Larval diapause in the tick *Ixodes ricinus* L. and its relation to external conditions. IV. Interactions between exogenous and endogenous factors in the control of larval diapause. *Entomologischeskoe Obozr.* 46, 760–768 [in Russian].

Belozerov, V.N. (1967b) Nymphal diapause in the tick *Ixodes ricinus* L. (Acarina, Ixodidae). II. Different types of nymphal diapause and peculiarities of their regulation in ticks in the Leningrad district. *Parazitologiya* 1, 279–287 [in Russian].

Belozerov, V.N. (1982) Diapause and biological rhythms in ticks. *Current Themes in Tropical Science* 1, 409–500.

Berger, R.S., Dukes, J.C. & Chow, Y.S. (1971) Demonstration of a sex pheromone in three species of hard ticks. *Journal of Medical Entomology. BP Bishop Museum Honolulu* 8, 84–86.

Bergström, J. (1979). Morphology of fossil arthropods as a guide to phylogenetic relationships. In: Gupta, A.P. (ed.), *Arthropod Phylogeny.* Van Nostrand Reinhold, New York, pp. 3–56.

Berlese, A. (1900) *Gli Acari Agrarii.* Firenze, 168pp.

Berlese, A. (1904) Illustrazione iconographica delgi acari mirmecofili. *Redia* 1, 299–474.

Berlese, A. (1906) Monographia del genere *Gamasus* Latr. *Redia* 3, 66–304.

Berlese, A. (1912) Trombidiidae. *Redia* 8, 1–291.

Bernini, F. (1991) Fossil Acarida. In: Simonetta, A. & Morris, S.C. (eds), *The Early Evolution of Metazoa and the Significance of Problematic Taxa.* Cambridge University Press, Cambridge, pp. 253–262.

Berry J.S. & Holtzer, T.O. (1990) Ambulatory dispersal behaviour of *Neoseiulus fallacis* (Acarina: Phytoseiidae) in relation to prey density and temperature. *Experimental & Applied Acarology* 8, 253–274.

Bertram, D.S. (1939) The structure of the capitulum in *Ornithodoros*: a contribution to the study of the feeding mechanisms of ticks. *Annals of Tropical Medicine and Parasitology* 33, 229–258.

Bhattacharyya, S.K. (1963) A revision of the British mites of the genus *Pergamasus* Berlese s. lat. (Acari: Mesostigmata). *Bulletin of the British Museum (Natural History)* (*Zool.*) 11, 131–242.

Binnington, K.C. (1972) The distribution and morphology of probable photoreceptors in eight species of ticks (Ixodoidea). *Zeitschrift fuer Parasitenikunde* 40, 321–332.

Binnington, K.C. (1978) Sequential changes in salivary gland structure during attachment and feeding of the cattle tick *Boophilus microplus. International*

Journal for Parasitology 8, 97–115.

Binnington, K.C. & Obenchain, F.D. (1982) Structure and function of the circulatory, nervous and neuroendocrine systems in ticks. *Current Themes in Tropical Science* 1, 351–398.

Binnington, K.C. & Tatchell, R.J. (1973) The nervous system and neurosecretory cells of *Boophilus microplus* (Acarina: Ixodidae). *Zeitschrift für Wissenschaftliche Zoologie, Leipzig* 185, 193–206.

Binns, E.S. (1972) *Arctoseius cetratus* (Sellnick) (Acarina: Ascidae) phoretic on mushroom sciarid flies. *Acarologia* 14, 350–356.

Binns, E.S. (1975a) A reassessment of the numerical distribution of watermites (*Arrenurus* spp.) on dragonflies in relation to parasitism and dispersal. *Acarologia* 17, 529–535.

Binns, E.R. (1975b) Negative binomial distribution of phoretic mites. *Entomologist's Monthly Magazine* 110, 223–226.

Binns, E.S. (1979) *Scutacarus baculitarsus* Mahunka (Acarina: Scutacaridae) phoretic on the mushroom fly *Megaselia halterata* (Wood). *Acarologia* 21, 91–107.

Binns, E.S. (1982) Phoresy as migration – some functional aspects of phoresy in mites. *Biological Review* 57, 571–620.

Blaszak, C. (1975) A revision of the family Zerconidae (Acari, Mesostigmata) (Systematic studies on family Zerconidae – I). *Acarologia* 17, 553–569.

Blaszak, C. (1979) Systematic studies on the family Zerconidae. 4. Asian Zerconidae (Acari Mesostigmata). *Acta Zoologica Cracoviensia* 24, 3–112.

Blauvelt, W.E. (1945) The internal anatomy of the common red spider mite (*Tetranychus telarius* Linn.). *Memoirs Cornell University Agricultural Experiment Station* No. 270, 1–35.

Blommers-Schlösser, R. & Blommers, L. (1975) Karyotypes of eight species of phytoseiid mites (Acarina: Mesostigmata) from Madagascar. *Genetica* 45, 145–148.

Boczek, J. (1974) Reproductive biology of *Tyrophagus putrescentiae* (Schr.) (Acari: Acaridae). *Proceedings of the 1st International Conference on Stored Products Entomology* pp. 154–159.

Boczek, J. & Griffiths, D.A. (1979) Spermatophore production and mating behaviour in stored product mites *Acarus siro* and *Lardoglyphus konoi*. In: Rodriguez, J.R. (ed.), *Recent Advances in Acarology*, Vol. 1. Academic Press, New York, pp. 279–284.

Boczek, J., Jura, C. & Krzysztofowicz, A. (1969) The comparison of the structure of the internal organs of postembryonic stages of *Acarus farris* (Oud.) with special reference to the hypopus. In: Evans, G.O. (ed.), *Proceedings of the 2nd International Congress of Acarology*. Akadémiai Kiadó, Budapest, pp. 265–271.

Bondarenko, N.V. (1958) Diapause peculiarities in *Tetranychus urticae* Koch. *Zoologicheskii Zhurnal* 37, 1012–1023 [in Russian].

Bonnet, A. (1907) Recherches sur l'anatomie comparée et le développement des ixodidés. *Annales de l'Université de Lyon* n.s. 1, fasc. 20, 180pp.

Booth, T.F. (1989) Wax lipid secretion and ultrastructural development in the egg-waxing (Gené's) organ in ixodid ticks. *Tissue & Cell* 21, 113–122.

Booth, T.F., Beadle, D.J. & Hart, R.J. (1984) The ultrastructure of Gené's organ in the cattle tick, *Boophilus microplus* Canestrini. In: Griffiths, D.A. & Bowman, C.E. (eds), *Acarology VI*, Vol. 1. Ellis Horwood, Chichester, pp. 261–267.

Börner, C. (1902) Arachnologische studien V. Die Mundbildung bei den milben. *Zoologischer Anzeiger* 26, 99–109.

Bostanian, N.J. & Morrison, F.O. (1973) Morphology and ultrastructure of sense organs in the two-spotted spider mite (Acarina: Tetranychidae). *Annals of the Entomological Society of America* 66, 379–383.

Boudreaux, H.B. (1979) Significance of intersegmental tendon system in arthropod phylogeny and a monophyletic classification of the Arthropoda. In: Gupta, A.P. (ed.), *Arthropod Phylogeny*. Van Nostrand Reinhold, New York, pp. 551–586.

Boudreaux, H.B. & Dosse, G. (1963) The usefulness of new taxonomic characters in females of the genus *Tetranychus* Dufour (Acari: Tetranychidae). *Acarologia* 5, 13–33.

Bourdeau, F.G. (1956) The gnathosoma of *Megalolaelaps ornata* (Acarina – Mesostigmata – Gamasides). *American Midland Naturalist* 55, 357–362.

Bowman, C.E. (1984) Some aspects of feeding and digestion in the soil predatory mite *Pergamasus longicornis* (Berlese) (Parasitidae: Mesostigmata). In: Griffiths, D.A. & Bowman, C.E. (eds), *Acarology VI*, Vol. 1. Ellis Horwood, Chichester, pp. 316–322.

Bowman, C.E. & Childs, M. (1982) Polysaccharides in astigmatid mites (Arthropoda: Acari). *Comparative Biochemistry and Physiology* 72B, 551–557.

Bregetova, N.G. (1956) Gamasid mites (Gamasoidea). *Opredelite po Faune SSR, Akademie Nauk, Leningrad* 61, 1–246 [in Russian].

Breucker, H. & Horstmann, E. (1972) Die Spermatogenese der Zecke *Ornithodorus moubata* (Murr.). *Zeitschrift für Zellforschung und Mikroskopische Anatomie, Berlin* 123, 18–46.

Brinton, L.P. & Burgdorfer, W. (1971) Fine structure of normal hemocytes in *Dermacentor andersoni* Stiles. *Journal of Parasitology* 57, 1110–1127.

Brinton, L.P. & Oliver, J.H. Jr. (1971) Fine structure of oogonial and oocyte development in *Dermacentor andersoni* Stiles (Acari: Ixodidae). *Journal of Parasitology* 57, 720–747.

Brinton, L.P., Burgdorfer, W. & Oliver, J.H. Jr. (1974) Histology and fine structure of spermatozoa and egg passage in the female tract of *Dermacentor andersoni* Stiles (Acari: Ixodidae). *Tissue & Cell* 6, 109–125.

Brody, A.R. (1970) Observations on the fine structure of the developing cuticle of a soil mite *Oppia coloradensis* (Acarina: Cryptostigmata). *Acarologia* 12, 421–431.

Brody, A.R. and Wharton, G.W. (1970) *Dermatophagoides farinae*: ultrastructure of the lateral opisthosomal dermal glands. *Transactions of the American Microscopical Society* 89, 499–513.

Brody, A.R., McGrath, J.C. & Wharton, G.W. (1972) *Dermatophagoides farinae*: The digestive system. *Journal of the New York Entomological Society* 80, 152–177.

Brody, A.R., McGrath, J.C. & Wharton, G.W. (1976) *Dermatophagoides farinae*: The supracoxal glands. *Journal of the New York Entomological Society* 84, 34–47.

Brown, J.R.C. (1952) The feeding organs of the adult common "chigger". *Journal of Morphology* 91, 15–52.

Brucker, E.A. (1900) Monographie de *Pediculoides ventricosus* Newport et pièces buccales des Acariens. *Bulletin Scientifique de la France et de la Belgique*, 35, 355–442.

Brust, G.E. & House G.J. (1988) A study of *Tyrophagus putrescentiae* (Acari: Acaridae) as a facultative predator of Southern Corn Rootworm eggs. *Experimental and Applied Acarology* 4, 335–344.

Bull, C.M. (1986) Ticks on reptiles, a different scale. In: Sauer, J.R. & Hair, J.A. (eds), *Morphology, Physiology and Behavioural Biology of Ticks.* Ellis Horwood, Chichester, pp. 391–405.

Bull, C.M. & Andrews, R.H. (1984) Two different mating signals used by female reptile ticks. In: Griffiths, D.A. & Bowman, C.E. (eds), *Acarology VI*, Vol. 1. Ellis Horwood, Chichester, pp. 427–430.

Burgdorfer, W. (1951) Analyse des Infektionsverlaufes bei *Ornithodoros moubata* (Murray) und der natürlichen Übertragung von *Spirochaeta duttoni. Acta Tropica* 8, 193–262.

Burgdorfer, W., Barbour, A.G., Hayes, S.F., Banach, J.L., Grunwaldt, E. & Davis, J.P. (1982) Lyme disease – a tick-borne spirochetosis. *Science [N.Y.]* 216, 1317–1319.

Butler, L. & Hunter, P.E. (1968) Redescription of *Megisthanus floridanus* with observations on its biology (Acarina: Megisthanidae). *Florida Entomologist* 51, 187–197.

Callani, G. & Mazzini, M. (1984) Fine structure of the egg shell of *Tyrophagus putrescentiae* (Schrank) (Acarina: Acaridae). *Acarologia* 25, 359–364.

Cameron, W.P.L. (1925) The fern mite (*Tarsonemus tepidariorum* Warburton). *Annals of Applied Biology* 12, 93–112.

Camicas, J.L. & Morel, P.C. (1977) Position systématique et classification des tiques (Acarida Ixodida). *Acarologia* 18, 410–420.

Camin, J.H. (1953a) Observations on the life history and sensory behavior of the snake mite. *Ophionyssus natricis* (Gervais) (Acarina: Macronyssidae). *Special Publication Chicago Academy of Sciences* No. 10, 1–75.

Camin, J.H. (1953b) A revision of the Cohort Trachytina Tragardh, 1938 with the description of *Dyscritaspis whartoni*, a new genus and species of polyaspid mite from tree holes. *Bulletin of the Chicago Academy of Sciences* 9, 335–385.

Camin, J.H. & Drenner, R.W. (1978) Climbing behavior and host-finding of larval rabbit ticks (*Haemaphysalis leporispalustris*). *Journal of Parasitology* 64, 905–909.

Camin, J.H. & Gorirossi, F. (1955) A revision of the suborder Mesostigmata (Acarina) based on comparative morphological data. *Special Publication Chicago Academy of Sciences* No. 11, 1–70.

Camin, J.H., Moss, W.W., Oliver, J.H. & Singer, G. (1967) Cloacaridae, a new family of cheyletoid mites from the cloaca of aquatic turtles (Acari: Acariformes: Eleutherogona). *Journal of Medical Entomology. BP Bishop Museum Honolulu* 4, 261–272.

Cancela da Fonseca, J.P. (1969) Le spermatophore de *Damaeus quadrihastatus* Märkel et Meyer (Acarina, Oribate). In: Evans, G.O. (ed.), *Proceedings of the*

2nd International Congress of Acarology. Akadémiai Kiadó, Budapest, pp. 227–232.

Canestrini, G. (1891) Abbozzo del sistema Acarologico. *Atti del Reale Instituto Veneto di Scienze* 38, 699–725.

Carpenter, P.D. & Greenberg, B. (1960) Factors in phoretic association between a mite and fly. *Science* [*N.Y.*] 132, 738–739.

Cassagne-Méjean, F. (1969) Sur les calyptostases des Hydrachnelles. In: Evans, G.O. (ed.), *Proceedings of the 2nd International Congress of Acarology*. Académiai Kiadó, Budapest. pp. 93–97.

Chinery, W.A. (1965) Studies on the various glands of the tick *Haemaphysalis spinigera* Neumann 1897, Part III. The salivary glands. *Acta Tropica* 22, 321–349.

Claparède, E. (1869) Studien an Acariden. *Zeitschrift für Wissenschaftliche Biologie, Leipzig* 18, 445–546.

Clifford, C.M., Kohls, G.M. & Sonenshine, D.E. (1964) The systematics of the subfamily Ornithodorinae (Acarina: Argasidae). 1. The genera and subgenera. *Annals of the Entomological Society of America* 57, 429–437.

Clift, A.D. & Larsson, S.F. (1987) Phoretic dispersal of *Brennandania lambi* (Kcrzal) (Acari: Tarsonemida: Pygmephoridae) by mushroom flies (Diptera: Sciaridae and Phoridae) in New South Wales, Australia. *Experimental and Applied Acarology* 3, 11–20.

Clift, A.D. & Toffolon, R.B. (1981) Biology, fungal host preferences and economic significance of two pygmephorid mites (Acari: Pygmephoridae) in cultivated mushrooms, N.S.W. Australia. *Mushroom Science* 11, 245–253.

Coineau, Y. (1970) A propos de l'oeil antérieur du naso des Caeculidae (Acariens, Prostigmates). *Acarologia* 11, 678–696.

Coineau, Y. (1974) Eléments pour une monographie morphologique, écologique et biologique des Caeculidae (Acariens). *Mémoirs de Museum National d'Histoire Naturelle, Paris* sér. A, 81, 1–299.

Coineau, Y. (1976) Les parades sexuelles des Saxidrominae Coineau 1974 (Acariens prostigmates, Adamystidae). *Acarologia* 18, 234–240.

Coineau, Y. (1979) Les Adamystidae, une étonnante famille d'acariens prostigmates primitifs. In: Piffl, E. (ed.), *Proceedings of the 4th International Conference of Acarology*. Akadémiai Kiadó, Budapest, pp. 431–435.

Coineau, Y. & Legendre, R. (1975) Sur un mode de régénération appendiculaire inédit chez les arthropodes: la régénération des pattes marcheuses chez les opilioacariens (Acari: Notostigmata). *Compte Rendu Hebdomadaire des Séances de l'Academie des Sciences, Paris* 280D, 41–43.

Coineau, Y & Van der Hammen, L. (1979) The postembryonic development of Opilioacarida, with notes on new taxa and on a general model for the evolution. In Piffl, E. (ed.), *Proceedings of the 4th International Congress of Acarology*. Akadémiai Kiadó, Budapest, pp. 437–441.

Compton, L. & Krantz, G.W. (1978) Mating behavior and related morphological specialization in the uropodine mite *Caminella perphora*. *Science* [*N.Y.*] 200, 1300–1301.

Cone, W.W. (1979) Pheromones of Tetranychidae. In: Rodriguez, J.R. (ed.), *Recent Advances in Acarology*, Vol. 2. Academic Press, New York, pp. 309–317.

Cone, W.W. (1985) Mating and communication. In: Helle, W. & Sabelis, M.W. (eds), *Spider Mites, their Biology, Natural Enemies and Control*, Vol. 1A. Elsevier, Amsterdam, pp. 243–251.

Cook, D.R. (1966) The water mites of Liberia, West Africa. *Memoirs of the American Entomological Institute* No. 6, 1–418.

Cook, D.R. (1967) Water mites from India. *Memoirs of the American Entomological Institute* No. 9, 1–411.

Cook, D.R. (1974) Water mite genera and subgenera. *Memoirs of the American Entomological Institute* No. 21, 1–860.

Cook, D.R. (1980) Studies on Neotropical water mites. *Memoirs of the American Entomological Institute* No. 31, 1–645.

Cooley, R.A. (1938) The genera *Dermacentor* and *Otocentor* (Ixodidae) in the United States, with studies on variation. *Bulletin of the National Institute of Health*, Washington, No. 15, 1–60.

Cooley, R.A. (1946) The genera *Boophilus*, *Rhipicephalus* and *Haemaphysalis* (Ixodidae) of the New World. *Bulletin of the National Institute of Health*, Washington, No. 187, 1–22.

Cooley, R.A. & Kohls, G.M. (1944) The Argasidae of North America, Central America and Cuba. *American Midland Naturalist Monographs* 1, 1–152.

Cooley, R.A. & Kohls, G.M. (1945) The genus *Ixodes* in North America. *Bulletin of the National Institute of Health*, Washington, No. 184, 1–246.

Coons, L.B. & Axtell, R.C. (1971) Circular organization in the synganglion of the mite *Macrocheles muscaedomesticae* (Acarina: Macrochelidae). An electron microscope study. *Zeitschrift für Zellforschung und Mikroskopische Anatomie, Berlin* 119, 309–320.

Coons, L.B. & Axtell, R.C. (1973) Sensory setae of the first tarsi and palps of the mite *Macrocheles muscaedomesticae*. *Annals of the Entomological Society of America*, 66, 539–544.

Coons, L.B. & Roshdy, M.A. (1973) Fine structure of the salivary glands of unfed *Dermacentor variabilis* (Say) (Ixodoidea, Ixodidae). *Journal of Parasitology* 59, 900–912.

Coons, L.B., Roshdy, M.A. & Axtell, R.C. (1974) Fine structure of the central nervous system of *Dermacentor variabilis* (Say), *Amblyomma americanum* (L.) and *Argas arboreus* Kaiser, Hoogstraal and Kohls (Ixodoidea). *Journal of Parasitology* 60, 687–698.

Coons, L.B., Tarnowski, B. & Ourth, D.D. (1982) *Rhipicephalus sanguineus*: localization of vitellogenin synthesis by immunological methods and electron microscopy. *Experimental Parasitology* 54, 331–339.

Costa, M. (1967) Notes on macrochelids associated with manure and coprid beetles in Israel. II. Three new species of *Macrocheles pisentii* complex, with notes on their biology. *Acarologia* 9, 304–329.

Costa, M. (1969) The association between mesostigmatic mites and coprid beetles. *Acarologia* 11, 411–428.

Crooker, A.R. & Cone, W.W. (1979) Structure of the reproductive system of the adult female two-spotted spider mite *Tetranychus urticae*. In: Rodriguez, J. (ed.), *Recent Advances in Acarology*, Vol. 2, Academic Press, New York, pp. 404–409.

Crooker, A.R., Drenth-Diephuis, L.J., Ferwerda, M.A. & Weyda, F. (1985)

Histological techniques. In: Helle, W. & Sabelis, M.W. (eds), *Spider Mites. Their Biology, Natural Enemies and Control*, Vol. 1A. Elsevier, Amsterdam, pp. 359–381.

Cross, E.A. (1965) The generic relationships of the family Pyemotidae (Acarina: Trombidiformes). *Kansas University Science Bulletin* 45, 29–275.

Cross, E.A. & Bohart, G.E. (1969) Phoretic behavior of four species of alkali bee mites as influenced by season and host sex. *Journal of the Kansas Entomological Society* 42, 195–219.

Cross, H.F. (1964) Observations on the formation of the feeding tube by *Trombicula splendens* larvae. *Acarologia* 6 (fasc. h.s.), 255–261.

Crowe, J.H. & Magnus, K.A. (1974) Studies on acarine cuticles. II. Plastron respiration and levitation in a water mite. *Comparative Biochemistry and Physiology* 49A, 301–310.

Cunliffe, F. (1952) Biology of the cockroach parasite, *Pimeliaphilus podapolipophagus* Tragardh, with discussion of the genera *Pimeliaphilus* and *Hirstiella* (Acarina, Pterygosomidae). *Proceedings of the Entomological Society of Washington*, 54 153–162.

Cunnington, A.M. (1985) Factors affecting oviposition and fecundity in the grain mite *Acarus siro* L. (Acarina: Acaridae), especially temperature and relative humidity. *Experimental and Applied Acarology* 1, 327–344.

Cusack, P.D., Evans, G.O. & Brennan, P.A. (1975) A survey of the mites of stored grain products in the Republic of Ireland. *Scientific Proceedings of the Royal Dublin Society* ser. B, 3, 273–329.

Czezuga, B. & Czerpak, R. (1968) The presence of carotenoids in *Eylais hamata* (Koenike, 1897) (Hydracarina, Arachnoidea). *Comparative Biochemistry and Physiology* 24, 37–46.

Dahme, V.E. & Popp, E. (1963) Todesfälle bei Seelöwen (*Zalophus californicus* Lesson) verursacht durch eine bisher unbekannte Milbe *Orthohalarachne letalis* Popp. *Berliner und Muencher Tierärtzliche Wochenschrift*, 76, 441–443.

Daniel, M. & Černý, V. (eds) (1971) *Key to the Fauna of the CSSR.*, Vol. 4. Československá Akademie Věd. Academia, Prague, pp. 307–603 [in Czech].

Davids, C. (1973) The water mite *Hydrachna conjecta* Koenike, 1895 (Acari, Hydrachnellae), bionomics and relation to species of Corixidae (Hemiptera). *Netherlands Journal of Zoology* 23, 363–429.

Davidson, J.A. (1958) A new species of lizard mite and a generic key to the Pterygosomidae. *Proceedings of the Entomological Society of Washington* 60, 5–79.

Davis, G.E. (1951) Parthenogenesis in the argasid tick *Ornithodoros moubata* (Murray, 1877). *Journal of Parasitology* 37, 99–101.

de Bruyne, M, Dicke, M. & Tjallingii, W.F. (1991) Receptor cell responses in the anterior tarsi of *Phytoseiulus persimilis* to volatile kairomone components. *Experimental and Applied Acarology* 13, 53–58.

De Geer, C. (1778) *Mémoires pour Servir à l'Histoire des Insectes*, Vol. 7. Stockholm.

De Jong, D., Morse, R.A. & Eickwort, G.C. (1982) Mite pests of honey bees. *Annual Review of Entomology* 27, 229–252.

De Jong, J.H., Lobbes, P.V. & Bolland, H.R. (1981) Karyotypes and sex determination in two species of laelapid mite (Acari: Gamasida). *Genetica* 55, 187–190.

Den Heyer, J. (1981) Systematics of the family Cunaxidae Thor, 1902 (Actinedida: Acarida). *Publications of the University of the North, S. Africa*, Ser. A. 24, 1–19.

Desch, C.E. (1988) The digestive system of *Demodex folliculorum* (Acari: Demodicidae) of man: a light and electron microscope study. In: Channa-Basavanna, G.P. & Viraktamath, C.A. (eds), *Progress in Acarology*, Vol. 1. Oxford & IBH Publishing, New Delhi, pp. 187–195.

Desch, C.E. & Nutting, W.B. (1972) *Demodex folliculorum* (Simon) and *D. brevis* Akbulatova of man: redescription and revaluation. *Journal of Parasitology* 58, 169–177.

Desch, C.E., O'Dea, J. & Nutting, W.B. (1971) The proctodeum – a new key character for demodicids (Demodicidae). *Acarologia* 12, 522–526.

Dicke, M. & Groeneveld, A. (1986) Hierarchical structure in kairomone preference of predatory mite *Amblyseius potentillae*: dietary component indispensable for diapause induction affects prey location behaviour. *Ecological Entomology* 11, 131–138.

Dicke, M., Sabelis, M.W. & De Jong, M. (1988) Analysis of prey preference in phytoseiid mites using an olfactometer, predation models and electrophoresis. *Experimental and Applied Acarology* 5, 225–241.

Dicke, M., van Beek, T.A., Posthumus, M.A., van Bokhoven, H. & de Groot, A.E. (1990) Isolation and identification of volatile kairomones that affect acarine predator-prey interactions: involvement of host plant in its production. *Journal of Chemical Ecology* 16, 381–396.

Diehl, P.A., Aeschlimann, A. & Obenchain, F.D. (1982) Tick reproduction: oogenesis and oviposition. *Current Themes in Tropical Science* 1, 277–350.

Dindal, D.L. (ed.) (1990) *Soil Biology Guide*. John Wiley, New York, 1349pp.

Dinsdale, D. (1974a) Feeding activity of a phthiracarid mite (Arachnida: Acari). *Journal of the Zoological Society of London* 174, 15–21.

Dinsdale, D. (1974b) The digestive activity of a phthiracarid mite mesenteron. *Journal of Insect Physiology* 20, 2247–2260.

Dinsdale, D. (1975) Excretion in an oribatid mite *Phthiracarus* sp. (Arachnida: Acari). *Journal of the Zoological Society of London* 177, 225–231.

Dittrich, V. (1971) Electronmicroscope studies of the respiratory mechanism of spider mite eggs. *Annals of the Entomological Society of America* 64, 1134–1143.

Dittrich, V. & Streibert, P. (1969) The respiratory system of spider mite eggs, *Zeitschrift für Angewandte Entomologie, Berlin* 63, 200–211.

Domrow, R. (1955) A second species of *Holothyrus* (Acarina: Holothyroidea) from Australia. *Proceedings of the Linnean Society of New South Wales* 79, 156–158.

Domrow, R. (1966) Some mite parasites of Australian birds. *Proceedings of the Linnean Society of New South Wales* 90, 190–217.

Donisthorpe, H. St. J.K. (1927) The guests of British ants: their habits and life-histories. (Chapter 14; Acarina, pp. 202–217). Routledge, London.

Dosse, G. (1959) Über den Kopulationvorgang bei Raubmilben aus Gattung *Typhlodromus* (Acari, Phytoseiidae). *Pflanzenschutzber Wien* 22, 125–133.

Douglas, J.R. (1943) The internal anatomy of *Dermacentor andersoni. University of California Publications in Entomology* 7, 207–272.

Dubinin, V.B. (1951) Feather mites (Analgesoidea). Part 2. Families Epidermoptidae and Freyanidae. In: *Fauna SSSR, Arachnoidea,* 6 (6). Akademia Nauk, Moscow, pp. 1–412 [in Russian].

Dubinin, V.B. (1956) Arachnoidea – feather mites (Analgesoidea). Part 3. P Pterolichidae. In: *Fauna SSSR, Arachnoidea,* 6 (7). Akademia Nauk, Moscow, pp. 1–813 [in Russian].

Dubinin, V.B. (1959) Subtype Chelicerophora W. Dubinin nom. nov. and its position in the system. I. Structural peculiarities of the Chelicerophora and their classification. *Zoologicheskii Zhurnal* 38, 1163–1189 [in Russian].

Dubinin, V.B. (1962) Class Acaromorpha: mites, or gnathosomic chelicerate arthropods. In: Rodendorf, B.B. (ed.), *Fundamentals of Palaeontology.* Academy of Sciences, USSR, Moscow, pp. 375–530 [in Russian].

Dugès, Ant. (1834) Recherches sur l'ordre des Acariens en général et la famille des Trombidiés en particulier. *Annales des Sciences Naturelles, Paris* (Sér. 2), 1, 5–46.

Dusbàbek, F., Šimek, P., Jegorov, A. & Tríska, J. (1991) Identification of xanthine and hypoxanthine as components of assembly pheromones in excreta of argasid ticks. *Experimental and Applied Acarology* 11, 307–316.

Easterbrook, M.A. (1984) The biology and control of the rust mites *Aculus schlechtendali* and *Epitrimerus piri* on apple and pear in England. In: Griffiths, D.A. & Bowman, C.E. (eds), *Acarology VI,* Vol. 2. Ellis Horwood, Chichester, pp. 797–803.

Ebermann, E. (1982) Fortpflanzungsbiologische Studien an Scutacariden (Acari, Trombidiformes). *Zoologische Jahrbuecher. Abteilung fuer Systematik, Oekologie und Geographie der Tiere* 109, 98–116.

Ebermann, E. (1991) Thanatosis or feigning death in mites of the family Scutacaridae. In: Schuster, R. & Murphy, P.W. (eds), *The Acari, Reproduction, Development and Life-History Strategies,* Chapman & Hall, London, pp. 399–401.

Edwards, M.A. (1975) The chaetotaxy of the pedipalps and legs of some larval ticks (Acari: Argasidae). *Transactions of the Zoological Society of London* 33 (1), 1–76.

Edwards, M.A. & Evans, G.O. (1967) Some observations on the chaetotaxy of the legs of larval Ixodidae (Acari: Metastigmata). *Journal of Natural History* 4, 595–601.

Egan, M.E. (1976) The chemosensory bases of host discrimination in a parasitic mite. *Journal of Comparative Physiology* 109, 69–89.

Ehara, S. (1960) Comparative studies on the internal anatomy of three Japanese trombidiform acarinids. *Journal of the Faculty of Science of Hokkaido University,* ser. 6, 14, 410–434.

Ehrhardt, P. & Voss, G. (1961) Die Carbohydrasen der Spinnmilbe *Tetranychus urticae* Koch (Acari. Trombidiformes, Tetranychidae). *Experientia* 17, 307.

Ehrnsberger, K. (1981) Ernahrungsbiologie der bodenbewohnenden Milbe *Rhagidia* (Trombidiformes). *Osnabrucker Naturwissenschaftliche Mitteilungen* 8, 127–132.

Eickwort, G.C. (1979) Mites associated with sweat bees (Halictidae). In:

Rodriguez, J.R. (ed.), *Recent Advances in Acarology*, Vol. 1. Academic Press, New York, pp. 575–581.

Ellis, B.J. & Obenchain, F.D. (1984) In vivo and in vitro production of ecdysteroids by nymphal *Amblyomma variegatum* ticks. In: Griffiths, D.A. & Bowman, C.E. (eds), *Acarology VI*, Vol. 1. Ellis Horwood, Chichester, pp. 400–404.

El Shoura, S.M. (1988) Fine structure of sight organs in the tick *Hyalomma (Hyalomma) dromedarii* (Ixodoidea: Ixodidae). *Experimental and Applied Acarology* 4, 109–116.

El Shoura, S.M. (1989) Fine structure of muscles in the tick *Hyalomma (Hyalomma) dromedarii* (Ixodoidea: Ixodidae). *Experimental and Applied Acarology* 7, 327–335.

Elzinga, R.J. (1982a) The genus *Coxequesoma* (Acari: Uropodina) and descriptions of four new species. *Acarologia* 23, 215–224.

Elzinga, R.J. (1982b) The genus *Antennequesoma* (Acari: Uropodina) and descriptions of four new species. *Acarologia* 23, 319–325.

Enigk, K. & Grittner, I. (1952) Die Exkretion der Zecken. *Zeitschrift fuer Tropenmedizin und Parasitologie* 4, 78–94.

Evans, G.O. (1954) On the genus *Asternolaelaps* Berl, 1923 (Acarina: Mesostigmata). *Entomologist's Monthly Magazine* 90, 88–90.

Evans, G.O. (1955a) A revision of the family Epicriidae (Acarina: Mesostigmata). *Bulletin of the British Museum (Natural History) [Zool.]* 3, 171–200.

Evans, G.O. (1955b) A review of the laelaptid paraphages of Myriapoda with descriptions of three new species (Acarina: Laelaptidae). *Parasitology* 45, 352–368.

Evans, G.O. (1955c) A collection of mesostigmatid mites from Alaska. *Bulletin of the British Museum (Natural History) [Zool.]* 2, 285–307.

Evans, G.O. (1958) Some mesostigmatid mites from a nest of social spiders in Uganda. *Annals and Magazine of Natural History* 13, 580–590.

Evans, G.O. (1963a) Observations on the chaetotaxy of the legs in the free-living Gamasina (Acari: Mesostigmata). *Bulletin of the British Museum (Natural History) [Zool.]* 10, 275–303.

Evans, G.O. (1963b) The systematic position of the *Gammaridacarus brevisternalis* Canaris (Acari: Mesostigmata). *Annals and Magazine of Natural History* (13), 5, 395–399.

Evans, G.O. (1963c) The genus *Neocypholaelaps* Vitzthum (Acari: Mesostigmata). *Annals and Magazine of Natural History* (13), 6, 209–230.

Evans, G.O. (1964) Some observations on the chaetotaxy of the pedipalpi in the Mesostigmata (Acari). *Annals and Magazine of Natural History* (13), 6 [1963], 513–527.

Evans, G.O. (1968) The external morphology of the postembryonic developmental stages of *Spinturnix myoti* Kol. (Acari: Mesostigmata). *Acarologia* 10, 589–608.

Evans, G.O. (1969) Observations on the ontogenetic development of the leg chaetotaxy of the tarsi of legs II–IV in the Mesostigmata (Acari). In: Evans, G.O. (ed.), *Proceedings of the 2nd International Congress of Acarology*. Akadémiai Kiadó, Budapest, pp. 195–200.

Evans, G.O. (1972) Leg chaetotaxy and the classification of the Uropodina (Acari: Mesostigmata). *Journal of Zoology [Lond.]* 167, 193–206.

Evans, G.O. (1982) Observations on the genus *Protogamasellus* Karg (Acari: Mesostigmata) with a description of a new species. *Acarologia* 23, 303–313.

Evans, G.O. (1984) Presidential address. In: Griffiths, D.A. & Bowman, C.E. (eds), *Acarology VI*, Vol. 1. Ellis Horwood, Chichester, pp. 1–7.

Evans, G.O. (1987) The status of the species of Phytoseiidae (Acari) described by Carl Wilmann. *Journal of Natural History* 21, 1461–1467.

Evans, G.O. & Browning, E. (1955) Techniques for the preparation of mites for study. *Annals and Magazine of Natural History* (12), 8, 631–635.

Evans, G.O. & Hyatt, K.K. (1963) Mites of the genus *Macrocheles* Latr, (Mesostigmata) associated with coprid beetles in the collections of the British Museum (Natural History). *Bulletin of the British Museum (Natural History)* [*Zool.*] 9, 325–401.

Evans, G.O. & Loots, G. (1975) Scanning electron microscope study of the structure of the hypostome of *Phityogamasus*, *Laelaps* and *Ornithonyssus* (Acari: Mesostigmata). *Journal of Zoology* [*Lond.*] 176, 425–436.

Evans, G.O. & Sheals, J.G. (1959) Three new mesostigmatic mites associated with millepedes in Indonesia. *Entomologische Berichten, Amsterdam* 19, 107–111.

Evans, G.O. & Till, W.M. (1965) Studies on the British Dermanyssidae (Acari: Mesostigmata). Part I. External morphology. *Bulletin of the British Museum (Natural History)* (*Zool.*) 12, 247–294.

Evans, G.O. & Till, W.M. (1966) Studies on the British Dermanyssidae (Acari: Mesostigmata). Part II. Classification. *Bulletin of the British Museum (Natural History)* (*Zool.*) 14, 107–370.

Evans, G.O. & Till, W.M. (1979) Mesostigmatic mites of Britain and Ireland (Chelicerata: Acari: Parasitiformes). An introduction to their external morphology and classification. *Transactions of the Zoological Society of London* 35, 139–270.

Evans, G.O., Sheals, J.G. & Macfarlane, D. (1961) *The Terrestrial Acari of the British Isles*, Vol. 1. British Museum (Nat. Hist.), London, 219pp.

Evans, G.O., Fain, A. & Bafort, J. (1963) Découverte du cycle évolutif du genre *Myialges* avec description d'une espèce nouvelle (Myialgidae: Sarcoptiformes). *Bulletin et Annales de la Société Royale Entomologique de Belgique, Bruxelles* 99, 486–500.

Everson, P.R. & Addicott, J.F. (1982) Mate selection by male mites in the absence of intersexual selection by females: a test of six hypotheses. *Canadian Journal of Zoology* 60, 2729–2736.

Ewing, H.E. (1922) Studies on the taxonomy and biology of the tarsonemid mites together with a note on the transportation of *Acarapis* (*Tarsonemus*) *woodi* Rennie (Acarina). *Canadian Entomologist* 54, 104–113.

Faasch, H. (1967) Beitrag zur Biologie der einheimischen Uropodiden *Uroobovella marginata* (C.L. Koch, 1839) und *Uropoda orbicularis* (O.F. Müller, 1776), und experimentelle Analyse ihres Phoresieverhaltens. *Zoologische Jahrbuecker. Abteilung fuer Systematik, Oekologie und Geographie der Tiere* 94, 521–608.

Fain, A. (1962) Les acariens parasites nasicoles des batraciens. Revision des Lawrencarinae Fain 1957. *Bulletin de l'Institut Royal des Sciences Naturelles de Belgique* 38, 1–69.

Fain, A. (1963a) La spermathéque et ses canaux adducteurs chez les acariens mésostigmatiques parasites des voies respiratoires. *Acarologia* 5, 463–479.

Fain, A. (1963b) Chaetotaxie et classification des Speleognathidae (Acarina: Trombidiformes). *Bulletin de l'Institut Royal des Sciences Naturelles de Belgique* 39, 1–80.

Fain, A. (1965a) Quelques aspects de l'endoparasitisme par les acariens. *Annales de Parasitologie Humaine et Comparée* 40, 317–327.

Fain, A. (1965b) A review of the family Epidermoptidae Trouessart parasitic on the skin of birds (Acarina: Sarcoptiformes). *Verhandelingen der Koninklÿke Akademie van Wetenshappen* 27 (84), pt. I, 1–176; pt. II (illustrations) 1–185.

Fain, A. (1965c) Les acariens nidicoles et detriticoles de la famille Pyroglyphidae Cunliffe (Sarcoptiformes). *Revue de Zoologie et de Botanique Africaines, Bruxelles* 72, 257–288.

Fain, A. (1965d) Les acariens producteurs de gale chez les edentes et les marsupiaux (Psoroptidae et Lobalgidae; Sarcoptiformes). *Bulletin de l'Institut Royal des Sciences Naturelles de Belgique* 41, 1–41.

Fain, A. (1966) Nouvelle description de *Dermatophagoides pteronyssinus* (Trouessart, 1879), importance de cet acarien en pathologie humaine (Psoroptidae: Sarcoptiformes). *Acarologia* 8, 302–327.

Fain, A. (1967a) Les hypopes parasites des tissues cellulaires des oiseaux (Hypodectidae, Sarcoptiformes). *Bulletin de l'Institut Royal des Sciences Naturelles de Belgique* 43, 1–139.

Fain, A. (1967b) Le genre *Dermatophagoides* Bogdanov, 1864. Son importance dans les allergies respiratoires et cutanées chez l'homme (Psoroptidae: Sarcoptiformes). *Acarologia* 9, 179–225.

Fain, A. (1968a) Acariens nidicoles et détriticoles en Afrique au Sud du Sahara. III. Espèces et genres nouveaux dans les sous-familles Labidophorinae et Grammolichinae (Glycyphagidae: Sarcoptiformes). *Acarologia* 10, 86–100.

Fain, A. (1968b) Un hypope de la famille Hypoderidae Murray, 1877 vivant sous la peau d'un rongeur (Hypoderidae: Sarcoptiformes). *Acarologia* 10, 111–115.

Fain, A. (1968c) Notes sur les acariens de la famille Cloacaridae Camin *et al.* parasites du cloaque et des tissues profonde des tortues (Cheyletoidea: Trombidiformes). *Bulletin de l'Institut Royal des Sciences Naturelles de Belgique* 44 (15), 1–33.

Fain, A. (1968d) Etude de la variabilité de *Sarcoptes scabiei* avec une révision des Sarcoptidae. *Acta Zoologica et Pathologica Antverpiensia* 47, 1–196.

Fain, A. (1969a) Adaptation to parasitism in mites. *Acarologia* 11, 429–449.

Fain, A. (1969b) Les deutonymphes hypopiales vivant en association phoretique sur les mammiferes (Acarina: Sarcoptiformes). *Bulletin de l'Institut Royal des Sciences Naturelles de Belgique* 45, 1–262.

Fain, A. (1970a). *Coreitarsonemus* un nouveau genre d'acariens parasitant la glande odoriférante des Hémiptères Coreidae (Tarsonemidae: Trombidiformes). *Revue de Zoologiques et de Botaniques Africaines, Bruxelles* 82, 315–324.

Fain, A. (1970b) Les Myocoptidae en Afrique au sud du Sahara (Acarina: Sarcoptiformes). *Annales Musée Royal de l'Afrique Centrale,* Tervuren 179, 1–27.

Fain, A. (1971) Evolution de certains groupes d'hypopes en fonction du parasitisme (Acarina: Sarcoptiformes). *Acarologia* 13, 171–175.

Fain, A. (1972a) Développement post-embryonaire chez les acariens de la sous-famille Speleognathidae (Ereynetidae: Trombidiformes). *Acarologia* 13, 607–614.

Fain, A. (1972b) Notes sur les acariens des familles Cheyletidae et Harpyrhynchidae producteurs de gale chez les oiseaux et mammiferes *Acta Zoologica et Pathologica Antverpiensia* 56, 37–59.

Fain, A. (1972c) Notes sur les hypopes des Saproglyphidae (Acarina: Sarcoptiformes). *Acarologia* 14, 225–249.

Fain, A. (1977) Notes sur les acariens nasicoles d'oiseaux du Cameroun. *Revue de Zoologique et de Botanique Africaines, Bruxelles* 91, 83–116.

Fain, A. (1978) Epidemiological problems of scabies. *International Journal of Dermatology* 17, 20–31.

Fain, A. (1979) Observations sur les genres *Sciurochirus, Aeromychirus* et *Tamiopsochirus* (Acari: Listrophoridae). *Acarologia* 20, 270–285.

Fain, A. (1982) Cinq espéces du genre *Schweibea* Oudemans, 1916 (Acari: Astigmata) dont trois nouvelles découvertes dans des sources du sous-sol de la ville de Vienne (Austriche) au cours des travaux du Metro. *Acarologia* 23, 359–371.

Fain, A. (1989) Notes on mites associated with Myriapoda 4. New taxa in the Heterozerconidae (Acari: Mesostigmata) *Bulletin de l'Institut Royal des Sciences Naturelles de Belgique Entomol.* 59, 145–156.

Fain, A. & Bafort, J. (1964) Les acariens de la famille Cytoditidae (Sarcoptiformes) description de sept espèces nouvelles. *Acarologia* 6, 504–528.

Fain, A. & Bafort, J. (1967) Cycle évolutif et morphologie des *Hypodectes* (*Hypodectoides*) *propus* (Nitzsch) acarien nidicole à deutonymphe parasite tissulaire des pigeons. *Bulletin de l'Academie Royale de Belgique Classe des Sciences, Bruxelles* 53, 501–533.

Fain, A. & Beaucournu, J.C. (1972) Observations sur le cycle évolutif de *Pelecanectes evansi* Fain et description d'une espèce nouvelle du genre *Phalacrodectes* Fain (Hypoderidae: Sarcoptiformes). *Acarologia* 13, 374–382.

Fain, A. & Herin, A. (1979) La prélarva chez les Astigmates (Acari). *Acarologia* 20, 566–571.

Fain, A. & Hyland, K. (1974) The listrophorid mites in North America. 2. The family Listrophoridae Mégnin and Trouessart (Acarina; Sarcoptiformes). *Bulletin de l'Institut Royal des Sciences Naturelles de Belgique* 50, 1–69.

Fain, A. & Johnston, D.E. (1975) A new algophagine mite, *Algophagopsis pneumatica* gen. n., sp. n. living in a river. *Bulletin et Annales de la Société Royale Entomologique de Belgique, Bruxelles* 111, 65–70.

Fain, A., Camerik, A.M., Lukoschus, F.S. & Kniest, F.M. (1980) Notes on the hypopi of *Fibulanoetus* Mahunka, 1973, an anoetid genus with a pilicolous clasping organ. *International Journal of Acarology* 6, 39–44.

Fain, A., Whitaker, J.O. Jr. & Smith, M.A. (1984) Fur mites of the genus *Schizocarpus* Trouessart, 1986 (Acari: Chirodiscidae) parasitic on the American beaver *Castor canadensis* in Indiana, U.S.A. *Bulletin et Annales de la Société Royale Entomologique de Belgique, Bruxelles,* 120, 211–239.

Fain, A., Guerin, B. & Hart, B.J. (1988) *Acariens et Allergies,* Allerbio, Belgium, 179pp.

Farish, D.J. & Axtell, R.C. (1971) Phoresy redefined and examined in *Macrocheles muscaedomesticae* (Acarina: Macrochelidae). *Acarologia* 13, 16–29.

Farkas, H. (1965) Spinnentiere. Eriophyidae (Gallmilben). *Tierwelt Mitteleurpas* 3 (3), 1–155.

Fashing, N.J. (1977) The evolutionary modification of dispersal in *Naiadacarus arboricola* Fashing, a mite restricted to water-filled treeholes (Acarina: Acaridae). *American Midland Naturalist* 95, 337–346.

Fashing, N.J. (1984) A possible osmoregulatory organ in the Algophagidae (Astigmata). In: Griffiths, D.A. & Bowman, C.E. (eds), *Acarology VI*, Vol. 1. Ellis Horwood, Chichester, pp. 310–315.

Fashing, N.J. (1988) Fine structure of the Claparède organs and genital papillae of *Naiadacarus arboricola* (Astigmata: Acaridae), an inhabitant of water-filled treeholes. In: ChannaBasavanna, G.P. & Viraktamath, C.A. (eds), *Progress in Acarology*, Vol. 1. Oxford & IBH Publishing, New Delhi, pp. 219–228.

Feider, Z. (1955) Arachnida, Acarina, Trombidoidea. *Fauna Republicii Populaire Romîne, Bucuresti* 5 (1), 1–187 [in Romanian].

Feider, Z. (1956) Nouveaux acariens parasites des insectes nuisables appartenant au genre *Phyllotreta. Analele Stiintifice de Universitătii A1. I. Cuza din Iasi.* N.S. Sect. 2, 2, 163–174 [in Romanian].

Feider, Z. (1965) Arachnida, Acaromorpha, Suprafamilia Ixoidoidea. *Fauna Republicii Populaire Romîne, Bucuresti* 5, (2), 1–404 [in Romanian].

Feider, Z. & Vasiliu, N. (1969) Revision critique de la famille Nicoletiellidae. In: Evans, G.O. (ed.), *Proceedings of the 2nd International Congress of Acarology*. Akadémiai Kiadó, Budapest, pp. 202–207.

Feiertag-Koppen, C.C.M. & Pijnacker, L.P. (1985) Oogenesis. In: Helle, W. & Sabelis, M.W. (eds), *Spider Mites, their Natural Enemies and Control*, Vol. 1A. Elsevier, Amsterdam, pp. 117–127.

Feldman-Muhsam, B. (1967) Spermatophore formation and sperm transfer in *Ornithodoros* ticks. *Science [N.Y.]* 156, 1252–1253.

Feldman-Muhsam, B. (1969) Components of the spermatophore, their formation and their biological function in argasid ticks. In: Evans, G.O. (ed.), *Proceedings of the 2nd International Congress of Acarology*. Akadémiai Kiadó, Budapest. pp. 357–362.

Feldman-Muhsam, B. (1986) Observations on the mating behaviour of ticks. In: Sauer, J.R. & Hair, J. (eds), *Morphology, Physiology and Behavioural Biology of Ticks*. Ellis Horwood, Chichester, pp. 217–232.

Feldman-Muhsam, B. & Borut, S. (1971) Copulation in ixodid ticks. *Journal of Parasitology* 57, 630–634.

Feldman-Muhsam, B. & Borut, S. (1984) Differences in the structure of the spermatophore between argasid and ixodid ticks. In: Griffiths, D.A. & Bowman, C.E. (eds), *Acarology VI*, Vol. 1. Ellis Horwood, Chichester, pp. 504–511.

Feldman-Muhsam, B. & Filshie, B.K. (1979) The ultrastructure of the prospermium of *Ornithodoros* ticks and its relation to sperm maturation and spermatelleosis. In: Fawcett, D.W. & Bedford, J.M. (eds), *The Spermatozoon*. Urbam & Schwarzenberg, Baltimore, pp. 355–369.

Feldman-Muhsam, B. & Havivi, Y. (1960) Accessory glands of Gené's organ in ticks. *Nature* [*Lond.*] 187, 964.

Filipponi, A. (1955) Sulla natura dell'associazione tra *Macrocheles muscaedomesticae* e *Musca domestica, Rivista de Parassitologia, Roma* 16, 83–102.

Filipponi, A. (1964) Experimental taxonomy applied to the Macrochelidae (Acari: Mesostigmata). *Acarologia* (fasc. hors sér.), 92–100.

Filipponi, A. & Francaviglia, G. (1964) Larviparita facoltativa in alcuni Macrochelidi (Acari; Mesostigmata) associati a Muscidi di interesse sanitario. *Parassitologia* 6, 99–113.

Filippova, N.A. (1966) Argasidae. *Fauna SSSR* (N.S.) No. 96, 1–225 [in Russian].

Filippova, N.A. (ed.) (1985) *Taiga Tick, Ixodes persulcatus Schulze* (Acarina: Ixodidae). Nauka Publ., Leningrad, 416pp.

Filshie, B.K. (1976) The structure and deposition of the epicuticle of the adult female cattle tick (*Boophilus microplus*). In: Hepburn, H.R. (ed.), *The Insect Integument.* Elsevier, Amsterdam, pp. 193–206.

Finnegan, S. (1931) On a new species of mite of the family Heterozerconidae parasitic on a snake. *Proceedings of the Zoological Society of London* 1349–1357.

Fleschner, C.A., Badgley, M.E., Richer, D.W. & Hall, J.C. (1956) Air drift of spider mites. *Journal of Economic Entomology* 49, 624–627.

Foelix, R.F. & Axtell, R.C. (1971) Fine structure of the tarsal sensilla in the tick *Amblyomma americanum* (L.) *Zeitschrift für Zellforschung und Mikroskopische Anatomie, Berlin* 114, 22–37.

Foelix, R.F. & Axtell, R.C. (1972) Ultrastructure of Haller's organ in the tick *Amblyomma americanum* (L.) *Zeitschrift für Zellforschung und Mikroskopische Anatomie, Berlin.* 124, 275–292.

Foelix, R.F. & Chu-Wang, I-Wu. (1972) Fine structural analysis of the palpal receptors in the tick *Amblyomma americanum* (L.) *Zeitschrift für Zellforschung und Mikroskopische Anatomie, Berlin* 129, 548–560.

Furman, D. (1977) Observations on the ontogeny of halarachnid mites (Acarina: Halarachnidae). *Journal of Parasitology* 63, 748–755.

Furman, D., Bonasch, H., Springsteen, R., Stiller, D. & Rahlmann, D.F. (1974) Studies on the biology of the lung mite, *Pneumonyssus simicola* Banks (Acarina: Halarachnidae) and diagnosis of infestation in macaques. *Laboratory Animal Science* 24, 622–629.

Furumizo, R.T. (1973) The biology and ecology of the house dust mite *Dermatophagoides farinae* Hughes, 1961 (Acarina: Pyroglyphidae). PhD Thesis, University of California, Riverside (quoted in Fain & Herin, 1979).

Gabby, S. & Warburg, M.R. (1976) The diversity of neurosecretory cell types in the cave tick *Ornithodoros tholozani. Journal of Morphology* 153, 371–386.

Gabe, M. (1955) Donnés histologiques sur le neuro-sécrétion chez les Arachnides. *Archives d'Anatomie Microscopique et de Morphologie Expérimentale, Paris* 44, 351–383.

Galun, R. & Kindler, S.H. (1968) Chemical basis of feeding in the tick *Ornithodoros tholozani. Journal of Insect Physiology,* 14, 1409–1421.

Garcia, R. (1969) Reaction of the winter tick, *Dermacentor albipictus* (Packard), to CO_2. *Journal of Medical Entomology* 6, 286.

Gaud, J. (1962) *Sarcoptiformes plumicoles* (Analgesoidea) parasites d'oiseaux de l'ile Rennell, Scientific Results of the Danish Rennell Expedition 1951 and the British Museum (Natural History) *Expedition, 1953* 4, 31–50.

Gaud, J. (1966) Nouvelles definitions de la famille Pterolichidae Mégnin et Trouessart et creation de genres nouveaux appartenant à cette famille. *Acarologia* 8, 115–128.

Gaud, J. & Atyeo, W.T. (1967) Genres nouveaux de la famille des Analgidae, Trouessart & Mégnin. *Acarologia* 9, 447–464.

Gaud, J. & Atyeo, W.T. (1975) Gabuciniidae, famille nouvelle de *Sarcoptiformes plumicoles. Acarologia* 16 (1974), 522–561.

Gaud, J. & Atyeo, W.T. (1978) Nouvelles superfamilies pour les acariens astigmates parasites d'oiseaux. *Acarologia* 19 (1977), 678–685.

Gaud, J. & Mouchet, J. (1959) *Acariens plumicoles* (Analgesoidea) parasites des oiseaux du Cameroun II. Analgesidae. *Annales de Parasitologie, Humaine et Comparée, Paris* 34, 149–208.

Gerhardt, U. & Kaestner, A. (1937) Araneae (part) in Kükenthal and Krumbach. *Handbuch der Zoologie*, Vol. 3, (2), 3, 394–496. Berlin & Leipzig.

Gerson, U. (1972) Mites of the genus *Ledermuelleria* associated with mosses in Canada. *Acarologia* 13, 319–343.

Gilyarov, M.S. (ed.) (1978) *Handbook for the Identification of Soil-Inhabiting Mites. Trombidiformes.* Zoological Institute of the Academy of Science, SSSR, Leningrad, 271pp [in Russian].

Gilyarov, M.S. & Bregetova, N.G. (eds) (1977) *Handbook for the Identification of Soil-Inhabiting Mites. Mesostigmata.* Zoological Institute of the Academy of Science, SSSR, Leningrad, 717pp [in Russian].

Gilyarov, M.S. & Krivolutsky, D.A. (eds) (1975) *Handbook for the Identification of Soil-Inhabiting Mites, Sarcoptiformes.* Zoological Institute of the Academy of Science, SSSR, Leningrad 491pp [in Russian].

Glinyanaya, E.I. (1972) The role of photoperiod in the reactivation of arthropods with summer and winter diapause. In: Goryshin, N.I. (ed.), *Problems of Photoperiodism and Diapause in Insects.* Leningrad State University, Leningrad, pp. 88–102 [in Russian].

Goddard, D.G. (1982) Feeding biology of free-living Acari at Signy Island, South Orkney Islands. *Bulletin of the British Antarctic Survey* No. 51, 290–293.

Gorgol, V.T. & Yastrebtsov, A.V. (1987) Muscular system of the mite *Cheyletus eruditus* (Trombidiformes, Cheyletidae). *Vestnik Zoologii* 60–67 [in Russian].

Gorirossi, F.E. (1950) The mouthparts of the adult female tropical rat mite, *Bdellonyssus bacoti* (Hirst, 1913) Fonseca, 1941 [=*Liponyssus bacoti* (Hirst)], with observations on the feeding mechanism. *Journal of Parasitology* 36, 301–317.

Gorirossi, F.E. (1955a) The anatomy of the feeding apparatus of *Uropoda agitans* Banks, 1908, a mesostigmatid mite. *American Midland Naturalist* 53, 145-155.

Gorirossi, F.E. (1955b) The gnathosoma of the Celaenopsina (Acarina – Mesostigmata). *American Midland Naturalist* 54, 153–167.

Gorirossi, F.E. (1955c) The anatomy of the gnathosoma of *Pergamasus vargasi* n. sp. (Acarina – Mesostigmata – Gamasides). *American Midland Naturalist* 54, 405–412.

Gorirossi, F.E. & Wharton, G.W. (1953) The anatomy of the feeding apparatus of *Megisthanus floridanus* Banks, 1904, a large mesostigmatid mite. *American Midland Naturalist* 50, 433–447.

Goroshchenko, Yu. L. (1965) Cytological investigations of certain properties of gametogenesis and karyologic testimony of systematic subdivision of argasid ticks. *Report submitted for Scientific Degree of Candidate of Biological Science*, Leningrad, 28pp [in Russian].

Gossel, P. (1935) Beiträge zur Kenntnis der Hautsinnesorgane und Hautdrusen der Cheliceraten und der Augen der Ixodiden. *Zeitschrift für Morphologie und Okologie der Tiere* 30, 177–205.

Graf, J-F. (1978) Copulation, nutrition et ponte chez *Ixodes ricinus* L. (Ixodoidea: Ixodidae) – 2e part. *Mitteilungen der Schweizerischen Entomologischen Gesellschaft, Lausanne* 51, 241–253.

Grandjean, F. (1933) Observations sur les organes respiratoires des Oribates [Acariens]. *Bulletin de la Société Entomologique de France, Paris* 38, 123–127.

Grandjean, F. (1934a) Les organes respiratoires secondaires des Oribates. *Annales de la Société Entomologique de France, Paris* 103, 109–146.

Grandjean, F. (1934b) Oribates de l'Afrique du Nord (2e série). *Bulletin de la Société d'Histoire Naturelle de l'Afrique du Nord Alger* 25, 235–252.

Grandjean, F. (1934c) Observations sur les Oribates (Arach. Acar.) (7e série). *Bulletin du Muséum d'Histoire Naturelle, Paris* 6, 423–431.

Grandjean, F. (1934d) La notation des poils gastronomiques et des poils dorsaux du propodosoma chez les Oribates (Acariens). *Bulletin de la Société Zoologique de France* 59, 12–44.

Grandjean, F. (1935a) Observations sur les Acariens (le série). *Bulletin du Muséum d'Histoire Naturelle, Paris* 7, 119–126.

Grandjean, F. (1935b) Observations sur les Acariens (2e série). *Bulletin du Muséum d'Histoire Naturelle, Paris* 7, 201–208.

Grandjean, F. (1935c) Les poils et les organes sensitifs portés par les pattes et le palp chez les Oribates. *Bulletin de la Société Zoologique de France* 60, 6–39.

Grandjean, F. (1936a) Un acarien synthétique: *Opilioacarus segmentatus* With. *Bulletin de la Société d'Histoire Naturelle de l'Afrique du Nord Alger* 27, 413–444.

Grandjean, F. (1936b) Le genre *Pachygnathus* Dugès (*Alycus* Koch) (Acariens). 1re Partie. *Bulletin du Muséum d'Histoire Naturelle, Paris* 8, 398–405.

Grandjean, F. (1937) Sur quelques caractères des Acaridae libres. *Bulletin de la Société Zoologique de France*, 62, 388–398.

Grandjean, F. (1938a) *Retetydeus* et les stigmates mandibulaires des Acariens Prostigmatiques. *Bulletin du Muséum d'Histoire Naturelle, Paris* 10, 279–286.

Grandjean, F. (1938b) Observations sur les Bdelles [Acariens]. *Annales de la Société Entomologique de France, Paris* 107, 1–24.

Grandjean, F. (1938c) Sur l'ontogenie des acariens. *Compte Rendu Hebdomadaire des Séances de l'Academie des Sciences, Paris* 206, 146–150.

Grandjean, F. (1938d) Observations sur les Acaridiae (2e série). *Bulletin de la Société Zoologique de France* 63, 278–288.

Grandjean, F. (1938e) Description d'une nouvelle prélarve et remarques sur la bouche des acariens. *Bulletin de la Société Zoologique de France* 63, 58–68.

Grandjean, F. (1939a) Les segments post-larvaires de l'hysterosoma chez les Oribates (Acariens). *Bulletin de la Société Zoologique de France* 64, 273-284.

Grandjean, F. (1939b) Quelques genres d'acariens appartenant au groupe des Endeostigmata. *Annales des Sciences Naturelles, Zoologie* 11 sér., 2, 3–122.

Grandjean, F. (1939c) La chaetotaxie des pattes chez les Acaridiae. *Bulletin de la Société Zoologique de France* 64, 50–60.

Grandjean, F. (1940a) Les poils et les organes sensitifs portés par les pattes et le palpe chez les Oribates II. *Bulletin de la Société Zoologique de France* 65, 32–44.

Grandjean, F. (1940b) Observations sur les Oribates (15e série). *Bulletin du Muséum d'Histoire Naturelle, Paris* 12, 332–339.

Grandjean, F. (1941a) L'ambulacre des Acariens (1e série). *Bulletin du Muséum d'Histoire Naturelle, Paris* 13, 422–429.

Grandjean, F. (1941b) Statistique sexuelle et parthenogenèse chez les Oribates (Acariens). *Compte Rendu Hebdomadaire des Séances de l'Academie des Sciences, Paris* 212, 463–476.

Grandjean, F. (1941c) La chaetotaxie comparée des pattes chez les Oribates (1e série). *Bulletin de la Société Zoologique de France* 66, 33–50.

Grandjean, F. (1942) Quelques genres d'acariens appartenant au groupe des Endeostigmata (2e série). Première Partie. *Annales des Sciences Naturelles, Zoologie* 11 sér., 4, 85–135.

Grandjean, F. (1943a) Le développement postlarvaire d'"*Anystis*" (Acarien). *Mémoires du Muséum d'Histoire Naturelles, Paris* 18, 33–77.

Grandjean, F. (1943b) Quelques genres d'acariens appartenant au groupe des Endeostigmata (2e série). Deuxième Partie. *Annales des Sciences Naturelles, Zoologie* 11 sér. 5, 1–59.

Grandjean, F. (1943c) L'ambulacre des Acariens (2e série). *Bulletin du Muséum d'Histoire Naturelle, Paris* 15, 303–310.

Grandjean, F. (1945) Observations sur les Acariens (8e série). *Bulletin du Muséum d'Histoire Naturelle, Paris* 17, 399–406.

Grandjean, F. (1946a) Au sujet de l'organe de Claparède, des eupathides multiples et des taenidies mandibulaires chez les acariens actinochitineux. *Archives des Sciences Physiques et Naturelles, Genève* 28, 63–87.

Grandjean, F. (1946b) Observations sur les Acariens (9e série). *Bulletin du Muséum d'Histoire Naturelle, Paris* 18, 337–344.

Grandjean, F. (1946c) Les Enarthronota (Acariens) – premiere série. *Annales des Sciences Naturelles, Zoologie* 8, 213–248.

Grandjean, F. (1947) L'origine de la pince mandibulaire chez les acariens actinochitineux. *Archives des Sciences Physiques et Naturelles, Genève* 29, 305–355.

Grandjean, F. (1948) Quelques caractères de Tétranyques. *Bulletin du Muséum d'Histoire Naturelle, Paris* 20, 517–524.

Grandjean, F. (1949a) Observation et conservation des très petits Arthropodes. *Bulletin du Muséum d'Histoire Naturelle, Paris* 21, 363–370.

Grandjean, F. (1949b) Les Enarthronota (Acariens) (2e série). *Annales des Sciences Naturelles, Zoologie,* 10, 29–58.

Grandjean, F. (1950) Observations éthologiques sur *Camisia segnis* (Herm.) et *Platynothrus peltifer* (Koch) (Acariens). *Bulletin du Muséum d'Histoire Naturelle, Paris* 22, 224–231.

Grandjean, F. (1951a) Sur le tégument des Oribates. *Bulletin du Muséum d'Histoire Naturelle, Paris* 23, 497–504.

Grandjean, F. (1951b) Les relations chronologiques entre ontogenèses et phylogenèses d'après les petits caractères discontinus des acariens. *Bulletin Biologique de la France et de la Belgique* 85, 269–292.

Grandjean, F. (1951c) Etude sur les Zetorchestidae *Memoires Muséum National d'Histoire Naturelle, Paris* sér. A, 4, 1–50.

Grandjean, F. (1951d). Observations sur les Oribates (23e série). *Bulletin du Muséum d'Histoire Naturelle, Paris* 23, 261–268.

Grandjean, F. (1952a) Au sujet de l'ectosquelette du podosoma chez les Oribates supérieurs et de sa terminologie. *Bulletin de la Société Zoologique de France* 77, 13–36.

Grandjean, F. (1952b) Le morcellement secondaire des tarses de *Tarsolarkus* sp. (Acarien). *Archives de Zoologie Experimentale et Generale.* 89, Notes et Revue No. 3, 113–123.

Grandjean, F. (1952c) Sur les articles des appendices chez les Acariens actinochitineux. *Compte Rendu Hebdomadaire des Séances de l'Academie des Sciences, Paris* 235, 560–564.

Grandjean, F. (1953) La coalescence fémorogénuale chez *Fusacarus.* (Acaridié, Acarien). *Bulletin du Muséum d'Histoire Naturelle, Paris* 25, 387–394.

Grandjean, F. (1954a) Observations sur les Oribates (31e serie). *Bulletin du Muséum d'Histoire Naturelle, Paris* 26, 582–589.

Grandjean, F. (1954b) Les deux sortes de temps et l'evolution. *Bulletin Scientifique de la France et de la Belgique* 88, 413–433.

Grandjean, F. (1954c) Sur les nombres d'articles aux appendices des acariens Actinochitineux. *Archives de Science, Genève* 7, 337–362.

Grandjean, F. (1954d) Etude sur les Palaeacaroides (Acariens, Oribates). *Memoires du Muséum d'Histoire Naturelles, Paris* ser. A, 7, 179–272.

Grandjean, F. (1954e) Essai de classification des Oribates (Acariens). *Bulletin de la Société Zoologique de France* 78, 421–446.

Grandjean, F (1955) Sur un acarien des Iles Kerguélen, *Podacrus auberti (Oribate). Memoires Muséum National d'Histoire Naturelle, Paris* 8A, 109–150.

Grandjean, F. (1956a) Observations sur les Oribates (34e série). *Bulletin du Muséum d'Histoire Naturelle, Paris* 28, 205–212.

Grandjean, F. (1956b) Caractères chitineux de l'ovipositeur, en structure normale chez les Oribates (Acariens). *Archives de Zoologie Expérimentale et Générale, Paris* 93, 96–106.

Grandjean, F. (1956c) Observations sur les Galumnidae, le série (Acariens, Oribates). *Revue Française d'Entomologie, Paris* 23, 137–146.

Grandjean, F. (1957a) L'evolution selon l'âge. *Archives de Sciences, Genève* 10, 477–526.

Grandjean, F. (1957b) Les stases du developpement ontogenetique chez *Balaustium florale* (Acarien, Erythroide). Première Partie. *Annales de la Société Entomologique de France, Paris* 15, 135–152.

Grandjean, F. (1957c) L'infracapitulum et la manducation chez les Oribates et d'autre Acariens. *Annales des Sciences Naturelles, Zoologie,* 19, 233–281.

Grandjean, F. (1958) Au sujet du naso et son oeil infère chez les Oribates et les Endeostigmata. *Bulletin du Muséum d'Histoire Naturelle, Paris* 30, 427–435.

Grandjean, F. (1962) Prélarves d'Oribates. *Acarologia* 4, 423–439.

Grandjean, F. (1967) Nouvelles observations sur les Oribates (5e série). *Acarologia* 9, 242–283.

Grandjean, F. (1969) Considerations sur le classement des Oribates leur division en 6 groupes majeur. *Acarologia* 11, 127–153.

Grandjean, F. (1970a) Stases. Actinopiline. Rappel de ma classification des Acariens en 3 groupes majeurs. Terminologie en soma. *Acarologia* 11, (1969), 796–827.

Grandjean, F. (1970b) Nouvelles observations sur les Oribates (7e série). *Acarologia* 12, 432–460.

Grandjean, O. & Aeschlimann, A. (1973) Contribution to the study of digestion in ticks: histology and fine structure of the mid-gut epithelium of *Ornithodoros moubata* Murray (Ixodoidea, Argasidae) *Acta Tropica*, 30, 193–212.

Grassé, P-P. (ed.) (1949) *Traité de Zoologie – Anatomie, Systematique, Biologie. Onychophores – Tardigrades – Arthropodes – Trilobitomorphes – Chelicerates,* Vol. 6. Masson et Cie, Paris, 979pp.

Gray, J.S. (1987) Mating and behavioural diapause in *Ixodes ricinus* L. *Experimental and Applied Acarology* 3, 61–71.

Gray, J.S. (1991) The development and seasonal activity of the tick *Ixodes ricinus*: a vector of Lyme borreliosis. *Revue of Medical and Veterinary Entomology* 79, 323–333.

Greenberg, B. (1953) New Labidostommidae with keys to the New World species (Acarina). *New York Entomological Society Journal* 60 (1952), 195–209.

Gregson, J.D. (1956) *The Ixodoidea of Canada.* Scientific Services, Entomology Division Canadian Department of Agriculture Publ. 930, 1–92.

Gregson, J.D. (1960a) Morphology and functioning of the mouthparts of *Dermacentor andersoni* Stiles. Part I. The feeding mechanism in relation to the tick. *Acta Tropica* 17, 48–72.

Gregson, J.D. (1960b) Morphology and functioning of the mouthparts of *Dermacentor andersoni* Stiles. Part II. The feeding mechanism in relation to the host. *Acta Tropica* 17, 72–79.

Gregson, J.D. (1967) Observations on the movement of fluids in the vicinity of the mouthparts of naturally feeding *Dermacentor andersoni* Stiles. *Parasitology* 57, 1–8.

Gridelet-de-Saint-Georges, D. (1976) Techniques d'extraction applicables à l'étude écologique des acariens des poussières de maison. Comparaison qualitative et quantitative de divers types de poussiéres. *Acarologia* 17, 693–708.

Griffiths, D.A. (1964a) A revision of the genus *Acarus* L. 1758 (Acaridae, Acarina). *Bulletin of the British Museum (Natural History) [Zool.]* 11, 415–464.

Griffiths, D.A. (1964b) Experimental studies on the systematics of the genus *Acarus* Linnaeus, 1758 (Sarcoptiformes: Acarina). *Acarologia* 6 (fasc, hors. sér.), 101–116.

Griffiths, D.A. (1966) Nutrition as a factor influencing hypopus formation in *Acarus siro* species complex (Acarina, Acaridae). *Journal of Stored Products Research* 1, 325–340.

Griffiths, D.A. (1969) The influence of dietary factors on hypopus formation in *Acarus immobilis* Griffiths (Acari, Acaridae). In: Evans, G.O. (ed.), *Proceed-*

ings of the 2nd International Congress of Acarology. Akadémiai Kiadó, Budapest, pp. 419–432.

Griffiths, D.A. (1977) A new family of astigmatid mites from Iles Crozet, sub-Antarctica: introducing a new concept relating to ontogenetic development of the idiosomal setae. *Journal of Zoology [Lond.]* 182, 291–308.

Griffiths, D.A. & Boczek, J. (1977) Spermatophores of some acaroid mites (Astigmata: Acarina). *International Journal of Insect Morphology and Embryology* 231–238.

Griffiths, D.A. & Bowman, C.E. (1981) World distribution of the mite *Varroa jacobsoni*, a parasite of honeybees. *Bee World* 62, 154–163.

Griffiths, D.A., Atyeo, W.T., Norton, R.A. & Lynch, C.A. (1990) The idiosomal chaetotaxy of astigmatid mites. *Journal of Zoology [Lond.]* 220, 1–32.

Guirgis, S.S. (1971) The subgenus *Persicargas* (Ixodoidea, Argasidae, *Argas*) 13. Histological studies on *A. (P.) arboreus* Kaiser, Hoogstraal & Kohls in Egypt. *Journal of Medical Entomology* 8, 648–667.

Haarløv, N. & Mørch, J. (1975) Interaction between *Ornithocheyletia hallae* Smiley 1970 (Acarina: Cheyletiellidae) and *Micromonospora chalcea* (Foulerton 1905) Ørskov 1923 (Streptomycetaceae, Actonomycetales) in skin of pigeon. *Acarologia* 17, 284–299.

Hackman, R.H. (1982) Structure and function in tick cuticle. *Annual Review of Entomology* 27, 75–95.

Haggart, D.A. & Davis, E.E. (1980) Ammonia-sensitive neurones on the first tarsi of the tick, *Rhipicephalus sanguineus*. *Journal of Insect Physiology* 26, 517–523.

Halaškova, V. (1970) Zerconidae of Czechoslovakia (Acari: Mesostigmata) *Acta Universitatis Carolinae (Biol.) Prague* 3, 175–352.

Hall, C.G. (1959) A dispersal mechanism in mites. *Journal of the Kansas Entomological Society* 32, 45–46.

Hamdy, B.H. (1973) Biochemical and physiological studies of certain ticks (Ixodoidea). Cycle of nitrogenous excretion of *Hyalomma dromedarii* Koch (Ixodidae). *Journal of Medical Entomology, BP Bishop Museum, Honolulu* 10, 345–348.

Hammer, F-L. (1804) In Hermann, J-F., *Mémoire apterologique*. Strasburg.

Hänel, H. (1986) Effect of juvenile hormone (III) from the host *Apis mellifera* (Insecta: Hymenoptera) on neurosecretion of the parasitic mite *Varroa jacobsoni* (Acari: Mesostigmata). *Experimental and Applied Acarology* 2, 257–271.

Harding, D.J.L. & Stuttard, R.A. (1974) Microarthropods. In: Dickinson, C.H. & Pugh, G.J.F. (eds), *Biology of Plant Litter Decomposition, Vol. 2*. Academic Press, London. pp. 489–532.

Hartl, D.L. & Brown, S.W. (1970) The origin of male haploid genetic systems and their expected sex ratio. *Theoretical Population Biology* 1, 165–190.

Heineman, R.L. & Hughes, R.D. (1969) The cytological basis for reproductive variability in the Anoetidae (Sarcoptiformes: Acari). *Chromosoma* 28, 346–356.

Helle, W. (1962) Genetic variability of photoperiodic response in an arrhentokous mite (*Tetranychus urticae*). *Entomology Experimental and Applied* 11, 101–113.

Helle, W. & Pijnacker, L.P. (1985) Parthenogenesis, chromosomes and sex. In: Helle, W. & Sabelis, M.W. (eds), *Spider Mites, their Biology, Natural Enemies and Control,* Vol. 1A. Elsevier, Amsterdam, pp. 129–139.

Helle, W. & Sabelis, M.W. (eds) (1985) *Spider Mites, their Biology, Natural Enemies and Control,* Vol. 1A and 1B. Elsevier, Amsterdam.

Helle, W. & Wysoki, M. (1983) The chromosomes and sex-determination in some actinotrichid taxa (Acari) with special reference to Eriophyidae. *International Journal of Acarology* 9, 67–71.

Helle, W., Bolland, H.R., Van Arendonk, R., de Beer, R. Schulten, G.C.M. & Russel, W.M. (1978) Genetic evidence for biparental males in haplo-diploid predatory mites (Acarina: Phytoseiidae). *Genetica* 49, 165–171.

Helle, W., Bolland, H.R., Jeurissen, S.H.M. & Van Seventer, G.A. (1984) Chromosome data on the Actinedida and Oribatida. In: Griffiths, D.A. & Bowman, C.E. (eds), *Acarology VI,* Vol. 1. Ellis Horwood, Chichester, pp. 449–454.

Henking, H. (1882) Beitrage zur Anatomie, Entwicklungsgeschichte und Biologie von *Trombidium fuliginosum. Zeitschrift für Wissenschaftliche Zoologie, Leipzig* 37, 553–663.

Hennig, W. (1950) *Grundzwüge einer Theorie der phylogenetischen Systematik.* Deutscher Zentralverlag, Berlin.

Hennig, W. (1966) *Phylogenetic Systematics.* University of Illinois Press, Urbana.

Herfs, A. (1926) Ökologisches Untersuchungen an *Pediculoides ventricosus* (Newp.) Berl. *Zoologica* 74, 1–68.

Hess, E. & Loftus, R. (1984) Warm and cold receptors of two sensilla on the foreleg tarsi of the tropical bont tick, *Amblyomma variegatum. Journal of Comparative Physiology* 155, 187–195.

Hess, E. & Vlimant, M. (1984) The distal tarsal slit sense organ (DTSSO), a new type of mechanoreceptor on the walking legs of the ixodid tick *Amblyomma variegatum* Fabricius 1794 (Ixodida: Metastriata). In: Griffiths, D.A. & Bowman, C.E. (eds), *Acarology VI,* Vol. 1. Ellis Horwood, Chichester, pp. 253–260.

Hess, E. & Vlimant, M. (1986) Leg sense organs of ticks. In: Sauer, J.R. & Hair, J.A. (eds), *Morphology, Physiology and Behavioural Biology of Ticks.* Ellis Horwood, Chichester, pp. 361–390.

Hessein, N.A. & Perring, T.M. (1986) Feeding habits of the Tydeidae with evidence of *Homeopronematus anconai* (Acari: Tydeidae) predation on *Aculops lycopersici* (Acari: Eriophyidae). *International Journal of Acarology* 12, 215–221.

Hessein, N.A. & Perring, T.M. (1988) *Homeopronematus anconi* (Baker) (Acari – Tydeidae) predation on citrus flat mite *Brevipalpus lewisi* McGregor (Acari: Tenuipalpidae). *International Journal of Acarology* 14, 89–90.

Hevers, J. (1978) Interspezifische Beziehungen zwischen *Unionicola* – larven (Hydrachnellae, Acari) und Chironomidae (Diptera, Insecta). *Verhandlungen der Gesellschaft für Ökologie* 7, 211–217.

Higgs, G.A., Vane, J.R., Hart, R.J., Potter, C. & Wilson, R.G. (1976) Prostaglandins in the saliva of the cattle tick *Boophilus microplus* (Canestrini) (Acari: Ixodidae). *Bulletin of Entomological Research* 66, 665–670.

Hinton, H.E. (1967) The structure of the spiracles of the cattle tick, *Boophilus microplus. Australian Journal of Zoology* 15, 941–945.

Hinton, H.E. (1971a) Some neglected phases in metamorphosis. *Proceedings of the Royal Entomological Society of London (C)* 35, 55–64.

Hinton, H.E. (1971b) Plastron respiration in the mite, *Platyseius italicus*. *Journal of Insect Physiology* 17, 1185–1199.

Hirschmann, W. (1957) Gangsystematik der Parasitiformes. Teil 1. Rumpfbehaarung und Rückenflächen. *Acarologie, Schriftenreihe vergl. Milbenkunde*, Fürth. Folge 1, pp. 1–20.

Hirschmann, W. (1971) A fossil mite of the genus *Dendrolaelaps* (Acarina, Mesostigmata, Digamasellidae) found in amber from Chiapas, Mexico. *University of California Publications on Entomology, Berkeley* 63, 69–70.

Hirschmann, W. (1975) Gangsystematik der Parasitiformes. Teil 206. Teilgänge und Stadien von 9 neuen *Macrodinychus* – Arten, Wiederbeschreibung von 2 bekannten *Macrodinychus* – Arten. *Acarologie, Schriftenreihe vergl. Milbenkunde*, Fürth. Folge 21, Teil 206, pp. 39–43.

Hirschmann, W. (1979) Stadiensystematik der Parasitiformes. Teil 1. Stadienfamilien und Stadiengattungen der Atrichopygidiina, erstelle im Vergleich zum Gangsystem Hirschmann 1979. *Acarologie, Schriftenreihe vergl. Milbenkunde*, Fürth. Folge, 26, Teil 1, pp. 57–68.

Hirschmann, W. (1991) Gangsystematik der Parasitiformes. Teil. 529–531. Weltweite revision der Ganggattung Sejus C.L. Koch 1836 (Trichopygidiina). *Acarologie, Schriftenreihe vergl. Milbenkunde*, Fürth. Folge 38, pp. 107–221.

Hirschmann, W. & Zirngiebl-Nicol. I. (1964) Das Gangsystem der Familie Uropodidae (Berlese, 1892). *Acarologie, Schriftenreihe vergl. Milbenkunde*, Fürth. Folge 6, Teil 7, pp. 1–22.

Hirst, S. (1917) On an apparently undescribed English saltorial mite (*Speleorchestes poduroides*). *Journal of Zoological Research* 2, 115–122.

Hirst, S. (1919) *Studies on Acari I. The genus* Demodex. British Museum (Natural History), London.

Hirst, S. (1920) *Arachnida and Myriapoda Injurious to Man.* British Museum (N.H.) Econ. Ser. No. 6, 59pp.

Hirst, S. (1921) Notes on parasitic Acari. *Journal of the Microscopical Club* 14, 229–236.

Hirst, S. (1922) *Mites Injurious to Domestic Animals* British Museum (N.H.) Econ. Ser., No. 13, 107pp.

Hirst, S. (1923) On some arachnid remains from Old Red Sandstone (Rhynie Chert Bed, Aberdeenshire). *Annals and Magazine of Natural History* ser. 9, 12, 455–474.

Hislop, R.G. & Jeppson, L.R. (1976) Morphology of the mouthparts of several species of phytophagous mites. *Annals of the Entomological Society of America* 69, 1125–1135.

Hislop, R.G. & Prokopy, R.J. (1981) Mite predator responses to prey and predator-emitted stimuli. *Journal of Chemical Ecology* 7, 895–904.

Hodson, W.E.H. (1948) *Narcissus Pests.* Ministry of Agriculture Bulletin No. 551. London.

Hoebel-Mävers, M. (1967) Funktionsanatomische Untersuchungen der Verdaungstrakt der Hornmilben (Oribatei). Dissertation. Tech. Hochschule Carolo-Wilhelmina zu Braunschweig, Germany.

Holmes, B.E. (1980) Reconsideration of some systematic concepts and terms. *Evolutionary Theory* 5, 35–87.

Holzmann, C. (1969) Die Familie Parasitidae Oudemans, 1901. *Acarologie Schriftenreihe vergl. Milbenkunde*, Fürth. Folge 13, 3–24; 25–55.

Hoogstraal, H. (1956) *African Ixodoidea, Vol. 1. Ticks of the Sudan*. U.S. Government Printing Office, 1101pp.

Hoogstraal, H. (1966) Ticks in relation to human diseases caused by viruses. *Annual Review of Entomology* 11, 261–308.

Hoogstraal, H. (1967) Ticks in relation to human diseases caused by *Rickettsia* species. *Annual Review of Entomology* 12, 377–420.

Houck, M.A. & OConnor, B.M. (1991) Ecological and evolutionary significance of phoresy in the Astigmata. *Annual Review of Entomology* 36, 611–636.

Howell, C.J., Neitz, A.W.H. & Potgieter, D.D.J. (1975) Some toxic, physical and chemical properties of the oral secretion of the sand tampan *Ornithodoros savignyi* Audouin (1827). *Onderstepoort Journal of Veterinary Research* 42, 99–102.

Hoy, M.A. (1975a) Effect of temperature and photoperiod on the induction of diapause in the mite, *Metaseiulus occidentalis*. *Journal of Insect Physiology* 21, 605–611.

Hoy, M.A. (1975b) Diapause in the mite *Metaseiulus occidentalis*: stages sensitive to photoperiodic induction. *Journal of Insect Physiology* 21, 745–751.

Hoy, M.A. (1985) Recent advances in genetics and genetic improvement of the Phytoseiidae. *Annual Review of Entomology* 30, 345–370.

Hoy, M.A. & Smilanick, J.M. (1979) A sex pheromone produced by immature and adult females of the predatory mite, *Metaseiulus occidentalis*. *Entomologia Experimentalis et Applicata* 26, 291–300.

Hoy, M.A., Groot, R. & Van de Baan, H.E. (1985) Influence of aerial dispersal on persistence and spread of pesticide-resistant *Metaseiulus occidentalis* in California almond orchards, *Entomologia Experimentalis et Applicata* 37, 17–31.

Hughes, A.M. (1976) *The Mites of Stored Food and Houses*. Technical Bulletin of the Ministry of Agriculture Fisheries and Food, No. 9. HMSO London, pp. 1–400.

Hughes, R.D. & Jackson, C.G. (1958) A review of the Anoetidae (Acari). *Virginia Journal of Science* 9, 5–198.

Hughes, T.E. (1949) The functional morphology of the mouthparts of *Liponyssus bacoti*. *Annals of Tropical Medicine and Parasitology* 43, 349–360.

Hughes, T.E. (1950a) The embryonic development of the mite *Tyroglyphus farinae* Linnaeus, 1758. *Proceedings of the Zoological Society of London* 119, 873–886.

Hughes, T.E. (1950b) Some stages of *Litomosoides carnii* in *Liponyssus bacoti*. *Annals of Tropical Medicine and Parasitology* 44, 285–290.

Hughes, T.E. (1950c) The physiology of the alimentary canal of *Tyroglyphus farinae*. *Quarterly Journal of Microscopical Science* 91, 45–61.

Hughes, T.E. (1952) The morphology of the gut in *Bdellonyssus bacoti* (Hirst, 1913). *Annals of Tropical Medicine and Parasitology* 46, 54–60.

Hughes, T.E. (1953) Functional morphology of the mouth parts of the mite *Anoetus sapromyzarum* Dufour, 1839, compared with those of the more typical Sarcoptiformes. *Proceedings of the Academy of Sciences, Amsterdam* 56C, 278–287.

Hughes, T.E. (1954a) The internal anatomy of the mite *Listrophorus leuckarti* (Pagenstecher, 1861). *Proceedings of the Zoological Society of London* 124, 239–256.

Hughes, T.E. (1954b) Some histological changes which occur in the gut epithelium of *Ixodes ricinus* females during gorging and up to oviposition. *Annals of Tropical Medicine and Parasitology* 48, 397–404.

Hughes, T.E. (1958) The respiratory system of the mite *Cheyletus eruditus* (Schrank, 1781). *Proceedings of the Zoological Society of London* 130, 231–239.

Hughes, T.E. (1959) *Mites or the Acari.* Athlone Press, London.

Hughes, T.E. (1964) Neurosecretion, ecdysis and hypopus formation in the Acaridei. *Acarologia* 6 (fasc. hors. sér.), 338–342.

Husband, R.W. (1980) Review of the genus *Podapolipus* (Acarina: Podapolipidae) with emphasis on species associated with tenebrionid beetles. *International Journal of Acarology* 6, 257–270.

Husband, R.W. (1984) *Dilopolipus, Panesthipolipus, Peripolipus* and *Stenopolipus,* new genera of Podapolipidae (Acarina) from the Indo-Australian region. *International Journal of Acarology* 10, 251–269.

Huxley, H.E. (1958) The contraction of muscle. *Scientific American* 199, 66–86.

Hyatt, K.H. (1959) Mesostigmatid mites associated with *Geotrupes stercorarius* (L.) (Col., Scarabaeidae). *Entomologist's Monthly Magazine* 95, 22–23.

Hyatt, K.H. (1980) Mites of the subfamily Parasitinae (Mesostigmata: Parasitidae) in the British Isles. *Bulletin of the British Museum (Natural History)* [*Zool.*] 38, 237–378.

Ifantidis, M.D. (1983) Ontogenesis of the mite *Varroa jacobsoni* Oud. in the worker and drone brood cells of the honey bee *Apis mellifera.* In: Cavalloro, R. (ed.). Varroa jacobsoni *Oud, affecting Honey Bees: Present Status and Needs.* A.A. Balkema, Rotterdam, pp. 37–39.

Immamura, T. (1952) Notes on the moulting of the adult of the water mite, *Arrenurus uchidai* n.sp. *Annotationes Zoologicae Japanenses, Tokyo* 25, 447–451.

Ioffe, I.D. (1963) Der Bau Nervenapparates von *Dermacentor pictus* Herm. (Chelicerata, Acarina). *Zoologicheskii Zhurnal* 42, 1472–1484 [in Russian].

Ioffe, I.D. (1965) Distribution of neurosecretory cells in the central nervous apparatus of *Dermacentor pictus* Herm. (Acarina, Chelicerata). *Dokladÿ Akademii Nauk SSSR, Leningrad* 154, 25–29 [in Russian].

Ivanov, V.P. (1979) Central nervous system. In: Balashov, Yu. S. (ed.), *An Atlas of Ixodid Tick Ultrastructure.* Nauka, Leningrad [in Russian]. (English translation in: *Miscellaneous Publications of the Entomological Society of America* 1983, 175–190.)

Ivanov, V.P. & Leonovich, S.A. (1979) Sensory organs. In: Balashov, Yu. S. (ed.), *An Atlas of Tick Ultrastructure.* Nauka, Leningrad [in Russian]. (English translation in: *Miscellaneous Publications of the Entomological Society of America* 1983, 191–202.)

Ivanov-Kazas, O.M. (1979) *Comparative Embryology of Invertebrates: Arthropods.* Nauka, Moscow, 223pp [in Russian].

Jack, K.M. (1962) Observations on the genus *Pterygosoma* (Acari: Pterygosomidae). *Parasitology* 52, 261–295.

Jacot, A.P. (1930) Oribatid mites of the subfamily Phthiracarinae of the northeastern United States. *Boston Society of Natural History* 39, 209–261.

Jagers op Akkerhuis, G., Sabelis, M.W. & Tjallingii, W.F. (1985) Ultrastructure of chemoreceptors on the pedipalp and first tarsi of *Phytoseiulus persimilis*. *Experimental and Applied Acarology* 1, 235–251.

Jakeman, L.A.R. (1961) The internal anatomy of the spiny rat mite, *Echinolaelaps echidninus* (Berlese). *Journal of Parasitology* 47, 329–349.

Jalil, M. & Rodriguez, J.G. (1970) Studies on the behaviour of *Macrocheles muscaedomesticae* (Acarina: Macrochelidae) with emphasis on its attraction to the house fly. *Annals of the Entomological Society of America* 63, 738-744.

Jenkin, P.M. & Hinton, H.E. (1966) Apolysis in arthropod moulting cycles. *Nature* [Lond.] 211, 871.

Jeppson, L.R., Keifer, H.H. & Baker, E.W. (1975) *Mites Injurious to Economic Plants*. University of California Press, Berkeley.

Jesiotr, L. & Suski, Z.W. (1970) A case of deuterotoky in the two-spotted spider mite *Tetranychus urticae* Koch (Acarina, Tetranychidae). *Bulletin de l'Academie Polonaise des Sciences (Ser. Sci. Biol.)* 18, 33–35.

Johnston, D.E. (1965) Comparative studies of the mouth-parts of the mites of the suborder Acaridei (Acari). Dissertation, Ohio State University.

Johnston, D.E. (1982) In: Parker, S.P. (ed.), *Synopsis and Classification of Living Organisms*, Vol. 2, Mesostigmata, pp. 112–116; Oribatida, pp. 145–146. McGraw-Hill, New York.

Johnston, D.E. & Bruce, W.A. (1965) *Tyrophagus neiswanderi*, a new acarid mite of agricultural importance. *Research Bulletin, Ohio Agricultural Experimental Station* No. 977, 1–17.

Johnston, D.E. & Wacker, R.R. (1967) Observations on the postembryonic development in *Eutrombicula splendens* (Acari: Acariformes). *Journal of Medical Entomology, BP Bishop Museum, Honolulu* 4, 306–310.

Johnson, D.T. & Croft, B.A. (1976) Laboratory study of the dispersal behaviour of *Amblyseius fallacis* (Acarina: Phytoseiidae). *Annals of the Entomological Society of America* 69, 1019–1023.

Johnson, D.T. & Croft, B.A. (1981) Dispersal of *Amblyseius fallacis* in an apple tree ecosystem. *Environmental Entomologist* 10, 313–319.

Jones, B.M. (1950a) Acarine growth: a new ecdysial mechanism. *Nature* [Lond.] 166, 908–909.

Jones, B.M. (1950b) The penetration of host tissue by the harvest mite, *Trombicula autumnalis* Shaw. *Parasitology* 40, 246–260.

Jones, J.C. (1962) Current concepts concerning insect hemocytes. *American Zoologist* 2, 209–246.

Juvara-Bals, I. & Baltac, M. (1977) Deux nouvelles espèces d'*Opilioacarus* (Acarina: Opilioacarida) de Cuba. In: *Resultats des expeditions biospéologiques cubano-roumaines á Cuba, 2.* Instituto de Spéologie "Emil Racovitza", Roumania, pp. 169–184.

Kaestner, A. (1968) *Invertebrate Zoology*, Vol 2, Arthropod relatives, Chelicerata, Myriapoda. [Translated and adapted by H.W. Levi and L.R. Levi.] Wiley-Interscience, New York.

Kahl, O. (1991) Lyme borreliosis – an ecological perspective of a tick-borne human disease. *Anzieger für Schädlingskunde, Berlin* 64, 3, 45–55.

Kahl, O. & Gray, J.S. (1991) Ticks, tick hosts and Lyme borreliosis. *EURAAC News Letter* 4 (1–2), 9–17.

Kahl, O. & Knülle, W. (1986) Water vapour uptake by engorged immature ixodid ticks. In: *Proceedings of the 3rd European Congress of Entomology*, Amsterdam, pp. 275–278.

Kahn, J. (1964) Cytotaxonomy of ticks. *Quarterly Journal of the Microscopical Society*, London 105, 123–137.

Kaiser, T. & Alberti, G. (1991) The fine structure of the lateral eyes of *Neocarus texanus* Chamberlin and Mulaik, 1942 (Opilioacarida, Acari, Arachnida, Chelicerata). *Protoplasma* 163, 19–33.

Kaltenrieder, M. (1990) Scototaxis and target perception in the camel tick *Hyalomma dromedarii*. *Experimental and Applied Acarology* 9, 267-278.

Karafiat, J. (1959) Systematik und Ökologie der Scutacaridae. In: Stammer, H-J. (ed.),*Beiträge zur Systematik und Ökologie Mitteleuropäischer Acarina*, Leipzig 1 (2), 627–712.

Karg, W. (1968) Ökologische Untersuchungen an Milben aus Komposterden im Freiland und unter Glas besonders im Hinblick auf die *Uroobovella marginata* C.L. Koch. *Archiv für Pflanzenschutz* 4, 93–122.

Karg, W. (1971) Acari (Acarina), Milben, Unterordnung Anactinochaeta (Parasitiformes). Die Freilebenden Gamasina (Gamasides), Raubmilben. In: *Die Tierwelt Deutschlands*, 59. Gustav Fischer Verlag, Jena, 475pp.

Karg, W. (1982) Diagnostik und Systematik der Raubmilben aus der Familie Phytoseiidae Berlese in Obstanlagen. *Zoologische Jahrbuecher. Abteilung für Systematik, Öekologie und Geographie der Tiere* 109, 188–210.

Karg, W. (1983) Verbeitung und Bedeutung von Raubmilben der Cohors Gamasina als Antagonisten von Nematoden. *Pedobiologia* 25, 419–432.

Karg, W. (1989) Acari (Acarina), Milben, Unterordnung Parasitiformes (Anactinochaeta) Uropodina Kramer, Schildkrotenmilben. In: *Die Tierwelt Deutschlands*, 67. Gustav Fischer Verlag, Jena, 203pp.

Karl, E. (1965) Untersuchungen zur Morphologie und Ökologie von Tarsonemiden gartnerischer Kulturpflanzen. II. *Hemitarsonemus latus* (Banks). *Tarsonemus confusus* Ewing. *T. talpae* Schaarschmidt, *T. setifer* Ewing, *T. smithi* Ewing und *Tarsonemoides belemnitoides* Weis-Fogh. *Biologisches Zentralblatt* 84, 331–357.

Kaufman, W.R. & Sauer, J.R. (1982) Ion and water balance in feeding ticks: mechanisms of tick excretion. *Current Themes Tropical Science* 1, 213--244.

Keifer, H.H. (1959) Eriophyid studies XXVI. *Bulletin of the California Department of Agriculture* 47 (4), 271–281.

Keifer, H.H. & Wilson, N.S. (1956) A new species of eriophyid mite responsible for the vection of peach mosaic virus. *Bulletin of the California Department of Agriculture* 44 (4), 145–146.

Keifer, H.H. Baker, E.W., Kono, T., Delfinado, M. & Styer, W.E. (1982) An illustrated guide to plant abnormalities caused by eriophyid mites in North America. *USDA Agriculture Handbook* No. 573, 1–178.

Keirans, J.E. & Clifford, C.M. (1975) *Nothoaspis reddelli*, new genus and new species (Ixodoidea: Argasidae). *Annals of the Entomological Society of America* 68, 81–85.

Keirans, J.E., Clifford C.M., Hoogstraal, H. & Easton, E.R. (1976) Discovery of *Nuttalliella namaqua* Bedford (Acarina: Ixodoidea: Nuttalliellidae) in Tanzania and redescription of the female based on scanning electron microscopy. *Annals of the Entomological Society of America* 69, 926–932.

Keirans, J.E., Clifford, C.M. & Reddell, J.R. (1977) Description of the immature stages of *Nothoaspis reddelli* (Ixodoidea: Argasidae). *Annals of the Entomological Society of America* 70, 591–595.

Kemp, D.H. & Tatchell, R.J. (1971) The mechanism of feeding and salivation in *Boophilus microplus* (Canestrini, 1877). *Zeitschrift für Parasitenkunde* 37, 55–69.

Kemp, D.H., Stone, B.F. & Binnington, K.C. (1982) Tick attachment and feeding: role of mouthparts, feeding apparatus, salivary gland secretions and the host response. *Current Themes in Tropical Science* 1, 119–168.

Kennedy, G.G. & Smitley, D.R. (1985) Dispersal. In: Helle, W. & Sabelis M.W. (eds), *Spider Mites, their Biology, Natural Enemies and Control*, Vol. 1A. Elsevier, Amsterdam, pp. 233–242.

Kennett, C.E., Flaherty, D.L. & Hoffman, R.W. (1979) Effect of wind-borne pollen on the population dynamics of *Amblyseius hibisci* (Acarina: Phytoseiidae). *Entomology Experimental and Applied* 28, 116–122.

Kethley, J.B. (1977) A review of the higher categories of the Trigynaspida (Acari: Parasitiformes). *International Journal of Acarology* 3, 129–149.

Kethley, J.B. (1978) *Narceolaelaps* n.g. (Acari: Laelapidae) with four new species parasitizing spiroboloid millipeds. *International Journal of Acarology*, 4, 195–210.

Kethley, J.B. (1982) In: Parker, S.P. (ed.), *Synopsis and Classification of Living Organisms*, Vol. 2, Prostigmata. McGraw-Hill, New York, pp. 117–145.

Khalil, G.M. & Shanbaky, N.M. (1976) The subgenus *Persicargas* (Ixodoidea – Argasidae: *Argas*). 21. The effect of some factors in the process of mating on egg development and oviposition in *A.* (*P.*) *arboreus*. *Experimental Parasitology* 39, 431–437.

Kielczewski, B. & Wiseniewski, J. (1977) Notes on the larval development of the mites of the genus *Trichouropoda* (Trichouropodini, Uropodinae). *Acarologia* 18, 107–409.

Kinn, D.W. (1972) A new species of *Holocercomegistus*, including some observations on the chaetotaxy of the pedipalpal and ambulatory appendages of the Cercomegistidae (Acarina). *Acarologia* 13, 258–265.

Kirkland, W.L. (1971) Ultrastructural changes in the nymphal salivary glands of the rabbit tick, *Haemaphysalis leporispalustris*, during feeding. *Journal of Insect Physiology* 17, 1933–1946.

Knop, N.F. (1985) Mating behaviour in the tydeid mite *Homeopronematus anconi* (Acari: Tydeidae). *Experimental and Applied Acarology* 1, 115–125.

Knülle, W. (1957) Morphologische und Entwicklungsgeschichtliche Untersuchungen zum Phylogenetischen System der Acari: Acariformes Zachv. I. Oribatei: Malaconothridae. *Mittelungen aus der Zoologischen Museum in Berlin* 33, 97–213.

Knülle, W. (1959) Morphologische und Entwicklungsgeschichtliche Untersuchungen zum Phylogenetischen System der Acari: Acariformes Zachv. II. Acaridiae. *Mittelungen aus der Zoologischen Museum in Berlin* 35, 347–418.

Knülle, W. (1984) Water vapour uptake in mites and insects: an ecophysiological and evolutionary perspective. In: Griffiths, D.A. & Bowman, C.E. (eds), *Acarology VI*, Vol. 1. Ellis Horwood, Chichester, pp. 71–82.

Knülle, W. (1987) Genetic variability and ecological adaptability of hypopus formation in a stored product mite. *Experimental and Applied Acarology* 3, 21–32.

Knülle W. & Rudolph, D. (1982) Humidity relationships and water balance of ticks. *Current Themes in Tropical Science* 1, 43–70.

Kohls, G.M., Sonenshine, D.E. & Clifford, C.M. (1965) The systematics of the subfamily Ornithodorinae (Acarina: Argasidae) 2. Identification of the larvae of the Western Hemisphere and descriptions of three new species. *Annals of the Entomological Society of America* 58, 331–364.

Kolata, D.R., Norstog, K.J. & Rhode, C.J. Jr. (1973) Ultrastructure of the egg envelopes of a uropodid mite. In: Daniel, M. & Rosický, B. (eds), *Proceedings of the 3rd International Congress of Acarology*. Academia, Prague, pp. 47–51.

Korn, W. (1982), Zur Fortpflanzung von *Poecilochirus carabi* G. u. R. Canestrini 1882 (syn. *P. necrophori* Vitzt.) und *P. austroasiaticus* Vitzthum 1930 (Gamasina, Eugamasidae). *Spixiana* 5, 261–288.

Kramer, P. (1877) Grundzüge zur systematik der Milben. *Archiv für Naturgeschichte*, Berlin 2, 215–247.

Krantz, G.W. (1974) *Phaulodinychus mitis* (Leonardi, 1899). An intertidal mite exhibiting plastron respiration. *Acarologia* 16, 11–20.

Krantz, G.W. (1978) *A Manual of Acarology*, 2nd. edn. Oregon State University Book Stores, Corvallis.

Krantz, G.W. & Baker, G.T. (1982) Observations on plastron mechanism of *Hydrozetes* sp. (Acari: Oribatida: Hydrozetidae). *Acarologia* 23, 273–277.

Krantz, G.W. & Lindquist, E.E. (1979) Evolution of phytophagous mites. *Annual Review of Entomology* 24, 121–158.

Krantz, G.W. & Redmond, B.L. (1987) Identification of glandular and poroidal idionotal systems in *Macrocheles perglaber* F & P. (Acari: Macrochelidae). *Experimental and Applied Acarology* 3, 243–253.

Krantz, G.W. & Redmond, B.L. (1988) On the structure and function of the cribrum, with special reference to *Macrocheles perglaber* (Gamasida: Macrochelidae). In: ChannaBasavanna, G.P. & Viraktamath, C.A. (eds), *Progress in Acarology*, Vol. 1, Oxford & IBH Publishing, New Delhi, etc., pp. 179–185.

Krantz, G.W. & Wernz, J.G. (1979) Sperm transfer in *Glyptholaspis americana*. In: Rodriguez, J.G. (ed.), *Recent Advances in Acarology*, Vol. 2. Academic Press, New York, etc., pp. 441–446.

Krczal, H. (1959) Systematik und Ökologie der Pyemotiden. In: Stammer, H-J. (ed.), *Beiträge zur Systematik und Ökologie Mitteleuropäischer Acarina* 1, Geest and Portig, Leipzig, pp. 385–625.

Krivolutsky, D.A. (1979) Some Mesozoic Acarina from the USSR. In: Piffl, E. (ed.), *Proceedings of the 4th International Congress of Acarology*. pp. 471–475 Académiai Kiadó, Budapest.

Krivolutsky, D.A. & Druk, A.Y. (1986) Fossil oribatids. *Annual Review of Entomology* 31, 533–545.

Krolak, J.M., Ownby, C.L. & Sauer, J.R. (1982) Alveolar structure of the salivary glands of the lone star tick, *Amblyomma americanum* (L.): unfed females. *Journal of Parasitology* 68, 61–82.

Krombein, K.V. (1961) Some symbiotic relations between saproglyphid mites and solitary vespid wasps (Acarina, Saproglyphidae and Hymenoptera, Vespidae). *Journal of the Washington Academy of Sciences* 51, 89–92.

Kuo, J.S., McCully, M.E. & Haggis, G.H. (1971) The fine structure of muscle attachments in an acarid mite, *Caloglyphus mycophagus* (Megnin) (Acarina). *Tissue & Cell* 3, 605–613.

Kuwahara, Y., Matsumoto, K. & Wada, Y. (1980) Pheromone study on acarid mites IV. Ciral: composition and function as an alarm pheromone and its secretory gland in four species of acarid mites. *Japanese Journal of Sanitary Zoology* 31, 73–80.

Lai-Fook, J. (1967) The structure of developing muscle insertions in insects. *Journal of Morphology* 123, 503–528.

Laing, J.E. & Knop, N.F. (1983) Potential use of predaceous mites other than Phytoseiidae for biological control of orchard pests. In: Hoy, M.A., Cunningham, G.L. & Knutson, L. (eds), *Biological Control of Pests by Mites*. University of California, Berkeley, pp. 28–35.

Lange, A.B. (1960) The prelarvae of mites of the order Acariformes and their peculiarities in the Paleacariformes. *Zoologicheskii Zhurnal* 39, 1819–1834 [in Russian].

Langenscheidt, M. (1973) Zur Wirkungsweise Sterilität erzeugenden Stoffen bei *Tetranychus urticae* (Acari: Tetranychidae). 1. Entwicklung der weiblichen Genitalorgane und Oogenese bei unbenhandelten Spinnmilben *Zeitschrift für Angewandte Entomologie, Berlin* 73, 103–106.

Lavoipierre, M.M.J. & Beck, A.J. (1967) Feeding mechanisms of *Chiroptonyssus robustipes* (Acarina: Macronyssidae) as observed in the transilluminated bat wing. *Experimental Parasitology* 20, 312–320.

Lavoipierre, M.M.J. & Riek, R.F. (1955) Observations on the feeding habits of argasid ticks and of the effect of their bites on laboratory animals with a note on the production of coxal fluid by several of the species studied. *Annals of Tropical Medicine and Parasitology* 49, 93–113.

Lawrence, R.F. (1935) The prostigmatic mites of South African lizards. *Parasitology* 28, 1–45.

Leal, W.S., Kuwahara, Y. & Suzuki, T. (1989) β-Acaridial, the sex pheromone of the acarid mite *Caloglyphus polyphyllae*. Pheromone study of acarid mites XXI. *Naturwissenschaften* 76, 332–333.

Lebrun, Ph. (1970) Écologie et biologie de *Nothrus palustris*. 3 iéme note. Cycle de vie. *Acarologia* 12, 193–207.

Leclerc, P. (1989) Considérations palaeobiogeographiques à propos la découverte en Thailande d'opiloacariens nouveaux (Acari: Notostigmata). *Compte Rendu des Séances de la Société de Biogeographie* 65 (4), 162–174.

Lee, D.C. (1974) Rhodacaridae (Acari: Mesostigmata) from near Adelaide, Australia. 3. Behaviour and development. *Acarologia* 16, 21–44.

Lee, D.C. (1981) Sarcoptiformes (Acari) of South Australian soils. 1. Notation. 2.

Bifemorata and Ptyctima (Crytostigmata). *Record of the South Australian Museum, Adelaide* 18, 199–222.

Lee, D.C. (1984) A modified classification for oribate mites. In: Griffiths, D.A. & Bowman, C.E. (eds), *Acarology VI*, Vol. 1. Ellis Horwood, Chichester, pp. 241–248.

Lee, D.C. (1987) Introductory study of advanced oribate mites (Acardia: Cryptostigmata) and a redescription of the only valid species of *Constrictobates* (Oripodoidea). *Record of the South Australian Museum*, Adelaide 21, 35–42.

Lee, D.C. & Southcott, R.V. (1979) Spiders and other arachnids of South Australia. Extract from *S.A. [South Australia] Year Book 1979*, pp. 1–15.

Lees, A.D. (1946) Chloride regulation and the function of coxal glands. *Parasitology* 37, 172–184.

Lees, A.D. (1947) Transpiration and structure of epicuticle in ticks. *Journal of Experimental Biology* 23, 379–410.

Lees, A.D. (1948) The sensory physiology of the sheep tick, *Ixodes ricinus* L. *Journal of Experimental Biology* 25, 145–207.

Lees, A.D. (1952) The role of cuticle growth in the feeding process of ticks. *Proceedings of the Zoological Society of London* 121, 759–772.

Lees, A.D. (1953a) Environmental factors controlling the evocation and termination of diapause in the fruit tree red spider mite *Metatetranychus ulmi* Koch (Acarina: Tetranychidae). *Annals of Applied Biology* 40, 449–486.

Lees, A.D. (1953b) The significance of light and dark phases in the photoperiodic control of diapause in *Metatetranychus ulmi* Koch. *Annals of Applied Biology* 40, 487–497.

Lees, A.D. (1961) On the structure of the egg shell in the mite *Petrobia latens* Müller (Acarina: Tetranychidae). *Journal of Insect Physiology* 6, 146–151.

Lees, A.D. & Beament, J.W.L. (1948) An egg waxing organ in ticks. *Quarterly Journal of Microscopical Science* 89, 291–332.

Legendre, R. (1978) Quelques progrès récents concernant l'anatomie des Araignées (systéme nerveux sympathique et appareil digestif). In: Merret, P. (ed.), *Arachnology, Seventh International Congress Symposium of the Zoological Society of London* No. 42, 379–388.

Lehtinen, P.T. (1980) A new species of Opilioacarida (Arachnida) from Venezuela. *Acta Biologica Venezuela, Caracas* 10, 205–214.

Lehtinen, P.T. (1981) New Holothyrina (Arachnida, Anactinotrichida) from New Guinea and South America. *Acarologia* 22, 3–13.

Liesering, R. (1960) Beitrag zum phytopathologischen Wirkungsmechanismus von *Tetranychus urticae* Koch. *Zeitschrift für Pflanzenkrankheiten, Pflanzenpathologie und Pflanzenschutz, Stuttgart* 67, 524–542.

Lindquist, E.E. (1963) A taxonomic revision of the genus *Hoploseius* Berlese (Acari: Blattisocidae). *Canadian Entomologist* 95, 1175–1185.

Lindquist, E.E. (1975) Associations between mites and other arthropods in forest floor habitats. *Canadian Entomologist* 107, 425–437.

Lindquist, E.E. (1976) Transfer of the Tarsochelidae to the Heterostigmata, and reassignment of the Tarsonemina and Heterostigmata to lower hierarchic status in the Prostigmata (Acari). *Canadian Entomologist* 108, 23–48.

Lindquist, E.E. (1983) Some thoughts on the potential for use of mites in biological control including a modified concept of "Parasitoids". In: Hoy, M.A.,

Cunningham, G.L. & Knutson, L. (eds), *Biological Control of Pests by Mites.* University of California, Berkeley, pp. 12–20.

Lindquist, E.E. (1984) Current theories on the evolution of major groups of Acari and on their relationships with other groups of Arachnida, with consequent implications for their classification. In: Griffiths, D.A. & Bowman, C.E. (eds), *Acarology VI,* Vol. 1. Ellis Horwood, Chichester, pp. 28–62.

Lindquist, E.E. (1985) External anatomy. In: Helle, W. & Sabelis, M.W. (eds), *Spider Mites, their Biology, Natural Enemies and Control.* Vol. 1A. Elsevier, Amsterdam, pp. 3–28.

Lindquist, E.E. (1986) The world genera of the Tarsonemidae (Acari: Heterostigmata): a morphological, phylogenetic, and systematic revision, with a reclassification of family-group taxa in the Heterostigmata. *Memoirs of the Entomological Society of Canada* No. 136, 1–517.

Lindquist, E.E. & Evans, G.O. (1965) Taxonomic concepts in the Ascidae, with a modified setal nomenclature for the idiosoma of the Gamasina (Acarina: Mesostigmata). *Memoirs of the Entomological Society of Canada* No. 47, 1–64.

Lindquist, E.E. & Kethley, J.B. (1975) The systematic position of the Heterocheylidae (Acariformes: Prostigmata). *Canadian Entomologist* 107, 887–898.

Lindquist, E.E. & Walter, D.E. (1989) *Antennoseius (Vitzthumia) janus* n. sp. (Acari: Ascidae), a mesostigmatic mite exhibiting adult female dimorphism. *Canadian Journal of Zoology* 67, 1291–1310.

Lister, A. (1984) Predation in an Antarctic micro-arthropod community. In: Griffiths, D.A. & Bowman, C.E. (eds), *Acarology VI,* Vol. 2. Ellis Horwood, Chichester, pp. 886–892.

Locke, M. (1974) The structure and function of the integument in insects. In: Rockstein, M. (ed.), *Physiology of Insecta,* 6. Academic Press, New York, pp. 123–213.

Lönnfors, F. (1930) Beiträge zur Morphologie der Analginen *Acta Zoologica Fennica* 8, 1–81.

Loots, G.C. & Ryke, P.A.J. (1966) A comparative quantitative study of the microarthropods in different types of pasture soil. *Zoologica Africana* 2, 167–192.

Loots, G.C. & Ryke, P.A.J. (1968) Two new genera of rhodacarid mites (Mesostigmata: Acari) from soil in the Ethiopian region. *Wetenskaplejke Bydraes van die Potchefstroomse Universiteit* No. 1, 1–16.

Lukoschus, F.S. & De Cock, A.W.A.M. (1971) Life cycle of *Melesodectes auricularis* Fain & Lukoschus (Glycyphagidae, Sarcoptiformes). *Tijdschrift voor Entomologie* 114, 173–183.

Lundblad, O. (1927) De Hydracarinen Schwedens I. *Zoologiska Bidrag från Uppsala* 11, 185–540.

Lundblad, O. (1929) Einiges über die Kopulation bei *Aturus scaber* und *Midea orbiculata. Zeitschrift für Morphologie und Ökologie der Tiere* 15, 474–480.

Lundblad, O. (1957) Zur Kenntnis sud- und mitteleuropäischer Hydrachnellen *Arkiv für Zoologi Stockholm* 10, 1–306.

Luxton, M. (1964) Some aspects of the biology of salt-marsh Acarina. Proceedings of the 1st International Congress of Acarology. *Acarology* 6 (fasc. hors sér.), 172–182.

Luxton, M. (1972) Studies on oribatid mites of a Danish beech wood soil, 1.

Nutritional biology. *Pedobiologia* 12, 434–463.

Luxton, M. (1981) Studies on the oribatid mites of a Danish beech wood soil. 4. Developmental biology. *Pedobiologia* 21, 312–340.

Luxton, M. (1982) Species of the genus *Crotonia* (Acari: Cryptostigmata) from New Zealand. *Zoological Journal of the Linnean Society of London* 76, 243–271.

Luxton, M. (1989) Mites of the family Hyadesidae (Acari: Astigmata) from New Zealand. *Zoological Journal of the Linnean Society of London* 95, 71–95.

McDaniel, B., Morihara, D. & Lewis, J.K. (1976) The family Teneriffiidae Thor, with a new species from Mexico. *Annals of the Entomological Society of America* 69, 527–537.

McEnroe, W.D. (1961) Guanine excretion by the two-spotted spider mite (*Tetranychus telarius*). *Annals of the Entomological Society of America* 54, 883–887.

McEnroe, W.D. (1963) The role of the digestive system in water balance of the two-spotted spider mite. In: Naegele, J.A. (ed.), *Advances in Acarology*, Vol. 1, pp. 225–231.

McEnroe, W.D. & Dronka, K. (1969) Eyes of the two-spotted spider mite. II. Behavioural analysis of the photoreceptors. *Annals of the Entomological Society of America* 62, 466–469.

McEnroe, W. & Dronka, K. (1971) Photobehavioural classes of the spider mite *Tetranychus urticae* (Acarina: Tetranychidae). *Entomology Experimental and Applied* 14, 420–424.

McLaughlin, P.J. (1968) Histochemical studies on the stylostome components and on feeding, digestion and excretion in the chigger *Eutrombicula splendens.* Unpublished PhD Thesis, University of Maryland, USA.

McMullen, H.L., Sauer, S.L. & Burton, R.L. (1976) Possible role of uptake of water vapour by ixodid tick salivary glands. *Journal of Insect Physiology* 22, 1281–1285.

MacQuitty, M. (1984) The feeding behaviour of two species of *Agauopsis* (Halacaroidea) from California. In: Griffiths, D.A. & Bowman, C.E. (eds), *Acarology VI*, Vol. 1. Ellis Horwood, Chichester, pp. 571–580.

Madge, D.S. (1964) The humidity reactions of mites. *Acarologia* 6, 566–591.

Mahunka, S. (1965) Identification key for the species of the family Scutacaridae (Acari: Tarsonemini). *Acta Zoologica Hungarica* 11, 353–401.

Mahunka, S. (1978) Schizoglyphidae fam n, and new taxa of Acaridae and Anoetidae (Acari: Acarida). *Acta Zoologica Hungarica* 24, 107–131.

Manton, S.M. (1977) *The Arthropoda. Habits, Functional Morphology and Evolution.* Clarendon Press, Oxford, 527pp.

Manton, S.M. (1978) Habits, functional morphology and the evolution of pycnogonids. *Zoological Journal of the Linnean Society of London* 63, 1–21.

Märkel, K. (1964) Die Euphthiracaridae Jacot, 1930 und ihre Gattungen (Acari, Oribatei). *Zoologische Verhundelingen, Leiden* 67, 4–78.

Martin, N.A. (1978) *Siteroptes* (*Siteroptoides*) species with *Pediculaster*-like phoretomorphs (Acari: Tarsonemida: Pygmephoridae) from New Zealand. *New Zealand Journal of Zoology* 5, 121–155.

Massee, A.M. (1928) The life-history of the black currant gall mite *Eriophyes ribis* (Westw.) Nal. *Bulletin of Entomological Research* 18, 297–309.

Mathur, S.N. & LeRoux, E.J. (1965) The musculature of the velvet mite *Allothrombium lerouxi. Annals of the Entomological Society of Quebec* 10, 33–61.

Mathur, S.N. & LeRoux, E.J. (1970) The reproductive organs of the velvet mite *Allothrombium lerouxi* (Trombidiformes: Trombidiidae). *Canadian Entomologist* 102, 144–157.

Megaw, M.W.J. & Robertson, H.A. (1974) Dopamine and noradrenaline in the salivary glands and brain of the tick, *Boophilus microplus*: effect of reserpine. *Experientia* 30, 1261–1262.

Mégnin, P. (1876) Mémoire sur l'organisation et la distribution zoologique des acariens de famille des Gamasides. *Journal de l'Anatomie et de la Physiologie Normales et Pathologiques des Animaux, Paris* 12, 316.

Meredith, J. & Kaufman, W.R. (1973) A proposed site of fluid secretion in the salivary gland of the ixodid tick, *Dermacentor andersoni. Parasitology* 67, 205–217.

Michael, A.D. (1884) *British Oribatidae*, Vol. 1. Ray Society, London.

Michael, A.D. (1888a). *British Oribatidae*, Vol. 2. Ray Society, London.

Michael, A.D. (1888b) The *Hypopus* question or the life-history of certain Acari. *Journal of the Linnean Society of London [Zool.]* 17, 371–394.

Michael, A.D. (1889) Observations on the special internal anatomy of *Uropoda krameri. Journal of the Royal Microscopical Society* 1889, 1–15.

Michael, A.D. (1890) On the variations of the female reproductive organs, especially the vestibule, in different species of Uropoda. *Journal of the Royal Microscopical Society* 1890, 142–152.

Michael, A.D. (1892) On the variations in the internal anatomy of the Gamasinae, especially in that of the genital organs and their mode of coition. *Transactions of the Linnean Society of London* (ser. 2), 5, 281–324.

Michael, A.D. (1896) The internal anatomy of *Bdella. Transactions of the Linnean Society of London* (ser. 2), 6, 477–528.

Michael, A.D. (1901) *British Tyroglyphidae*, Vol. 1. Ray Society, London.

Michael, A.D. (1903) *British Tyroglyphidae*, Vol. 2. Ray Society, London.

Michalska, K. & Boczek, J. (1990) Sexual behaviour of males attracted to quiescent deutonymphs in the Eriophyoidea (Acari). *Abstracts of the VIII International Congress of Acarology*, České Budějovice, p. 79.

Michener, C.D. (1946) The taxonomy and bionomics of some Panamanian trombidiid mites (Acarina). *Annals of the Entomological Society of America* 39, 349–380.

Micherdzinski, W. (1969) *Die Familie Parasitidae Oudemans 1901 (Acarina, Mesostigmata)*. Kraków (Państwowe Wuydawnictwo Naukowe), 690pp.

Millot, J. (1949) Classe des Arachnides (Arachnida). I. Morphologie générale et anatomie interne. In: Grassé, P-P. (ed.), *Traité de Zoologie*, Vol. 6. Masson et Cie., Paris, pp. 263–319.

Mills, L.R. (1973) Morphology of glands and ducts in the two-spotted spider mite, *Tetranychus urticae* Koch, 1836. *Acarologia* 15 (1973), 649–658.

Mills, L.R. (1974a) Structure of dorsal setae in the two-spotted spider mite, *Tetranychus urticae* Koch, 1836. *Acarologia* 15 (1973), 649-658.

Mills, L.R. (1974b) Structure of the visual system of the two-spotted spider mite, *Tetranychus urticae. Journal of Insect Physiology* 20, 795–808.

Mischke, U. (1981) Die Ultrastruktur der Lateralaugen und die Medianauges der Süssewassermilbe *Hydryphantes ruber* (Acarina: Parasitengona). *Entomologia Generalis* 7, 141–156.

Mitchell, R. (1957a) Locomotor adaptations of the family Hydryphantidae (Hydrachnellae, Acari). *Abhandlungen des Naturwissenschaftlichen Vereins zu Bremen* 35, 75–100.

Mitchell, R. (1957b) The mating behavior of pionid water-mites. *American Midland Naturalist* 58, 360–366.

Mitchell, R. (1957c) Major evolutionary lines in water mites. *Systematic Zoology* 6, 137–148.

Mitchell, R. (1962a) The musculature of a trombiculid mite, *Blankaartia ascoscutellaris* (Walch.). *Annals of the Entomological Society of America* 55, 106–119.

Mitchell, R. (1962b) The structure and evolution of water mite mouthparts. *Journal of Morphology* 110, 41–59.

Mitchell, R. (1964) The anatomy of an adult chigger mite, *Blankaartia ascoscutellaris*. *Journal of Morphology* 114, 373–391.

Mitchell, R. (1967) Host exploitation of two closely related water mites. *Evolution* [Lancaster, Pa.] 21, 59–75.

Mitchell, R. (1968) Site selection by larval water mites parasitic on the damselfly, *Cercion hieroglyphicum* Brauer. *Ecology* 49, 40–47.

Mitchell, R. (1970) Evolution of the blind gut in trombiculid mites. *Journal of Natural History* 4, 221–229.

Mitchell, R. (1972) The tracheae of water mites. *Journal of Morphology* 136, 327–336.

Mitchell, R. & Nadchatram, M. (1969) Schizeckenosy: the substitute for defecation in chigger mites. *Journal of Natural History* 3, 121–124.

Moorhouse, D.E. (1969) The attachment of some ixodid ticks to their natural hosts. In: Evans, G.O. (ed.), *Proceedings of the 2nd International Congress of Acarology*. Akadémiai Kiadó, Budapest, pp. 319–327.

Moorhouse, D.E. (1975) Studies on the feeding of larval *Argas persicus* Oken. *Zeitschrift für Parasitenkunde* 48, 65–71.

Moorhouse, D.E. & Tatchell, R.J. (1966) The feeding processes of the cattle tick *Boophilus microplus* (Canestrini): A study of host-parasite relations. *Parasitology* 56, 623–632.

Moritz, M. (1976) Revision der europäischen Gattung und Arten der Familie Brachychthoniidae (Acari, Oribatei). Teil 1. *Liochthonius* v.d. Hammen, *Verachthonius* nov. gen. und *Paraliochthonius* nov. gen. Teil 2. *Mixochthonius* Niedbala, 1972. *Neobrachychthonius* nov. gen., *Synchthonius* v.d. Hammen, 1959, *Poecilochthonius* Balogh, 1943, *Brachychthonius* Berlese, 1910, *Brachychochthonius* Jacot, 1938. *Mitteilungen aus dem Zoologische Musem in Berlin* 52, 27–136, 227–319.

Moser, J.C. & Cross, E.A. (1975) Phoretomorph: a new phoretic phase unique to the Pyemotidae. *Annals of the Entomological Society of America* 68, 1775–1798.

Moss, W.W. (1960) Description and mating behaviour of *Allothrombium lerouxi* new species (Acarina: Trombidiidae), a predator of small arthropods in Quebec apple orchards. *Canadian Entomologist* 92, 898–905.

Moss, W.W. (1962) Studies on the morphology of the trombidiid mite *Allothrombium lerouxi* Moss. *Acarologia* 4, 315–345.

Mothes, U. & Seitz, K-A. (1980) Licht- und elektronmikroskopische Untersuchungen zur Funktionsmorphologie von *Tetranychus urticae* (Acari: Tetranychidae). I Exkretionssysteme. *Zoologische Jahrbuecher. Abteilung für Anatomie und Ontogenie der Tiere* 104, 500–529.

Mothes, U. & Seitz, K-A. (1981a) The transformation of male sex cells of *Tetranychus urticae* (Acari: Tetranychidae) during passage from the testis to the oocytes: an electron microscopic study. *International Journal of Invertebrate Reproduction* 4, 81–94.

Mothes, U. & Seitz, K-A. (1981b) Fine structure and function of the prosomal glands of the two-spotted spider mite, *Tetranychus urticae* (Acari, Tetranychidae). *Cell Tissue Research* 221, 339–349.

Mothes, U. & Seitz, K-A. (1981c) Functional microscopic anatomy of the digestive system of *Tetranychus urticae* (Acari: Tetranychidae). *Acarologia* 22, 257-270.

Mothes, U. & Seitz, K-A. (1982) Action of the microbial metabolite and chitin synthesis inhibitor Nikkomycin on the mite *Tetranychus urticae* (Acari: Tetranychidae); an electron microscope study. *Pesticide Science* 13, 426–441.

Muraoka, M. & Ishibashi, N. (1976) Nematode-feeding mites and their feeding behaviour. *Applied Entomology and Zoology, Tokyo* 11, 1–7.

Murley, M.R. (1951) Seeds of Cruciferae of northeastern North America. *American Midland Naturalist* 46, 1–81.

Murphy, P.W. (1955) Ecology of the fauna of forest soils. In: Kevan, D.K. McE. (ed.), *Soil Zoology*. Butterworth Scientific, London, pp. 99–124.

Murray, R.A. & Solomon, M.G. (1978) A rapid technique for analysing predators by electrophoresis. *Annals of Applied Biology* 90, 7–10.

My-Yen, L.T., Wada, Y., Matsumato, K. & Kuwahara, Y. (1980) Pheromone study of acarid mites. VI. Demonstration and isolation of an aggregation pheromone in *Lardoglyphus konoi* Sasa et Asanuma. *Japanese Journal of Sanitary Zoology* 31, 249–254.

Nagelkerke, C.J. & Sabelis, M.W. (1991) Precise sex-ratio control in the pseudo-arrheno tokous phytoseiid mite, *Typhlodromus occidentalis*, Nesbitt. In: Schuster, R. & Murphy, P.W. (eds), *The Acari. Reproduction, Development and Life-History Strategies.* Chapman & Hall, London, etc., pp. 193–207.

Nalepa, A. (1884) Die Anatomie der Tyroglyphen. *Sitzungsberichte der Akamemie der Wissenschaften, Wien* 90, 197–228.

Nalepa, A. (1885) Die Anatomie der Tyroglyphen. *Sitzungsberichte der Akamemie der Wissenschaften, Wien* 92, 116–167.

Nalepa, A. (1887) Die Anatomie der Phytopten. *Sitzungsberichte der Akamemie der Wissenschaften, Wien* 96, 1–51.

Nathanson, M.E. (1967) Comparative fine structure of sclerotized and unsclerotized integument of the rabbit tick, *Haemaphysalis leporispalustris* (Ixodidae). *Annals of the Entomological Society of America* 60, 1125–1135.

Naudo, M.H. (1963) Acariens notostigmata de l'Angola. *Das Publicações Culturais da Companhia de Diamantes de Angola, Lisbon* No. 63, 13–24.

Nault, L.R. & Styer, W.E. (1969) The dispersal of *Aceria tulipae* and three other grass-infesting eriophyid mites in Ohio. *Annals of the Entomological Society of America* 62, 1446–1455.

Needham, G.R. & Teel, P.D. (1986) Water balance by ticks between bloodmeals. In: Sauer, J.R. & Hair J.A. (eds), *Morphology, Physiology and Behavioural Biology of Ticks.* Ellis Horwood, Chichester, pp. 100–151.

Nelson-Rees, W.A., Hoy, M.A. & Roush, R.T. (1980) Heterochromatinization, chromatin elimination and haploidization in the parahaploid mite *Metaseiulus occidentalis. Chromosoma* 77, 262–276.

Nevill, E.M. (1964) The role of carbon dioxide as a stimulant and attractant to the sand tampan, *Ornithodoros savignyi* (Andouin). *Onderstepoort Journal of Veterinary Research* 31, 59–68.

Neville, A.C. (1975) *Biology of Arthropod Cuticle.* Springer Verlag, Heidelberg.

Newell, I.M. (1947a) A systematic and ecological study of the Halacaridae of eastern North America. *Bulletin of the Bingham Oceanographic Collection, Yale* 10 (3), 1–232.

Newell, I.M. (1947b) Studies on the morphology and systematics of the family Halarachnidae Oudemans 1906 (Acari, Parasitoidea). *Bulletin of the Bingham Oceanographic Collection, Yale* 10 (4), 235–266.

Newell, I.M. (1957) Studies on the Johnstonianidae (Acari, Parasitengona). *Pacific Science* 11, 396–466.

Newell, I.M. (1963) Feeding habits in the genus *Balaustium* (Acarina, Erythraeidae) with special reference to attacks on man. *Journal of Parasitology* 49, 498–502.

Newell, I.M. & Tevis, L. Jr. (1960) *Angelothrombium pandorae* n.g., n.sp. (Acari, Trombidiidae), and notes on the biology of the giant red velvet mites. *Annals of the Entomological Society of America* 53, 293–304.

Newstead, R. & Duvall, H.M. (1918) Bionomic, morphological and economic report of acarids of stored grain and flour. *Royal Society Reports of the Grain Pests (War) Committee* No. 2, 1–48.

Niedbala, W. (1976) Brachychthoniidae Polski (Acari, Oribatei). *Monografie Fauny Polski.* 6, 1–144 [in Polish].

Nikonov, G.I. (1958) Mouthparts of the tick *Ixodes persulcatus. Trudy̆ Izhevskago Gosudarstvennogo Meditsinskogo Instituta, Izhevsk* 17, 262–273 [in Russian].

Nordenskiöld, E. (1908) Zur anatomie und histologie von *Ixodes reduvius. Zoologische Jahrbuecher, Abteilung für Anatomie und Ontogenie der Tiere* 25, 637–674.

Norton, R.A. (1975) Elliptochthoniidae, a new mite family (Acari; Oribatei) from mineral soil in California *New York Entomological Society Journal* 83, 209–216.

Norton, R.A. (1977) An example of phoretomorphs in the mite family Scutacaridae. *Journal of the Georgia Entomology Society* 12, 185–186.

Norton, R.A. (1980) Observations on phoresy by oribatid mites. *International Journal of Acarology* 6, 121–130.

Norton, R.A. (1984) Monophyletic groups in the Enarthronota. In: Griffiths, D.A. & Bowman, C.E. (eds), *Acarology VI,* Vol. 1. Ellis Horwood, Chichester, pp. 233–240.

Norton, R.A. (1985) Aspects of the biology and systematics of soil arachnids, particularly saprophagous and mycophagous mites. *Quaestiones Entomologicae* 21, 523–541.

Norton, R.A. (1988) Parthenogenesis in Nothridae and related groups. In: Channa Basavanna, G.P. & Viraktamath, C.A. (eds), *Progress in Acarology*, Vol. 1. Oxford & IBH, New Delhi, etc., pp. 271–277.

Norton, R.A. (1990) Acarina: Oribatida. In: Dindal, D.L. (ed.), *Soil Biology Guide*. John Wiley, New York & Chichester, pp. 779–803.

Norton, R.A. & Ide, G.S. (1974) *Scutacarus baculitarsus agaricus* n. subsp. (Acarina: Scutacaridae) from commercial mushroom houses with notes on phoretic behaviour. *Journal of the Kansas Entomological Society* 47, 527–534.

Norton, R.A. & Palmer, S.C. (1991) The distribution, mechanisms and evolutionary significance of parthenogenesis in oribatid mites. In: Schuster, R. & Murphy, P.W. (eds), *The Acari. Reproduction, Development and Life-History Strategies*. Chapman & Hall, London, pp. 107–136.

Norton, R.A., Bonamo, P.M., Grierson, J.D. & Shear, W.A. (1988) Fossil mites from the Devonian of New York State. In: ChannaBasavanna, G.P. & Viraktamath, C.A. (eds), *Progress in Acarology*, Vol. 1. Oxford & IBH, New Delhi, etc., pp. 271–277.

Nucifora, A. (1963) Osservazioni sull riproduzione di *Hemitarsonemus latus* (Banks) (Acarina, Tarsonemidae). *Atti della Accademia Nazionale Italiana di Entomologia Bologna* 10, 142–153.

Nuttall, G.H.F. (1911) Types of parasitisim in ticks, illustrated by a diagram together with some remarks upon longevity in ticks. *Parasitology* 4, 175–182.

Nuttall, G.H.F. & Warburton, C. (1911) *Ticks. A Monograph of the Ixodoidea*, Part II. The genus *Ixodes*. Cambridge University Press, pp. 105–348.

Nuttall, G.H.F. & Warburton, C. (1915) *Ticks. A Monograph of the Ixodoidea*, Part III. The genus *Haemaphysalis*. Cambridge University Press, pp. 349–550.

Nuttall, G.H.F., Cooper, W.F. & Smedley, R.D. (1905) The buccal apparatus of a tick (*Haemaphysalis punctata* C. and F.). *Report of the 75th Meeting of the British Association, S. Africa*, pp. 439–441.

Nuttall, G.H.F., Cooper, W.F. & Robinson, L.E. (1908a) On the structure of the spiracles of a tick – "*Haemaphysalis punctata*", Canestrini and Fanzago. *Parasitology* 1, 347–351.

Nuttall, G.H.F., Warburton, C., Cooper, W.F. & Robinson, L.E. (1908b) *Ticks. A Monograph of the Ixodoidea*, Part I. Argasidae. Cambridge University Press, pp. 1–104.

Nuzzaci, G. (1979a) Studies on the structure and function of mouth parts of eriophyid mites. In: Rodriguez, J.G. (ed.), *Recent Advances in Acarology*, Vol. 2. Academic Press, New York, etc., pp. 411–415.

Nuzzaci, G. (1979b) Contributo alla conscenza dello gnatosoma delgi Eriofidi (Acarina: Eriophyoidea). *Entomologia* [Bari] 15, 73–101.

Nuzzaci, G. & de Lillo, E. (1989) Contributo all conoscenza dello gnatosoma degli Acari Tenuipalpidi (Tetranychoidea: Tenuipalpidae). *Entomologica* [Bari] 24, 5–32.

Nuzzaci, G. & de Lillo, E. (1991) Fine structure and function of the mouthparts involved in the feeding mechanisms in *Cenopalpus pulcher* (Canestrini & Fanzago). In: Schuster, R. & Murphy, P.W. (eds), *The Acari. Reproduction,*

Development and Life-History Strategies. Chapman & Hall, London, pp. 367–376.

Nuzzaci, G. & Solinas, M. (1984) An investigation into sperm formation, transfer, storage and utilization in eriophyid mites. In: Griffiths, D.A. & Bowman, C.E. (eds), *Acarology VI*, Vol. 1. Ellis Horwood, Chichester, pp. 491–503.

Obenchain, F.D. (1974) Structure and anatomical relationships of the synganglion in the American dog tick, *Dermacentor variabilis* (Acari: Ixodidae). *Journal of Morphology* 142, 205–224.

Obenchain, F.D. & Oliver, J.H. (1975) Neurosecretory system of the American dog tick, *Dermacentor variabilis* (Acari: Ixodidae). 2. Distribution of secretory cell types, axonal pathways and putative neurohemal–neuroendocrine associations; comparative, histological and anatomical implications. *Journal of Morphology* 145, 269–294.

Obenchain, F.D. & Oliver, J.H. (1976) Peripheral nervous system of the ticks, *Amblyomma tuberculatum* Marx and *Argas radiatus* Raillet (Acari: Ixodoidea). *Journal of Parasitology* 62, 811–817.

Oboussier, H. (1939) Beiträge zur Biologie und Anatomie der Wohnmilben. *Zeitschrift für Angewandte Entomologie* 26, 253–296.

OConnor, B.M. (1979) A review of the family Heterocoptidae. In: Rodriguez, J.G. (ed.), *Recent Advances in Acarology*. Academic Press, New York, pp. 429–433.

OConnor, B.M. (1982a) Astigmata. In: Parker, S.P. (ed.), *Synopsis and Classification of Living Organisms*, Vol. 2. McGraw-Hill, New York, pp. 146–169.

OConnor, B.M. (1982b) Evolutionary ecology of astigmatic mites. *Annual Review of Entomology* 27, 385–409.

OConnor, B.M. (1984) Phylogenetic relationships among higher taxa in the Acariformes, with particular reference to the Astigmata. In: Griffiths, D.A. & Bowman, C.E. (eds), *Acarology VI*, Vol. 1. Ellis Horwood, Chichester, pp. 19–27.

O'Dowd, D.J. & Wilson, M.F. (1989) Leaf domatia and mites on Australian plants: ecological and evolutionary implications. *Biological Journal of the Linnean Society* 37, 191–236.

Oldfield, G.N., Hobza, R.F. & Wilson, N.S. (1970) Discovery and characterization of spermatophores in the Eriophyoidea (Acari). *Annals of the Entomological Society of America* 63, 520–526.

Oliver, J.H. Jr. (1971) Parthenogenesis in mites and ticks (Arachnida: Acari). *American Zoologist* 11, 283–299.

Oliver, J.H. Jr. (1977) Cytogenetics of mites and ticks. *Annual Revue of Entomology* 22, 407–429.

Oliver, J.H. Jr. (1982) Tick reproduction: sperm development and cytogenetics. *Current Themes in Tropical Science* 1, 245–275.

Oliver, J.H. Jr. & Brinton, L.P. (1973) Cytogenetics of ticks (Acari: Ixodoidea). 7. Spermatogenesis in the Pacific coast tick, *Dermacentor occidentalis* Marx (Ixodidae). *Journal of Parasitology*. 58 (1972), 365–379.

Oliver, J.H. Jr. & Delfin, E.D. (1967) Gynandromorphism in *Dermacentor occidentalis* (Acari: Ixodidae). *Annals of the Entomological Society of America* 60, 1119–1121.

Oliver, J.H. Jr. & Krantz, G.W. (1963) *Macrocheles rodriguezi*, a new species of mite from Kansas (Acarina: Macrochelidae) with notes on its life cycle and behavior. *Acarologia* 5, 519–525.

Olivier, P.G. & Ryke, P.A.J. (1965) Seasonal fluctuations of the mesofauna in soil under kikuyu grass. *Memórias do Instituto de Investigação Científica de Moçambique* 7A, 235–279.

Otieno, D.A., Hassanali, A., Obenchain, F.D., Sternberg, A. & Galun, R. (1985) Identification of guanine as an assembly pheromone in ticks. *Insect Science Application* 6, 667–670.

Oudemans, A.C. (1906) Nieuwe classificatie der Acari. *Entomologische Berichtew, Amsterdam* 2, 43–46.

Oudemans, A.C. (1914) Acarologisches aus Maulwurfsnestern. *Archiv für Naturgeschichte, Berlin* 79A (1913), Heft, 10, 1–69.

Oudemans, A.C. (1917) Notizen über Acari, 25 Reihe (Trombidiidae, Oribatidae, Phthiracaridae). *Archiv für Naturgeschichte, Berlin* 82A (1916), Heft 6, 1–84.

Oudemans, A.C. (1923) Studie over de sedert 1877 ontworpen Systemen der Acari; Nieuwe classificatie; Phylogenetische Beschouwingen. *Tijdschrift voor Entomologie* 66, 49–85.

Oudemans, A.C. (1928) Fauna Buruana: Acari. *Teubia* 7 (Suppl. No. 2), 37–100.

Oudemans, A.C. (1931a) Over zijne nieuwste ontdekkingen over de ligging der stigmata bij eenige Acari. *Tijdschrift voor Entomologie* 74, Verslagen, XIX–XXVI.

Oudemans, A.C. (1931b) Acarologische Aanteekeningen CXI. *Entomologische Berichtew Amsterdam* 8, 312–331.

Oudemans, A.C. (1936a) *Kritisch Historisch Oversicht der Acarologie. Derde Gedeelte, 1805–1850.* Vol. A. E.J. Brill, Leiden, 430pp.

Oudemans, A.C. (1936b) Neues über Anystidae (Acari). *Archiv für Naturgeschichte, Berlin* N.F. 5, (3), 364–446.

Oudemans, A.C. (1937) Mededeelingen. *Tijdschrift voor Entomologie* 80, Verslagen, IV–XVI.

Parker, S.P. (ed.) (1982), *Synopsis and Classification of Living Organisms,* Vol. 2. McGraw-Hill, New York.

Pater de Kroot, I., Van Brouswijk, J.E.M.H. & Fain, A. (1979) Redescription and biology of the mite *Acanthophthirius polonicus* Hartlinger, 1978, parasitic on the pond bat, *Myotis dasycneme* (Bioe, 1825) (Prostigmata: Myobiidae). *International Journal of Acarology* 5, 291–298.

Pauly, F. (1952) Die "Copula" der Oribatiden. *Naturwissenschaften* 39, 572–573.

Pauly, F. (1956) Zur Biologie einiger Belbiden (Oribatei, Moosmilben) und zur Funktion ihrer pseudostigmatischen Organe. *Zoologische Jahrbuecher. Abteilung für Systematik, Ökologie und Geographie der Tiere* 84, 275–328.

Penman, D.R. & Cone, W.W. (1974a) Structure of cuticular lyrifissures in *Tetranychus urticae. Annals of the Entomological Society of America* 67, 1–4.

Penman, D.R. & Cone, W.W. (1974b) Role of web, tactile stimuli, and female sex pheromone in attraction of male two-spotted spider mites to quiescent female deutonymphs. *Annals of the Entomological Society of America* 67, 179–182.

Philip, C.B. (1963) Ticks as purveyors of animal ailments: a review of pertinent data and recent contributions. In: Neagele, J.A. (ed.), *Advances in Acarology,*

Vol. 1. Academic Press, New York, pp. 285–325.

Phillis, W.A. & Comroy, H.L. (1977) The microanatomy of the eye of *Amblyomma americanum* (Acari: Ixodidae) and resultant implications of its structure. *Journal of Medical Entomology, BP Bishop Museum, Honolulu* 13, 686–698.

Piersig, R. (1895) Beiträge zur Systematik und entwicklungsgeschichte der Süsswassermilben. *Zoologischer Anzeiger* 18, 19–25.

Piesman, J. (1989) Transmission of Lyme disease spirochetes. *Experimental and Applied Acarology* 7, 71–80.

Piffl, E. (1991) A new interpretation of the epimeral theory of Grandjean. In: Schuster, R. & Murphy, P.W. (eds), *The Acari. Reproduction, Development and Life-History Strategies.* Chapman & Hall, London, pp. 353–354.

Pomerantsev, B.I. (1948) Fundamental directions of Ixodoidea (Acarina) evolution. *Parasitologicheskii Sbornik, Leningrad* 10, 5–19 [in Russian].

Pomerantsev, B.I. (1950) *Arachnoidea, Ticks (Ixodidae), Fauna of the U.S.S.R.*, N.S., No. 41, 4, pp. 1–224. Academy of Sciences, Moscow [in Russian].

Porres, M.A., McMurtry, J.A. & Marsh, R.B. (1975) Investigations of leaf sap feeding by three species of phytoseiid mites by labelling with radioactive phosphoric acid ($H_3{}^{32}PO_4$). *Annals of the Entomological Society of America* 68, 871–872.

Pospelova-Shtrom, M.V. (1946) On the systematics of the Argasidae (with descriptions of two new subfamilies, three new tribes and one new genus). *Meditsinskaya Parazitologiya i Parazitarhŷe Bolezni, Moskva* 15 (3), 47–58 [in Russian].

Pospelova-Shtrom, M.V. (1969) On the system of classification of ticks of the family Argasidae, Can., 1890 *Acarologia* 11, 1–22.

Prasad, V. (1967) Biology of the predatory mite *Phytoseiulus macropilis* in Hawaii (Acarina: Phytoseiidae). *Annals of the Entomological Society of America* 60, 905–908.

Prasad, V. & Cook, D.R. (1972) The taxonomy of water mite larvae. *Memoirs of the American Entomological Institute* 18, 1–326.

Prasse, J. (1967a) Zur Anatomie und Histologie der Acaridae mit besonderer Berüchsichtigung von *Caloglyphus berlesei* (Michael 1903) und *C. michaeli* (Oudemans 1924). I. Das Darmsystem. *Wissenschaftliche Zeitschrift der Universität, Halle* 16, 789–812.

Prasse, J. (1967b) Zur Anatomie und Histologie der Acaridae mit besonderer Berücksichtigung von *Caloglyphus berlesei* (Michael 1903) und *C. michaeli* (Oudemans 1924). II Das Exkretionssystem und das Bindegewebe. *Wissenschaftliche Zeitschrift der Universität, Halle* 16, 963–970.

Prasse, J. (1968a) Untersuchungen über Oogenese, Befruchtung, Eifurchung und spermatogenese bei *Caloglyphus berlesei* Michael 1903 und *C. michaeli* Oudemans 1924 (Acari, Acaridae). *Biologisches Zentralblatt* 87, 757–775.

Prasse, J. (1968b) Zur Anatomie und Histologie der Acaridae mit besonderer Berücksichtigung von *Caloglyphus berlesei* (Michael 1903) und *C. michaeli* (Oudemans 1924). III. Die Drüsen und drüsenähnlichen Gebilde der Podocephalkanal. *Wissenschaftliche Zeitschrift der Universität, Halle* 17, 629–646.

Prasse, J. (1970) Zur Anatomie und Histologie der Acaridae mit besonderer Berücksichtigung von *Caloglyphus berlesei* (Michael 1903) und *C. michaeli*

(Oudemans 1924). IV. Das Genitalsystem. *Wissenschaftliche Zeitschrift der Universität, Halle* 19, 93–116.

Pritchard, A.E. & Baker, E.W. (1952) A guide to the spider mites of deciduous fruit trees. *Hilgardia* 21, 253–287.

Pritchard, A.E. & Baker, E.W. (1955) A revision of the spider mite family Tetranychidae. *Memoirs of the Pacific Coast Entomological Society, San Francisco* 2, 1–472.

Pritchard, A.E. & Baker, E.W. (1958) The false spider mites (Acarina: Tenuipalpidae). *University of California Publications on Entomology, Berkeley* 14, 175–274.

Pugh, P.J.A. & King, P.E. (1985) Feeding in intertidal Acari. *Journal of Experimental Marine Biology and Ecology* 94, 269–280.

Pugh, P.J.A., King, P.E. & Fordy, M.R. (1987a) The structure and probable function of the peritreme in intertidal Gamasina (Acarina; Mesostigmata). *Zoological Journal of the Linnean Society* 89, 393–407.

Pugh, P.J.A., King, P.E. & Fordy, M.R. (1987b) Structural features associated with respiration in some intertidal Uropodina (Acarina: Mesostigmata). *Journal of Zoology* [Lond.] 211, 107–120.

Pugh, P.J.A., King, P.E. & Fordy, M.R. (1987c) A comparison of the structure and function of the cerotegument in two species of Cryptostigmata (Acarina). *Journal of Natural History* 21, 603–616.

Pugh, P.J.A., King, P.E. & Fordy, M.R. (1988) The spiracle of *Ixodes ricinus* (L.) (Ixodidae: Metastigmata: Acarina) – a passive diffusion barrier for water vapour. *Zoological Journal of the Linnean Society* 93, 113–131.

Pugh, P.J.A., King, P.E. & Fordy, M.R. (1990) Spiracular transpiration in ticks: a passive diffusion barrier in three species of Ixodidae. *Journal of Natural History* 24, 1529–1547.

Pugh, P.J.A., Evans, G.O., Fordy, M.R. & King, P.E. (1991) The functional morphology of the respiratory system of the Holothyrida (=Tetrastigmata) Acari: Anactinotrichida. *Journal of Zoology* [Lond.] 225, 153–172.

Pulpán, J. & Verner, P.H. (1965) Control of tyroglyphid mites in stored grain by the predatory mite *Cheyletus eruditus* (Schrank). *Canadian Journal of Zoology* 43, 417–432.

Purvis, G. (1982) The soil arthropod fauna (Acari and Collembola) of a coastal locality in southeast Ireland. *Journal of Life Sciences Royal Dublin Society* 3, 379–396.

Rack, G. (1967) Untersuchungen uber die Biologie von *Dolichocybe* Krantz, 1957 und Beschreibung von zwei neuen Arten (Acarina, Pyemotidae). *Mitteilungen aus der Hamburgischen Zoologischen Museum und Institut, Hamburg* 64, 29–42.

Radford, C.D. (1950) Systematic check list of mite genera and type species. *International Union of Biological Sciences* Ser C, No. 1, 1–232.

Radinovsky, S. (1965) The biology and ecology of granary mites of the Pacific Northwest, 3. Life history and development of *Leiodinychus krameri* (Acarina: Uropodidae). *Annals of the Entomological Society of America* 58, 259–267.

Radovsky, F.J. (1967) The Macronyssidae and Laelapidae (Acarina, Mesostigmata) parasitic on bats. *University of California Publications in Entomology* 46, 1-288.

Radovsky, F.J. (1969) Adaptive radiation in parasitic Mesostigmata. *Acarologia* 11, 450–483.

Rafferty, D.E. & Gray, J.S. (1987) The feeding behaviour of *Psoroptes* spp. on rabbits and sheep. *Journal of Parasitology* 73, 901–906.

Raikhel, A.S. (1979a) The intestine. In: Balashov, Yu. S. (ed.), *An Atlas of Tick Ultrastructure.* Nauka, Leningrad [in Russian]. (Translation in: *Miscellaneous Publications of the Entomological Society of America* (1983), 59–97.)

Raikhel, A.S. (1979b) Excretory system. In: Balashov, Yu. S. (ed.), *An Atlas of Ixodid Tick Ultrastructure.* Nauka. Leningrad [in Russian]. (Translation in: *Miscellaneous Publications of the Entomological Society of America* (1983), 129–146.)

Raw, F. (1957) Origin of the chelicerates. *Journal of Paleontology* 31, 139–192.

Regenfuss, H. (1973) Beinreduktion und Verlagerundes Kopulationsapparates in der Milbenfamilie Podapolipidae, ein Beispiel fur verhaltensgesteuerte Evolution morphologischer Strukturen. *Zeitschrift für Zoologische Systematik und Evolutionforschung* 11, 173–195.

Regev, S. & Cone, W.W. (1975) Evidence of farnesol as a male sex attractant of the two-spotted spider mite, *Tetranychus urticae* Koch (Acarina: Tetranychidae). *Environmental Entomology* 4, 307–311.

Remane, A. (1952) *Die grundlagen des natürlichen Systems, der vergleichenden Anatomie und der Phylogenerik; Theoretische Morphologie und Systematik I.* Akademische Verlagsgesellschaft, Geest u. Portig, Leipzig.

Remane, A., Storch, V. & Welsch, U. (1976) *Systematische Zoologie – Stamme des Tierreiches.* G. Fisher Verlag, Stuttgart (cited in Alberti, 1979).

Reuter, E. (1909) Zur Morphologie und Ontogenie der Acariden mit besonderer Berucksichtigung von *Pediculopsis graminum. Acta Societatis Scientiarum Fennicae* 36, 1–288.

Rhode, C.J. (1964) Some techniques in the preparation of stained whole mounts and serial sections of mite embryos and adults. *Acarologia* 6, fasc. h. s., 208-214.

Robaux, P. (1974) Recherches sur le développment et la biologie des acariens 'Thrombidiidae'. *Memoires Muséum National d'Histoire Naturelle, Paris* Sér. A. 85, 1–186.

Robaux, P. (1977) Observations sur quelques Actinedida (=Prostigmates) du sol d'Amérique du Nord. VI. Sur deux espèces nouvelles de Labidostommidae (Acari). *Acarologia* 18, 442–461.

Robinson, I. (1953) The hypopus of *Hericia hericia* (Kramer), Acarina, Tyroglyphidae. *Proceedings of the Zoological Society of London* 123, 267–272.

Robinson, L.E. (1926) *Ticks. A monograph of the Ixodoidea, Pt. IV The genus* Amblyomma. Cambridge University Press, pp. 1–302.

Robinson, L.E. & Davidson, J. (1913a) The anatomy of *Argas persicus* (Oken, 1818). Part I. *Parasitology* 6, 20–48.

Robinson, L.E. & Davidson, J. (1913b) The anatomy of *Argas persicus* (Oken, 1818). Part II. *Parasitology* 6, 217–256.

Robinson, L.E. & Davidson, J. (1914) The anatomy of *Argas persicus* (Oken, 1818). Part III. *Parasitology* 6, 382–424.

Rock, G.C., Monroe, R.J. & Yeargan, D.R. (1976) Demonstration of a sex pheromone in the predaceous mite *Neoseiulus fallacis. Environmental Entomology* 5, 264–266.

Rockett, C.L. (1980) Nematode predation by oribatid mites. *International Journal of Acarology* 6, 219–224.

Rockett, C.L. & Woodring, J.P. (1972) Comparative studies of acarine limb regeneration, apolysis and ecdysis. *Journal of Insect Physiology* 18, 2319–2336.

Roesler, R. (1934) Histologische, Physiologische und Serologische Untersuchungen über die Verdauung bei der Zeckengattung *Ixodes* Latr. *Zeitschrift für Morphologie und Ökologie der Tiere* 28, 297–317.

Romeis, B. (1968) *Mikroskopische Technik*, 16 Aufl. Oldenbourg Verlag., München, 757pp.

Roshdy, M.A. (1961) Comparative internal anatomy of subgenera of *Argas* ticks (Ixodoidea. Argasidae). 1. Subgenus *Carios*: *Argas vespertilionis* (Latreille, 1802). *Journal of Parasitology* 47, 987–994.

Roshdy, M.A. (1972) The genus *Persicargas* (Ixodoidea, Argasidae, *Argas*) 15. Histology and histochemistry of the salivary glands of *A.* (*P.*) *persicus* (Oken) *Journal of Medical Entomology, BP Bishop Museum, Honolulu* 9, 143–148.

Roshdy, M.A. (1974) Structure of the nymphal spiracle and its formation in the pharate adult *Haemaphysalis (Kaiseriana) longicornis* Neumann (Ixodoidea: Ixodidae). *Zeitschrift für Parasitenkunde* 44, 1–14.

Roshdy, M.A. & Hefnawy, T. (1973) The functional morphology of *Haemaphysalis* spiracles (Ixodoidea: Ixodidae). *Zeitschrift für Parasitenkunde* 42, 1–10.

Roshdy, M.A., Foelix, R.F. & Axtell, R.C. (1972) The subgenus *Persicargas* (Ixodoidea: Argasidae: *Argas*). 16. Fine structure of Haller's organ and associated tarsal setae of adult *A. (P.) arboreus* Kaiser, Hoogstraal and Kohls. *Journal of Parasitology* 58, 805–816.

Roshdy, M.A., Banaja, A.A. & Wassef, H.Y. (1982) The subgenus *Persicargas* (Ixodoidea, Argasidae, *Argas*). 34. Larval respiratory system structure and spiracle formation in pharate nymph *Argas (P.) arboreus*. *Journal of Medical Entomology, BP Bishop Museum, Honolulu* 19, 665–670.

Roshdy, M.A., Hoogstraal, H., Banaka, A.A. & El-Shoura, S.M. (1983) *Nuttalliella namaqua* (Ixodoidea: Nuttalliellidae); spiracle structure and surface morphology. *Zeitschrift für Parasitenkunde* 69, 817–821.

Royce, L.G. & Krantz (1989) Observations on pollen processing by *Pneumolaelaps longanalis* (Acari: Laelapidae), a mite associate of bumblebees. *Experimental and Applied Acarology* 7, 161–165.

Rudolph, D. & Knülle, W. (1974) Site and mechanism of water vapour uptake from the atmosphere in ixodid ticks. *Nature* [Lond.] 249, 84–85.

Rudolph, D. & Knülle, W. (1979) Mechanisms contributing to water balance in non-feeding ticks and theoretical implications. In: Rodriguez, J.G. (ed.), *Recent Advances in Acarology*. Vol. 1. Academic Press, London, pp. 375–383.

Rudzinska, M.A., Spielman, A., Lewengrub, S., Piesman, J. & Karakashian, S. (1982) Penetration of the peritrophic membrane of the tick by *Babesia microti*. *Cell & Tissue Research* 221, 471–481.

Ruser, M.Z. (1933) Beiträge zur Kenntnis der Chitins und der Muskultur der Zecken (Ixodidae). *Zeitschrift für Morphologie und Ökologie der Tiere* 27, 199–261.

Sabelis, M.W. (1981) Biological control of two-spotted spider mites using phytoseiid predators. Part 1. Modelling the predatory-prey interaction at the individual level. *Agricultural Research Reports* 910. Centre for Agricultural Publishing and Documentation, Wageningen, The Netherlands, 242pp.

Sabelis, M.W. & Dicke, M. (1985) Long range dispersal and searching behaviour. In: Helle, W. & Sabelis, M.W. (eds), *Spider Mites, their Biology, Natural Enemies and Control*, Vol. 1B. Elsevier, Amsterdam, pp. 141–160.

Sabelis, M.W., Afman, B.P. & Slim, P.J. (1984) Location of distant spider mite colonies by *Phytoseiulus persimilis*: localization and extraction of a kairomone. In: Griffiths, D.A. & Bowman, C.E. (eds), *Acarology VI*, Vol. 1. Ellis Horwood, Chichester, pp. 431–440.

Sachs, H. (1951) Zur Morphologie von *Acarapis*. 1. Bau und funktion der Mundwerkzeuge der Tracheenilbe *Acarapis woodi* Rennie, 1921. *Zeitschrift für Bienenforschung* 1, 103–112.

Saito, Y. (1985) Life types of spider mites. In: Helle, W. & Sabelis, M.W. (eds), *Spider Mites, their Biology, Natural Enemies and Control*, Vol. 1A. Elsevier, Amsterdam, pp. 253–264.

Samšiňák, K. (1971) Die auf *Carabus*-arten (Coleoptera, Adephaga) der Palaearktischen Region lebenden Milben der Unterordnung Acariformes (Acari); ihre Taxonomie und Bedeutung für die Lösung zoogeographischer, entwicklungsge-schichtlicher und parasitophyletischer Fragen. *Entomologische Abhandlunden und Berichte aus der Staatlichen Museum für Tierkunde in Dresden* [1970–71], 145–234.

Samšiňák, K. (1977) *Rosensteinia hileri* sp.n., and a revision of the taxonomy and status of the family Rosensteiniidae (Acari: Sarcoptiformes). *Acta Entomologica Bohemoslovaca Praha* 74, 419–425.

Samšiňák, K. & Dusbàbek, F. (1971) Mesostigmata. In: Daniel, M. & Černý, V. (eds), *Key to the Fauna of the CSSR*, Vol. 4. Akademia, Prague, pp. 313–352.

Sardar, M.A. & Murphy, P.W. (1987) Feeding tests of grassland soil-inhabiting gamasine predators. *Acarologia* 28, 117–121.

Sasa, M. (1964) Special problems of mites in stored food and drugs in Japan. *Acarologia* 6, 390–391.

Savory, T.H. (1977) *Arachnida*. Academic Press, London, New York, etc.

Schaarschmidt, L. (1959) Systematik und Oekologie der Tarsonemiden. In: Stammer, H-J. (ed.), *Beiträge zur Systematik und Ökologie Mitteleuropäischer Acarina* 1, Abschn. 5, 713–823. Akademische Verlags., Geest & Portig, Leipzig.

Schaller, F. (1954) Die indirekte Spermatophorenübertrangnung und ihre Probleme.*Forschungen und Fortschritte, Korrespondenzblatt der Deutschen Wissenschaft und Technik, Berlin* 28, 321–326.

Schaller, F. (1971) Indirect sperm transfer by soil arthropods. *Annual Review of Entomology* 16, 407–446.

Schaller, F. (1979) Significance of sperm transfer and formation of spermatophores in arthropod phylogeny. In: Gupta, A.P. (ed.), *Arthropod Phylogeny*. Van Nostrand Reinhold, New York, pp. 587–608.

Schaub, R. von (1888) Über die Anatomie und Histologie von *Hydrodroma* (C.L. Koch). *Sitzungsberichte der Akademie der Wissenschaften, Wien* 97, 98–151.

Scheucher, R. (1957) Systematik und Ökologieder deutschen Anoetinen. In: Stammer, H-J. (ed.), *Beiträge zur Systematik und Ökologie Mitteleuro-*

päischer Acarina 1, Abschn. 2, 233–384, Akademische Verlags., Geest & Portig, Leipzig.

Schlein, Y. & Gunders, A.E. (1981) Pheromone of *Ornithodoros* spp. (Argasidae) in the coxal fluid of female ticks. *Parasitology* 82, 467–471.

Schliesske, J. (1984) Effect of photoperiod and temperature on the development and reproduction of the gall mite *Aculus fockeui* (Nalepa & Trouessart) (Acari: Eriophyoidea) under laboratory conditions. In: Griffiths, D.A. & Bowman, C.E. (eds), *Acarology VI*, Vol. 2. Ellis Horwood, Chichester, pp. 804–808.

Schmidt, U. (1935) Beiträge zur Anatomie und Histologie der Hydracrinen, besonders von *Diplodontus despiciens* O.F. Müller. *Zeitschrift für Morphologie und Ökologie der Tiere* 30, 99–176.

Schulten, G.G.M. (1985) Pseudo-arrhenotoky. In: Helle, W. & Sabelis, M.W. (eds), *Spider Mites, their Biology, Natural Enemies and Control*, Vol. 18. Elsevier, Amsterdam, pp. 67–71.

Schulze, P. (1942) Über die Hautsinnesorgane der Zecken besonders über eine bisher unbekannte Art von Arthropoden-Sinnesorganen, die Krobylophoren. *Zeitschrift für Morphologie und Ökologie der Tiere* 39, 379–419.

Schuster, I.J. & Schuster, R. (1970) Indirekte Spermaübertrangung bei Tydeidae (Acari, Trombidiformes). *Naturwissenschaften* 57, 256.

Schuster, R. (1956) Der Anteil der Oribatiden an den Zersetzungsvorgangen im Boden. *Zeitschrift für Morphologie und Ökologie der Tiere* 45, 1–33.

Schuster, R. (1962) Nachweis eines Paarungszeremoniells bei den Hornmilben. *Naturwissenschaften* 49, 502.

Schuster, R & Pötsch, H. (1988) Another record of an active prelarva in mites. In: ChannaBasavanna, G.P. & Viraktamath, C.A. (eds), *Progress in Acarology*, Vol. 1. Oxford & IBH, New Delhi, pp. 261–265.

Schuster, R. & Schuster, I.J. (1966) Über das Fortpflanzungsverhalten von Anystiden-Männchen. *Naturwissenschaften* 53, 162–163.

Schuster, R. & Schuster, I.J. (1977) Ernährungs – und fortplanzungsbiologische Studieren an der Milben-familie Nanorchestidae (Acari, Trombidiformes). *Zoologischer Anzeiger.* 199, 89-94.

Schuster, R.O. & Summers, F.M. (1978) Mites of the family Diarthrophallidae. *International Journal of Acarology* 4, 279–385.

Schwarz, H.H. & Müller, J.K. (1992). The dispersal behaviour of the phoretic mite *Poecilochirus carabi* (Mesostigmata, Parasitidae): adaptation to the breeding biology of its carrier *Necrophorus vespilloides* (Coleoptera, Silphidae). *Oecologia* 89, 487–493.

Segenbusch, H.G. (1977) Review of oribatid mite – anoplocephalan tapeworm relationships (Acari; Oribatei: Cestoda; Anoplocephalidae). In: Dindal, D.L. (ed.), *Biology of Oribatid Mites.* State University, New York, pp. 87–102.

Sellnick, M. (1929) Hornmilben, Oribatei. *Tierwelt Mitteleuropas* 3 (9), 1–42.

Sellnick, M. (1944) *Zercon* C.L. Koch. *Acari-Blätter für Milbenkunde, Königsberg* 5, 30–41.

Sellnick, M. (1960) Hornmilben, Oribatei. *Tierwelt Mitteleuropas* 3, Ergänzung, 45–134.

Sellnick, M. & Forsslund, K-H. (1955) Die Camisiidae Schwedens, *Arkiv für Zoologi* 8, 473–530.

Seniczak, S. & Stefaniak, O. (1978) The microflora of the alimentary canal of *Oppia nitens* (Acarina, Oribatei). *Pedobiologia* 18, 110–119.

Severino, G., Oliver, J.H. Jr. & Pound, J.M. (1984) Synganglial and neurosecretory morphology of the chicken mite, *Dermanyssus gallinae* (De Geer) (Mesostigmata: Dermanyssidae). *Journal of Morphology* 181, 49–68.

Sharov, A.G. (1966) *Basic Arthropodan Stock.* Pergamon Press, Oxford.

Shcherbak, G.I. & Akimov, I.A. (1979) The importance of "scleronodules" in the systematics of the family Rhodacaridae Oudemans, 1902. In: Piffl, E. (ed), *Proceedings of the 4th International Congress of Acarology* Akadémiai Kiadó, Budapest, pp. 467–470.

Shevtchenko, V. & Silvere, A.P. (1958) The feeding organs of the four-legged mites (Acarina, Eriophyoidea). *Academy of Sciences of the Estonian SSSR Institute of Experimental Biology* 3, 248–264 [in Russian].

Shih, C.T., Poe, S.L. & Comroy, H.L. (1976) Biology, life table and intrinsic rate of increase of *Tetranychus urticae. Annals of the Entomological Society of America* 69, 362–364.

Shultz, J.W. (1989) The morphology of the locomotor appendages in Arachnida: Evolutionary trends and phylogenetic implications. *Zoological Journal of the Linnean Society* 97, 1–56.

Silvere, A.P. & Setjein-Margolina, V. (1976) Tetrapodili – Chetyreknogie Kleshci. *Academy of Sciences of the Estonian SSSR Institute of Experimental Biology* 167pp. [in Russian].

Sinclair, A.N. & Kirkwood, A.C. (1983) Feeding behaviour in *Psoroptes ovis. Veterinary Record* 15, 65.

Sitnikova, L.G. (1978) The main evolutionary trends of the Acari and the problem of their monophyletism. *Entomologicheskoe Obozrenie* 57, 431–457.

Siuda, K. (1982) The differentiation of nymphal *Argas (Argas) polonicus* Siuda, Hoogstraal, Clifford and Wassef, 1979 (Acarina, Ixodides, Argasidae). *Wiadomosci Parazytologiczne, Warszawa* 28, 51–55 [in Polish].

Sixl, W. & Sixl-Voigt, B. (1974) Ein Beitrag zur Klärung des Feinaufbaues der Innenstrukturen bei Stigmen von Zecken (*Haemaphysalis inermis*). *Mitteilungen der Abteilung für Zoologie Landesmuseums* 3, 35–39.

Sixl, W., Dengg, E. & Waltinger, H. (1971) Rasterelektronoptische untersuchungen bei Zecken. III. Die stigmen von *Ixodes ricinus* Linné, *Ixodes canisuga* Johnston, *Ixodes redikorsevi* Olenev, *Dermacentor marginatus* Sulzer, *Argas reflexus* Latreille, *Ornithodoros papillipes* Birula. *Archives des Sciences, Genève* 24, 403–407.

Skaife, S.H. (1952) The yellow-banded carpenter bee, *Mesotrichia caffra* Linn., and its symbiotic mite, *Dinogamasus braunsi* Vitzthum. *Journal of the Entomological Society of Southern Africa* 15, 63–76.

Smallman, B.N. & Shunter, C.A. (1972) Authentication of the cholinergic system in the cattle tick, *Boophilus microplus. Insect Biochemistry* 2, 67–77.

Smiley, R.L. & Knutson, L. (1983) Aspects of taxonomic research and services relative to mites as biological control agents. In: Hoy. M.A., Cunningham, G.L. & Knutson, L. (eds), *Biological Control of Pests by Mites.* University of California, Berkeley, pp. 148–154.

Smith, B.P. (1983) The potential of mites as biological control agents of mosquitoes. In: Hoy, M.A., Cunningham, G.L. & Knutson, L. (eds), *Biological*

Control of Pests by Mites. University of California, Berkeley, pp. 79–85.

Smith, D.S. (1968) *Insect Cells; their Structure and Function.* Oliver & Boyd, Edinburgh, 372pp.

Smith, I.M. (1976) A study of the systematics of the water mite family Pionidae (Prostigmata: Parasitengona). *Memoirs of the Entomological Society of Canada* 1–249.

Smith Meyer, M.K.P. (1974) A revision of the Tetranychidae (Acari) of Africa with a key to genera of the world. *Entomology Memoir of the Department of Agricultural Technical Services, Republic of South Africa* No. 36, 1–292.

Smith Meyer, M.K.P. (1979) The Tenuipalpidae (Acari) of Africa with keys to the world fauna. *Entomology Memoir of the Department of Agricultural Technical Services, Republic of South Africa* No. 50, 1–135.

Smith Meyer, M.K.P. & Ueckermann, E.A. (1987) A taxonomic study of some Anystidae (Acari: Prostigmata). *Entomology Memoir of the Department of Agriculture and Fisheries, Republic of South Africa* No. 68, 1–37.

Sneath, P.H.A. & Sokal, R.R. (1973) *Numerical Taxonomy.* W.H. Freeman, San Francisco. 573pp.

Snodgrass, R.E. (1938) Evolution of the Annelida, Onychophora and Arthropoda. *Smithsonian, Miscellaneous Collection* 97, no. 6, 1–159.

Snodgrass, R.E. (1948). The feeding organs of the Arachnida, including mites and ticks *Smithsonian, Miscellaneous Collection* 110, no. 10, 1–93.

Soar, C.D. & Williamson, W. (1925) *The British Hydracarina,* Vol. 1. Ray Society, London.

Sokolov, I. (1977) The protective envelopes of the eggs of the Hydrachnellae. *Zoologischer Anzeiger* 198, 36–46.

Solomon, K.R., Mango, C.K.A. & Obenchain, F.D. (1982) Endocrine mechanisms in ticks: effects of insect hormones and their mimics on development and reproduction. *Current Themes in Tropical Science* 1, 399–428.

Sonenshine, D.E. (1984) Pheromones of Acari and their potential use in control strategies. In: Griffiths, D.A. & Bowman, C.E. (eds), *Acarology VI,* Vol. 1. Ellis Horwood, Chichester, pp. 100–108.

Sonenshine, D.E. (1985) Pheromones and other semiochemicals of the Acari. *Annual Review of Entomology* 30, 1–28.

Sonenshine, D.E. (1986) Tick pheromones: an overview. In: Sauer, J.R. & Hair, J.A. (eds), *Morphology, Physiology and Behavioural Biology of Ticks.* Ellis Horwood, Chichester, pp. 342–360.

Sonenshine, D.E. (1988) In: Monath, T.P. (ed.), *The Arboviruses: Epidemiology and Ecology,* Vol. 1. CRC Press, Boca Raton, Florida, pp. 219–243.

Sonenshine, D.E. & Gregson, J.D. (1970) A contribution to the internal anatomy and histology of the bat tick *Ornithodoros kelleyi* Cooley and Kohls, 1941. I. The alimentary system with notes on the food channel in *Ornithodoros denmarki* Sonenshine & Clifford, 1965. *Journal of Medical Entomology, BP Bishop Museum, Honolulu* 7, 46–64.

Sonenshine, D.E., Gainsburg, D.M., Rosenthal, M.D. & Silverstein, R.M. (1981) The sex pheromone glands of *Dermacentor variabilis* (Say) and *Dermacentor andersoni* Stiles sex pheromone stored in neutral lipid. *Journal of Chemical Ecology* 7, 345–357.

Sonenshine, D.E., Silverstein, R.M. & Rechav, Y. (1982) Tick pheromone mechanisms. *Current Themes in Tropical Science* 1, 549–468.

Sonenshine, D.E., Homscher, P.J., Carson, K.A. & Wang, V.D. (1984) Evidence of the role of the cheliceral digits in the perception of genital sex pheromones during mating in the American dog tick, *Dermacentor variabilis* (Acari: Ixodidae). *Journal of Medical Entomology, BP Bishop Museum, Honolulu* 21, 296–306.

Southcott, R.V. (1961) Studies on the systematics and biology of the Erythraeoidea (Acarina), with a critical revision of the genera and subfamilies. *Australian Journal of Zoology* 9, 367–610.

Southcott, R.V. (1963) The Smarididae (Acarina) of North and Central America and some other countries. *Transactions of the Royal Society of South Australia* 89, 159–245.

Southcott, R.V. (1986) Studies on the taxonomy and biology of the subfamily Trombidiinae (Acarina; Trombidiidae) with a critical revision of the genera. *Australian Journal of Zoology* Suppl. ser., 123, 116pp.

Spieksma, F.Th.M. (1967) The house dust mite *Dermatophagoides pteronyssinus* (Trouessart, 1897), producer of house dust allergen (Acari: Psoroptidae). Thesis, cited by Fain & Herin (1979).

Stanley, J. (1931) Studies on the musculature system and mouthparts of *Laelaps echidninus* Berl. *Annals of the Entomological Society of America* 24, 1–10.

Starovir, I.S. (1982) Functional digestive histology in *Amblyseius reductus* (Parasitiformes: Phytoseiidae). *Vestnik Zoologii* 1982 (1), 69-74 [in Russian].

Stefaniak, O. & Seniczak, S. (1976) The microflora of the alimentary canal of *Achiptera coleoptrata* (Acarina, Oribatei). *Pedobiologia* 16, 185–194.

Stefaniak, O. & Seniczak, S. (1981) The effect of fungal diet on the development of *Oppia nitens* (Acari, Oribatei). *Pedobiologia* 21, 202–210.

Sternlicht, M. & Griffiths, D.A. (1974) The emission of spermatophores and fine structure of adult *Eriophyes sheldoni* (Acarina, Eriophyoidea). *Bulletin of Entomological Research* 63, 561–565.

Storch, V. & Ruhberg, H. (1977) Zur Enstehung der Spermatophore von *Opisthopatus cinctipes* Purcell, 1899 (Onychophora, Peripatopsidae). *Zoomorphologie* 87, 263–276.

Storms, J.J.H. (1971) Some physiological effects of spider mite infestations on bean plants. *Netherlands Journal of Plant Pathology* 77, 154–167.

Strandtmann, R.W. (1971) The eupodoid mites of Alaska (Acarina: Prostigmata). *Pacific Insects* 13, 75–118.

Stratil, H.U. & Knülle, W. (1985) Die Induktion des Hypopusstadiums bei der im Lagergetreide vorkommenden Milbe *Glycyphagus destructor* (Schrank, 1781) (Astigmata, Glycyphagidae). *Zeitschrift für Angewandte Entomologie* 99, 350–365.

Strenzke, K. (1946) Zur Fortflanzung der Moosmilben. *Microkosmos* 38, 177–180.

Strenzke, K. (1954) *Nematalycus nematoides* n.gen., n.sp. (Acarina: Trombidiformes) aus dem Grundwasser der Algerischen Küste. *Vie et Milieu* 4, 638–647.

Strenzke, K. (1963) Entwicklung und Verwandtschaftsbeziehungen der Oribatidengattung. *Senkenbergiana Biologica Frankfurt A.M.* 44, 231–255.

Strube, H.G.R. & Flechtmann, C.H.W. (1985) Study on the peritreme of the female of *Varroa jacobsoni* Oud., 1904 (Acari: Mesostigmata). *Experimental and Applied Acarology* 1, 87–89.

Stunkard, H.W. (1937) The life-cycle of *Moniezia expansa. Science* 86, 312.

Stunkard, H.W. (1941) Studies on the life-history of the anoplocephaline cestodes of hares and rabbits. *Journal of Parasitology* 27, 299–325.

Summers, F.M. (1962) The genus *Stigmaeus* (Acarina: Stigmaeidae). *Hilgardia* 33, 491–537.

Summers, F.M. & Price, D.W. (1970) Review of the mite family Cheyletidae. *University of California Publications of Entomology, Berkeley* 6, 1–153.

Summers, F.M. & Schlinger, E.I. (1955) Mites of the family Caligonellidae (Acarina). *Hilgardia* 23, 529–561.

Summers, F.M. & Witt, R.L. (1972) Nesting behaviour of *Cheyletus eruditus* (Acarina: Cheyletidae) *Pan-Pacific Entomologist* 48, 261–269.

Summers, F.M. & Witt, R.L. (1973) Oviposition and mating tendencies of *Cheyletus malaccensis* (Acarina: Cheyletidae). *Florida Entomologist* 56, 277–285.

Summers, F.M., Gonzales, R.H. & Witt, R.L. (1973) The mouthparts of *Bryobia rubrioculus* (Sch.) (Acarina: Tetranychidae). *Proceedings of the Entomological Society of Washington* 75, 96–111.

Suski, Z.W. (1972) Tarsonemid mites on apple trees in Poland. X. Laboratory studies on the biology of certain species of the family Tarsonemidae (Acarina, Heterostigmata). *Zeszty Problemowe Postepow Nauk Rolniczych, Warsawa* 129, 111–137.

Suski, Z.W. (1989) Occurrence of the respiratory apparatus in the eggs of some mite species (Acarina). *Polskie Pismo Entomologiczne* 59, 311–318.

Suski, Z.W. & Naegele, J.A. (1963) Light response in the two-spotted spider mite. 2. Behaviour of the "sedentary and dispersal" phases. In: Naegele, J.A. (ed.), *Advances in Acarology* Vol. 1. Cornell Press, Ithaca, New York, pp. 445–453.

Taberly, G. (1957) Observations sur les spermatophores et leur transfert chez les Oribates. *Bulletin de la Société Zoologique de France* 82, 139–145.

Taberly, G. (1987a) Recherches sur la parthénogenèse thélytoque de deux espèces d'acariens oribates: *Trhypochthonius tectorum* Berlese et *Platynothrus peltifer* (Koch). II. Étude anatomique, histologique et cytologique des femmelles parthénogenètique. 1re partie. *Acarologia* 28, 285–293.

Taberly, G. (1987b) Recherches sur la parthénogenèse thélytoque de deux espèces d'acariens oribates: *Trhypochthonius tectorum* Berlese et *Platynothrus peltifer* (Koch). III. Étude anatomique, histologique et cytologique des femelles parthénogenètiques. 2eme partie. *Acarologia* 28, 389–403.

Taberly, G. (1988) Recherches sur la parthénogenèse thélytoque de deux espèces oribates: *Trhypochthonius tectorum* Berlese et *Platynothrus peltifer* (Koch). IV. Observations sur les males ataviques. *Acarologia* 29, 95–107.

Tanigoshi, L.K. (1982) Advances in knowledge of Phytoseiidae. In: Hoy, M.A. (ed.), Recent advances in the knowledge of the Phytoseiidae. University of California, Division of Agriculture, Scientific Publication 3284, 1–22.

Tanigoshi, L.K. & Davis, R.W. (1978) An ultrastructural study of *Tetranychus mcdanieli* feeding injury to the leaves of 'Red Delicious' apple. *International Journal of Acarology* 4, 47–56.

Tatchell, R.J. (1964) Digestion in the tick, *Argas persicus* Oken. *Parasitology* 54, 423–440.

Tatchell, R.J. (1967) Salivary secretion in the cattle tick as a means of water elimination. *Nature* [Lond.] 213, 940–941.

Tatchell, R.J. (1969a) The significance of host-parasite relationships in the feeding of the cattle tick *Boophilus microplus*. In: Evans, G.O. (ed.), *Proceedings of the 2nd International Congress of Acarology*. Akadémiai Kiadó, Budapest, pp. 341–345.

Tatchell, R.J. (1969b) The ionic regulatory role of salivary secretion of the cattle tick, *Boophilus microplus*. *Journal of Insect Physiology* 15, 1421–1430.

Tatchell, R.J. & Binnington, K.C. (1973) An active constituent of the saliva of the cattle tick *Boophilus microplus*. In: Daniel, M & Rosický, B. (eds), *Proceedings of the 3rd International Congress of Acarology*. Academia, Prague, pp. 745–748.

Tauber, M.J., Tauber, C.A. & Masaki, S. (1986) *Seasonal Adaptations of Insects*. Oxford University Press, New York, 411pp.

Theodor, O. & Costa, M. (1960) New species and new records of Argasidae from Israel. Observations on the rudimentray scutum and the respiratory system of the larvae of the Argasidae. *Parasitology* 50, 365–386.

Theron, P.D. (1976) New species of the genus *Terpnacarus* Grandjean (Acari: Terpnacaridae) with notes on the biology of one species. *Journal of the Entomological Society of Southern Africa* 39, 132–141.

Theron, P.D. (1979) The functional morphology of the gnathosoma of some liquid and solid feeders in the Trombidiformes, Cryptostigmata and Astigmata (Acarina). In: Piffl, E. (ed.), *Proceedings of the 4th International Congress of Acarology*. Akadémiai Kiadó, Budapest, pp. 575–579.

Theron, P.D. & Ryke, P.A.J. (1969) The family Nanorchestidae Grandjean (Acari: Prostigmata) with descriptions of new species from South African soils. *Journal of the Entomological Society of Southern Africa* 32, 31–60.

Theron, P.D. & Ryke, P.A.J. (1975a) Three new species of the family Sphaerolichidae (Acari: Endeostigmata) from South Africa. *Acarologia* 17, 220–235.

Theron, P.D. & Ryke, P.A.J. (1975b) Five new species of the family Lordalycidae (Acari: Endeostigmata) from South Africa. *Acarologia* 17, 631–635.

Thomae, H. (1926) Beiträge zur Anatomie der Halacariden. *Zoologische Jahrbuecher. Abteilung für Anatomie und Ontogenie der Tiere* 47, 155–190.

Thomas, R.H. & Zeh, D.W. (1984) Sperm transfer and utilization strategies in arachnids. In: Smith, R.L. (ed.), *Sperm Competition and the Evolution of Animal Mating Systems*. Academic Press, Orlando, pp. 179–221.

Thon, K. (1905a) Neue Luftorgane bei Milben. *Zoologischer Anzeiger* 28, 587–594.

Thon, K. (1905b) Über die Drusen der Holothyriden. *Sitzungsberichte der Königlichen Bömischen Gesellschaft der Wissenschaften* 4, 1–41.

Thon, K. (1906) Die aussere Morphologie und die Systematik der Holothyriden. *Zoologische Jahrbuecher. Abteilung für Systematik, Ökologie und Geographie der Tiere* 23, 677–724.

Thor, S. (1904). Recherches sur l'anatomie comparée des acariens prostigmatiques. *Annales des Sciences Naturelles, Zoologie* 19, 1–190.

Thor, S. (1931) Acarina. Bdellidae, Nicoletiellidae, Cryptognathidae. *Das Tier-*

reich, Lief. 56, 1–87. Walter de Gruyter, Berlin & Leipzig.

Thor, S. (1933) Acarina. Tydeidae, Ereynetidae. *Das Tierreich*, Lief. 60, 1–84.

Thor, S. & Willmann, C. (1947) Acarina, Trombidiidae. *Das Tierreich*, Lief, 71B, 187–541.

Thorpe, W.H. (1950) Plastron respiration in aquatic insects. *Biological Review* 25, 344–390.

Thurm, U. (1964) Mechanoreceptors in the cuticle of the honey bee: Fine structure and stimulus mechanism. *Science* [New York] 145, 1063–1065.

Till, W.M. (1961) A contribution to the anatomy and histology of the brown ear tick *Rhipicephalus appendiculatus* Neumann. *Memoirs of the Entomological Society of Southern Africa* No. 6, 1–124.

Timms, S., Ferro, D.N. & Emberson, R.M. (1980) Selective advantage of pleomorphic male *Sancassania berlesei* (Michael) (Acari: Acaridae). *International Journal of Acarology* 6, 97–102.

Timms, S., Ferro, D.N. & Emberson, R.M. (1982) Andropolymorphism and its heritability in *Sancassania berlesei* (Michael). *Acarologia* 22, 385–390.

Tomalski, M.D., Kutney, R., Bruce, W.A., Brown, M.B., Blum, M.S. & Travis, J. (1989) Purification and characterization of insect toxins derived from the mite *Pyemotes tritici*. *Toxicon* 27 (10), 1151–1167.

Tomczy, A. & Kropczynska, D. (1985) Effects on host plant. In: Helle, W. & Sabelis, M.W. (eds), *Spider Mites. their Biology, Natural Enemies and Control.* Elsevier, Amsterdam, pp. 317–329.

Trägårdh, I. (1942) Microgyniina, a new group of Mesostigmata. *Entomologisk Tidskrift*, 63, 120–133.

Travé, J. (1959) Dimorphisme sexuel chez *Pirnodus detectidens* Grandjean (Acariens-Oribates). Notes écologiques et ethologiques. *Vie et Milieu* 9, 454-468.

Travé, J. (1970) Les stases immatures du genre *Neoribates* (Parakalumnidae, Oribates). Parakalumnidae et Galumnidae. *Acarologia* 12, 208–215.

Travé, J. (1976) Les prélarves d'acariens. Mise au point et données récentes. *Revue d'Ecologie et de Biologie du Sol* 13, 161–171.

Travé, J. (1983) Observations morphologiques sur les Holothyrides. 1re partie. Le corps de *Thonius braueri* (Thon, 1906). *Acarologia* 24, 333–341.

Travé, J. (1986) Les taenidies respiratoires des Oribates. *Acarologia* 27, 85–94.

Treat, A.E. (1975) *Mites of Moths and Butterflies.* Comstock Publishing Associates, Ithaca & London.

True, G.H. (1932) Studies on the anatomy of the parjaroello tick, *Ornithodorus coriaceus* Koch. 1. The alimentary canal, *University of California Publications of Entomology, Berkeley* 6, 21–48.

Tsvileneva, V.A. (1964) The nervous structure of the ixodid ganglion. *Zoologische Jahrbuecher. Abteilung für Anatomie und Öntogenie der Tiere* 81, 579–602.

Türk, E. & Türk, F. (1957) Systematik und Ökologie der Tyroglyphiden Mitteleuropas. In: Stammer, H-J. (ed.), *Beiträge zur Systematik und Ökologie Mitteleuropischer Acarina*, Vol. 1 (1), 3–231. Geest and Portig, Leipzig.

Usher, M.B. & Bowring, M.F.B. (1984) Laboratory studies on predation by the Antarctic mite *Gamasellus racovitzai* (Acari: Mesostigmata). *Oecologia* 62, 245–249.

Van Bronswijk, J.E.M.H. & de Kreek, E.J. (1976) *Cheyletiella* (Acari: Cheyletiellidae) of dog, cat and domesticated rabbit, a review. *Journal of Medical*

Entomology, BP Bishop Museum, Honolulu 13, 315–327.

Van der Geest, L.P.S. (1985) Aspects of physiology. In: Helle, W. & Sabelis, M.W. (eds), *Spider Mites, their Biology, Natural Enemies and Control*, Vol. 1A. Elsevier, Amsterdam, pp. 171–184.

Van der Hammen, L. (1952) The Oribatei (Acari) of the Netherlands. *Zoologische Verhandelingen, Leiden* 17, 1–139.

Van der Hammen, L. (1960) *Fortuynia marina* nov. gen, nov. spec., an oribatid mite from the intertidal zone in Netherlands New Guinea. *Zoologische Mededeelingen, Leiden* 37, 1–9.

Van der Hammen, L. (1961) Description of *Holothyrus grandjeani* nov. spec., and notes on the classification of the Acari. *Nova Guinea, Zoology* 9, 173–194.

Van der Hammen, L. (1963a) Description of *Fortuynia yunkeri* nov. spec., and notes on the Fortuynidae nov. fam. (Acarida, Oribatei). *Acarologia* 5, 152–167.

Van der Hammen, L. (1963b) The oribatid family Phthiracaridae. I. Introduction and redescription of *Hoplophthiracarus pavidus* (Berlese). *Acarologia* 5, 306–317.

Van der Hammen, L. (1964) The morphology of *Glyptholaspis confusa* (Foà, 1900) (Acarida, Gamasina). *Zoologische Verhandelingen, Leiden*, 71, 1–56.

Van der Hammen, L. (1966) Studies on the Opilioacarida (Arachnida). I. Description of *Opilioacarus texanus* (Camberlin & Mulaik) and revised classification of genera. *Zoologische Verhandelingen, Leiden* 86, 1–80.

Van der Hammen, L. (1968a) Stray notes on Acarida. *Zoologische Mededeelingen, Leiden* 42, 261–280.

Van der Hammen, L. (1968b) The gnathosoma of *Hermannia convexa* (C.L. Koch) (Acarida: Oribatina). *Zoologische Verhandelingen, Leiden* 94, 1–45.

Van der Hammen, L. (1968c) Studies of Opilioacarida (Arachnida) II. Redescription of *Paracarus hexophthalmus* (Redikorzev). *Zoologische Mededeelingen Leiden* 43, 57–76.

Van der Hammen, L. (1970) La segmentation primitive des Acariens. *Acarologia* 12, 3–10.

Van der Hammen L. (1972) A revised classification of the mites (Arachnidea; Acarida) with diagnoses, a key and notes on phylogeny. *Zoologische Mededeelingen, Leiden* 47, 273–292.

Van der Hammen, L. (1973) Classification and phylogeny of mites. In: Daniel, M. & Rosický, B. (eds), *Proceedings of the 3rd International Congress of Acarology*. Academia, Prague, pp. 275–282.

Van der Hammen, L. (1975) L'evolution des Acariens, et les modèles de l'evolution des Arachnides. *Acarologia* 16, 377–381.

Van der Hammen, L. (1977a) A new classification of Chelicerata. *Zoologische Mededeelingen, Leiden* 51, 307–319.

Van der Hammen, L. (1977b) Studies on the Opilioacarida (Arachnidea). IV. The genera *Panchaetes* Naudo and *Salfacarus* nov. gen. *Zoologische Mededeelingen, Leiden* 51, 43–78.

Van der Hammen, L. (1978) The evolution of the chelicerate life-cycle. *Acta Biotheoretica* 27, 44–60.

Van der Hammen, L. (1979) Comparative studies in Chelicerata, I. The Cryptognomae (Ricinulei, Architarbi and Anactinotrichida). *Zoologische Verhande-*

lingen, Leiden 174, 1–62.

Van der Hammen, L. (1980). *Glossary of Acarological Terms. 1. General Terminology.* W. Junk, The Hague, 244pp.

Van der Hammen, L. (1983) New notes on the Holothyrida (anactinotrichid mites). *Zoologische Verhandelingen, Leiden* 207, 1–48.

Van der Hammen, L. (1989) *An Introduction to Comparative Arachnology.* SPB Academic Publishing, The Hague, 576pp.

Van Eyndhoven, G.L. (1964) *Cheyletomorpha leipidopterorum* (Shaw, 1794) (=*Ch. venutissima*) (Acari, Cheyletidae) on Lepidoptera. *Beaufortia* 11, 53–60.

Van Houten, Y.M. (1989) Photoperiodic control of adult diapause in the predacious mite, *Amblyseius potentillae*: repeated diapause induction and termination. *Physiological Entomology* 14, 341–348.

Van Houten, Y.M., Overmeer, W.P.J., Van Zon, A.Q. & Veerman, A. (1988) Thermoperiodic induction of diapause in the predacious mite, *Amblyseius potentillae. Journal of Insect Physiology* 34, 285–290.

Van Impe, G. (1985) Contribution à la conception de stratégies de contrôle de l'acarien tisserand commun, *Tetranychus urticae* (Acari: Tetranychidae). Unpublished Thesis, Université Catholique de Louvain, Belgium, 382pp.

Veerman, A. (1974) Carotenoid metabolism in *Tetranychus urticae* Koch (Acari: Tetranychidae). *Comparative Biochemistry and Physiology* 47B, 101–116.

Veerman, A. (1977a) Aspects of induction of diapause in a laboratory strain of the mite *Tetranychus urticae. Journal of Insect Physology* 23, 703–711.

Veerman, A. (1977b) Photoperiodic termination of diapause in spider mites. *Nature* [Lond.] 266, 526–527.

Veerman, A. (1985) Diapause. In: Helle, W. & Sabelis, M.W. (eds), *Spider Mites, their Biology, Natural Enemies and Control,* Vol. 1A. Elsevier, Amsterdam, pp. 279–316.

Veerman, A. (1991) Physiological aspects of diapause in plant-inhabiting mites. In: Schuster, R. & Murphy, P.W. (eds), *The Acari. Reproduction, Development and Life-History Strategies.* Chapman & Hall, London, pp. 245–265.

Veerman, A. & Vaz Nunes, M. (1987) Analysis of the operation of the photoperiodic counter provides evidence for hourglass time measurement in the spider mite *Tetranychus urticae. Journal of Comparative Physiology* A, 160, 421–430.

Vercammen-Grandjean, P.H. (1965) *Iguanacarus,* a new subgenus of chigger mites from the nasal fossae of the marine iguana in the Galapagos Islands, with a revision of the genus *Vatacarus* Southcott (Acarina: Trombiculidae). *Acarologia* Suppl., 7, 266–274.

Vercammen-Grandjean, P.H. (1976) Les organes des Claparède et les papilles génitales de certains acariens – sont-ils des organes respiratoires? *Acarologia* 17, 1975, 624–630.

Vermeulen, N.M.J., Gothe, R., Senekal, A.C. & Neitz, A.W.H. (1986) Investigations into the function and chemical composition of the porose areas secretion of *Rhipicephalus evertsi evertsi* during oviposition. *Oonderstepoort Journal of Veterinary Research* 83, 147–152.

Viets, K. (1925) Wassermilben. Hydracarina. *Tierwelt Mitteleuropas* 3, 1–157.

Vistorin, H.E. (1978) Fortpflanzung und Entwicklung der Nicoletiellidae (Labidostomidae); Acari: Trombidiformes. *Zoologische Jahrbuecher. Abteilung für Systematik, Ökologie und Geographie der Tiere* 105, 462–473.

Vistorin, H.E. (1980) Ernährungsbiologie und Anatomie des verdauungstraktes der Nicoletiellidae (Acari: Trombidiformes). *Acarologia* 21, 204–215.

Vistorin-Thesis, G. (1976) Morphologisch-taxonomische Studien an der Mibenfauna Calyptostomidae (Acari: Trombidiformes). *Sitzungsberichte der Österreichischen Akademie der Wissenschaften* (Abt. 1), 185, 55–89.

Vistorin-Theis, G. (1978) Anatomische untersuchungen an Calyptostomiden (Acari: Trombidiformes). *Acarologia* 19, 242–257.

Vitzthum, H.G. (1930) Das Atmungssytem von *Allothrombium meridionale. Zoologischer Anzeiger* 91, 217–220.

Vitzthum H.G. (1931) Acari. In: Kükenthal u. Krombach, *Handbuch der Zoologie*, 3, Hälfte 2, Lief. 1, 1–160. Walter de Gruyter, Berlin & Leipzig.

Vitzthum, H.G. (1940–43) Acarina. In: *Bronn's Klassen und Ordnungen des Tierreiches*, 5, Abt. 4, Buch 5, Lief. 1–7, 1–1011. Leipzig.

Volgin, V.I. (1969) Mites of the family Cheyletidae from the world fauna. *Opredeliteli po Faune SSSR, Izdavalmÿe Zoologicheskim Muzeum Akademi Nauk, Leningrad* 101, 1–431 [in Russian].

Volkonsky, M. (1940) *Podapolipus diander* n. sp. acarien hétérostigmatique parasite du criquet migrateur (*Locusta migratoria* L.). *Archives de l'Institut Pasteur d'Algerie* 18, 321–340.

Wachmann, E. (1975) Feinstruktur der Lateralaugen einer räuberischen Milbe *Microcaeculus* (Acari: Prostigmata: Caeculidae). *Entomologische Germanica, Berlin* 1, 300–307.

Waladde, S.M. (1982) Tip-recording from ixodid tick olfactory sensilla: responses to odours. *Journal of Comparative Physiology* 148, 411–418.

Waladde, S.M. & Rice, M.J. (1977) The sensory nervous system of the adult cattle tick, *Boophilus microplus* (Canestrini), Ixodidae. Part 3. Ultrastructure and electrophysiology of cheliceral receptors. *Journal of the Australian Entomological Society* 16, 442–453.

Waladde, S.M. & Rice, M.J. (1982) The sensory basis of tick feeding behaviour. *Current Themes in Tropical Science* 1, 71–118.

Waladde, S.M., Kokwaro, E.D. & Chimtawi, M. (1981) A cold receptor on the tick, *Rhipicephalus appendiculatus*: electrophysiological and ultrastructural observations. *Insect Science Application* 1, 191–196.

Wallace, M.M.H. (1974) An attempt to extend the biological control of *Sminthurus viridis* (Collembola) to new areas in Australia by introducing a predatory mite *Neomolgus capillatus* (Bdellidae). *Australian Journal of Zoology* 22, 519–529.

Wallwork, J.A. (1965) A leaf-boring galumnoid mite (Acari: Cryptostigmata) from Uruguay. *Acarologia* 7, 758–764.

Wallwork, J.A. (1967) Acari. In: Burges, A. & Raw, F. (eds), *Soil Biology*. Academic Press, London, pp. 363–395.

Walter, C. (1920) Die Bedeutung der Apodermata in der Epimorphose der Hydracarina. *Festschr. zu Feier des 60 Geburtstages von F. Zschotte, Basel*, 24, 14pp.

Walter, D.E. (1988) Predation and mycophagy by endostigmatid mites (Acariformes: Prostigmata). *Experimental and Applied Acarology* 4, 159–166.

Walter, D.E. & Kaplan, D.T. (1991) Observations on *Coleoscirus simplex* (Acarina: Prostigmata), a predatory mite that colonizes greenhouse cultures of rootknot nematode (*Melidogyne* spp.), and a review of feeding behaviour in the Cunaxidae. *Experimental and Applied Acarology* 12, 47–59.

Walter, D.E. & Lindquist, E.E. (1989) Life history and behaviour of mites in the genus *Lasioseius* (Acari: Mesostigmata: Ascidae) from grassland soils in Colorado, with taxonomic notes and description of a new genus. *Canadian Journal of Zoology* 67, 2797–2813.

Walter, D.E. & Oliver, J.H. Jr. (1989) *Geolaelaps oreithyiae* n. sp. (Acari: Laelapidae), a thelytokous predator of arthropods and nematodes, and a discussion of clonal reproduction in the Mesostigmata. *Acarologia* 30, 293–303.

Walter, D.E., Hudgens, R.A. & Freckman, D.W. (1986) Consumption of nematodes by fungivorous mites, *Tyrophagus* spp. (Acarina: Astigmata: Acaridae) *Oecologia* 70, 357–361.

Walzl, M.G. (1987) The cheliceral sense organs of the adult oribatid mite *Hermannia gibba* (C.L. Koch) (Actinotrichida: Acari). Abstract in *European Cell Biology* Suppl. 18, Vol. 43, no. 98.

Walzl, M.G. & Waitzbauer, J. (1980) Kritisch-Punkt-Trocknung von Milben: eine modifizierte Präparationsmethode. *Mikroskopie* 36, 164–168.

Warburton, C. & Embleton; A.L. (1902) The life-history of the black currant gall-mite, *Eriophyes (Phytoptus) ribis*, Westwood. *Journal of the Linnean Society, London* 23, 366–378.

Warren, E. (1941) On the genital system and modes of reproduction and dispersal in certain gamasid mites. *Annals of the Natal Museum* 10, 95–126.

Webb, J.R. (1979) Host locating behaviour of nymphal *Ornithodoros concanensis* (Acarina: Argasidae). *Journal of Medical Entomology, BP Bishop Museum, Honolulu* 16, 437–447.

Weis-Fogh, T. (1970) Structure and formation of insect cuticle. *Symposia of the Royal Entomological Society of London* 5, 165–195.

Welbourn, W.C. (1983) Potential use of trombidioid and erythraeoid mites as biological control agents of insect pests. In: Hoy, M.A., Cunningham, G.L. & Knutson, L. (eds), *Biological Control of Pests by Mites.* University of California Press, Berkeley, pp. 103–140.

Wernz, J.G. & Krantz, G.W. (1976) Studies on the function of the tritosternum in selected Gamasida (Acari). *Canadian Journal of Zoology* 54, 202–213.

Weyda, F. (1980) Reproductive system and oogenesis in active females of *Tetranychus urticae* (Acari, Tetranychidae). *Acta Entomologica Bohemoslovaca, Praha* 77, 375–377.

Weygoldt, P. (1966) Vergleichende Untersuchungen zur Fortpflanzungsbiologie der Pseudoscorpione. Beobachtungen über das Verhalten die Samenübertragungsweisen und die Spermatoophoren einiger einheimscher Arten. *Zeitschrift für Morphologie und Ökologie der Tiere* 56, 39–92.

Weygoldt, P. & Paulus, H.F. (1979a) Untersuchungen zur Morphologie, Taxonomie und Phylogenie der Chlicerata. 1. Morphologische Untersuchungen. *Zeitschrift für Zoologische Systematik und Evolutionforschung* 17, 85–116.

Weygoldt, P. & Paulus, H.F. (1979b) Untersuchungen zur Morphologie, Taxonomie und Phylogenie der Chelicerata. 2. Cladogramme und die Entfaltung der

Chelicerata. *Zeitschrift für Zoologische Systematik und Evolutionforschung* 17, 177–200.

Wharton, G.W. (1976) House dust mites. *Journal of Medical Entomology, BP Bishop Museum, Honolulu* 12, 577–621.

Wharton, G.W. & Fuller, H.S. (1952) A manual of chiggers. *Memoirs of the Entomological Society of Washington* 4, 184pp.

Wharton, G.W., Parish, W. & Johnston, D.E. (1968) Observations on the fine structure of the cuticle of the spiny rat mite, *Laelaps echnidina* (Acari: Mesostigmata). *Acarologia*, 10, 207–214.

Wharton, G.W., Duke, K.M. & Epstein, H.M. (1979) Water and the physiology of house mites. In: Rodriguez, J.G. (ed.), *Recent Advances in Acarology*. Academic Press, New York, etc., pp. 325–335.

Wharton, R.H. & Utech, K.W.B. (1969) The engorgement and dropping of *Boophilus microplus* (Canestrini). In: Evans, G.O. (ed.), *Proceedings of the 2nd International Congress of Acarology*. Akadémiai Kiadó, Budapest, pp. 347–348.

Wiesmann, R. (1968) Untersuchungen über die Verdauungsvoränge bei der gemeinen Spinnmilbe, *Tetranychus urticae* Koch. *Zeitschrift für Angewandte Entomologie* 61, 457–465.

Wigglesworth, V.B. (1961) *The Principles of Insect Physiology*, 5th edn. Methuen, London.

Wiles, P.R. (1984) Watermite respiratory systems. *Acarologia* 25, 27–31.

Wiley, E.O. (1981) *Phylogenetics: the Theory and Practice of Phylogenetic Systematics*. John Wiley, New York, etc.

Willmann, C. (1931) Moosmilben oder Oribatiden. *Tierwelt Deutschlands* 22, 79–200. Fischer, Jena.

Winkler, W. (1888) Die Anatomie der Gamasiden. *Arbeiten aus der Zoologischen Instituten der Universität, Wien* 7, 317–354.

Wiśniewski, J. (1979) Zur Kenntnis der Uropodiden-Fauna Polens. *Acarologie, Schriftenreihe, Vergl. Milbenkd* [Fürth] Folge 26, Teil 339, 68–74.

Witalinski, W. (1975) Spermatogenesis in a freeliving mite, *Pergamasus viator* Halas. (Mesostigmata, Parasitidae). 1. Fine structure of spermatozoa. *Zeitschrift für Mikroskopische – Anatomische Forschung, Leipzig* 89, 1–17.

Witalinski, W. (1979a) Fine structure of the respiratory system in mites from the family Parasitidae. *Acarologia* 21, 330–339.

Witalinski, W. (1979b) Fine structure of the spermatozoa in the mite *Parasitus niveus* (Mesostigmata, Acari). *International Journal of Invertebrate Reproduction* 1, 141–149.

Witalinski, W. (1986) Egg shells in mites. 1. A comparative ultrastructural study of vitelline envelope formation. *Cell & Tissue Research* 244, 209–214.

Witalinski, W. (1987) Egg shells in mites: cytological aspects of vitelline envelope and chorion formation in *Pergamasus barbarus* Berlese (Gamasida, Pergamasidae). *International Journal of Acarology* 13, 189–196.

Witalinski, W. (1988) Egg-shells in mites. Vitelline envelope and chorion in a water mite. *Limnochares aquatica* L. (Acari, Limnocharidae). *Journal of Zoology* [Lond.] 214, 285–294.

Witalinski, W., Szlendak, E. & Boczek, J. (1990) Anatomy and ultrastructure of the reproductive systems of *Acarus siro*. *Experimental and Applied Acarology* 10, 1–31.

With, C. (1904) The Notostigmata, a new suborder of Acari. *Videnskabelige Meddelelser fra Dansk Naturhistorisk Foreningi Kjøbenhaus* 137–192.

Witte, H. (1975a) Funktionsanatomie der Genitalorgane und Fortpflanzungsverhalten bei den Männchen der Erythraeidae (Acari, Trombidiformes). *Zeitschrift für Morphologie und Ökologie der Tiere* 80, 137–180.

Witte, H. (1975b) Funktionsanatomie des Weiblichen Genitaltraktes und Oogenese bei Erythraeiden (Acari, Trombidiformes). *Zoologische Beiträge* 21, 247–277.

Witte, H. (1978) Die postembryonal Entwicklung und die funktionelle Anatomie des Gnathosoma in der Milbenfamilie Erythraeidae (Acari, Prostigmata). *Zoomorphologie* 91, 157–189.

Witte, H. (1984) The evolution of the mechanisms of reproduction in the Parasitengonae (Acarina: Prostigmata). In: Griffiths, D.A. & Bowman, C.E. (eds), *Acarology VI*. Ellis Horwood, Chichester, pp. 470–478.

Witte, H. (1991) Indirect sperm transfer in prostigmatic mites from a phylogenetic viewpoint. In: Schuster, R. & Murphy, P.W. (eds), *The Acari. Reproductive Strategies, Development and Life-History Strategies*. Chapman & Hall, London, pp. 137–176.

Womersley, H. (1961a) Some Acarina from Australia and New Guinea paraphagic upon millipedes and cockroaches and on beetles of the family Passalidae. *Transactions of the Royal Society of South Australia* 84, 11–26.

Womersley, H. (1961b) On the family Diarthrophallidae (Acarina-Mesostigmata-Monogynaspida) with particular reference to the genus *Passalobia* Lombardini 1926. *Transactions of the Royal Society of South Australia* 84, 26–44.

Woodring, J.P. (1969) Preliminary observations on moulting and limb regeneration in the mite *Caloglyphus boharti*. *Journal of Insect Physiology* 15, 1719–1728.

Woodring, J.P. (1970) Comparative morphology, homologies and functions of the male system in oribatid mites (Arachnida: Acari). *Journal of Morphology* 132, 425–452.

Woodring, J.P. (1973) Comparative morphology, functions and homologies of the coxal glands in oribatid mites (Arachnida: Acari). *Journal of Morphology* 139, 407–430.

Woodring, J.P. & Carter, S.C. (1974) Internal and external morphology of the deutonymph of *Caloglyphus boharti* (Arachnida: Acari). *Journal of Morphology* 144, 275–295.

Woodring, J.P. & Cook, E.F. (1962a) The internal anatomy, reproductive physiology, and moulting process of *Ceratozetes cisalpinus* (Acarina: Oribatei). *Annals of the Entomological Society of America* 55, 164–181.

Woodring, J.P. & Cook, E.F. (1962b) The biology of *Ceratozetes cisalpinus* Berlese, *Scheloribates laevigatus* Koch and *Oppia neerlandica* Oudemans with descriptions of all stages. *Acarologia* 4, 101–137.

Woodring, J.P. & Galbraith, C.A. (1976) The anatomy of the adult uropodid *Fuscuropoda agitans* (Arachnida: Acari), with comparative observations on other Acari. *Journal of Morphology* 150, 19–58.

Woodroffe, G.E. (1953) An ecological study of insects and mites in the nests of certain birds in Britain. *Bulletin of Entomological Research* 44, 739–772.

Woolley, T.A. (1972). Scanning electron microscopy of the respiratory apparatus of ticks. *Transactions of the American Microscopical Society* 91, 348–363.

Woolley, T.A. (1988) *Acarology, Mites and Human Welfare.* John Wiley, New York.

Wraith, D.G. Cunnington, A.M. & Seymour, W.M. (1979) The role and allergenic importance of storage mites in house dust and other environments. *Clinical Allergy* 9. 545–561.

Wright, F.C., Riner, J.C. & Guillot, F.S. (1983) Cross-mating studies with *Psoroptes ovis* (Hering) and *Psoroptes cuniculi* Delafond (Acarina; Psoroptidae). *Journal of Parasitology* 69, 696–700.

Wright, J.E. (1969) Hormonal termination of larval diapause in *Dermacentor albipictus. Science* [New York] 163, 390–391.

Wright, K.A. & Newell, I.M. (1964) Some observations on the fine structure of the midgut of the mite *Anystis* sp. *Annals of the Entomological Society of America* 57, 684–693.

Wuest, J., El Said, A., Swiderski, Z. & Aeschlimann, A. (1978) Morphology of the spermatid and spermatozoon of *Amblyomma hebraeum* Koch (Acarina, Ixodidae). *Zeitschrift für Parasitenkunde* 55, 91–99.

Wysoki, M. (1985) Karyotyping. In: Helle, W. & Sabelis, M.W. (eds), *Spider Mites, their Biology, Natural Enemies and Control,* Vol. 1B. Elsevier, Amsterdam, pp. 191–196.

Yalvac, S. (1939) Histologische Untersuchungen über die Entwicklung des Zeckenadultus in der Nymph. *Zeitschrift für Morphologie und Ökologie der Tiere* 35, 535–585.

Yoshikura, M. (1975) Comparative embryology and phylogeny of the Arachnida. *Kumamoto Journal of Science* (Biology) 12, 71–142.

Young, J.H. (1959) The morphology of *Haemogamasus ambulans* (Thorell) with emphasis on the reproductive system (Acarina: Haemogamasidae). Thesis, University of California, Berkeley.

Young, J.H. (1968a) The morphology of *Haemogamasus ambulans.* 1. Alimentary canal. *Journal of the Kansas Entomological Society* 41, 101–107.

Young, J.H. (1968b) The morphology of *Haemogamasus ambulans.* 2. Reproductive system. *Journal of the Kansas Entomological Society* 41, 532–543.

Young, J.H. (1970) The muscles and endosternum of *Haemogamasus ambulans. Canadian Entomologist* 102, 157–163.

Yunker, C. (1955) A proposed classification of the Acaridiae (Acarina: Sarcoptiformes). *Proceedings of the Helminthological Society of Washington* 22, 98–105.

Zacharda, M. (1980) Soil mites of the family Rhagidiidae (Actinedida: Eupodoidea). *Acta Universitatis Carolinae (Biol.) Prague* 1978, 489–785.

Zachvatkin, A.A. (1941) *Fauna of USSR, Arachnoidea,* Vol. 6. No. 1 Tyroglyphoidea [in Russian]. (English translation by American Institute of Biological Sciences Washington, 1959.)

Zachvatkin, A.A. (1952) The division of the Acarina into orders and their position in the system of the Chelicerata. *Parazitologicheskii Sbornik* 14, 5–46 [in Russian].

Zdárková, E. (1967) Stored food mites in Czechoslovakia. *Journal of Stored Products Research* 3, 155–171.

Zebrowski, G. (1926) A preliminary report on the morphology of the American dog tick. *Transactions of the American Entomological Society* 51, 331–369.

Zinkler, D. (1971) Vergleichende Untersuchungen zum Wirkungsspektrum der Carbohydrasen laubstreubewohnender Oribatiden. *Verhandlungen der Deutschen Gesellschaft* 65, 149–153.

Zukowski, K. (1964) Investigations into the embryonic development of *Pergamasus brevicornis* Berl. (Parasitiformis, Mesostigmata). *Zoologica Poloniae* 14, 247–268.

Zukowski, K. (1966) The development of the fourth pair of walking legs during ontogenesis of some Gamasida. *Zoologica Poloniae* 16, 31–46.

Zung, J.L., Lewengrub, S., Rudzinski, M.A., Spielman, A., Telford III, S.R. & Piesman, J. (1989) Fine structural evidence for the penetration of the Lyme disease spirochete *Borrelia burgdorferi* through the gut and salivary tissues of *Ixodes dammini*. *Canadian Journal of Zoology* 67, 1737–1748.

Index to taxa

Comprising index to genera and also to suprageneric taxa in Chapter 12.

Page numbers in **bold** type denote illustrations.

Subject index

Page numbers in **bold** type denote illustrations.